LIBRARY
16 OCT 1979
INST. OF ANIMAL PHYS.

The
Nutrient Requirements
of
Farm Livestock

No. 4
COMPOSITION OF BRITISH FEEDINGSTUFFS
TECHNICAL REVIEW AND TABLES

AGRICULTURAL RESEARCH COUNCIL
LONDON
1976

This work was compiled for
the
A.R.C. Technical Committee on the Nutrient Requirements of Farm Livestock
by
DR I. LEITCH & MR A.W. BOYNE
with the assistance of
MRS G.F. GARTON

Published by the
AGRICULTURAL RESEARCH COUNCIL
and obtainable from
HER MAJESTY'S STATIONERY OFFICE
At the following addresses
49 High Holburn, London W.C.1.
423 Oxford Street, London W.1.
13A Castle Street, Edinburgh 2
109 St. Mary Street, Cardiff
Brazennose Street, Manchester 2
50 Fairfax Street, Bristol 1
35 Smallbrook Street, Birmingham 5
80 Chichester Street, Belfast 1
or through any bookseller

ISBN 0 7084 0027 2

FOREWORD

By Sir David Cuthbertson, C.B.E., M.D., D.Sc., F.R.C.P. (Edin.)., F.R.S.E.
Chairman of the Agricultural Research Council
Technical Committee
on the Nutrient Requirements of Farm Livestock (1959-72).

The Technical Committee which the Agricultural Research Council set up in 1959 was charged with the task of reappraising the accepted standards for feeding of farm livestock. Having embarked on its immediate task of reviewing the literature on nutrient requirements of various classes of stock and of publishing its recommendations it then decided that to complete the basis for calculating input to meet these needs it was necessary to review the literature on the composition of feeding stuffs since most of the available information was old and some of it, particularly on the energy side, dated from Kellner. Dr. Isabella Leitch, O.B.E., M.A., D.Sc., LL.D., who is mainly responsible for the major part of this work, namely the review of the literature, and Mr. A. W. Boyne, B.Sc., F.R.S.E., who organised most of the collection and reduction of data summarized in Part Two, volunteered to conduct this considerable operation and this is their report.

Until recently, successive editions of Bulletin 48, "Rations for Livestock", written first by Professor Woodman in 1921, were the reference works in this country on the nutritive requirements and practical feeding of farm animals. Partly as a result of other publications of this A.R.C. Technical Committee, the Agricultural Development and Advisory Service has published an advisory paper, No. 11, dealing with the nutrient allowance and composition of feeding stuffs for ruminants, and Bulletin 48 has been withdrawn.
The report which follows differs from these publications in several ways. It consists of two parts, the first of which derives from a survey of the literature and contains information on the many aspects of nature and nurture which influence feed composition. For different classes of feeding stuffs there are instanced varietal differences and variations due to region, season, fertilizer treatment, stage of growth and methods of conservation. A feature of this part is that the source reference of every value quoted in the extensive tables is made available to the reader who may thus pursue his own studies on these aspects of feed composition which interest him.

The data summarized in Part Two derive in the first place from the records of analyses made by the National Agricultural Advisory Service, the Scottish Colleges of Agriculture and the Macaulay Institute for Soil Research. Representative firms in the feed processing industry were shown these data, and made their own records available to us from which conspicuous gaps in the collation might be filled. Some unusual feeding stuffs in occasional use in the U.K. are also listed. In this part, each mean is accompanied, wherever practical, by the range and the standard deviation to indicate the spread of individual values encountered.

The Agricultural Research Council is greatly indebted to Mrs. G.A. Garton, B.Sc., and Mr. T.D. Bell, B.Sc., M.I. Inf. Sc., of the Commonwealth Bureau of Nutrition for help in the production of this book and to Dr. J.A.B. Smith, C.B.E., who scrutinized text and tables.

FOREWORD

CONTENTS

PREFACE

The Material from Published Reports: Part One

Our terms of reference were that the search of literature should be for data on the composition of feeds used in Britain, Northern Ireland and Eire, and of origin not earlier than about 1930. The nature and the scope of the material that would be collected were unknown and procedures had to be devised as the collection grew.

The journals scrutinised were:

Agriculture, London: 1939-40, *46* to 1965, *72*
Agriculture in Northern Ireland: 1945-46, *20* to 1962-63, *37*
Analyst: only from references
Animal Production: 1959, *1* to 1965, *7*
British Journal of Nutrition: 1947, *1* to 1964, *18*
British Poultry Science: 1960, *1* to 1965, *6*
Economic Proceedings of the Royal Dublin Society: only from references
Empire Journal of Experimental Agriculture: 1933, *1* to 1964, *32;* 1965, *1*
Journal of Agricultural Science: 1931, *21* to 1965, *65*
Journal of the Bath and West and Southern Counties Association: only from references
Journal of the British Grassland Society: 1946, *1* to 1965, *20*
Journal of Dairy Research: 1934, *5* to 1965, *32*, No. 1
Journal of the Department of Agriculture, Dublin: 1940, *37* to 1965, *61: later*
Journal of the Department of Agriculture and Fisheries, Dublin: 1966, *62*
Journal of the Marine Biological Association of the United Kingdom: only from references
Journal of the Ministry of Agriculture: 1932-33, *39* to 1938-39, *45: later* Agriculture, London
Journal of the National Institute of Agricultural Botany: 1929-34, *3* to 1965, *10*
Journal of the Science of Food and Agriculture: 1950, *1* to 1965, *16*
Journal of the Society of Chemical Industry, London: 1933, *52* to 1949, *68*
Manx Journal of Agriculture: 1952, *6* to 1961, *15*
Nature, London: only from references
Scottish Agriculture: 1944, *24* to 1964-65, *44*
Transactions of the Royal Highland and Agricultural Society of Scotland: 1930, *42* to 1962, *7*
Welsh Journal of Agriculture: 1932, *8* to 1945, *18* (ceased)

A large number of reports were seen also.

In addition to published reports in those journals, unpublished information on the composition of hays was contributed by Mr. G. Alderman, of silages by Dr. P. McDonald and of cereals by Mr. D.E. Morgan. Unpublished details of the composition of cereals analysed at the laboratories of the Research Association of British Flour Millers were given us by courtesy of the Director and Mr. N.L. Kent who was helpful with explanations and advice on several occasions. Mr. F. Raymond and Dr. D. G. Armstrong both supplied extra information and calculations from their records of published work.

The criteria evolved for the inclusion of data were that the material analysed should be used as a feed for one or other farm animal, and that, where required, there should be a satisfactory description of it. The second of those criteria had sometimes to be interpreted leniently but, for example, the data were usually omitted where the feeding stuffs were analysed as components of rations without any description other than a name such as "hay", or where the analyses presented were means for a number of samples so used, in one instance, 59 samples; or the results were presented only as graphs (Black, 1948, Ref. 60).

The amount of information for a single feed varied widely. In some cases with a full description of variety and treatment there would be complete details of proximate constituents, two or more major elements and some trace elements, and one vitamin or more. At the opposite extreme only a single constituent was recorded, most often crude protein.

For grasses in general a picture of change of composition with season and stage of growth could be built up and that pattern was adopted as standard. Where there was abundant detailed information data averaged for a number of cuts were rejected, but where there was little information available, all of it was included. In a few instances apparent discrepancies have been found in the figures given. For instance, in the presentation of analyses of leaf, stem and whole plant, (references 509 and 214) the values for whole plant do not always lie between those for leaf and stem. The failure of agreement may be due to estimation of composition from different sub-samples. In all such cases the figures as published are quoted, with or without comment.

For grains the evident differences found were varietal and regional. Composition is therefore tabulated for varieties separately and special studies are presented of regional differences in composition. The change in composition of oats from North to South of Great Britain is shown in Table 1.20 and Fig. 0.1 and the composition of wheat grown in the Fens is compared with that from other areas in Table 1.30.

The pattern of presentation of data from the literature is therefore not always the same, but has been varied to give as much useful information as possible.

Field Data: Part Two

When the work on tables of feed composition was initiated it was decided to supplement the more detailed and specific information obtained from published reports by summaries of the many analyses of feed composition carried out at the National Agricultural Advisory Service centres in England and Wales, at the Colleges of Agriculture in Scotland, and at the Macaulay Institute for Soil Research.

With the willing co-operation of these centres data were obtained spanning the years 1947-63. A large number of transient part-time workers spent many hours categorising the types of feeds and feeding stuffs and extracting the information on composition contained in the original record books.

After extraction from the original records the data were punched on to paper tape for processing on the Elliot 803 computer of Aberdeen University. The information was summarised in the form of mean, standard deviation (SD) and range estimated on a dry matter (DM) basis apart from dry matter itself and pH. The numbers of samples of different kinds of materials ranged from one to over a thousand, and the conventions were adopted of estimating SD for 20 or more samples and of replacing the maximum and minimum by the upper and lower deciles if 60 or more samples were recorded.

When tabulation of these data was completed it was found that many feeds were not represented. In order to fill the obvious gaps in our tables the co-operation of the feed-processing industry was sought and several organizations made their laboratory records available to Mr. A.H. Sim who was employed by us to visit these organizations and extract from their records information on up to 60 samples of each class of "missing" feed. These have subsequently been analysed and incorporated in our tables following the conventions outlined above.

The organizations whose records were utilized and to whom we would here express our indebtedness were at the time:

Aynsome Laboratories Ltd, Grange-over-Sands, Lancs
J. Bibby (Agriculture) Ltd, Wirral, Cheshire
BOCM Silcock Ltd, London
Thos. Borthwick (Glasgow) Ltd
Cerebos (Agriculture) Ltd, London
Crosfields and Calthrop Ltd, Liverpool
Eastern Counties Farmers Ltd, Ipswich
Quaker Oats Ltd, Southall
Research and Advisory Services (Agric.) Ltd, Poole
Salamon and Seaber, London
Scottish Co-operative Wholesale Society Ltd, Edinburgh
Scottish Grain Distillers Ltd, Edinburgh
Scottish Malt Distillers, Elgin
R. Silcock and Sons Ltd, Liverpool
Silcock and Lever Feeds Ltd, Birkenhead
Spillers Ltd, London

The purpose of giving the measures of variability and the numbers of samples included in the estimates is to indicate the reliance which may be placed on the means, and also to guide the potential user in the preparation of a diet. Many gaps still exist in the tables of field data and we hope that future analytical activities will help to fill them.

Terminology

Protein nitrogen

The first difficulty is that of the structure of proteins. The convention is that "crude protein" is nitrogen x 6.25, or that all protein contains 16% nitrogen. But feed proteins vary in nitrogen content from 15.7% for milk through 16.8 for rice to 18.9 for oilseeds; the corresponding factors are 6.35, 5.95 and 5.30. For oilseeds, important sources of protein, crude protein as N x 6.25 overestimates protein content by 18%.

It is true also that the nitrogen content of body protein is not uniformly 16%. The main interest of animal husbandry is in "retained" nutrients, but the capacity of the animal body to store protein is small, and the body has many different nitrogen-containing substances of which a considerable proportion is scleroprotein, of lower nitrogen content, so that the mean may be nearer 15 than 16%.

The practical implications of these differences in structure have not been considered in detail. For the present it has been assumed that the errors, which will to some extent cancel each other, will not be of much importance and that the usual convention should continue. Crude protein, throughout the tables, is N x 6.25.

Metabolism

Research has long been in progress on the changes in the components of feedingstuffs during digestion, especially digestion in the ruminant. It is clear that the assumption that the end products are closely representative of the original components is, in some cases, chemically speaking, nonsense. It was shown many years ago, and since amply confirmed, that diet protein and non-protein nitrogen are in the main converted in the rumen to the protein of microorganisms, and the nitrogenous compounds of faeces are for the most part the unabsorbed end products of the metabolism of microorganisms and their undigested structural proteins *plus* the unabsorbed residues of nitrogenous substances lost to the intestine in enzymes, mucus and shed epithelium.

It has been shown further that the lipids that leave the rumen are mostly fatty acids derived from microbial hydrolysis of diet lipids, together with the structural lipids of rumen microorganisms; the unsaturated acids undergo partial or complete hydrogenation in the rumen. The end products in faeces consist mainly of the structural lipids of the bacteria of the lower gut, with a small proportion of unabsorbed free fatty acids, some in the form of soaps (Garton, 1969 and private communication). To what extent these changes might invalidate the assumption that absorbed lipid is adequately represented by the ether extract of feed *less* that of faeces is not known. And indeed, ether may not extract all the lipid of either feed or faeces.

To cover the uncertainties it has become usual to speak of "apparently absorbed" protein and lipid. In the interests of brevity we have left that saving qualification to be understood.

Metabolisable energy

Metabolisable energy is taken to be the heat of combustion of feed less the combined heats of combustion of methane, faeces and urine, either measured by calorimeter or estimated by the use of accepted heats of combustion of the several components. What is measured in the conventional procedure is the energy available for accumulation in growth, for storage as fat, or for external production, and expended in the translocations and conversions involved in *all* the shifts of tissue components, productive and non-productive, including absorption and excretion. Since it is reasonable to suppose non-productive energy expenditure above maintenance to be directly related to production, internal or external, the present convention seems to cover the practical issues, and so to be acceptable as defined.

Methods of Analysis: Effect on Results

It has not been possible in this study to take account of methods of analysis for two reasons, namely that the method used is not always stated and that methods described by the same name are not always identical in practice. For that reason, the following summary is included of the results of the Survey made by the Agricultural Research Council (1963) which gives a picture of the variation to be expected within a range of research stations such as have supplied all the data assembled here (Table 0.1).

Included in the summary are results for 10 uniform materials (apple leaves and tomato leaves have been omitted) supplied to each co-operating centre to be analysed for N, P, K, Ca, Mg, S, Fe, Mn, B, Cu, Mo, Zn, Co, Ni, V, Pb, Sn, Ti, Ba, Sr.

The samples were issued in two series and the figures in the summary table under No. of Centres refer to the number of centres producing analyses in each series. The number following the name of material is the ARC sample number; numbers 1—6 belong to the first series, numbers 14—19 to the second. Hence, for instance, 6 centres contributed estimates of N in materials 2, 3, 4, 5 and 6; 5 centres in materials 14, 15, 17, 18 and 19. Results are reproduced only for those elements estimated at more than one centre.

The Arrangement of Information

It was not possible to treat the data for cereals and fodder crops in the same way. Interest in cereal grains derives from variety, year, location, and manurial treatment; in fodder crops also from variety, year and fertiliser treatment, but even more from stage of growth. The arrangement of Tables has been planned to show as clearly as possible the effects of all those variables.

Cereals

During the sorting of the data certain inferences appeared which were thought worthy of comment. These comments form the first part of the Introduction. The full data are shown in extended tables. Data for the grain, milling offals and distillers' and brewers' residues are tabulated as concentrates; the cereal plant as fresh fodder, hay or straw separately with other fodders. Rye appears only as fodder.

Fodder crops

The heading gives the Latin name, English name, variety and strain, or English name only, or trade name. Individual named varieties and strains of each species are treated separately; then mixtures of named varieties, followed by samples without designation of variety or strain. All the data are in terms of dry matter (DM).

Most of the material is arranged in four columns which show:

(1) The nature of the data as shown in the sequence below.

(2) Analytical data by month (in three subdivisions), or year, or both as may be possible, or in a single statement. When a single analysis represents samples taken over a space of time the entry is made at about the mid-point of that time. When no date is given, the entry is made about mid-year. There are explanatory notes. In studies designed to compare primary growth and regrowth the status of the crop is indicated.

(3) Descriptive and explanatory notes.

TABLE 0.1 AGRICULTURAL RESEARCH COUNCIL SURVEY (1963)

Units	Element and Methods		No. of Centres	Pasture herbage (2)	Pasture herbage (17)	Wheat grain (3)	Wheat straw (4)	Oat grain (15)	Broccoli leaves (5)	Cauliflower leaves (6)	Kale leaves (18)	Cabbage leaves (19)	Lucerne (14)
% of DM	Nitrogen :	Mean	6 + 5	1.31	2.09	1.31	0.47	1.87	4.54	4.20	3.56	2.86	2.52
		Range		1.18 – 1.39	2.00 – 2.17	1.61 – 1.97	0.45 – 0.52	1.76 – 2.00	4.26 – 4.89	4.05 – 4.42	3.34 – 3.70	2.84 – 2.89	2.34 – 2.60
	Phosphorus :	Mean	6 + 5	0.23	0.25	0.29	0.07	0.39	0.44	0.38	0.34	0.61	0.26
		Range		0.21 – 0.24	0.25 – 0.26	0.24 – 0.32	0.06 – 0.08	0.36 – 0.41	0.42 – 0.46	0.36 – 0.40	0.33 – 0.35	0.60 – 0.63	0.25 – 0.26
	Potassium (Colorimetry, Titration Turbidimetry) :	Mean	4 for 1st 5 1 for 2nd 5	2.12	2.01	0.40	1.05	0.48	2.64	3.25	2.19	3.69	2.03
		Range		2.10 – 2.14		0.37 – 0.46	0.97 – 1.12		2.54 – 2.91	3.16 – 3.34			
	Potassium :	Mean	5 + 5	2.06	2.14	0.40	0.99	0.54	2.49	3.25	2.29	3.91	2.14
	(Spectrochemical)	Range		1.98 – 2.14	2.05 – 2.25	0.38 – 0.45	0.90 – 1.08	0.49 – 0.56	2.42 – 2.54	3.08 – 3.38	2.25 – 2.34	3.81 – 3.98	2.11 – 2.26
	Calcium :	Mean	6 + 4	0.54	0.98	0.07	0.33	0.12	2.73	3.28	3.39	0.92	1.69
	(Chemical)	Range		0.51 – 0.61	0.97 – 1.00	0.04 – 0.10	0.31 – 0.36	0.11 – 0.13	2.64 – 2.81	3.18 – 3.45	3.27 – 3.55	0.83 – 0.99	1.62 – 1.86
	Calcium :	Mean	4 + 5	0.49	0.94	0.07	0.30	0.14	2.66	3.29	3.25	0.81	1.63
	(Spectrochemical)	Range		0.36 – 0.56	0.85 – 0.99	0.05 – 0.12	0.25 – 0.34	0.10 – 0.18	2.22 – 2.95	3.21 – 3.37	3.14 – 3.36	0.69 – 0.88	1.60 – 1.68
	Magnesium :	Mean	9 + 7	0.13	0.22	0.10	0.08	0.12	0.26	0.27	0.13	0.13	0.15
		Range		0.11 – 0.15	0.20 – 0.26	0.08 – 0.11	0.06 – 0.09	0.11 – 0.13	0.24 – 0.31	0.24 – 0.29	0. 10 – 0.17	0. 10 – 0.15	0.14 – 0.18
	Sulphur :	Mean	5 + 2 or 1	0.19	–	0.14	0.21	0.13	1.11	1.12	–	–	–
		Range		0.18 – 0.21	0.19 & 0.09	0.13 – 0.16	0.17 – 0.23	–	0.94 – 1.40	1.04 – 1.26	1.38 & 0.98	1.01 & 0.57	0.20 & 0.17
μg/g of DM	Iron :	Mean	8 + 7	51	73	39	103	96	147	370	146	98	341
		Range		38 – 59	59 – 81	32 – 51	78 – 111	83 – 105	107 – 167	279 – 409	138 – 153	78 – 116	302 – 365
	Manganese :	Mean	6 + 4	109	73	43	48	67	53	22	39	49	38
		Range		101 – 125	71 – 79	38 – 46	44 – 51	63 – 70	49 – 55	19 – 23	37 – 41	45 – 55	35 – 43
	Boron :	Mean	6 + 8	20	13	9	12	2	45	38	38 | 28	22	35
		Range		10 – 29	11 – 15	2 – 16	4 – 20	1 – 4	37 – 51	29 – 50	29 – 49 | 20 – 31	14 – 25	28 – 41
	Copper :	Mean	5 + 4	5.0	5.8	7.5	4.2	4.9	4.8	7.4	3.5	4.9	18.6
		Range		3.5 – 6.6	5.4 – 6.5	6.5 – 9.1	3.0 – 5.5	4.7 – 5.0	4.0 – 6.2	5.9 – 10.0	3.1 – 3.8	4.7 – 5.1	16.6 – 20.3
	Molybdenum :	Mean	5 + 4 or 3	0.78	0.98	0.06	0.22	0.15	0.15	0.51	0.79	0.30	0.27
		Range		0.66 – 1.0	0.94 – 1.04	0.05 – 0.06	0.16 – 0.31	0.12 – 0.18	0.08 – 0.30	0.47 – 0.54	0.76 – 0.81	0.30 – 0.31	0.19 – 0.48
	Zinc :	Mean	4 + 1	33	40	42	18	43	55	62	33	60	20
		Range		26 – 39		39 – 44	13 – 23		49 – 61	52 – 69			

(4) References. Each entry or group of entries has a journal or report reference. The author index has been numbered in alphabetical order and those numbers are given as references except in the Preface and Introduction where author(s) and date are given.

In the main tables of composition data are arranged in sequence as follows, as many items as may occur. Some items appear only for silage.

Digestibility of dry matter (DM) or organic matter (OM) or both, per cent.
Proximate constituents: Crude protein, ether extract, nitrogen-free extract, crude fibre and ash, in that order.
Amino acids: alphabetically.
Amines and volatile bases.
Organic acids and pH.
Carbohydrate fractions.
Major elements: Ca, P, Mg, K, Na, Cl, S.
Trace elements: B, Ba, Co, Cr, Cu, Fe, I, Mn, Mo, Ni, Pb, Ru, Se, Sn, Sr, Ti, V, Zn.
Vitamins: vitamin A or carotene; vitamin D; tocopherol; B vitamins (thiamin, riboflavin, nicotinic acid, pantothenic acid, pyridoxine, biotin); vitamin C.
Miscellaneous: e.g. myoinositol, choline.

In the tables in which the estimates of metabolisable energy are collected the arrangement is: in the first line proximate constituents; in the second line digested organic components as reported or as computed from coefficients of digestibility; and in the third line metabolisable energy in kcal or kcal and kJ, per g dry matter (DM), organic matter (OM) or digested organic matter (DOM).

There is little information from which metabolisable energy could be calculated for the cultivated grasses and other fodder crops. The information collected on metabolisable energy has not been reproduced in the main tables of analytical data, because it is more easily found in Tables 1.73 to 1.160, or in the tables in the Introduction (Tables 0.28 to 0.30) prepared to show comparisons of metabolisable energy computed with different forumlae, or different coefficients of energy value. Cross references are provided.

Amino acids

This note has been prepared in response to a request from the Chairman of the Technical Committee.

The data collected on the Composition of British Feedingstuffs include little information on amino acids and what there is relates chiefly to animal products. The sparse data assembled in Part I are expressed in 7 different ways and usually without the information necessary to reduce the data to a standard form, so that a summary table was judged to be worthless.

Also, the assembly covers literature only to 1966 and much work has been done since then. Those in search of a guide to the amino acids of individual feedingstuffs may refer to the most recent compilation, produced for the United Nations Food and Agriculture Organization (1970). In one table the publication details the amino acid composition of a large number of feedingstuffs; in another the ranges of values are shown. The source of each item of information is shown also. Some of the ranges are wide; they probably reflect, at least in part, difficulties of analytical techniques.

Data on amino acids, at least as encountered up to 1966, fall into two classes. Those that appear in agricultural journals relate to the total protein in feed. Those in the more technical journals are not usually in terms readily adapted to practical needs; they are concerned with the physico-chemical character and amino acid composition of individual proteins, the synthesis of "protoplasmic" and "reserve" proteins and their possible interconversions; not with the whole protein as a component of feed. For instance, the work of Folkes and Yemm (1956) describes in detail the isolation from "barley grains" of amide N, hordein (prolamine), hordenin (glutelin), globulin and albumin and the amino acid composition, mostly by microbiological methods, of each of these 4 proteins. No information is given about the relative proportions of the proteins from which the amino acid composition of the grain could be computed. The main immediate interest of the data is in the lysine content of the two reserve proteins: lysine N as per cent of protein N, hordein 0.8, hordenin 4.8, which suggests that, to improve the feeding value of barley in which lysine is the limiting amino acid, plant breeders might look for varieties high in hordenin and low in hordein.

In a recent paper in which he repeats his earlier results Folkes (1970) says that hordenin, by N content, does vary in different varieties from about 30 to 40 per cent of total protein in grain, so that some selection for high hordenin might be possible; but again his main interest is in the physiological processes involved in protein synthesis.

The protein and amino acid contents of cereals have, in recent years, received a great deal of speculative attention because most human diets rely on a cereal or cereals as chief source of energy, and in developing countries malnutrition appears to be associated with lack of protein, or insufficiency of limiting amino acids. The same problem is common in animal production. There is the constant need to balance cereal rations with protein or amino acids from other sources, usually more costly than cereals. So the question is asked: can plant breeders produce cereals of higher protein content or higher content of limiting amino acids. It appears from what was said above about the proteins of barley that raising protein content may aggravate amino acid imbalance if, for instance, increase of protein means more hordein of lower lysine content, as happens when protein content is raised by high nitrogen applications (Folkes, 1970).

The hope of improvement of cereals by appropriate hybridisation is encouraged by the discovery in America of a gene in maize, opaque—2, which reduces the zein and raises the glutelin content of the grain, so increasing the content of both lysine and tryptophan. Maize with the gene opaque—2 is of high nutritional value but its total protein content, kernel yield and kernel quality are low. The aim of plant breeders is therefore to combine the opaque—2 character with genes for high protein and high yield, if that should be found to be possible. The investigation is still in the exploratory stage.

A variety of barley of similar promise has been identified. It is described by a United Nations Panel of Experts (1971) as "from Ethiopia called 'hi-proly'" and as having twice as much lysine in grain as do ordinary varieties of barley. Mossberg (1970) of the Weibullsholm Plant Breeding Institute, Karlskrona describes 4 trials, 2 in America and 2 at Svalov (Sweden), in which the barley CI3947 (Hi-proly) averaged 16.3 per cent protein in dry matter of grain and 4.2 per cent lysine "in protein", compared with controls: 11.6 and 3.5 for 3 commercial varieties, 10.7 and 3.9 for 3 low protein varieties and 15.0 and 3.4 for 5 high protein varieties. In percentage protein in dry matter CI3947 equals or exceeds the other high protein varieties tested (though there are varieties with even higher protein contents, up to 20 per cent, in the World List) and has roughly 20 per cent more lysine in protein than present commercial varieties. It will presumably have more of other basic amino acids also. Mossberg says nothing about other characters of CI3947, or conditions of growth and cultivation.

Assuming that the genes that determine protein content, or more likely the content of individual proteins, can be identified and controlled, a breeding programme to produce hybrid varieties with all the qualities necessary for commercial exploitation may take 6 or 7 years (Whitehouse, 1970) or even 10 to 15 years (O'Sullivan, 1965), and its initiation would depend on the commercial prospects. The malting industries prefer low protein barley (O'Sullivan, 1965) and so would offer no market for a new high protein variety. The incentive to produce would depend on the relatively small needs of the Middle East populations that rely on barley as staple cereal, and the amount used in animal production. The question of the possible demand for animal feeding of a hypothetical high protein or high lysine barley, in competition with other sources of protein or lysine, and the premium necessary to encourage production, are discussed by Carpenter (1968; 1970).

There are methods by which better utilisation of existing varieties might be attained; for instance by grinding and air classification into protein rich and protein poor fractions. Cheap new fungal sources of lysine might render breeding programmes unprofitable. And methods of harvesting, drying and storing grain require study and improvement.

Carpenter, K.J. and Taylor, J.H. The economic value of a hypothetical high-lysine barley. *Proc. Nutrition Soc.*, 1968, **27**, 1A.

Carpenter, K.J. Nutritional considerations in attempts to change the chemical composition of crops. *Proc. Nutrition Soc.*, 1970, **29**, 3–12.

FAO: Food policy and food science service, Nutrition Division. Amino acid content of foods and biological data on proteins. FAO nutr. Stud. No. 24, Rome, 1970.

Folkes, B.F. and Yemm, E.W. The amino acid content of the proteins of barley grains. *Biochem. J.*, 1956, **62**, 4–11.

Folkes, B.F. The physiology of the synthesis of amino acids and their movement into the seed proteins of plants. *Proc. Nutrition Soc.*, 1970, **29**, 12–20

Mossberg, R. Practical problems concerning the marketing of cereals with improved protein value. *Proc. Nutrition Soc.*, 1970, **29**, 39–48.

O'Sullivan, T.J.R. Report of the cereal station, Ballinacurra, 1962. *J. Dept. Agric. Dublin*, 1965, **61**, 165–208.

United Nations: Panel of Experts. Report on the protein problem confronting developing countries. [Annex II] U.N. Publication E.71.II.A.17, New York, 1971.

Whitehouse, R.N.H. The prospects of breeding barley, wheat and oats to meet special requirements in human and animal nutrition. *Proc. Nutrition Soc.*, 1970, **29**, 31–39.

PART ONE

A Survey of the Literature of Experimental Studies

INTRODUCTION

Concentrates

In the compounding of rations for animal feeding, certain assumptions are taken for granted; for instance that the digestibility of dry matter of grain will be about 80 per cent; that the protein of plants is ill balanced and requires supplementation with animal protein or specific amino acids; that ruminants are well and pigs ill adapted to the digestion of fibre. The main interest in a compilation of data on the composition of concentrates lies in differences in crude composition between different grains and, for instance within cereals, between varieties of the same grain. Recently there has been increased interest in the amino acid composition of plant proteins, but up to the completion of this assembly, the total amount of information on the amino acids of feedingstuffs was small (see Preface). And indeed the interest of farmers has been and still is more in yield and price of grain than in its composition.

Information in the Concentrate section of Part I is therefore, for the most part, on crude composition as described by the Weende system of analysis, with some details of major and trace elements and occasional estimates of vitamin content. There is even less on the details of carbohydrate composition than on amino acids.

Oat: Whole grain

There is more information on the oat than on any other grain and more on oat kernel than on the whole grain. Detailed analyses of the data have been made to show differences in composition by variety, place and year. For instance, Table 1.9 shows the oil content of 10 Spring and 4 Winter varieties. The Winter varieties have more oil than do the Spring varieties: ranges 3.9–5.6 for Spring and 5.7–7.5 per cent of dry matter for Winter varieties.

Tables 1.5 and 1.8 show the effect on crude protein of extra nitrogen fertiliser, of undersowing different grasses and of state of maturity at harvest. In Table 1.5, extra nitrogen consistently raised the N content of the grain, by almost 9 per cent on average, without otherwise changing the composition of the grain. The chief effect from the farmers' point of view would be on yield. Table 1.7 shows the variation of crude protein in 5 varieties over 4 years. The variety Unique appears only in 1949 with the highest crude protein of the series. Year-to-year differences are definite, 1947 and 1949 better for all varieties than 1946 and 1948, but the varieties do not follow the same sequence of protein contents in each year, or even in the two better or the two worse years.

The meagre data on carbohydrate fractions (Table 1.10) show nothing of interest except the wide ranges of available carbohydrate in each year in the Welsh study, in which the highest values are 47, 55 and 60 per cent above the lowest in the three years. The major elements (Table 1.11) show wider ranges, of over 100 per cent, and some of the trace elements (Table 1.12) still greater variation, especially manganese, zinc and cobalt.

In Table 1.13 is set out in detail a summary of a study of kernel weight, crude protein and thiamin in 6 popular varieties of oat grown in different parts of Scotland, 4 grown also in Wales, and 17 other varieties grown with 5 of the 6 popular varieties in one area only. There was no great varietal or geographic difference, but there was a significant positive correlation of 0.42 between crude protein and thiamin. The only other information about vitamins in whole oat grain is for tocopherols in one variety, Yielder, which varied from 17 to 32 μg/g dry matter, with 26 per cent in the α form and 63 per cent in the β form.

There was little information on digestibility. With such data as could be used metabolisable energy was computed with Axelsson's coefficients for concentrates, and is compared in Table 1.219, with other estimates from metabolism experiments or computed with other formulae. Huskless oats, as might be expected, have a higher value of dry matter and organic matter than the ordinary oat.

Oat kernel

General survey. Most of the information about the composition of oat kernels is for nitrogen and oil in dry matter. The accumulated data, covering 91 varieties and 11 years from 1951 to 1961, grown in Scotland, in England and Wales, or in both regions, are presented in Tables 1.14, 1.15, 1.16 and 1.17, the varieties arranged in alphabetical order. The tables distinguish Spring and Winter oats (S and W), popular varieties (P), varieties of no commercial importance (N), varieties restricted to certain regions (R), those grown only on limited acreage (L) and finally for the 10 restricted varieties the areas where they are grown are shown in footnotes. The individual annual values represent anything from a single sample to 30 or more samples from as many as 17 centres.

It was not possible to give detailed references in the tables to the sources of information. References to published data are included in the bibliography; extended information was supplied by some authors.

Type and Region. The data for Scotland refer to Spring oats with one exception, Maris Quest (Ab/26/284) grown on limited acreage in 1961; the data for England and Wales include both Spring and Winter varieties.

In all, 17 varieties of Winter oats were recorded. The distributions show that 14 of these 17 had less than 2.4 per cent of nitrogen in dry matter. Of the Spring varieties 16 were below and 24 above 2.4 per cent. All 17 Winter varieties had more than 7 per cent oil in dry matter; only 10 out of 40 Spring varieties had more than 7 per cent oil in England. Those findings are in contrast to 29 Spring and 1 Winter variety with more than 7 per cent oil out of 47 varieties grown in Scotland.

Fig. 0.1 REGIONAL TREND IN OIL CONTENT OF OAT KERNELS

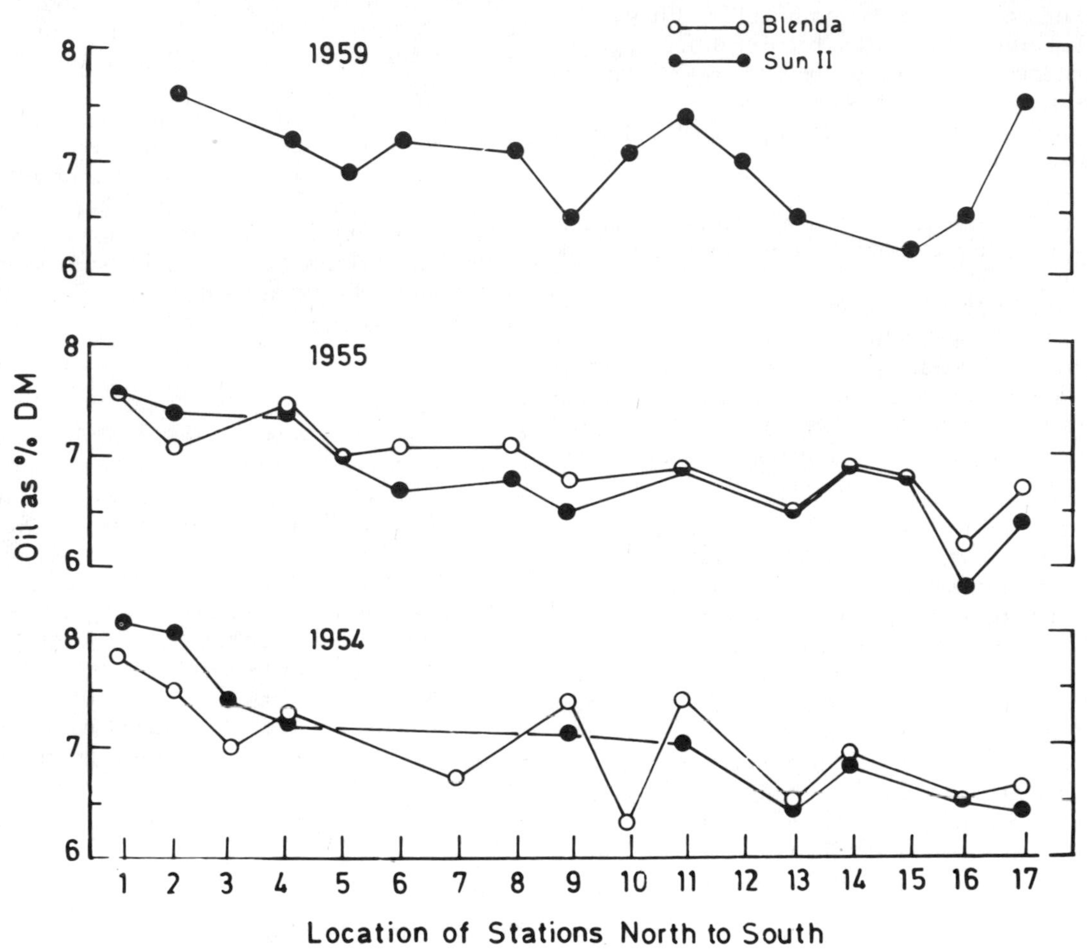

Key to Stations:

1. Aberdeen, Craibstone.
2. Edinburgh, Boghall.
3. Ayr, Auchincruive.
4. Newcastle, Cockle Park.
5. Winmarleigh.
6. Headley Hall.
7. Askham Bryan.
8. Harper Adams.
9. Sutton Bonington
10. Sprowston.
11. Trawscoed.
12. Rosemaund.
13. Cambridge.
14. Connington.
15. Wye.
16. Sparsholt.
17. Seale-Hayne.

According to information (November, 1966) from Dr. N.L. Kent of the Cereals Research Station, St. Albans, the latest relevant to the information assembled here, over the whole of England, Scotland and Wales, Astor, Blenda, Condor, Forward and Sun II accounted for about 95 per cent of the total of Spring oats grown.

Again according to Dr. Kent's information, Padarn and Powys accounted for about four-fifths of the total of Winter oats grown and Peniarth for about another tenth. S147 and S172 were the only others grown in any quantity.

The chances are, therefore, that odd samples, such as unnamed commercial samples, will belong to one, or a mixture, of those varieties which have been treated separately in Table 1.18 which summarises the extended information in Tables 1.14 to 1.17.

Because of the obvious differences with region, further analysis was made of Spring varieties grown in both England and Wales and Scotland. Of 13 varieties so listed 7 had nitrogen contents higher by more than 0.1 per cent in England and 2 had less nitrogen. Eleven of the 13 varieties had less oil in England than in Scotland.

The effect of region on composition is further illustrated in Table 1.20. Two of the most popular varieties, Blenda and Sun II are shown, Blenda in 2 years at 12 and 13 centres, Sun II in 3 years at 10, 13, and 13 centres. The values for oil are graphed in Fig. 0.1. The trend, is downwards from north to south, the whole decline from about 7.8 per cent in dry matter in north-east Scotland to about 6.5 per cent in Somerset.

Hutchinson and Martin (1955) from analyses of the kernel of Victory oats (Table 1.21) thought there were real differences between oats grown in Scotland and in England. The differences for nitrogen are small and those for oil less than for the more recently popular varieties Blenda and Sun II. Dent and Boyd (1964-66) report findings over the 5 years 1957 to 1961 for three varieties grown in the uplands of south-west Scotland, for one of which, Blenda, comparisons can be made with other data. Blenda gave 5-year averages for nitrogen and oil in whole grain of 1.66 and 6.4 with ranges 1.26–1.86 and 5.5–6.8 per cent, values which, on the assumption that the values in Table 1.1 are applicable, are roughly equivalent to 2.13 and 8.5 with ranges 1.62–2.39 and 7.3–9.0 in kernel. Those values do not differ greatly from the values for Blenda grown in Scotland shown in Table 1.14 with ranges 1.96–2.28 and 7.0–7.7 per cent of dry matter. Morgan (1967) more recently has found low values in whole grain of oats grown in Wales and fed on the farms, as low as 1.2 per cent nitrogen(approx. 1.5 in kernel) and 0.9 per cent ether extract (approx. 1.2 in kernel) in dry matter of whole grain, or, excluding the abnormally low value 0.9, the range was still from 1.9 to 8.0. There is no information to show whether the difference in oil content as recorded between Blenda in the Scottish South-West Uplands and an unknown variety in Wales is varietal or environmental.

Year-to-year variations. The trend of year-to-year variation as shown by those Spring varieties of oats that have records for at least three of the five years 1951 to 1955 is shown in Table 1.19. The data are for Scotland only. To give all the varieties included representation in each year, missing values were computed by the usual least squares procedure. The 3 popular varieties in the series, Blenda, Forward and Sun II are represented in each year.

There were highly significant differences between years and still more highly significant differences between varieties. The range in nitrogen between varieties was 11.5 per cent and that for oil was 5.3 per cent. It seems clear that popularity is determined not by percentage content of N or oil in the kernels but presumably by qualities of yield and resistance to disease.

There is a broad inverse correlation between N and oil in the kernels, but with clear exceptions. Forward, for instance is low in both N and oil; Castleton Potato is well above average in N and highest of all in oil.

Typical deviations are shown in Table 1.22 in condensed form in terms of means for the 5 years as percentages of the general means.

Morgan (1967) has found similar differences in the whole grain of 171 samples grown in Wales, 82 in 1961, 39 in 1962 and 50 in 1963. His means and ranges for the 3 years are included in Table 1.3 which shows the proximate composition of whole grain. Crude protein had a coefficient of variation of 13.4 per cent and oil of 20.2 per cent. The coefficient of variation of metabolisable energy for poultry, Table 1.219, was only 7.2 per cent so that digestible crude protein and digestible oil must have varied inversely.

Table 1.23 gives means and ranges for nitrogen in 23 varieties where complete analyses were not made. Samples came from 5 to 50 centres representing trials by the NIAB, NAAS, Scottish Colleges of Agriculture and Scottish Seed Certification Scheme.

Table 1.24 reports a single study of the amino acids of oat kernels, means only, without distinction of variety.

Table 1.25 shows the proximate composition of two grades of oatmeal, milled kernel, their available carbohydrate and calcium content.

Table 1.26 presents a varietal study of the oil content of varieties grown together at different centres and the free fatty acid content of the dry kernel.

Table 1.27 from the same study as that of whole grain (Table 1.13), shows the thiamin content at different stages of development of the kernel from soon after shooting to harvest, and its distribution in parts of the ripe grain. There is a rapid fall at first, as starch accumulates.

Oat Offals

Oat offals Table 1.28 exhibit the wide variation that would be expected from unstandardised products. The husk shows no advantage for the pig from grinding and is not inferior in energy value of dry matter or organic matter to some samples of Oat Dust. Those low grade feeds have also a low energy value of digested organic matter (DOM) in kcal (kJ)/g: feed meal 3.8 (15.9), husk 3.7 (15.5), straw 3.9 (16.3), compared with whole grain, 4.1 (17.2) or 4.2 (17.6). But Oat Dust, with a higher oil content, rises to 4.5 (18.8) (white) and 4.4 (18.4) (brown) kcal(kJ)/g digested organic matter.

Wheat: Whole Grain

Our information about wheat is less extensive than that about oats and much of what we have for proximate constituents is for nitrogen, or crude protein, only. A few more detailed statements are shown in Table 1.29. Table 1.30 shows the crude protein contents of 27 varieties grown at NIAB centres in England, in 1, 3 or 4 years, under normal and intensive manuring and in strip trials, and 9 varieties in 1, 2 or 3 years in Fen trials. These show a variation with variety, year and fertiliser treatment, but the greatest difference is between wheats grown in the Fen trials and the same varieties in the other trials. The summary table shows that, on the average of 7 varieties, crude protein in the Fen trials exceeded the mean of the other trials by 30 per cent.

There is little information on the energy value of the whole wheat grain and what there is is for poultry (Table 1.220). Carpenter and Clegg (1956) assessed two varieties, Guardsman and Hybrid 46 at 3.69 (15.4) and 3.71 (15.5) kcal (kJ)/g dry matter. Lewis and Morgan (1963) assembled 5 estimates of metabolisable energy and from them stated a "compromise value" of 3.3 (13.8) at 90 per cent dry matter content, or 3.67 (15.4) kcal (kJ)/g dry matter. From estimates of digestibility by poultry in one study, metabolisable energy has been computed with Axelsson's coefficients for concentrates and is 3.33 (13.9) kcal (kJ)/g dry matter, and 3.84 (16.1) kcal (kJ)/g digested organic matter. Per g dry matter Axelsson's value is about 10 per cent less than that found in recent experiments. Two samples with digestibility coefficients for sheep give computed values of 3 (12) kcal (kJ)/g dry matter or 3.8 (15.9)/g digested organic matter. In comparison with the whole oat grain the higher crude protein and higher nitrogen-free extract more than make up for the lower content of oil.

Table 1.31 summarises the results of 8 studies with 11 varieties at different centres and with differences in methods of cultivation and application of fertiliser. The effects of additional nitrogen as ammonium sulphate and Nitrochalk applied to the soil and ammonium nitrate or urea as a leaf spray varied widely, from none to nearly 20 per cent increase. The experiments were elaborate and the data are insufficient for a detailed analysis of associations between variety, year, fertiliser and procedure. Hence the simplified summary in terms of no extra nitrogen and additional nitrogen.

In experiments with Winter wheat, Table 1.32, application of phosphate and potash fertilisers had no effect on crude protein in dry matter of the whole grain but, in each of 3 years and on the average of 3 varieties over 4 years, Nitrochalk raised crude protein content, the increase rising with level of application from 125 to 500 kg per hectare. The same general effect of application of nitrogen was reproduced in extensive trials in Eire in 1962 and 1963 (Table 1.33); the magnitude of the effect varied with variety, area and year, with area possibly the most important.

Table 1.34 is from a study of 11 amino acids in a mixture of Canadian and English ground whole wheat, in the bran and in the germ. The percentage content of each amino acid was highest in germ (not reproduced); eight amino acids were higher in whole grain and 3 were higher in bran. Table 1.34 includes also carbohydrate fractions.

Major elements: Studies of the major elements in whole wheat grain are summarised in Table 1.35. The report of the Agricultural Research Council Special Survey (1963) (Preface Table 0.1) which was concerned with comparison of methods used at different research centres, shows the very wide range of values for the same sample that may be due to difference in methods or differences in use of the same method. The widest range among major elements is for calcium, 0.04 to 0.12 per cent of dry matter at 10 centres, and it is disconcerting to find that, in a report (Chambers, 1953) giving means and ranges for 11 fertiliser treatments at one centre, the range for calcium was only 0.04 to 0.05 per cent of dry matter. It seems clear that only camparisons made at one centre can be taken as sure to be valid. The study that deals with the application of lime to the wheat crop shows little change in the composition of the grain.

Trace elements: The ARC Special Survey shows a wide range of variation also for some of the trace elements, especially iron and boron (Table 1.36). A study of iron and manganese in 15 varieties at 7 centres, (Table 1.37), shows a still wider range for both elements. It is not possible to say whether the major part of the variation is due to centre or to variety.

A study from Eire (Table 1.36) shows little effect of lime on the manganese content of 4 varieties.

Vitamins: There are two studies of tocopherols, each of one variety. In Atle α-tocopherol is one-third of the total, in Yeoman, one-half (Table 1.38).

There are also two studies of thiamin content. The first, from NIAB centres (Table 1.39) records 17 varieties, 6 of Spring wheat, under normal fertiliser treatment, with means and ranges in μg per g, for Winter wheat 3.78 and 3.06 to 4.32, and for Spring wheat 4.57 and 3.90 to 5.67. There are also 5 varieties given intensive fertiliser treatment, with a mean of 3.59, 2 varieties in strip trials, mean 4.12, and 3 varieties in Fen trials, with a mean of 5.15.

Comparing those 3 varieties in the Fens and elsewhere, the Fen samples lead by 46 per cent, a difference even greater than that for nitrogen in the same trials.

Table 1.40 compares the nitrogen and thiamin contents of 11 varieties over 3 years. In each year Rivet had the least thiamin content. On the average of the 3 years, Bersée had most; it was highest in 1949, with 4.54 μg per g; equal with Vilmorin 27 in 1948 with 4.44 μg per g and, with 4.13 μg per g in 1947, was second to Vilmorin 27 with 4.26 μg per g.

Wheat Offals and Mixtures

Table 1.41 outlines the mode of derivation of wheat offals and their approximate average composition. Since most wheat bread at present is baked from flour of about 70 per cent extraction, there will be available as feed about 20 per cent by weight of fine wheatfeed and 10 per cent of bran.

Table 1.42 gives the composition of the main classes of wheat offals in terms of proximate composition, major and trace elements, tocopherols and thiamin. Table 1.43 shows an unpublished study of the fibre content of wheatfeeds. In another study the proximate composition is reported (Table 1.44) of 2 mixtures of wheat and maize and 1 of wheat and oats.

Barley

Table 1.45 summarises data for the proximate composition of barley, whole grain. The digestibility of the whole grain of barley has been measured with sheep and poultry and the metabolisable energy has been estimated in *ad hoc* experiments and computed from digested nutrients with Axelsson's coefficients for concentrates. The values for metabolisable energy so estimated vary between 2.98 (12.5) and 3.27 (13.7) kcal (kJ)/g dry matter, but are higher, between 3.09 (12.9) and 3.59 (15.0), when calculated from digested nutrients by Bolton's formula. Per g digested organic matter metabolisable energy by Axelsson's coefficients is 3.8 (15.9) or 3.9 (16.3) kcal (kJ) (Table 1.221).

Studies of the nitrogen content of the whole grain are summarised in Tables 1.46—1.48. They deal with varietal differences, the effect of fertilisers in terms of rate, time and method of application, the type of crop undersown, differences in response to nitrogen on farms of different type, and the effect of weed control on nitrogen content.

It is evident, incidentally, that uncontrollable variations due to "year" occur. For instance, in Table 1.48 the response to Nitrochalk was positive in 36 of 39 trials and 12 of 13 farms in 1955/56/57, but in 1954 with a wet growing season the average on each of 4 farms and 10 out of 12 responses was negative. Those studies are from South-East Scotland. Table 1.49 summarises a number of studies of the nitrogen content of whole grain of barley of different strains in Eire, over the 10 years 1939 to 1948. The year 1944 stands out with a mean nitrogen content of 1.74 per cent of dry matter for 3 varieties of Spratt Archer of which one was the reference barley, Spratt Archer 37 No. 3. It had a nitrogen content in 1944 of 1.68 per cent, a mean for the other 9 years represented of 1.42 per cent (range 1.34—1.52) and the lowest value of the series, 1.34 per cent in 1941.

It seems that there has been more interest in the nitrogen content of barley and the effect on it of fertilisers, than in that of other cereal grains, because of the inverse relation between crude protein and malting quality. David (1964) stresses the importance of low nitrogen in barley from the Maltster's point of view. O'Sullivan (1962) was more specific. "The nitrogen content of malting barley should be low. High nitrogen leads to high malting loss and to difficulty in making malt. From the brewer's point of view, high nitrogen is objectionable in that each 0.1% increase in nitrogen will decrease the yield of extract by about 1%.....". In Table 1.51 there is undoubtedly a highly significant inverse correlation between nitrogen, per cent of dry matter, and IGE (dry); that is between nitrogen and malting quality. Examination of the data in detail shows that the major part of the association stems from differences between centres. Indeed, centre is more important than either variety or nitrogen application, and nitrogen is more important than variety.

There is now a new focus of interest in high protein barley for barley beef, but no relevant information, apart from the varietal differences recorded, was found in this survey. Probably a new series of hybridisations and selections will be reported in the near future.

The mineral content, major and trace elements, of whole grain from named and unspecified varieties is shown in Tables 1.52 and 1.53 with the effect of fertilisers, usually lime and potash.

Other information is for the proximate constituents only of barley bran (Table 1.54).

Distillers' Barley and Maize, and Brewers' Grains and other residues

A substantial part of the information in Table 1.55 comes from the distillers and relates to the form in which residues are now sold for feed. Instead of selling wet or dried spent grain, the distilleries find it more profitable to dry the solubles with the grain. Some, but not all, of the combined products have a higher protein content than the spent grains. There is one set of data for 6 amino acids, and some information on the contents of calcium, phosphorus, sodium, chlorine and B vitamins.

All the information there is on brewers' grains comes from the Report of a Ministry of Agriculture, NAAS, Conference (1963). It includes crude protein, crude fibre, calcium and phosphorus.

For other residues there are single analyses of malt kiln offals and malt parings, and two of distillers' dreg meal, from one of which a metabolisable energy value of 5 kcal (21 kJ)/g digested organic matter has been estimated.

Maize

Studies are of whole grain and meal from plants home-grown or imported (Tables 1.56 and 1.57). For both, metabolisable energy for poultry, kcal (kJ)/g dry matter, (Table 1.222) has been estimated at 3.9 (16.3), but computed from digested nutrients by Axelsson's coefficients the value is 3.6 (15.1) for whole grain for poultry and 3.7 (15.5) kcal (kJ) for degermed cooked meal for pigs. Corresponding values in kcal (kJ)/g digested organic matter are 3.9 (16.3) and 3.8 (15.9). There are several single estimates of the proximate constituents, more important major elements and B vitamins, more than one estimate of the tocopherols, or tocopherols and tryptophan.

There are analyses of flaked maize and of maize gluten feed and germ meal (Tables 1.58 and 1.59) the last two estimated to have low metabolisable energy values for poultry in terms of dry matter: 2.1 (8.8) and 2.5 (10.5) kcal (kJ) for gluten feed and 2.4 (10.0) for germ meal. The last two values, computed with Axelsson's coefficients, are low because of the relatively low digestibility. Values per gram digested organic matter are 4.1 (17.2) and 4.9 (20.5).

(For maize distillers' grains see Table 1.55). The limited information for millet and rice is shown in Tables 1.60 and 1.61.

Other grains and seeds

Legumes: There are two studies of soya bean meal (Table 1.62), of which one has measurements of urease activity, one has estimates of lysine, total and available, and methionine, one of cystine, and both of methionine. There are also two studies of groundnut meal, one with lysine and methionine and one with cystine and methionine. In addition there is a study of the proximate constituents of lupin seed.

Oilseeds: It was possible to estimate metabolisable energy of extracted palm kernel meal and sunflower seed, Hungarian Striped. Per g dry matter, organic matter and digested organic matter the values are in kcal (kJ):palm kernel 2.47 (10.3), 2.58 (10.8) and 3.75 (15.7); and sunflower seed, unextracted, 3.03 (12.7), 3.12 (13.0) and 6.82 (28.5). There are a few other analyses (Table 1.63) of these and of cottonseed meal and linseed with the residue flax chaff.

Miscellaneous: There are studies (Table 1.64) of proximate constituents of cocoa bean and the husk residue of the bean; a more extensive study of tapioca *(Cassava),* sago pith meal (*Metroxylon*) and of horse chestnut meal, the residue after removal of glucose. It appears that the residue had no digestible crude protein for sheep but about 25 per cent of dry matter as crude protein is digested by pigs so that the metabolisable energy was higher for pigs than for sheep. The pig digested none of the crude protein or crude fibre in sago pith meal so that its metabolisable energy in terms of dry matter or organic matter is low (Table 1.223).

Grass seed: There is a single series of analyses (Table 1.65) of seed from the 7 most cultivated grasses and from 10 occasionally cultivated or components of natural pasture. The uncultivated had more protein, more ether extract and more ash, but less nitrogen-free extract. The cultivated had more nitrogen-free extract and potassium.

Leaf Protein Extract

There is a study (Table 1.66) of N in dry protein preparations and the 3 amino acids, arginine, histidine and lysine in protein extracted from leaf of cocksfoot S143, perennial ryegrass S23 with S101, meadow fescue S53 and timothy S48. The grasses were grown in heavily fertilised plots and cut at intervals at a height of 20 cm. N in the dry preparations varied from 7.9 to 15.8 per cent.

A second study reports percentage crude protein and Ca and P in preparations from the leaf of Italian ryegrass, grass and white clover and lucerne and grass, the extracted protein steam coagulated and dried by different processes. A third study reports proximate constituents and the major elements Ca, P, K, Na and Cl and carotene. Neither of those gives N in the dry preparations.

Animal Products:

Milk

Since the Milk Marketing Boards publish annually extensive reports on the composition of milk as produced for human consumption, no such information is included here. Only some analyses of preparations of milk, modified for use in the early weaning of animals, are presented (Table 1.224).

Meat and Meat Residues

There are analyses of meat meal, whalemeat meal and meat and bone meal of grades that differ chiefly in fat content (Table 1.68, 1.69, 1.70). According chiefly to fat content metabolisable energy per g digested organic matter rises from 4.6 to 5.5 kcal (19.2—23.0 kJ).

Fishmeal

There is much more information for fishmeal, means and ranges for proximate constituents of roughly 50 samples of meal from white fish and a few samples of processed herring meal (Table 1.71). The white fishmeal data include a study of B vitamins of which thiamin, riboflavin, pantothenic acid and pyridoxine show large differences between fishing seasons, but nicotinic acid and cobalamin show none.

There is one extensive study of 18 amino acids in processed herring meal and one of 17 in Peruvian fishmeal; some data for lysine in both white and herring meal and 3 studies of methionine in herring meal.

There are several estimates of Ca, P and Cl in fishmeal, but we have not found one of I or any other trace element.

Residues

In addition to some or all of the proximate constituents there are estimates of lysine and methionine in blood meal, lysine total and available, cystine, methionine and tryptophan in whale solubles; 18 amino acids in fish solubles and in a preparation of defatted herring used in fish cakes and meals, (Table 1.72). There are estimates of five B vitamins including B_{12} in whale liver, of riboflavin, pantothenic acid and vitamin B_{12} in whale solubles and of vitamins A and D in shark-liver oil. Thirteen trace elements are recorded in whale solubles.

Metabolisable energy has been estimated for the several modified milks with a range from 4.1 kcal (17.2 kJ) per g of digested organic matter for separated milk spray-dried, to 5.5 kcal (23.0kJ) for the same milk with added oil. Estimates for white fish meal vary from 2 to 3 kcal (8 to 12 kJ)/g dry matter (Table 1.225).

Fodders

A perpetually recurring problem of animal husbandry is the estimation of the nutritive value of herbage, and that means estimation at different stages of growth, so that the best use may be made of it. Empirical knowledge of differences, a sort of clinical judgement, has been replaced by knowledge of the chemical composition of the herbage and recent history is that of the testing of one chemical criterion after another: protein, fibre, digestibility and now metabolisable energy.

Central to the development of that succession of criteria, and essential to the understanding of the problem, are differences in composition of pasture plants and, even more important, the changes that take place in all of them during growth to maturity. For that reason it has not been thought sufficient to present the data gathered from research papers in the form of averages with standard deviations, and so to waste the refinements of analysis on which so much effort has been expended. Hence, in Part 1 of this report, the data for composition of plants are set out to show the seasonal changes in composition. In the Tables for the cultivated grasses recorded it will be possible to trace the changes from early growth to maturity and, where cutting sequences have been devised to simulate grazing or to provide material for conservation, to trace also the changes that occur in regrowth.

In Tables 1.219 to 1.239 are summarised all the studies found for the time interval under review here in which full proximate analysis and digestibility of the organic components of the herbage were reported. The Tables show proximate analysis, digested nutrients, and metabolisable energy, as given in the original report, or, in the great majority of cases, computed by use of Axelsson's coefficients.

An important part of the work of the National Agricultural Advisory Service (NAAS) was to provide information to the farmer on the composition of herbage and its possible feeding value. Since, at present, such guidance must be based on proximate analysis, possibly aided by inspection of the material, the data in the Tables referred to in the last paragraph have been examined to see whether the proximate constituents themselves give worthwhile information on metabolisable energy.

The several components of the problem are briefly discussed below.

Determinants of quality

There are generic differences in composition between pasture plants, between the grasses and clover and lucerne; species differences, between the ryegrasses, cocksfoot, timothy and the fescues; and varietal or strain differences, between early and late varieties, between diploid and tetraploid strains. Examples of all those differences will be found in the main tables. It is accepted that, in general, ryegrasses are more digestible than cocksfoot, timothy, or the fescues and are for that reason to be preferred, other things such as yield and pattern of growth being equally acceptable. It is accepted that, at the same protein content, grasses are more digestible than lucerne, but not more digestible than clover. It will be shown that those views take insufficient account of pattern of growth and that they are true only in certain situations.

The pattern of growth. In brief, the very young plant is "all leaf" and the cellular structures are loosely put together and readily disintegrated. As the plant grows and elongates, more and more supporting structures are required and they in turn become more and more rigid. In the young plant the weight of dry matter in leaf, or leaf lamina, bears a high ratio to that of stem, and leaf lamina contains more protein than does stem. The digestibility of the young plant is higher than at any later stage and, by uncritical deduction, crude protein content came to be regarded as a criterion of digestibility, and so of quality. But, in 260 trials at Hurley, sheep given different cuts of herbage with 2.5 per cent nitrogen in the organic matter digested between 53 and 83 per cent of the organic matter. Further, the application of extra nitrogen to pasture may, at the same time, raise the nitrogen content of the herbage and reduce the digestibilities of dry matter and organic matter.

The reasons lie in the complex of chemical changes that accompany growth. In the very young plant, during first growth, the crude protein content falls continuously, but soluble carbohydrate rises to a maximum at ear emergence. In the immature plant, leaf sheath and stem are at least as digestible as leaf (Bland and Dent, 1964). Up to ear emergence loss of protein is balanced, or more than balanced, by rise of soluble carbohydrate. After ear emergence digestibility depends on the rate at which changes occur in structural carbohydrates. For instance, in four cuts of perennial ryegrass S23, given to sheep at approximately maintenance level, the digested cellulose, araban, glucan and galactan fell from between 92 and 97 per cent in the first cut to between 69 and 75 per cent in the fourth, but digested xylan and uronic acids fell from 92 to 52 per cent and pectin from 72 to 35 per cent. The xylan and uronic acids together accounted for 11 per cent of dry matter in Cut 1 and 21 per cent in Cut 4. The combined effects of differences in digestibility and changes in the proportions, and in the chemical composition, of leaf blade, stem, leaf sheath and inflorescence on digestibility are summed up by Terry and Tilley (1964) as follows: before ear emergence digestibility of the whole plant falls at the rate of 0.2 per cent per day and after ear emergence at the rate of 0.5 per cent per day, in ryegrass and cocksfoot. The rate of decrease of 0.5 per cent in the whole plant is made up of 0.2 per cent in leaf blade, 0.8 per cent in stem and 0.4 per cent in leaf sheath. In timothy and tall fescue the fall to ear emergence is less steep because stem elongates earlier. Johnston and Waite (1965) have reported the changes in lignification in different parts of the plant; leaf, sheath, stem and head.

Sometimes it is impossible to explain the recorded differences in composition. Bland and Dent (1964) make no comment on the striking differences, common to all the six varieties grown in their Auchincruive study of animal preference, between the composition of the grasses in 1961 and 1962 (Tables 1.73, 1.74 and 1.75). In particular, in per cent of dry matter in the September samples, crude protein averaged 29.6 in 1961 and 15.6 in 1962: soluble carbohydrates averaged 2.9 in 1961 and 6.2 in 1962. Assuming the treatment of the plots to have been the same in both years, with alternate grazing and cutting to maintain all "in a vegetative state" and application of Nitrochalk between the Spring and June and between the June and September grazings, the question remains whether the differences were due to ageing of the crop, sown in 1960, or to a cold Spring and slow growth in 1962.

It was, no doubt, inevitable that the view should become current that leafy varieties of grasses must be more digestible than stemmy. But for instance in cocksfoot the most digestible varieties are the stemmy European, Roskilde II and Kammekes, and the home variety Scotia; the less digestible S37 and S143 were selected for leafiness (Bland and Dent, 1964; Minson *et al*, 1964; Raymond, 1964; Terry and Tilley, 1964). And again, the tetraploid strain of ryegrass, Reveille, is more digestible than the diploid S24 at the same stage of growth. Their chemical composition is set out in Tables 1.73—1.75. Soluble carbohydrate is high and fibre low in Reveille.

When species or varieties are compared the conclusions drawn will depend on whether the comparisons are made at the same stage of development, or at the same date. In the studies made by Armstrong and his colleagues (Waite, Johnston and Armstrong, 1964; Armstrong, 1964; Armstrong, Blaxter and Waite, 1964) when ryegrass S23 and S24 are compared at the same date the early S24 is in flower with a digestibility of organic matter of 80 per cent while the later S23 is young and leafy and the advantage is with S23 with a digestibility of 86 per cent. But if the comparison is with S23 in flower, with digestibility of organic matter 79 per cent, the advantage is with the earlier species. In that series, timothy S48 was less digestible at all stages than cocksfoot or ryegrass and had a higher content in dry matter of xylan and uronic acids.

In lucerne, the digestibility of the whole plant depends closely on the ratio of leaf to stem. The stem, like that of grasses at early stages, has more soluble carbohydrate than the leaf. The changes in structural carbohydrates are such that digestible hexosans may provide less digestible dry matter than crude fibre over a wide range of stages of development (Terry and Tilley, 1964). In white clover, part of its special value in pasture lies in the persistence of high digestibility late in the grazing season, associated presumably with slow changes in the structural carbohydrates (Harkess, 1963).

It was perhaps inevitable also from the early studies of chemical composition and digestibility that high quality should come to be associated with "earliness". Dent and Aldrich (1963) made a wide study of the relation of digestibility to heading date in 26 varieties of ryegrass, 19 of timothy, 14 of cocksfoot and 11 of meadow fescue. There were 8 monthly cuts and 4 hay cuts of each species, all at two centres (here called C and T) giving 1680 samples in all. The results forcefully illustrate the complexity of the relations and the difficulties involved in trying to reduce them to any system. For instance, taking the first 3 and the 6th cut of each species, all varieties averaged together, of the 16 cuts digestibility and crude protein gave significant positive correlations in only 8, digestibility and soluble carbohydrate in only 4; digestibility and crude fibre gave significantly negative correlations in only 6 of the 16 cuts.

Again, when the digestibilities of the 26 varieties of perennial ryegrass were ranked in a decile system the disagreement was such that two varieties each ranked 1 at one centre and 10 at the other. The maximum difference in digestibility was 4 per cent at centre C and 3.7 per cent at centre T.

Those and other findings in the study raise the question of what may be taken as a difference in digestibility of practical significance. The authors say: "...taking the year as a whole, ryegrass was about 1% higher than timothy and 2–3% higher than cocksfoot." But the range of digestibilities within the 4 cuts of 26 varieties of ryegrass was 5.7 per cent at the second cut; for timothy the maximum difference was 4.3 per cent at the 3rd cut, for cocksfoot 4.2 per cent at the 3rd cut and for meadow fescue 3.9 per cent at the 2nd cut. More helpful is the picture of the different species at the first hay cut, 10 days after heading, i.e. after the date at which it was judged that 50 per cent of the heads had emerged. It is clear that at centre T ryegrass was, in general, superior to cocksfoot at dates when both species were heading. It is clear also that, at centre C, early flowering varieties of ryegrass were 2 – 3 per cent better than cocksfoot varieties that headed at the same time. Late flowering ryegrass varieties were less digestible than the earlier cocksfoot, perhaps a little better than meadow fescue that headed a little earlier, or the earlier timothy, but little, if any, better than the *latest* timothy varieties. Generalisations in comparison of species seem to call for careful statement of the conditions with appropriate qualifications. *What appears to be established is* that very small differences in digestibility, here between species, are regarded as important. If that is so, then it in turn emphasises the importance of varietal differences within species.

To avoid the changes associated with maturation, cutting systems have been devised which give a leafy regrowth with relatively high and uniform protein content and digestibility. But neither the protein content nor the digestibility of regrowth quite equals that of early primary growth. Again there are species with varietal differences. During successive cuts, ryegrass varieties maintain their high quality better than do varieties of cocksfoot so that the gap between them widens.

Such systems of cutting are designed to imitate grazing but grazing is complicated in that the animal selects either some parts of the sward in preference to others, or the top of the plant, so that the part left is not of the same composition as the part eaten. For those reasons the part eaten may not be of the same composition as samples cut by hand or by machine. The magnitude of such differences will depend on the botanical composition of the sward, and, in a mixed sward, on the stages of maturity attained at a given cut, or grazing period, by its individual components. An animal that bites off the upper leafy part of the plant (top growth) will get relatively more protein, at least at the later stages of maturity; it may also do better than the one that is less selective or is restricted to bottom growth, but will not always do so. Grasses are chosen for mixed swards to give a yield as uniform as possible over a grazing season and that involves changes in the dominant species and in type of plant growth and composition over the season.

In spite of those complexities, Corbett, Langlands and Reid (1963) devised relatively simple equations relating percentage digestibility of organic matter in cut herbage to the date of cutting for the year in which their experiment was made. For obvious reasons it would be unwise to assume that they would be of general application.

McDonald (1956) has an important observation on the relative values of Spring and Autumn grass. Grasses cut at a similar stage of maturity, with similar protein and fibre contents, respectively 11.1 and 26.8 per cent of dry matter in Spring (June) and 12.8 and 27.1 per cent in Autumn (September) grass, and dried, showed little difference in digestibility. In particular, digestibility of crude protein was 63.6 per cent in Spring and 67.9 per cent in Autumn. Many examples of those relations will be seen in the tables of grasses and legumes. One taken at random of ryegrass S24 shows the fall of nutritive value during primary growth to maturity and the effect of regular cutting (Fig. 0.2).

Fig 0.2 LOLIUM PERENNE (perennial ryegrass) : S24

Organic matter digestibility % and crude protein % of DM measured at different times.

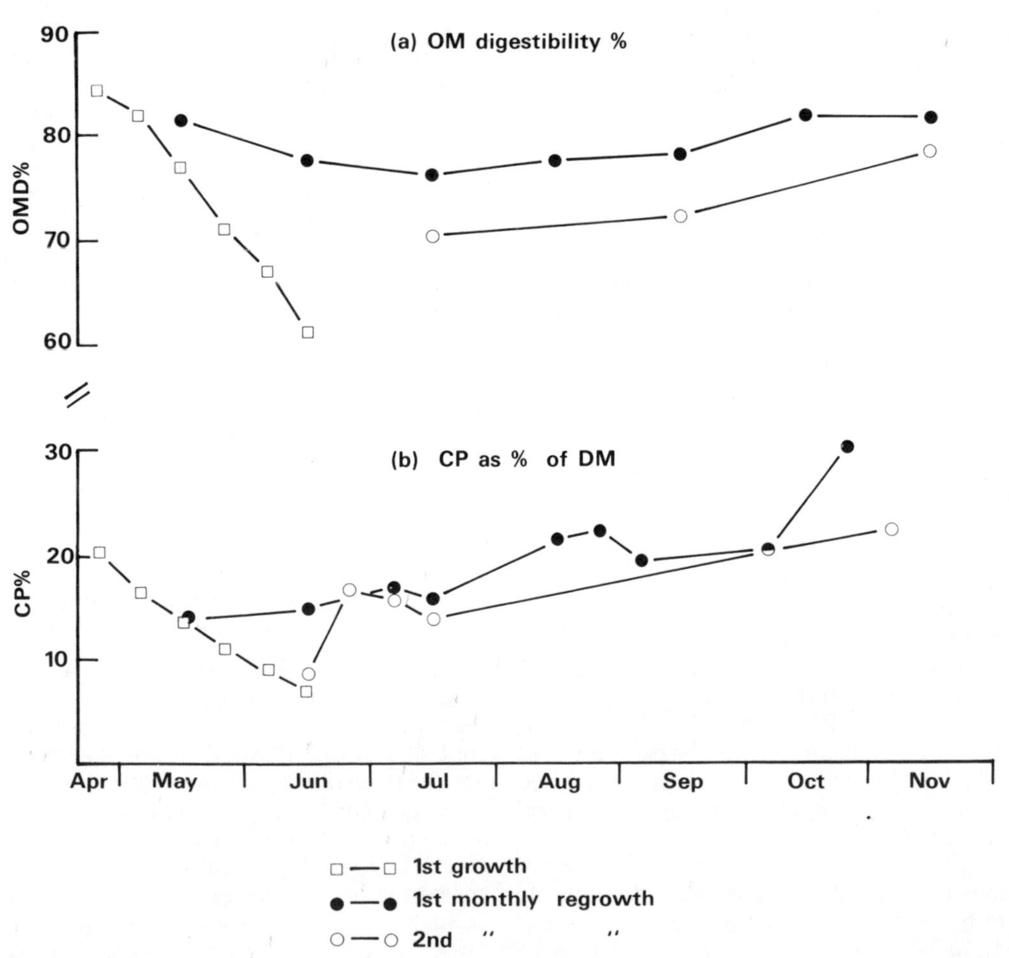

It might be thought that year to year and seasonal (in the sense of "good" and "bad", "early" and "late") variations would be explicable in the same way, an early or good season giving condensation of growth stages and a late or bad season extension. But there is a certain contradiction there. Condensation of growth would give early maturity, but also early deterioration of digestibility. Raymond (personal communication) explains regional differences on the same basis, saying that the composition and digestibility of the same plant at the same stage of growth will be "the same at centres as widely dispersed as, say, Cambridge and Aberystwyth". That conclusion obviously precludes any significant effective difference in soil type. If, with that proviso, it is true of forage plants, it is not true, as has been shown elsewhere in this report, for all parts of the Gramineae. See composition of the oat grain, p.9 of Introduction.

Raymond and his colleagues in their search for a criterion of quality of herbage tested also the hypothesis that digestibility was so closely related to crude fibre content that the one could be predicted from the other. Their conclusions are shown under Digestibility, p.33. The correlation was not found to be such as to be applicable to estimate the value of individual feeds.

than organic matter, soluble carbohydrate, cellulose, hemicellulose, and organic acids in 3 of the 4 grasses and only ether solubles, pectin, and lignin were less digestible than protein. The 4th grass differed only in detail. Non-protein nitrogen (NPN) was less well digested than protein in 2 grasses, equally well in 1 and better in 1. Without labouring the argument, these findings at least suggest that a change in protein as the plant matures accompanies, or even precedes, the grosser changes in structural carbohydrates with extensive lignification and deposit of encrusting substances, as described by Baker and Harriss (1947).

ap Griffith and Jones (1965) found that the technique of estimating fibre affected the coefficient of correlation between digestibility and crude fibre. In a study of 32 samples of herbage of known *in vivo* and *in vitro* digestibility, when digestion with pepsin and hydrochloric acid preceded the usual normal acid digestion, a correlation of 0.969 was obtained. In a subsequent study the correlation was less as herbage matured, suggesting a change in some structural component other than fibre, possibly xylan (Bland and Dent 1964).

Dent and Aldrich (1963) found positive correlations between digestibility of dry matter on the one hand and content of protein and of soluble carbohydrates on the other, and negative correlations between digestibility of dry matter and fibre content; the level of significance varied greatly. But crude protein and soluble carbohydrate contents were negatively correlated in 12 of the 15 samples of 4 grasses described by Dent and Aldrich so that the broad pattern might be described as depending on protein *or* soluble carbohydrate, which is outstandingly high in young grass. The means for each grass at the first and August cuts are shown in the table above.

	Crude protein %DM		Soluble carbohydrate %DM	
	1st cut	August cut	1st cut	August cut
Perennial ryegrass	17.0	22.0	31.3	7.1
Timothy	17.8	22.3	22.9	4.3
Cocksfoot	18.7	23.9	20.6	4.0
Meadow fescue	19.5	21.9	21.2	5.7

The dependence of the digestibility of dry matter, or organic matter, on the fibre content of forage is held to have two components: (1) as a plant matures changes occur in the structural carbohydrates, the chief components of crude fibre, which render them less digestible and (2) crude fibre as it increases is thought to depress the digestibiltiy of other components, protein in particular. It seems unlikely that the fall of digestibility of, say, crude protein is to be attributed simply to the rise of crude fibre and its reduced digestibility. The content of crude protein falls as fibre rises and there is no reason to suppose that the composition and the physico-chemical state of the residual protein is the same as that at younger stages of growth or indeed that any component continues identical during development to maturity. We have already referred to the changes in composition recorded by Armstrong and his colleagues in a series of 12 cuts of 4 grasses in primary growth. If the digestibility of protein at the later, or latest cuts (4th for ryegrass S23, 2nd for ryegrass S24, 3rd for cocksfoot S37 and timothy S48) be compared with the digestibilities of the carbohydrate components, it will be seen that protein was *less* digestible

In agreement Terry and Tilley (1964) say: "Before ear emergence, therefore, increases in soluble-carbohydrate content partially compensated for decreases in crude-protein content and helped to maintain pepsin-digestible contents at a constant level; . . ."

Drying and Ensiling. It is accepted that fodder crops lose digestibility when dried. The degree of change depends on the method. Grass "barn dried" or "artificially dried" usually ranks higher than grass dried in the field, and tripod-dried grass higher than grass dried on the ground.

Apart from mechanical loss of the more fragile parts, the difference is, no doubt, due to speed of drying with control of fermentative and putrefactive change. Commercial drying is subject to faults of overheating. At best, commercially dried grass compares with experimentally barn dried; technical faults apart it usually ranks with good hay.

Drying alters the magnitude but not the sense of the differences in composition due to type of plant, stage of growth, primary growth and regrowth, and appears to affect all fodder crops in the same way. In the main tables comparisons will be found between fresh and dried samples of the same crops, their composition and digestibility. Some of the most important studies were made with artifically dried grass, stored frozen, since it is difficult or impossible with fresh herbage to ensure uniformity of supply through a long feeding experiment.

Differences in the techniques of making silage add complications to the study. The intake by the animal varies widely and depends on moisture content of the silage and its "palatability", and judgement of quality must depend to a large extent on intake.

McDonald (1956) made experimental silages, in different states of preservation, in small experimental silos and, for comparison, hay on tripods and cured in the ordinary field way. Silage was made from both Spring and Autumn grass, with and without molasses, and, in 2 experiments, from wilted and unwilted herbage. Wilted herbage gave silage with slightly more soluble carbohydrate than unwilted, and slightly more digestible in 1 of the 2 experiments. Molasses had little effect on preservation except to give a little more lactic acid in the silage; it slightly increased digestion of crude protein. In terms of conservation of nutrients and digestibility, silage ranked higher than tripod hay and tripod hay above field cured.

Harris and Raymond (1963) compared the digestibilities of pure stands of grasses, cut at different stages of growth, cold stored, barn dried, and ensiled by different methods, "cold" and "hot", unwilted, wilted, or lacerated, and the intakes of grass, hay and silage as g dry matter per kg live weight $^{0.73}$. The grasses were ryegrass S24 and H1, cocksfoot S37 and Germinal, meadow fescue S215 and timothy S48. The dry matter of silage was estimated by toluene distillation to avoid loss of volatile substances in drying.

Summing up, Harris and Raymond say that, for the 15 silages studied, there were negative correlations between intake and digestibility, and intake and moisture, ammonia, and butyric acid in the silages, so that, for silage, high digestibility is not a reliable criterion of quality as it is for fresh grass and hay. It appears from the tables that both digestibility and intake of timothy, as both "cold" and "hot" silage, were much higher than those of ryegrass H1; and that cold stored grass and both wilted and unwilted silage of timothy gave higher intakes, though lower digestibility, than Germinal cocksfoot, H1 ryegrass and S215 meadow fescue.

Dent and Harris (1963) compared the digestibilities of silages of Early, Medium and Late maturing maize *in vivo* with sheep and *in vitro* at two centres, the Grassland Research Station and the National Institute of Agricultural Botany. There was little or no difference *in vivo*. *In vitro* at both centres the late maturing maize was slightly better than the other two, 76.9 and 75.3 per cent, compared with 72.4 and 71.6 per cent. Later Harris (1965) compared the digestibilities by sheep of the same types of maize, fresh and ensiled, with their protein contents and pH. There is not enough information to separate the possible effects of type and composition.

Differences in the technique of making silage make little difference to its composition and digestibility. The variants in technique by 1) addition of carbohydrate, usually as molasses, sometimes as cereal, 2) addition of acids of several kinds, and 3) treatment of the crop by chopping, lacerating or crushing, or 4) by wilting, all are designed to limit butyric acid fermentation and ammonia production. In the trials quoted here and from the examples in the main tables, wilting appears to be the most important of those treatments, but the evidence is that the composition of the crop itself is more important than any of the treatments. As with fresh and dried fodder, type of plant and stage of growth are of prime importance. Unlike fresh and dried fodder, it seems that, for the best silage, preference should be for grasses, and among these for varieties or strains of low protein and high soluble carbohydrate content.

The effect of fertilisers on composition

In the main tables will be found many examples of the effects of fertilisers on the composition of the plants listed. For this commentary some have been selected that are concerned specifically with the effect of a fertiliser on one component, or more, of the crop studied, or with the improvement of poor pasture, two effects that are not always separable. The interpretation of the data describing the action of a particular fertiliser in a given series is often made difficult by lack of precision in the description of the samples analysed. For instance, the effect of nitrogenous fertilisers on the nitrogen content of the plant may be positive or negative, with the same grass, at what purports to be the same stage of growth, say "early flowering". Such terms as "early flowering", "heading date", appear to be elastic, or to involve so large an element of judgement as sometimes to be of little value. The position may be further complicated by effects of fertilisers on the botanical composition of the sward itself, by encouragement of one species at the expense of another, by allowing the appearance of legumes or causing their elimination. For instance, agrostis and red fescue may replace sown grasses in potassium deficiency. (Hemingway, 1961 (b)) and clover may disappear when sulphate of ammonia is applied (Hood (1957); Hemingway (1961(b)); Reid (1962)). Reports seldom detail changes in the flora, but the possibility of significant changes suggests that the interpretation of changes in the composition of samples of unsorted herbage should be regarded as changes in the composition of the sward, not necessarily of the species sown, or described at the beginning of an experiment.

The practical importance of the experimental studies reviewed here depends to some extent on practice in the use of fertilisers. The study of Boyd, Church and Hills (1963) shows that the rate of application of nitrogen on grassland in 1957-58 varied from less than 12.5 to over 125 kg per hectare with a modal value of about 44 kg. With a similar range the modal value for phosphorus was about 19 kg P_2O_5 and for potassium 44 kg K_2O. Much of the experimental work has been with applications above those modal values, some above the upper limits of the ranges, and no doubt the trends shown should be interpreted in that light. Applications for the improvement of poor pasture and to correct the balance of trace elements may be more directly interpreted.

Nitrogenous Fertilisers

Effect on soluble carbohydrate. A series of studies at the Welsh Plant Breeding Station, Aberystwyth, planned with reference to the suggestion that pasture improvement has been associated with an increase of disorders of metabolism in grazing stock, has shown conclusively that nitrogen fertilisers depress the content of water-soluble carbohydrate in pasture plants, so confirming some earlier observations. For reference to the first 4 papers in the series see the fifth (Jones, ap Griffith and Walters, 1965) which gave the following results in well controlled experiments.

In perennial ryegrass S23 and cocksfoot S143 in pure stand extra nitrogen, 88 kg per hectare as nitrate of soda 1 month before sampling, did not affect the growth pattern of either grass, but the soluble carbohydrate of each sample cut between 5 April (early growth) and 13 June (hay stage) was reduced; the average reduction was of the order of 50 per cent.

To compare fertilisers, ryegrasses S23 and S24, cocksfoot S143 and timothy S48 were given extra nitrogen, 88 kg per hectare, as nitrate of soda, sulphate of ammonia, or urea, 1 month before the first cut and after each cut except the last, and were cut monthly from mid-April to the end of September, with these results:

Soluble carbohydrate: per cent of dry matter
Means of 3 field replicates and 7 cuts

	No extra N	Nitrate of Soda	Sulphate of Ammonia	Urea
Ryegrass S23	16.0	10.0	10.2	11.9
Ryegrass S24	13.8	7.2	10.4	12.3
Cocksfoot S143	9.2	5.3	4.8	5.2
Timothy S48	12.1	7.2	6.2	6.9

The seasonal changes are shown in Tables 1.74, 1.88, 1.89, 1.114.

In ryegrass S23 and cocksfoot S143, total sugars and fructosan were estimated in each cut. Both were reduced by nitrogen, fructosan relatively more than total sugars. Means and ranges were:

	Total sugars: % of dry matter		Fructosan: % of dry matter	
	Without N	With N	Without N	With N
Ryegrass S23	7.3	6.0	7.6	4.3
	(5.0-8.7)	(5.2-6.7)	(2.0-13.2)	(1.3-8.8)
Cocksfoot S143	5.6	4.5	2.9	0.6
	(4.6-7.2)	(4.2-5.0)	(0.7-5.3)	(0.3-1.6)

Finally, the effect of Nitrochalk at 4 different levels, 125, 250, 500 and 1000 kg per hectare, applied either a) 1 month before each sampling date, or b) all at once 1 month before the first cut, was investigated with perennial ryegrass S24 (Table 1.89). Eight samples, all of primary growth, were taken between March 26 and June 13. In both tests, soluble carbohydrate was increasingly depressed as level of nitrogen application rose, slightly less in the b) than in the a) test. Those changes were superimposed on an approximately normal pattern of growth and composition. It seems likely that the reduction of soluble carbohydrate accounts for a significant part of the reduction of digestibility which has been found to follow application of nitrogenous fertiliser.

Effect on crude protein content. In 1946 and 1947 Hughes and Evans (1951) studied the effect of sulphate of ammonia on a run-down sheep ley. There were 160 plots each of 1/40 acre (roughly 100 m²) in groups of 4, 1 control and 3 dressed with sulphate of ammonia, 125, 250 or 500 kg/ha. The plots were dressed once, in a monthly sequence from February to November. Half of each was cut for silage in May, June and September; and half for hay in June and aftermath in September.

The effect of nitrogen application on crude protein in the first and second silage cuts was not consistent in relation to either interval between application and cutting, or level of application. In general, the higher the level of application of nitrogen the less crude protein in dry matter. In the September silage cut all the crude protein values were higher and they were highest with the largest nitrogen dressings. Protein in the hay cut was less than in the silage cuts and was no more consistent. In aftermath all the protein values were high, as in the September silage cut. Protein content was least after application of nitrogen in May and at a minimum with the highest application. Protein was highest after dressing in August and at a maximum with the highest rate of application. In general depression of the nitrogen of the crop was associated with reduction of clover in the herbage, but the relation was not always apparent.

Experiments with cultivated pasture have been chiefly with Nitrochalk in amounts varying from 250 to 1250 kg per hectare. In illustration, there are two studies, each with a sward of perennial ryegrass S23 and white clover S100. In the first (Hood, 1957), Nitrochalk was applied at the rate of 250, 500 or 750 kg per hectare to a 3rd year ley.

At the two lower rates, applied 'early', i.e., at the start of the season and soon after each cut, Nitrochalk reduced crude protein in the herbage; applied 'late', i.e., a week or two before cutting, it raised percentage nitrogen. Applied at the highest rate early or late, Nitrochalk raised percentage nitrogen in the crop. Means and ranges per cent of dry matter were:

No Nitrochalk: 16.9 (13.1–19.3)

Lower nitrogen levels:

applied 'early', 15.0 (12.5–16.6); 16.5 (15.2–17.4)
applied 'late', 17.7 (14.8–19.0); 20.4 (18.3–21.8)

Higher nitrogen levels:

applied 'early', 19.2 (17.9–20.4)
applied 'late', 22.4 (20.1–24.1)

In the second study (Reid, 1962), with ryegrass S23 and white clover S100, 1250 kg per hectare of Nitrochalk was applied, 250 kg in Spring and after 4 of 5 cuts, all between mid-May and mid-October; or 250–437 kg after the 2nd, 3rd and 4th cuts only. The experiment was repeated for 5 years, 1957-61. In each year, Nitrochalk reduced the mean crude protein of mixed herbage, per cent of dry matter. Means and ranges were: no Nitrochalk 16.8 (15.3–17.8): and the experimental herbages 14.8 (13.2–16.9) and 15.8 (14.6–17.3). In both trials the fall of percentage nitrogen was associated with reduction of clover in the sward.

Combined fertilisers: N.P.K. Similar results are seen in the experiments with routine fertilisers providing nitrogen, phosphorus and potassium.

In a plot study designed to simulate pasture cut for silage (Walker, Edwards, Cavell and Rose, 1952a) plots at 7 centres in 1950 and 12 centres in 1951 were given sulphate of ammonia, or superphosphate, or sulphate of potash, each at two levels, or nitrogen with phosphate or

Table 0.2 Effect of ammonium sulphate on crude protein and calcium: per cent of dry matter

Cut	None CP	None Ca	N1 CP	N1 Ca	N3 CP	N3 Ca
Perennial ryegrass, meadow fescue, timothy and meadow foxtail, alsike and N.Z. white clover: Hawkstone						
1	11.3	0.44	11.0	0.41	13.6	0.45
2	7.3	0.56	7.9	0.49	6.5	0.51
3	15.4	1.05	13.6	0.71	13.4	0.84
Cocksfoot, ryegrass and clover (S100 white): Weston						
1	13.0	0.76	12.0	0.61	12.6	0.47
2	15.2	1.04	10.6	0.54	9.4	0.54
3	26.8	1.59	16.4	0.79	24.8	1.52
Timothy, Italian ryegrass, lucerne and clover: Kinver						
1	11.7	0.74	11.0	0.54	11.1	0.36
2	15.8	1.76	11.2	0.91	10.8	0.99
3	21.9	1.74	19.5	1.24	20.6	1.58
Timothy, meadow fescue, cocksfoot and clover: Leighton						
1	13.5	0.46	13.6	0.36	16.3	0.38
2	15.7	0.91	14.2	0.60	13.7	0.64
3	20.9	0.99	16.1	0.46	16.9	0.63

potash. Since neither phosphate nor potash had a significant effect on crude protein, only the effect of nitrogen is reported. Nitrogen reduced crude protein, per cent of dry matter, the reduction greatest where initially there was most clover in the sward. The effects of the fertilisers on the percentage content in dry matter of calcium, phosphorus and potassium are recorded in a second report (Walker *et al.*, 1952 b). The results depended on the effect of the fertiliser on the ratio of legume to grass in the sward. With nitrogen, on average, calcium was reduced and phosphorus was reduced in the absence of superphosphate. Without nitrogen, both phosphate and potash raised calcium content as well as the respective contents of phosphorus and potassium in dry matter.

In a further plot study of pasture as cut for commercial grass drying (Walker, Edwards, Cavell and Rose, 1953) material from 6 different centres, all in the West Midlands, is described. Four sets of data, for typical swards, are reproduced here *. The Table 0.2 shows the effect of nitrogen as 250 kg per hectare (N1) of ammonium sulphate, or three times that amount (N3), on crude protein and calcium content of the crop as harvested.

Table 0.3 Effect of NPK fertiliser at different levels on phosphorus content of herbage: per cent of dry matter

Fertiliser	No nitrogen			Nitrogen N1			Nitrogen N3	
Cut	None	PK	3PK	N1	PK	3PK	N3	3PK
	Perennial ryegrass, meadow fescue, timothy and meadow foxtail, alsike and N.Z. white clover: Hawkstone							
1	0.27	0.29	0.35	0.26	0.35	0.35	0.29	0.38
2	0.26	0.27	0.26	0.24	0.25	0.27	0.27	0.27
3	0.42	0.45	0.44	0.40	0.46	0.44	0.40	0.43
	Cocksfoot, ryegrass and clover (S100 white): Weston							
1	0.25	0.29	0.30	0.26	0.29	0.30	0.29	0.31
2	0.29	0.31	0.30	0.24	0.32	0.26	0.22	0.27
3	0.45	0.51	0.53	0.38	0.44	0.31	0.42	0.44
	Timothy, Italian ryegrass, lucerne and clover: Kinver							
1	0.24	0.26	0.25	0.20	0.21	0.26	0.28	0.27
2	0.19	0.20	0.20	0.15	0.20	0.19	0.16	0.18
3	0.30	0.37	0.37	0.28	0.38	0.34	0.34	0.34
	Timothy, meadow fescue, cocksfoot and clover: Leighton							
1	0.30	0.33	0.25	0.31	0.34	0.25	0.33	0.37
2	0.30	0.30	0.33	0.29	0.32	0.32	0.28	0.37
3	0.43	0.48	0.46	0.38	0.40	0.49	0.40	0.47

*Footnote: If these results had been reproduced in the main tables those for the different areas would have been widely separated because of the different seeds mixtures used. It was thought that it was more effective to concentrate them here.

No simple summary can be made of the effect of nitrogen application on crude protein. On average in the first cut N3 gave the highest protein content, by a small margin. In both second and third cuts no nitrogen gave the highest crude protein. The net effect of nitrogen is compounded of a slight effect on the crude protein of clover and a greater effect on that of grass; a great decrease of clover in the sward; and the effect of stage of growth (cut) which, in the Weston and Kinver swards, was greater than that of the fertiliser. In general, nitrogen reduced percentage calcium in the sward; in the first cut the higher dressing had the greater effect. The result was dependent on the proportion of legumes in the sward. As with crude protein, the change of composition with stage of growth obscures that due to application of nitrogen.

The Tables 0.3 and 0.4 show the effects of nitrogen, phosphate and potash on the phosphorus and potassium contents of the herbage as cut. Single dressings were of nitrogen 250 kg per hectare sulphate of ammonia, phosphorus 250 kg per hectare superphosphate and potash 125 kg per hectare sulphate of potash. The treble dressings were of the same amounts in either 3 applications or all at once. The results of the different methods of application are not distinguished. Nitrogen greatly reduced the yield of legumes in the herbage.

Table 0.4 Effect of NPK fertiliser at different levels on potassium content of herbage: per cent of dry matter

Fertiliser		No Nitrogen			Nitrogen Nl		Nitrogen N3	
Cut	None	PK	3PK	N1	PK	3PK	N3	3PK
	Perennial ryegrass, meadow fescue, timothy and meadow foxtail, alsike and N.Z. white clover: Hawkstone							
1	2.39	2.48	2.74	2.10	2.24	2.71	1.79	2.69
2	1.60	1.89	1.98	1.55	1.74	1.93	1.49	1.81
3	2.56	2.86	2.70	2.14	3.30	2.91	2.42	2.81
	Cocksfoot, ryegrass and clover (S100 white): Weston							
1	2.09	2.35	2.47	2.07	2.32	2.47	2.41	2.58
2	1.94	2.35	2.27	1.83	2.45	2.50	1.95	2.00
3	2.96	3.10	3.49	2.49	3.12	3.06	2.65	2.91
	Timothy, Italian ryegrass, lucerne and clover: Kinver							
1	2.51	2.72	2.61	2.53	2.77	2.58	2.61	2.80
2	1.69	1.93	1.86	1.73	1.93	2.03	1.66	1.68
3	2.07	2.85	2.41	2.61	2.78	2.58	2.58	2.66
	Timothy, meadow fescue, cocksfoot and clover: Leighton							
1	2.38	2.43	2.57	2.39	2.68	2.77	2.24	2.67
2	2.02	2.21	2.29	2.06	2.36	2.40	1.88	2.50
3	2.38	3.04	2.78	2.17	2.69	2.34	2.51	2.70

On average nitrogen slightly reduced phosphorus and potassium in the swards, phosphorus raised the phosphorus and potassium raised the potassium content. The third cut, with or without nitrogen, had the highest phosphorus content, but potassium content lacked uniformity of response to fertiliser or stage of growth.

The effects of nitrogenous fertiliser on a mixed sward are seen in more detail in Hemingway's plot studies (1961; 1962) in which grasses and clover were hand separated for analysis. The sward was predominantly of ryegrass, cocksfoot, timothy and wild white clover and four cuts were taken for silage in each of three successive years, the first in late May and the last in early September. Fertilisers were in kg per hectare: 500 sodium chloride, 250 magnesium sulphate, 375 ammonium sulphate, 375 superphosphate and 250 muriate of potash. All were applied in Spring, nitrogen also after each cut. In the third year (1959) potassium values had fallen as low as 0.04 per cent in the plots not given potash, there was potassium deficiency and an invasion of *Agrostis tenuis* and *Festuca rubra,* replacing the sown grasses. In the plots given ammonium sulphate, clover disappeared after the first year. Both those changes in the flora must have affected the changes in the composition of the herbage. The analysis of grass and clover separately makes clear the effect of suppression of legumes, but the invasion of red fescue may have modified the composition of the grass portion of the herbage. The data are presented *in extenso* in the tables for mixed grasses and white clover.

In brief summary, the results of the analysis of grasses suggest that season, i.e. number of cut, had more effect on calcium content than any of the fertilisers. Magnesium content was least when potash was applied. Potassium showed no effect of salt and was conspicuously low in the third year except where potash was applied. Later cuts had more sodium than early; sodium tended to vary inversely with potassium and potash depressed sodium.

Copper, molybdenum and iron rose in the later cuts. Nitrogen raised copper but the effect was less than that of cut. Nitrogen depressed molybdenum. Phosphorus raised iron. Manganese was raised by nitrogen and phosphorus together and depressed by potash; it showed no consistent change with cut.

In clover, calcium fell from year to year and in 1959 was less than half what it was in 1957. In each year it fell with cut. There was little difference between fertilisers.

Potassium in clover showed no consistent change with year or cut. Compared with clover without fertiliser, potash increased potassium content from a mean of 1.75 to 2.64 per cent of dry matter, a rise of 51 per cent.

Magnesium fell with cut in 1958 but not in the other two years; it was less in the third year, especially with sodium chloride as fertiliser. Magnesium sulphate raised magnesium content by only 7 per cent; there was no effect of the other fertilisers.

Sodium fell with cut except with ammonium sulphate in the first year. Thereafter there was a marked rise, greatest in the third year, except in the plots given potash where there was a great reduction. Sodium chloride raised the sodium content of clover by 47 per cent and potash reduced it to 48 per cent of that without fertiliser.

Of the trace elements, without added nitrogen copper rose with cut from about 11 μg per g dry matter to about 14, according to Hemingway, but the effect was almost entirely in the first year where the rise was from a mean of 10.5 in the first cut to 16.7 μg per g in the fourth. With added nitrogen in the first year the rise was bigger, from 11 to 19 μg per g. Superphosphate and potash had little or no effect.

Clover contained more molybdenum than grass in the first two years but not in the third. There was little change with cut. Potash reduced molybdenum: 1.6 μg per g was the mean for the three years compared with 2.6 without fertiliser and 2.1 with phosphate.

Table 0.5

Effect of ammonium sulphate on sulphur of herbage of sward of mixed ryegrasses, timothy, cocksfoot and clovers
SS = sulphate sulphur, TS = total sulphur: per cent of dry matter

		Ammonium sulphate: kg per hectare in single (1) or divided (2) applications								
		0	250	500 (1)	500 (2)	750	1000 (1)	1000 (2)	1500 (1)	1500 (2)
1950	SS	0.138	0.187	0.193	0.209	0.200	0.182	0.178	0.145	0.161
	TS	0.252	0.348	0.356	0.341	0.357	0.333	0.347	0.357	0.358
1953	SS	0.132	0.255	0.257	0.262	0.296	0.255	0.282	0.230	0.270
	TS	0.261	0.378	0.385	0.389	0.437	0.424	0.416	0.437	0.424

Manganese rose with cut when no fertiliser was given, but not consistently with phosphate or potash. Phosphate and potash raised manganese in the first two years but not in the third when the control level was higher. There was no effect on the mean.

Iron rose with cut in 1957 and 1958 in the absence of nitrogen, but not in 1959 when all the values were lower. Phosphate slightly raised and potash reduced iron.

Hemingway summarises the results as showing that the content of those four elements in herbage is determined chiefly by the relative amounts of grass and clover when the samples are cut, and the effect of nitrogen. "From the viewpoint of animal health, the seasonal changes in copper content are probably of greatest interest . . .Heavy applications of nitrogen, which may suppress clover, will themselves increase grass copper . . .reduce the molybdenum level . . . and greatly increase the manganese content."

The effect of ammonium sulphate on the sulphur content of herbage was recorded (Jones, 1960) on four swards of different components in 1950 the first, and 1953 the fourth, year of treatment. In 1950, cuts were taken between the 23rd May and the 14th June; in 1953 between the 20th and 27th June. The results for two of the swards are shown in Tables 0.5, 0.6. Data for the other two are presented in Table 1.93.

Table 0.6 Effect of ammonium sulphate on sulphur of herbage of timothy S48, meadow fescue S215, clover white S100 and wild white clover

SS = sulphate sulphur, TS = total sulphur: per cent of dry matter

		Ammonium sulphate: kg per hectare in single (1) or divided (2) applications								
		0	250	500 (1)	500 (2)	750	1000 (1)	1100 (2)	1500 (1)	1500 (2)
1950	SS	0.091	0.131	0.163	0.136	0.172	0.188	0.159	0.166	0.165
	TS	0.163	0.255	0.308	0.270	0.277	0.336	0.277	0.312	0.287
1953	SS	0.124	0.198	0.236	0.225	0.250	0.253	0.257	0.216	0.270
	TS	0.257	0.317	0.431	0.390	0.487	0.470	0.405	0.475	0.483

Points of interest are the difference in behaviour of the two swards and of the two fractions, organic and inorganic sulphur. Sulphate sulphur seems to reach a plateau at the lowest levels of fertiliser, a lower plateau in the timothy and fescue sward than in the ryegrass and cocksfoot sward. But organic sulphur rises progressively with level of fertiliser, in timothy and fescue more rapidly and to a higher level than in the ryegrass sward. Since it is unlikely that crude protein would rise at similar rates, it would be interesting to know with what the sulphur is coupled. According to Jones "the change in the grass to clover ratio from 1950 to 1953 has had little effect on the sulphate levels, probably as a result of an abundant supply of soil sulphate. There is, however, a slight increase in the non-sulphate sulphur in the control plots which might be expected from an increase in the clover content of the swards".

The improvement of poor pasture

Procedures are of three main types: the liming of acid soils; ploughing, application of the usual fertilisers and reseeding of previously uncultivated or old pastures; and the correction of localised deficiencies or excesses of individual mineral elements.

Fagan, Jones, Williams and Davies (1940) describe the effects of liming on the composition of the herbage on acidic soils in 7 centres in Merionethshire. Results are summarised in Table 0.7.

As the result of application of lime, crude protein rose in all except the herbage on one of the less acid soils, No.6. Calcium rose in every case. Phosphorus and potassium rose in the pastures from highly acid soils; herbage from the others showed little change in phosphorus, and potassium fell.

In a repeat survey of three of the farms, three to six years later, without further treatment, the improvements were well maintained (Fagan, Jones, Williams and Davies, 1945).

At the opposite extreme of soil type there are plot studies (Norman, 1956) of the effect of phosphorus and potash fertilizers on the crude protein and phosphorus content of pasture on chalk. The flora was mainly of red fescue *(Festuca rubra),* creeping bent *(Agrostis stolonifera)* and tall oat grass *(Arrhenatherum elatius),* with some other grasses, herbs and legumes. All plots were given in kg per hectare 125 muriate of potash and 250 x 2 Nitrochalk annually, and from 125 to 1125 superphosphate, all at once or in three parts, annually in 1952, 1953 and 1954. The plots were grazed by sheep.

Table 0.7 Effect of liming on composition of herbage on acidic soils: means of 3 cuts

Centre		Fertiliser			Percentage increase	
		None	Ground Lime	Ground Limestone	Lime	Limestone
1	kg/hectare	–	156	280		
	CP % of DM	10.8	13.8	13.4	28	24
	P ”	0.12	0.19	0.17	58	42
	Ca ”	0.26	0.47	0.56	81	115
	K ”	1.70	2.23	1.97	31	16
2	kg/hectare	–	187	325		
	CP % of DM	10.8	15.9	13.3	47	23
	P ”	0.29	0.35	0.40	21	38
	Ca ”	0.55	0.63	0.86	15	56
	K ”	1.29	1.15	1.99	–	54
3	kg/hectare	–	187	325		
	CP % of DM	12.8	16.2	15.8	26	23
	P ”	0.22	0.26	0.34	18	55
	Ca ”	0.29	0.57	0.65	97	124
	K ”	1.99	2.22	2.54	12	28
4	kg/hectare	–	206	362		
	CP % of DM	13.7	16.8	15.0	23	10
	P ”	0.21	0.27	0.21	29	–
	Ca ”	0.26	0.69	0.71	166	173
	K ”	1.90	2.28	1.99	20	5
5	kg/hectare	–	250	437		
	CP % of DM	18.4	21.6	20.8	17	13
	P ”	0.35	0.29	0.30	–	–
	Ca ”	0.53	0.68	0.74	28	40
	K ”	2.14	1.92	1.78	–	–
6	kg/hectare	–	250	437		
	CP % of DM	15.9	14.8	15.1	–	–
	P ”	0.26	0.24	0.28	–	8
	Ca ”	0.62	0.83	0.84	34	35
	K ”	2.11	1.78	1.99	–	–
7	kg/hectare	–	175	313		
	CP % of DM	13.2	15.3	13.7	16	4
	P ”	0.24	0.27	0.27	12	12
	Ca ”	0.58	0.69	0.76	19	31
	K ”	1.46	1.39	1.07	–	–

There was little difference between the different levels of phosphate and the results are summarised for the weighted means of 5 cuts in 1952 and 1953 and 6 cuts in 1954 (Table 0.8).

The fertilizers raised phosphorus content in each year and crude protein in the second and third years. The flora was not greatly changed. *Dactylis glomerata, Festuca rubra, Arrhenatherum elatius, Poa trivialis, Holcus lanatus, Lolium perenne* and *Trifolium repens* increased their percentage of cover; others were unaffected or decreased.

Hypomagnesaemia. Field, McCallum and Butler (1958) made a study of permanent pasture to which different fertilisers had been applied. The fields were grazed by sheep and hypomagnesaemia was present on fields 6 and 7 a week before sampling and on field 8 at the time of collection and occurred regularly on that field. (Table 0.9).

Table 0.8 Effect of P and K fertilisers on pasture on chalk

Per cent of DM	1952 CP	1952 P	1953 CP	1953 P	1954 CP	1954 P
Control	12.8	0.23	12.8	0.18	13.9	0.23
Treated	12.8	0.27	14.6	0.25	15.7	0.28

Similar values for Mg in cultivated pasture on which hypomagnesaemia occurred in grazing dairy cows are given by Line *et al* (1958). Sampled between the 24th April and 9th May, and between the 9th and 14th May, 1956, the values, per cent of dry matter, were 0.135 and 0.182; between the 1st and 18th April, 1957, 0.199 and 0.183. At the same time crude protein was between 18.8 and 23.5 per cent. See also Butler *et al* 1963.

hypomagnesaemia. The results tabulated show the effect of magnesian limestone (Mg carbonate 44.3, Ca carbonate 46.7, silica 4.5 per cent) on the composition and quality of the silage (Table 0.11).

Each of the values tabulated above is the mean for three samples of the silage, which differed slightly in quality.

Table 0.9 Comparison of the effects of different fertilisers on Mg, Ca and CP of permanent pasture grazed by sheep

Field	Fertiliser	Time of sample	Mg (Per cent of DM)	Ca (Per cent of DM)	CP (Per cent of DM)
1	Nitrochalk	May	0.094	0.548	11.6
3	Compound	June	0.173	0.783	21.9
6	Potato	May	0.105	0.737	16.6
			0.109	0.711	
7	Potato	May	0.118	0.683	16.4
			0.142	0.645	
8	Sulph. ammon. pot. superphos.	May	0.155	0.650	19.4
			0.147	0.663	
8	” ” ” ”	May	0.169	0.765	20.4
			0.175	0.757	

Hunt, Alexander and Rutherford (1964) showed in a plot experiment that nitrogenous fertiliser and magnesium sulphate improved the Mg content of pasture and potash reduced it. The effect of magnesium sulphate in Kieserite (17.4%K) on the magnesium content is shown in Table 0.10.

The sward was of perennial ryegrass and red and white clover; clover, per cent of fresh weight, at the four sampling dates was: 0.7, 1.3, 0.8 and 1.3.

Ibbotson (1963) experimented with the addition of magnesian limestone or magnesium oxide to herbage as it was ensiled, as a possible measure to prevent

The conclusion was that addition of magnesian limestone at the time of ensiling had little effect on the quality of the silage and greatly increased both calcium and magnesium content.

Aphosphorosis in Ireland. An extensive study was made in 1947 and 1948, in the midlands of Ireland, on the composition of pastures on which aphosphorosis occurred in cattle (Sheehy, O'Donovan, Day and Curran, 1948; Curran, 1949).

In the first study of composition in Co. Offaly, moor and upland pastures on which aphosphorosis occurred were compared with good permanent pasture in the same county and with "prime" pasture in Meath and Dublin. Means are shown in Table 0.12.

Table 0.10 Effect of magnesium sulphate on magnesium content of pasture

Date of sampling:	April 21	April 28	May 5	May 12
DM% of fresh	19.6	19.2	16.9	20.3
CP% of DM	18.9	15.7	13.5	11.5
	Magnesium: per cent of DM			
Fertiliser				
None	0.150	0.147	0.136	0.137
500 kg per hectare	0.180	0.168	0.156	0.156
1500 kg per hectare	0.212	0.204	0.185	0.177

Table 0.11

Magnesian limestone as additive to silage
per cent of dry matter

Type of silage	CP	Ca	Mg	pH	Quality
Grass: untreated	9.1	0.51	0.14	4.00	Excellent
treated	10.5	1.48	0.43	4.10	Good
" untreated	9.1	1.06	0.38	3.82	Satisfactory
treated	8.8	3.06	1.12	4.20	Good
Oats, barley, vetches:					
untreated	11.2	1.08	0.27	3.95	Satisfactory
treated	12.7	1.72	0.71	5.18	Good
Grass and clover:					
untreated	15.1	0.88	0.37	4.83	Good
treated	13.7	1.37	0.93	5.23	Satisfactory
Grass and clover:					
untreated	15.3	0.85	0.19	4.35	Satisfactory
treated	11.1	3.08	1.06	5.40	Satisfactory

The poor pastures, in addition to P, appear to lack Cl and protein, but not Ca or nitrogen-free extract. Digestibility of organic matter was not studied.

The effect on phosphorus content of application of superphosphate to poor pastures was studied in plot experiments at 8 centres. The maximum application of 1125 kg per hectare raised the phosphorus content only to 0.332 per cent of dry matter, still below that of good pasture.

In the second study (Curran, 1949), a total of 178 upland and moor pastures were sampled, with a range for phosphorus content from 0.070 to 0.288 per cent of dry matter. Six pastures treated with superphosphate gave values between 0.218 and 0.332 and six good upland pastures between 0.345 and 0.485 per cent of dry matter.

Analyses of individual species from plots, untreated and treated with superphosphate, showed considerable differences between species in the same plot and between the responses of species to application of phosphate. Yorkshire Fog and Crested Dogstail from initial values of 0.16 and 0.11 per cent of dry matter each increased phosphorus content by 76 per cent; wild white clover from 0.20 increased by 74 per cent and red clover from 0.13 by 83 per cent, self heal from 0.20 by 65 per cent, a plantain from 0.13 by 85 per cent, but sedges, two samples with 0.20 and 0.15 per cent initially, increased by only 33 and 26 per cent. Previous studies of natural moor pastures, or "sedge pastures", showed low phosphorus and surprisingly high calcium contents. Means and ranges of P and Ca for 17 samples in August, per cent of dry matter, were: 0.11 (range 0.09–0.15) and 0.71 (range 0.57–0.79); and in November 0.13 (range 0.08–0.18) and 0.63 (range 0.51–0.79) (Sheehy *et al*, 1948). Samples from different Counties gave similar values for phosphorus and crude protein values between 8.0 and 16.6 per cent. Application of superphosphate in each County gave phosphorus content from 0.22 to 0.33 and crude protein from 15.9 to 20.8 per cent (Curran, 1949).

Analyses of sedges in Mid-Wales gave a similar low phosphorus content but calcium was low also. (Table 0.13).

Table 0.12

Aphosphorosis in cattle: composition of pasture
Means: per cent of DM

		CP	EE	NFE	CF	Ash	Ca	P	Cl
Offaly:	moor	11.1	2.0	53.5	26.1	7.3	0.86	0.13	0.40
	upland	10.8	1.9	51.6	25.3	10.4	1.22	0.15	0.64
	good permanent	19.8	2.0	45.4	21.8	11.0	0.86	0.40	0.93
Meath:	"prime"	21.3	1.6	42.3	23.4	11.4	0.66	0.52	1.33
Dublin:	"prime"	22.6	2.2	42.0	21.2	12.0	0.59	0.49	1.35

Table 0.13 Composition of sedges (*Carex* sp.) in Mid-Wales (Fagan and Watkins, 1932)

CP	EE	NFE	CF	Ash	Ca	P	K	Cl
7.4	4.8	53.3	30.2	4.3	0.14	0.13	1.53	0.31

Manurial trials by Castle and Drysdale (1962) show how difficult it may be to judge the true causes, or indeed the occurrence, of change in the composition of herbage. They began with a sward of meadow fescue S53 and white clover S100, the clover showing ground cover varying from 9 to 47 per cent. At the end of 3 years there was an invasion of 19–30 per cent of other grasses and between 29 and 36 per cent of bare ground. Cuts were taken 5 times in each year at 15–20 cm height, leafy growth, and there were 4 replicates of the plots.

The main interest lies in the low levels of sodium in the first year in all the manured samples; range 0.07 to 0.21, with 4 less than 0.13 per cent of dry matter. But still lower values are reported by ap Griffith, Jones and Walters (1965) in a comparison of species and varieties of grasses. For instance, 8 samples of *Festuca pratensis* S215 gave a range from 0.025 to 0.076, mean 0.044; and 8 of *F. arundinacea* S170 from 0.062 to 0.154, mean 0.097, all but 1 below the estimated deficiency level of 0.13.

In all 7 grasses were analysed over the season, with the following results for sodium content of those other than the Fescues shown above.

		% of dry matter	
Phleum pratense S48	:	0.018 to 0.055;	mean 0.030
Dactylis glomerata S37	:	0.205 to 0.301;	mean 0.246
Lolium perenne S24	:	0.131 to 0.357;	mean 0.269
,, ,, S23	:	0.126 to 0.228;	mean 0.189
Lolium multiflorum S22	:	0.060 to 0.138;	mean 0.103

The analyses are of pure swards on replicated plots, close together and uniformly treated with 375 kg per hectare of NPK (12:12:18) and 375 kg per hectare of Nitrochalk. The fertilisers are complicating, but species differences do seem sufficiently large to affect the sodium content of a mixed sward.

Teart pastures. Analyses of Somerset pastures (Ferguson, Lewis and Watson, 1943) over 4 years show the difference in molybdenum content between normal grazings and those on which "teart" occurred. In 1937, 11 samples of normal grazing from 7 farms gave a mean of 5.5 and a range of 3–12 μg per g dry matter; in 1938, 31 samples from 4 farms gave a mean and range of 4.4 and 4–5 μg per g. At the same times, 24 samples from 8 teart farms averaged 35 with a range of 14–52 μg per g in 1937 and in the next 9 samples from 4 farms gave a mean and range of range of 44 and 18–59. The seasonal variation on one farm is shown in Table 0.14.

Table 0.14 Molybdenum in teart herbage: μg per g DM

	Apr.	May	June	July	Aug.	Sept.	Oct.	Nov.	Dec.
1937	32	34	26	48		80			29 (frosted)
1938		42	49		66		79		
1939	49	42		53	93	66			
,,	45	66		87	108	102			
1940 *	74		62		68				

* Extraordinarily dry season.

Values for Cu in both teart and normal pasture were high, 7.6–17.8 and 8.0–15.0 μg per g.

Since the problem of excess molybdenum has now been resolved, it seems unnecessary to deal more fully with the older literature.

Swayback. Before the role of copper in myelination had been clearly described, Shearer and McDougall (1944) reported on the copper and lead contents of pastures grazed by sheep and subject to the occurrence of swayback in lambs. The following means and ranges, converted to μg per g dry matter, were found:

Table 0.15 Swayback: copper and lead in pasture

	Samples	Copper (μg/g DM) Mean	Copper (μg/g DM) Range	Lead (μg/g DM) Mean	Lead (μg/g DM) Range
Northumberland, old permanent pasture	6	14.6	11.4 - 19.6	14.0	9.0 - 21.0
Bristol area	8	10.7	4.4 - 18.0	18.0	7.0 - 27.0
Derbyshire, old pasture	1	22.2		7.0	
" ploughed and reseeded	2	16.6, 16.6		23.0, 24.2	
" moor plants: sheep's fescue and bilberry *(Festuca ovina* and *Vaccinium myrtillus)*	1	21.8		91.1	
Buckinghamshire	4	7.7	4.9 - 10.8		
	2			11.0, 11.0	

Since Shearer and McDougall worked on swayback, the estimation of Cu in pasture has become almost a routine. A great deal of information will be found in the main tables. Also, it is now established that swayback is due to deficiency of copper but, within the time span covered here, no comparative study of copper in normal and affected pasture appeared.

Sodium. A survey was made of estimates of sodium in pasture with a view to assessing the possible incidence of deficiency, for which purpose 0.13 per cent of dry matter was taken as the lower limit of normality. A more detailed survey of the data adds information on the possible part of fertilisers in inducing some of the deficiencies.

A pasture cut regularly for hay, chief components ryegrass, agrostis and *Poa trivialis,* and treated with basic slag, lime, sulphate of ammonia and Nitrochalk in successive years, was sampled 17 times between the 21st of May and the 31st of December. There was 1 sample with 0.07 per cent, that is less than the postulated normal minimum. The range of the other 16 samples was 0.13 to 0.50 per cent, with both the lowest and the highest values in November samples (Ferguson, 1931).

In another study (Stewart and Holmes, 1953) a ryegrass, timothy and clover sward was given superphosphate or muriate of potash or both, each with or without Nitrochalk. In the first series of plots, as shown in the Table 0.16, no nitrogen was given; in the second, Nitrochalk was given at the rate of 750 kg per hectare in March and after each cut. Muriate of potash was at a rate equivalent to 297 kg K_20 per hectare in 1949 and 1950 and 440 kg in 1951; superphosphate at a rate equivalent to 132 kg P_20_5 per hectare in 1949 and 1950 and 198 kg in 1951. There were 4 cuts without nitrogen and 5 cuts with nitrogen. Averages are given.

Table 0.16 Effect of fertilisers on sodium content of mixed sward
Sodium per cent of dry matter: means of 4 or 5 cuts

	Fertilisers: None	K	P	PK
No nitrogen				
1949	1.03	0.35	0.65	0.34
1950	0.97	0.23	0.68	0.30
1951	1.08	0.11	0.69	0.19
Nitrogen as Nitrochalk				
1949	1.17	0.96	1.31	0.94
1950	1.05	0.62	1.33	0.73
1951	0.58	0.28	1.07	0.25

Marked changes occurred in the botanical composition of the sward. The sward deteriorated in the plots without potash, whether or not they were given nitrogen. The sown species suffered replacement by *Agrostis tenuis, Festuca rubra* and *Holcus lanatus*. Potash gave distinct improvement with increase of clover; potash and phosphate together produced a clover dominant sward. Whatever the effect of those changes, nitrogen raised and potash depressed sodium, except in the third year with N alone, and even when the levels of phosphorus and potassium were raised.

Those results are confirmed and extended by Hemingway's (1961, 1962) analyses of a sward of grasses and clover, detailed above. The changes in sodium content were large and consistent, and occurred both in grass and clover. In the grasses in each year regardless of fertiliser, sodium content rose as the season advanced. Magnesium sulphate and superphosphate had little effect; sodium chloride gave a mean increase of about 50 per cent and ammonium sulphate roughly doubled the control value. Potash reduced sodium, on the average to half the control value and to much less in the first cut. The values for the first cut in each year and for the second cut in the first year were below the assumed normal minimum.

In clover the results differed only in detail. In the first year there was no trend with cut, i.e. with advance of season, but in the second and third years sodium content rose with number of cut. In the plots treated with ammonium sulphate clover disappeared after the first year, and such disappearance, though it does not complicate the present results since grass and clover were hand separated and analysed separately, is certainly a complication in similar fertiliser trials where the herbage is analysed as sampled.

In the first year the sodium content of the clover rose by 50 per cent with sodium chloride, by 30 per cent with ammonium sulphate, and fell by 30 per cent with potash. Magnesium and phosphate had no effect, and the same was true in the next two years. On the average of the three years sodium chloride raised sodium content by 38 per cent and potash reduced it to 36 per cent of the control value. All but 1 of the 8 values for the second and third years were below the deficiency level of 0.13 per cent.

In three studies (Woodman, Evans and Norman, 1933; 1934; Woodman and Eden, 1935) the sodium content of lucerne varied widely without obvious trend. In the variety Provence, without fertiliser, it varied between 0.12 and 0.30 per cent in whole plant, between 0.007 and 0.35 in leaf alone. In Hungarian, it varied between 0.02 and 0.30 per cent. In Provence, on a site that had had an application of chalk at one time and of potash more recently, the range did not differ: 0.04 to 0.27. Of the 36 values recorded 14 were below 0.13 per cent of dry matter.

The sodium content of moor plants varies within very wide limits. Some have high sodium contents and, for that reason, their invasion of cultivated pasture may be important. The following examples illustrate the range.

Table 0.17 Sodium content of moor plants: per cent of dry matter

Calluna vulgaris: Heather 24 samples: mean 0.058, range 0.037 – 0.083	(Thomas, Escritt and Trinder, 1945)
Agrostis setacea: Heath bent 6 samples: mean 0.10, range 0.07 – 0.13	(Dougall, 1954)
Achillea millefolium: Yarrow 3 samples: 0.02, 0.05, 0.05	(Thomas, *et al.*, 1952)
Plantago lanceolata: Ribwort plantain 6 samples: mean 0.70, range 0.36 – 0.86	(Thomas and Thompson, 1948)
Ranunculus spp: Buttercup 6 samples: mean 0.22, range 0.16 – 0.24	(Thomas and Thompson, 1948)
Rumex acetosella: Sheep's sorrel 6 samples: mean 0.24, range 0.18 – 0.31	(Thomas and Thompson, 1948)
Rhinanthus Christa-galli: Yellow rattle 6 samples: mean 0.30, range 0.16 – 0.36	(Thomas and Thompson, 1948)
Cynosurus cristatus: Crested dogstail 6 samples: mean 0.24, range 0.19 – 0.30	(Thomas and Thompson, 1948)
Holcus lanatus: Yorkshire fog (Cut 1) 6 samples: mean 0.16, range 0.12 – 0.23	(Thomas and Thompson, 1948)
Holcus lanatus: Yorkshire fog (Cut 2) 6 samples: mean 0.19, range 0.13 – 0.23	(Thomas and Thompson, 1948)
Festuca rubra: Red fescue 6 samples: mean 0.41, range 0.35 – 0.48	(Thomas and Thompson, 1948)

Clearly, any considerable invasion of red fescue into a sward lacking potash, as described above, would alter the sodium content of the sward and the change might be ascribed to other causes.

Digestibility

Since neither the crude fibre nor the crude protein content of herbage was found to provide a satisfactory index of quality, attention turned to digestibility, which had the added importance that voluntary intake of herbage is related to digestibility. But the digestibility of any given sample of forage depends on the stage of growth at which it is harvested and the subsequent treatment. Also, it differs in detail from one experimental animal to another and, to some extent, with the level of feeding in the same animal.

Composition of herbage. More attention has been given to the relation of digestibility of herbage to its fibre content than to other constituents. We are indebted to Mr. F. Raymond (letter, 1964) for a retabulation of records of digestibility of organic matter and crude fibre in 90 grasses* and to Mr. G. Alderman (1966) for the details of a study of the composition and digestibility of 47 hays. These will serve to illustrate the relation of digestibility to crude fibre and other components of herbage. The terminology is that of Raymond or Alderman and differs from that used elsewhere in this survey.

The regression of digestibility of organic matter (OMD) on crude fibre (CF) in Raymond's 90 grasses gives the following equation:

$$OMD=108.1-1.53\ CF$$

with a residual standard deviation of ±3.1 per cent. The coefficient of correlation was about 0.8, sufficient to show that, in general, digestibility falls as crude fibre rises, but of little predictive value for the individual sample.

In the study of 47 hays Alderman and his colleagues related the digestible organic matter as per cent of dry matter to crude fibre by the equation:

$$DOMD=98.6-1.26\ CF$$

with a residual standard deviation of 4.6 and a coefficient of variation of 8.2 per cent. The inclusion of crude protein in the equation made little difference:

$$DOMD=86.2+0.73CP-1.10\ CF$$

with a residual standard deviation of 4.4 and a coefficient of variation of 7.9 per cent.

Alderman gives also the following table of correlations between digestible organic matter per cent of dry matter, and the several components estimated by him and his colleagues.

Table 0.18 Correlation coefficients with digestible organic matter as per cent of dry matter (DOMD)

Component		Coefficient
Crude protein		0.439
Crude fibre		-0.646
Soluble carbohydrate	Deriaz *	0.199
Pentosan	Deriaz *	-0.399
Hexosan	Deriaz *	-0.458
Lignin	Deriaz *	-0.768
Trichloroacetic acid fibre		-0.492
N-acid fibre, 1 hour		-0.585
N-acid fibre, 3 hour		-0.675
Pepsin HCl digestible crude protein		0.603
Modified acid detergent fibre		-0.724

* Deriaz, R.E. (1961) J. Sci. Food Agric., **12**, 152.

Only the correlation with soluble carbohydrate was not significant. The closest correlation was with lignin, but, for individual samples, prediction on that basis would be of little value.

Obviously none of these coefficients is of such predictive value as to give confidence for individual samples of herbage. There is also the disadvantage attached to the estimation of digestibility by animal experiment that the *in vivo* technique is both laborious and expensive c.f. Blaxter, Graham and Wainman (1956); Greenhalgh, Reid and McDonald (1966).

An *in vitro* technique to replace it was therefore devised by Tilley, Terry, Deriaz and Outen and described in the 16th Annual Report of the Grassland Research Institute, 1962-63, pp. 64-65.

Digestibility in vitro. Since then a great deal of work has been done with the *in vitro* method as described or slightly modified. The technique consists of incubation of a 0.5 g sample of grass, or other material, first with rumen liquor at pH 6.7 to 6.9 and then with pepsin and hydrochloric acid. The pH during incubation with rumen liquor must be accurately controlled.

*Digestibilities of cocksfoot S37, ryegrass S23 and S24 in the main tables against Reference 357 are not those published, but are as recalculated.

The relation of digestibility *in vitro* to that *in vivo* has been studied at several centres. Tilley and Terry (1963) give the following equation for 130 samples of grass:

$$y = 0.99x - 1.01 \text{ (S.E. } \pm 2.31) \text{ .... (a)}$$

where y is percentage digestibility of dry matter *in vivo* and x is percentage digestibility of dry matter *in vitro*. For 18 samples of lucerne and clover the corresponding equation was:

$$y = 0.85x + 8.37 \text{ (S.E. } \pm 1.42) \text{ ....(b)}$$

The range of digestibilities was from 55.0 to 77.9 per cent *in vivo* and over that range the equations were not significantly different.

A cooperative study at 2 centres of the use of the method with fodder crops, maize, rape, radish, cabbage and kale, showed excellent agreement between the Grassland Research Institute and the National Institute of Agricultural Botany, and a similar relation between *in vitro* and *in vivo* estimates, the *in vitro* slightly higher with 2 exceptions, rape and the stem of thousand-headed kale, which were better digested *in vivo* (Dent, 1963).

Raymond (1963) gives another equation relating the digestibilities:

$$x = 0.97y + 0.38 \text{ ....(c)}$$

where x is percentage digestibility of dry matter *in vivo* and y percentage digestibility of dry matter *in vitro*. The range of digestibilities *in vivo* in that series was about 46 to 83 per cent. The similarity of the results with the three equations over the common range is illustrated by the Table 0.19.

Alderman and his colleagues give the following equation relating digestible organic matter per cent of dry matter *in vivo* (DOMD) to that *in vitro* (IVD):

$$\text{DOMD} = 0.80 \text{ IVD} + 11.8$$

with RSD 3.0 and coefficient of variation 5.4 per cent. The correlation between the two estimates was 0.866. Raymond has found an overall correlation of 0.97 between the two methods applied to samples of herbage (letter, 1964).

Alderman presents equations involving *in vitro* digestibility and either crude protein or crude fibre, which predict *in vivo* digestibility with SD of 3.0 in each case. Since the inclusion of either crude protein or crude fibre does not give any more information than the equation with *in vitro* digestibility alone, these equations are not presented here.

The Experimental Animal

Sheep: The animal most often used in digestibility trials is the sheep. The question whether age affects digestibility has been investigated at the Grassland Research Station in a series of 9 experiments (Raymond, Harris and Kemp, 1954). Male sheep of the same breeding, Halfbred x Suffolk, were fed in proportion to their maintenance requirements as assessed by Woodman, Evans and Eden (1937) according to this formula:

$$\text{Maintenance in lb SE per week} = \frac{\text{LW(lb)}}{20} + 4$$

All the sheep used were born in February or March and, up to 12 months old they were classed as lambs, from 12 to 24 months as one-year, and so on. The same animals were used in more than one experiment; in experiments 1 to 7, only 38 sheep in 72 trials. Feeds were cut and stored frozen in amounts sufficient for a trial. The results for digestibility of dry matter are collected here, commencing with Experiment 9 with the same 6 "lambs" over a year. Born, as the general description says, in February or March, they were 7 or 8 months old at the beginning of the experiment. (Table 0.20).

Table 0.19 — *In vivo* digestibility predicted from *in vitro* digestibility by equations (a), (b) and (c) in the text

In vitro digestibility: per cent	*In vivo* digestibility: per cent		
	(a)	(b)	(c)
55	53.4	55.1	53.7
65	63.3	63.6	63.4
75	73.1	72.1	73.1
80	78.2	76.4	78.0

Table 0.20 Effect of age on digestion by sheep over a year

Month of test	Digestibility of dry matter per cent: Mean	Range
Cocksfoot and clover: cut August, stored frozen		
October	79.0	76.3 - 81.1
January	78.0	76.8 - 79.8
March	79.0	77.5 - 80.1
May	78.4	77.5 - 79.7
July	79.4	78.6 - 80.2
September	80.0	78.7 - 81.5

The comment on those results was: " . . . there was a general upward trend with time and superimposed on this was a seasonal fluctuation. However, neither of these effects was consistent from sheep to sheep . . . it is not possible to predict that these effects will occur in other similar experiments".

Table 0.21 Effect of age on digestion by sheep of different ages

Expt.	Age	lamb	1 year	2 years	3 years	4 years	Number in group
		Digestibility of dry matter per cent					
1	Ryegrass and clover: cut July, stored frozen	67.3	69.4				2
2	Ryegrass and clover: cut June, stored frozen	71.9	73.3	75.1			3
3	Grass and clover: cut June, hay	63.4	63.3	63.5			3
4	Grass and clover: cut July, dried	67.3	68.7	69.4			3
5	Lucerne and grass: cut July, stored frozen, chopped	59.0	62.0	62.9	64.5		3
6	Cocksfoot and clover: cut May, stored frozen	76.4	76.2	77.8	78.0	77.8	3 or 4
7	Ryegrass and clover: cut May, fresh			78.0	78.6		3
	Ryegrass and clover: cut May, frozen			78.5	78.0		3
8	Ryegrass and clover: cut June, frozen						
	September 1949	75.4					4*
	April 1950	73.7					3
	August 1950	75.6					3

*Only 1 of the lambs was tested in all three trials. That lamb and two tested twice showed a fall of digestibility with time.

The following table of regression coefficients of digestibility upon age in years is given.

Table 0.22 Regression of digestibility on age of sheep

Experiment	Total data	Lambs, 1-year, 2-year only
1	2.05	–
2	1.60	1.60
3	0.05	0.05
4	1.10	1.10
5	1.76	2.00
6	0.47	0.70
7	0.05	–
Average regression coefficients (unweighted)		
Including (5)	1.01	1.09
Excluding (5)	0.89	0.86

"The average rise in digestibility found was about 1 unit for each year's increase in the age of the sheep; a similar rise of 1 unit was found in Experiment 9 for the same six sheep fed in October 1951 and September 1952. ...While the differences found with sheep are small, they are likely to be of importance as more precise digestibility data are required...While the present results do not justify the correction of digestibility for age, it seems advisable to be aware of the trend indicated".

It is suggested also that "the digestive efficiency of sheep decreases during the winter and rises during the following summer" with possible relation to fluctuations in rumen flora. But, assuming that digestibility increases with age, there is no evidence on which to assess the pattern of change, or how it may differ with change of diet.

Sheep and Cattle compared. Harkess (1963) compared the digestibility of mixed herbage by wether sheep and dairy cows, when intake was "slightly below voluntary intake". Ages are not stated (Table 0.23).

There was no significant difference between sheep and cow with any of the feeds.

Sheep and Rabbit compared. Of less interest to the farmer but of some importance to the technical investigator is the comparison of sheep and rabbits reported by Watson and Horton (1936). Digestibility of the dry matter of three grasses by sheep averaged 77.7, 73.1 and 72.2 per cent; by rabbits, 63.4, 63.0 and 62.8 per cent. The details of the difference are of some interest. The rabbits digested crude protein as well as did the sheep, but, on the average, only 46 per cent of ether extract compared with 52 per cent, 66 per cent of nitrogen-free extract against 78 per cent and 49 per cent of crude fibre compared with 80 per cent. In Watson and Godden's (1935) trials with a single sample of a mixed pasture, artificially dried, the corresponding percentage means for rabbit and sheep were: ether extract 26 and 52, nitrogen-free extract 56 and 81, crude fibre 26 and 75; but the rabbits were relatively inefficient also in digestion of crude protein with a mean coefficient of 62 compared with 77 for sheep (See also Metabolisable Energy,p.38). Clearly, if such results are typical, rabbits can not replace sheep in the estimation of the nutritive values of forage crops in farm practice.

Pigs and poultry compared with sheep

Examples will be found in the main tables and in tables 0.28 to 0.33 of computed metabolisable energy, of the composition of feeds and their digestibility by pigs and

Table 0.23 The digestibility per cent of the organic matter of three herbages (Modified from Table 7 of the original)

State of herbage	Animals per group	Collection (days)	Sheep	Cow	SE of difference
Fresh	3	12	81.4	81.7	± 0.70
Barn dried	3	16	70.6	71.5	± 1.03
Field cured	6	20	53.6	55.4	± 0.88

poultry. In some examples pig and sheep are compared. It is generally accepted, and the tabulated results are in agreement, that pigs and birds make little use of crude fibre. They digest protein and nitrogen-free extract at least as efficiently as do sheep. The proportion of ether extract digested by all of the animals varies widely, the variation probably due to differences in the chemical make-up of the ether extract. Oils and fats are well digested by all three and indeed an excess of oil or fat may interfere with the digestion of other components of the feed.

Effect of level of feeding on digestibility. Raymond, Harris and Kemp (1955) report 6 experiments of which the first was a cross-over trial with two sheep and was considered technically unsatisfactory. The remaining 5, each with 2 groups of comparable sheep fed on herbage stored frozen, at 2 different levels of intake, (H high, L low) are summarised in Table 0.24.

Table 0.24 Digestibility: effect of level of feeding

Feed	Wethers No. H	Wethers No. L	Age m.	Dry matter Intake: g/day H	Dry matter Intake: g/day L	Digestibility of DM: per cent H	Digestibility of DM: per cent L	Intake L per cent of H
Timothy	3	3	24	1357	1127	70.7	72.3	83
Ryegrass and Clover	3	3	18	1583	1303	69.4	72.1	82
Timothy and Fescue	4	4	42	1700	1300	73.1	73.4	76
Timothy and Fescue	6	6	12	653	455	72.0	73.3	70
Ryegrass and Clover	6	6	24	1085	827	75.9	76.2	76

The mean percentage digestibility of dry matter in the 5 experiments was 72.7 at high and 73.8 at low intake. Statistical tests indicated that the percentage digested at high intake was in general less than that at low intake. The difference is small, about 1 per cent, but was thought to be important on the grounds that digestibility estimated at maintenance level would over-estimate digestibility at production levels of intake.

Raymond, Minson and Harris (1959) in a second trial tested the effect of level of feeding on the digestibility of herbage stored frozen. There were 6 tests. In each, one group was fed at such a level that very little feed was refused (high level) and the other at approximately two-thirds of the high level. The results are summarised as follows in Table 0.25.

The authors conclude again that "the more herbage sheep consume, the less efficiently do they digest it"; the difference was attributed to more rapid passage of the feed through stomach and intestine.

Experiments by Harkess (1963) were quoted above in which sheep and cows were compared at approximately maximum voluntary intake. Comparisons were made also with both species between digestibilities of organic matter at that level and at half voluntary intake of barn dried hay, and 60 per cent of maximum voluntary intake of fresh herbage. The results are reproduced in Table 0.26. Digestibility by both species was greater with restricted feed, but the difference was significant only for sheep.

Campling, Freer and Balch (1963) in their series of studies of factors affecting the voluntary intake of feed by cows have one in which long hay and the same hay, ground and pelleted, are compared for digestibility at two levels (Table 0.27).

The digestibility of dry matter, organic matter and all components of long hay, except crude protein, was reduced by restriction of intake. Grinding and pelleting, which had little effect on proximate composition, greatly reduced digestibility at both levels of intake. For the ground and pelleted hay, restriction raised digestibility of dry matter and organic matter by raising the digestibility of nitrogen-free extract and crude fibre; that of ether extract was reduced.

Dent and Harris (1963) compared the digestibilities of silages of early, medium and late maturing maize *in vivo*

Table 0.25 Digestibility: effect of level of feeding

Feed	Wethers No. H	Wethers No. L	Age m.	Dry Matter Intake: g/day H	Dry Matter Intake: g/day L	Digestibility of DM: per cent H	Digestibility of DM: per cent L	Significance of difference
Ryegrass	12	12	23	1220	800	76.6 ± 0.39	78.3 ± 0.23	xxx
Ryegrass	11	12	15	1430	1020	73.5 ± 0.27	75.0 ± 0.34	xx
Ryegrass and Clover	12	12	18	950	640	70.1 ± 0.24	70.6 ± 0.27	ns
Permanent pasture	12	12	27	1510	1040	66.3 ± 0.32	67.5 ± 0.38	x
" "	5	6	17	1140	690	62.1 ± 0.78	64.1 ± 0.67	ns
Fibre residue +	7	7	14	900	540	60.3 ± 0.54	62.3 ± 0.44	x

+ Fibrous residue after extraction of protein

Table 0.26 Daily dry matter intake and digestibility of organic matter

	Barn dried hay		Fresh herbage	
	Intake: g	Digestibility: per cent	Intake: g	Digestibility: per cent
Sheep, high	1 360	70.6	1 483	81.4
Sheep, low	680	74.5	907	83.6
Cows, high	10 106	71.5	10 533	81.7
Cows, low	5 058	73.6	6 305	83.3

Table 0.27 Effect of processing on digestibility of hay

Digestibility per cent

Intake: g per day	D.M.	O.M.	C.P.	E.E.	N.F.E.	C.F.	Ash
			Long Hay				
9253	61.8	63.6	30.6	44.6	65.6	66.2	41.1
3946	59.0	61.8	36.5	36.6	63.4	65.6	37.3
			Hay, ground and pelleted				
9435	42.8	42.9	29.4	54.0	49.0	31.6	42.0
3946	46.0	47.7	29.4	45.8	54.8	38.8	22.7

with sheep and *in vitro* at two centres, the Grassland Research Station and the National Institute of Agricultural Botany. There was little or no difference *in vivo*. *In vitro* at both centres the late maturing maize was slightly better than the other two, 76.9 and 75.3 per cent, compared with 72.4 and 71.6 per cent. Later Harris (1965) compared the digestibilities by sheep of the same three types of maize, fresh and ensiled, with their protein contents and pH. There is not enough information to separate the possible effects of type and composition.

A substantial contribution towards solving the problems of digestibility has been made by Waite, Johnston and Armstrong (1964). Twelve cuts of 4 grasses were fed at different levels, from well below to roughly twice maintenance level, or near maximum voluntary intake. Results for the effect of level of feeding are presented for one grass only, ryegrass S23, at the lowest level "well below that needed for maintenance" and the highest "close to maximum voluntary intake", each compared on a percentage basis with coefficients of digestibility about maintenance level. In general, under-feeding increased and high level feeding reduced digestibility. In detail, differences due to underfeeding were greatest for ether soluble components and lignin and, of hemicellulose polysaccharides, greatest for xylan in the most mature cut. Differences due to high feeding were noticeable only in the glucan, galactan and uronic acid components, and only in the most mature cut when digestibility of organic matter had fallen from 86 to 62 per cent. "The decrease in digestibility of the organic matter at a feeding level approximately twice that required for maintenance was small even in the most mature grass, and as a factor governing digestibility it is clear that the stage of growth of the grass was much more important".

Metabolisable energy

It has been shown that content of crude protein is not a good index of quality of fodder, partly because the digestibility of crude protein varies within wide limits and partly because crude proteins of approximately equal digestibility do not always have the same biological value. It has been shown also that, although digestibility of dry matter tends to vary inversely with crude fibre, crude fibre is of little predictive value for quality. Nor is digestibility of dry matter or organic matter always a reliable criterion of feeding value since the nature of the digested nutrients varies with type of plant, stage of growth and treatment of the crop. Of the digested organic matter, a much higher proportion comes from the crude fibre of timothy than from that of ryegrass, and a much higher proportion from the crude protein of cocksfoot than from the crude protein of ryegrass. Since the energy values of the digested nutrients, for both maintenance and production, differ *inter se*, the sum of digested dry matter or digested organic matter in grams cannot accurately represent the energy value of a feeding stuff.

Those comments refer specifically to fodder plants but the argument applies to other plants also. The problem is most urgent for grassland products.

Before the assembly of data on the composition of British feeding stuffs began, the Technical Committee had decided that metabolisable energy was at once the most informative characteristic and the least subject to technical error (Agric. Res. Council, 1965). For those reasons the Technical Committee recommended that, if possible, estimates of metabolisable energy should be included in the tables to be compiled. From the data assembled a special study has been made of the estimation of metabolisable energy from the proximate constituents. The study is analytical and includes all material presented in sufficient detail.

It was further suggested by the Technical Committee that Axelsson's (1941) coefficients for the metabolisable energy of digested nutrients should be used. Before embarking on a long series of calculations of metabolisable energy, it seemed to us to be desirable to search for possible series in which metabolisable energy computed with Axelsson's coefficients for digested nutrients could be compared with metabolisable energy measured in metabolism experiments in which losses in urine and as methane were measured, or from nitrogen and carbon balances, or by direct calorimetry. No comparison exactly of the sort hoped for has been possible. The nearest approach was found in the studies of Armstrong and his colleagues (1964) of 12 cuts of 4 grasses in primary growth, the samples dried artificially and stored frozen. Proximate analyses in the conventional form are not published but, with Axelsson's coefficients for individual carbohydrates and some minor assumptions, a comparison was possible.

For the calculation of metabolisable energy in that series the following assumptions were made:

that the metabolisable energy of digested organic acids was 2.4 kcal (10.0 kJ) per g (FAO, 1947)

that the metabolisable energy of the "not isolated" fraction, since it was present "in the water extract", was the same as that of nitrogen-free extract, 3.7 kcal (15.5 kJ) per g

that the metabolisable energy of digested cellulose was 3.76 (15.7) and of digested hemicellulose 3.7 kcal (15.5 kJ) per g.

The comparison is shown below in summary; the extended calculations are in Table 1.227.

In view of the assumptions made for the calculations and of those regarding the identity of fractions of feed and faeces made in the estimation of "apparent digestibility", on which the detailed comparison depends, the excellence of the agreement between the Axelsson and the calorimetric estimate, and not the difference, is surprising. There is no suggestion here, as appears later, that Axelsson's coefficients under-estimate metabolisable energy. Possibly the chemical details of nitrogen-free extract, crude fibre, and "carbohydrate fractions" might repay further study.

All other relevant work of Dr. Blaxter and his team at the Hannah Dairy Research Institute (Blaxter, 1964) is expressed in terms of energy and so cannot be used for such comparisons.

Three other series of studies were examined in detail, those of Walker and Hepburn (1954-55) on 24 hays and (1956) on 24 silages, and that of Alderman and his

Table 0.28

Metabolisable energy of 12 cuts of 4 grasses, primary growth

Metabolisable energy: kcal or kJ per g dry matter

		Original estimate		With Axelsson's coefficients	
		kcal	kJ	kcal	kJ
Ryegrass S23	(1)	3.06	12.80	3.18	13.31
	(2)	2.75	11.51	2.98	12.47
	(3)	2.78	11.63	2.85	11.92
	(4)	2.37	9.92	2.28	9.54
Ryegrass S24	(1)	3.23	13.51	3.21	13.43
	(2)	2.86	11.97	2.86	11.97
Cocksfoot S37	(1)	3.09	12.93	3.11	13.01
	(2)	2.78	11.63	2.82	11.80
	(3)	2.55	10.67	2.61	10.92
Timothy S48	(1)	2.71	11.34	2.77	11.59
	(2)	2.48	10.38	2.51	10.50
	(3)	2.08	8.70	2.19	9.16
	Mean	2.73	11.42	2.78	11.63

colleagues (unpublished, 1966) on 47 hays. By manipulation of Walker and Hepburn's tables metabolisable energy in kcal per g of dry matter was computed and compared with estimates for which Axelsson's coefficients were applied to the proximate analyses. The comparisons are shown in Tables 0.31 and 0.32. The agreement is not good. Part at least of the difference probably lies in the formula used by Walker and Hepburn to compute loss of methane:

$$\text{Methane in g} = 2.41x + 9.80$$

where x = digested carbohydrate in units of 100 g and 1 g methane = 13.34 kcal, which underestimates the loss of energy in methane, as Walker and Hepburn say in the silage paper and as judged by recent direct estimation. It seems likely that Walker and Hepburn overestimated metabolisable energy. The extended data are in Table 1.229 and Table 1.231. References 516 (Hay) and 517 (Silage).

Alderman estimated digested energy by bomb calorimetry of feed and faeces and metabolisable energy by Armstrong's formula:

$$\text{ME} = 0.81 \times \text{digested energy}$$

In comparison with the estimates on that basis, Axelsson's coefficients applied to digested nutrients consistently underestimate metabolisable energy. The results are shown in Table 0.33 (extended data Table 1.230). In view of the much better agreement of Axelsson with the calorimetric estimates in the Armstrong series of dried grasses, it might be suggested that the formula derived from Armstrong's data is not suited for use with hays. The same suggestion arises from the table of frequency distribution of metabolisable energy per gram digested organic matter (see below, Table 0.34), where the difference between Armstrong's grasses and the other artificially dried grasses and the hays, Alderman's and the rest, is so great.

To test the measure of agreement that might be found with different coefficients, estimates were made for Walker and Hepburn's and Alderman's hays with Nehring's (1967) coefficients. The comparisons are shown in Tables 0.31 and 0.33. The coefficients are shown below.

Table 0.29

Metabolisable energy: coefficients

Metabolisable energy: kcal or kJ per gram digested nutrients

Axelsson (1941)	CP		EE		NFE		CF	
	kcal	kJ	kcal	kJ	kcal	kJ	kcal	kJ
Coarse fodders	4.3	18.0	7.8	32.6	3.7	15.5	2.9	12.1
Concentrates	4.5	18.8			3.7	15.5	2.9	12.1
Cereals			8.3	34.7				
Oilseeds			8.8	36.8				
Animal products			9.3	38.9				

Carbohydrates: kcal per gram	kcal	kJ
Polysaccharides	3.76	15.73
Trisaccharides	3.62	15.15
Disaccharides	3.56	14.90
Monosaccharides	3.38	14.14

FAO Committee on Calorie Conversion Factors (Washington, 1947)

Available organic acids: 2.45 kcal (10.25 kJ) per gram

Nehring (1967)		CP	EE	NFE	CF
Cattle *	kcal	3.63	8.17	3.87	3.06
	kJ	15.19	34.18	16.19	12.80
Pig	kcal	4.50	8.63	4.17	4.00
	kJ	18.83	36.11	17.45	16.74

*No equation is given for sheep. We have assumed that there is no significant difference between cattle and sheep in their use of the same feed.

In the comparisons made of the Axelsson and Nehring coefficients, since the digestibilities were with sheep, the occasion did not arise to use Nehring's coefficients for the pig.

With Nehring's coefficients the mean metabolisable energy of Walker and Hepburn's hays is 4.4 per cent below the mean of the original estimates, nearer than with Axelsson's coefficients which gave a mean difference of 7 per cent. For Alderman's series the original and Nehring estimates are generally in agreement except for a few values above about 2.40 kcal per g, where Armstrong's formula gives the higher estimate. Again, since there is a question of the applicability of the Armstrong formula, derived from pure stands of grass, artificially dried and stored frozen, to hays of mixed origin and varying quality, the comparison, like that of the Walker and Hepburn series, serves only to show that Nehring's coefficients give a slightly higher estimate of metabolisable energy than do Axelsson's. Nehring's coefficients cannot be applied to Armstrong's data, as Axelsson's particular coefficients were.

For those reasons we decided that the difference between estimates with Axelsson's coefficients and those of Nehring were not of a magnitude to warrant repetition of all the calculations already made for the tables of metabolisable energy. In those tables, all the values have been computed with Axelsson's coefficients.

Metabolisable energy per gram of dry matter. The simplest first approximation to metabolisable energy value would be with a formula based on proximate analysis only, without estimation of digestibility and without aid of calorimeter other than that embodied in accepted heats of combustion of proximate components of feed. The assembled computations of metabolisable energy were therefore examined to see whether such a formula could be found.

Table 0.30 Classification of feeds, metabolisable energy and prediction equations

Types of feed for which equations were obtained predicting ME from proximate constituents, all on a DM basis

Type	No. of observations	Mean ME kcal and kJ per g DM	SD as CV%	SD as CV% after prediction
Pasture, fresh	87	kcal 2.50 kJ 10.46	7.2	4.5
Pasture, artificially dried	30	kcal 2.33 kJ 9.75	13.2	7.3
Pasture, silage	148	kcal 2.40 kJ 10.04	10.5	7.9
Hays, grass with or without clover	96	kcal 2.07 kJ 8.66	10.2	5.7
Kale	15	kcal 2.72 kJ 11.38	12.9	7.4

Equations predicting ME as kcal and kJ per g DM from proximate constituents as percentage of DM

Material	Equation	Residual SD	CV %
Pasture, fresh	kcal ME = 3.16 – 0.027 CF	± 0.14	4.5
	kJ 13.22 – 0.113 CF	± 0.59	
Pasture, artificially dried	kcal ME = 2.57 + 0.15 EE – 0.026 CF	± 0.17	7.3
	kJ 10.75 + 0.63 EE – 0.109 CF	± 0.71	
Pasture, silage	kcal ME = 2.94 + 0.049 EE – 0.026 CF	± 0.19	7.9
	kJ 12.30 + 0.205 EE – 0.109 CF	± 0.79	
Hays	kcal ME = 3.04 + 0.023 CP – 0.038 CF	± 0.12	5.7
	kJ 12.72 + 0.096 CP – 0.159 CF	± 0.50	
Kale	kcal ME = 0.80 + 0.034 NFE	± 0.20	7.4
	kJ 3.35 + 0.142 NFE	± 0.84	

It was clear from a superficial examination of the data that it was out of the question to attempt a single prediction of metabolisable energy for all classes of feeds from information on proximate constituents. Separate equations were sought for those types of feed for which a reasonable number of observations with ruminants was available. The subdivisions into types and some other information are shown in Table 0.30.

The regression analyses were based on crude protein, ether extract, crude fibre, nitrogen-free extract by difference, and ash. For each regression equation only those variables were used which contributed significantly to information on metabolisable energy estimated according to Axelsson from the digested organic constituents. This does not mean that the variables excluded were not related to metabolisable energy. For fresh pasture, for example, there was a positive correlation between metabolisable energy and crude protein, but that between metabolisable energy and crude fibre was larger. There was also a negative correlation between crude fibre and crude protein such that, when the variability of metabolisable energy associated with crude fibre was accounted for, the inclusion of crude protein gave no additional information.

A point of special interest is that, for artificially dried pasture and silage, ether extract is of more importance than crude protein, which is the chief positive determinant of metabolisable energy of hay. The equation for kale is based on few observations.

These equations are not presented here as definitive, but to indicate that this would appear to be a useful approach for providing equations to estimate ME from the relatively easily obtained measures of proximate composition. The method we would suggest is that the range of feeds and feeding stuffs should be divided into a number of narrow classes, and that an adequate number of representative samples be analysed from each class (say 30 samples). Metabolism trials with these materials should then be made to provide the metabolisable energy data necessary for the prediction equations.

Metabolisable energy per g digested organic matter. It has been suggested as an alternative to Armstrong's formula that the metabolisable energy of forages may be estimated without great error as 3.6 kcal (15.1 kJ) per g digested organic matter. Not that that formula, any more than the expression of metabolisable energy as a percentage of digested energy, solves the problem of a quick evaluation of energy value. A quick evaluation should rest on proximate analysis and not require animal experiment or the estimation of *in vivo* digestibility. In view of the different energy values attributed to the several components in a proximate analysis it was thought of interest to examine the frequency distribution of metabolisable energy per g digested organic matter in the 603 samples of feeds for which we have computed values. The result is shown in Table 0.34

It will be seen that 3.6 kcal (15.1 kJ) per g digested organic matter in Axelsson's terms is not applicable as a mean value to either artificially dried grass or hay. The Armstrong series of grasses all have values above 3.6 (range 3.7–4.0) (15.5–16.7) and there is only 1 of Alderman's hays with a value as high as 3.6 (range 3.3–3.6) (13.8–15.1). Over the range of fodders the scatter is wide and does not suggest any simple, generally applicable formula.

To sum up, it appears that there is no satisfactory basis on which a quick assessment of feeding value may be made from chemical composition. We suggest that further work is required on typical feeding stuffs in which detailed analysis, digestibility trials and calorimetric measurements are made together. In our opinion, at this stage, the emphasis should be on detail. An attempt should be made to assess the energy values of different components of herbage at different stages of growth, in extension of the work of Waite, Johnston and Armstrong. It is unlikely that any further search for a short cut will be more successful than those that have been made. Further progress to simplification, we believe, must come through detail.

Table 0.31 Metabolisable Energy: 24 hays estimated in 3 ways

Walker and Hepburn (1954-55)

From bomb calorimetry of hay, faeces and urine: methane estimated with equation of Swift *et al:* J. Animal Sci., 1948, 7, 475

$$E = 2.41x + 9.80$$

where E = methane in g, x = digested carbohydrate in units of 100 g and energy value of methane is 13.34 kcal per g.

per g DM

	Walker and Hepburn		Axelsson		Nehring	
	kcal	kJ	kcal	kJ	kcal	kJ
1A	2.31	9.67	2.12	8.87	2.19	9.16
B	2.87	12.01	2.89	12.09	2.96	12.38
C	2.34	9.79	2.25	9.41	2.30	9.62
2A	2.13	8.91	1.97	8.24	1.98	8.28
B	2.40	10.04	2.21	9.25	2.26	9.46
C	2.16	9.04	1.95	8.16	1.99	8.33
3A	2.16	9.04	1.98	8.28	2.04	8.54
B	2.60	10.88	2.37	9.92	2.43	10.17
C	2.40	10.04	2.17	9.08	2.24	9.37
4A	2.27	9.50	1.97	8.24	1.98	8.28
B	2.06	8.62	1.88	7.87	1.92	8.03
C	2.39	10.00	2.16	9.04	2.19	9.16
5A	2.06	8.62	1.95	8.16	1.99	8.33
B	2.10	8.79	2.01	8.41	2.05	8.58
C	2.16	9.04	2.06	8.62	2.11	8.83
6A	2.20	9.20	2.07	8.66	2.14	8.95
B	1.94	8.12	1.84	7.70	1.89	7.91
C	2.02	8.45	1.85	7.74	1.91	7.99
7A	1.99	8.33	1.89	7.91	1.94	8.12
B	2.14	8.95	2.05	8.58	2.12	8.87
C	2.11	8.83	1.97	8.24	2.03	8.49
8A	2.16	9.04	1.97	8.24	2.03	8.49
B	2.78	11.63	2.76	11.55	2.82	11.80
C	2.09	8.74	1.98	8.28	2.04	8.54
Mean	2.24	9.37	2.10	8.79	2.15	9.00
Range kcal	1.94 – 2.87		1.84 – 2.89		1.89 – 2.96	
kJ	8.12 – 12.01		7.70 – 12.09		7.91 – 12.38	

Table 0.32 Metabolisable Energy of 24 silages
Walker and Hepburn, (1956)

Estimation as shown for 24 hays

per g DM

	Walker and Hepburn		Axelsson	
	kcal	kJ	kcal	kJ
1A	2.60	10.88	2.59	10.84
B	2.81	11.76	2.76	11.55
C	2.62	10.96	2.39	10.00
2A	2.59	10.84	2.46	10.29
B	2.61	10.92	2.39	10.00
C	2.74	11.46	2.54	10.63
3A	2.29	9.58	2.17	9.08
B	2.36	9.87	2.07	8.66
C	2.35	9.83	2.06	8.62
4A	2.73	11.42	2.45	10.25
B	2.87	12.01	2.75	11.51
C	2.61	10.92	2.37	9.92
5A	2.76	11.55	2.57	10.75
B	2.69	11.25	2.56	10.71
C	2.38	9.96	2.22	9.29
6A	2.83	11.84	2.49	10.42
B	2.70	11.30	2.47	10.33
C	2.74	11.46	2.40	10.04
7A	2.95	12.34	2.79	11.67
B	2.96	12.38	2.76	11.55
C	2.48	10.38	2.29	9.58
8A	2.81	11.76	2.58	10.79
B	2.61	10.92	2.50	10.46
C	2.81	11.76	2.69	11.25
9x*	2.78	11.63	2.69	11.25
y*	2.03	8.49	1.82	7.61
Silage range kcal	2.29 – 2.96		2.06 – 2.79	
kJ	9.58 – 12.38		8.62 – 11.67	

* 9x and y are two "reference hays".

Table 0.33 Metabolisable Energy of 47 hays computed in 3 ways

Alderman *et al* (1966)

ME as given compared with that computed by Axelsson's and Nehring's factors

ME (Alderman) = 81% of digestible energy, determined.

kcal or kJ/g DM

Identification number	Alderman	Axelsson	Nehring	Identification number	Alderman	Axelsson	Nehring
764/65	1.37 5.73	1.31 5.48	1.35 5.65	743	2.03 8.49	1.96 8.20	2.01 8.41
765	1.73 7.24	1.64 6.86	1.70 7.11	747	1.88 7.87	1.81 7.57	1.84 7.70
798	2.02 8.45	1.92 8.03	1.96 8.20	797	2.15 9.00	2.05 8.58	2.10 8.79
606/65A	1.62 6.78	1.66 6.95	1.71 7.15	802	1.75 7.32	1.64 6.86	1.68 7.03
591	1.96 8.20	1.90 7.95	1.97 8.24	803	2.15 9.00	1.97 8.24	2.02 8.45
592	2.46 10.29	2.36 9.87	2.41 10.08	815	2.41 10.08	2.26 9.46	2.31 9.67
593	2.04 8.54	1.99 8.33	2.06 8.62	6000	2.20 9.20	2.12 8.87	2.18 9.12
603	2.11 8.83	2.05 8.58	2.09 8.74	6001	2.09 8.74	2.01 8.41	2.08 8.70
604	2.18 9.12	2.15 9.00	2.20 9.20	6053	1.97 8.24	1.96 8.20	2.02 8.45
605	2.32 9.71	2.27 9.50	2.31 9.67	6056	2.19 9.16	2.14 8.95	2.21 9.25
631	1.95 8.16	1.93 8.08	2.00 8.37	6057	2.17 9.08	2.11 8.83	2.18 9.12
635	2.06 8.62	1.97 8.24	2.05 8.58	6058	2.26 9.46	2.15 9.00	2.20 9.20
654	1.79 7.49	1.85 7.74	1.90 7.95	545/66	2.03 8.49	1.94 8.12	1.99 8.33
658	1.94 8.12	1.85 7.74	1.90 7.95	549	2.04 8.54	1.99 8.33	2.05 8.58
659	1.75 7.32	1.57 6.57	1.61 6.74	550	2.16 9.04	2.06 8.62	2.12 8.87
676	2.17 9.08	2.04 8.54	2.09 8.74	557	1.53 6.40	1.52 6.36	1.54 6.44
680	2.63 11.00	2.42 10.13	2.45 10.25	592	1.72 7.20	1.71 7.15	1.77 7.41
700	2.24 9.37	2.17 9.08	2.19 9.16	502	2.19 9.16	2.09 8.74	2.15 9.00
702	2.29 9.58	2.15 9.00	2.18 9.12	630	1.83 7.66	1.78 7.45	1.84 7.70
715	2.75 11.51	2.60 10.88	2.61 10.92	632	1.88 7.87	1.83 7.66	1.87 7.82
719	1.82 7.61	1.78 7.45	1.83 7.66	633	1.80 7.53	1.72 7.20	1.76 7.36
723	2.06 8.62	1.97 8.24	2.03 8.49				
727	2.14 8.95	2.03 8.49	2.09 8.74	No. 47			
731	1.99 8.33	1.82 7.61	1.88 7.87	Mean	2.04 8.54	1.96 8.20	2.01 8.41
735	1.99 8.33	1.90 7.95	1.96 8.20	Range	1.37–2.75 5.73–11.51	1.31–2.60 5.48–10.88	1.35–2.61 5.65–10.92
739	2.05 8.58	2.04 8.54	2.12 8.87	Nehring % of Alderman = 98.5 Axelsson % of Alderman = 96.1			

Table 0.34 Frequency distribution of Metabolisable Energy: kcal/g DOM above, kJ below

kcal/	2.9-	3.0-	3.1-	3.2-	3.3-	3.4-	3.5-	3.6-	3.7-	3.8-	3.9-	4.0-	4.1-	4.2-	4.3-	4.4-	4.5-	4.6-	4.7-	4.8-	4.9-	5.0-	5.1-	5.2-	5.3-	5.4-	5.5-	5.6-	5.7-	5.8-	5.9-	7.0-	7.1-	Number of Samples
Animal products																																		
Beef tallow																																1	1	2
Milk, whole																											1	1	1					3
Milk, separated												1	1																					2
" " with lactose											1																							1
" " with oil																								1	1	2	1							5
Meat meal, oil extracted																		1																1
" " oil expelled																						1												1
" " oil drained																											1							1
Whale meat meal, oil expelled																		1																1
White fish meal																	1																	1
Plants and plant products																																		
Cereals, miscellaneous concentrates and residues									8	6	10	13	11	6	2	2	1				1	1										1		62
Cereals as fodder	1	2	1		1	3	2																											10
Fresh grasses							18	36	23	8	2																							87
Fresh Lucerne								2	7																									9
Grass, artificially dried (Waite *et al*)									2	4	4	2																						12
Grass ± legume, artificially dried						2	8	9	11	5	1																							36
Hay: Alderman					3	23	20	1																										47
Walker and Hepburn						1	8	11	4																									24
General (excluding above)					2	24	34	17	2																									79
Silage: Walker and Hepburn						1		2	1	6	9	2	2	1																				24
General (excluding above)						2	24	39	32	16	5																							118
Uncultivated plants: Grasses and rushes						4	11	2																										17
Bracken: fresh hay, silage							1	6	3																									10
Heather							1	8																										9
Kale									9	2	1																							12
Roots					1			5	4		1	3	1																					15
Seaweed							1	2		2																								5
Miscellaneous fodders				3	1		1	3	1																									9
kJ	12.1-	12.6-	13.0-	13.4-	13.8-	14.2-	14.6-	15.1-	15.5-	15.9-	16.3-	16.7-	17.2-	17.6-	18.0-	18.4-	18.8-	19.2-	19.7-	20.1-	20.5-	20.9-	21.3-	21.8-	22.2-	22.6-	23.0-	23.4-	23.8-	24.3-	24.7-	29.3-	29.7-	

Tables

Table 1.1 OAT, WHOLE GRAIN and its FRACTIONS: Composition as % of DM "Best averages" values
Reference: 324

	Crude protein	Ether extract	Nitrogen-free extract	Crude fibre	Ash
Whole grain	11.6	5.3	69.8	10.4	2.9
Kernel	14.9	7.0	74.6	1.4	2.1
Oatmeal	14.2	7.5	75.0	1.2	2.1
Rolled oats	14.7	7.6	74.7	1.0	2.0
Oat flour	15.5	7.9	73.4	1.1	2.0
Husk	1.4	0.4	55.9	37.8	4.5

Table 1.2 OAT, WHOLE GRAIN: Percentage digestibility of DM and OM

Description of sample	Digestibility DM	Digestibility OM	Notes Experimental animal	Reference
Whole grain				
var. Victory		69.8	Leghorn Cockerels	366
Huskless Oat		86.2		
Commercial sample	53.8		Light Sussex Cockerels	495
Ground whole grain				
Farm ground	67.8	69.1	Pig	579
Sussex ground	64.4	65.8		
Coarse ground	69.4	71.2	Pig	114
Medium ground	68.8	70.7	Digestibility refers to rations of which oats formed 85.5%	
Fine ground	69.6	71.4		
Crushed oats	54.7	56.7	Pig	579
	72.1	73.5	Ruminant	572

Table 1.3 OAT, WHOLE GRAIN: Proximate analysis as % of DM by variety, place of growth and year

Variety	CP	EE	NFE	CF	Ash	Place of growth	Year	Notes	Reference
Bell	12.7	5.1	69.7	9.6	2.9	Scotland	1932		
Early Miller	10.6	5.9	68.8	11.9	2.8		1933		
Elder	10.6	5.1	70.4	10.2	3.7		1932	Mean of samples from several areas	440
Potato	13.3	4.0	71.9	7.9	2.9		1932		
	12.2	4.5	70.0	10.5	2.9		1933		
Sandy	13.1	5.0	67.2	11.2	3.5		1932		
Golden Rain II	9.6	6.0	69.5	11.7	3.2	Mid-Wales	1932-34		
Golden Rain X Radnorshire Sprig, 266	9.8	5.9	69.5	11.5	3.3		1932-34		
Golden Rain X Red Algerian 258	10.4	5.9	69.8	10.5	3.3		,,		
Golden Rain X Yielder, 254	11.0	5.9	69.6	10.7	2.9		,,		
Marvellous	9.6	5.9	69.4	12.0	3.1		,,	Mean for 5 altitudes and 3 years	18
Record	9.5	6.4	68.5	12.7	2.9		,,		
Record X Radnorshire Sprig 275	10.1	5.7	68.0	13.2	2.9		,,		
Record X Radnorshire Sprig, 274	9.7	5.8	71.1	10.7	2.8		,,		
Star	9.7	5.5	71.8	9.8	3.2		,,		
Victory	9.3	5.4	69.9	11.9	3.5		,,		
Victory X Red Algerian, 289	9.6	5.8	70.3	11.2	3.0				
Ceirch Llwyd Cwta	14.1	6.6					1934	Mean of 5 altitudes	
	13.1-14.8	6.3-6.9						Range of 5 altitudes	
Ceirch Llwyd 578	13.4	6.7					,,		
	12.7-14.5	6.4-7.2							
Engelbrecht	11.7	5.7							
	10.9-11.9	4.6-6.6							
Ceirch du Bach	11.3	7.1	66.9	11.3	3.4	Aberystwyth	1939	Cut 19th August	
	11.6	7.0	66.1	11.3	4.0		,,	Cut 5th September	
Golden Rain	10.6	6.2	69.0	10.1	4.1		,,	Cut 21st August	211
	11.2	6.4	67.5	11.4	3.5		,,	Cut 5th September	
Haidd Garw	11.6	4.8	76.0	4.6	3.0		,,	Cut 2nd August	
	11.2	5.0	76.4	4.8	2.5		,,	Cut 29th August	
Blenda	10.3	5.6	70.1	11.1	2.9	West of England and Wales	1955	3 upland centres	
	10.2	5.5	71.0	10.5	2.8		,,	1 lowland centre	
	11.9	4.3	68.6	12.2	3.0		1956	4 upland centres	163
	11.0	4.4	70.5	10.7	3.4		,,	2 lowland centres	
	13.0	5.2	68.0	11.1	2.7		1957	4 upland centres	
	11.8	5.0	69.8	10.4	3.0		,,	2 lowland centres	
Maldwyn	10.6	5.3	67.9	13.4	2.8		1955	3 upland centres	
	10.9	5.5	69.2	11.6	2.8		,,	1 lowland centre	
	12.2	4.0	66.3	14.7	2.8		1956	4 upland centres	
	11.7	4.2	65.7	14.3	4.1		,,	2 lowland centres	
	14.0	5.0	65.4	12.9	2.7		1957	4 upland centres	
	11.9	5.1	67.8	12.0	3.2		,,	2 lowland centres	

Radnorshire Sprig	11.6	5.9	66.3	13.2	3.0		1955	3 upland centres	
	10.9	5.0	69.5	11.9	2.7		"	1 lowland centre	
	12.9	4.0	66.0	14.1	3.0		1956	4 upland centres	
	13.2	4.0	67.1	12.1	3.6		"	2 lowland centres	
	14.8	5.2	64.7	12.4	2.9		1957	4 upland centres	
	12.7	5.0	67.1	11.5	3.7		"	2 lowland centres	
Blenda	11.6	5.5	69.2	10.9	2.8	Scotland, South West	1957		
	7.9	6.2	68.2	15.1	2.6	Scotland, Uplands	1958		
	10.2	6.7	70.6	9.4	3.1		1959		167
	10.9	6.8	70.8	8.9	2.6		1960		
	11.3	6.6	67.5	12.2	2.4		1961		
Maldwyn	13.3	5.4	65.7	12.7	2.9		1957		
	8.1	6.2	67.0	16.2	2.5		1958		
	11.6	6.2	67.6	11.4	3.1		1959		
	11.3	6.5	69.0	10.3	2.9		1960		
	12.2	6.2	64.8	14.2	2.6		1961		
Ayrshire Potato	14.0	5.9	65.0	10.9	3.2		1957		
	9.6	6.7	66.4	14.6	2.7		1958		
	12.0	7.6	67.2	10.0	3.3		1959		
	11.8	8.0	68.0	9.4	2.8		1960		
	12.1	7.3	64.9	13.0	2.7		1961		
Maelor	12.7	6.8	67.6	10.4	2.6		1961		
Commercial samples	9.2	7.3	68.3	11.5	3.7				92
	10.6	7.4	68.3	10.5	3.3				563
	13.8	4.4	65.7	13.0	3.1				90
	12.0	4.9	68.6	10.8	3.8				568
	12.3	6.0	66.0	12.1	3.5				568
Sample unspecified	12.4	4.1	69.0	11.1	3.5				40
Sample unspecified	7.6	1.8	73.5	14.5	2.6				495
Varieties grown in Wales	9.8	5.2	68.9	13.0	3.1	Wales	1961	82 samples as used	
	7.2-12.6	1.9-8.0	63.8-74.2	9.0-17.9	2.4-4.1			on farms	
	11.0	5.5	68.5	12.1	2.9		1962	39 samples as used	370
	8.9-13.2	3.6-7.6	63.9-71.1	10.0-15.5	2.2-3.6			on farms	
	11.9	4.9	68.1	12.0	3.1		1963	50 samples as used	
	9.5-14.5	0.9-6.8	64.1-72.2	8.0-15.8	2.2-4.1			on farms	

Table 1.4 OAT, WHOLE GRAIN: Proximate analysis as % of DM. Selected sample for which metabolisable energy computed
For digested nutrients and metabolisable energy see table 1.219

	CP	EE	NFE	CF	Ash	Notes	Reference
Whole grain							
Victory	9.25	5.78	70.38	10.60	3.98		366
						Means and ranges for:	
Varieties grown in Wales:	9.8	5.2	68.9	13.0	3.1	82 samples in 1961	
	(7.2-12.6)	(1.9-8.0)	(63.8-74.2)	(9.0-17.9)	(2.4-4.1)		
	11.0	5.5	68.5	12.1	2.9	39 samples in 1962	370
	(8.9-13.2)	(3.6-7.6)	(63.9-71.1)	(10.0-15.5)	(2.2-3.6)		
	11.9	4.9	68.1	12.0	3.1	50 samples in 1963	
	(9.5-14.5)	(0.9-6.8)	(64.1-72.2)	(8.0-15.8)	(2.2-4.1)		
Huskless oat	14.14	5.40	76.50	1.93	2.01		366
Ground whole grain							
Farm ground	7.2	5.2	73.3	9.4	4.6	Australian oats	
	9.1	2.7	77.7	8.6	2.0	Scottish oats	119
Sussex ground	8.6	4.3	74.1	10.2	2.8	Sample unspecified	
Farm ground	12.90	5.38	66.92	11.52	3.28		579
Sussex ground	13.32	6.47	62.97	13.07	4.17		
Coarse ground	13.1	6.0	66.8	11.5	2.6		
Medium ground	12.9	6.1	67.3	11.1	2.6		114
Fine ground	12.6	6.4	68.0	10.4	2.6		
Crushed oats							
bulk sample	11.18	5.49	66.85	12.74	3.74		579

Table 1.5 OAT, WHOLE GRAIN: Proximate analysis as % of DM. Effect of extra nitrogen fertiliser and state of maturity at harvest

Reference: 162

Variety	Year	State at harvest	CP N_0	CP N_1	EE N_0	EE N_1	NFE N_0	NFE N_1	CF N_0	CF N_1	Ash N_0	Ash N_1	Notes
Eagle	1952	Binder ripe	10.9	12.3	5.6	5.6	69.5	68.1	10.6	10.6	3.4	3.4	N_0: no extra fertiliser
		Combine ripe	11.1	12.5	5.4	5.3	69.0	67.4	11.1	11.2	3.4	3.6	N_1: 250 kg/ha Nitrochalk
	1953	Binder ripe	12.6	13.5	5.6	5.5	68.0	67.0	10.3	10.7	3.5	3.3	
		Combine ripe	12.4	13.5	5.3	5.2	68.8	67.5	10.1	10.3	3.4	3.5	
	1954	Binder ripe	12.5	13.5	5.6	5.4	66.8	65.5	11.0	11.5	4.1	4.1	
		Combine ripe	12.4	13.4	5.3	5.0	67.5	65.8	10.7	11.3	4.1	4.5	
Milford	1952	Binder ripe	11.2	12.3	5.1	4.8	69.6	68.3	10.8	11.2	3.3	3.4	
		Combine ripe	11.5	12.6	4.8	4.7	69.4	68.4	11.0	10.9	3.3	3.4	
	1953	Binder ripe	12.9	13.7	5.2	5.1	68.8	68.1	10.0	10.0	3.1	3.1	
		Combine ripe	12.7	13.7	5.0	4.9	69.5	68.3	9.7	10.0	3.1	3.1	
	1954	Binder ripe	12.6	13.4	5.2	4.9	67.2	66.5	11.2	11.9	3.1	3.3	
		Combine ripe	12.4	13.2	4.9	4.8	68.6	67.1	10.9	11.5	3.2	3.4	
Onward	1952	Binder ripe	10.6	11.6	4.0	3.9	68.9	67.8	13.4	13.6	3.1	3.1	
		Combine ripe	10.8	11.9	4.0	3.5	68.3	66.8	13.8	14.4	3.1	3.4	
	1953	Binder ripe	11.6	12.4	4.2	3.8	68.0	67.5	13.1	13.3	3.1	3.0	
		Combine ripe	11.3	12.6	3.9	4.0	69.6	68.0	12.3	12.4	2.9	3.0	
	1954	Binder ripe	11.4	12.5	4.0	3.8	67.5	66.4	14.1	14.2	3.0	3.1	
		Combine ripe	11.4	12.5	3.8	3.6	67.7	66.3	14.0	14.4	3.1	3.2	
	Means	Binder ripe	11.8	12.8	4.9	4.7	68.4	67.3	11.6	11.9	3.3	3.3	
		Combine ripe	11.8	12.9	4.7	4.6	68.7	67.2	11.5	11.8	3.3	3.5	

Table 1.6 OAT, WHOLE GRAIN: Nitrogen as % of DM

Variety		Notes	Ref
Black Supreme	1.87	14 Varieties of spring oats grown in England	306
Black Tartarian	2.10		
Blenda	1.88		
Eagle	1.82		
Early Miller	2.02		
Golden Rain II	1.91		
Maldwyn	1.90		
Marvellous	1.86		
Milford	1.93		
Onward	1.78		
Star	1.81		
Sun II	1.88		
Victory	1.87		
S 84	1.98		
Black Winter	1.98	6 Varieties of winter oats grown in England.	
Grey Winter	1.82		
Picton	1.88		
S 81	1.92		
S 147	1.88		
S 172	1.95		
Unspecified	1.87 (1.76-2.00)	Mean and range of 5 estimates on the same sample	

Table 1.7 OAT, WHOLE GRAIN: Crude protein as % of DM. Year to year variation, Hertfordshire, England. Reference: 244

Variety	1946	1947	1948	1949	Notes
Black Winter	13.7	17.0	15.4	16.3	Means of 3 or 4 trials
Grey Winter	12.8	15.1	13.5	16.3	
Picton	13.4	15.4	13.4	17.6	
S 147	13.3	17.0	14.0	16.6	
S 172	14.4	16.6	14.1	16.5	
Unique	–	–	–	20.2	

Table 1.8 OAT, WHOLE GRAIN: Effect on crude protein of extra nitrogen fertiliser

	Crude protein as % of DM			Notes	Ref.
	N: Normal practice	N: applied 1	2		
Spring oat: 6 varieties Milford, Star, Maldwyn Victory, Sun II and Onward,	11.3	12.4	13.6	N applied: Nitrochalk 1) 250 kg/ha in April 2) same with 250 kg/ha in June 54 samples from each of 3 centres. Varietal differences less than between centres. Lowest mean for 3 N levels 10.1, highest, 14.8%	243
Early Miller	8.0	9.2		N applied: 125 kg/ha ammonium sulphate	459
Milford S 225	12.3	11.8		Undersown: ryegrass and clover	135
	12.3	12.5		Undersown: cocksfoot and clover	
	11.8	12.1		Undersown: timothy and clover	
	11.9	12.8		Undersown: a mixture of 3 above	
				N applied: 375 kg/ha Nitrochalk	

Table 1.9 OAT, WHOLE GRAIN: Oil content (petroleum extract) as % of DM
Reference: 305

Variety		Notes
Black Supreme	3.9	10 varieties of spring oats grown together at 22 centres.
Black Tartarian	4.2	
Eagle	5.5	
Early Miller	5.6	
Golden Rain II	5.1	
Marvellous	4.8	
Onward	4.1	
Star	4.8	
Victory	4.7	
S 84	4.9	
Grey Winter	7.5	4 varieties of winter oats grown together at 46 centres.
Picton	6.9	
S 147	5.7	
S 172	7.0	

Table 1.10 OAT, WHOLE GRAIN: Carbohydrate fractions as % of DM

		Notes	Ref.
Available carbohydrate by Bolton's (1960) method (reference 84)		Varieties grown in Wales	370
	46.8 (37.5-55.0)	82 in 1961	
	47.4 (36.3-56.5)	39 in 1962	
	46.4 (33.3-53.8)	50 in 1963	
Starch		Australian oats	
	41.3	Farm ground	119
		Scottish oats	
	55.7	Farm ground	
		Unspecified	
	45.8	Sussex ground	
	47.9	Unspecified	120
Sugars		Australian oats	
	3.0	Farm ground	119
		Scottish oats	
	1.7	Farm ground	
		Unspecified	
	1.6	Sussex ground	
	1.4	Unspecified	120

Table 1.11 OAT, WHOLE GRAIN: Major elements as % of DM

	Variety		Notes	Ref.
Calcium	Ceirch Llwyd Cwta	0.12 0.10-0.14	Mean and range of 5 samples from different altitudes	18
	Ceirch Llywd 578	0.13 0.10-0.18	Mid-Wales 1934	
	Engelbrecht	0.09 0.08-0.10		
	Golden RainII	0.09	Mid-Wales 1932-34	
	Golden RainX Radnorshire Sprig 226	0.11	Mean for 5 altitudes in 3 years	
	Golden Rain X Red Algerian 258	0.10		
	Golden Rain X Yielder 254	0.08		
	Marvellous	0.08		
	Record	0.08		
	Record X Radnorshire Sprig 275	0.11		
	Record X Radnorshire Sprig 274	0.09		
	Star	0.10		
	Victory	0.09		
	Victory X Red Algerian 289	0.08		
	Commercial sample	0.13		92
	Varieties grown in Wales	0.11 0.07-0.18	Mean and range of 171 samples in 3 years 1961-63	371
	Sample unspecified	0.11		278
	Sample unspecified	0.13 0.10-0.18	Mean and range for 8 estimates on same sample	1
	Ceirch du Bach	0.11 0.12	Harvested at normal stage Top growth (cut when 30 cm high)	211
	Golden Rain	0.11 0.11	Harvested at normal stage Top growth (cut when 30 cm high)	
	Haidd Garw	0.08 0.08	Harvested at normal stage Top growth (cut when 30 cm high)	
Phosphorus	Ceirch Llwyd Cwta	0.44 0.34-0.48	See calcium same reference	18
	Ceirch Llwyd 578	0.47 0.39-0.62		
	Engelbrecht	0.34 0.29-0.38		
	Golden Rain II	0.36		
	Golden Rain X Radnorshire Sprig 226	0.37		
	Golden Rain X Red Algerian 258	0.34		
	Golden Rain X Yielder 254	0.28		
	Marvellous	0.35		
	Record	0.36		
	Record X Radnorshire Sprig 275	0.37		
	Record X Radnorshire Sprig 274	0.33		
	Star	0.36		
	Victory	0.37		
	Victory X Red Algerian 289	0.34		
	Commercial Sample	0.55		92
	Varieties grown in Wales	0.38 0.29-0.59	See calcium same reference	371
	Sample unspecified	0.40		278
	Sample unspecified	0.39 0.36-0.41	Mean and range for 5 estimates on same sample	1
	Ceirch du Bach	0.34 0.38	See calcium same reference	211
	Golden Rain	0.34 0.36		
	Haidd Garw	0.46 0.47		

continued/

Table 1.11 OAT, WHOLE GRAIN: Major elements as % of DM (continued)

	Variety		Notes	Ref.
Magnesium	Unspecified	0.12		278
	Varieties grown in Wales	0.13 0.10-0.18	See calcium same reference	371
	Unspecified	0.12 0.11-0.13	Mean and range for 7 estimates on same sample	1
Potassium	Golden Rain X Radnorshire Sprig 266 Golden Rain X Red Algerian 258 Golden Rain X Yielder 254 Marvellous Record Record X Radnorshire Sprig 275 Record X Radnorshire Sprig 274 Star Victory Victory X Red Algerian 289	0.58 0.55 0.61 0.59 0.55 0.60 0.52 0.58 0.37 0.61	See calcium same reference	18
	Ceirch du Bach Golden Rain Haidd Garw	1.00 1.05 0.98 1.00 1.13 0.94	See calcium same reference	211
	Sample unspecified	0.54		278
	Sample unspecified	0.53 0.48-0 56	Mean and range for 6 estimates on same sample	1
	Varieties grown in Wales	0.47 0.31-0.65	See calcium same reference	371
Sodium	Varieties grown in Wales	0.02 0.004-0.06	See calcium same reference	371
Chlorine	Golden Rain X Red Algerian 258 Golden Rain X Yielder 254 Marvellous Record Record X Radnorshire Sprig 275 Record X Radnorshire Sprig 274 Star Victory Victory X Red Algerian 289	0.10 0.12 0.17 0.10 0.15 0.07 0.09 0.09 0.11	See calcium same reference	18
	Varieties grown in Wales	0.09 0.04-0.18	See calcium same reference	371
Sulphur, total	Sample unspecified	0.13	1 centre only	1

Table 1.12 OAT, WHOLE GRAIN: Trace elements as $\mu g/g$ DM

	Varieties		Notes	Ref.
Barium	Sample unspecified	3.7	One estimate only	1
Boron	Sample unspecified	2 (1-4)	Mean and range for 8 estimates on same sample	1
	Sample unspecified " "	4.4 1.5	Spectrographic Colorimetric	450
Cobalt	Varieties grown in Wales	0.05 0.02-0.17	Mean and range for 163 samples	371
	Sample unspecified	0.04	One estimate only	1
Copper	Sample unspecified	4.9 4.7-5.0	Mean and range for 3 estimates on same sample	1
	Varieties grown in Wales	4.7 3.0-8.2	Mean and range for 171 samples in 3 years 1961-63	371
Iron	Sample unspecified	96 83-105	Mean and range for 7 estimates on same sample	1
Lead	Sample unspecified	<1	One estimate only	1
Manganese	Sample unspecified	67 63-70	Mean and range for 4 estimates on same sample	1
	Varieties grown in Wales	45 22-79	171 samples in 3 years 1961-63	371
Molybdenum	Sample unspecified	0.15 0.12-0.18	Mean and range for 3 estimates on same sample	1
Nickel	Sample unspecified	2.10	One estimate only	1
Strontium	Sample unspecified	2.1	One estimate only	1
Tin	Sample unspecified	<1	One estimate only	1
Titanium	Sample unspecified	2.8	One estimate only	1
Vanadium	Sample unspecified	0.03	One estimate only	1
Zinc	Sample unspecified	43	One estimate only	1
	Varieties grown in Wales	37 21-70	Mean and range for 171 samples in 3 years 1961-63	371

Table 1.13 OAT, WHOLE GRAIN: Tocopherol and Thiamin

Tocopherols. Reference: 100

Variety	Total µg/g DM	% of total α form	β Form	
Yielder	17-32	26	63	

Thiamin, Kernel, crude protein and thiamin contents of whole grain. Reference: 280

Variety	Kernel content %	Crude protein % DM	Thiamin µg/g DM	Notes
Abundance	75.9	12.4	4.8	Scotland and Wales (1944)
Bell	77.8	12.1	4.4	All samples not otherwise marked grown
Black Tartarian	72.4	12.8	3.3	in one area, South-East Scotland
Castleton Potato	77.0	12.9	5.6	
	75.1	12.4	4.4	Means and ranges of 7 samples from
	70.0-77.5	11.3-12.9	3.8-5.6	7 different areas, Scotland
Castleton Sandy	77.6	14 8	6.3	
Early Miller	78.6	13.1	4.4	
Golden Rain	75.1	12.1	4.8	
Gordon	73.2	12.2	5.1	
Grey Winter	76.7	11.8	5.1	
Marvellous	73.4	11.0	4.0	Means and ranges of 13 samples from
	71.4-75.8	9.3-12.4	3.4-3.3	10 different areas, Scotland
Onward	68.2	10.7	4.4	
	70.9	9.5	3.9	Means and ranges of 21 samples from
	68.2-73.1	7.5-11.5	3.1-4.8	16 different areas, Scotland
	68.1	10.0	3.8	Means of 4 samples grown in Wales
Potato	75.7	13.0	4.8	
Pure Line Potato	74.6	12.8	4.4	
Quality	76.3	12.8	3.8	
Record	72.1	10.2	4.2	
Sandy	74.7	13.8	5.5	
Scots Berlie	78.0	14.9	6 3	
Small Welsh	75.6	16.3	5.9	
Star	73.5	11.3	4.8	
	75.6	11.1	4 0	Means and ranges of 28 samples from
	73.5-78.2	9.2-13.6	3.1-5.2	15 different areas, Scotland
	73 5	10 4	4.0	Means of 4 samples grown in Wales
Tam Finlay	75.0	13 2	5.1	
Victory	74.1	11.2	4.1	
	76.0	10.4	3 8	Means and ranges of 22 samples from
	73.1-76.8	8.8-12.4	3.1-4.9	14 different areas, Scotland
	72.5	11.7	4.3	Means of 4 samples grown in Wales
Yielder	67.8	11.3	3.8	
	69.2	11.5	3.8	Means and ranges of 25 samples from
	66.5-72.9	10.0-13.0	3.1-4.2	16 different areas, Scotland
	66.7	11 9	3.8	Means of 4 samples grown in Wales
Tiree Oat *(Avena strigosa)*	74.2	15.2	5.7	

Distribution of thiamin within the grain

Distribution by weight of fractions and thiamin (µg/g DM) in each fraction

	Total %	Total µg/g	Husk %	Husk µg/g	Embryo %	Embryo µg/g	Scutellum %	Scutellum µg/g	Bran %	Bran µg/g	Endosperm %	Endosperm µg/g
Early Miller	(98.9)	4.6	24	0	1.4	63	1.5	83	8	27	64	0.5
Star	(99.5)	4.1	26	0	0.9	42	1.6	70	9	28	62	0.2

Footnote: There was no appreciable varietal or geographical difference but a significant positive correlation (r = 0.42) between crude protein and thiamin.

E

Table 1.14 OAT KERNEL: Nitrogen as % of DM by variety and year. SCOTLAND

Variety		1951	1952	1953	1954	1955	1956	1957	1958	1959	1960	1961	Mean
Aa 724	SN		1.95	3.12	2.46	2.30							2.46
Aa 726	SN		2.01										2.01
Aa 729	SN	2.68	2.50	2.60									2.59
Aa 732	SN		2.51										2.51
Albyn Donside	SN		2.76	2.85		2.80							2.80
Albyn Express	SN	2.38	2.35	2.32									2.35
Ardee Potato	SN			2.58									2.58
Astor	SP										2.60	2.31	2.46
Awnless Potato	SN	2.54											2.54
Ayr Ally	SR		2.09	2.51									2.30
Ayr Bounty	SR[1]	2.86	2.83										2.85
Ayr Commando	SR[1]	2.30	2.26	2.41	2.11	2.36							2.29
Ayr Everest	SN				2.20	2.19							2.20
Ayr Line	SR[1]		2.37			2.22							2.30
Ayr Line Potato	SR[1]		2.04	2.64	2.12	2.75							2.39
Bambu	SN	2.16	1.70										1.93
Blenda	SP	2.08	2.05	2.08	1.96	2.28							2.09
Castleton Potato	SR[2]	2.55	2.54	2.58									2.56
Craig's Afterlea	SN	2.70	2.62	2.88									2.73
Eagle	SN	2.12		2.09									2.11
Early Miller	SN	2.48	2.18	2.35									2.34
Express	SN			1.80									1.80
Forward	SP	2.12	2.19	2.17	2.11	2.24							2.17
Garton's 401	SN				2.12								2.12
Harvester	SN					2.55							2.55
Lightning	SN				2.04								2.04
Maris Quest (AB/26/284)	WL											2.40	2.40
Marne	SN			1.68	1.89	2.09	1.86						1.88
Marvellous	SN	2.20	2.14										2.17
Max	SN						1.96						1.96
Milford	SL				1.96	2.03							2.00
Minor	SN	2.29	2.19	2.09	1.92								2.12
Onward	SL	2.14	2.07	2.12	2.08								2.10
Orion III	SN			2.67									2.67
Palu	SN			1.80	1.99	2.32	2.06						2.04
Pendek	SN					1.94	1.87						1.91
Primus	SN			2.01									2.01
R 30	SN		2.39	2.39									2.39
Reagle	SN		2.23	2.35	2.09	2.31							2.25
Rex	SN			1.63	1.88	2.06	1.90						1.87
Sandy	SL		2.69	2.73									2.71
Sisu	SN						1.75						1.75
Star	SL	2.13	1.94										2.04
Sun II	SP	2.14	2.08	2.11	1.96	2.23	1.89	2.04					2.06
Supreme	SR		2.31	2.60									2.46
Victory	SN		1.82	2.70	2.15	2.30							2.24
Yielder	SR	2.64	2.26	2.55									2.48

S = Spring W = Winter N = No commercial importance P = Popular varieties R = Restricted to certain areas L = Limited acreage
1 – Caithness 2 – West Coast of Scotland.

OAT KERNELS (Addendum) Varieties under Test, 1966

		Kernel as % of whole grain	N as % of DM	Oil as % DM
Mostyn	S	75	2.2	5.7
Nina (W. 16414)	S	76.5	2.1	5.8
Tarpan (Ngh 5823)	S	74	2.5	7.1

Table 1.15 OAT KERNEL: Nitrogen as % of DM by variety and year. ENGLAND and WALES

Variety		1951	1952	1953	1954	1955	1956	1957	1958	1959	1960	1961	Mean
Angus	SN								2.45	2.45			2.45
Barnwell	WN							2.33	2.18				2.26
Black Tartarian	SN			2.70									2.70
Blanche de Wattines	SN							2.67					2.67
Blenda	SP			2.15	2.02	2.12							2.10
CA 148 No. 4	WN							2.30					2.30
CA 148 No. 5	WN							2.31					2.31
Civena	SN							2.71					2.71
Condor	SP								2.19	2.25	2.28		2.24
Clinton	SN				3.43	2.81							3.12
Craig's Afterlea	SN			2.86									2.86
Deva	SN							2.72					2.72
E 2295	WN								2.10				2.10
E 2296	WN										2.28		2.28
Eagle	SN			2.17	2.08	2.13							2.13
Early Miller	SN			2.71	2.40								2.56
Feltwell	WN							2.27	2.10				2.19
Flamande Deprez	SN							2.68					2.68
Flamingskrone	SN										2.59	2.38	2.49
Golden Rain II	SN			2.13									2.13
Grey Winter	WN			2.34	2.30								2.32
Karin (MGH 462)	SN											2.33	2.33
Linda (SV 01760)	SN											2.35	2.35
Maelor	SN								2.66	2.51			2.59
Maldwyn (S 221)	SN			2.19	2.17	2.28							2.21
Manod	SR[1]								2.56	2.93			2.75
Marino	SN										2.66	2.30	2.48
Marvellous	SN			2.18									2.18
Max	SN									2.33	2.80		2.57
Milford	SL			2.30	2.03	2.29				2.43		2.37	2.28
Milns Prophet (32/78/4)	WN										2.30	2.30	2.30
Nestor	SN							2.73					2.73
Onward	SL			2.19	2.10								2.15
Opus	SN				2.02	2.17							2.10
Orion III	SN				2.45	2.51							2.48
Padarn	WP									2.18	2.38		2.28
Pendek	SN							2.69					2.69
Peniarth (E 2297)	WP										2.31	2.38	2.35
Pennant	WN							2.24	2.13				2.19
Penrhyn	WR[2]									2.23			2.23
Phoenix	SN									2.38	2.67		2.53
Picton	WN			2.29	2.40	2.27		2.55					2.38
Powys	WP							2.51		2.33	2.41	2.51	2.44
Primus II	SN				2.41	2.50							2.46
S 81	SN			1.87									1.87
S 84	SN			2.45	2.29								2.37
S 147	WR[3]			2.14	2.34	2.28		2.49	2.11	2.43	2.36	2.34	2.31
S 172	WR[4]			2.30	2.37	2.41		2.62		2.37		2.52	2.43
Sun II	SP			2.02	2.05	2.20		2.78	2.42	2.54	2.75	2.39	2.39
Supreme	SR[1]			2.27	2.13								2.20
Victory	SN			2.24									2.24
Vigor	SN								2.54	2.46			2.50
Vollbringer	SN										2.65	2.31	2.48
W 16004	SN							2.43					2.43
Zandster	SN							2.75					2.75
32/256/3	WN											2.56	2.56
4757/6/2/19	SN							3.00					3.00

S = Spring W = Winter N = No commercial importance P = Popular varieties R = Restricted to certain areas L = Limited acreage
1 – Wales, Isle of Man 2 – Surrey 3 – Anglesey, Randor, Derby, Worcestershire, Devon, Somerset, Sussex 4 – Randor, Nottingham, Oxford, Wiltshire, Rutland, Lincoln, Hampshire, Surrey, Sussex.

Table 1.16 OAT KERNEL: Oil as % of DM by variety and year. SCOTLAND

Variety		1951	1952	1953	1954	1955	1956	1957	1958	1959	1960	1961	Mean
Aa 724	SN		7.25	6.74	7.30	7.15							7.11
Aa 726	SN		8.93										8.93
Aa 729	SN	7.00	7.40	6.88									7.09
Aa 732	SN		6.58										6.58
Albyn Donside	SN		6.45	5.94		6.45							6.28
Albyn Express	SN	6.70	7.42	7.22									7.11
Ardee Potato	SN			8.47									8.47
Astor	SP										7.00	7.60	7.30
Awnless Potato	SN	10.60											10.60
Ayr Ally	SR		8.04	6.86									7.45
Ayr Bounty	SR[1]	7.40	7.05										7.23
Ayr Commando	SR[1]	4.80	4.81	4.36	4.95	5.03							4.79
Ayr Everest	SN				6.07	6.23							6.15
Ayr Line	SR[1]				8.78			7.90					8.34
Ayr Line Potato	SR[1]		8.53	7.75	8.71	8.20							8.30
Bambu	SN	5.50	6.20										5.85
Blenda	SP	7.00	7.41	7.11	7.69	7.47							7.34
Castleton Potato	SR[2]	9.30	8.81	8.29									8.80
Craig's Afterlea	SN	6.90	7.10	6.62									6.87
Eagle	SN	7.50		7.70									7.60
Early Miller	SN	8.00	7.95	7.97									7.97
Express	SN			7.55									7.55
Forward	SP	5.80	6.07	5.97	6.15	6.41							6.08
Garton's 401	SN				7.30								7.30
Harvester	SN					6.60							6.60
Lightning	SN				7.56								7.56
Maris Quest (AB/26/284)	WL											7.80	7.80
Marne	SN			7.98	7.70	7.78	7.77						7.81
Marvellous	SN	7.00	6.78										6.89
Max	SN						7.47						7.47
Milford	SL				7.08	7.03							7.06
Minor	SN	7.50	7.76	7.94	8.20								7.85
Onward	SL	5.80	6.24	6.00	6.16								6.05
Orion III	SN			6.30									6.30
Palu	SN			8.37	7.94	7.70	7.50						7.88
Pendek	SN					8.70	8.20						8.45
Primus	SN			6.24									6.24
R 30	SN		10.03	9.04									9.54
Reagle	SN		6.51	6.44	6.78	7.06							6.70
Rex	SN			8.10	7.66	7.57	7.43						7.69
Sandy	SL		8.88	8.04									8.46
Sisu	SN						7.75						7.75
Star	SL	6.60	6.85										6.73
Sun II	SP	7.20	7.43	7.36	7.95	7.52	7.65						7.52
Supreme	SR		5.57	5.10									5.34
Victory	SN		6.96	6.09	6.87	7.22							6.79
Yielder	SR	5.80	6.16	6.44									6.13

S = Spring W = Winter N = No commercial importance P = Popular varieties R = Restricted to certain areas L = Limited acreage
1 – Caithness 2 – West Coast of Scotland.

Table 1.17 OAT KERNEL: Oil as of % DM by variety and year. ENGLAND and WALES

Variety		1951	1952	1953	1954	1955	1956	1957	1958	1959	1960	1961	Mean
Angus	SN								8.30				8.30
Barnwell	WN							8.80	9.00				8.90
Black Tartarian	SN			5.75									5.75
Blanche de Wattines	SN							6.30					6.30
Blenda	SP			6.48	6.90	6.90							6.76
CA 148 No.4	WN							9.40					9.40
CA 148 No.5	WN							9.40					9.40
Civena	SN							6.10					6.10
Condor	SP								7.50	7.50	7.40		7.47
Clinton	SN				5.08	5.80							5.44
Craig's Afterlea	SN			6.40									6.40
Deva	SN							5.60					5.60
E 2295	WN								7.00				7.00
E 2296	WN										8.60		8.60
Eagle	SN			7.13	7.43	7.40							7.32
Early Miller	SN			6.93	7.19								7.06
Feltwell	WN							9.40	9.60				9.50
Flamande Desprez	SN							6.40					6.40
Flamingskrone	SN										7.10	7.30	7.20
Golden Rain II	SN			6.88									6.88
Grey Winter	WN			8.76	9.74								9.25
Karin (MGH 462)	SN											7.40	7.40
Linda (SV 01760)	SN											6.80	6.80
Maelor	SN								6.40	6.80			6.60
Maldwyn (S 221)	SN			6.85	7.00	7.00							6.95
Manod	SR[1]								7.40	7.00			7.20
Marino	SN										6.70	7.50	7.10
Marvellous	SN			6.32									6.32
Max	SN									7.20	6.50		6.85
Milford	SL			6.11	6.71	6.40				6.80		6.30	6.46
Milns Prophet (32/78/4)	WN										8.50	8.80	8.65
Nestor	SN							6.50					6.50
Onward	SL			5.83	5.70								5.77
Opus	SN				7.38	7.30							7.34
Orion III	SN				6.28	6.60							6.44
Padarn	WP									8.00	7.90		7.95
Pendek	SN							6.96					6.96
Peniarth (E 2297)	WP										8.60	8.10	8.35
Pennant	WN							8.90	8.80				8.85
Penrhyn	WR[2]									9.00			9.00
Phoenix	SN									7.70	6.80		7.25
Picton	WN			8.18	8.87	8.80		8.20					8.51
Powys	WP							8.10		8.60	8.20	9.10	8.50
Primus II	SN				6.02	6.10							6.06
S 81	SN			6.65									6.65
S 84	SN			6.75	6.82								6.79
S 147	WR[3]			6.98	7.47	7.40		7.00	7.30	7.40	7.50	7.70	7.34
S 172	WR[4]			8.81	9.30	9.00		8.50		9.40		9.10	9.02
Sun II	SP			6.51	6.86	6.80		5.80	7.10	7.00	6.70	7.20	6.75
Supreme	SR[1]			4.97	5.42								5.20
Victory	SN			6.18									6.18
Vigor	SN								5.70	6.00			5.85
Vollbringer	SN										6.30	6.60	6.45
W 16004	SN							4.60					4.60
Zandster	SN							6.10					6.10
32/256/3	WN											9.50	9.50
4757/6/2/19	SN							6.00					6.00

S = Spring W = Winter N = No commercial importance P = Popular varieties R = Restricted to certain areas L = Limited acreage
1 – Wales, Isle of Man 2 – Surrey 3 – Anglesey, Randor, Derby, Worcestershire, Devon, Somerset, Sussex. 4 – Randor, Nottingham, Oxford, Wiltshire, Rutland, Lincoln, Hampshire, Surrey, Sussex.

Table 1.18 OAT KERNEL: Nitrogen and Oil as % of DM. (Summary Table)

		Nitrogen		Oil	
		Mean[1]	Range[2]	Mean[1]	Range[2]
WINTER OAT (England only)					
Popular varieties: Peniarth, Padarn and *Powys*					
Varieties	3	2.36	2.28-2.44	8.57	8.35-8.85
Averages by year	8		2.18-2.51		7.9-9.1
Varieties restricted by locality or acreage					
Varieties	3	2.32	2.23-2.43	8.45	7.34-9.02
Averages by year	15		2.11-2.62		6.98-9.4
All other varieties tested					
Varieties	11	2.29	2.10-2.56	8.87	7.00-9.50
Averages by year	19		2.10-2.62		7.0-9.6
SPRING OAT (England)					
Popular varieties: Blenda, Condor and *Sun II*					
Varieties	3	2.24	2.10-2.39	6.99	6.75-7.45
Averages by year	14		2 02-2.78		5.8-7.5
Varieties restricted by locality or acreage					
Varieties	4	2.35	2.15-2.75	6.56	5.77-7.20
Averages by year	11		2.02-2.78		5.8-7.5
All other varieties tested					
Varieties	33	2.50	1.87-3.12	6.54	4.60-8.30
Averages by year	50		1.87-3.43		4.6-8.3
SPRING OAT (Scotland)					
Popular varieties: Astor, Blenda, Forward and *Sun II*					
Varieties	4	2.20	2.06-2.46	7.06	6.08-7.34
Averages by year	19		1.89-2.60		5.8-7.95
Varieties restricted by locality or acreage					
Varieties	12	2.37	2.00-2.85	7.06	4.79-8.80
Averages by year	33		1.94-2.86		4.36-9.3
All other varieties tested					
Varieties	29	2.24	1.75-2.80	7.40	5.85-10.60
Averages by year	65		1.63-3.12		5.03-10.60

1 The mean is the mean of all averages by year

2 The range for varieties is that of general means for the several varieties; the range for averages by year is that of individual annual means. It includes year-to-year variation.

Table 1.19 OAT KERNEL: (Scotland)

Nitrogen and Oil as % of DM: 20 varieties in 5 years with means and analysis of variance. Values in parentheses have been fitted.

Nitrogen

Variety	1951	1952	1953	1954	1955	Means
Aa 724	(2.51)	1.95	3.12	2.46	2.30	2.47
Aa 729	2.68	2.50	2.60	(2.50)	(2.73)	2.60
Albyn Donside	(2.82)	2.76	2.85	(2.68)	2.80	2.78
Albyn Express	2.38	2.35	2.32	(2.26)	(2.49)	2.36
Ayr Commando	2.30	2.26	2.41	2.11	2.36	2.29
Ayr Line Potato	(2.44)	2.04	2.64	2.12	2.75	2.40
Blenda	2.08	2.05	2.08	1.96	2.28	2.09
Castleton Potato	2.55	2.54	2.58	(2.47)	(2.69)	2.57
Craig's Afterlea	2.70	2.62	2.88	(2.64)	(2.87)	2.74
Early Miller	2.48	2.18	2.35	(2.25)	(2.48)	2.35
Forward	2.12	2.19	2.17	2.11	2.24	2.17
Marne	(1.90)	(1.73)	1.68	1.89	2.09	1.86
Minor	2.29	2.19	2.09	1.92	(2.28)	2.16
Onward	2.14	2.07	2.12	2.08	(2.26)	2.14
Palu	(2.05)	(1.88)	1.80	1.99	2.32	2.01
Reagle	(2.29)	2.23	2.35	2.09	2.31	2.26
Rex	(1.87)	(1.70)	1.63	1.88	2.06	1.83
Sun II	2.14	2.08	2.11	1.96	2.23	2.10
Victory	(2.29)	1.82	2.70	2.15	2.30	2.25
Yielder	2.64	2.26	2.55	(2.39)	(2.62)	2.49
Means	2.33	2.17	2.35	2.19	2.42	2.29

Analysis of variance

	DF	SOS	MS	VR
Varieties	19	6.64	0.35	10.70
Years	4	0.93	0.23	7.09
Residual	50	1.63	0.03	

Oil

Variety	1951	1952	1953	1954	1955	Means
Aa 724	(6.86)	7.25	6.74	7.30	7.15	7.06
Aa 729	7.00	7.40	6.88	(7.36)	(7.31)	7.19
Albyn Donside	(6.07)	6.45	5.94	(6.44)	6.45	6.27
Albyn Express	6.70	7.42	7.22	(7.38)	(7.33)	7.21
Ayr Commando	4.80	4.81	4.36	4.95	5.03	4.79
Ayr Line Potato	(8.05)	8.53	7.75	8.71	8.20	8.25
Blenda	7.00	7.41	7.11	7.69	7.47	7.34
Castleton Potato	9.30	8.81	8.29	(9.07)	(9.02)	8.90
Craig's Afterlea	6.90	7.10	6.62	(7.14)	(7.09)	6.97
Early Miller	8.00	7.95	7.97	(8.24)	(8.19)	8.07
Forward	5.80	6.07	5.97	6.15	6.41	6.08
Marne	(7.57)	(7.84)	7.98	7.70	7.78	7.77
Minor	7.50	7.76	7.94	8.20	(8.00)	7.88
Onward	5.80	6.24	6.00	6.16	(6.20)	6.08
Palu	(7.76)	(8.02)	8.37	7.94	7.70	7.96
Reagle	(6.45)	6.51	6.44	6.78	7.06	6.65
Rex	(7.53)	(7.79)	8.10	7.66	7.57	7.73
Sun II	7.20	7.43	7.36	7.95	7.52	7.49
Victory	(6.53)	6.96	6.09	6.87	7.22	6.74
Yielder	5.80	6.16	6.44	(6.40)	(6.35)	6.23
Means	6.93	7.19	6.97	7.30	7.25	7.13

Analysis of variance

	DF	SOS	MS	VR
Varieties	19	85.56	4.50	61.49
Years	4	2.26	0.56	7.71
Residual	50	3.66	0.07	

Table 1.20 OAT KERNEL: Regional trend in nitrogen and oil contents as % of DM of two popular varieties, Blenda and Sun II

(For Graph see fig 0.1)

Regions North to South	Blenda 1954		Blenda 1955		Sun II 1954		Sun II 1955		Sun II 1959	
	N	Oil	N	Oil	N	Oil	N	Oil	N	Oil
Aberdeen, Craibstone	2.10	7.8	2.32	7.6	2.25	8.1	2.42	7.6		
Edinburgh, Boghall	1.91	7.5	2.17	7.1	1.67	8.0	2.18	7.4	2.10	7.6
Ayr, Auchincruive	2.03	7.0			2.10	7.4				
Newcastle, Cockle Park	1.82	7.3	1.95	7.5	1.99	7.2	1.90	7.4	2.72	7.2
Winmarleigh			2.10	7.0			2.06	7.0	2.61	6.9
Headley Hall			2.23	7.1			2.36	6.7	2.56	7.2
Askham Bryan	2.21	6.7								
Harper Adams			2.22	7.1			2.27	6.8	2.40	7.1
Sutton Bonington	1.52	7.4	2.09	6.8	1.60	7.1	2.13	6.5	2.65	6.5
Sprowston	2.07	6.3							2.98	7.1
Trawscoed	1.75	7.4	2.32	6.9	1.85	7.0	2.12	6.9	2.05	7.4
Rosemaund									2.61	7.0
Cambridge	2.09	6.5	2.30	6.5	2.19	6.4	2.24	6.5	2.71	6.5
Connington	1.91	6.8	1.93	6.9	2.00	6.9	2.03	6.9		
Wye			1.98	6.8			2.05	6.8	2.65	6.2
Sparsholt	2.54	6.5	2.34	6.2	2.58	6.5	3.09	5.8	2.44	6.5
Seale-Hayne	2.38	6.6	1.90	6.7	2.31	6.4	1.96	6.4	2.53	7.5

Table 1.21 OAT KERNEL: Nitrogen and oil as % of DM of VICTORY oat kernel grown in England and in Scotland
Reference: 305: 306

	Oats grown in: England	Scotland
Nitrogen		
No. of samples	127	92
Mean and SD	2.43±0.34	2.07±0.22
Range	1.72 - 3.44	1.69 - 2.72
Oil		
No. of samples	100	74
Mean and SD	6.2 ± 0.41	6.5 ± 0.34
Range	5.2 - 7.0	5.9 - 7.3

Table 1.22 OAT KERNEL: Nitrogen and oil as % of DM of 3 popular and other selected varieties. Percentage deviation of the means of all varieties.

Variety	N % of general mean	Oil % of general mean
Blenda	91	103
Forward	95	85
Sun II	92	105
Albyn Donside (highest N)	121	88
Ayr Commando (lowest oil)	100	67
Castleton Potato (highest oil)	112	125
Rex (lowest N)	80	108

Table 1.23 OAT KERNEL: Nitrogen as % of DM of 23 varieties at several centres
Reference: 306

Variety	Mean	Range	Number of centres	Notes
S 172	2.50	1.99 - 3.17	50	All samples from NIAB, NAAS,
	2.57	1.99 - 3.17	20	Scottish Colleges of Agriculture and
	2.39	2.05 - 2.80	11	Scottish Seed Certification Scheme.
S 147	2.37	1.71 - 3.25	50	
	2.50	1.89 - 3.25	20	
	2.27	1.71 - 2.62	11	
S 81	2.40	1.82 - 2.98	11	
S 84	2.64	1.79 - 3.31	22	
Black Supreme	2.53	1.92 - 3.18	22	
Black Tartarian	2.82	2.00 - 3.83	22	
Black Winter	2.58	1.91 - 3.25	20	
Blenda	2.38			In 13 comparisons with Victory: q.v.
"	2.15			In 26 comparisons with Sun II: q.v.
Clinton	3.10	2.66 - 3.64	5	In these comparisons (see also below)
Eagle	2.34	1.64 - 3.14	22	ranges are not given for the varieties
Early Miller	2.62	2.05 - 3.25	22	separately.
Grey Winter	2.26	1.77 - 3.14	50	
" "	2.34	1.82 - 3.14	20	
Golden Rain II	2.50	1.96 - 3.11	22	
Maldwyn	2.50			In 43 comparisons with Victory: q.v.
Marvellous	2.46	1.81 - 3.17	22	
Milford	2.51			In 45 comparisons with Victory: q.v.
Onward	2.43	1.77 - 3.07	22	
Orion III	2.76	2.44 - 3.18	5	
Picton	2.34	1.85 - 3.06	50	
	2.39	1.86 - 3.06	20	
Primus	2.81	2.41 - 3.27	5	
Star	2.36	1.69 - 2.95	22	
Sun II	2.18			In 26 comparisons with Blenda
	2.40			In 39 comparisons with Victory: q.v.
Victory	2.43	1.72 - 3.44		127 samples grown in England
	2.07	1.69 - 2.72		92 samples grown in Scotland
	2.45	1.72 - 3.09	22	
	2.42			In 43 comparisons with Maldwyn
	2.51			In 45 comparisons with Milford
	2.41			In 39 comparisons with Sun II
	2.41			In 13 comparisons with Blenda

Table 1.24 OAT KERNEL: Amino acids as % of DM

Means for from 4 to 15 samples of 10 varieties: microbiological assay.
Reference: 269

Amino acid	Number of samples analysed	Mean	Range
Alanine	6	0.90	0.69 - 1.15
Arginine	15	0.99	0.71 - 1.26
Aspartic Acid	10	0.56	0.28 - 0.75
Cystine	13	0.27	0.17 - 0.38
Glutamic acid	10	2.89	2.07 - 3.84
Glycine	10	0.63	0.45 - 0.84
Histidine	12	0.23	0.14 - 0.33
*Iso*leucine	11	0.80	0.50 - 1.05
Leucine	11	0.87	0.63 - 1.12
Lysine	10	0.51	0.32 - 0.69
Methionine	13	0.23	0.14 - 0.31
Phenylalanine	15	0.72	0.55 - 0.95
Proline	13	0.87	0.63 - 1.17
Serine	6	0.69	0.52 - 0.97
Threonine	11	0.54	0.44 - 0.69
Tryptophan	12	0.20	0.18 - 0.25
Tyrosine	15	0.50	0.32 - 0.74
Valine	10	0.99	0.77 - 1.28

Table 1.25 OAT KERNEL MILLED: Oatmeal

Reference: 347

	Crude protein	Ether extract	Nitrogen-free extract	Crude fibre	Ash	Notes
Proximate constituents %DM	14.76	8.48	73.85	1.00	1.91	Coarse ground, "pinhead", oatmeal.
	10.85	9.15	77.18	1.09	1.73	Medium ground oatmeal.
Carbohydrate fractions %DM	Available carbohydrate					
	68.2					Coarse ground, "pinhead", oatmeal.
	68.2					Medium ground oatmeal.
Major elements %DM	Calcium					
	0.053					Coarse ground, "pinhead", oatmeal.
	0.059					Medium ground oatmeal.

Table 1.26 OAT KERNEL: Oil content (light petroleum extract) and free fatty acids as % of DM
Reference: 305

Variety	Oil Mean	Oil Range	Free fatty acids	Notes
Black Supreme	5.4	4.5 - 6.4	0.33	10 varieties of spring oats grown together at 22 centres
Black Tartarian	5.8	4.9 - 6.6	0.35	
Eagle	7.5	6.5 - 8.2	0.37	
Early Miller	7.4	6.4 - 8.3	0.39	
Golden Rain II	6.8	6.2 - 7.5	0.37	
Marvellous	6.5	5.6 - 7.2	0.37	
Onward	5.7	4.7 - 6.7	0.32	
Star	6.4	5.6 - 7.0	0.35	
Victory	6.3	5.4 - 7.0	0.33	
S 84	6.6	5.8 - 7.4	0.38	
Grey Winter	9.6	8.1 - 11.1	0.37	4 varieties of winter oats grown together at 46 centres
Picton	8.8	7.3 - 9.9	0.39	
S 147	7.4	6.2 - 8.6	0.33	
S 172	9.2	7.8 - 10.2	0.39	
Victory	6.2	5.2 - 7.0		100 samples grown in England
	6.5	5.9 - 7.3		74 samples grown in Scotland

Table 1.27 OAT KERNEL: Thiamin content as μg/g by stage of growth and distribution in the kernel.
Reference: 280

Stage of growth	July 17	July 24	July 31	August 14	August 31	September 6 (harvest)
Marvellous	8.7	7.9	7.3	5.3	5.9	5.8
Star	11.6	9.4	7.7	5.1	-	4.4
		Distribution in kernel				
		Fractions of kernel				
	Whole Kernel	Embryo	Scutellum	Bran	Endosperm	
Early Miller	6.2	63	83	27	0.5	
Star	5.5	42	70	28	0.2	

Table 1.28 OAT OFFALS

OAT FEEDMEAL

Proximate constituents %DM	CP	EE	NFE	CF	Ash	Notes	Reference
	1.9 1.9 3.1 5.6	0.9 0.7 1.4 2.1	57.0 57.4 57.8 58.5	35.7 36.0 31.9 29.9	4.5 4.0 5.8 3.9		435
	4.2 (1.4-8.7)	1.8 (0.5-3.7)	55.6	34.1 (23.6-41.0)	4.3	Means and ranges for 24 samples from 18 mills	436
	6.8 5.4	3.4 2.3	61.1 60.3	23.8 28.1	4.9 3.9		501
	3.4	1.7	63.1	28.8	3.0		120
	8.5	4.8	67.3	16.6	2.8		112
	9.4	5.7	55.7	24.7	4.5	Digestibility of organic matter 60.5 For digested nutrients and metabolisable energy see table 1.219	137
Carbohydrate fractions %DM							
Starch	1.9 3.0 7.8						435
	6.2 (1.1-16.7) 3.0						436
Sugar	10.3 0.7						120
Major elements %DM							
Calcium Phosphorus Sodium Chlorine	0.09 0.32 0.08 0.12						112

OAT DUST

Proximate constituents %DM	CP	EE	NFE	CF	Ash	Notes	Reference
White (Heavy)	16.9 17.5 17.5	10.3 9.6 9.5	59.0 63.8 64.5	8.8 5.0 4.3	5.0 4.1 4.2		435
Brown (Light)	9.4 9.4 10.6	3.5 3.6 4.7	52.4 53.9 55.4	27.8 27.2 22.0	6.9 5.9 7.3		
Cyclone	8.8 10.6 10.6	4.0 5.8 5.6	54.9 60.7 63.1	25.5 17.9 15.7	6.8 5.0 5.0		
Polisher Brown screenings Unspecified	15.6 8.8 13.8	10.0 5.4 9.1	65.5 59.5 63.8	5.1 20.7 9.7	3.8 5.7 3.7		
White Brown (scree)	14.9 16.9 13.1	9.3 9.1 7.0	55.2 52.8 50.8	16.0 16.6 23.5	4.6 4.6 5.6	For digested nutrients and metabolisable energy see table 1.219	137

Table 1.28 OAT OFFALS (continued)
OAT DUST (continued)

	CP	EE	NFE	CP	Ash	Notes	Reference
Carbohydrate fractions %DM							
Starch	26.2						137
White (Heavy)	39.3						
	36.3						
Brown (Light)	3.8						
	7.9						
	7.9						
Cyclone	9.3						
	22.0						
	23.7						
Polisher	38.4						
Brown Screenings	15.0						
Unspecified	36.6						
Vitamins							
Thiamin μg/g							
Unspecified type	2.4					Mean of 4 samples	280

MEAL SEEDS OR SIDS

Proximate constituents %DM	CP	EE	NFE	CF	Ash	Notes	Reference
	6.2	2.9	65.9	21.8	3.2		435
	10.0	4.4	66.5	16.4	2.7		
	8.7	4.0	65.8	18.2	3.3		
Carbohydrate fractions %DM							
Starch	23.8						
	32.4						
	27.7						
Vitamins							
Thiamin μg/g	1.6					Mean of 4 samples	280

OAT HUSKS, HULLS

Digestibility %	DM	OM				Notes	Reference
Coarsely ground	69.0	70.9					
Medium ground	68.5	70.2				Digestibility by pig	114
Fine ground	68.2	70.0					
Proximate constituents %DM	CP	EE	NFE	CF	Ash	Notes	Reference
	1.5	2.6	52.5	39.9	3.5		136
Coarsely ground	2.5	0.9	59.3	33.0	4.3	For digested nutrients	114
Medium ground	2.2	1.0	58.8	34.4	3.6	and metabolisable energy	
Fine ground	2.2	0.9	58.9	34.6	3.4	see table 1.219	
	3.8	1.6	59.2	31.5	3.9		435
	1.2	0.3	56.4	38.1	4.0		
	1.2	0.4	54.8	39.3	4.3		
	1.2	0.4	57.2	35.8	5.4		
	1.5	0.4	54.8	38.8	4.5		436
	–	–	–	37.3 (34.8 – 39.3)	–	Mean and range for 11 varieties, each grown at 7 centres.	437
Amino acids N as % of total N							
Glutamic acid	2.2						269
Histidine	0.9						
Lysine	2.1						
Proline	3.8						

continued

Table 1.28 OAT OFFALS (continued)
OAT HUSKS (continued)

	CP	EE	NFE	CP	Ash				Notes	Reference
Carbohydrate fractions %DM										
Cellulose	33.4									136
Lignin	13.6									
Starch	5.6									435
	1.0									
	0.8									
	0.8									
Starch	0.9									436
Cellulose % of CF+NFE	39									
Hemicellulose										
+ pentosan "	29									
Lignin "	16									
Vitamins	July 24th	July 31st	Aug. 7th	Aug. 14th	Aug. 21st	Aug. 26th	Aug. 31st	Sept. 6th	Sampling date 1944/45	280
Thiamin μg/g										
Marvellous	2.4	2.2	1.4	0.8	0.3	0.2	0.2	–		
Star	2.1	1.6	1.0	0.7	0.5	0.2	–	0.1		

OAT CHAFF

Proximate constituents %DM	CP	EE	NFE	CF	Ash	Notes	Reference
	9.1	1.2	50.9	29.7	9.2		80

Table 1.29 WHEAT, WHOLE GRAIN: For metabolisable energy see table 1.220

						Reference
Digestibility %	DM	OM				
		88				82
	77.2					495
Proximate constituents %DM	CP	EE	NFE	CF	Ash	
Guardsman	13.8	1.4	81.2	1.7	1.9	119
Hybrid 46	13.2	2.0	80.9	1.8	2.2	
Unspecified	10.2	1.9	83.9	2.1	1.9	538
	10.4	2.0	83.4	2.4	1.8	
	11.5	2.1	82.2	2.5	1.7	82
Whole	8.9	1.0	85.0	3.0	2.1	495
Crushed	11.2	2.2	80.6	4.0	2.0	94, 92
Rolled	11.8	1.5	83.3	1.7	1.7	506

Table 1.30 WHEAT, WHOLE GRAIN: DM and crude protein as % of DM of 27 varieties in different manurial and soil trials
NIAB Centres, England
Reference: 233, 234, 235

Variety	Intensive manuring								Normal manuring							
	1935		1937		1938		1939		1935		1937		1938		1939	
	DM	CP	DM	CP	DM	CP	DM	CP	DM	CP	DM	CP	DM	CP	DM	CP
198 (20 C)																
55/10/2/C																
162/8/1/E	85.68	11.34							85.17	10.67						
162/55/1	85.48	11.23							85.18	11.14						
190/101			84.0	11.16	83.0	11.07					82.6	10.87	82.8	10.84		
202(47B)					83.0	10.18	84.3	9.95					82.5	10.19	84.7	9.07
April Bearded																
Atle																
Blanka																
Brown's 15/100																
Desprez 80 (Joncquien)			83.0	10.61	83.5	10.76	85.3	9.97			82.7	10.65	82.6	10.17	85.1	9.88
Diamond II																
Extra Kolben II																
Holdfast							84.3	10.98							84.8	10.24
Juliana			83.6	10.47	82.4	10.68					83.3	10.09	82.3	9.85		
Little Joss																
Little Tich			83.4	11.17							83.2	11.00				
Redman																
Red Marvel																
Ritchies'																
Squarehead's II																
Squarehead's Master	85.45	12.37	83.4	12.00	83.3	11.32	85.0	11.36	84.73	12.15	83.6	11.14	82.9	10.48	85.3	10.55
Steel	85.19	10.94							85.24	10.74						
W 81																
W 70A	85.81	11.98							85.60	13.01						
The Warden							84.3	9.50							85.1	8.67
Yeoman II	85.90	11.32	83.2	11.00	82.9	11.66	83.7	10.57	85.40	11.65	82.8	10.36	82.3	10.35	83.8	10.24
	85.6	11.53	83.4	11.07	83.0	10.94	84.5	10.39	85.2	11.56	83.0	10.68	82.6	10.31	84.8	9.77

Mean CP, intensive manuring 10.98
Range 9.50 - 12.37

Mean CP, normal manuring 10.58
Range 8.67 - 13.01

To show the "Fen" effect, mean CP has been calculated for all varieties with data in other subdivisions.

N manuring

Variety	Intensive	Normal	Strip	Fen trials
198(20C)			10.75	14.21
190/101	11.12	10.85		14.40
202(47B)	10.06	9.63		13.15
Desprez 80	10.45	10.23		13.10
Holdfast	10.98	10.24		14.16
Juliana	10.57	9.97		12.84
Yeoman II	11.14	10.65	10.76	14.27
Mean		10.53		13.73
Difference				30%

Table 1.30 WHEAT, WHOLE GRAIN: DM and crude protein as % of DM of 27 varieties in manurial and soil trials (continued)

Strip trial								Fen trial								
1935		1937		1938		1939		1935		1937		1938		1939		Mean
DM	CP	DM	CP	DM	CP	DM	CP	DM	CP	DM	CP	DM	CP	DM	CP	CP
				83.4	10.70	85.3	10.80			83.7	14.21					11.90
				84 2	10.81											10.81
																11.00
85.83	9.46															10.61
												83.1	14.40			11.67
										84.3	12.54	82.9	12.81	84.3	14.10	11.26
84.47	12.11					84.4	11.19									11.65
				81.1	10.09	84.9	10.84									10.46
				81.8	9.03	85.6	10.49									9.76
84.92	10.10															10.10
												82.6	13.40	84.7	12.80	11.03
						82.8	10.52									10.52
				81.9	9.71	84.3	12.60									11.15
										84.0	14.16					11.79
												82.5	12.84			10.79
										88.3	15.42					15.42
																11.08
85.98	10.64	83.6	10.60													10.62
				82.2	10.43	85.4	10.29									10.36
		83.5	10.82													10.82
												82.4	15.04			15.04
86.37	10.40															
85.12	10.62			83.4	10.68	85.2	11.46									11.21
																10.84
84.81	11.59															11.59
																12.49
																9.08
85.68	11.39	84.0	10.14							82.6	14.04	82.4	14.43	83.5	14.35	11.65
85.4	10.79	83.7	10.52	82.6	10.21	84.7	11.02			84.6	14.07	82.6	13.82	84.2	13.75	11.30 11.36

Mean CP in Strip trials 10.63
Range 9.03 - 12.60

Mean CP in Fen trials 13.88
Range 12.54 - 15.42

F

Table 1.31 WHEAT, WHOLE GRAIN: Nitrogen content by Variety and Year: effect of N fertiliser

Variety		1934		1935		1936		1950		1951		1952		1953		1954	
		No N	+N	No N	+N	No N	+N	No N	+N	No N	+N	No N	+N	No N	+N	No N	+N
Atle	(Area 1)															2.16	2.20
																+ 1.9%	
	(Area 2)															2.08	2.17
																+ 4.3%	
Cappelle	(Area 1)																
	(Area 2)																
Heine 7																	
Hybrid 46	(Area1)																
	(Area 2)													2.34	2.36		
														+ <1%			
Jubilegem										1.88	1.89						
Minister																	
Svenno																	
Staring												1.96	2.01				
												+ 2.6%					
Wilma								2.54	2.55								
								+ <1%									
Victor	(Area 1)	2.13	2.23	1.85	1.96	1.90	2.02										
		+ 4.7%		+ 5.9%		+ 6.3%											
	(Area 2)	2.98	3.13	1.80	1.82	1.73	1.68										
		+ 5.0%		+ 1.1%		– 2.9%											
Yeoman												1.73	2.00	1.88	[1] 2.14 [2] 2.25		
												+ 15.6%		[1] +13.8% [2] +19.7%			

1955	1956	1957	1958	1959	1960	Notes	Ref.
No N +N	No N +N	No N +N	No N +N	No N +N	No N +N		
2.13 2.10 –1.4%						Area 1: Northaw, 1954, 1955 Area 2: Tilsworth, 1954 31 or 62 kg N/ha as ammonium sulphate, combine-drilled or broadcast on seedbed, half of plots topdressed with 500 kg/ha Nitrochalk.	544
1.37 1.30 – 5.1%		(Area 2)	1.97 [1]2.12 [2]2.31 [1]+8% [2]+17%	1.90 [1]2.04 [2]2.19 [1]+7% [2]+15%	1.78 [1]1.90 [2]2.05 [1]+7% [2]+15%	1958: 75 or 150 kg N/ha 1959, 1960: 62 or 125 kg N/ha all as sulphate of ammonia	73 547
1.51 1.65 + 9.3% ...	1.98 2.19 +10.6%	1.82 2.01 +10.4%				19 kg N/ha as Nitrochalk in autumn and 2 or 3 top dressings totalling 94 or 375 kg/ha	543
1.69 1.82 +7.7% 2.19 2.21 +<1%	2.06 2.24 +8.7% 2.22 2.25 + 1.3%	1.98 2.18 +10.1%				Area 1: Bedfordshire and Hertfordshire Fertilisers as for Heine. Area 2: (1953) Cambridge (1955,1956) Topdressed 187 kg N/ha	543 264 263
							264
1.52 1.69 +11.2%	2.01 2.21 +10.0%	1.82 2.05 +12.6%				As Heine	543
	2.21 [1]2.31 [2]2.49 [3]3.08 [1]+4.5% [2]+12.7%					Northaw, 1956 : [1]31 kg/ha as ammonium sulphate [2]62 kg/ha .. .. [3]125 kg/ha	544
						125 kg/ha as sulphate of ammonia	264
						125 kg/ha as sulphate of ammonia	264
						Area 1: Rothamsted 38 kg N/ha as sulphate of ammonia Area 2: Woburn 50 kg N/ha as sulphate of ammonia	527
						[1]62 kg N/ha [2]125 kg N/ha Nitrochalk on soil, or leaf spray of ammonium nitrate or urea	496

Table 1.32 WHEAT, WHOLE GRAIN: Effect of nitrogen fertiliser on crude protein as % of DM

	Normal	1	2	3	4	Notes	Reference
Winter wheat							
Holdfast	11.8	12.5	14.4			Means for 14 centres. 1949-52. N level 1) 250 kg Nitrochalk/ha in April and 2) additional 250 kg/ha in June.	242
Atle	11.8	12.4	14.0				
Little Joss	11.7	12.2	14.4				
Hybrid 46	11.6	12.0	13.7				
Vilmorin 27	11.3	12.0	13.4				
Scandia	11.2	12.1	13.7				
Bersee	11.2	11.6	12.7				
N. 59	10.8	11.5	13.1				
Pilot	10.0	11.0	12.6				
Capelle	10.8	11.2	11.2	11.7	12.0	Means of four years	281
Heine VII	11.8	11.6	12.7	12.9	13.3	Means of two years	
Minister	9.7	10.2	10.4	11.1	11.9	One year. N 125, 250, 375 or 500 kg/ha as Nitrochalk	
Capelle-Desprez	10.8	11.2	11.8	12.2		Means for 3 or 4 trials in 3 years. N 50.5, 101 or 146 kg/ha as Nitrochalk	354

Table 1.33 WHEAT, WHOLE GRAIN: Effect of N fertiliser on crude protein as % of DM. Spring wheat variety trials by area, Eire 1962, 1963.

Area	1			2			3			4a			4b			Notes	Ref.
8 varieties in 3 areas	11.4			9.6			11.5									*percentage of DM assumed. Stated as protein%	
Extreme range by variety				8.3 – 12.2												Varieties: Atle, Atson, Carpo, Drott, Jufy I, Karn II, Koga II, Svenno	239
	N_1	N_2	N_3	N_1	N_2	N_3	N_1	N_2	N_3	N_1	N_2	N_3	N_1	N_2	N_3		
Trial 1 Atle	9.8	9.7	9.8	11.8	12.1	12.1	10.2	11.6	12.3	10.1	10.7	10.7	9.9	9.8	9.9	3 trials; 4 areas, 1 with 2 soil types	395
5 test wheats	8.8	9.2	9.5	11.7	12.5	12.8	10.3	10.9	11.6	10.0	10.2	10.7	9.1	9.6	9.9	Atle, reference wheat	
Trial 2 Atle	9.1	9.5	10.3	11.8	12.1	12.6	10.8	11.0	11.3	10.4	10.5	11.4	10.4	10.9	10.9	Trial 1: AP76, Ring, Els, SVO1311, SVO1312	
6 AP hybrids	9.4	9.4	9.7	11.6	11.8	12.3	11.0	11.3	11.7	10.4	10.8	11.2	9.8	10.3	10.8	Trial 2: 6 AP hybrids	
Trial 3 Atle	9.4	9.8	-	11.3	11.5	-	10.4	11.6	-	10.6	10.9	-	-	-	-	Trial 3: 2 7W hybrids 3 N levels (fertiliser not stated) in Trials 1 and 2	
2 7W hybrids	9.2	9.6	-	11.1	11.9	-	10.8	11.4	-	10.2	10.8	-	-	-	-	2 N levels in Trial 3: in kg/ha N_1 N_2 N_3 0 32.5 65 0 52	
Trial 1 Atle	11.1	11.4	11.8	12.0	12.0	11.8	11.2	11.0	11.3							2 trials: 3 areas Atle reference wheat	396
6 test wheats	11.4	11.3	11.7	11.9	11.8	12.1	10.9	11.4	11.8							Trial 1: Ring, Els and 4 SVO hybrids	
Trial 2 Atle	12.1	12.1	12.7	11.0	11.6	11.8	10.7	12.9	11.6							Trial 2: Quern, 2 8W hybrids, 3 AP hybrids	
6 test wheats	12.0	12.2	12.8	11.4	11.6	11.9	11.3	11.6	11.8							3 N levels: kg/ha N_1 N_2 N_3 0 32.5 65	

Table 1.34 WHEAT, WHOLE GRAIN: Amino acids and Carbohydrate fractions

Amino Acids as % of DM		Reference
Arginine	0.66	57
Cystine	0.31	
Histidine	0.29	
Isoleucine	1.22	
Leucine	1.45	
Lysine	0.43	
Methionine	0.23	
Phenylalanine	0.64	
Threonine	0.49	
Tryptophan	0.18	
Valine	0.71	

Carbohydrate fractions as % of DM with digestibility, where recorded

	Starch	Sugars	Pentosan	Cellulose	Lignin		Reference
Var. Guardsman	67.2	3.1					119
Var. Hybrid 46	67.5	3.1					
Unspecified	68.4	1.7	6.7	2.3	1.0		82
Digestibility %	100	100	32.6	0	0		
Unspecified	96.6		39.8	48.2		Poultry, intact birds	495
Digestibility %	98.7		33.3	43.4		Poultry, birds without caeca	

Table 1.35 WHEAT, WHOLE GRAIN: Major elements as % of DM

		Notes	Ref.
Calcium			
Unspecified	0.04		92
	Lo Li		
Var. Brons	0.17 0.17	Lo: no extra lime Li: lime supplying CaO 3750 kg/ha	
var. Ella Fylgia Karn	0.20 0.14 0.17 0.15 0.16 0.16		181
Unspecified	0.042 0.04 - 0.05	Mean and range of 9 treatments	134
Unspecified	0.07 0.04 - 0.12	Mean and range of same sample at 8 centres.	1
Phosphorus			
Crushed grain	0.18		92
	Lo Li		
var. Brons Ella Fylgia Karn	0.28 0.28 0.27 0.29 0.28 0.27 0.30 0.28	Lo: no extra lime Li: lime supplying CaO 3750 kg/ha	181
Unspecified	0.29 0.24 - 0.32	Mean and range of same sample at 6 centres.	1
Magnesium			
Unspecified	0.12 0.11 - 0.13	Mean and range of 9 treatments	134
Unspecified	.09 0.08 - 0.11	Mean and range of same sample at 9 centres.	1
Potassium			
Unspecified	0.50 0.48 - 0.53	Mean and range of 9 treatments	134
	Lo Li		
Var. Brons Ella Fylgia Karn	0.20 0.21 0.20 0.23 0.23 0.21 0.23 0.25	Lo: no extra lime Li: lime supplying CaO 3750 kg/ha	181
Unspecified	0.40 0.37 - 0.46	Mean and range of same sample at 9 centres	1
Sodium	Lo Li		
Var. Brons Ella Fylgia Karn	0.05 0.05 0.05 0.04 0.06 0.04 0.06 0.04	Lo: no extra lime Li: lime supplying 3750 kg/ha	181
Unspecified	0.003 0.003 - 0.004	Mean and range of 9 treatments	134
Sulphur			
Unspecified	0.14 0.13 - 0.16	Mean and range of same sample at 5 centres.	1

Table 1.36 WHEAT, WHOLE GRAIN: Trace elements as μg/g DM

		Notes	Ref.
Iron Unspecified	39 32 – 51	Mean and range for same sample at 8 centres See also Table 1.37	1
Manganese	Lo Li		
Var. Brons	170 150	Lo; no extra lime	
Ella	120 140	Li; lime supplying CaO	181
Fylgia	170 160	3750 kg/ha	
Karn	180 130		
15 varieties	32.3 15.0 – 67.0	Mean and range at 7 centres See also Table 1.37	251
Unspecified	F_0 61 F_9 59 48-70	Fo: no fertiliser F9: mean and range of 9 fertilisers; highest value with farmyard manure	551
Unspecified	41 38-44	Mean and range for same sample at 6 centres	1
Copper Unspecified	F_0 10 F_9 9 7-12	FO: no fertiliser F9: mean and range of 9 fertilisers: highest values with farmyard manure	551
Unspecified	7.5 6.5 - 9.1	Mean and range for same sample at 4 centres	1
Molybdenum Unspecified	F_0 0.06 F_9 0.06 0.03-0.09	FO: no fertiliser F9: mean and range of 9 fertilisers: highest values with farmyard manure	551
Unspecified	0.05 0.05-0.06	Mean and range for same sample at 4 centres	1
Zinc Unspecified	F_0 45 F_9 50 42-67	FO: no fertiliser F9: mean and range of 9 fertilisers: highest values with farmyard manure	551
Unspecified	42 39-44	Mean and range for same sample at 4 centres	1
Boron Unspecified	1.4 2	Spectrographic Colorimetric	450
Unspecified	8.6 2-16	Mean and range for same sample at 5 centres	1
Cobalt Unspecified	>0.02	1 centre only	
Nickel ,,	0.44		
Vanadium ,,	>0.03		
Lead ,,	<0.6		
Tin ,,	0.5		1
Titanium ,,	1.0		
Barium ,,	3.0		
Strontium ,,	1.3		

Table 1.37 WHEAT, WHOLE GRAIN: Iron and Manganese as $\mu g/g$ DM 7 Centres, Scotland and England, 1947.

Reference: 251

Variety	Fe		Mn	
	Mean	Range	Mean	Range
Bersée	39.4	33.5 - 48.5	32.8	17.3 - 63.5
Holdfast	40.5	32.3 - 49.7	38.6	26.6 - 67.0
Jubilégem	36.6	32.3 - 46.2	33.0	20.8 - 56.6
Juliana	40.0	32.3 - 47.3	32.3	19.6 - 54.3
Little Joss	46.7	34.6 - 56.6	35.7	24.2 - 60.0
Redman	40.8	32.3 - 50.8	32.0	18.5 - 57.7
Rivet	40.1	32.3 - 50.8	30.5	19.6 - 50.8
Squarehead II	44.9	33.5 - 70.4	35.4	21.9 - 52.0
Squarehead's Master	39.1	32.3 - 48.5	32.4	20.8 - 52.0
Steadfast	39.4	30.0 - 49.7	31.6	16.2 - 46.2
Victor	38.5	30.0 - 45.0	26.9	17.3 - 37.0
Vilmoran 27	38.1	31.2 - 47.3	30.7	16.2 - 58.9
Warden	35.7	27.7 - 48.5	27.4	15.0 - 37.0
Wilma	32.8	25.4 - 40.4	32.7	18.5 - 46.2
Yeoman	38.1	30.0 - 46.2	33.0	17.3 - 54.3
Mean and range of 15 varieties at 7 centres	39.4	25.4 - 70.4	32.3	15.0 - 67.0

Table 1.38 WHEAT, WHOLE GRAIN: Tocopherols as $\mu g/g$ DM

Variety	Total	Fractions as % of total				Reference
		α	β	ϵ	ζ	
Yeoman	34 to 39	50	20			100
Atle	42.7	33.6	12.0	44.4	10.0	248

Table 1.39 WHEAT, WHOLE GRAIN: Thiamin content as μg/g DM, 11 varieties, 4 treatments

Reference: 367

Variety	Manurial treatment		Strip trial	Fen trial
	Normal	Intensive		
Airing II	3.99			
Chevalier	3.93			
Desprez 80	3.06	3.99		4.62
Hold fast	4.32			
Iron red	3.45			
Scandia	4.20			
Squarehead's Master	4.29	2.34	3.99	
The Warden	3.75	4.29		
Yeoman II	3.66	3.81		5.43
198 (20C)			4.26	
202 (47B)	3.12	3.51		5.40
Mean	(10) 3.78	(5) 3.59	(2) 4.12	(3) 5.15
April Bearded	4.89			
Atle	4.20			
Blanka	4.38			
Extra Kolben II	3.90			
Hybrid 29	4.38			
Red Marvel	5.67			
Mean	4.57			

Table 1.40 WHEAT, WHOLE GRAIN: N as % of DM and Thiamin as μg/g DM, means of 6 centres

Reference: 252

Variety	1947		1948		1949		Mean Thiamin
	N	Thiamin	N	Thiamin	N	Thiamin	
Bersée	2.08	4.13	1.98	4.44	1.89	4.54	4.37
Holdfast	2.14	3.40	2.09	4.09	2.05	4.02	3.84
Jubilégem	1.89	3.36	1.91	4.20	1.85	3.95	3.84
Juliana	1.91	3.57	1.95	3.95	1.87	4.09	3.87
Little Joss	2.05	3.61	2.11	4.02	2.07	4.20	3.94
Redman	2.09	3.81	2.02	4.00	2.02	4.00	3.94
Rivet	2.00	2.98	2.03	3.02	1.89	3.05	3.02
Squarehead's Master	2.10	3.64	2.13	4.23	2.01	4.20	4.02
Squarehead II	2.20	3.81	2.06	4.30	2.07	4.30	4.14
Steadfast	1.92	3.50	1.89	3.92	1.83	4.09	3.84
Vilmorin 27	2.03	4.26	2.11	4.44	2.00	4.26	4.32
	Means 2.04	3.64	2.03	4.06	1.96	4.06	3.92
	Range	2.98-4.26		3.02-4.44		3.05-4.54	

Highest thiamin on average, Bersée
Lowest " in each year, Rivet

Table 1.41 WHEAT OFFALS: Composition of Flour and Milling offals at various extraction rates
(Expressed on 13 % moisture basis)

Reference: 312

Material	Yield %	Protein % (N x 5.7)	Oil %	Ash %	Fibre %	Thiamin µg/g	Nicotinic acid, µg/g
Wheat:	100	12.4	2.0	1.5	2.0	3.87	57
Flour:							
85% extraction	85	12.5	1.5	0.92	0.33	3.42	–
80% extraction	80.5	12.0	1.4	0.72	0.20	2.67	19
70% extraction	70	11.4	1.2	0.44	0.10	0.7	10
Fine Wheatfeed: (shorts)							
85% extraction	10	12.6	4.7	5.1	10.6	6.0	–
80% extraction	12.5	14.3	4.7	4.7	8.4	10.4	191
70% extraction	20	15.4	4.7	3.5	5.2	14.0	113
Bran:							
85% extraction	5	11.1	3.7	6.1	13.5	4.6	–
80% extraction	7	12.4	3.9	5.9	11.1	5.0	302
70% extraction	10	13.0	3.5	5.1	8.9	6.0	232

Table 1.42 WHEAT OFFALS: For metabolisable energy see table 1.220

Digestibility %	DM	OM	Digestibility by:	Ref.
Broad bran		39.4	Poultry	257
Coarse bran	59.8	63.6	Sheep	563
” ”	53.9	56.5	Pig	
Milled coarse bran	57.6	59.8	Pig	
Coarse bran		44.2	Poultry	260
” . ”		64.1	Rabbit	
Fine bran	63.9	66.6	Sheep	563
” ”	65.7	67.9	Pig	
” ”		54.8	Poultry	260
Middlings (coarse) (straight-run pollards)		63.2	Poultry	257
Middlings (fine)		84.6	Poultry	
Middlings (coarse) (fine wheatfeed)	67.4	70.6	Sheep	563
” ”	69.0	71.2	Pig	
Middlings (fine)		75.6	Poultry	83
Wheat and barley offals (10% barley)				
Fine bran		40.9	Poultry	260
Coarse miller's offals	54.7	57.9	Sheep	563
” ” ”	45.6	47.9	Pig	
Fine miller's offals	63.2	66.1	Sheep	
” ” ”	56.1	58.5	Pig	

Proximate constituents as %DM	CP	EE	NFE	CF	Ash	Ref.
Broad bran	16.6	4.2	60.6	11.9	6.7	257
Coarse bran	17.3	4.9	60.0	11.9	5.9	563
Coarse bran milled	17.2	5.1	59.9	11.9	5.9	
Coarse bran	17.4	5.8	58.2	12.9	5.7	260
	15.6	4.9	60.6	11.8	7.1	119
Fine bran	17.5	5.1	61.5	10.8	5.1	563
	17.3	4.7	63.6	9.3	5.1	260
Fine bran (10% barley)	16.0	5.0	61.6	11.3	6.1	260
Miller's coarse offals (10% barley)	14.1	4.4	59.1	15.5	6.9	563
Miller's fine offals (10% barley)	16.2	5.0	61.2	11.9	5.7	
Middlings (coarse) (Straight-run pollards)	18.9	5.0	64.7	7.1	4.3	257
Middlings, fine	9.7	2.6	83.4	2.2	2.1	
Middlings (Fine wheatfeed) (3 samples)	16.8	5.0	62.2	10.2	5.8	196
Middlings, coarse (Fine wheatfeed)	18.8	5.8	62.2	8.6	4.6	563
Middlings, fine	20.2	5.9	65.5	4.7	3.7	83
Fine parings	17.6	5.0	67.4	6.1	3.9	119
French pollards	18.6	4.3	62.9	8.4	5.8	

Table 1.42 WHEAT OFFALS: (continued)

Bran		Reference
Amino acids %DM		
Arginine	0.99	57
Cystine	0.25	
Histidine	0.28	
*Iso*leucine	0.74	
Leucine	1.08	
Lysine	0.64	
Methionine	0.18	
Phenylalanine	0.46	
Threonine	0.47	
Tryptophan	0.30	
Valine	0.67	
Carbohydrate fractions %DM		
Starch	38.3	119
Sugars	9.3	
Major elements %DM		
Phosphorus	0.15	368
Phytate phosphorus	0.13	
Trace elements		
Iron μg/g DM	124	
Vitamins		
Tocopherols in Yeoman variety		
total μg/g DM	69 to 87	100
Tocopherols in α form % of total	14	
" in β form % of total	6	
Thiamin μg/g	4.8	368
Nicotinic acid μg/g	250	
Riboflavin μg/g	5.0	
Middlings and similar forms		
Fine parings		
Carbohydrate fractions %DM		
Starch	38.3	119
Sugars	9.3	
French pollards		
Carbohydrate fractions %DM		
Starch	22.4	
Sugars	9.2	
Fine wheatfeed (middlings)		
Vitamins		
(Average of 3 samples)		
Tocopherols, total μg/g	104	196
α form % of total	23	
β form % of total	21	
ε form % of total	42	

Table 1.43 WHEAT OFFALS: Fibre as % of DM

Reference: 268
Courtesy of Research Association of British Flour-Millers

	Fine Wheatfeed			Coarse Wheatfeed			Weatings		
	Samples	Mean	Range	Samples	Mean	Range	Samples	Mean	Range
April 1960 – Sept, 1960	16	6.62	4.8–7.3	64	10.40	8.7–11.4	41	6.13	3.7–7.9
Sept. 1960 – March 1961	23	6.50	5.1–7.5	67	10.37	8.6–11.8	46	6.17	4.9–7.5
April 1961 – Sept. 1961	22	6.25	4.7–7.5	60	10.14	8.4–11.6	44	5.96	4.9–7.2
Oct. 1961 – March 1962	24	6.17	4.5–8.0	57	10.10	7.6–12.4	42	6.26	3.6–8.4
							Superfine Weatings		
May 1958 – Sept. 1958							12	5.05	3.8–6.2
Oct. 1958 – March 1959							5	4.96	4.2–6.5
March 1959–Sept. 1959							5	4.54	4.2–5.0
Oct. 1959–March 1960							2	4.9	4.7–5.0

Percentage distribution of samples over the range of fibre contents

Fibre %	Fine wheatfeed		Coarse wheatfeed		Weatings	
	1960	1961	1960	1961	1960	1961
	4 Apr. '60 to 31 Mar. '61	1 Apr. '61 to 31 Mar. '62	4 Apr. '60 to 31 Mar. '61	1 Apr. '61 to 31 Mar. '62	4 Apr. '60 to 31 Mar. '61	1 Apr. '61 to 31 Mar. '62
3.5 - 4.0		2.2		0.9	2.3	1.2
4.0 - 4.5	2.6	13.0		1.7		1.2
4.5 - 5.0	10.3	8.7	1.5	4.3	2.3	3.5
5.0 - 5.5	7.7	17.4	5.3	8.5	11.5	17.4
5.5 - 6.0	25.6	23.9	21.4	25.6	27.6	25.6
6.0 - 6.5	30.8	13.0	33.6	32.5	28.7	26.7
6.5 - 7.0	23.1	13.0	25.2	23.1	19.5	18.6
7.0 - 7.5		8.7	12.2	1.7	4.6	3.5
7.5 - 8.0			0.8	0.9	3.5	1.2
8.0 - 8.5				0.9		1.2
	No. of samples 39	46	131	117	87	86

Table 1.44 CEREALS: Mixed

		Reference
Wheat + Maize (67:33 mixture) Grain		
Proximate constituents %DM		
Crude protein	12.1	206
Ether extract	2.9	
Nitrogen-free extract	80.1	
Crude fibre	2.9	
Ash	2.0	
Major elements %DM		
Calcium	0.11	
Phosphorus	0.46	
Sodium	0.05	
Chlorine	0.07	
Wheat + Maize, kibbled (50:50 mixture) Grain		
Proximate constituents %DM		
Crude protein	12.1	
Ether extract	1.7	
Nitrogen-free extract	83.1	
Crude fibre	1.6	
Ash	1.5	
Major elements %DM		
Calcium	0.11	
Phosphorus	0.32	
Sodium	0.04	
Chlorine	0.06	
Wheat + Oats (50:50 mixture) Grain		
Proximate constituents %DM		
Crude protein	13.5	
	12.8	
Ether extract	4.6	
Nitrogen-free extract	71.5	
Crude fibre	7.5	
	5.8	
Ash	2.9	
Major elements %DM		
Calcium	0.1	
Phosphorus	0.4	
Sodium	0.04	
Chlorine	0.07	

Table 1.45 BARLEY: WHOLE GRAIN

For digested nutrients and metabolisable energy see table 1.221

						Notes	Ref.
Digestibility: %	DM	OM					
		88				Sheep	538
		78				Poultry	82
	68.7					Poultry	495
Proximate constituents %DM	CP	EE	NFE	CF	Ash		
Unspecified	9.8	1.5	81.4	4.6	2.7		538
	11.3	1.9	79.5	4.2	3.1		
	12.9	2.5	75.5	6.3	2.8		82
	11.0	1.4	78.7	5.7	3.2		119
	11.4	1.1	74.3	5.6	7.6		
82	9.6	1.9	80.2	5.9	2.4	1961 (82 samples)	370
	(6.6-12.6)	(1.5-2.5)	(76.3-83.3)	(4.0-7.3)	(1.8-3.1)		
39	11.3	2.0	79.3	5.0	2.4	1962 (39 samples)	
	(8.5-13.7)	(1.4-2.6)	(75.9-82.3)	(4.0-6.6)	(2.0-2.8)		
58	12.0	1.8	78.1	5.5	2.6	1963 (58 samples)	
	(9.2-15.3)	(1.1-3.2)	(74.2-80.7)	(3.8-7.1)	(1.7-4.2)		
69	11.8	2.1	78.1	5.7	2.3	Means and ranges: 1936-1965 (69 samples)	Collected from 30 reports *
	(6.6-15.0)	(0.8-3.9)	(73.8-87.5)	(3.2-7.7)	(2.0-7.6)		

*82, 90, 92, 93, 94, 95, 97, 119, 120, 199, 200, 201, 202, 203, 495, 538, 557, 558, 559, 560, 561, 562, 565, 566, 569, 571, 573, 574, 575, 576.

Table 1.46 BARLEY: WHOLE GRAIN: Nitrogen content as % of DM

Variety trials

Variety	1933	1934	1935	1936	Notes	Ref.
Beavens 35/7		1.42	1.42	1.43		
Golden Archer	1.42	1.44				
Kenia		1.88				
New Cross		1.49	1.48	1.50	Six NIAB centres	335
Plumage Archer 1924		1.42	1.48	1.48	8 varieties over 4 years	
Spratt Archer	1.42	1.46	1.42	1.41		
Victory		1.83				
35/513			1.48	1.50		
Carlsberg II	1.47					
Drost	1.49					
Herta	1.60					
Ingrid	1.55				Scotland	524
Maythorpe	1.54				Means only of 3 years	
Proctor	1.60				1957-1960	
Provost	1.62					
Ymer	1.55					
B8(8)	1.74				E. Scotland. Normal fertiliser	459
	1.59				Extra 125 kg ammonium sulphate/ha in May	

Effect of rate and method of application of fertiliser and other treatments

Variety	Area	1930		1931		1932		Notes	Ref.
		N	I	N	I	N	I		
Plumage Archer 1924	Cambridge	1.56	1.62	1.58	1.52	1.56	1.58	Extra N as ammonium sulphate or sodium nitrate	16
	Good Easter	1.40	1.50	1.45	1.38	1.31	1.38	N: normal rate	
	Cannington	1.37	1.33	1.34	1.38	1.39	1.41	I: intensive	
	Long Sutton	1.37	1.36	1.47	1.48	1.52	1.61		
	Newport	1.46	1.43	1.43	1.37	1.31	1.48		
	Sprowston	1.37	1.41	1.28	-	1.46	1.58		
	Mean	1.42	1.44	1.42	1.43	1.42	1.51		
Spratt Archer	Cambridge	1.54	1.63	1.58	1.54	1.58	1.66		
	Good Easter	1.32	1.33	1.38	1.53	1.38	1.32		
	Cannington	1.23	1.28	1.34	1.38	1.43	1.39		
	Long Sutton	1.38	1.38	1.34	1.50	1.54	1.64		
	Newport	1.44	1.50	1.52	1.41	1.32	1.48		
	Sprowston	1.34	1.44	1.32	-	1.45	1.61		
	Mean	1.38	1.43	1.41	1.47	1.45	1.52		

continued

Table 1.46 BARLEY: WHOLE GRAIN: Nitrogen content as % of DM (continued)

Effect of time of application of nitrogen

	1936	1937	Notes	Reference
Spratt Archer				
	N_0 1.50	N_0 1.66	1 centre only	303
Preflowering	N_1 1.46	N_1 1.61	N_0: no extra nitrogen	
At or after flowering	N_1 1.76	N_1 2.02	N_1: 125 kg/ha as nitrate of soda	
Before and after		N_1 1.91		
Spratt				
		N_0 1.75		
Preflowering		N_1 1.68		
At or after flowering		N_1 2.05		
Before and after		N_1 1.96		

Effect of level and method of application of nitrogen

	N_0	N_1	N_2	Notes	Reference
	1.50			N_0: no extra nitrogen	
Broadcast		1.51	1.59	N_1: 31 kg/ha	544
Drilled		1.47	1.58	N_2: 62 kg/ha	
Topdressed		1.59	1.74	As sulphate of ammonia alone or with Nitrochalk top dressed For details see table 1.47	

Effect of level and method of application of nitrogen

	1957	1958	1959	Notes	Reference
	N_0 1.91	N_0 1.58	N_0 1.67	N_0: no extra nitrogen	545
Broadcast	N_1 2.00	N_1 1.57	N_1 1.71	N_1: N_2, N_3 : 37.5, 75.0,	
Drilled	2.03	1.56	1.75	112.5 kg/ha as sulphate of ammonia on seedbed.	
Drilled and topdressed			1.89	Nitrochalk topdressing 37.5 kg N/ha	
Broadcast	N_2 2.14	N_2 1.69	N_2 1.87		
Drilled	2.15	1.69	1.98	Varieties: Proctor at 11 centres,	
Drilled and topdressed	2.14	1.68	2.02	Herta at 2	
Broadcast	N_3 2.26	N_3 1.79	N_3 2.04	Rika at 1	
Drilled	2.30	1.81	2.25	Ingrid at 1	
Drilled and topdressed	2.35	1.82	2.20		

Effect of type of farm on response to nitrogen

		Response to extra N			Notes	Reference
	N_0	N_1	N_2	N_4	N_0: no extra nitrogen N_1, N_2, N_4: 125, 250 or 500 kg/ha as sulphate of ammonia	
Mean for varieties	1.51	0.00	0.04	0.26	Arable 1 year ley in rotation	282
Ymer, Freja, Earl	1.54	0.07	0.08	0.22	Ley more than 1 year but longer in crop than ley	
Proctor and Domen	1.65	-0.03	0.16	0.39	Cropping and ley of approx, equal duration. For details see table 1.48	

Effect of potassium and nitrogen application

	1958	1959	Notes	Reference
Variety Proctor	1.70 N_0, 1.84 N_1, 2.08 N_2 } K_0	1.85 N_0, 1.93 N_1, 2.05 N_2 } K_0	N applied as ammonium sulphate K as granulated PK compound fertiliser	546
	1.76 N_0, 1.84 N_1, 2.06 N_2 } K_1	1.91 N_0, 1.99 N_1, 2.12 N_2 } K_1	N_0: no extra nitrogen N_1, N_2: 50, 100 kg/ha K_0: no extra potassium K_1, K_2: 56, 112 kg/ha of K_2O	
	1.71 N_0, 1.88 N_1, 2.02 N_2 } K_2	1.93 N_0, 1.96 N_1, 2.12 N_2 } K_2		

Effect of nitrogen application and undersown crop

	N_0	N_1	N_2	Notes	Reference
				N_0: no extra N : N_1 and N_2: 187 or 375 kg Nitrochalk/ha	
Variety				Undersown crops: lucerne, broad red clover or trefoil; cocksfoot or Italian ryegrass	
Abed Kenia	1.77	1.84	1.89	None	309
	1.78	1.83	1.99	Legume	
	1.76	1.79	1.88	Legume and grass	
	1.74	1.79	1.85	Grass	

continued

Table 1.46 BARLEY: WHOLE GRAIN: Nitrogen content as % of DM (continued)
Effect of weed control by spraying with sulphuric acid

Variety	H_2SO_4			Notes	Reference
Plumage Archer	None	1.41			71
	9.2%		1.48		
	13.8%		1.46		
	18.4%		1.41		
Golden Archer	None	1.30			
	13.8%		1.43		
	18.4%		1.43		
Spratt Archer	None	1.34			
	9.2%		1.41		
	13.8%		1.41		
	18.4%		1.46		
	9.2%		1.43		
	9.2%		1.37		
	Mean	1.35	1.43		

Table 1.47 BARLEY: WHOLE GRAIN: Nitrogen content as % of DM. Effect of level and manner of application of fertiliser nitrogen.
Reference: 544

Date	Variety	Area 1	Area 2	Area 3	Area 4	Area 5	Notes
1954	Proctor	1.42 1.41= 1.36≠ - 1.48† 1.40+ -					Nitrogen applications are indicated thus: = 31 kg/ha as Amm. Sulphate: Broadcast ≠ 31 kg/ha as Amm. Sulphate: Drilled † 62 kg/ha as Amm. Sulphate: Broadcast + 62 kg/ha as Amm. Sulphate: Drilled * 31 kg/ha as Nitrochalk Topdressed ** 62 kg/ha as Nitrochalk Topdressed No mark indicates no application
1954	Plumage Archer 1 area same as Area 3 for Proctor			1.31 1.20= 1.21≠ - 1.30† 1.24+ -			
1955	Proctor		1.39 1.47= 1.37≠ 1.35* 1.45† 1.49+ 1.53**	1.41 1.50= 1.45≠ 1.53* 1.50† 1.60+ 1.71**			
1956	Proctor	1.39 1.29= 1.36≠ 1.58* 1.32† 1.46+ 1.63**		1.51 1.65= 1.50≠ 1.56* 1.72† 1.74+ 1.65**	1.56 1.64= 1.62≠ 1.63* 1.74† 1.74+ 1.96 **	1.71 1.75= 1.74≠ 1.78* 1.90† 1.77+ 1.89**	
1956	Herta	1.60 1.51= 1.56≠ 1.63* 1.64† 1.58+ 1.79*	1.68 1.68= 1.55≠ 1.69* 1.81† 1.73+ 1.80*				

Summary in Table 1.46

Table 1.48 BARLEY: WHOLE GRAIN: Nitrogen content as % of DM. Effect of type of farm, fertiliser and year.
Reference: 282

Date	Variety	Ymer					Fertiliser
	Farm	1	2	3	4		Kg N/ha as Amm. sulphate
		1.33	1.33	1.63	1.68		None
		1.34	1.42	1.73	1.61		125
		1.40	1.42	1.75	1.82		250
		1.70	1.79	1.88	1.90		500
1954	Farm	23	24	25	26		Kg N/ha as Nitrochalk
		1.60	1.95	1.73	2.10		None
		1.55	1.82	1.67	2.14		250 at 1st tiller
		1.59	1.83	1.63	2.06		250 at end of tillering
		1.48	2.11	1.59	2.05		250 at ear emergence
	Variety	Ymer				Freja	
	Farm	5	6	7	8	9	Kg N/ha as Amm. sulphate
		1.41	1.27	1.36	1.28	1.38	None
		1.40	1.31	1.63	1.30	1.33	125
		1.33	1.30	1.51	1.33	1.47	250
		1.65	1.54	1.42	1.42	1.80	500
1955							Kg N/ha as Nitrochalk
		1.31	1.46	1.57	1.24	1.35	None
		1.39	1.65	1.40	1.33	1.60	250 at 1st tiller
		1.57	1.49	1.39	1.37	1.60	250 at end of tillering
		1.36	1.49	1.37	1.25	1.70	250 at ear emergence
	Variety	Ymer			Freja		
	Farm	10	11	12	14	16	Kg N/ha as Amm. sulphate
		1.55	1.56	1.95	1.68	1.85	None
		1.41	1.52	1.91	1.68	1.82	125
		1.47	1.60	2.01	1.73	2.18	250
		1.73	1.76	2.08	1.95	2.41	500
1956							Kg N/ha as Nitrochalk
		1.44	1.49	1.93		2.11	None
		1.73	1.57	2.06		2.34	250 at 1st tiller
		1.45	1.61	1.95		2.27	250 at end of tillering
		1.47	1.77	2.02		2.31	250 at ear emergence
	Variety	Ymer		Earl	Proctor	Domen	
	Farm	17	18	19	20	21	Kg N/ha as Amm. sulphate
		1.96	1.62	1.57	1.39	1.66	None
		1.66	1.72	1.68	1.48	1.76	125
		1.78	1.95	1.60	1.49	1.82	250
		2.10	2.01	1.96	1.75	1.84	500
1957							Kg N/ha as Nitrochalk
		1.53	1.77	1.55	1.49		None
		1.97	2.01	1.79	1.71		250 at 1st tiller
		1.85	1.99	1.72	1.66		250 at end of tillering
		1.69	2.03	2.06	1.83		250 at ear emergence

Summary in Table 1.46.

Table 1.49 BARLEY: WHOLE GRAIN: Nitrogen as % of DM. Eire: by variety and year.

Variety	1939	1940	1941	1942	1943	1944	1945	1946	1947	1948	References
Archer			1.38	1.42							173, 174
Balder										1.40	180
Beavens 54/12/3							1.43	1.34 (9)			177, 178
D.S.K. Binder									1.50	1.43	179, 180
Golden Archer 1 x Hybrid 4B1				1.47							174
Golden Archer 2	1.56	1.42	1.37								171, 172, 173
Hume's Archer No. 1									1.47	1.33	179, 180
Kenia								1.65 (6)			178
Spratt-Archer 37 No. 3	1.52	1.41	1.34	1.36	1.37	1.68	1.48	1.43 (9)	1.48	1.36	171, 172, 173 174, 175, 176 177, 179, 180
Spratt-Archer 37 No. 3 H9	1.50										171
Spratt-Archer 37/6 No. 7	1.56	1.46									171, 172
Spratt-Archer 37 No. 3 H9 x Golden Archer 2 No. 2						1.76		1.49 (9)			176, 178
Spratt-Archer 37/9 x Golden Archer 2/1								1.42 (9)			178
Spratt-Archer 37 No. 3 x Golden Archer 2/2							1.44				177
Spratt-Archer 37/3 H9 x Hybrid 4B1 No. 2					1.51						175
Spratt-Archer 37 No. 4 x July Six-rowed 16/2		1.46									172
Spratt-Archer 37 No. 3 x Victory 2					1.44	1.79					175, 176
Ymer									1.46		179
Means of 18 varieties each grown at 10 (except where number given in brackets) centres over 10 years	1.53	1.44	1.36	1.42	1.44	1.74	1.45	1.47	1.48	1.38	

Table 1.50 MALTING BARLEY: Nitrogen as % of DM and Malt extract. Variety trials. Eire 1961.
Reference: 397

Area \ Variety	Proctor		Hunter		Volla		Wisa	
	N	IGE*	N	IGE	N	IGE	N	IGE
1	1.26	140.9	1.26	143.2	1.20	146.8	1.24	144.7
2	1.36	140.2	1.29	141.0	1.59	140.9	1.44	142.1
3	1.24	143.4	1.23	142.6	1.25	147.0	1.26	145.4
4	1.30	142.9	1.36	140.1	1.34	142.7	1.48	139.1
5	1.32	143.4	1.24	143.5	1.28	145.1	1.24	145.0
6	1.30	143.4	1.34	141.0	1.43	142.5	1.38	141.1
7	1.65	135.0	1.79	133.0	1.82	133.3	1.83	129.0
8	1.31	137.9	1.32	140.1	1.34	141.7	1.32	143.1
9	1.63	140.2	1.59	140.6	1.68	141.6	1.64	138.9
10	1.23	144.0	1.18	145.8	1.26	145.2	1.26	144.0
Mean	1.36	141.1	1.36	141.1	1.42	142.7	1.41	141.2

*Institute of Brewing Extract; approximates brewing practice. See comment, p. 13.

continued

Table 1.50 MALTING BARLEY: Nitrogen as % of DM and Malt extract. Variety trials. Eire 1961.
Reference: 397 (continued)

	N	Malt extract: IGE (dry)	
Hunter 24 test barleys	1.36 1.43 (1.33-1.50)	141.2 140.3 (137.8-143.1)	Small scale trial. Hunter, reference barley; 24 test barleys. Neutron Beorna and 23 hybrids
Hunter 5 test barleys	1.60 1.61 (1.56-1.68)	135.6 135.8 (134.3-136.6)	Drill strip trial, 1. Neutron Beorna and 4 hybrids
Proctor 5 test barleys	1.40 1.33 (1.27-1.48)	142.8 (139-143.6)	Drill strip trial, 2. Proctor reference barley, 5 hybrids
		Malt extract % DM	
Hunter 7 test barleys	1.35 1.40 (1.35-1.42)	81.4 81.9	European Brewery Convention trial, 1961

Table 1.51 MALTING BARLEY: Nitrogen as % of DM and Malt extract. Variety trials. Eire 1962: Effect of N fertiliser.
Reference: 397

Area	1		2		3		4		5		6		Notes
Fertiliser	N_0	N_1	N_0	N_1	N_0	N_1	N_0	N_1	N_0	N_1	N_0	N_1	Proctor and Hunter reference barleys 2 levels of N as ammonium sulphate N_0 N_1 0 125 kg/ha at seeding 6 test barleys: Beorna and 5 hybrids
	N												
Proctor	1.2	1.2	1.7	1.7	1.5	1.6	1.5	1.8	1.5	1.6	1.7	1.7	
Hunter	1.2	1.2	1.6	1.7	1.5	1.6	1.5	1.6	1.4	1.5	1.7	1.8	
Beorna and 5 hybrids	1.2	1.2	1.7	1.8	1.5	1.6	1.5	1.7	1.5	1.5	1.7	1.8	
	IGE (dry)												
Proctor	140.6	141.0	132.2	132.6	132.0	134.5	141.7	135.8	140.4	138.6	133.2	136.8	
Hunter	143.6	141.2	134.4	133.0	139.4	137.0	141.2	137.4	141.0	141.4	132.9	133.4	
Beorna and 5 hybrids	142.3	142.3	135.0	132.0	138.0	136.3	141.7	138.5	140.8	139.9	135.7	134.8	

Area	1	2	3	4	5		Notes
	N						Proctor and Hunter, reference barleys 3 hybrids: 48/9, 50/2, 92/21
Proctor	1.4	1.6	1.4	1.5	1.5		
Hunter	1.3	1.5	1.4	1.4	1.4		
3 Hybrids	1.3	1.6	1.4	1.4	1.4		
	IGE (dry)						
Proctor	143.3	140.5	145.4	143.5	144.2		
Hunter	144.7	139.3	146.4	142.7	144.0		
3 Hybrids	143.9	142.0	145.3	142.8	143.5		

Trial	2			4			Notes
Fertiliser	N_0	N_1	N_2	N_0	N_1	N_2	4 trials of which 1 and 3 without extra fertiliser Only 2 and 4 reported here. 3 extra N levels: kg/ha N_0 N_1 N_2 0 25 37.5 Test barleys, 8 CS hybrids (CS hybrids are bred at the Cereal Station, Ballinacurra)
	N						
Hunter	1.4	1.4	1.5	1.4	1.5	1.7	
Proctor	1.4	1.5	1.5	1.4	1.6	1.6	
8 CS hybrids	1.4	1.4	1.4	1.4	1.5	1.6	
	IGE (dry)						
Hunter	144.5	142.4	141.6	142.8	141.2	139.3	
Proctor	143.9	140.2	141.4	141.0	138.9	137.7	
8 CS hybrids	142.1	140.0	140.7	142.0	138.9	139.0	

Table 1.52 BARLEY: WHOLE GRAIN: Major elements as % of DM

		Notes	Reference
Calcium		Pot experiments in washed silica sand and granulated peat. K supplied as sulphate 50,500 or 5000 μg/g at base and top.	
Spratt Archer	0.05 0.03 0.16	Low potassium Medium potassium High potassium	522
Unspecified Unspecified	F_0 0.04 F_1 0.04 F_2 0.05 0.05 0.04	F_0: no extra calcium F_1: 3750 kg CaO/ha F_2: 3750 kg CaO and 3750 kg magnesian limestone/ha 1946 1947	434
Unspecified (meal)	0.07		95
Freja Kenia Ymer No. 5	F_0 0.19 F_1 0.18 F_0 0.18 F_1 0.16 F_0 0.19 F_1 0.16 F_0 0.13 F_1 0.17	F_0: no extra calcium F_1: 3750 kg CaO/ha	181
Unspecified	0.08		92
Unspecified	0.08		575
Unspecified	0.08 (0.05 - 0.16)	Mean and range of 179 samples grown in Wales	371
Phosphorus Spratt Archer	0.37 0.25 0.42	Low potassium Medium potassium High potassium	522
Unspecified Unspecified	F_0 0.49 0.53 F_1 0.51 0.54 0.35 0.36	Area 1 Area 2 Area 1 Area 2 F_0: no extra calcium F_1: 3750 kg/ha CaO 1946 1947	434
Unspecified (meal)	0.39		95
Freja Kenia Ymer No. 5	F_0 0.27 F_1 0.29 F_0 0.27 F_1 0.30 F_0 0.29 F_1 0.32 F_0 0.28 F_1 0.34	F_0: no extra calcium F_1: 3750 kg/ha CaO	181
Unspecitied	0.44		575
All varieties 1852-1937	0.39	Rothamsted survey	446
Unspecified	0.42		133
Unspecified	0.38 (0.26 - 0.52)	Mean and range of 179 samples grown in Wales	371

continued

Table 1.52 BARLEY: WHOLE GRAIN: Major elements as % of DM (continued)

Magnesium Spratt Archer	0.04 0.06 0.05	Low potassium Medium potassium High potassium	522
Unspecified	F_0 0.13 F_1 0.16 F_0 0.16 F_1 0.16 F_2 0.15	Area 1 Area 2 F_0: no extra calcium F_1: 3750 kg/ha CaO F_2: 3750 kg/ha CaO and 3750 kg/ha magnesian limestone	434
Unspecified	0.12 (0.11 - 0.14) 0.12 (0.10 - 0.17)	Means and ranges for 4 samples in 2 years 1946 and 1947	434
Unspecified	0.12 (0.09 - 0.16)	Mean and range for 179 samples grown in Wales	371
Potassium Spratt Archer	1.01 1.01 2.15	Low potassium Medium potassium High potassium	522
Unspecified	F_0 0.62 F_1 0.66 F_0 0.63 F_1 0.64	Area 1 Area 2 F_0: no extra calcium F_1: 3750 kg/ha CaO	434
Unspecified	0.56 0.62	1946 1947	434
Freja Kenia Ymer No. 5	F_0 0.29 F_1 0.32 F_0 0.28 F_1 0.29 F_0 0.29 F_1 0.30 F_0 0.24 F_1 0.30	F_0: no extra calcium F_1: 3750 kg/ha CaO	181
All varieties (1852-1937)	0.50	Rothamsted survey	446
Unspecified	0.53		133
Unspecified	0.49 (0.35 - 0.63)	Mean and range of 179 samples grown in Wales	371
Sodium Spratt Archer	0.12 0.10 0.08	Low potassium Medium potassium High potassium	522
Freja Kenia Ymer No. 5	F_0 0.06 F_1 0.06 F_0 0.06 F_1 0.06 F_0 0.07 F_1 0.06 F_0 0.05 F_1 0.06	F_0: no extra calcium F_1: 3750 kg/ha CaO	181
Unspecified	0.02 (0.006 - 0.04)	Mean and range of 179 samples grown in Wales	371
Chlorine Unspecified	0.08		92
Unspecified	0.08		575
Unspecified	0.14 (0.08 - 0.22)		371
Sulphur, Total Spratt Archer	0.42 0.29 0.26	Low potassium Medium potassium High potassium	522

Table 1.53 BARLEY: WHOLE GRAIN: Trace elements and Tocopherol as μg/g DM

		Notes	Reference
Cobalt			
Unspecified	0.07 (0.02 - 0.18)	Mean and range of 174 samples grown in Wales	371
Copper			
Unspecified	F_0 8 F_1 10 F_2 8 F_1+F_2 11 F_3 9 F_1+F_3 13 F_2+F_3 9 F_1+F_2+F_3 9 F_4 12 F_1+F_2+F_3+F_4 13	F_0: no extra fertiliser F_1: N 56 kg/ha as Nitrochalk F_2: P_2O_5 62 kg/ha as Superphosphate F_3: K_2O 125 kg/ha as potassium sulphate F_4: Farmyard manure	551
Unspecified	6.6 (3.5 - 19.8)	Mean and range of 179 samples grown in Wales	371
Manganese			
Unspecified	F_0 22 F_1 17 F_2 26 F_1+F_2 17 F_3 22 F_1+F_3 44 F_2+F_3 26 F_1+F_2+F_3 39 F_4 26 F_1+F_2+F_3+F_4 35	As for Copper	551
Freja	F_0 180 F_1 210	F_0: no extra calcium F_1: 3750 kg/ha CaO	181
Kenia	F_0 200 F_1 180		
Ymer	F_0 160 F_1 240		
No. 5	F_0 140 F_1 210		
Unspecified	16 (5 - 47)	Mean and range of 179 samples grown in Wales	371
Molybdenum			
Unspecified	F_0 0.21 F_1 0.24 F_2 0.12 F_1+F_2 0.19 F_3 0.11 F_1+F_3 0.21 F_2+F_3 0.12 F_1+F_2+F_3 0.18 F_4 0.18 F_1+F_2+F_3+F_4 0.21	As for Copper	551
Zinc			
Unspecified	F_0 37 F_1 40 F_2 30 F_1+F_2 33 F_3 34 F_1+F_3 42 F_2+F_3 34 F_1+F_2+F_3 37 F_4 32 F_1+F_2+F_3+F_4 49	As for Copper	551
Unspecified	37 (19 - 77)	Mean and range of 179 samples grown in Wales	371

Tocopherols

	Total	Fractions as % of total						Reference
		α	β	γ	ϵ	ζ	η	
Proctor	45	21.0	9.3	trace	26.5	43.2	trace	248
Bére	56 - 71	10	67	0				100

Table 1.54 BARLEY BRAN: Proximate constituent as % of DM

		Ref.
Crude protein	11.3	
Ether extract	3.7	
Nitrogen free extract	63.4	507
Crude fibre	17.2	
Ash	4.4	

Table 1.55 DISTILLERS' AND BREWERS' GRAINS AND OTHER RESIDUES

Proximate constituents as % of DM

	CP	EE	NFE	CF	Ash	Carbo-hydrates	Notes	Ref.
Draff: distillers' wet grains	17.5	7.8	44.7	17.5	12.5			72
Light grains (Bourbon)	30.0	9.4	44.0	13.3	3.3		Bourbon whisky distillers' grains = maize distillers' dried grains. Light grains are thoroughly washed and dried. Scotch malt and Scotch grain distillers' grains = barley distillers' dried grains.	88
Grains with solubles (Bourbon)	29.6	9.7	45.0	10.5	5.2			
Solubles (Bourbon)	28.2	9.6	45.7	4.1	12.4			
Light grains (Scotch malt)	19.8	8.4	50.1	18.6	3.1			
Grains with solubles (Scotch malt)	25.8	3.5	52.6	12.0	6.1			
Solubles (Scotch malt)	30.6	0.2	60.3	2.2	6.7			
Light grains (Scotch grain)	20.8	4.6	52.5	18.8	3.3			
Grains with solubles (Scotch grain)	19.7	4.5	61.0	10.0	4.8			
Solubles (Scotch grain)	16.9	4.0	70.6	3.0	5.5			
Dark maize grains	22.9	7.5	55.8	9.5	4.3	45.1	*Dried maize grains* are normally dried to 5% moisture in a once-through drying operation which reduces the moisture content from approx. 60% to 5%. Production of *Scotaferm* is a complex process. It is dried distillers' solubles. *Dark grains* are a combination of distillers solubles and wet maize grains, but the definition varies from distillery to distillery. *Dried barley grains* and *Scotasol* are spray dried distillers solubles obtained from the malted grain process.	593
Dried maize grains	21.7	8.1	51.4	17.0	1.8	40.3		
Barley grains and solubles	21.0	7.7	54.6	13.2	3.5	46.6		
Scotaferm and dried grain solubles	29.0	8.4	47.3	2.3	13.0	43.2		
Scotasol and malt solubles	25.5	0.2	49.9	0.2	24.2	45.6		
Dark grains — Sample no.								
1A	20.6	5.6	58.1	11.2	4.5		Dark grains is wet draff mixed with the syrup from the first distillation, potale; the mixture is dried and sold as "dark grains".	592
1B	20.7	5.5	57.6	11.4	4.8			
2	22.1	4.0	60.4	8.6	4.9			
3	20.9	5.8	58.5	11.3	3.5			
4	20.9	5.6	59.3	11.6	2.6			
5	21.8	3.7	61.7	8.4	4.4			
6	23.0	3.0	61.4	7.8	4.8			
7	21.6	5.6	57.7	11.0	4.1			
Malt distillers' dried solubles	28.2	0.2	53.5	-	18.1		A typical commercial sample	591
Distillers' dreg: wet	49.3	20.0	22.5	6.3	1.9		Dreg is the liquid drawn off to leave draff	500
Distillers' dreg meal	49.7	14.7	28.7	5.2	1.7		For digested nutrients and metabolisable energy see table 1.221	528
Brewers' grains	18.6 (17.9-19.4)	-	-	17.3 (14.5-19.0)			Freshly delivered on farm	594
	20.3 (15.9-22.9)			19.7 (16.4-21.5)			Stored	
	22.4	6.8	46.6	20.1	4.1			501
Grains: dried	22.2	6.1	48.5	19.6	3.6		Not further described	501
	20.5	8.4	50.9	17.6	2.6			502
	23.9	7.3	48.8	16.7	3.3			505

Major elements as % of DM

	Ca	P	Na	Cl		Notes	Ref.
Dark maize grains	0.25	0.53	0.08	0.12		See under Proximate constituents, same reference	593
Dried maize grains	0.08	0.12	0.04	0.06			
Barley grains and solubles	-	-	0.08	0.12			
Scotaferm and dried grain solubles	4.05	1.15	0.16	0.24			
Scotasol and malt solubles	7.01	1.60	0.36	0.54			
Malt distillers' dried solubles	4.6	-	-	-		A typical commercial sample	591
Brewers' grains	0.55 (0.42-0.70)	0.74 (0.43-1.02)				Freshly delivered on farm	594
	0.65 (0.50-0.78)	0.82 (0.52-1.22)				Stored	

continued

Table 1.55 DISTILLERS' AND BREWERS' GRAINS AND OTHER RESIDUES (continued)

		Vitamins as $\mu g/g$						
	Thiamin	Ribo-flavin	Nicotinic acid	Pantoth-enic acid	Pyrid-oxine	Biotin	Notes	
Malt distillers' dried solubles	1.5	21	510	67	19	0.7	A typical commercial sample contained also: Inositol 9.9 and Choline 2.0 mg/g DM	591
Light grains (Bourbon)		4	43	3.5			See under Proximate constituents same reference	88
Grains with solubles (Bourbon)		8	70	11				
Solubles (Bourbon)		28	120	23.5				
Light grains (Scotch malt)		1	25	0.7				
Grains with solubles (Scotch malt)		15	360	12				
Solubles (Scotch malt)		28	700	24				
Light grains (Scotch grain)		1.2	70	5				
Grains with solubles (Scotch grain)		10	112	12				
Solubles (Scotch grain)		19.5	164	19				

Table 1.56 MAIZE: WHOLE GRAIN

Proximate constituents %DM	CP	EE	NFE	CF	Ash	Soluble carbo-hydrate	Notes	Ref.
	10.0					66.7	1 sample	255
	10.4					67.5	Mean of 2 samples	
	10.6	5.6	80.2	2.1	1.4			120
	9.6	4.5	82.0	2.8	1.1		Plate Argentinian, white	330
	11.6	4.2	81.2	1.6	1.3		For digested nutrients and metabolisable energy see table 1.222	82
Carbohydrate fractions % DM	Sugar	Starch	Pentosan	Cellu-lose	Lignin			
	1.16 (100)	72.5 (100)	5.07 (38.1)	1.72 (9.6)	0.48 (14.6)		Digestibility by poultry in brackets where recorded	82
	1.3	62.4						120
Major elements % DM	Ca	P						
	0.01	0.30					1 sample	255
	0 01	0.32					Mean of 2 samples	

Vitamins Tocopherols $\mu g/g$ DM	Total	Fractions as % of total					Notes	Ref.
		α	γ	δ	ζ	η		
	16.0							330
	108	5.2	87.2	trace	2.3	5.3	Sweet corn: John Innes Hybrid	248

B Vitamins $\mu g/g$ DM	Thiamin	Ribo-flavin	Pyrido-xine	Calcium panto-thenate	Nico-tinic acid	Folic acid	Biotin	Vitamin B12	Notes	Ref.
	4.6	0.43	3.4	3.4	(1) 0.3 (2) 17.7	70	20	<0.1	Plate Argentinian white 1: free 2: bound Tryptophan, mg/g DM 0.85	330

Table 1.57 MAIZE: MEAL

Proximate constituents % DM	CP	EE	NFE	CF	Ash	Notes	Ref.
	10.6 10.9	5.0 5.0	81.1 80.7	2.1 2.0	1.2 1.4		500
	11.6	6.2	77.8	2.8	1.6		579
	10.7	5.8	79.3	2.4	1 8		584
	10.7 11.1	4.8 4.7	80.9 80.1	1.9 2.3	1.7 1.8		560
	10.9 10.9	4.5 4.3	80.4 80.6	2.5 2.4	1.7 1.8		563
	10.8	7.3	74.8	5.0	2.1		570
	10.6 10.9	5.2 5.0	80.7 80.5	2.0 2.0	1.5 1.6		97
	10.5	5.0	80.8	2.2	1.5		234
	10.3	3.7	82.4	2.0	1.6	Yellow maize, For metabolisable energy see table 1.222	119
	10.5	1.7	85.4	1.4	1.0	Degermed, cooked. For digested nutrients and metabolisable energy table 1.222	555
Carbohydrate fractions % DM	Sugars	Starch					
	1.9	69.3					119
Major elements % DM	Ca	P	Mg	K	Na		
	0.01 0.01	0.36 0.19	0.11 0.08	0.33 0.31	0.02 0.007	Degermed, cooked.	555
Vitamins Tocopherols μg/g	Total	% of total					
		α	γ				
	42	10	90				100
Nicotinic acid μg/g	Free	Bound					
	0.3	18 0				Plate Argentinian, Yellow. Tryptophan, mg/g: 1.08	331
	0.3	14.4				Mexican. white. Tryptophan mg/g: 1.00	

Table 1.58 MAIZE: FLAKED

Proximate constituents % DM	CP	EE	NFE	CF	Ash	Notes	Ref.
	10.7	3.9	81.5	2.4	1.5		54
	12.0	2.9	82.9	1.3	0.9		93
	13.8	4.2	77.6	2.6	1.8		92
	9.9 (9.5-10.9)	3.7 (3.2-4.9)	83.0 (80.2-83.8)	2.1 (1.8-2.4)	1.3 (1.2-1.6)	Means and ranges for 8 samples	25
	10.1	3.0	83.4	2.1	1.4		53
	8.3	1.6	-	1.6	-		151
	10.0	2.0	85.1	1.7	1.2		98
Major elements % DM	Ca	P					
	0.04	0.03					54
	0.02	0.21					92
	0.04	0.35					151
Tocopherols µg/g DM	Total	% of total					
		a	γ				
	11	10	90				100

Table 1.59 MAIZE: PREPARED FEEDS: Proximate constituents and carbohydrate fractions as % of DM

	CP	EE	NFE	CF	Ash	Sugars	Starch	Pentosan	Cellulose	Lignin	Notes	Ref.
Maize gluten feed	25.6	1.1	62.0	10.2	1.1	0.9	15.7				For metabolisable energy for poultry see table 1.222	119
	27.9	2.1	56.3	7.8	5.9	0.9 (100)	18.4 (100)	16.2 (10.3)	8.6 (nil)	0.9 (-ve)	Digestibility % by poultry in brackets where recorded. For digested nutrients and metabolisable energy see table 1.222	83
Maize germ meal	21.0	12.6	47.2	16.5	2.7	14.1 (100)		23.3 (5.6)	13.5 (nil)	2.8 (11.5)	Digestibility %, by poultry in brackets where recorded. For digested nutrients and metabolisable energy see table 1.222	83

Table 1.60 MILLET

Digestibility of OM%						Notes	References
Dura (white ground) grain			86.4				85
Proximate constituents %DM	CP	EE	NFE	CF	Ash		
Dura (white ground) grain	10.30	3.25	82.39	1.30	2.76		
Sorghum grain	11.2	2.8	82.4	1.8	1.8		119
Sorghum meal	12.8 11.6	3.7 3.7	79.4 81.7	2.6 1.4	1.5 1.6	2 samples, tested as pig-feed	97

Table 1.61 RICE

Proximate constituents %DM	CP	EE	NFE	CF	Ash	Ref.
Grain, unspecified	8.1	0.25	90.84	0.38	0.43	503
Meal	12.9	17.52	52.88	7.32	9.38	504
Offals, bran	11.7	16.69	56.30	6.31	9.00	501

Major elements %DM	Ca	P	Mg	Ref.
var. *Ceylon Country* (unpolished and parboiled)	0.022	0.35	0.077	144
Milchard No. 2 (polished and parboiled)	0.021	0.24	0.009	
Europe No. 1 (polished and raw)	0.013	0.17	0.008	
Europe No. 2 (polished and raw)	0.014	0.18	0.007	
Small Mills Special (polished and raw)	0.029	0.19	0.010	

Table 1.62 SOYA BEAN MEAL, GROUNDNUT MEAL and LUPINS:
Proximate constituents as % of DM and amino acids as % of crude protein

	Proximate constituents					Amino acids				
	CP	EE	NFE	CF	Ash	Lysine	Cystine	Methio-nine	Notes	Ref.
Soya bean meal	47.0					4.4		1.1	Mean of 8 samples. Available lysine	
						6.2			Total lysine. 1 sample only	110
Sample No. 1	45.3	0.5	43.2	4.8	6.2		1.8	1.6	Sample No. 1 urease activity: 8.6	256
2	47.0	1.1	39.1	6.4	6.4		2.0	1.8	2 urease activity: 10.0	
22	44.5	0.5	43.9	5.0	6.1		1.7	1.8	22 urease activity: 1.9. Expelled at 192°F before extraction and lightly toasted (15 min. at 212°F)	
23	45.0	0.4	42.9	5.4	6.3		1.7	1.9	23 urease activity: 1.1. Expelled at 223°F before extraction, dried but not toasted	
24	43.7	0.3	44.4	5.6	6.0		1.9	1.8	24 urease activity: 1.1. Extracted only and lightly toasted (30 min at 212°F) Urease activity determined by the method of Croston, Smith and Cowan (1955) and expressed in the units they recommend.	
Groundnut meal	50.5					2.4		(C) 0.6	Mean of 6 samples. Available lysine	
						2.9		(M) 0.7	Total lysine. 2 samples only	110
									C: chemical estimation, M: microbiological estimation } 1 sample	
	48.2	1.8	36.3	9.7	4.0		2.0	1.4	Sample No. 2	256
	48.7	1.2	42.0	3.7	4.4		1.5	1.1	Sample No. 16	
Bitter blue lupin, seed	32.4	6.1	44.6	14.0	2.9				Alkaloid as lupanine: 2.2%	394
Sweet yellow lupin, seed	47.3	6.6	26.6	14.6	4.9				Alkaloid as lupanine: 0.5%	

Table 1.63 OTHER OILSEEDS

	Proximate constituents % DM					Carbohydrate fractions %DM		Vitamins μg/g DM	Notes	Ref.
	CP	EE	NFE	CF	Ash	Cellulose	Lignin	Total Tocopherols		
Palm kernel Extracted meal	20.2	2.3	53.8	19.4	4.3				Digestibility of DM% 65.9 Pig " "OM% 68 8 For digested nutrients and metabolisable energy see table 1.223	565
	20.3	1.4	63.1	10.7	4.5					399
Palm kernel not extracted	7.7	51.1	30.9	8.5	1.8					501
Palm kernel cake								4		100
Palm kernel extracted cake	18.2	1.4						0.5	Composition stated as % and mg/100g. No DM given	338
Sunflower seed Hungarian striped	19.3	28.3	20.6	27.8	4.0				Trials with pullets. Digestibility of OM 49.6%. For digested nutrients and metabolisable energy see table 1.223	259
Cottonseed meal	42.1	8.1	34.7	8.2	6.9					399
Linseed Feeding linseed	26.1 (22.3-28.0)	28.9 (26.6-31.4)	29.2 (27.1-30.2)	10.7 (7.8-13.8)	5.1 (4.6-10.7)				Mean and range of 5 samples	136
Flax chaff	9.2 (7.0-11.0)	5.7 (2.5-9.6)	40.2 (33.3-47.9)	36.9 (34.0-41.2)	8.0 (7.2-8.7)				Mean and range of 7 samples, commercial	
Flax chaff (1)	10.4	6.3	37.4	37.8	8.1				Two samples each used in 2 digestibility trials with sheep. For digested nutrients and metabolisable energy see table 1.223	
(2)	11.4	7.5	37.4	38.0	8.6					

Table 1.64 OTHER CONCENTRATES

	Proximate constituents %DM						Notes	Ref.
	CP	EE	NFE	CF	Ash			
Cocoa beans: cocoa cake meal	26.1 26.7	1.2 3.2	55.7 53.2	7.2 8.9	7.8 8.0		Untreated: 2.5-2.7% theobromine Extracted: 0.45% theobromine	94
Cocoa offals: husks	17.1	14.8	48.1	14.3	5.7			508
Cassava: Tapioca root flour	2.1	0.5	92.2	2.8	2.4		Root with rind or bark removed before grinding Digestibility of DM (2 pigs) 97.5;96.2% Digestibility of OM (2 pigs) 97.7;96.8% For digested nutrients and metabolisable energy see table 1.223	582
Tapioca	1.7	0.4					As stated: DM not given Total tocopherols, 0.11 mg/100g	338
Tapioca flour	1.5	0.5	91.7	3.7	2.6			54
Tapioca starch	0.12		99.9					80
	Major elements %DM							
	Ca	P	Mg	K	Na	Cl		
Tapioca flour	0.18	0.13	0.09	0.98	0.015	0.01		582
	0.15	0.10						54
	Proximate constituents %DM							
	CP	EE	NFE	CF	Ash			
Sago pith meal *Metroxylon: Sagus*	1.9	0.4	88.8	5.0	3.9		Digestibility % (2 pigs): DM 69.5;71.3 OM 72.4;74.0	582
		Major elements %DM					For digested nutrients and metabolisable energy see table 1.223	
	Ca	P	Mg	K	Na	Cl		
	0.27	0.06	0.09	0.51	0.21	0.33		
Horse chestnut *Aesculus hippocastanum*	Proximate constituents %DM							
	CP	EE	NFE	CF	Ash			
Meal, alcohol extracted Meal, water extracted Meal residue	7.35 7.73 11.17	7.38 7.76 6.89	74.60 73.62 52.90	7.99 8.48 22.88	2.68 2.41 6.16		Meal prepared from artificially dried nuts for production of glucose. Residue: pressed residue from alcohol extracted meal after removal of glucose	567
Meal, alcohol extracted Meal, water extracted	7.82 7.65	6.44 6.61	75.66 75.67	7.49 7.44	2.59 2.63			568
	Digestibility %							
	2 sheep		2 pigs					
	DM	OM	DM	OM				567
Meal, alcohol extracted Meal, water extracted Meal residue	56.8 54.6 31.3	58.3 55.8 33.8	60.7 60.7 30.8	62.4 62.4 33.5			For digested nutrients and metabolisable energy see table 1.223	
	Major elements %DM			Trace elements %DM				
	Ca	P	Cl		Fe			
Meal, alcohol extracted Meal, water extracted Meal residue (a)	0.24 0.41 1.29	0.12 0.17 0.30	0.045 0.035 0.06		0.45 0.39 1.76		(a) Values raised by added neutraliser	
Meal residue (b)	0.13						(b) In untreated residue	

Table 1.64 OTHER CONCENTRATES (continued)

	Proximate constituents %DM							
	CP	EE	NFE	CF	Ash	pH	Notes	Reference
Apple pomace: fresh	6.5	2.0	65.9	23.5	2.1	3.2		462
Apple pomace: fresh pectin extracted	9.4	3.6	51.2	33.4	2.4	3.6		
Apple pomace: silage	7.8	3.4	58.9	25.3	4.6	3.8		
Apple pomace: silage pectin extracted	10.9	4.0	48.0	35.3	1.8	3.1		
Molasses	13.5		79.3		7.2			568
	13.6		78.8		7.6			570
	25.0 25.9						Ammoniated: acidified with acetic acid Ammoniated: acidified with phosphoric acid	53
Penicillium felt, dried from deep-culture method	43.0	3.2	27.4	10.3	16.1		For proximate constituents and metabolisable energy of dried felt from surface–culture see table 1.223	569
	Major elements %DM							
	Ca	P	Cl					
from surface–culture method	0.61	2.57	0.14					
from deep–culture method	0.45	4.05	0.02					
	Proximate constituents %DM							
	CP	EE	NFE	CF	Ash			
Yeast, dried	52.3	0.3	35.8		11.6		For proximate constituents and metabolisable energy of other yeasts see table 1.223	95
	56.4	0.5	38.9		4.2			120
Yeast dried *(Candida arborea)*	55.5	5.1			8.5			58
(Torulopsis utilis)	58.9	5.7			10.0			
(Zygosaccharomyces lactis)	58.3	6.3			9.2			
Yeast, brewers', dried	52.3	2.4			6.7			92
	41.2	0.4	52.2	0.2	10.4		Mean of 3 commercial samples	199
	46.4	0.3	42.9		10.4			201
	45 5						Lysine, available, g/16 g N: 5.1	110
	47.2	0.5					Total tocopherols, mg/100 g DM: 0.08	338
Yeast, dried *(Torula)*	40.0						Lysine, available, g/16 g N: 5.4	110
Yeast, dried residue*	38.6	4.5	42.1		14.8			95
	Major elements %DM							
	Ca	P	Cl					
Yeast, dried	0.22	2.51	0.02					569
	0.30	2.56						95
Yeast, brewers', dried	0.15	1.20						92
	0.12	2.27	0.01					574
Yeast, dried residue*	0.05	1.90						95

*Residue from preparation of a proprietary yeast extract

continued

Table 1.64 OTHER CONCENTRATES (continued)

	Vitamins µg/g DM						
	Thiamin	Ribo-flavin	Nicotinic acid	Folic acid	Biotin		
Yeast, dried *(Candida arborea)*	15–25	30–45	150–300	10–14	3		58
(Torulopsis utilis)	15–25	45–60	350–450	6–7	3		
(Zygosaccharomyces lactis)	25–30	50–65	125–180	6–7	3		

Table 1.65 GRASS SEED: Proximate composition and major elements as % of DM. Commercial samples

Reference: 153

	Proximate constituents					Major elements		
	CP	EE	NFE	CF	Ash	Ca	P	K
Dactylis glomerata Cocksfoot	17.7	8.5	52.5	15.8	5.5	0.2	0.5	0.8
Festuca elatior Tall fescue	14.8	2.7	61.8	14.1	6.6	0.3	0.4	0.8
Festuca ovina Sheep's fescue	17.0	3.5	54.8	18.6	6.1	0.3	0.4	0.6
Festuca pratensis Meadow fescue	13.5	2.5	68.3	12.3	3.4	0.2	0.4	0.9
Lolium italicum Italian ryegrass	12.3	2.6	69.7	10.9	4.5	0.2	0.4	0.7
Lolium perenne Perennial ryegrass	12.4	2.9	67.5	12.3	4.9	0.2	0.4	0.8
Phleum pratense Timothy	18.1	4.8	68.4	5.9	2.8	0.2	0.4	0.7
Agrostis canina Bent grass	20.0	13.1	46.2	10.8	9.9	0.4	0.6	0.8
Agrostis tenuis Brown top	19.8	12.7	52.4	8.3	6.8	0.3	0.6	0.4
Alopecurus pratensis Meadow foxtail	19.7	9.7	50.9	14.1	5.6	0.2	0.5	0.7
Anthoxanthum odoratum Sweet vernal	26.9	15.6	43.9	8.1	5.5	0.2	0.6	0.6
Arrhenatherum avenaceum Tall oat grass	14.3	7.5	60.0	12.6	5.6	0.2	0.4	0.5
Avena flavescens Golden oat grass	20.0	14.9	43.2	14.9	7.0	0.3	0.4	0.8
Cynosurus cristatus Crested dogstail	15.7	6.5	58.4	13.3	6.1	0.2	0.4	0.6
Festuca rubra Red fescue	15.0	3.4	62.7	14.4	4.5	0.2	0.4	0.7
Poa pratensis Smooth stalked meadow grass	13.6	5.4	54.7	16.8	9.5	0.2	0.4	0.5
Poa trivialis Rough stalked meadow grass	17.1	3.9	61.5	10.3	7.2	0.4	0.4	0.8

Table 1.66 EXTRACTED PROTEIN: LEAF PROTEIN EXTRACT

Cocksfoot S143: extracted protein

	April	May	June	July	August	Sept.	Oct.	Notes	Ref.
N % in dry protein (1948)		14.2			7.9			Half plots grown until September, half cut at 8" (late April - early May) and then at 8" each time until autumn. 250 kg/ha superphosphate + potassium chloride in February. 375kg/ha Nitrochalk in April and after each cut.	510
(1951)				13.7	15.0	13.4			
(1952)	14.8								
Amino acids g per 16 g N									
Arginine (1948)		6.8			7.6				
(1951)				8.1	7.5	8.6			
(1952)	7.7								
Histidine (1948)		1.3			1.3				
(1951)				1.2	1.2	1.5			
(1952)	1.7								
Lysine (1948)		4.7			6.3				
(1951)				5.0	5.3	4.9			
(1952)	4.9								

Timothy S48: extracted protein

	April	May	June	July	August	Sept.	Oct.	Notes	Ref.
N % in dry protein (1948)		12.6			13.3			As above	510
(1950)						13.8			
(1951)				13.4		15.0			
(1952)	15.8								
Amino acids g per 16 g N									
Arginine (1948)		7.5			6.5				
(1950)						6.2			
(1951)				7.5		6.9			
(1952)	6.8								
Histidine (1948)		1.5			1.7				
(1950)						1.3			
(1951)				1.6		1.3			
(1952)	1.7								
Lysine (1948)		6.5			5.3				
(1950)						5.0			
(1951)				5.5		6.1			
(1952)	5.9								

Meadow fescue S53: extracted protein

	April	May	June	July	August	Sept.	Oct.	Notes	Ref.
N % in dry protein (1948)	15.8	14.9				8.0		As above	510
(1950)						14.0			
(1951)				13.4		15.2			
Amino acids g per 16 g N									
Arginine (1948)	6.1	7.6				7.1			
(1950)						8.0			
(1951)				6.6		6.6			
Histidine (1948)	1.5	1.5				1.2			
(1950)						1.4			
(1951)				1.3		1.6			
Lysine (1948)	6.0	6.9				5.8			
(1950)						6.4			
(1951)				6.0		6.4			

Perennial ryegrass S23 and S101: extracted protein

	April	May	June	July	August	Sept.	Oct.	Notes	Ref.
N % in dry protein (1948)	14.5				11.4			As above	510
(1950)						14.5			
(1951)			15.5		14.9				
(1952)	15.4								
Amino acids g per 16 g N									
Arginine (1948)	6.6				7.8				
(1950)						7.2			
(1951)			7.9		8.2				
(1952)	7.8								
Histidine (1948)	1.6				1.5				
(1950)						1.5			
(1951)			1.1		1.5				
(1952)	1.8								
Lysine (1948)	5.3				6.2				
(1950)						5.5			
(1951)			6.3		6.0				
(1952)	6.4								

continued

Table 1.66 EXTRACTED PROTEIN: LEAF PROTEIN EXTRACT (continued)
Italian Ryegrass: leaf protein extract

	May	June	July	August	Sept.	Oct.	Nov.	Dec.	Notes	Ref.
Crude protein % DM			(1953)	(1) 40.0					(1) frozen, roller dried. (2) frozen, oven dried. (3) " " ", extracted with 90% ethyl alcohol. (4) frozen, oven dried, extracted with 85% acetone. (5) frozen, oven dried, extracted with CCl_4. (6) frozen, oven dried, washed, oven dried. (7) (pH 4.5) frozen, washed, oven dried. (8) (pH 8.2) frozen, washed, oven dried. (9) frozen, oven dried, alkali soluble, oven dried. Fresh herbage macerated in fixed-hammer mill. Juice expressed in screw expeller. Protein coagulated by steam heating. Curd dried by methods described above.	142
		1954	(2) 34.5							
		1954	(3) 47.5							
		1954	(4) 42.6							
		1954	(5) 44.6							
		1954	(6) 36.1							
		1954	(7) 34.3							
		1954	(8) 37.4							
		1954	(9) 48.9							
Major elements % DM Calcium			(1953)	(1) 1.2						
		1954	(2) 1.3							
		1954	(3) 1.9							
		1954	(4) 1.9							
		1954	(5) 1.7							
		1954	(6) 1.4							
		1954	(7) 1.0							
		1954	(8) 2.0							
		1954	(9) trace							
Phosphorus			(1953)	(1) 0.5						
		1954	(2) 0.6							
		1954	(3) 0.6							
		1954	(4) 0.6							
		1954	(5) 0.6							
		1954	(6) 0.7							
		1954	(7) 0.6							
		1954	(8) 1.0							
		1954	(9) 0.08							

Grass and White Clover: leaf protein extract

	June	July	August	Sept.	Oct.	Nov.	Dec.	Notes	Ref.
Crude protein % DM			37.7					52% clover, 33% grass on DM basis. Fresh herbage macerated in fixed-hammer mill. Juice expressed in screw expeller. Protein coagulated by steam-heating. Curd dried by freezing followed by roller-drying.	142
Major elements % DM Calcium			1.8						
Phosphorus			0.6						

Lucerne, unspecified: leaf protein extract

	Mar.	Apr.	May	June	July	August	Sept.	October	Nov.	Dec.	Notes	Ref.
Crude protein % DM					1953	(1) 48.4		(2) 35.6			(1) frozen, roller dried. (2) roller dried, whole juice. (3) roller dried. (4) roller dried, washed, oven dried (5) oven dried, washed, oven dried. (6) frozen, washed, oven dried. (7) frozen, oven dried. Fresh herbage macerated in fixed-hammer mill, juice expressed in screw expeller. Protein coagulated by steam heating. Curd dried by methods described above.	142
								(3) 43.4				
								(4) 52.3				
			1954	(5) 58.1		(6) 60.4						
						(7) 47.5						
Major elements % DM Calcium					1953	(1) 2.4		(2) 3.1				
								(3) 2.3				
								(4) 1.9				
			1954	(5) 0.7		(6) 3.5						
						(7) 6.0						
Phosphorus					1953	(1) 0.6		(2) 0.5				
								(3) 0.5				
								(4) 0.6				
			1954	(5) 0.3		(6) 1.6						
						(7) 2.3						

Table 1.66 EXTRACTED PROTEIN: LEAF PROTEIN EXTRACT (continued)
Lucerne and Grass: leaf protein extract

	April	May	June	July	Aug.	Sept.	Notes	Ref.
Crude protein % DM			(1) 42.8 (2) 42.5 (3) 44.7 (4) 46.0				(1) frozen, oven dried (2) frozen, roller dried (3) frozen, freeze dried (4) frozen, roller dried, washed, oven dried.	142
Major elements % DM Calcium			(1) 2.0 (2) 1.7 (3) 1.8 (4) 1.3				54.7% lucerne, 45.3% grass on DM basis. Fresh herbage macerated in fixed-hammer mill, juice expressed in screw expeller. Protein coagulated by steam heating. Curd dried by methods described.	
Phosphorus			(1) 0.8 (2) 0.5 (3) 0.5 (4) 0.5					

Lucerne, Cocksfoot and Meadow Fescue: leaf protein extract

	June	July	August	September	Oct.	Nov.	Dec.	Notes	Ref.
Proximate constituents % DM									
Crude protein				43.2				Cut 4 September. Used in chicken feeding trial.	299
Ether extract				7.2					
Nitrogen-free extract				28.2					
Crude fibre				1.6					
Ash				19.8					
Major elements % DM									
Calcium				2.4					
Phosphorus				0.4					
Potassium				2.6					
Sodium				0.58					
Chlorine				0.9					
Vitamins									
Carotene mg/kg DM				202					

Table 1.67 'CYTOPLASMIC PROTEIN' PREPARATIONS FROM GRASSES AND LUCERNE

Cocksfoot S37:

Amino acids: N% of total protein N*		Notes	Reference
Alanine	6.7	Cut 12 August 1958, regrowth after cutting on 9 July 1958. *excluding cystine and tryptophan (insufficient sample)	553
Arginine	12.9		
Aspartic acid	6.3		
Glutamic acid	6.4		
Glycine	6.9		
Histidine	3.6		
Isoleucine	3.4		
Leucine	5.7		
Lysine	6.6		
Methionine	1.1		
Phenylalanine	3.6		
Proline	3.9		
Serine	4.0		
Threonine	3.4		
Tyrosine	2.2		
Valine	5.1		
Total N recovered	95.8		

Perennial Rygrass S23:

Amino acids: N% of total protein N*		Notes	Reference
Alanine	7.0	Cut 27 May, 1958 *excluding cystine and tryptophan (insufficient sample)	553
Arginine	13.7		
Aspartic acid	6.6		
Glutamic acid	6.9		
Glycine	6.9		
Histidine	4.0		
Isoleucine	3.7		
Leucine	6.3		
Lysine	8.4		
Methionine	1.2		
Phenylalanine	3.3		
Proline	4.0		
Serine	3.4		
Threonine	3.9		
Tyrosine	2.4		
Valine	5.5		
Cystine	0.9	Determined separately on special hydrolysates	
Tryptophan	2.1		
Total N recovered	98.8		

Table 1.67 'CYTOPLASMIC PROTEIN' PREPARATIONS FROM GRASSES AND LUCERNE (continued)

Italian Ryegrass S22:

Amino acids: N% of total protein N*		Notes	Reference
Alanine	6.5		
Arginine	11.5		
Aspartic acid	7.1		
Glutamic acid	6.7	Cut 12 August 1958, regrowth after cutting on 9 July 1958	
Glycine	6.4	*excluding cystine and tryptophan (insufficient sample)	
Histidine	4.0		553
Isoleucine	3.6		
Leucine	6.2		
Lysine	8.2		
Methionine	1.0		
Phenylalanine	3.3		
Proline	3.8		
Serine	4.4		
Threonine	3.8		
Tyrosine	2.4		
Valine	5.6		
Cystine	0.7	Determined separately on special hydrolysates	
Tryptophan	2.0		
Total N recovered	95.3		

Meadow Fescue S53:

Amino acids: N% of total protein N*		Notes	Reference
Alanine	6.8		
Arginine	12.1		
Aspartic acid	7.5		
Glutamic acid	6.5	Cut 12 August 1958, regrowth after cutting on 9 July 1958	553
Glycine	6.9	* excluding cystine and tryptophan (insufficient sample)	
Histidine	4.0		
Isoleucine	3.8		
Leucine	6.2		
Lysine	8.2		
Methionine	1.1		
Phenylalanine	3.4		
Proline	4.1		
Serine	4.4		
Threonine	4.0		
Tyrosine	2.6		
Valine	5.5		
Total N recovered	97.8		

continued

Table 1.67 'CYTOPLASMIC PROTEIN' PREPARATIONS FROM GRASSES AND LUCERNE (continued)

Timothy S48:

Amino acids N% of total protein N*		Notes	Reference
Alanine	6.9	Cut 12 Agusut 1958, regrowth after cutting on 9 July 1958 *excluding cystine and tryptophan (insufficient sample)	553
Arginine	11.3		
Aspartic acid	6.9		
Glutamic acid	6.2		
Glycine	6.4		
Histidine	4.3		
Isoleucine	3.6		
Leucine	6.2		
Lysine	8.4		
Methionine	1.0		
Phenylalanine	3.2		
Proline	3.8		
Serine	4.2		
Threonine	4.2		
Tyrosine	2.4		
Valine	5.6		
Total N recovered	95.8		

Lucerne var. Du Puits:

Amino acids: N% of total protein N*	April	May	June	Notes	Reference
Alanine	6.3	6.0	5.8	*excluding cystine and tryptophan (insufficient sample)	553
Arginine	14.4	13.7	14.0		
Aspartic acid	6.5	6.6	6.8		
Glutamic acid	6.9	6.5	6.2		
Glycine	6.4	6.0	6.6		
Histidine	4.4	4.2	4.2		
Isoleucine	3.8	3.7	3.6		
Leucine	6.1	6.4	6.4		
Lysine	8.5	7.6	7.6		
Methionine	1.1	0.9	1.0		
Phenylalanine	3.3	3.2	3.4		
Proline	4.1	3.6	3.7		
Serine	3.5	3.5	3.5		
Threonine	3.2	3.7	3.6		
Tyrosine	2.3	2.4	2.5		
Valine	5.2	4.9	5.0		
Cystine	0.8	0.8	0.9	Determined separately on special hydrolysates	
Tryptophan	1.9	2.0	1.7		
Total N recovered	96.8	94.9	96.1		

Table 1.68 MEAT MEAL

		Notes	References
Digestibility %		*In vivo* with 2 pigs. For metabolisable energy see table 1.225	
DM	85.4 88.0 75.3	A grade A meat meal: fat merely drained off B grade B meat meal: fat expelled by pressure only C grade C meal: fat expelled and extracted Approx. 10% of bone added to raw material	557
OM	88.3 93.1 84.8	A B C	
Proximate constituents %DM Crude protein	69.3 74.8 74.8	A 66.4 B 71.6 } CP = N x 100/16.7 C 71.7	
	60.8		93
	66.4	Used in pig feeding trials	94
	53.3 (41.4-79.4)	Mean and range of 19 samples	110
	67.0	Mean of 3 samples	256
	65.8 (64.2-67.3) 79.8	Mean and range of 6 samples, dry rendered 1 sample, wet rendered	247
Ether extract	18.8 11.1 3.2	A B C	557
	16.8		93
	15.1		94
	7.6 (2.0-17.4)	Mean and range of 19 samples	110
	13.4	Mean of 3 samples	256
	16.0 (10.7-25.7) 16.1	See crude protein, same reference	247
Nitrogen free extract	17.5	Given as "carbohydrates"	93
Ash	12.0 14.9 20.9	A B C	557
	4.9		93
	17.4		94

continued

Table 1.68 MEAT MEAL (continued)

		Notes	References
Proximate constituents %DM Ash (continued)	26 (7-42)	Mean and range of 19 samples	110
	14.0	Mean of 3 samples	256
	17.2 (5.2-23.1) 3.4	See crude protein, same reference	247
Amino acids: g/16 g N Cystine	1.3	Mean of 3 samples	256
Lysine	3.7 (1.9-4.6) 3.7 (1.2-4.9) 6.2 (2.1-7.8)	(1) Dinitrofluorobenzene method (2) Methyl chloroformate modification of (1) Available (1) } Mean and range of 19 samples Available (2) } Total (mean and range of 8 samples)	110
	5.0 4.8 5.6	Assay (1) Mean of 3 samples } dry rendered Assay (2) Mean of 4 samples } Assay (2) 1 sample, wet rendered	247
Methionine	1.0 (0.2-1.6) 1.2 (0.9-1.7)	Mean and range of 14 samples, Chemical estimation Mean and range of 5 samples, microbiological estimation	110
	1.6	Mean of 3 samples	256
Major elements %DM Calcium	2.8 3.3 5.3	A B C	557
	6.7 (3.0-15.1)	Mean and range of 6 samples	110
	4.9 (0.8-8.1) 0.6	See crude protein, same reference	247
Phosphorus	1.7 2.0 2.7	A B C	557
	2.9 (0.7-6.0)	Mean and range of 6 samples	110
	2.4 (0.3-3.6) 0.1	See crude protein, same reference	247

Table 1.68 MEAT MEAL (continued)

		Notes	References
Major elements %DM (cont'd)			
Potassium	0.7 0.6 0.7	A B C	557
Sodium	1.11 1.02 1.42 1.49 1.25 1.21	A { in Na_20 / in NaCl B { in Na_20 / in NaCl C { in Na_20 / in NaCl salt by extraction	
Chlorine	0.7 1.6 1.0 2.3 0.8 1.9	A { as Cl_2 / in NaCl B { as Cl_2 / in NaCl C { as Cl_2 / in NaCl salt by extraction	

Table 1.69 WHALEMEAT MEAL

		Notes	References
Digestibility %		*In vivo* with 2 pigs	
		For metabolisable energy see table 1.225	
DM	87.87	2 samples. Cooked, oil expelled and	
OM	87.80	steam dried	574
Proximate constituents % DM			
Crude protein	83.8	Blue whale vacuum dried 80°C	
	85.5	Fin whale, normal dried 145°C	
		Commercial samples used in feeding experiments	258
	92.93	See digestibility, same reference	574
		No. of samples — *Sample origin* — *Meal type*	
	90.7	3 — British floating factory — Good quality	
	92.8	2 — South African land station — Variable quality	408
	60.9	1 — Australian — Meat and Bone	
	19.4	1 — British floating factory — Mainly bone	
	85.2	Pure meat, blue whale } Factory ship products, vacuum dried	122
	93.8	Grax alone	
	88.1	Grax + meat	
	89.7	Av. production meal	
	88.6	Pure meat, fin whale } Land station products, flame dried	
	91.7	Av. production meal	
		Grax meal: prepared by separation and drying of the suspended solids resulting from cooking whale tissues, rich in fat, in rotary cookers	
	76.7 (57.3-88.3)	15 samples, commercial	110
	91.4	2 samples, commercial, floating stations	410
	99.6	1 sample, (time post mortem short)	
	96.9	1 sample, (time post mortem medium)	
	93.5	1 sample, (time post mortem long)	
	89.1	2 samples, commercial land stations. S. Georgia	
Ether extract	10.2 7.3	See crude protein, same reference	258
	3.66	See digestibility, same reference	574
	4.0 5.0 14.9 15.6	See crude protein, same reference	408
	8.6 3.5 6.0 6.0 4.4 3.8	See crude protein same reference	122

Table 1.69 WHALEMEAT MEAL (continued)

		Notes	References
Proximate constituents %DM Ether extract (cont'd)	4.0 (2.2-9.7)	12 samples, commercial	110
	4.7 4.0 4.6 6.8 7.1	See crude protein, same reference	410
Ash	6.7 3.8	See crude protein, same reference	258
	2.18	See digestibility, same reference	574
	2.9 3.2 23.1 61.2	See crude protein, same reference	408
	3.9 2.7 5.3 5.4 3.5 3.4	See crude protein, same reference	122
	9.9 (2-29)	15 samples, commercial	110
	7.3 2.4 2.4 2.2 4.9	See crude protein, same reference	410
Amino acids %DM Arginine	4.4 (3.7-5.0) 5.0	5 commercial samples, floating stations 2 commercial samples, land stations, S. Georgia	410
Cystine	1.0 0.5 0.7 0.9 0.9 0.7	See crude protein, same reference	122
	0.9 (0 5-1.1) 0.6	See arginine, same reference	410
Histidine	2.1 (1.8-2.5) 2.3	See arginine. same reference	410
*Iso*leucine	4.4 4.7	1 commercial sample, floating stations 2 " " land stations, S. Georgia	410

continued

Table 1.69 WHALEMEAT MEAL (continued)

			Notes	References
Amino acids (cont'd)				
Leucine		7.2 (6.1-8.1) 6.8	See arginine, same reference	410
Lysine		10.3 11.0 10.4 9.1 9.2 9.0	See crude protein, same reference	122
	g/16 g N	5.8 (4.0-7.3) 6.2 (3.3-8.1) 9.0 (8.1-9.7)	Available, 12 samples, method (1) " 12 " " (2) Total, 5 samples (1) Dinitrofluorobenzene method (2) Methyl chloroformate modification of (1)	110
		7.0 (6.6-7.8) 6.3	See arginine, same reference	410
Methionine		2.8 3.1 3.4 3.3 3.6 3.6	See crude protein, same reference	122
	g/16 g N	1.8 (1.6-2.0) 2.5 (2.1-2.8)	9 samples, chemical estimation 8 samples, microbiological estimation	110
		1.5 (1.1-2.1) 1.7	See arginine, same reference	410
Phenylalanine		4.5 (3.3-5.3) 3.2	See arginine, same reference	410
Threonine		6.4 (5.7-7.0) 5.2	See arginine, same reference	410
Tryptophan		1.0 1.1 1.0 - 0.8 -	See crude protein, same reference	122
		0.9 (0.8-1.0) 1.0	See arginine, same reference	410

Table 1.69 WHALEMEAT MEAL (continued)

		Notes	References
Amino acids (cont'd) Tyrosine	2.1 (1.8-2.3) 2.2	See arginine, same reference	410
Valine	2.6 (2.4-2.8) 2.5	See arginine, same reference	410
Major elements %DM Calcium	0.05 0.6	see crude protein, same reference	258
	19.13* 46.70*	*Sample origin* — *Meal type* Australian — Meat and bone British floating factory. — Mainly bone *Given as Ca phosphate, % of ash	408
	0.04	See digestibility, same reference	574
	0.4 0.04 1.0 1.1 0.5 0.5	See crude protein, same reference	122
	2.3 0.8	2 commercial samples, floating stations 2 " " land stations, S. Georgia	410
Phosphorus	0.7 0.6	See crude protein, same reference	258
	8.76* 21.38*	See calcium, same reference *Given as P_2O_5, % of ash	408
	0.4	See digestibility, same reference	574
	0.7 0.4 0.9 0.7 0.8 0.7	See crude protein, same reference	122
	1.0 0.3 0.4 0.5 0.6	See crude protein, same reference	410
Sodium	0.06 0.07	2 samples, floating stations 2 samples, land stations, S. Georgia Given as NaCl. See also chlorine	410

continued

Table 1.69 WHALEMEAT MEAL (continued)

		Notes	References
Major elements %DM (cont'd) Chlorine	0.02	See digestibility, same reference	574
	0.7 0.4 0.7 0.7 0.07 0.2	See crude protein, same reference	122
	0.09 0.1	2 samples, floating stations 2 samples, land stations, S. Georgia Given as NaCl, see also sodium	410
Vitamins µg/g DM Thiamin	1.4	2 samples	411
Riboflavin	7.8	2 samples	411
	5.5 6.1 1.7 0.4	See crude protein, same reference	408
	6.6 0.9 6.6 - 4.4 -	See crude protein, same reference	122
	4.0 (1.6-6.7)	6 samples, commercial	110
	4.1 (3.3-4.8) 4.0	See arginine, same reference	410
Nicotinic acid	102.9	2 samples	411
	121.4 126.3 51.1 15.8	See crude protein, same reference	408
	91.5 (64.6-137.7)	6 samples, commercial	110
	81.1 (57.5-88.8) 114.4	See arginine, same reference	410
Pantothenic acid	3.0	2 samples	411
	9.3 8.9 3.4 1.3	See crude protein, same reference	408

Table 1.69 WHALEMEAT MEAL (continued)

			Notes	References
Vitamins Pantothenic acid (cont'd)		5.5 4.4 6.0 4.1 7.7 3.8	See crude protein, same reference	122
		8.7 (4.5-16.2)	6 samples, commercial	110
		6.2* (3.3-9.1) 2.4	See arginine, same reference *Given as Ca d-pantothenate	410
Pyridoxine		9.0	2 samples	411
		7.2 5.5 1.3 1.0	*No. of samples* 1 See crude protein, 1 same reference	408
		2.6 (1.3-4.0) 2.4	See arginine, same reference	410
Vitamin B_{12}		0.05	2 samples	411
		0.07 0.11 0.07 0.01	See crude protein, same reference	408
		0.09 0.08 0.11 0.07 0.08 0.09	See crude protein, same reference	122
		0.03 (0.02-0.04) 0.07	See arginine, same reference	410
Miscellaneous Choline	mg/g DM	4.3 (2.9-5.2) 3.2	See arginine, same reference	410
Myoinositol	μg/g DM	280 134 62 18	See crude protein, same reference	408

Table 1.70 MEAT AND BONE MEAL

		Notes	Reference
Proximate constituents %DM Crude protein	49.7		92
	47.9 52.7 59.7 69.6	*Meal grade:* 1 45 % protein 2 50 % protein 3 55 % protein 4 64 % protein Commercial samples	409
	28.5 59.9 34.4 87.4	Raw materials for the manufacture of meat and bone meal. Commercial samples A. processed "bone" } before defatting B. processed "meat" } C. defatted "bone" D. defatted "meat" Selected for analysis as high in bone or meat	192
	59.7 (54.0-63.4) 56.7 56.7 (51.2-62.0)	A Mean and range of 6 samples dry rendered B 1 sample, wet rendered C Mean and range of 5 samples, process unknown	247
Ether extract	9.8		92
	3.9 4.5 5.0 4.1	1 2 See crude protein, same reference 3 4	409
	15.2 31.4	A See crude protein, same reference B	192
	13.0 (9.9-16.1) 12.9 14.3 (13.1-15.9)	A. See crude protein, same reference B C	247
Ash	33.6		92
	47.2 41.2 32.1 21.3	1 2 See crude protein, same reference 3 4	409
	53.3 7.3 63.7 10.9	A B See crude protein, same reference C D	192
	25.1 (19.6-34.5) 29.2 29.7 (23.1-36.3)	A See crude protein, same reference B C	247

Table 1.70 MEAT AND BONE MEAL (continued)

				Notes	Reference
Amino acids					
Alanine	g/16 g N	9.1 8.0	C D	See crude protein, same reference	192
Arginine	% DM	2.1 2.5 3.7 4.7	1 2 3 4	See crude protein, same reference	409
	g/16 g N	7.3 6.6	C D	See crude protein, same reference	192
Aspartic acid	g/16 g N	6.5 7.1	C D		
Cystine	% DM	0.5 0.5 0.5 0.3	1 2 3 4	See crude protein, same reference	409
Glutamic acid	g/16 g N	10.1 10.8	C D	See crude protein, same reference	192
Glycine	g/16 g N	19.8 15.7	C D		
Histidine	% DM	1.3 1.6 1.7 2.0	1 2 3 4	See crude protein, same reference	409
	g/16 g N	0.9 1.5	C D	See crude protein, same reference	192
Hydroxylysine	g/16 g N	0.7 0.5	C D	See crude protein same reference	192
Hydroxyproline	g/16 g N	10.3 7.2	C D		
*Iso*leucine	g/16 g N	1.7 2.5	C D		
Leucine	% DM	2.3 2.7 3.1 4.4	1 2 3 4	See crude protein, same reference	409
	g/16 g N	4.0 5.4	C D	See crude protein, same reference	192

continued

Table 1.70 MEAT AND BONE MEAL (continued)

				Notes	References
Amino acids (cont'd)					
Lysine	% DM	4.2 4.9 4.9 6.9	1 2 3 4	See crude protein, same reference	409
	g/16 g N	4.0 4.8	C D	See crude protein, same reference	192
		5.5 (4.8-5.9) 5.2 4.8 (4.1-5.2)	A B C	See crude protein, same reference	247
Methionine	% DM	0.8 0.9 1.3 1.1	1 2 3 4	See crude protein, same reference	409
	g/16 g N	0.5 1.0	C D	See crude protein, same reference	192
Phenylalanine	% DM	2.5 2.9 2.9 2.9	1 2 3 4	See crude protein, same reference	409
	g/16 g N	2.6 3.1	C D	See crude protein, same reference	192
Proline	g/16 g N	11.7 9.9	C D	See crude protein, same reference	192
Serine	g/16 g N	3.7 4.0	C D		
Threonine	% DM	1.9 2.1 2.2 2.6	1 2 3 4	See crude protein, same reference	409
	g/16 g N	2.5 3.3	C D	See crude protein, same reference	192
Tryptophan	% DM	0.2 0.2 0.3 0.4	1 2 3 4	See crude protein, same reference	409
Tyrosine	% DM	1.3 1.5 1.3 1.1	1 2 3 4		
	g/16 g N	1.0 1.8	C D	See crude protein, same reference	192

Table 1.70 MEAT AND BONE MEAL (continued)

				Notes	References
Amino acids (cont'd) Valine	% DM	2.7 2.1 1.9 2.4	1 2 3 4	See crude protein, same reference	409
	g/16 g N	3.1 3.9	C D	See crude protein, same reference	192
Major elements %DM Calcium		9.0			92
		14.3 11.5 8.9 5.9	1 2 3 4	See crude protein, same reference	409
		8.9 (6.4-13.5) 9.4 10.4 (7.3-12.9)	A B C	See crude protein, same reference	247
Phosphorus		7.9			92
		7.5 5.9 4.4 3.0	1 2 3 4	See crude protein, same reference	409
		9.5 0.7	A B	See crude protein, same reference	192
		4.7 (2.1-9.9) 4.5 4.7 (3.5-5.9)	A B C	See crude protein, same reference	247
Sodium		0.18 0.46 0.74 0.84	1 2 3 4	See crude protein, same reference Given as NaCl. See also Cl.	409
Chlorine		0.28 0.71 1.14 1.30	1 2 3 4	Given as NaCl. See also Na	
Vitamins µg/g DM Riboflavin		2.7 3.4 4.4 5.7	1 2 3 4		
Nicotinic acid		17.1 33.3 42.6 75.3	1 2 3 4		

continued

Table 1.70 MEAT AND BONE MEAL (continued)

		Notes	References
Vitamins μg/g DM (cont'd)			
Pantothenic acid	3.8 3.1 3.9 3.2	1 2 3 4	
Pyridoxine	1.0 1.2 1.5 1.3	1 2 3 4	
Vitamin B_{12}	0.01 0.02 0.04 0.04	1 2 3 4	
Miscellaneous mg/g DM			
Choline	1.1 1.7 2.2 2.8	1 2 3 4	

Table 1.71 FISHMEAL

		Notes	Reference
White Fish			
Digestibility %			
DM	75.14 71.99	With 2 pigs, each shown	574
OM	92.94 89.91	" " " "	
		For metabolisable energy see table 1.225	
Proximate constituents %DM			
Crude protein	69.7 64.2 - 78.8	Mean and range of 55 samples used in feeding trials.	25, 41, 91, 95, 96, 118, 120,
Ether extract	4.1 1.1 - 7.0	Mean and range of 52 samples used in feeding trials.	199, 200, 201, 203, 338, 558,
Nitrogen-free extract	1 3 0.1 - 5.8	Mean and range of 38 samples used in feeding trials.	560, 561, 562, 563, 565, 566,
Ash	25.0 19.0 - 30.2	Mean and range of 51 samples used in feeding trials.	568, 569, 571, 574, 575, 576
Amino acids g/16 g N			
Lysine	6.4 6.1 - 6.7	Mean and range of 4 samples used in hatchability trials	118
Major elements %DM			
Calcium	7.0 6.0 - 8.2	Mean and range of 7 samples used in feeding trials.	41, 91, 95, 96, 569, 575,
Phosphorus	4.0 3.4 - 4.5	Mean and range of 7 samples used in feeding trials.	"
Sodium	0.76 0.39 - 1.11	Mean and range of 9 samples used in feeding trials.	41, 91, 96, 118, 558,
Chlorine	1.10 0.57 - 1.69	Mean and range of 11 samples used in feeding trials.	41, 91, 96, 118, 558, 569, 575
Vitamins			
Tocopherol mg/100 g DM	0.42	Total tocopherol	338
B vitamins μg/g DM		*Fishing season:*	
Thiamin	4.3 2.8	Nov. – Dec. May – June	412
Riboflavin	14.9 6.6	Nov. – Dec. May – June	
Nicotinic acid	50.9 48.7	Nov. – Dec. May – June	
Pantothenic acid	13.0 6.2	Nov. – Dec. May – June	
Pyridoxine	12.4 3.0	Nov. – Dec. May – June	
Cobalamin	0.13 0.13	Nov. – Dec. May – June	
Miscellaneous μg/g DM			
Myoinositol	256	May – June	
Choline	4460	May – June	
Herring			
Proximate constituents %DM			
Crude protein	63.2	Alkali reduction meal	121
	70.0	'Cook and Press'	
	72.0	Norwegian, 2 samples	110
	70.6	Danish	
	80.8 79.6 80.8	Press cake, Press cake meal, Wholemeal.	81
	85.2	Commercial, poor quality	
	66.0	Meal	123
Ether extract	0.7	Alkali-reduction meal	121
	7.1	Cook and Press'	
Ash	25.9	Alkali-reduction meal	121
	11.9	'Cook and Press'	
	10.6 10.5 11.2	Press cake, Press cake meal, Wholemeal.	81
	10.2	Commercial, poor quality	
	17.6	Meal	123

continued

Table 1.71 FISHMEAL (continued)

		Notes	Reference
Herring (continued) *Amino acids* g/16 g N			
Alanine	7.7 7.5 7.5 7.0	Press cake, Press cake meal, Wholemeal Commercial, poor quality	81
Arginine	8.2 8.1 7.9 7.1	As above	
Aspartic acid	9.9 9.8 9.1 8.8	As above	
Cystine	1.6 1.6 1.3 1.0	As above	
Glutamic acid	12.1 12.0 11.8 10.6	As above	
Glycine	5.4 5.7 6.3 5.6	As above	
Histidine	2.0 2.1 1.8 1.5	As above	
*Iso*leucine	6.8 6.7 6.1 5.6	As above	
Leucine	7.5 7.6 6.9 6.3	As above	
Lysine	9.1 9.1 8.2 5.7	As above	
	6.4	Norwegian herring	110
	5.5	Meal	123
Methionine	2.7 2.7 2.5 2.2	As above	81
	1.9	Norwegian, 2 samples	110
	2.1	Danish	
	1.6	Meal	123
Phenylalanine	3.9 3.9 3.6 3.3	As above	81
Proline	4.3 4.3 4.6 4.1	As above	
Serine	4.5 4.6 4.2 4.1	As above	
Threonine	4.1 4.2 4.0 3.5	As above	
Tryptophan	0.8 0.8 0.7 0.5	As above	
Tyrosine	3.3 3.3 2.9 2.6	As above	
Valine	5.7 5.9 5.3 4.7	As above	
Major elements %DM			
Calcium	0.2 2.7	Alkali-reduction meal 'Cook and Press'	121
Phosphorus	0.9 1.3	Alkali-reduction meal 'Cook and Press'	
Sodium	8.1 1.1	Alkali-reduction meal 'Cook and Press'	
Chlorine	12.5 1.8	Alkali-reduction meal 'Cook and Press'	
Pilchard, South Africa *Proximate constituents*			
Crude protein %DM	65.4		110
Amino acids g/16 g N			
Methionine	2.0		

Table 1.71 FISHMEAL (continued)

		Notes	Reference
Peruvian fishmeal			
Proximate constituents			
Crude protein % of DM	72.8	2 samples from the same importing agency	407
Ether extract	3.6		
Ash	17.3		
Amino acids g/16 g N			
Alanine	7.4		
Arginine	6.2		
Cystine	1.1		
Glutamic acid	14.2		
Glycine	6.5		
Histidine	2.7		
*Iso*leucine	5.5		
Leucine	8.8		
Lysine	8.5		
Methionine	3.0		
Phenylalanine	4.7		
Proline	4.8		
Serine	4.2		
Threonine	4.9		
Tryptophan	1.2		
Tyrosine	3.7		
Valine	6.4		
Major elements %DM			
Calcium	4.6		
Phosphorus	2.7		
Vitamins			
α-tocopherol mg/100g DM	2.3		100

Table 1.72 ANIMAL PRODUCTS: Residues

Blood meal

		Notes	Reference
Proximate constituents %DM			
Crude protein	95.9	Commercial sample used in feeding experiments	566
	87.3		110
Ether extract	0.9	As above same reference	566
Ash	3.2	As above same reference	680
Amino acids g/16 g N			
Lysine	(1) 6.7 (2) 7.8 8.6	Available (1) Dinitrofluorobenzene method " (2) Methylchloroformate modification of (1) Total	110
Methionine	1.1	Chemical estimation	

Slaughterhouse waste

		Notes	Ref.
Proximate constituents % DM			
Crude protein	48.4	Dried. Commercial samples used in feeding experiments	560
Ether extract	17.5		
Nitrogen-free extract	5.9		
Crude fibre	1.5		
Ash	26.7		

Whale liver meal

		Notes	Ref.
Vitamins µg/g DM			
Thiamin	2.8	Mean of 3 samples. Assayed microbiologically.	411
Riboflavin	84.7		
Nicotinic acid	213.9		
Pantothenic acid	38.9		
Pyridoxine	10.2		
Vitamin B_{12}	0.5		

Table 1.72 ANIMAL PRODUCTS: Residues (continued)
Whale solubles

		Notes	Ref.
Proximate constituents % DM			
Crude protein	89.6		200
True protein	22.2		201
	85.0	Average production 1950–51	122
	86.2	" " 1951–52	
	85.5		110
True protein	24.8		
	78.1		
True protein	34.4		
Ether extract	0.5		200
	1.6	Average production 1950–51	201
	1.6	Average production 1951–52	
Nitrogen-free extract	3.8	Average production 1950–51	122
	3.6	Average production 1951–52	
Ash	9.6	Average production 1951–52	
	8.6	Average production	
Amino acids % DM			
L-lysine	9.8	Average production 1950–51	122
	9.0	Average production 1951–52	
L-cystine	< 0.2	Average production 1950–51	
	< 0.2	Average production 1951–52	
L-methionine	3.0	Average production 1950–51	
	3.6	Average production 1951–52	
DL-tyrptophan	–	Average production 1950–51	
	0.8	Average production 1951–52	
Available lysine g/16 g N	3.4		110
	4.3		
Major elements % DM			
Calcium	0.2		200
			201
	0.12	Average production 1950–51	122
	0.08	Average production 1951–52	200
Phosphorus	0.7		201
	0.42	Average production 1950–51	122
	0.46	Average production 1951–52	
Chlorine	2.2		200
			201
	4.6	Average production 1950–51	122
	3.1	Average production 1951–52	
Trace elements μg/g Ash			
Copper	120	Average production 1950–51	122
Zinc	1880	N.B. Values per g. Ash	
Molybdenum	<10		
Manganese	62		
Iron	1500		
Cobalt	<10		
Nickel	54		
Lead	140		
Tin	855		
Vanadium	<10		
Titanium	<10		
Chromium	<10		
Silver	<10		
Vitamins μg/g DM			
Riboflavin	1.3	Average production 1950–51	
Pantothenic acid	0.8	Average production 1950–51	
	1.8	Average production 1951–52	
Vitamin B_{12}	0.06	Average production 1950–51	
	0.04	Average production 1951–52	

continued

Table 1.72 ANIMAL PRODUCTS: Residues (continued)
Fish solubles

		Notes	Ref.
Proximate constituents % DM			
Crude protein	69.1	Fish not stated	201
	76.4	Herring	
	71.6	Dried fish solubles.	110
	27.3	A Press water (defatted, from press cake before meal production) condensed to approx 30% DM	81
	25.1	B Condensed solubles (press water condensed in triple condenser) approx 30% DM. Herring samples taken in January.	
Ether extract	5.6	Fish not stated	201
	5.3	Herring	
Nitrogen-free extract	5.1	Fish not stated	
	2.2	Herring	
Ash	20.2	Fish not stated	
	16.1	Herring	
	5.2	A	81
	5.4	B	
Amino acids g/16 g N			
Alanine	7.3	A	
	7.9	B	
Arginine	5.4	A	81
	5.2	B	
Aspartic acid	5.0	A	
	4.9	B	
Cystine	0.42	A	81
	0.08	B	
Glutamic acid	7.7	A	
	7.5	B	
Glycine	10.2	A	
	10.9	B	
Histidine	1.2	A	81
	0.7	B	
*Iso*leucine	2.0	A	
	1.9	B	
Leucine	3.3	A	
	3.2	B	
Lysine	(1) 5.3	(1) See Bloodmeal. same reference	110
	(2) 3.6	(2)	
	4.6	A	81
	4.3	B	
Methionine	1.1	Chemical estimation	110
	1.3	A	81
	1.4	B	
Phenylalanine	1.6	A	81
	1.5	B	
Proline	4.6	A	81
	4.8	B	
Serine	2.9	A	
	–	B	
Threonine	2.2	A	
	2.1	B	
Tryptophan	0.16	A	
	0.10	B	81
Tyrosine	0.7	A	81
	0.7	B	
Valine	2.6	A	
	2.5	B	
Major elements % DM			
Calcium	0.2	Fish not stated	201
	0.1	Herring	
Phosphorus	1.4	Fish not stated	
	1.5	Herring	
Chlorine	2.7	Fish not stated	
	3.4	Herring	

Table 1.72 ANIMAL PRODUCTS: Residues (continued)

Fish Waste

		Notes	Ref.
Proximate constituents % DM			
Crude protein	76.2	Fresh (defatted) herring.	81
Ash	10.5	Samples taken in January. Used in production of various cakes, meals and solubles.	
Amino acids g/16 g N			
Alanine	76.		
Arginine	7.1		
Aspartic acid	9.4		
Cystine	1.4		
Glutamic acid	11.4		
Glycine	6.3		
Histidine	1.9		
*Iso*leucine	6.2		
Leucine	7.1		
Lysine	8.3		
Methionine	2.6		
Phenylalanine	3.6		
Proline	4.2		
Serine	4.1		
Threonine	4.1		
Tryptophan	0.8		
Tyrosine	3.0		
Valine	5.4		

Table 1.73 Cocksfoot S37

	January			February			March			April			May			June		
Digestibility of OM%												78.0	76.3	74.0	67.9	61.0	60.0 76 2*	56.2
Crude protein %DM												24.5	22.6	17.3	13.6	9.7	8.3 18.6*	6.8
Digestibility of OM%												78.5	76.3	71.8 77.5*	65.6	59.7	55.7 71.4*	
Crude protein %DM												26.3	17.2	13.7 22.8*	10.8	10.1	7.4 18.4*	
Digestibility of OM%														N_0 N_1	71.9 69.8	65.4 61.0		
Crude protein %DM														N_0 N_1	10.1 15.9	7.6 9.8		
Digestibility of OM%																		
Crude protein %DM																		
Digestibility of DM%										1961 1962	69.8	77.6 73.6 75.5 73.3	69.9 73.5 L SS	68.2 72.6 69.4 72.5	64.1 69.9 70.6 75.2		72.8	73.8
										W LB LS S Dd I		78 79 81 69	76 79 78 86 65		66 70 65 65 63 73	64 70 63 58 62 64		
of OM%												[1]88		[2]77	[3]70			
Proximate constituents %DM Crude protein		16.5						15.4										
						(1947) (1948) (1949) (1950)		16.3 12.3 15.4										
	No N1 16.5 N2 18.9							13.1 15.2 13.7										
	17.4																	
	G 22.8 G 22.2 F 11.7 F 12.3																	
	W G F	12.8 20.3 8.8				12.8 22.4 9.5												
														(1948) (1949) (1950)		10.4	7.9	
				(1951) (1952) (1953) (1954)				23.5 19.0 16.0			26.7 25.2 22.3			19.7 23.4 19.8				
				W G F	19.4 29.6 14.0	19.6 26.8 13.5												
											15.0					17.4		
														15.3				
										(1961) (1962)		29.1 21.0					16.1*	21.7*
												18.8 D 84		13.8 D 76	11.1 D 76			

July			August			September			October			November			December			Notes	Reference
	48.1 73.0* 65.3**			74.2*			72.2* 66.0**			72.4*			71.0**					1958 *Monthly regrowth **Two-monthly regrowth	357
	6.2 17.2* 9.8**			25.9*			21.4* 16.1**			26.1*			26.2**						
	52.1 69.0* 70.8**			71.8*			71.4* 65.7**			69.6*								1959	
	6.6 20.4* 20.9**			20.9*			23.8* 17.3**			26.9*									
																		Nitrogen application. 1958 N_0 : none N_1 : 57 kg/ha. Fertiliser not specified	
								N_1 N_2	65.2 66.7									1959 N_1 : 38 kg/ha N_2 : 192 kg/ha Fertiliser not specified	
								N_1 N_2	10.4 18.4										
							65.3 66.0											Digestibility *in vitro* L : leaf SS : sheath and stem	75
53 66 58 44 56 55			72* (calc.) 73* 61*						75* (calc.) 73* 63*									Digestibility *in vitro* *Regrowth W : whole plant LB : leaf blade LS : leaf sheath S : stem Dd : dead I : inflorescence	465
																		By sheep: [1]Late leafy [2]40% heads emerged [3]90% heads emerged	514
													19.4			16.5		Nov./Dec. 1946 Jan./March 1947	293
				16.9 14.8 15.1			14.7 12.9 13.9									16.1 14.0 18.2			205
													20.9			15.1 16.4 18.2		N_0 : no fertiliser N_1 : sulphate of ammonia 250 kg/ha N_2 : sulphate of ammonia 500 kg/ha Nov. and Dec. 1946 Jan. and March 1947	296
											16.1					13.3		Sown broadcast 1948	
																20.0 18.8 12.4 11.0		Sown in drills. Broadcast. Sown in drills. Broadcast. G: green F: frostburnt	
											11.9 12.9 8.6				12.5 16.2 8.4		11.0	W : whole plant G : green F : frostburnt Jan./Feb. 1951 Oct./early Dec. 1952 Late Dec. 1953	
		12.4			10.6			18.7 14.6											439
										16.1 17.4 12.6						15.1 17.0 14.0		Pre-grazing samples Sown 27/4/51	249
																		Possibly some *Poa* and clover contamination. W : whole plant G : green F : frostburnt Early Feb. 1956. Middle Feb.1957	337
		13.6				17.0													141
	10.4*						30.6* 14.6*											*Regrowth Values for April - May represent means of 3 Springs grazings. Later values followed topping and application of fertiliser.	75
												F	10.9					53% frostburnt	355
																		D : digestibility %	514

continued

K

Table 1.73 Cocksfoot S37 (continued)

	January			February			March			April			May			June		
Proximate constituents (Contd.) Crude protein													L LS S H	21.5 8.1 8.9	22.7 7.4 6.3		17.9 7.8 3.7 13.4	14.6 5.8 3.0 14.0
										(1958) (1959) (1958) (1959) (1958) (1959)		24.9 25.8	22.7 17.3	17.3 13.3 22.8*	13.6 10.8	9.8 10.1	8.3 18.6* 18.4*	6.8
Ether extract												8.6 D66		6.6 D58	5.2 D55			
Crude fibre														21.9				
										(1961) (1962)		20.6 21.6					24.7*	
Ash											11.8					8.2		
												6.7 D60		6.5 D62	6.0 D60			
Carbohydrate fractions % DM Soluble carbohydrate											15.0					9.7		
														11.3				
									(1961) (1962) (1961) (1962)		11.8	8.1 16.1 10.7 15.9	10.8 15.9	13.6 12.2	17.8 12.6		16.2* 8.0*	
									W LB LS S Dd I			9 9 12 5	17 15 19		12 10 13 19	9 6 11 15 2		
												12.0 D100		5.5 D100	5.9 D100			
Hexose														4.0				
Sucrose														4.6				
Fructosan														2.6				
Pectin												2.4 D54		2.0 D42	2.2 D44			
Cellulose												18.0 D89		23.8 D85	27.2 D79			
Xylan												6.4 D88		11.3 D79	12.8 D65			
Araban												1.7 D92		2.4 D87	2.3 D82			
Glucan												1.8 D94		1.7 D92	1.8 D76			
Galactan												0.7 D89		0.8 D85	0.8 D77			
Aldobiouronics												4.1 D75		4.9 D71	5.4 D64			
Total hemicelluloses												14.7 D85		21.1 D79	23.1 D71			
Lignin											5.9					6.0		
												4.0 D37		5.6 D33	6.5 D19			
												L LS S H		3.8 4.3 3.9	3.6 5.7 5.2		5.1 6.8 8.1 9.9	6.3 7.9 10.2 9.2

July			August			September			October			November			December			Notes	References
14.7 5.2 2.5 11.4	12.3 5.2 2.3 10.8	3.9 1.7 10.7		5.8 2.9 1.5 9.7														L: leaf LS: leaf sheath S: stem H: head	311
6.8 17.2*	6.2 20.4* 9.8** 20.9**		25.9*	20.9*	21.4*	23.8* 16.1**	17.3**	26.1*	26.9*			26.2**						* Monthly Regrowth ** Two-monthly Regrowth	542
																		D: digestibility %	514
	30.2*																	* Regrowth	
24.2*							24.1* 29.2*											See crude protein Same reference	75
		8.6				10.5													141
																		D: digestibility %	514
		13.3				10.1													141
	4.4*																	* Regrowth	75
							3.1* 5.7*											Auchincruive Cambridge	
12 6 10 14 3 6			6* 5*						9* 7* 3*									* Regrowth W: whole plant LB: leaf blade LS: leaf sheath S: stem Dd: dead I: inflorescence	465
																		D: digestibility %	514
																			75
																		D: digestibility %	514
																			514
		5.2				5.1													141
																		D: digestibility%	514
6.6 8.3 10.4	6.8 8.8 10.8	12.4 9.3		10.1 13.0														L: leaf LS: leaf sheath S: stem H: head	311

continued

Table 1.73 Cocksfoot S37 (continued)

	January			February			March			April			May			June		
Major elements % DM																		
Calcium							(1958)					0.83	0.72	0.62	0.60	0.51	0.51	0.46
							(1959)					0.89	0.69	0.72	0.66	0.60		
							(1964)			0.53		0.52		0.56	0.44	0.28		0.33
							(1965)				0.37	0.35		0.28	0.27	0.27		0.20
							(1958)							0.64*			0.76*	
							(1959)					0.93*		0.81*			1.01*	
							(1958)							0.59**				
							(1959)							0.69**				
Phosphorus							(1958)					0.41	0.38	0.34	0.27	0.25	0.25	0.22
							(1959)					0.40	0.35	0.28	0.25	0.23		
							(1964)			0.36		0.33		0.29	0.24	0.19		0.18
							(1965)				0.48	0.44		0.34	0.26	0.25		0.17
							(1958)							0.35*			0.39*	
							(1959)					0.43*		0.36*			0.25*	
							(1958)							0.34**				
							(1959)							0.27**				
Magnesium							(1958)					0.15	0.13	0.13	0.14	0.12	0.12	0.11
							(1959)					0.15	0.10	0.11	0.11	0.10		
							(1964)			0.20		0.17		0.16	0.14	0.13		0.13
							(1965)				0.22	0.22		0.14	0.11	0.18		0.11
							(1958)							0.13*			0.19*	
							(1959)					0.16*		0.14*			0.19*	
							(1958)							0.13**				
							(1959)							0.10**				
							(1959)	0.14			0.15			0.16			0.16	
							(1960)	0.14			0.18			0.19			0.18	
Potassium							(1958)					3.61	3.33	3.45	2.48	2.24	2.36	2.18
							(1959)					3.96	2.49	1.96	1.78	1.94		
							(1964)			2.96		4.20		3.40	2.28	1.72		1.76
							(1965)				3.60	3.18		2.35	1.91	1.40		1.15
							(1958)							3.61*			3.77*	
							(1959)					3.11*		2.39*			1.45*	
							(1958)							3.56**				
							(1959)							1.93**				
								2.4			3.0	2.9		2.8	2.3	2.1		1.9
Sodium							(1958)					0.152	0.205	0.137	0.300	0.232	0.152	0.137
							(1959)					0.259	0.279	0.203	0.191	0.203		
							(1964)			0.14		0.31		0.38	0.34	0.27		0.29
							(1965)				0.46	0.62		0.49	0.28	0.36		0.37
							(1958)							0.102*			0.104*	
							(1959)					0.281*		0.190*			0.176*	
							(1958)							0.108**				
							(1959)							0.174**				
								0.2			0.2	0.2		0.3	0.3	0.2		0.2
Chlorine							(1958)					1.18	1.08	1.04	0.98	0.85	0.91	0.92
							(1959)					1.45	1.31	1.11	1.02	1.09		
							(1958)							1.07*			1.21*	
							(1959)					1.52*		1.01*			0.98*	
							(1958)							1.02**				
							(1959)							1.10**				
Sulphur							(1958)					0.42	0.39	0.31	0.28	0.25	0.22	0.22
							(1959)					0.40	0.33	0.32	0.30	0.27		
							(1958)							0.33*			0.50*	
							(1959)					0.43*		0.46*			0.41*	
							(1958)							0.30**				
							(1959)							0.31**				
Trace elements µg/g DM																		
Iodine							(1966)							0.32				
Vitamins µg/g DM																		
Carotene																		320

July			August			September			October			November			December			Notes	References
0.60 0.75*	0.47 1.25* 0.72** 1.27**		0.91*	1.24*	1.02*	1.39* 0.87**	1.39**	1.24*	1.87*			1.10**						*Monthly regrowth **Two-monthly regrowth	542
0.15 0.39*	0.21 0.24* 0.30** 0.21**		0.37*	0.31*	0.38*	0.31* 0.34**	0.23**	0.39*	0.29*			0.39*							
0.11 0.16*	0.12 0.21* 0.16** 0.22**		0.25*	0.26*	0.24*	0.29* 0.19**	0.28**	0.30*	0.34*			0.25**							
	0.20 0.20			0.22 0.19			0.20 0.22			0.22			0.22					Cut when about 10cm high. A dressing of magnesian limestone, 5 tonnes/ha, raised Mg content by 12% in 1959 and by 13% in 1960. Magnesium sulphate had little effect and carboniferous limestone reduced Mg in herbage	322
1.20 3.61*	1.79 1.58* 2.30** 1.25**		3.34*	1.73*	2.92*	1.16* 2.08**	0.94**	2.51*	0.82*			2.00**							542
																			254
0.133 0.086*	0.133 0.181* 0.221** 0.285**		0.157*	0.195*	0.238*	0.194* 0.471**	0.311**	0.338*	0.207*			0.556**						*Monthly regrowth **Two-monthly regrowth	542
																			254
0.81 1.17*	0.79 0.96* 0.82** 0.84**		0.73*	0.94*	0.53*	0.71* 0.65**	0.80**	0.52*	0.75*			0.58**							542
0.26 0.50*	0.23 0.42* 0.38** 0.39**		0.54*	0.45*	0.51*	0.50* 0.43**	0.52**	0.52*	0.74*			0.52**						S as stated Presumably total sulphur	
										0.32								Average values	
																		Date uncertain	209

Table 1.74 Cocksfoot S143

		January			February			March			April			May			June		
Digestibility of DM%										(1961) (1962)		67.8	77.1 75.5 75.2 67.0	75.5 66.1	70.6 71.2 L SS	68.7 71.5 73.2 75.2	69.4 72.1	72.1 73.6	
Proximate constituents %DM Crude protein	F L S	6.5 24.6 15.8	(86.5)																
	F W L S		5.9 10.9 15.7 9.1	(58.1)			6.1 8.5 15.1 9.7	(69.6)		6.4 8.3 15.9 9.8	(72.6)								
	G F	16.0 7.5	(53)			15.1 7.2 (70)													
													(1951) (1952) L S	20.9 22.9 13.8		25.6 29.1 18.5		8.0 24.4 25.8 15.9	
		19.2																	
	G F G F	22.7 10.7 23.3 13.4																	
										(1948) (1949) (1950)							11.1	8.6	
													20.0			10.7			
												18.7					11.6		
															H L	14.4 17.5			
												(1958) (1959) (1960) (1958) (1959) (1960)		20.5 17.3 18.1 24.6 18.3 19.9			14.7 12.2 11.5 17.2 12.0 13.9		
													(1961) 31.4 (1962) 23.7		15.4			17.0*	21.9*
Ether extract													7.0			5.3			
Crude fibre															22.8				
													(1961) 20.7 (1962) 21.3					24.8*	24.5*
Ash													12.5			8.9			
												10.9					8.9		
														H L		5.7 9.6			

July			August			September			October			November			December			Notes	References
							64.6 68.6											Digestibility *in vitro* L : Leaf SS : Sheath and stem	75
																		F : frostburnt (% in brackets) L : leaf S : stem	293
											4.6 (25.3) 14.3 13.1 5.0					4.9 (50.4) 11.5 13.9 3.3		F : frostburnt (% in brackets) W : whole plant L : leaf S stem	294
																		G : green F : frostburnt (% in brackets)	204
20.2 26.4 28.0 15.8		22.9 24.9 26.4 15.5		24.7	29.2 23.0 23.5 14.8		30.6	 18.5 19.4 10.6										L : Leaf S : Stem	509
											15.3					14.2		Sown broadcast 1948	296
																		Sown in 2ft. drills Broadcast	
		 12.1			11.4				<19.6 14.4										439
4.5							2.5											Study of primary growth Average height of grass at date shown: 12, 26, 82, 87 cm.	511
		17.6				9.9													141
																		H : head L : leaf	236
17.4 18.4 14.8 15.1 18.2 20.0			15.3 19.2 16.7 13.9 17.1 16.7			20.1 18.1 19.9 19.1 21.2 20.6			23.6 21.7 28.5 27.9 28.6				 26.8					*Low manuring:* 500 kg/ha/annum superphosphate 250 kg/ha/annum muriate of potash No Nitrochalk *High manuring:* 750 kg/ha/annum superphosphate 500 kg/ha/annum muriate of potash 1500 kg/ha/annum Nitrochalk given in 6 applications of 250 kg over 6 months Clover S184 at the rate of 0.55 kg/ha was sown with 32.2 kg/ha of grass seed, but made no apparent contribution to crude protein.	270
	10.9*						29.5* 13.9*											*Regrowth Values for April-May represent means for 3 Spring grazings. Later values follow topping and application of fertiliser.	75
3.1							2.4												511
	30.0*																	*Regrowth	75
							24.4* 29.5*												
7.3							4.9												511
		10.5				10.9													141
																			236

continued

Table 1.74 Cocksfoot S143 (continued)

	January			February			March			April			May			June		
Carbohydrate fractions %DM																		
Soluble carbohydrate												12.0			14.1			
											13.6					12.3		
														8.3				
											(1961)	6.2					13.9*	
											(1962)	12.2						7.2*
										(1961)	10.4	9.5	9.6	13.1	14.8			
										(1962)		13.3	11.1	11.0	13.7			
										N_0	8.1			11.6		9.0		
										N_1	4.7			4.6		6.5		
										N_2	5.0			3.3		4.3		
										N_3	5.5			3.6		4.6		
Total soluble sugars											(1951)						18.6	
											(1952)		9.1		5.0		5.6	
												L	6.5		4.6		5.5	
												S	12.6		3.3		6.2	
Hexose											(1951)						2.0	
											(1952)		2.1		0.9		1.1	
												L	1.3		0.5		0.8	
												S	4.2		1.5		2.8	
														3.1				
Sucrose											(1951)						4.5	
											(1952)		4.3		3.5		3.9	
												L	4.4		3.9		4.1	
												S	3.3		1.6		2.7	
														4.2				
Fructosan											(1951)						10.7	
											(1952)		2.4		0.5		0.4	
												L	0.5		0.2		0.4	
												S	4.4		0.1		0.5	
														1.1				
Lignin												4.8			5.1			
											6.9					6.1		

July			August			September			October			November			December			Notes	Reference
11.6							0.5												511
		7.7				9.8													141
	4.0*																	*Regrowth See Crude protein, same reference Auchincruive	75
							2.5* 6.4*											Cambridge	
12.9 8.3 6.4 7.6			7.4 4.0 3.7 3.5			8.5 4.5 5.2 5.8		6.8 4.4 5.9 5.8										N_0 : no fertiliser N_1 : sodium nitrate N_2 : ammonium sulphate N_3 : urea To supply 88 kg N/ha before the first cut and after each cut except the last.	314
5.9 2.6 2.6 2.8		7.7 2.0 2.0 2.0		5.5	4.0 2.7 2.5 5.2		4.4	 5.8 5.2 10.6										L : leaf S : stem	509
0.9 0.8 0.6 1.8		1.5 0.9 0.8 1.2		1.0	0.7 0.3 0.1 2.9		0.5	 1.1 0.9 2.7											
																			75
3.6 1.6 1.9 1.0		5.0 1.0 1.0 0.8		3.4	2.8 2.1 2.1 1.8		3.5	 3.3 3.4 2.8											509
																			75
0.8 0.0 0.0 0.0		0.9 0.1 0.1 0.0		0.9	0.3 0.1 0.1 0.4		0.2	 1.0 0.6 4.4											509
																			75
8.9							14.6												511
		6.2				4.5													141

continued

Table 1.74 Cocksfoot 143 (continued)

	January			February			March			April			May			June		
Major elements %DM Calcium														H L S	0.1 0.6 0.3			
Phosphorus														H L S	0.4 0.3 0.3			
Magnesium														H L	0.1 0.2			
Potassium														H L S	1.8 2.6 2.7			
Trace elements µg/g DM Barium														H L S	2.0 8.6 9.3			
Boron														H L S	8.0 10.0 4.0			
Chromium														H L S	0.2 0.3 0.2			
Cobalt														H L S	0.05 0.03 0.04			
Copper														H L S	7.2 7.1 5.4			
Iodine												N_0 N_1	0.6* 0.2*					0.3* 0.3*
Iron														H L S	44.0 67.0 21.0			
Lead														H L S	0.6 1.9 0.5			
Manganese														H L S	37 105 103			
Molybdenum														H L S	0.3 0.8 0.2			
Nickel														H L S	3.2 1.0 0.9			
Strontium														H L S	1.6 8.6 4.9			
Tin														H L S	0.1 0.2 0.2			
Titanium														H L S	1.7 3.3 1.6			
Vanadium														H L S	0.11 0.23 0.05			
Zinc														H L S	35 23 14			
Vitamins µg/gDM α-tocopherol											261	288	341	340	176	108		

July			August			September			October			November			December			Notes	Reference
																		H : head L : leaf S : stem	236
							0.5* 0.6*											N_0 : no fertiliser N_1 : monthly application of 500 kg ammonium sulphate/ha *Regrowth	542
																			236
313 – 362 174 – 242 13 – 18				40		29												Average of cuts at 20-25cm high, presumably regrowth The same, dried Hay cut at 30cm. Losses of tocopherol approach 90%	100

Table 1.75 Cocksfoot: named varieties
Digestibility and part analysis

Variety: S26	January			February			March			April			May			June		
Digestibility of DM %												77.0					73 2	
									(1961)		69.3	75.2	73.8	70.6	66 8			
												74.8						74 2
												70.0	72.1	71.1	70.3			
									(1962)					L	71.5	69.0		
														S	75.3	70.6		
Proximate constituents %DM Crude protein																		
	(1)		16.1															
	(2)		11.0															
	(3)	14.4																
													(1958)	17.0				
									(1961)			29.6					19.5*	
									(1962)			23.9						21.4*
Ether extract																		
Crude fibre																		
														23.0				
									(1961)			20.8					25 1*	
									(1962)			21.0						24.5*
Carbohydrate fractions %DM Soluble carbohydrate									(1958)					8.3				
									(1961)			7.0					11.7*	
									(1962)			12.7						7 6*
									(1961)		9.3	9.5	9.3	11.2	14.5			
									(1962)			13.9	12.4	10.6	13.2			
Hexose									(1958)					4.4				
Sucrose									(1958)					3.5				
Fructosan									(1958)					0.4				
Vitamins µg/g DM Carotene																	(1937)	320

July			August			September			October			November			December			Notes	Reference
							66.8 68.4											Digestibility *in vitro* (L) leaf (S) sheath and stem	75
	N_0 N_1 N_2 N_3	14.9 15.6 18.8 14.7																Means of 6 cuts, May to October N_0 : no fertiliser N_1 : Nitrochalk, 250 kg/ha N_2 : Nitrochalk, 500 kg/ha N_3 : sown with clover	288
																		(1) 1954-55 Mean of 3 cuts taken on 19/11, 24/1, 2/3 (2) 1955-56 Mean of 4 cuts taken on 8/11, 12/12, 23/1, 5/3 (3) 1956-57 Mean of 4 cuts taken on 13/11, 13/12, 17/1, 19/2	240
															G F	17.2 11.5		G : green F : frost burn	
	10.7*																	*Regrowth Values for April - May represent means of 3 Spring grazings. Later values followed topping. and application of fertilisers	75
							27.9* 15.5*												
	N_0 N_1 N_2 N_3	5.4 5.5 5.7 5.4																See Crude protein, same reference	288
															G F	2.6 2.2		G : green F : frost burn	240
	N_0 N_1 N_2 N_3	25.4 25.9 25.6 25.8																See Crude protein, same reference	288
															G F	28.1 30.8		G : green F : frost burn	240
	30.8*																	See Crude protein, same reference	75
							25.6* 29.2*												
	3.7*																		
							3.4* 6.4*											Auchincruive Cambridge	
																		Headley Hall	
																		Date uncertain	209

continued

Table 1.75 Cocksfoot: named varieties (continued)

	January			February			March			April			May			June		
Variety: Roskilde II *Digestibility of* DM%										(1961)	72.1 (1962)	77.9 78.0 76.1 76.6 (L) (S)	73.9 76.1 75.2 77.2	71.0 73.3 70.2 72.3	67.0 70.5		72.8	75.7
Proximate constituents %DM Crude protein														14.7				
											(1961) (1962)	29.3 20.1					16.6*	20.9*
Crude fibre														22.8				
											(1961)	21.2 22.0					24.0*	24.9*
Carbohydrate fractions %DM Soluble carbohydrate														10.0				
										(1961) (1962) (1961) (1962)	11.7	7.6 16.3 10.8 15.7	9.6 13.1	12.2 12.5	16.7 14.0		17.3*	9.3*
Hexose														4.2				
Sucrose														4.7				
Fructosan														1.2				
Variety: Scotia *Digestibility of* DM%										(1961)	75.3 (1962)	77.9 79.3 77.5 75.6 (L) (S)	76.4 75.7 75.5 76.8	72.1 74.4 71.5 75.3	68.0 73.0		73.4	76.1
Proximate constituents %DM Crude protein														14.8				
											(1961) (1962)	29.5 21.9					16.5*	22.3*
Crude fibre														20.7				
											(1961) (1962)	20.2 19.7					23.0*	23.2*
Carbohydrate fractions %DM Soluble carbohydrate														12.1				
										(1961) (1962) (1961) (1962)	13.7	8.4 18.4 16.1 17.0	14.0 16.6	15.3 14.6	18.1 15.8		19.1*	8.9
Hexose														5.2				
Sucrose														3.8				
Fructosan														3.2				
Variety: Van Kammekes *Digestibility of* DM%										(1961)	74.4 (1962)	77.1 75.4 76.4 75.4 (L) (S)	72.6 75.8 70.5 75.9	68.9 72 1 69.7 73.5	63.4 71.0		73 8	74.7
Proximate constituents %DM Crude protein													(1958)	14.9				
											(1961) (1962)	29.9 21.1					18.5*	22.3*
Crude fibre													(1958)	23.6				
											(1961) (1962)	19.5 21.3					24.6*	23.8*
Carbohydrate fractions %DM Soluble carbohydrate													(1958)	10.4				
										(1961) (1962) (1961) (1962)	11.2	9.2 16.7 12.2 17.1	11.5 13.8	14.6 11.8	20.1 15.2		16.8*	9.0*
Hexose													(1958)	4.9				
Sucrose													(1958)	4.8				
Fructosan													(1958)	0.7				
Variety: Germinal *Digestibility of* DM%												75.1	73.3		56.7 69.4*	57.4	65.8*	
Digestibility of OM%												77.0	75.2		58.0 71.1*	58.4	68.0*	
Crude protein %DM												19.4	15.0		10.0 18.1*	7.5	18.8*	

July			August			September			October			November			December			Notes	Reference
							65.3 69.6											Digestibility *in vitro* Leaf (L) Sheath and stem (S)	75
	10.8*																	*Regrowth Values for April/May represent means for 3 spring grazings. Later values follow topping and application of fertiliser	
							30.7* 15.1*												
	30.2*																		
							23.6* 29.0*												
	4.5*																		
							3.0* 6.4*											} Auchincruive } Cambridge	
																		} Headley Hall	
							66.5 71.6											Digestibility *in vitro* Leaf (L) Sheath and stem (S)	
	10.1*																	*Regrowth Values for April/May represent means of 3 spring grazings. Later values follow topping and application of fertiliser.	
							29.1* 17.9*												
	30.6*																		
							23.6* 27.3*												
	4.3*																		
							2.8* 5.4*											} Auchincruive } Cambridge	
																		} Headley Hall	
							68.4 69.7											Digestibility *in vitro* Leaf (L) Sheath and stem (S)	
	12.3*																	*Regrowth Values for April/May represent means for 3 spring grazings. Later values follow topping and application of fertiliser.	
							30.0* 16.5*												
	30.0*																		
							22.8* 28.1*												
	4.3*																		
							2.8 6.6											} Auchincruive } Cambridge	
																		} Headley Hall	
	69.9*			68.4*		65.5*			69.7*									Digestibility *in vivo* *Regrowth	356
	72.6*			70.1*		71.2*			73.5*										
	21.3*			16.9*		21.3*			23.1*										

Table 1.76 Cocksfoot: named varieties
Incomplete analysis

	May			June			July			August			Notes	Reference
Variety: Garton's *Proximate constituents* %DM														
Crude protein		15.4						9.7*					*Regrowth	75
Crude fibre		23.7						31.8*						
Carbohydrate fractions %DM														
Soluble carbohydrate		9.1						4.1*						
Hexose		3.2												
Sucrose		4.4												
Fructosan		1.4												
Variety: Glasnevin *Proximate constituents* %DM														
Crude protein		15.5						10.5*						
Crude fibre		23.2						29.6*						
Carbohydrate fractions %DM														
Soluble carbohydrate		8.9						4.2*						
Hexose		3.8												
Sucrose		3.9												
Fructosan		1.3												
Variety: Hercules *Proximate constituents* %DM														
Crude protein		16.3						11.5*					*Regrowth Considerably grazed	
Crude fibre		22.3						30.0*						
Carbohydrate fractions %DM														
Soluble carbohydrate		11.9						4.8*						
Hexose		5.4												
Sucrose		4.8												
Fructosan		1.6												
Variety: Lodi 22 *Proximate constituents* %DM														
Crude protein		14.2						12.3*						
Crude fibre		24.7						31.0*						
Carbohydrate fractions %DM														
Soluble carbohydrate		10.9						3.8*						
Hexose		6.0												
Sucrose		3.7												
Fructosan		1.1												
Variety: Minerva *Proximate constituents* %DM														
Crude protein		15.4						10.7*					*Regrowth Slightly grazed	
Crude fibre		23.7						31.0*						
Carbohydrate fractions %DM														
Soluble carbohydrate		1.0						3.8*						
Hexose		4.1												
Sucrose		4.8												
Fructosan		1.1												
Variety: Mommersteeg *Proximate constituents* %DM														
Crude protein		17.4						10.4*						
Crude fibre		22.9						31.0*						
Carbohydrate fractions %DM														
Soluble carbohydrate		9.8						3.6*						
Hexose		4.0												
Sucrose		4.4												
Fructosan		1.4												
Variety: Pajbjerg Milka II *Proximate constituents* %DM														
Crude protein		14.9						10.3*					*Regrowth Moderately grazed	
Crude fibre		21.1						29.6*						
Carbohydrate fractions %DM														
Soluble carbohydrate		11.6						4.5*						
Hexose		4.7												
Sucrose		3.7												
Fructosan		3.2												
Variety: Souche II *Proximate constituents* %DM														
Crude protein		15.6						11.0*					*Regrowth Slightly grazed	
Crude fibre		22.7						30.0*						
Carbohydrate fractions %DM														
Soluble carbohydrate		8.6						4.0*						
Hexose		3.7												
Sucrose		4.0												
Fructosan		1.0												

Table 1.76 Cocksfoot: named varieties
Incomplete analysis (continued)

	May			June			July			August			September			October			Notes	Reference
14 named varieties *Proximate constituents* %DM Crude protein	18.7	21.5		21.3 12.0 9.5			 21.4* 10.2*			23.9			17.4			21.4			For the individual varieties only yields and digestibilities of dry matter are given. Cuts in June, for hay, and July, 1st aftermath (bracketed) are from different centres *Regrowth	166
Crude fibre	19.7	22.1		18.9 30.8 28.2			 24.4* 32.1*			23.9			25.9			23.7				
Carbohydrate fractions %DM Soluble carbohydrate	20.6	12.1		17.2 11.2 16.4			 5.8* 10.9*			4.0			6.2			7.2				
	April			May			June			July			August			September				
Variety: Danish *Proximate constituents* %DM Crude protein					15.6			12.2			6.0			9.0*			19.7*		*Regrowth Grazed regrowth in September	460
Major elements %DM																				
Calcium (1964)	0.4		0.6		0.6	0.4	0.4		0.3											542
(1965)		0.2	0.2		0.1	0.1	0.1		0.1											
Phosphorus (1964)	0.4		0.4		0.3	0.3	0.2		0.2											
(1965)		0.4	0.4		0.3	0.3	0.2		0.2											
Magnesium (1964)	0.1		0.1		0.1	0.1	0.1		0.1											
(1965)		0.2	0.2		0.1	0.1	0.1		0.1											
Potassium (1964)	3.5		4.5		2.6	2.2	2.0		1.7											
(1965)		3.5	3.5		2.4	1.8	1.6		1.2											
Sodium (1964)	0.1		0.2		0.2	0.3	0.1		0.3											
(1965)		0.4	0.6		0.4	0.6	0.3		0.2											
Vitamins μg/g DM Carotene					382			166			74			167*			321*		*Regrowth Grazed regrowth in September	460

Table 1.77 Cocksfoot: Crude protein only

		May	June	July	August	September	October	November	Notes	Reference
Variety: Adefa II *Crude protein* %DM	(1958)	18.1	12.7	16.5	15.1	19.5	23.4		*Low manuring* 500 kg/ha/annum superphosphate 250 kg/ha/annum muriate of potash. No Nitrochalk.	270
	(1959)	14.7	10.7	18.3	20.1	18.5				
	(1960)	15.7	10.8	15.0	17.5	21.2	23.4			
	(1958)	20.2	15.3	14.7	14.2	17.9	28.2		*High manuring* 750 kg/ha/annum superphosphate 500 kg/ha/annum muriate of potash 1500 kg/ha/annum Nitrochalk given in 6 applications of 250kg over 6 months. Clover S184 at the rate of 0.55 kg/ha was sown with 23.7 to 35.6 kg/ha of grass seeds of different varieties. but made no apparent contribution to crude protein.	
	(1959)	15.1	10.6	17.0	16.4	20.7	27.0	28.1		
	(1960)	16.9	12.6	18.7	15.5	20.4	27.9			
Variety: Barenza *Crude protein* %DM	(1958)	19.6	12.2	17.6	14.1	19.2	24.2		"Low" manuring	
	(1959)	15.5	13.4	18.5	20.5	17.6				
	(1960)	15.7	9.7	15.5	17.5	21.5	23.8			
	(1958)	22.6	14.4	15.2	13.9	18.2	29.2		"High" manuring	
	(1959)	16.5	10.8	17.1	16.9	20.6	24.8	28.4		
	(1960)	17.7	12.6	19.1	15.6	20.4	28.0			
Variety: Daeno II *Crude protein* %DM	(1958)	16.8	14.4	15.7	14.1	18.6	23.5		"Low" manuring	
	(1959)	13.5	14.3	18.4	19.7	18.5				
	(1960)	14.8	11.7	15.4	17.2	21.0	24.9			
	(1958)	19.4	17.2	14.5	14.2	18.5	29.4		"High" manuring	
	(1959)	14.1	11.4	16.2	16.7	20.8	24.7	27.7		
	(1960)	16.4	14.4	18.8	14.8	18.8	28.0			
Variety: Frode *Crude protein* %DM	(1958)	18.1	12.0	16.8	13.9	18.7	23.8		"Low" manuring	
	(1959)	15.4	11.9	19.3	20.0	18.6				
	(1960)	16.6	10.3	15.2	17.0	20.2	24.5			
	(1958)	21.1	15.4	15.1	13.8	18.5	29.3		"High" manuring	
	(1959)	15.3	10.9	17.1	15.7	19.9	26.7	26.2		
	(1960)	16.9	13.0	17.7	14.7	19.8	27.2			
Variety: Gullaker *Crude protein* %DM	(1958)	18.4	12.5	15.6	13.2	17.9	22.7		"Low" manuring	
	(1959)	13.2	13.5	17.7	19.1	18.5				
	(1960)	14.8	11.0	14.9	16.6	20.5	23.9			
	(1958)	20.3	16.4	14.3	13.7	18.4	30.1		"High" manuring	
	(1959)	14.7	9.8	15.7	15.3	19.6	28.0	27.8		
	(1960)	15.9	13.3	17.2	14.7	19.0	27.6			
Variety: Late Roskilde II *Crude protein* %DM	(1958)	17.2	13.7	15.7	13.9	18.4	23.2		"Low" manuring	
	(1959)	14.1	10.7	18.1	19.4	18.6				
	(1960)	15.4	11.9	15.1	17.1	20.4	23.7			
	(1958)	20.6	15.7	14.2	13.4	18.2	28.6		"High" manuring	
	(1959)	15.6	10.4	16.4	16.0	20.3	26.4	26.8		
	(1960)	16.7	13.1	17.9	15.0	19.7	26.7			

Table 1.77 Cocksfoot: Crude protein only (continued)

		May	June	July	August	September	October	November	Notes	Reference
Variety: New Zealand 'Grasslands" *Crude protein* %DM	(1958)	18.5	14.4	16.6	15.0	19.2	22.1		"Low" manuring	270
	(1959)	14.8	13.9	17.7	19.7	18.9				
	(1960)	15.5	13.0	15.2	16.9	21.0	22.7			
	(1958)	22.2	17.9	15.3	14.7	19.2	29.9		"High" manuring	
	(1959)	16.8	11.7	17.0	17.3	21.2	26.8	27.5		
	(1960)	17.1	15.7	18.7	15.6	20.9	27.2			
Variety: Pajbjerg II *Crude protein* %DM	(1958)	17.0	13.6	15.4	13.4	18.1	22.6		"Low" manuring	
	(1959)	12.7	11.6	17.8	18.6	18.2				
	(1960)	15.3	11.6	15.1	16.8	20.1	24.3			
	(1958)	19.2	16.4	14.0	14.0	18.6	28.9		"High" manuring	
	(1959)	13.8	10.7	15.7	15.7	20.1	28.1	28.8		
	(1960)	16.2	13.3	18.2	15.3	19.9	27.6			
Variety: Potomac *Crude protein* %DM	(1958)	17.0	14.5	15.2	13.9	18.9	23.2		"Low" manuring	
	(1959)	13.5	14.5	17.9	20.5	18.4				
	(1960)	15.2	10.7	15.5	15.7	20.8	24.7			
	(1958)	19.8	15.5	14.4	13.5	19.0	28.5		"High" manuring	
	(1959)	14.5	10.9	16.3	16.7	20.5	26.0	[illegible]2		
	(1960)	16.6	14.5	18.0	15.7	20.0	28.5			
Variety: Tardus II *Crude protein* %DM	(1958)	18.5	13.9	16.1	14.3	19.3	23.1		"Low" manuring	
	(1959)	15.0	11.1	19.4	20.4	18.6				
	(1960)	17.1	11.8	15.1	16.8	20.3	24.1			
	(1958)	20.5	16.1	14.5	13.8	18.4	28.2		"High" manuring	
	(1959)	16.1	10.4	17.2	15.7	20.0	27.7	27.2		
	(1960)	17.6	14.0	18.4	15.3	19.7	28.2			
Variety: Trifolium II *Crude protein* %DM	(1958)	17.1	13.9	15.0	14.3	18.6	22.5		"Low" manuring	
	(1959)	12.7	12.4	20.4	19.5	18.3				
	(1960)	15.1	11.4	14.8	16.0	20.9	24.5			
	(1958)	20.4	17.1	13.9	14.0	18.6	28.3		"High" manuring	
	(1959)	14.6	11.4	16.6	17.9	20.1	27.7	27.2		
	(1960)	16.0	14.7	18.0	15.2	19.5	28.1			

		January	February	March	October	November	December	Notes	Reference
Variety: U.S.A. *Crude protein* %DM	N_0	17.3		15.5		–	17.2	N_0: No fertiliser	296
	N_1	18.2		15.5		–	18.8	N_1: Ammonium sulphate 250 kg/ha	
	N_2	18.5		17.7		20.9	19.1	N_2: Ammonium sulphate 500 kg/ha	
								Nov./Dec. 1946 Jan./March 1947	

Table 1.78 Cocksfoot S37, S143 and White Clover S100

	January			February			March			April			May			June		
Proximate constituents % DM Crude protein												23.6	20.1 20.3	17.7		19.9	17.9 19.4	16.5
									24.5		21.8	16.1	14.3		10.8			
												23.0	17.0	18.4	19.1			
Ether extract									4.5		4.4	3.4	4.1		3.1			
Nitrogen-free extract									40.6		44.3	51.6	51.0		45.5			
Crude fibre									20.1		20.3	19.7	21.9		33.1			
Ash									10.3		9.2	9.2	8.7		7.5			
Major elements % DM Calcium										(1952)	N_1 0.37 N_2 0.39 N_3 0.87		0.26 0.33 0.65					
													(1954)	N_4 0.37 N_5 0.39 N_6 0.49 N_7 0.41				
												(1955)	N_8 N_9	0.33 0.42				
Phosphorus										(1952)	N_1 0.80 N_2 1.02 N_3 0.91		0.67 0.99 0.85					
													(1954)	N_4 0.45 N_5 0.46 N_6 0.43 N_7 0.43				
												(1955)	N_8 N_9	0.51 0.47				

July			August			September			October			November			December			Notes	Reference
	14.6	14.8 13.7			18.1 15.6				24.2	24.5 24.3								Pre-grazing samples. Caged samples taken at end of grazing period. Clover ca. 12% DM basis	124
20.6	15.0		12.6		12.4	12.1							13.8					Pre-grazing samples. Strip grazed by dairy cows. Clover ca. 10% DM basis.	
	13.3	11.7				15.9	15.0	12.6			17.3								
3.6	4.1		3.6		3.6	3.4							3.4						
38.3	40.3		42.4		44.4	41.6							40.8						
26.6	29.4		30.4		28.9	30.6							28.8						
10.9	11.2		11.0		10.7	12.3							13.2						
																		1952 (17/4 – 29/4) (29/4 – 10/5) In kg/ha N_1 : 312 Superphosphate, 3125 magnesite and (twice) 375 ammonium sulphate N_2 : Superphosphate and ammonium sulphate as N_1	443
																		1954 (14/5 – 27/5) N_3 : Superphosphate as above and 500 potassium sulphate N_4 : 250 Superphosphate, 375 ammonium sulphate (magnesite in 1952)	
																		1955 (11/5 – 20/5) N_5 : Superphosphate + ammonium sulphate as N_4 (No magnesite in 1952)	
																		N_6 : 250 Superphosphate N_7 : 250 Superphosphate and 500 potassium sulphate N_8 : 290 Mixed fertiliser and 562 (in two lots) ammonium sulphate	
																		N_9 : 312 Mixed fertiliser	

continued

Table 1.78 Cocksfoot S37, S143 and White Clover S100 (continued)

			March			April	May			June			Notes	Reference
***Major elements* (continued)**														443
Magnesium	(1952)	N_1				0.27	0.23							
		N_2				0.16	0.15							
		N_3				0.14	0 14							
	(1954)	N_4							0.28					
		N_5							0.13					
		N_6							0.12					
		N_7							0.10					
	(1955)	N_8						0.11						
		N_9						0.10						
Potassium	(1952)	N_1				1.77	1.60							
		N_2				1.44	1.73							
		N_3				1.74	1.75							
	(1954)	N_4							2.32					
		N_5							2.44					
		N_6							2.19					
		N_7							2.92					
	(1955)	N_8						2.99						
		N_9						2.73						
Sodium	(1952)	N_1				0.09	0.10							
		N_2				0.14	0.15							
		N_3				0.08	0.07							
	(1954)	N_4							0.36					
		N_5							0.31					
		N_6							0.24					
		N_7							0.09					
	(1955)	N_8						0.25						
		N_9						0.16						
Chlorine	(1952)	N_1				0.65	0.54							
		N_2				0.38	0.26							
		N_3				0.52	0.67							
	(1954)	N_4							0.67					
		N_5							0.49					
		N_6							0.50					
		N_7							0.64					
	(1955)	N_8						1.23						
		N_9						1.04						
Sulphur (total)	(1952)	N_1				0.48	0.26							
		N_2				0.50	0.51							
		N_3				0.41	0.40							
	(1954)	N_4							0.54					
		N_5							0.53					
		N_6							0.34					
		N_7							0.35					
	(1955)	N_8						0.44						
		N_9						0.29						

Table 1.78 Cocksfoot S37, S143 and White Clover S100 (continued)

(With Wild White Clover)	April	May	June	July	Notes	Reference
Major elements %DM Sulphate sulphur		N_0 0.11	0.12	0.18	N_0 : no fertiliser	318
		N_1 0.17	0.22	0.28	N_1 : ammonium sulphate 250 kg/ha	
		N_2 0.19	0.27	0.25	N_2 : ammonium sulphate 2 x 250 kg/ha	
		N_3 0.17	0.28	0.25	N_3 : ammonium sulphate 500 kg/ha	
		N_4 0.19	0.25	0.18	N_4 : ammonium sulphate 4 x 500 kg/ha	
		N_5 0.17	0.27	0.27	N_5 : ammonium sulphate 750 kg/ha	
		N_6 0.19	0.27	0.23	N_6 : ammonium sulphate 6 x 250 kg/ha	
		N_7 0.17	0.27	0.23	N_7 : ammonium sulphate 1000 kg/ha	
		N_8 0.16	0.22	0.22	N_8 : ammonium sulphate 1500 kg/ha	
Total sulphur		N_0 0.18	0.26	0.27	First cuts 23/5–14/6/1950, Second cut 24/7/1950 } 1st year of fertiliser treatment	
		N_1 0.27	0.40	0.38		
		N_2 0.30	0.42	0.36		
		N_3 0.31	0.43	0.38		
		N_4 0.34	0.39	0.41	First cuts 20/6–27/6/1953 } 4th year of fertiliser treatment	
		N_5 0.31	0.42	0.44		
		N_6 0.32	0.45	0.37		
		N_7 0.36	0.48	0.41		
		N_8 0.32	0.43	0.41	Weighted mean values 24/7/1950 Mean values only 23/5–14/6/1950 and 20/6–27/6/1953	

See also Introduction p24.

Table 1.79 Cocksfoot S37 and clover: Crude protein %DM

	January			April			May			June			July			November			Notes	Reference
Crude protein %DM Cocksfoot S37 and White Clover S100					29.2	26.7		29.5	20.3			15.5	14.0	14.8					First cuts taken on 15/4, 21/4, 28/4/1951 Second cuts taken on 12/5, 22/5, 30/5/1951 Third cuts taken on 27/6, 5/7, 11/7/1951 250 kg/ha Nitrochalk given after first and second cuts.	452
									13.6 14.7 14.3										Fortnightly grazing Monthly grazing Cut for hay and aftermath } For previous 3 years no fertiliser Clover 16%–28% of ground cover. Sown 1945. Cut 30/5/1949	297
									11.5 11.2 10.3										Fortnightly grazing Monthly grazing Cut for hay and aftermath } For previous 3 years Fertiliser – 750 kg/ha ammonium sulphate in 3 equal dressings. Clover 3%–7% of ground cover. Sown 1945. Cut 30/5/1949	
Cocksfoot S37 and White Clover S100, S184								(1947) (1948) (1949)	16.0 20.0 19.3										Mean of several cuts before grazing Grazing samples Grazed15 May to late June and early Nov. to end of Dec. Yearly averages given	295
Cocksfoot S37 and White Clover unnamed	N_0	15.5														N_0 N_1	18.2 19.4		N_0: no fertiliser N_1: 250 kg/ha ammonium sulphate DM basis assumed	296

Table 1.80 Cocksfoot S37 and Lucerne

	April			May			June			July			August			October			Notes	Reference
Crude protein % DM Cocksfoot S37 and Lucerne White Seal	(1950)	25.9				14.9	15.9					20.7 17.5			25.3 26.1 21.8				Sown 17/5/1949 40-60% lucerne throughout	152
	(1952)					15.3 15.6 17.6	13.8						19.7	19.4 20.0		21.9 21.8 20.5				
												*12.7							Sown 17/5/1949 *Two cuts taken on 25-29/5/1952and 29/7 or 13-14/8/1952 Mean values only 25% lucerne	157
Major elements %DM Calcium												(1) 1.6 (2) 1.5	(3) 1.6						(1) Means for 18/4, 9/6. 21/7, 29/8/1950 and 31/5, 10/8, 1/10/1951 (2) Means for 31/5, 28/7, 31/8/1950 and 31/5, 10/8, 3/10/1951 (3) Means for 8/6, 31/8/1950 and 30/5, 2/8, 1/10/1951 Sown 17/5/1949. 40-60% lucerne throughout.	152
												*1.0							Sown 17/5/1949 *Two cuts taken on 25-29/5/1952 and 29/7 or 13-14/8/1952. Mean values only 25% lucerne.	157
Phosphorus												(1) 0.3 (2) 0.3	(3) 0.3						(1) Means for 18/4, 9/6, 21/7, 29/8/1950 and 31/5. 10/8, 1/10/1951 (2) Means for 31/5, 28/7, 31/8/1950 and 31/5, 10/8, 3/10/1951 (3) Means for 8/6, 31/8/1950 and 30/5, 2/8, 1/10/1951 Sown 17/5/1949. 40-60% lucerne throughout	152
												*0.3							Sown 17/5/1949. *Two cuts taken on 25-29/51 1952 and 29/7 or 13-14/8/1952 Mean values only. 25% lucerne.	157
Vitamins μg/g DM β-carotene												(1) 288 (3) 231	(6) 231	(2) 213 (4) 216	(5) 208				(1) Means for 18/4, 9/6, 21/7, 29/8/1950 (2) Means for 31/5, 10/8, 1/10/1951 (3) Means for 31/5, 28/7, 31/8/1950 (4) Means for 31/5, 10/8, 3/10/1951. (5) Means for 8/6, 31/8/1950 (6) Means for 30/5, 2/8, 1/10/1951 Sown 17/5/1949. 40-60% lucerne throughout.	152
Crude protein % DM Cocksfoot S37 and Lucerne AF1																	16.0		Sown in rows 26/6/1956 4 replicates – mean only. Lucerne 13% contrib. DM	159
Cocksfoot S37 and Lucerne Du Puits																	15.3		Sown in rows 26/6/1956 4 replicates – mean only. Lucerne 19% DM basis	
Cocksfoot S37 and Lucerne, unnamed												(1947) (1948) (1949)	13.7* 12.7* 12.4*						*Regrowth 1st regrowth 22/7 to 26/8 2nd regrowth 9/10 to 11/11 Mean of several cuts during the grazing year.	295

Table 1.81 Cocksfoot S26 and others with clover: Crude protein %DM

	January			February			March			April			May			June		
Cocksfoot S26, S143, Italian ryegrass, some White Clover S100								(1945) (1946)	N_0 N_1 N_2 N_3		24.1	14.5 11.7	12.5	13.6		14.1 9.8 12.0	N_0 N_1 N_2 N_3 12.4	
					N_0 N_1 N_2 N_0 N_1 N_2	13.8 12.6 13.9 15.1 17.8 18.3	(56) (42) (49) (44) (48) (49)											
Cocksfoot S26 and White Clover S100																		
Cocksfoot, Danish and White Clover S100																		
Cocksfoot, Scotia and White Clover S100																		
Cocksfoot, U.S.A. and White Clover. unspecified																		

July			August			September			October			November			December			Notes	Reference
15.0. 18.9 17.0 14.5 17.0 18.8 16.8	21.4	17.2 19.9	16.5 15.9	14.7 16.1 18.0 17.4	15.8	19.4 20.0			22.5 21.6 17.2 16.9	18.0	23.2 20.8 18.7							N_0 : no fertiliser N_1 : 437 kg/ha Nitrochalk in 1 dressing N_2 : 437 kg/ha Nitrochalk in 2 dressings N_3 : 437 kg/ha Nitrochalk in 3 dressings Some white clover. S100, late in season. The experiments simulated grass production for drying. Cuts were made preflowering, at a height of 20–30 cm. Presumably all except the first were of regrowth	284
																14.1 14.2 17.2 18.6 18.6 19.5	(59) (66) (59) (60) (30) (41)	N_0 : no fertiliser N_1 : 375 kg/ha Nitrochalk N_2 : 750 kg/ha Nitrochalk Figure in brackets is % burn by weight.	241
		N_0 17.8 N_1 14.5 N_2 16.8																N_0 : no fertiliser N_1 : 250 kg/ha Nitrochalk after each cut. N_2 : 250 kg/ha Nitrochalk after first and last cuts only	288
		N_0 18.0 N_1 14.4 N_2 16.5																26–59% clover 13–27% other grasses Grass leafy. Clover not flowering	
		N_0 17.7 N_1 16.4 N_2 18.2																May–Oct., 1951-1953. Weighted means only. No replicates 4–5 cuts per year	
												N_0 N_1	20.7 21.0					N_0 : no fertiliser N_1 : 250 kg N/ha as ammonium sulphate 3 year old DM basis assumed	296

Table 1.82 Cocksfoot unnamed

	January			February			March			April			May			June		
Proximate constituents %DM Crude protein																WM WP WD LM LP LD SM SP SD	8.1 8.1 9.1 8.9 8.9 11.1 6.3 7.2 6.1	
													N_0 N_1 N_0 N_1	15.2 19.1 19.0 19.4			10.5 11.2	
											14.9		14.4		8.5	6.7		
														15.9				
														15.8				
												8.1						
														20.0				
Ether extract											2.8		2.3		1.5	1.4		
														3.7				
Nitrogen-free extract											50.6		49.6		51.2	48.3		
														41.3				
Crude fibre											21.2		24.4		30.7	35.4		
														24.3				
												31.5						
													25.4					
Ash													N_0 N_1 N_0 N_1	12.2 10.6 9.0 9.3			6.6 6.2	
											10.5		9.3		8.1	8.2		
														12.2				
														9.6				

July			August			September			October			November			December			Notes	Reference
11.8† 11.6 11.9 12.1 12.6 13.5 8.9 8.8 10.7					8.4* 8.6* 10.0* 8.9* 9.5* 11.2* 6.1* 6.3* 6.6*													L: Leaf; W: Whole plant; S: Stem; M: sown in mixture with timothy. meadow foxtail and fine leaved fescue. P: pure plots D: pure drills † : Hay cut taken 19/6 Mean of 4 cuts taken on 5/5, 4/6, 5/7 and 28/9 * Regrowth	214
									N_0 N_2 N_0 N_2	12.9 11.4 11.4 11.4								N_0 : no fertiliser N_1 : Superphosphate 250 kg/ha, Kainit 500 kg/ha N_2 : As above and ammonium sulphate and sodium nitrate equivalent to 110 kg N/ha May 1931 June/Oct. 1932	215
5.8																			9
																		Grass at leafy stage. Date uncertain.	14
																		20 to 25 cm high. Date uncertain Third cut. Leafy	512
																			494
																		For metabolisable energy see table 1.226	351
1.1																			9
																			351
49.7																			9
																			351
35.2																			9
																		Grass at leafy stage. Date uncertain	14
																			494
																			351
									N_0 N_2 N_0 N_2	9.4 10.2 7.7 8.1								See crude protein, same reference	215
8.2																			9
																			512
																			351

continued

Table 1.82 Cocksfoot unnamed (continued)

	January			February			March			April			May			June		
Carbohydrate fractions %DM																		
Soluble carbohydrate														6.6				
Total sugars								9.5 9.5										
														4.6				
Glucose													0.9		0.9	1.2		
														0.9				
Fructose													1.5		1.3	2.1		
														1.1				
Sucrose													3.4		3.7	2.6		
														1.8				
Fructosan													2.0		3.9	4.9		
								4.7 5.4										
														1.6				
Pectin														4.4				
Cellulose											35.2		36.7		40.6	41.2		
														21.8				
														27.4				
Xylan														12.0				
Araban														3.2				
Galactan														1.2				
Total hemicelluloses														16.4				
Hemicellulose polysaccharide														14.9				
Hemicellulose uronic anhydride														2.4				
Lignin											5.5		4.9		6.1	8.3		
														6.0				
														5.9				

July			August			September			October			November			December			Notes	Reference
																		20 to 25 cm high. Date uncertain. Third cut. Leafy	512
																		Fresh Dried, oven, 80°, 4 h	313
																			351
																			333
																			351
																			333
																			351
																			333
																			351
																			333
																		Fresh Dried, oven, 80°, 4 h	313
																			351
																		20 to 25 cm high. Date uncertain Third cut. Leafy	512
43.4																			9
																		20 to 25 cm high. Date uncertain Third cut. Leafy	512
																			351
																		20 to 25 cm high. Date uncertain Third cut. Leafy	512
9.7																			9
																		20 to 25 cm high.Date uncertain Third cut. Leafy	512
																			351

continued

Table 1.82 Cocksfoot unnamed (continued)

	January			February			March			April			May			June		
Major elements %DM Calcium																WM WP WD LM LP LD SM SP SD	0.6 0.7 0.6 0.8 0.8 0.6 0.4 0.4 0.3	
													N_0 N_1 N_0 N_1	0.4 0.4 0.3 0.4			0.5 0.4	
											0.7		0.5		0.3		0.4	
														0.54				
Phosphorus																WM WP WD LM LP LD SM SP SD	0.3 0.2 0.3 0.3 0.3 0.3 0.3 0.4 0.3	
													N_0 N_1 N_0 N_1	0.4 0.4 0.2 0.3			0.2 0.2	
											0.3		0.3		0.3		0.2	
														0.32				
											0.3		0.2		0.2		0.2	
Magnesium													N_0 N_1	0.1 0.1				
Potassium																WM WP WD LM LP LD SM SP SD	3.2 3.2 2.5 2.8 3.6 2.6 3.6 3.5 2.4	
													N_0 N_1 N_0 N_1	2.3 2.4 2.1 2.7			1.7 1.4	
											2.8		2.6		2.2		2.1	
Sodium											0.15		0.13		0.13		0.11	
Chlorine																WM WP WD LM LP LD SM SP SD	1.0 0.8 0.7 0.9 0.8 0.6 1.2 0.9 0.8	
													N_0 N_1 N_0 N_1	1.5 1.1 0.9 0.9			0.8 0.9	
											0.3		0.3		0.3		0.3	

July			August			September			October			November			December			Notes	Reference
0.8† 0.7 0.6 0.8 0.9 0.7 0.5 0.5 0.4					0.7* 0.7* 0.6* 0.8* 0.7* 0.6* 0.4* 0.5* 0.4*													See crude protein, same reference	214
									N_0 N_2 N_0 N_2	0.6 0.7 0.5 0.6								See crude protein, same reference	215
	0.3																		490
																		Grass at leafy stage. Date uncertain	14
0.4† 0.4 0.3 0.4 0.3 0.3 0.5 0.4 0.4					0.3* 0.3* 0.3* 0.3* 0.3* 0.2* 0.4* 0.4* 0.3*													See crude protein, same reference	214
									N_0 N_2 N_0 N_2	0.4 0.4 0.2 0.3									215
	0.2																		490
																		Grass at leafy stage. Date uncertain	14
	0.3																		490
																		N_0 : No N N_1 : Ammonium sulphate	442
3.3† 3.2 3.3 3.2 3.3 3.7 3.3 3.3 3.5					2.9* 2.8* 2.8* 2.8* 2.9* 2.5* 3.2* 3.2* 3.2*													See crude protein, same reference	214
									N_0 N_2 N_0 N_2	1.3 1.5 1.3 1.5								See crude protein, same reference	215
	1.8																		490
	0.10																		490
1.0† 1.0 0.8 0.8 1.2 0.9 1.3 1.2 1.3					0.9* 1.0* 1.0* 0.9* 1.7* 0.7* 1.2* 1.6* 1.5*													See crude protein, same reference	214
									N_0 N_2 N_0 N_2	0.9 1.2 0.8 0.9									215
	0.3																		490

continued

Table 1.82 Cocksfoot unnamed (continued)

	January			February			March			April			May			June		
Trace elements µg/g DM Barium																	6 7	
Boron																		
Copper											7.0		11.4		7.3		15.0	
																  N_0 N_1 N_0 N_1	4.3 4.4 3.5 2.8 4.0 4.1	
													N_0 N_1	4.6 6.0				
Chromium																	0.19 0.20	
Cobalt											0.20		0.07		0.09		0.21	
																  N_0 N_1 N_0 N_1	0.13 0.73 0.06 0.59 0.05 0.68	
		0.16 to 0.22																
Iron											321		156		122		167	
																	50 43	
												95						
Lead																		
Manganese											61		50		48		36	
																	70 182	
													N_0 N_1	17 22				
Molybdenum																	1.2 1.4	
Nickel																	0.9 2.4	
Selenium																		
Strontium																	7 7	
Titanium																	3.8 2.8	
Vanadium																	0.08 0.07	
Zinc																	22 18	
Vitamins µg/g DM Carotene								597				311			203			138
Tocopherol											115			95			120	
Thiamin																		
Riboflavin																		
Nicotinic acid																		
Pantothenic acid																		
Pyridoxine																		
Biotin																		

July			August			September			October			November			December			Notes	References
																		Soil well drained Soil poorly drained	359
															3.5 to 5.5				542
	9.1																		490
																		Soil well drained Soil poorly drained N_0 : no fertiliser N_1 : topdressed $CuSO_4.5H_2O$, 22 kg/ha in 1954 and 1955 Sampled 1955	359
																		N_0 : no fertiliser N_1 : 22kg/ha $CuSO_4$ and 33kg/ha $MnSO_4$ Date uncertain	441
																		Soil well drained Soil poorly drained	359
	0.11																		490
																		Soil well drained Soil poorly drained N_0 : no fertiliser N_1 : topdressed $CoSO_4.7H_2O$, 1.65 kg/ha in 1954 and 1955 Sampled 1955	359
																			542
	232																		490
																		Soil well drained Soil poorly drained	359
																		Preflowering	494
																12.2 to 22.7		Samples December and January	542
	32																		490
																		Soil well drained Soil poorly drained	359
																		N_0 : no fertiliser N_1 : $CuSO_4$ 22 kg/ha and $MnSO_4$ 33 kg/ha Date uncertain	441
																		Soil well drained Soil poorly drained	359
															2.6 to 3.3				542
																		Soil well drained Soil poorly drained	359
																0.4 to 0.7		See Lead, same reference	542
																		Soil well drained Soil poorly drained	359
																		Soil well drained Soil poorly drained	
																		Soil well drained Soil poorly drained	
																		Soil well drained Soil poorly drained	
		141		149			292*			438*								*Regrowth	362
	135			200			115						122					Grazeable. Almost all leaf	86
									1.6									Preflowering stage	493
									9.2										
									32.7										
									4.7										
									5.5										
									0.1										

Table 1.83 Cocksfoot and Ryegrass unnamed

		May			June			July			August			Notes	Reference
Proximate constituents %DM															
Crude protein	W(1)	16.3	12.1			16.9	12.3	9.4	10.4		18.0	20.0		Second year ley grazed by dairy cows. Average composition: Cocksfoot 66%, perennial ryegrass 23%, *Poa* spp. 3%, *Agrostis* spp. 1%, clover 4%, weeds 3%. Sampled before grazing: (1) Close folding: 13 strips, 3 grazed in early May 4 in late June. (2) Rotational grazing. Sampled after grazing: (3) Same strips as in (1) W: whole plant L: leaf only S: stem only Data for leaf L(1) and stem S(1) do not correspond exactly with the values for whole plant W(1). The early May value is for 1 only of the 3 strips grazed. The corresponding W(1) value for crude protein was 14.7%	513
	L(1)	17.1				18.8	14.1		11.3			23.1			
	S(1)	9.9				10.6	7.2		5.1			13.6			
	W(2)	18.1	13.3			15.5	12.6	13.6	10.8			21.7			
	W(3)	13.5	9.9			11.5	9.9	6.1	7.6		13.7	13.2			
Ether extract	W(1)	4.9	4.1			4.3	3.8	8.0	3.3		4.6	5.3			
	L(1)	5.4				5.1	6.1		5.3			6.6			
	S(1)	2.5				2.8	2.3		2.1			4.0			
	W(2)	5.5	4.6			4.5	3.9	4.1	4.2			5.3			
	W(3)	3.9	3.1			3.7	3.0	2.5	2.5		3.7	3.8			
Nitrogen-free extract	W(1)	48.5	53.2			44.6	49.3	48.4	52.8		42.4	41.2			
	L(1)	47.2				42.0	43.8		46.7			36.5			
	S(1)	55.3				47.0	52.5		57.2			45.9			
	W(2)	45.6	50.9			44.8	49.6	44.5	50.4			38.7			
	W(3)	47.5	51.6			47.3	50.3	53.5	54.3		47.1	47.6			
Crude fibre	W(1)	21.1	22.1			25.1	26.7	29.9	26.2		24.6	24.7			
	L(1)	20.9				22.9	25.7		26.4			23.8			
	S(1)	23.1				29.0	30.9		29.2			26.0			
	W(2)	20.3	21.2			25.3	25.4	26.3	25.6			24.4			
	W(3)	23.8	25.4			27.6	27.9	29.2	27.9		25.3	26.2			
Ash	W(1)	9.2	8.5			9.1	7.9	8.3	7.3		10.4	8.8			
	L(1)	9.4				11.2	10.3		10.3			10.0			
	S(1)	9.2				10.6	7.1		6.4			10.5			
	W(2)	10.5	10.0			9.9	8.6	11.5	9.0			9.9			
	W(3)	11.3	10.0			9.9	8.8	8.7	7.7		10.2	9.2			
Carbohydrate fractions %DM															
Lignin	W(1)	2.5	2.1			3.4	3.8	4.2	4.7		3.4	3.0			
	W(3)	3.3	3.4			3.4	4.0	4.3	4.8		3.6	–			
Major elements %DM															
Calcium	W(1)	0.53	0.50			0.51	0.57	0.54	0.83		0.76	0.71			
	L(1)	0.56				0.53	0.65		0.83			0.71			
	S(1)	0.26				0.28	0.29		0.31			0.40			
	W(3)	0.69	0.59			0.58	0.65	0.59	0.71		0.82	0.76			
Phosphorus	W(1)	0.39	0.25			0.32	0.31	0.30	0.27		0.28	0.30			
	L(1)	0.31				0.30	0.31		0.41			0.41			
	S(1)	0.31				0.33	0.19		0.30			0.39			
	W(3)	0.34	0.23			0.23	0.29	0.25	0.26		0.31	0.28			
Vitamins µg g DM															
Carotene															
(Total carotenoids)	W(1)	242	191			272	191	139	95		239	294			
	W(3)	124	93			140	50	26	46		98	99			

Table 1.84 Cocksfoot and white clover, unnamed. Digestibility only

		May	June	July	August	Notes	Reference
Digestibility of DM%	(1949)				72.0	Digestibility trials with 3, 4 or 6 sheep	429
	(1950)			68.1	70.8		
					69.6	These values are not stated to apply to cuts of the same crop	
					68.3		
					67.5		
	(1951)	78.7	73.1		79.5		
		76.2			74.5		
Digestibility of OM%	(1949)				74.4		
	(1950)			70.2	72.5		
					71.4		
					70.8		
					69.3		
	(1951)	81.5	75.6		81.5		
		80.0			77.3		

Table 1.85 Cocksfoot and clover, unnamed. Crude protein %DM

		Notes	Reference
Crude protein %DM			
Cocksfoot and clover	O 21.6 F 22.5	O = ordinary grazing F = close-folding	414
	O 16.0 F 17.9	Sown 1945. Pre-grazing samples. At end of period 3, plots mown and raked	
	O 22.6 F 22.9	Samples taken from 19/4 to 14/8/1949; no exact dates given	
	O 18.1 F 17.3		
	O 16.6 F 16.2	O = ordinary grazing F = close-folding	413
	O 18.2 F 16.5	Second year sward.	
	O 17.5 F 17.0	Pre-grazing samples. No exact dates given.	
	O 17.1 F 17.5		
		Range only for 47 sites. Swards sampled in 1st half of December and 1st half of January to test effect of different resting periods. N fertiliser applied to sward at time of resting (type not stated). N_0 = No fertiliser N_1 = 57 kg N/ha N_2 = 114 kg N/ha Resting periods :	31
	N_0 8-12	2nd half July	
	N_0 12-18	1st half August	
	N_0 12-16	2nd half August	
	N_0 11-26	1st half September	
	N_0 10-23	2nd half September	
	N_1 13-14	2nd half July	
	N_1 13-22	1st half August	
	N_1 13-16	2nd half August	
	N_1 14-29	1st half September	
	N_1 12-21	2nd half September	
	N_2 13-14	2nd half July	
	N_2 13-24	1st half August	
	N_2 14-20	2nd half August	
	N_2 20-28	1st half September	
	N_2 18-27	2nd half September	
	14.5 16.7 22.3 22.0	A ley in its third year. Four grazing periods between April 29th and September 1957: "zero-grazing"	291
Cocksfoot with White and Red Clover	N_0 17.4 EN_2 15.7 EN_4 16.9 EN_6 20.4 LN_2 18.8 LN_4 21.8 LN_6 24.1	The second year of a cocksfoot, timothy, meadow fescue, red clover and white clover ley which had been grazed by dairy cows throughout 1950. Cocksfoot and S100 dominant. 4 replicate plots. 5 cuts between May and September: means only given E : fertiliser applied immediately after cutting (early) L : fertiliser applied 1-2 weeks before cutting (late) N_0 : no fertiliser N_2 : 250 kg/ha Nitrochalk N_4 : 500 kg/ha Nitrochalk N_6 : 750 kg/ha Nitrochalk	290
	N_0 18.2 EN_2 16.6 EN_4 17.4 EN_6 18.8 LN_2 19.0 LN_4 21.0 LN_6 22.5	The first year of a ley similar to that used in 1951. Red clover yield was high. Fertiliser treatment as above 4 cuts : means only	

Table 1.86 Cocksfoot with other grasses

		Notes	Reference
Cocksfoot and Timothy, with some Clover, Ryegrass and *Poa:*			
Proximate constituents %DM		Aftermath, mature and stemmy, cut for silage in September	401
Crude protein	13.4		
Ether extract	4.1	For composition of silage see table 1.133, same reference	
Nitrogen-free extract	45.4		
Crude fibre	28.0		
Ash	8.9		
Major elements %DM			
Calcium	0.71		
Phosphorus	0.27		

Table 1.87 Cocksfoot and Lucerne, unnamed

	May			June			July			August			Notes	References
Proximate constituents %DM Crude protein							19.8 16.8 17.4 16.8 15.2						Wilted for up to 5 h: at loading drier at building tripods at building pikes at building "fence" at loading drier; second batch	416
						16.4 16.1							Cut for hay. 3 replicates. Means only. Large proportion of lucerne.	378
						10.7							Lucerne predominant. Date uncertain	377
Ether extract						2.4 2.4								378
						2.5								377
Nitrogen-free extract						42.4 42.0								378
						48.9								377
Crude fibre						30.6 30.9								378
						30.1								377
Ash						8.2 8.6								378
						7.8								377
Vitamins μg/g DM Carotene							133 65 62 100 (estimated) 120						Wilted for up to 5 h: as above for crude protein	416
Vitamin D iu/g DM (air-dry material)	Light petroleum extract 1954 0.32 (0.28 – 0.37) 1955 0.04 (0.02 – 0.05) 1956						Light petroleum extract of saponifiable material 0.04 (0.02 – 0.06) 0.04 (0.03 – 0.06)						Mean values with their true fiducial limits at P = 0.95 for groups of 8 rats. Dates not given.	277

Table 1.88 Perennial Ryegrass S23

	January			February			March			April			May			June		
Digestibility of OM%													1958{	80.0	79.8	77.8	72.9 86.7*	62.5
												(1959){	83.2	84.9	79.7	78.7	75.2*	62.7
																		N_0 N_1 N_2
														86	83	79		62
Proximate constituents %DM Crude protein													(1952) (F)		14.1	(1951)(F)8.3	(F)	14.4
	W G B	15.6 19.4 12.5			W G B	15.7 20.3 13.7												
												20.9					6.1	
														7.5				
											17.2					10.1*		
													D	18.4 14.8	15.2 11.8	13.8 10.0		9.6 5.6
												(1958) (1959)	 18.9	20.4 19.1	14.7 15.7	14.7 14.0 (1958)14.8* (1959)17.2*	11.7	10.1 9.6
Carbohydrate fractions %DM Total carbohydrate											47.9					65.6*		
Water soluble carbohydrate												15.8					27.1	
														29.5				
											19.7					31.2*		
										N_0 N_1	11.2 6.5			21.4 7.8		23.1 12.4		
Soluble carbohydrate†														13.8	11.8	(F)11.3		(S)10.6
Total soluble sugars														(1952)(F)	17.9	(1951)(F)25.5		(F)21.6
Hexoses														(1952)(F)	2.9	(1951)(F)1.9		(F) 1.7
														4.6	6.3	(F)6.7		(S) 4.0
Sucrose														(1952)(F)	5.0	(1951)(F)5.4		(F) 5.4
														3.2	2.7	(F)2.8		(S) 2.8
Fructosan														(1952)(F)	9.0	(1951)(F)16.1		(F)12.7
														6.0	2.8	(F)1.8		(S) 3.8
Pectin														3.1				
													D	2.4 1.7	2.1 1.3	(F)2.2 (F)1.0		(S) 2.2 (S) 0.8
Cellulose														20.2				
													D	21.3 19.6	22.1 19.7	(F)23.9 (F)20.8		(S)26.7 (S)19.5
Xylan Araban Glucan Galactan Aldobiouronics Total hemicelluloses													 D	6.9 2.1 1.9 0.7 4.2 15.8 14.7	8.9 2.2 2.2 0.7 4.9 18.9 15.9	(F)10.1 (F) 2.3 (F) 2.0 (F) 0.7 (F) 4.3 (F)19.4 (F)15.3		(S) 14.4 (S) 2.5 (S) 1.7 (S) 0.7 (S) 6.4 (S) 25.7 (S) 14.4
Hemicellulose polysaccharide														12.3				
Uronic anhydride														1.7				
Lignin												3.0					3.3	
														3.5				
											6.1					4.5*		
													D	2.7 0.6	3.6 0.8	(F) 4.3 (F) 0.8		(S) 7.3 (S) 0.0

July			August			September			October			November			December			Notes	Reference
63.1 65.2 63.1 64.7	79.0* 67.4** 78.0* 75.6**			79.5* 78.8*			79.1* 79.2** 76.6* 72.4**			78.2* 80.0*			74.7* 78.5** 81 8* 79.8**					*Regrowth, monthly **Regrowth, 2-monthly N_0 : no extra nitrogen N_1 : 59 kg N/ha as a "chalk based" fertiliser N_2 : 98 kg N/ha as above Values recalculated to usual form from Mr. Raymond.	357
																		In vivo with sheep at a level of feeding close to maintenance	514
(NF) (NF)	16.9* 26.4*	24.9* 23.0*		26.1*	25.3*	 16.5*	24.3*				 20.3*							*Regrowth (F) Flower (NF) No flower	509
									W G B	14.5 17.5 11.6		W G B	16.2 20.2 13.4		W16.2 G20.3 B12.8			W : whole sample G : green B : frost burnt	296
(F)	3.4						3.1											Primary growth. Height at cutting in cm: 10, 25, 51–76, 60–90 (seed shed).	511
				(S)	3.4													(S) Seed	512
		17.5*				14.0*												*Regrowth	141
																		D : see under Digestibility of OM same reference. Digested nutrients calculated from specific digestibilities including those of protein and non-protein N	514
9.9 16.9* (1958) (1959)	 19.6* 9.1** 18.5**		25.3*	 20.2*	23.2*	 19.9* 17.2**	 14.8**	22.0*	 20.3*		21.8*	 29.9* 20.2** 24.1**						*Regrowth, monthly **Regrowth, 2-monthly	542
		54.7*				58.1*												(F) Flower	141
(F)	17.7					(S)	4.2											(NF) No flower	511
					(S) 4.6													(S) Seed	512
		16.5*				27.2*												*Regrowth	141
19.1 20.0			13.3 9.0			12.9 10.4		10.7 8.9										N_0 : no extra nitrogen N_1 : extra 90 kg N/ha 90 kg/ha	314
																			514
(NF) (NF)	9.1* 4.6*	6.0* 12.4*		7.2*	7.1*	 11.6*	5.8*				 12.1*							D : See crude protein same reference	509
(NF) (NF)	1.0* 1.3*	1.5* 2.6*		1.9*	1.4*	 1.7*	1.0*				 1.6*							Soluble carbohydrate, hexose, sucrose, fructosan, completely digested.	509
																			514
(NF) (NF)	3.5* 2.3*	2.8* 6.7*		4.5*	4.7*	 4.7*	3.9*				 6.9*							†For definition of Soluble Carbohydrate see reference 512	509
																			514
(NF) (NF)	4.6* 0.9*	1.4* 2.6*		0.5*	0.7*	 4.0*	1.1*				 3.0*								509
																			514
					(S) 4.1														512
																			514
					(S) 33.8														512
																			514
																			514
					(S). 21.5														512
					(S) 4.2														512
(F)	7.2					(S)	10.0												511
					(S) 11.3														512
		5.5*				2.8*													141
																			514

continued

Table 1.88 Perennial Ryegrass S23 (continued)

	January			February			March			April			May			June		
Major elements %DM																		
Calcium													(1958)	0.78	0.69	0.67	0.62	0.58
												(1959)	0.84	0.99	0.92	0.89		0.81
													(1958)	0.77*			0.61*	
													(1959)	0.99*			1.14*	
													(1958)	0.80**				
														(1959)	0.92**			
									(1964)	0.46		0.60		0.79	0.61	0.35		0.39
										(1965)	0.45	0.45		0.54	0.49	0.43		0.42
Phosphorus													(1958)	0.36	0.30	0.30	0.28	0.26
												(1959)	0.33	0.31	0.33	0.28		0.20
													(1958)	0.37*			0.37*	
													(1959)	0.31*			0.31*	
													(1958)	0.36**				
														(1959)	0.33**			
									(1964)	0.30		0.34		0.36	0.26	0.22		0.19
										(1965)	0.48	0.40		0.36	0.32	0.25		0.23
Magnesium													(1958)	0.118	0.108	0.109	0.117	0.114
												(1959)	0.102	0.115	0.126	0.126		0.123
													(1958)	0.120*			0.097*	
													(1959)	0.115*			0.219*	
													(1958)	0.119**				
														(1959)	0.126**			
									(1964)	0.13		0.14		0.17	0.11	0.11		0.11
										(1965)	0.18	0.17		0.14	0.17	0.15		0.11
Potassium								(1962)	1.93		2.54	2.24	2.27	2.28	1.88	1.67		1.63
													(1958)	3.30	2.38	2.36	2.16	1.81
												(1959)	2.43	1.86	2.54	1.92		1.23
													(1958)	3.38*			2.85*	
													(1959)	1.86*			1.67*	
													(1958)	3.37**				
														(1959)	2.54**			
									(1964)	2.64		4.10		3.45	2.48	1.92		1.52
										(1965)	3.60	3.45		2.40	2.60	1.55		1.34
Sodium								(1962)	0.18		0.21	0.23	0.22	0.20	0.18	0.17		0.13
													(1958)	0.116	0.204	0.173	0.190	0.212
												(1959)	0.185	0.197	0.103	0 264		0.186
													(1958)	0.091*			0.063*	
													(1959)	0.197*			0.139*	
													(1958)	0.108**				
														(1959)	0.103**			
									(1964)	0.17		0.24		0.55	0.46	0.34		0.34
										(1965)	0.42	0.58		0.66	0.38	0.47		0.37
Chlorine													(1958)	0.98	0.92	0.87	0.80	0.80
												(1959)	1.03	1.01	0.96	1.02		0.75
													(1958)	0.91*			1.14*	
													(1959)	1.01*			0.90*	
													(1958)	0.98**				
														(1959)	0.96**			
Sulphur													(1958)	0.40	0.33	0.32	0.31	0.27
												(1959)	0.40	0.37	0.38	0.36		0.39
													(1958)	0.40*			0.41*	
													(1959)	0.37*			0.50*	
													(1958)	0.41**				
														(1959)	0.38**			
Trace elements μg/g DM																		
Iodine												N_0	0.27*					0.29*
												N_1	0.17*					0.26*
Vitamins μg/gDM																		
α-tocopherol											352	288	250	179	129	115	127	

July	August	September	October	November	December	Notes	Reference
0.52 0.75* 1.37* 0.58** 1.29**	0.89* 1.21* 0.88*	1.23* 0.80** 1.20** 0.89*	1.18* 0.99*	2.06* 0.87** .0.90**		*Regrowth, monthly **Regrowth, 2-monthly N_0 : no extra nitrogen N_1 : 500 kg/ha monthly of ammonium sulphate Experiments 1958 and 1959 at Grassland Research Institute, 1964 and 1965 at Welsh Plant Breeding Station	542
0.27 0.41* 0.26* 0.26** 0.24**	0.40* 0.36* 0.42*	0.33* 0.41** 0.26** 0.39*	0.33* 0.39*	0.42* 0.39** 0.31**			
0.106 0.115* 0.239* 0.099** 0.268**	0.164* 0.246* 0.161*	0.263* 0.135** 0.258** 0.148*	0.263* 0.155*	0.216* 0.152** 0.215**			
							254
1.74 3.13* 1.61* 2.08** 1.38**	3.10* 2.07* 3.68*	1.81* 2.87** 1.34** 3.19*	1.68* 2.77*	2.34* 2.65** 1.55**			542
							254
0.206 0.057* 0.144 0.098** 0.276**	0.073* 0.139* 0.106*	0.194* 0.302** 0.366** 0.171*	0.232* 0.142*	0.229* 0.295** 0.420**			542
0.76 1.32* 1.07* 0.92** 0.90**	1.00* 0.92* 0.87*	0.72* 0.85** 0.78** 0.69*	0.69* 0.73*	0.83* 0.58** 0.78**			
0.26 0.54* 0.52* 0.37** 0.51**	0.61* 0.65* 0.56*	0.68* 0.53** 0.63** 0.54*	0.76* 0.72*	0.68* 0.61** 0.60**		S as stated. Presumably total sulphur.	
		0.46* 0.26*					
82 30–40* 187 – 235*	16	28				Primary growth to maturity Average of cuts at 20–25 cm high, presumably regrowth. The same, dried	100

Table 1.89 Perennial Ryegrass S24

	January			February			March			April			May			June		
Digestibility of DM%														75.1	74.3		70.6	
											(F) (S)	82.3 75.0	81.5 75.6	71.1 64.6				
											(WP) (LB) (LS)	83 83 87	(S) (I)	83 83 82 85 83	71 79 72 71 75		69 79 68 64 71	
Digestibility of OM%											(1958)	85.0 83.2	83.9 82.6	80.7 79.6 76.6	81.4 72.8	70.9 65.9	60.2 79.6*	
											(1959)	85.2 84.8 85.0	81.5 81.8 79.9	80.0 72.4 81.8*	66.8	66.7	62.5 76.3*	
													(1958)	{ N_0 { N_1	76.7 74.4		61.1 60.2	
														(1959)	{ N_0 { N_2	65.7 66.7		
											(1)	88	(2)80					
														(L) (LS) (S) (I)	83 85 75 77 79	(1)77 (1)74	80 62 64 59 64	73 56 54
Proximate constituents %DM Crude protein								18.2		23.1		18.1						
													(L+S)	29.1		25.2	25.6 (L) (S)	25.0 26.6 16.6
												17.8		7.5				(F)
													(1957)	N_0 N_1 N_2	10.9 15.3 18.8			(1956)
											15.4					10.0*		
											(1)	17.7	(2)	15.5				
														(L) (LS) (S) (H)	20.0 8.0 5.7	11.7 4.7 3.1 9.7	12.0 5.3 3.0 12.1	9.9 4.9 3.1 10.2
											(1958) (1959)	21.1 20.3	19.1 14.8 (1959)	16.9 11.3 14.2*	13.4 9.7	8.9 10.2 (1958)	7.1 7.9 16.1* 14.9*	

July			August			September			October			November			December			Notes	Reference
																		In vitro	423
																		In vivo F : fresh S : silage	267
																		In vitro WP : whole plant LB : leaf blade LS : leaf sheath S : stem I : inflorescence	465
	78.9* 66.8**		78.8*				79.8* 74.6**			77.6*			77.4**					*In vivo* *Regrowth, monthly **Regrowth, 2-monthly	357
	74.4* 74.7**		77.4*				76.8* 70.1**			86.4*			81.7* 79.7**						
																		N_0 : no extra N N_1 : 58 kg N/ha as a "chalk based" fertiliser	
																		N_2 : 97 kg N/ha as above	
																		In vivo with sheep at a level of feeding close to maintenance (1) Late leafy (2) 20% heads emerged	514
56 63	(2)66 (2)51 (2)54	63 48	(3)41 (3)44 (3)63															*In vitro* (1) Head fully emerged (2) Seed setting (3) Seed dispersal L : leaf LS : leaf sheath S : stem I : inflorescence	311
																			30
	20.3 22.6 13.5	15.6 18.2 10.0		12.5 15.9 8.7	13.0 16.7 8.6	8.3 (L+S) 10.6 5.5	14.8* (L) (S)	10.2 18.3* 13.4 20.7* 6.3 10.9*	10.2 15.9 6.8	12.7* 13.7* 9.7*	14.8* 16.4* 9.4*		14.6* 16.5* 9.3*					(L+S) leaf + stem (L) leaf only (S) stem only *Regrowth	345
3.7						(S)	3.7											(F) flowering (S) seed shedding	511
												W G F	15.7 19.5 9.4					Winter grazing W : whole sample G : green F : frost burnt	337
N_0 N_1 N_2	10.2 11.0 13.6		(1956)	N_0 N_1 N_2	14.2 17.8 19.8													N_0 : no extra nitrogen N_1 : 3x92 kg N/ha as ammonium nitrate N_2 : 3x185 kg N/ha " "	415
		17.1*				12.1*												*Regrowth	141
																		(1) Late leafy (2) 20% heads emerged	514
10.9 5.1 2.7 9.7	10.3 4.3 2.4 9.2	7.1 2.8 2.2 8.5	2.7 2.2 8.2															(L) leaf (LS) leaf sheath (S) stem (H) head	311
16 9* (1958) (1959)	16.1* 8.8**	16.9**		23.3* 20.5*	22.4*	19.7* 16.4**	14.0**		20.4* 21.3*		30.9*	19.9** 24.6**						*Regrowth, monthly **Regrowth, 2-monthly	542

continued

Table 1.89 Perennial Ryegrass S24 (continued)

	January			February			March			April			May			June		
Carbohydrate fractions %DM Total carbohydrate											47.1					61.7*		
Soluble carbohydrate (Sugars)												17.7		31.5				
											21.7					27.4*		
												17.2	17.5	18.2	20.6	19.3	20.9	
												18.4				(3)	13.5	
Water-soluble carbohydrate	(1953)	Herbage (u) (c) Stubble (u) (c)		3.6 7.1 6.2 7.3		(1953)	3.8 5.2 7.5 8.2											
										N_0 $N_e(0)$	15.6 7.3			19.2 7.0		20.5 13.4		
								$N_e(1)$ $N_e(2)$ $N_e(3)$ $N_e(4)$	20.1 16.7 14.2 12.3	20.0 17.8 13.6 11.1	24.4 21.9 17.5 13.3		35.8 31.5 28.5 20.2	38.9 34.2 27.7 20.9	31.2 28.9 22.2 16.2	31.5 29.9 26.9 21.0	26.3 24.6 22.0 18.5	
Sucrose														6.2 5.1 5.4 5.2				
	(1953)	Herbage (u) (c) Stubble (u) (c)		1.1 1.9 1.1 0.9		(1953)	1.4 2.2 0.8 0.8											
														8.5		12.2	3.2	
Glucose														1.3 1.3 1.9 1.6				
	(1953)	Herbage (u) (c) Stubble (u) (c)		0.5 1.4 0.6 1.0		(1953)	0.7 0.7 1.2 0.9											
														0.4		0.7	0.3	
Fructose														1.7 1.7 2.9 1.4				
	(1953)	Herbage (u) (c) Stubble (u) (c)		0.5 1.0 0.6 1.0		(1953)	0.5 0.8 0.8 0.9											
														0.4		0.7	1.5	
Fructosan														9.6 9.2 5.2 3.5				
	(1953)	Herbage (u) (c) Stubble (u) (c)		1.5 2.8 4.0 4.4		(1953)	1.3 1.6 4.8 5.7											
														0.6		2.5	2.0	
Other oligosaccharides																2.5	1.4	
Pentosan Hexosan												16.2 17.6	19.0 20.7	20.6 23.9	8.9 23.8	12.2 26.4	15.1 29.2	
Xylan Araban Glucan Galactan Aldobiouronics Total hemicelluloses											D	6.9 1.5 1.9 0.6 4.2 15.1 13.1	(3) (3) (3) (3) (3) (3) D	9.5 1.9 1.8 0.6 4.9 18.7 14.8				
Cellulose														8.0		16.5		
											D	18.5 17.4	(3) D	20.9 18.0				
Pectin											D	2.2 1.6	(3) D	2.3 1.4				
Lignin																		
												3.6		3.5				(3)
											3.7					4.9*		
												2.7	3.1	4.5	5.0	7.2	8.0	
											D	3.1 1.3	(3) D	4.1 0.8				
																W (3) L (3) Sh (3) St (3) H (3)	7.4 4.4 6.5 7.7 8.3	(4) (4) (4) (4) (4)

July			August			September			October			November			December			Notes	Reference
		52.1*				51.9*												*Regrowth	141
(3) 20.6							(4) 1.8											(3) Flowering (4) Seed	511
		14.9*				25.6*													141
																			183
																		(3) Flowering	514
												(1952)	5.7 6.2 13.5 15.6					u :uncut, primary growth c :cut 4 times before sampling Stubble: material left above ground when herbage removed plus stem bases and rhizomes below surface.	30
9.0 17.8			12.6 9.1			11.6 8.3		7.8 7.0										N_0 : no extra nitrogen N_e : extra nitrogen N_e (0) Mean for 3 different fertilisers sodium nitrate, ammonium sulphate and urea N_e (1)(2)(3)(4) Effect of 4 levels of Nitrochalk, 125, 250, 500 and 1000 kg/ha	314
																		As for Glucose	589
												(1952)	2.1 1.8 2.2 2.4					See Soluble Carbohydrate, same reference	30
							2.5*	2.3*		3.6*	5.6*		5.1*						345
																		Fresh grass Wilted 2 h " 24 h "' 8 days	589
												(1952)	1.0 1.1 1.1 1.4					See Soluble Carbohydrate, same reference	30
							0.4*	0.7*		0.8*	0.5*		0.5*						345
																		As for Glucose	589
												(1952)	0.8 0.9 1.1 1.6					See Soluble Carbohydrate, same reference	30
							0.5*	1.1*		1.1*	0.7*		0.8*						345
																		As for Glucose	589
												(1952)	1.8 2.4 9.1 10.3					See Soluble Carbohydrate, same reference	30
							2.0*	5.4*		5.1*	6.0*		8.9*						345
							4.9*	2.6*		3.3*	3.2*		2.1*						
																			183
																		(3) Flowering D : see under digestibility of OM, same reference. Digested nutrients calculated from specific digestibilities. (Soluble carbohydrates completely digested)	514
										16.7*			18.9*						345
																		(3) Flowering D : see above, same reference	514
										2.8*			4.6*						345
8.2						(4)	12.4											(3) Flowering (4) Seed	511
		5.3*				3.2*													141
																			183
																		(3) Flowering D : see above, same reference	514
7.9 4.8 7.0 8.7 7.2		(4) (4) (4) (4) (4)	8.6 4.9 8.4 9.6 7.2															W : whole plant L : leaf Sh : sheath St : stem H : head (3) Flowering (4) Seed	311

continued

Table 1.89 Perennial Ryegrass S24 (continued)

	January			February			March			April			May			June		
Major elements %DM																		
Calcium												(1958) 0.80	0.78	0.66	0.59	0.51	0.47	
												(1959) 0.88	0.74	0.72	0.69	0.71	0.68	
														(1958) 0.65*			0.77*	
												(1959) 0.94*		1.82*			1.02*	
														(1958) 0.71**				
														(1959) 0.75**				
										(1964) 0.49		0.57		0.62	0.53	0.40		0.43
											(1965) 0.42	0.42		0.42	0.29	0.39		0.35
Phosphorus												(1958) 0.42	0.41	0.35	0.29	0.26	0.23	
												(1959) 0.42	0.35	0.29	0.25	0.24	0.22	
														(1958) 0.39*			0.37*	
												(1959) 0.46*		0.37*			0.26*	
														(1958) 0.35**				
														(1959) 0.27**				
										(1964) 0.30		0.36		0.23	0.23	0.16		0.17
											(1965) 0.48	0.36		0.29	0.25	0.22		0.22
Magnesium												(1958) 0.126	0.113	0.115	0.103	0.096	0.095	
												(1959) 0.096	0.086	0.098	0.111	0.118	0.119	
														(1958) 0.109*			0.138*	
												(1959) 0.102*		0.128*			0.236*	
														(1958) 0.120**				
														(1959) 0.110**				
										(1964) 0.14		0.16		0.12	0.10	0.11		0.12
											(1965) 0.17	0.15		0.12	0.11	0.13		0.11
Potassium									1.75		2.39	2.12		2.19	1.66	1.45		1.16
												(1958) 3.00	2.73	2.95	2.10	1.75	1.54	
												(1959) 3.28	2.48	1.71	1.77	1.63	1.39	
														(1958) 3.12*			2.91*	
												(1959) 3.52*		2.24*			1.38*	
														(1958) 3.00**				
														(1959) 1.57**				
										(1964) 2.72		3.60		2.50	1.56	1.12		1.20
											(1965) 3.70	2.65		1.90	1.40	1.20		1.25
Sodium									0.22		0.36	0.34		0.32	0.25	0.21		0.13
												(1958) 0.141	0.276	0.140	0.317	0.224	0.130	
												(1959) 0.167	0.272	0.237	0.264	0.128	0.211	
														(1958) 0.118*			0.087*	
												(1959) 0.120*		0.115*			0.162*	
														(1958) 0.134**				
														(1959) 0.207**				
										(1964) 0.25		0.65		0.65	0.63	0.42		0.42
											(1965) 0.62	0.66		0.58	0.47	0.28		0.27
Chlorine												(1958) 1.02	1.15	1.19	1.14	0.99	0.93	
												(1959) 1.47	1.22	1.01	0.99	0.82	0.92	
														(1958) 1.21*			1.42*	
												(1959) 1.44*		0.86*			0.81*	
														(1958) 1.16**				
														(1959) 0.98**				
Sulphur												(1958) 0.42	0.37	0.33	0.28	0.28	0.30	
												(1959) 0.40	0.34	0.30	0.31	0.29	0.31	
														(1958) 0.33*			0.44*	
												(1959) 0.39*		0.39*			0.39*	
														(1958) 0.34**				
														(1959) 0.28**				
Trace elements µg/g DM																		
Iodine												N_0	0.20					0.30
												N_1	0.16					0.30

July			August			September			October			November			December			Notes	Reference
0.80*	1.28* 0.71**	1.32**		0.92* 1.36*	0.88*	1.22* 0.94**	1.37**		1.06* 1.24*		1.08*	1.10* 1.16**						*Regrowth, monthly **Regrowth, 2-monthly	542
0.39*	0.24* 0.30**	0.22**		0.43* 0.37*	0.45*	0.33* 0.42**	0.26**		0.46* 0.32*		0.41*	0.41** 0.32**						Experiments 1958 and 1959 at Grassland Research Institute 1964 and 1965 at Welsh Plant Breeding Station	
0.130*	0.275* 0.120**	0.273**		0.181* 0.274*	0.166*	0.283* 0.151**	0.285**		0.173* 0.305*		0.231*	0.207** 0.246**							
																			254
3.04*	1.54* 1.89**	1.34**		3.49* 2.34*	3.77*	1.86* 2.75**	1.41**		3.15* 1.59*		2.52*	2.76** 1.67**							542
																			254
0.062*	0.159* 0.097**	0.265**		0.081* 0.159*	0.109*	0.228* 0.352**	0.327**		0.186* 0.273*		0.303*	0.325** 0.475**							542
1.26*	0.95* 1.01**	0.89**		0.93* ·1.05*	0.82*	0.85* 0.84**	0.79**		0.60* 0.72*		0.93*	0.52** 0.87**							
0.49*	0.46* 0.44**	0.49**		0.57* 0.65*	0.60*	0.63* 0.47**	0.63**		0.58* 0.69*		0.59*	0.60** 0.55**						S as stated, presumably total	
							0.35 0.27											N_0 : no extra nitrogen N_1 : 500 kg/ha monthly of ammonium sulphate	

Table 1.90 Perennial Ryegrass, S23 and S24

		April	May		June		July	August	September	October	November	Notes	Reference
Crude protein %DM													
(1955)	N_0		14.0	14.4	11.2	11.6	11.8	10.4	14.7	17.5		N_0 : No Nitrochalk	464
	N_0		13.9	12.4	13.2	11.7	11.7	11.0	16.0	19.0		N_1 : Nitrochalk, 750 kg/ha, once.	
	N_1		13.5	13.4	17.0	16.0	12.6	13.6	19.4	23.3		N_2 : Nitrochalk, 750 kg/ha, twice.	
	N_3		16.3	14.2	16.8	18.2	14.5	13.0	20.8	27.8		N_3 : Nitrochalk, 1500 kg/ha, once.	
(1956)	N_0	19.3	14.1	10.0	14.6		14.0	14.9	17.8	20.9		N_4 : Nitrochalk, 1500 kg/ha, twice.	
	N_2	20.7	21.0	14.1	15.3		16.8	18.4	21.6	20.6	20.4		
							14.5						
	N_4	24.7	23.5	16.4	16.8		19.9	20.9	19.3	23.5	23.7		
							14.6						
	N_3	22.0	19.9	18.6	16.5		18.2	10.2	24.0	24.3	22.2		
								22.0					

Table 1.91 Perennial Ryegrass, other named varieties

S101	April	May	June	July	Notes	Reference
Proximate constituents %DM Crude protein				N_0 12.1 N_1 16.4 N_2 17.6 N_{0c} 14.6	N_0 : no extra nitrogen N_1 : Nitrochalk 250 kg/ha N_2 : Nitrochalk 500 kg/ha N_{0c} : no extra nitrogen, sown with clover Means of cuts May to October	288
Ether extract				N_0 3.6 N_1 4.9 N_2 5.3 N_{0c} 4.5		
Crude fibre				N_0 25.1 N_1 22.5 N_2 23.2 N_{0c} 23.7		
Carotene μg/g DM	Values for year 1937: Fresh, 290		Artificially dried: 230			209

S101 and N.Z.	April	May	June	July	Notes	Reference
Digestibility of DM%		74.6			Barn dried. Digestibility by sheep	265
Digestibility of OM%		76.7				

New Zealand H1	September	October	November	December	Notes	Reference
Crude protein %DM			W 13.6 G 16.7 F 12.4		W : whole sample G : green only F : frost burnt	337

Ayrshire	May	June	July	August	Notes	Reference
Crude protein %DM	8.1	7.4	4.1	9.5*	Dates assumed from stage of growth. May, preflowering; June, mid-flowering or heading; July, mature; August, *regrowth	460
Vitamins μg/g DM Carotene	88	64	15	135		

continued

Table 1.91 Perennial Ryegrass, other named varieties (continued)

Irish	May	June		July	August	Notes	Reference
Major elements %DM							
Calcium		H	0.23			Dates not given but assumed from stage of growth. Sampled at early maturity.	236
		L	0.87				
		S	0.30				
Phosphorus		H	0.42			H : head	
		L	0.32			L : leaf	
		S	0.27			S : stem	
Magnesium		H	0.13				
		L	0.17				
		S	0.09				
Potassium		H	1.7				
		L	2.3				
		S	1.7				
Trace elements μg/g DM							
Barium		H	6.1				
		L	18.0				
		S	6.7				
Boron		H	8				
		L	9				
		S	6				
Chromium		H	0.16				
		L	0.35				
		S	0.15				
Cobalt		H	0.04				
		L	0.04				
		S	0.03				
Copper		H	4.5				
		L	5.0				
		S	4.0				
Iron		H	39				
		L	101				
		S	22				
Lead		H	0.8				
		L	1.9				
		S	1.1				
Manganese		H	18				
		L	41				
		S	56				
Molybdenum		H	0.29				
		L	0.47				
		S	0.18				
Nickel		H	2.6				
		L	0.8				
		S	0.8				
Strontium		H	3.3				
		L	15.0				
		S	4.4				
Tin		H	0.14				
		L	0.21				
		S	0.18				
Titanium		H	1.7				
		L	9.1				
		S	1.4				
Vanadium		H	0.07				
		L	0.27				
		S	0.04				
Zinc		H	37				
		L	20				
		S	13				

Table 1.91 Perennial Ryegrass, other named varieties (continued)

26 named varieties	May			June			July			August			September			October			Notes	Reference
Digestibility of DM%	78.4	76.5*		78.5*						72.0*			70.2*			74.9*			Digestibility *in vitro* 1)Plots cut to simulate rotational grazing; presumably all cuts except the first were regrowth. *Regrowth For the individual varieties only yields and digestibilities of dry matter are given.	166
Proximate constituents %DM Crude protein	17.0	20.5*		19 6* 9.6} 6.6}			15.3*} 9.6*}			22.0*										
Crude fibre	17.4	21.7*		18.8* 30.4} 29.6}			27.0*} 27.4*}			24.3*										
Carbohydrate fractions %DM Soluble carbohydrate	31.3	15.8*		21.2* 17.6} 26 3}			11.1*} 22.6*}			7.1*			11.4*			14.1*			2)Cuts in June, hay; and July, 1st aftermath, (bracketed) are from different centres.	

Table 1.92 Perennial Ryegrass, unnamed varieties

	January			February			March			April			May			June		
Proximate constituents %DM Crude protein													N_0 N_1 N_2	16.2 16.5 16.8				
Ash													N_0 N_1 N_2	10.5 11.6 13.0				
Crude protein Ether extract Nitrogen-free extract Crude fibre Ash											14.3 2.5 53.1 19.9 10.2		11.7 1.7 57.3 20.7 8.7		7.0 1.1 60.6 24.2 7.2	6.6 1.5 54.6 30.9 6.5		
Crude protein														7.0				
Amino acids % of crude protein Cystine Lysine Methionine, DL form L form Tryptophan Tyrosine														0.77 3.38 1.41 1.25 1.75 4.15				
Carbohydrate fractions %DM Sugars							(1961)12.7						(1960)12.4					
Sucrose Glucose Fructose Fructosan													3.6 1.2 1.8 6.8		7.4 1.4 2.6 10.2	4.0 1.3 1.3 12.1		
							(1961) 4.7						(1960) 6.5					
Cellulose Lignin											34.1 4.7		35.5 3.7		35.0 4.5	42.3 6.9		
Major elements %DM Calcium													N_0 N_1 N_2	0.54 0.66 0.70				
												0.67		0.60			0.36	
													Mean of 5 samples 1948–49, 0.46					
														0.54				
Phosphorus													N_0 N_1 N_2	0.25 0.28 0.26				
												0.30		0.31			0.25	
													Mean of 5 samples 1948–49, 0.26					
														0.39				
Magnesium												0.31		0.21			0.16	
													Mean of 5 samples 1948–49, 0.11					
Potassium													N_0 N_1 N_2	1.78 2.02 2.57				
												2.31		2.25			1.96	
Sodium												0.19		0.13			0.13	
Chlorine													N_0 N_1 N_2	1.14 1.35 1.77				
												0.64		0.53			0.48	
													Mean of 5 samples 1948–49, 0.58					
Trace elements µg/g DM Boron																		
Cobalt														0.17			0.15	
Copper												15.2		7.1			4.8	
													Mean of 5 samples 1948–49, 4.3					
Iron												333		301			241	
													Mean of 5 samples 1948–49, 97					
Lead																		
Manganese												31.1		22.4			20.3	
													Mean of 5 samples 1948–49, 23					
Molybdenum																		
Selenium																		
Vitamins µg/g DM Carotene									430			260			180			90
Tocopherols													300 to 400, all α - tocopherol 187 to 235, all α - tocopherol					

July			August			September			October			November			December			Notes	Reference
																		N_0 : no fertiliser N_1 : superphosphate 250 kg/ha; Kainit 500 kg/ha N_2 : as above with ammonium sulphate and sodium nitrate equivalent to 110 kg N/ha	215
5.2 1.5 53.7 32.7 7.0																		April and early May: preflowering. Late May: early flower. June: late flower. July: seeding.	9
																		Cut at early flowering stage, height 22–30 cm.	12
																			313
																			333
																			313
46.2 12.7																		See Crude protein, same reference	9
																		See Crude protein, same reference	215
	0.36			0.34														Dates uncertain. April: young leaf. May: late leaf. June: mid-flowering. July: full maturity. August: in decline.	490
																			20
																		Grass at leafy stage, no date given	14
																		See Crude protein, same reference	215
	0.25			0.18														See Calcium, same reference	490
																			20
																		Grass at leafy stage. no date given	14
	0.25			0.21														See Calcium, same reference	490
																			20
																		See Crude protein. same reference	215
	1.93			1.44														See Calcium, same reference	490
	0.10			0.15															
																		See Crude protein, same reference	215
	0.53			0.39														See Calcium, same reference	490
																			20
								7.4 to 11.8											542
	0.15			0.14														See Calcium, same reference	490
														0.14 to 0.19					542
	6.5			9.1														See Calcium, same reference	490
																			20
	181			203														See Calcium, same reference	490
																			20
														6.5 to 15.0					542
	17.9			16.3														See Calcium, same reference	490
																			20
								1.36 to 1.56											542
														0.6 to 0.7					542
		140		190			400*			370*								*Regrowth	362
																		Fresh, at height 20–25cm Artificially dried	100

Table 1.93 Perennial Ryegrass with other grasses and clover

	Year		May	July	August	September	Notes	Reference
Perennial Ryegrass, Cocksfoot and Timothy from a sward with Wild White Clover *Major elements* %DM Calcium	(1957)	F_0	0.45	0.63	0.69	0.74	Grasses hand separated from a sward with wild white clover F_0 : no fertiliser F_1 : NaCl 500 kg/ha F_2 : $MgSO_4$ 250 kg/ha F_3 : ammonium sulphate 375 kg/ha F_4 : Superphosphate 375 kg/ha F_5 : Muriate of potash 250 kg/ha Cut at silage stage each time first cut in late May, last cut in early September. Dates of second and third cuts not stated: assumed to be early July and early August. Fertiliser applied in spring each year, except N, which was applied after each cut.	274
		F_1	0.44	0.67	0.69	0.70		
		F_2	0.43	0.56	0.65	0.73		
		F_3	0.49	0.66	0.69	0.82		
		F_4	0.44	0.61	0.71	0.75		
		F_5	0.43	0.66	0.65	0.74		
	(1958)	F_0	0.31	0.28	0.42	0.39		
		F_1	0.30	0.28	0.40	0.39		
		F_2	0.30	0.27	0.41	0.37		
		F_3	0.34	0.32	0.47	0.37		
		F_4	0.32	0.31	0.41	0.37		
		F_5	0.30	0.28	0.40	0.38		
	(1959)	F_0	0.25	0.26	0.33	0.36		
		F_1	0.22	0.23	0.28	0.31		
		F_2	0.24	0.25	0.32	0.36		
		F_3	0.26	0.28	0.34	0.36		
		F_4	0.27	0.28	0.33	0.36		
		F_5	0.21	0.23	0.28	0.33		
Magnesium	(1957)	F_0	0.09	0.13	0.18	0.18		
		F_1	0.08	0.13	0.18	0.16		
		F_2	0.09	0.13	0.18	0.18		
		F_3	0.09	0.15	0.20	0.22		
		F_4	0.09	0.13	0.18	0.18		
		F_5	0.09	0.11	0.16	0.16		
	(1958)	F_0	0.10	0.10	0.15	0.16		
		F_1	0.09	0.10	0.12	0.15		
		F_2	0.11	0.11	0.15	0.17		
		F_3	0.12	0.13	0.19	0.20		
		F_4	0.09	0.09	0.15	0.15		
		F_5	0.08	0.09	0.12	0.13		
	(1959)	F_0	0.14	0.17	0.21	0.22		
		F_1	0.12	0.14	0.18	0.18		
		F_2	0.15	0.18	0.23	0.26		
		F_3	0.15	0.21	0.26	0.27		
		F_4	0.13	0.17	0.21	0.22		
		F_5	0.13	0.13	0.18	0.20		
Potassium	(1957)	F_0	1.63	1.96	2.43	2.09		
		F_1	1.64	1.98	2.39	2.05		
		F_2	1.62	2.01	2.39	2.18		
		F_3	1.69	1.81	1.68	1.38		
		F_4	1.50	1.93	2.47	1.98		
		F_5	1.87	2.20	3.02	2.62		
	(1958)	F_0	2.01	1.87	1.85	1.90		
		F_1	2.02	1.70	1.74	1.69		
		F_2	2.09	2.03	1.88	1.97		
		F_3	1.88	1.62	1.04	0.72		
		F_4	1.87	1.83	1.88	1.74		
		F_5	2.79	2.76	2.62	2.71		
	(1959)	F_0	1.60	1.45	1.67	1.45		
		F_1	1.62	1.43	1.83	1.57		
		F_2	1.62	1.42	1.67	1.45		
		F_3	1.21	0.93	0.38	0.04		
		F_4	1.56	1.43	1.57	1.30		
		F_5	2.54	2.25	2.56	2.26		
Sodium	(1957)	F_0	0.144	0.224	0.371	0.446		
		F_1	0.214	0.329	0.507	0.481		
		F_2	0.138	0.182	0.321	0.378		
		F_3	0.231	0.472	0.803	0.838		
		F_4	0.139	0.203	0.355	0.407		
		F_5	0.082	0.127	0.208	0.288		
	(1958)	F_0	0.380	0.434	0.495	0.514		
		F_1	0.518	0.683	0.742	0.761		
		F_2	0.361	0.407	0.512	0.496		
		F_3	0.693	0.962	1.132	1.028		
		F_4	0.429	0.438	0.531	0.487		
		F_5	0.094	0.140	0.259	0.278		
	(1959)	F_0	0.416	0.416	0.510	0.537		
		F_1	0.691	0.573	0.779	0.898		
		F_2	0.381	0.390	0.471	0.556		
		F_3	0.764	0.840	1.078	0.975		
		F_4	0.461	0.434	0.505	0.560		
		F_5	0.102	0.193	0.271	0.295		

Table 1.93 Perennial Ryegrass with other grasses and clover (continued)

	Year		May	July	August	September	Notes	Reference
Perennial Ryegrass, Cocksfoot and Timothy from a sward with Wild White Clover (continued) *Trace elements* μg/g DM Copper	(1957)	F_0	3.8	5.8	10.1	9.8	See Calcium	275
		F_1	3.9	4.8	9.3	8.0		
		F_2	3.1	5.6	9.9	8.8		
		F_3	4.6	8.5	13.8	9.9		
		F_4	4.0	7.3	11.6	9.5		
		F_5	5.3	6.1	9.1	9.5		
	(1958)	F_0	5.9	5.9	6.5	8.3		
		F_1	5.0	5.8	6.8	8.3		
		F_2	5.0	7.5	6.5	8.4		
		F_3	7.9	9.0	11.0	10.6		
		F_4	7.8	5.9	8.5	9.4		
		F_5	7.0	7.8	8.5	8.8		
	(1959)	F_0	4.4	5.5	6.3	6.8		
		F_1	4.1	4.6	5.9	6.1		
		F_2	3.8	4.9	6.3	7.5		
		F_3	9.3	7.8	9.8	11.9		
		F_4	5.5	7.8	8.6	10.5		
		F_5	4.8	5.8	8.0	10.8		
Iron	(1957)	F_0	128	146	225	285		
		F_1	118	100	255	228		
		F_2	101	129	200	290		
		F_3	128	134	228	248		
		F_4	136	151	255	233		
		F_5	150	130	220	260		
	(1958)	F_0	63	74	136	183		
		F_1	70	46	140	142		
		F_2	50	93	103	121		
		F_3	101	102	123	143		
		F_4	85	173	150	167		
		F_5	89	84	104	171		
	(1959)	F_0	74	83	66	104		
		F_1	69	65	59	170		
		F_2	54	73	60	110		
		F_3	60	68	92	155		
		F_4	68	76	83	220		
		F_5	76	65	73	145		
Manganese	(1957)	F_0	18	14	26	15		
		F_1	35	28	39	25		
		F_2	23	39	43	19		
		F_3	35	59	77	70		
		F_4	21	42	68	45		
		F_5	20	18	49	38		
	(1958)	F_0	16	16	24	32		
		F_1	25	25	40	49		
		F_2	30	35	49	37		
		F_3	61	84	122	70		
		F_4	30	51	74	80		
		F_5	37	40	80	74		
	(1959)	F_0	16	17	26	25		
		F_1	30	30	50	43		
		F_2	33	56	52	42		
		F_3	59	87	94	133		
		F_4	59	70	70	84		
		F_5	54	54	77	70		
Molybdenum	(1957)	F_0	1.0	2.2	2.5	2.0		
		F_1	1.1	1.6	2.8	1.7		
		F_2	0.9	1.8	2.8	2.3		
		F_3	0.5	0.6	0.5	0.6		
		F_4	0.7	0.8	1.3	0.9		
		F_5	0.7	0.9	1.4	1.0		
	(1958)	F_0	1.2	2.0	2.2	2.6		
		F_1	0.9	1.6	2.1	2.2		
		F_2	0.9	1.4	2.2	2.2		
		F_3	0.6	0.7	0.9	1.2		
		F_4	0.6	0.7	0.9	1.2		
		F_5	0.6	0.9	0.9	1.4		
	(1959)	F_0	1.5	1.7	2.2	2.1		
		F_1	1.1	1.8	1.4	1.5		
		F_2	1.0	1.3	1.8	1.9		
		F_3	0.5	0.5	0.5	0.5		
		F_4	0.5	0.5	0.8	0.9		
		F_5	0.7	0.7	0.7	0.9		

		May	June	August	September	Notes	Reference
Ryegrass S24 and White Clover S100 *Digestibility of* OM%	(1)	75.2				Digestibility by sheep (1) Ryegrass 2 parts; clover 1 (2) Ryegrass 1 part; clover 2 (Clover alone 82.0; ryegrass alone 74.4)	265
	(2)	79.4					
Perennial Ryegrass and Broad Red Clover *Digestibility of* DM%			71.2			Digestibility by sheep	537
Digestibility of OM%			73.5				

continued

Table 1.93 Perennial Ryegrass with other grasses and clover (continued)

	May			June			July			August			September			October			Notes	Reference
Perennial Ryegrass S23 and White Clover S100 *Proximate constituents* %DM Crude protein		N_0 N_1	17.1 12.4																N_0 : No nitrogen fertiliser; clover ground cover 34% N_1 : 750 kg/ha ammonium sulphate; clover ground cover 18% Cut for silage	297
	14.6	12.9		17.0	12.9	13.5			14.5	14.7	10.5	13.7			20.0	18.9	17.1		Strip grazed by dairy cows. Sampled before grazing.	124
(1949)	(1) (2) (3)	20.7 15.6 20.2	16.7 12.6 14.7			15.8 10.9 14.3	16.1 11.4 14.1								15.7	21.2			Rotational grazing for dairy cows. (1) Sampled before grazing. (2) Residue after grazing. (3) Caged samples at end of grazing period.	
(1950)	(1) (2)	13.8 9.5			15.0 10.9	12.7 10.6	14.3 10.6		14.5 10.4	14.7 12.7	10.5 9.3	13.7 10.5			20.0 11.6	18.9 10.1	17.0 12.2		Strip grazing for dairy cows. (1) Sampled at start of grazing. (2) After grazing.	
(1951)		(1) (2)	13.7 10.7		11.3 10.2	13.0 9.1	12.5 8.6		10.2 10.3	13.0 10.1		18.5 13.5				18.6 14.6	19.5 13.9	18.7 13.3	As 1950. Clover about 10%.	
	(1)	21.4 (2) (3)	20.0 19.6 16.0	18.0															(1) Proportionately bulked samples from 16 plots before grazing. (2) Mean of 16 plots. Range 16.3 – 22.5 before grazing. (3) Mean of 16 plots. Range 12.1 – 20.9 after grazing.	143
		(1) (2)	20.3 19.3		15.8 18.2	16.8 17.2	18.6 16.2		18.9 20.1	17.7 17.1	17.8 17.2		17.2 20.1	22.3 17.7					(1) Free range grazing with calves. (2) Strip grazing with calves. Some mowing or sheep grazing to maintain an even sward.	4
(1947)		(1) (2)	15.3 15.1			16.5 15.6			18.7 20.8		24.2 23.0					21.8 23.1			All plots had Nitrochalk 375 kg/ha after each cut. (1) Weighted means for nitrogen alone and nitrogen + phosphorus (2) Weighted means for extra calcium and calcium + phosphorus (3) Weighted means for extra potassium and K + P Proportion of clover: 1947, 31%; 1948, 8%; 1949, 13%.	286
(1947)		(1) (2)	11.6 10.3					13.7 13.9			18.3 19.3						19.6 20.6			
(1948) (1) (3)	12.5 13.4			11.1 11.3				11.9 11.2				19.4 22.5					22.1 22.1			
(1949)	(1) (3)	12.7 12.4			12.1 11.4				13.7 12.7						16.8 20.3					
									N_0 19.6 N_1 18.6 N_2 21.0 N_0 18.3 N_1 19.4 N_2 20.5										Four or 5 cuts. N_0 : No nitrogen fertiliser. N_1 : Nitrochalk 250 kg/ha after each cut. N_2 : Nitrochalk 250 kg/ha after first and last cuts only. Weighted means for cuts. Proportion of clover (DM) in different cuts 26 – 59%.	288
Ether extract	3.3	3.2		3.2	2.7	2.9			2.5	3.5	2.8	3.5			3.3	3.6	3.2		Strip grazed by dairy cows Sampled before grazing.	124
		(2) (3)	3.1 2.7																(2) Mean of 16 plots. Range 2.8 – 3.5 before grazing (3) Mean of 16 plots. Range 2.4 – 3.3 after grazing.	143

continued

Table 1.93 Perennial Ryegrass with other grasses and clover (continued)

	May			June			July			August			September			October			Notes	Reference
Perennial Ryegrass S23 and White Clover S100 *Proximate constituents* %DM Nitrogen-free extract (cont'd)	52.1	52.6		47.0	48.8	50.7			47.5	47.6	49.8	48.3			40.4	42.0	44.1		Strip grazed by dairy cows. Sampled before grazing.	124
		(2) (3)	47.3 47.5																(2) Mean of 16 plots. Range 45.3 – 50.1 before grazing (3) Mean of 16 plots. Range 45.0 – 50.3 after grazing	143
Crude fibre	21.6	22.9		24.3	27.7	25.4			25.2	24.6	24.0	24.1			20.8	23.2	22.8		Strip grazed by dairy cows. Sampled before grazing	124
		(2) (3)	19.9 22.8																(2) Mean of 16 plots. Range 18.7 – 21.8 before grazing (3) Mean of 16 plots. Range 20.2 – 25.4 after grazing	143
Ash	8.4	8.4		8.5	7.9	7.6			10.3	9.6	12.9	10.4			15.5	12.3	12.9		Strip grazed by dairy cows. Sampled before grazing.	124
		(2) (3)	10.0 11.0																(2) Mean of 16 plots. Range 9.5 – 11.3 before grazing (3) Mean of 16 plots Range 9.5 – 12.0 after grazing	143
Carbohydrate fractions %DM Soluble carbohydrate								(1957)11.0					16.7							141
Lignin								(1957) 5.6					3.8							

continued

Table 1.93 Perennial Ryegrass with other grasses and clover (continued)

	January			February			March			April			May			June		
Perennial Ryegrass S24, S23 and White Clover (unnamed) Crude Protein %DM											N_0 N_0 N_1 N_2	15.2 16.0 16.9 17.0			15.0 15.4 14.8 15.6		19.2 19.2 19.2 16.5	14.2 14.5 17.4 18 3
														N_0 N_1 N_2 N_3	17.2 16.5 14.8 16.3			
Perennial Ryegrass S24, S101 and White Clover S100 Crude protein %DM										N_0 N_1 N_1 N_1	25.4	13.3 14.8	12.2	13.9*		15.0* 9.5* 11.4*	N_0 N_1 N_1 N_1 10.6*	20.9 18.4

July			August			September			October			November			December			Notes	Reference
		12.5				14.9		17.7										Pre-grazing samples N_0 : No fertiliser N_1 : Nitrochalk 750 kg/ha N_2 : Nitrochalk 1500 kg/ha Samples in late April had 12-23%, in late May 10-20% clover. In early June N_0 plots had 40% clover, those with N fertiliser 12-15%. In July clover was 20-21% on control plots, 13-16% with N. Later clover averaged 6-11% and 2-4%.	464
		12.6				15.7		18.6											
		13.2				18.8		20.8											
		14.0				19.6		23.8											
		15.3																N_0 : No fertiliser N_1 : Nitrochalk 750 kg/ha since 1955 N_2 : Nitrochalk 1500 kg/ha since 1955 N_3 : Nitrochalk 1500 kg/ha for one year only (1957) Samples in late May had 12% clover, in late July 37.4%.	
		12.3																	
		11.7																	
		12.4																	
14.2			15.8*				19.1*		20.1*									N_0 : No fertiliser N_1 : Nitrochalk 437 kg/ha *Regrowth The experiments simulated grass production for drying. All cuts were made preflowering at a height of 20 to 30 cm. It is assumed that all cuts after the first were of regrowth	284
	15.1*					16.5*			21.4*	(1945)									
	19.9*			14.0*					17.7*										
15.6		16.9*			15.0*				16.8*										
	17.9*					21.8*					22.1*								
		15.0*						20.4*											
	17.9*				17.2*						21.7*	(1946)							
	16.5*				17.7*						20.9*								

continued

Table 1.93 Perennial Ryegrass with other grasses and clover (continued)

	May			June			July			August			Notes	Reference
Perennial Ryegrass S23 and White Clover S100, S184 Crude protein %DM							(1947) (1948) (1949)	13.5 19.3 12.1					Grazed from the end of May to 22 July and late September to the beginning of November Averages only given for samples taken over the grazing periods in each year	295
							(1947) (1948) (1949)	16.5 19.5 14.0					Grazed between 22/4 and 14/7 and between 17/9 and 22/10. Averages only given for samples taken over the grazing periods in each year.	
Perennial Ryegrass S23, S101, White Clover S100 and Wild White Clover														
Major elements														
Sulphur %DM	(0)	(TS)		0.21		0.29							(0) No fertiliser (1) Sulphate of ammonia 250 kg/ha (2) Sulphate of ammonia 500 kg/ha (3) Sulphate of ammonia 750 kg/ha (4) Sulphate of ammonia 1000 kg/ha (5) Sulphate of ammonia 1500 kg/ha (6) Sulphate of ammonia 2 x 250 kg/ha (7) Sulphate of ammonia 4 x 250 kg/ha (8) Sulphate of ammonia 6 x 250 kg/ha (TS) Total sulphur. (SS) Sulphate sulphur Samples late May to mid-June taken in first year of fertiliser treatment, 1950; in late June, taken in fourth year of treatment, 1953. See introduction p	318
		(SS)		0.12		0.15								
	(1)	(TS)		0.29		0.41								
		(SS)		0.20		0.29								
	(2)	(TS)		0.33		0.46								
		(SS)		0.21		0.32								
	(3)	(TS)		0.33		0.48								
		(SS)		0.22		0.36								
	(4)	(TS)		0.38		0.50								
		(SS)		0.23		0.35								
	(5)	(TS)		0.31		0.49								
		(SS)		0.18		0.34								
	(6)	(TS)		0.29		0.41								
		(SS)		0 18		0.31								
	(7)	(TS)		0.31		0.47								
		(SS)		0.21		0.35								
	(8)	(TS)		0.31		0.46								
		(SS)		0.20		0.33								
Permanent pasture: Perennial Ryegrass 40%, *Poa* spp. 20% *Agrostis* spp. 18%, Yorkshire Fog 8%, other grasses 4%, Clover (unspec.) 3%, weeds 7%														
Proximate constituents %DM														
Crude protein	(1)	12.8	11 0	9.2	7.7			11.1	9.3	6.7			Permanent pasture grazed by dairy cows. (1) Close folding: values for 10 strips before grazing. 2 grazed in late May. mid July and late July. (2) Close folding: values for 5 selected strips. whole plant (W). leaf (L) and stem (S) separately. before grazing. (3) Close folding: values for 10 strips. as (1) after grazing. (4) Rotational grazing: values for 2 paddocks in late May and early June and the same paddocks in July. before grazing. (5) Same paddocks after grazing.	513
	(2)	(W)	11.3		7.7			11.1	9.3	6.7				
		(L)	14.8		10.0			12.8	13.6	9.7				
		(S)	7.9		5.7			6.3	6.5	3.8				
	(3)	9.6	7.8	7.3	6.4			8.1	7.3	5.6				
	(4)		13.4	9.6				10 4	11 2					
	(5)		10.7	7.5				8.8	8.0					
Ether extract	(1)	4.2	3.6	3.4	2.7			3.9	3.6	2.5				
	(2)	(W)	3.6		2.7			4.1	3.4	2.5				
		(L)	5.5		4.5			5.2	5.0	4.1				
		(S)	2.3		2.3			2.2	1.9	1.7				
	(3)	3.5	2.6	2.8	2.2			3.0	2.6	2.3				
	(4)		4.7	4.3				3.1	3.4					
Nitrogen-free extract	(1)	52.9	53.6	55.4	56.8			52.8	51.9	55.3				
	(2)	(W)	53.2		56.8			53.0	52.2	55.3				
		(L)	48.2		51.4			52.9	48.1	50.2				
		(S)	58.0		58.9			62.2	56.1	60.1				
	(3)	54.7	56.6	57.5	57.4			55.7	54.4	58.6				
	(4)		49.0	53.3				52.1	52.1					
Crude fibre	(1)	21.5	23.6	24.7	26.3			24.2	27.6	28.9				
	(2)	(W)	23.3		26.3			24.7	27.5	28.9				
		(L)	22.1		24.8			20.6	24.0	27.4				
		(S)	25.3		28.2			23.9	29.7	28.9				
	(3)	23.6	24.4	24.9	27.3			25.8	28.2	27.2				
	(4)		23.2	24.0				24.2	24.3					
Ash	(1)	8.6	8.2	7.3	6.5			8.0	7.6	6.6				
	(2)	(W)	8.6		6.5			7.1	7.6	6.6				
		(L)	9.4		9.3			8.5	9.3	8.6				
		(S)	6.5		4.9			5.4	5.8	5.5				
	(3)	8.6	8.6	7.5	6.7			7.4	7.5	6.3				
	(4)		9.7	8.8				10.2	9.0					
Carbohydrate fractions %DM														
Lignin	(1)	2.8	2.7	3.6	3.4			3.9	5.3	5.8				
	(3)	3.2	3.2	4.5	4.3			4.6	5.6	5.0				
Major elements %DM														
Calcium	(1)	0.59	0.57	0.49	0.43			0.67	0.55	0.52				
	(2)	(W)	0.61		0.43			0.63	0.55	0.52				
		(L)	0.76		0.72			0.69	0.64	0.57				
		(S)	0.23		0.19			0.22	0.23	0 21				
	(3)	0.71	0.56	0.47	0.42			0.53	0.58	0.45				
Phosphorus	(1)	0.36	0.36	0.33	0.24			0.25	0.21	0.16				
	(2)	(W)	0.35		0.24			0.26	0.20	0.16				
		(L)	0.31		0.18			0.23	0.24	0.32				
		(S)	0.16		0.13			0.18	0.19	0.18				
	(3)	0.34	0.27	0.28	0.23			0.24	0.18	0.21				
Vitamins μg/g DM														
Carotene	(1)	214	172	144	98			153	69	30				
	(3)	103	73	88	46			54	33	22				

continued

Table 1.93 Perennial Ryegrass with other grasses and clover (continued)

	May		Notes	Reference
Perennial Ryegrass, Italian Ryegrass, Cocksfoot, Timothy, Red Clover, *Poa*, White Clover and weeds *Proximate constituents* %DM			Main components of herbage, per cent: Perennial ryegrass 49 Cocksfoot 12 Red clover 21 Means and ranges of 7 samples, cut between 22 and 30 May. For composition of the silages see table 1.134 same reference	185
Crude protein	15.1 (13.9 – 17.6)			
Ether extract	2.3 (1.8 – 2.6)			
Nitrogen-free extract	50.7 (48.8 – 53.0)			
Crude fibre	24.1 (22.4 – 25.2)			
Ash	7.9 (7.6 – 8.3)			
	June			
Perennial Ryegrass with a little clover *Proximate constituents* %DM Crude protein	O F O F O F	18.5 17.2 16.9 19.8 16.8 14.8	O: ordinary grazing; F: close-folding } over 3 periods of grazing on plots between April and August. No exact dates given	414
	May			
Carbohydrate fractions %DM Glucose Fructose Sucrose Galactose Arabinose Xylose Fructosan Oligosaccharides Total soluble carbohydrates excluding oligosaccharides	 present	 6.7 4.5 5.2 nil nil nil 8.4 24.8	Cut for silage For composition of silage see table 1.134 same reference	589
	August			
Perennial Ryegrass, Cocksfoot, Timothy, Clover and Herbs *Proximate constituents* %DM Crude protein	M1 M2 M3	14.5 15.2 14.7	Mean of 5 samples taken between July and October from caged areas on 3 plots as follows: M1 : 50/50 Herbs/Grass and Clover. M2 : 35/65 Herbs/Grass and Clover. M3 : 20/80 Herbs/Grass and Clover. Herbs include chicory, plantain, yarrow, sainfoin, burnet. See also below reference 487	486
Ether extract	M1 M2 M3	2.7 2.6 2.7		
Nitrogen-free extract	M1 M2 M3	49.8 49.2 49.4		
Crude fibre	M1 M2 M3	22.5 22.8 23.4		
Ash	M1 M2 M3	10.4 10.2 9.8		
Carbohydrate fractions %DM Total sugars	M1 M2 M3	5.4 5.2 5.8		
Cellulose	M1 M2 M3	30.2 32.8 33.1		
Lignin	M1 M2 M3	6.2 5.8 4.9		

continued

Table 1.93 Perennial Ryegrass with other grasses and clover (continued)

	August		Notes	Reference
Perennial Ryegrass, Cocksfoot, Timothy, Clover and Herbs				
Major elements %DM				
Calcium	M1	0.83		486
	M2	0.71		
	M3	0.60		
Phosphorus	M1	0.33		
	M2	0.33		
	M3	0.30		
Magnesium	M1	0.23		
	M2	0.23		
	M3	0.22		
Potassium	M1	3.11		
	M2	3.13		
	M3	3.25		
Sodium	M1	0.14		
	M2	0.14		
	M3	0.13		
Chlorine	M1	1.11		
	M2	1.09		
	M3	1.03		
Trace elements μg/g DM				
Cobalt	M1	0.15		
	M2	0.16		
	M3	0.14		
Copper	M1	10.6		
	M2	10.7		
	M3	8.7		
Iron	M1	296		
	M2	326		
	M3	288		
Manganese	M1	195		
	M2	216		
	M3	235		
Vitamins μg/g DM				
Carotene	M1	234		
	M2	249		
	M3	259		

Table 1.93 Perennial Ryegrass with other grasses and clover (continued)

		September			Notes	Reference
Perennial Ryegrass, Cocksfoot, Timothy, clover and herbs						
Digestibility of OM%	M1	62.2			Aftermath cut 7 September from same plots as above (486) after a hay cut on 9 July. An additional plot without herbs sown: M4	487
	M2	59.5				
	M3	59.5				
	M4	57.5				
Proximate constituents %DM						
Crude protein	M1	10.7				
	M2	11.5				
	M3	11.6				
	M4	12.1				
Ether extract	M1	2.3				
	M2	2.7				
	M3	2.6				
	M4	2.8				
Nitrogen-free extract	M1	52.0				
	M2	50.2				
	M3	48.6				
	M4	47.2				
Crude fibre	M1	25.2				
	M2	25.1				
	M3	27.1				
	M4	27.5				
Ash	M1	9.8				
	M2	10.4				
	M3	10.0				
	M4	10.5				
Major elements %DM						
Calcium	M1	1.04				
	M2	1.06				
	M3	0.85				
	M4	0.68				
Phosphorus	M1	0.28				
	M2	0.27				
	M3	0.26				
	M4	0.27				
Magnesium	M1	0.24				
	M2	0.25				
	M3	0.26				
	M4	0.22				
Potassium	M1	2.38				
	M2	2.32				
	M3	2.51				
	M4	2.76				
Sodium	M1	0.300				
	M2	0.293				
	M3	0.414				
	M4	0.321				
Chlorine	M1	0.91				
	M2	1.01				
	M3	0.88				
	M4	0.84				
Trace elements µg/g DM						
Cobalt	M1	0.05				
	M2	0.16				
	M3	0.03				
	M4	0.15				
Copper	M1	12.7				
	M2	12.5				
	M3	10.2				
	M4	8.0				
Iron	M1	318				
	M2	257				
	M3	292				
	M4	296				
Manganese	M1	278				
	M2	251				
	M3	308				
	M4	303				

continued

Table 1.93 Perennial Ryegrass with other grasses and clover (continued)

	May			June			July			August			September			October			November			December			Notes	Reference
Perennial Ryegrass 45%, *Agrostis* spp. 25%, *Poa trivialis* 14%, Yorkshire Fog, *Festuca ovina*, *Alopecurus pratensis*, Wild White Clover.																										
Proximate constituents %DM																										
Crude protein			18.9	13.8	16.4		16.9	15.8	15.9		17.6	15.3		12.7	12.7	13.9		20.5	20.3	14.1		19.1	22.4	21.8	Pasture land. Given potassic basic slag 875 kg/ha in Dec. 1928, hydrated lime in Jan. 1929 and sulphate of ammonia 125 kg/ha in Feb. and June and the same of Nitrochalk in Nov.	222
Major elements %DM Calcium			0.8	0.7	0.6		0.6	0.7	0.7		0.7	0.6		0.6	0.7	0.6		0.5	0.5	0.5		0.5	0.4	0.4		
Phosphorus			0.3	0.3	0.3		0.2	0.2	0.2		0.3	0.2		0.2	0.2	0.2		0.3	0.3	0.3		0.4	0.6	0.4		
Potassium			2.1	2.4	2.7		2.6	2.8	2.1		2.4	2.5		2.4	2.5	1.5		2.4	2.6	1.6		2.4	2.0	1.9		
Sodium			0.2	0.3	0.4		0.1	0.2	0.2		0.1	0.2		0.3	0.2	0.1		0.2	0.5	0.1		0.2	0.2	0.4		
Chlorine			1.2	1.2	1.5		1.6	1.7	1.4		1.8	1.6		1.6	1.8	0.7		1.5	1.6	0.8		1.0	1.0	1.2	Flowerheads removed in early July; grazed by sheep end of Aug. and early Nov.	

Table 1.93 Perennial Ryegrass with other grasses and clover (continued)

	July	Notes	Reference
Perennial Ryegrass, unnamed, Timothy and White Clover S100		Perennial Ryegrass dominant. With NK and NPK there was almost no clover or weeds. With N alone or NP there was no clover and there was an invasion of *Agrostis tenuis, Festuca rubra* and *Holcus lanatus.* With K alone clover was moderate, and KP gave a clover dominant ryegrass sward.	
Proximate constituents %DM			
Crude protein	(1949) 13.4 (1950) 14.2 (1951) 12.9 } $N_0P_0K_0$	$N_0P_0K_0$: No additional fertiliser	463
	(1949) 23.2 (1950) 23.5 (1951) 21.9 } NP_0K_0	NP_0K_0 : Nitrochalk, 750 kg/ha in March and after each cut.	
	(1949) 12.1 (1950) 13.9 (1951) 13.8 } N_0PK_0	N_0PK_0 : Superphosphate equivalent to 132 kg P_2O_5/ha in 1949-50 and 198 kg/ha in 1951	
	(1949) 23.1 (1950) 23.6 (1951) 23.1 } NPK_0	NPK_0 : Nitrochalk and superphosphate, no potash	
	(1949) 13.3 (1950) 13.2 (1951) 15.4 } N_0P_0K	N_0P_0K : No Nitrochalk or superphosphate. Muriate of potash equivalent to 297 kg K_2O/ha in 1949-50 and 440 kg/ha in 1951	
	(1949) 20.3 (1950) 20.5 (1951) 19.6 } NP_0K	NP_0K : Nitrochalk and potash but no superphosphate	
	(1949) 13.7 (1950) 14.6 (1951) 16.6 } N_0PK	N_0PK : No Nitrochalk Superphosphate and potash given	
	(1949) 20.9 (1950) 20.6 (1951) 19.7 } NPK	NPK : Nitrochalk, superphosphate and potash given Averages of 4 cuts (without N) or 5 (with N) over each season, at height of 20-28 cm.	
Major elements %DM			
Calcium	(1949) 0.67 (1950) 0.64 (1951) 0.64 } $N_0P_0K_0$		
	(1949) 1.09 (1950) 0.97 (1951) 0.88 } NP_0K_0		
	(1949) 0.66 (1950) 0.59 (1951) 0.59 } N_0PK_0		
	(1949) 1.07 (1950) 1.01 (1951) 0.99 } NPK_0		
	(1949) 0.71 (1950) 0.61 (1951) 0.71 } N_0P_0K		
	(1949) 0.94 (1950) 0.80 (1951) 0.64 } NP_0K		
	(1949) 0.68 (1950) 0.55 (1951) 0.95 } N_0PK		
	(1949) 0.93 (1950) 0.78 (1951) 0.71 } NPK		

continued

Table 1.93 Perennial Ryegrass with other grasses and clover (continued)

	July		Notes	Reference
Major Elements %DM (contd.)				
Phosphorus	(1949) 0.37	$N_0P_0K_0$		463
	(1950) 0.38			
	(1951) 0.35			
	(1949) 0.33	NP_0K_0		
	(1950) 0.44			
	(1951) 0.32			
	(1949) 0.38	N_0PK_0		
	(1950) 0.38			
	(1951) 0.38			
	(1949) 0.35	NPK_0		
	(1950) 0.39			
	(1951) 0.42			
	(1949) 0.33	N_0P_0K		
	(1950) 0.33			
	(1951) 0.33			
	(1949) 0.31	NP_0K		
	(1950) 0.34			
	(1951) 0.31			
	(1949) 0.39	N_0PK		
	(1950) 0.39			
	(1951) 0.39			
	(1949) 0.35	NPK		
	(1950) 0.36			
	(1951) 0.35			
Magnesium	(1949) 0.24	$N_0P_0K_0$		
	(1950) 0.30			
	(1951) 0.33			
	(1949) 0.33	NP_0K_0		
	(1950) 0.43			
	(1951) 0.32			
	(1949) 0.19	N_0PK_0		
	(1950) 0.26			
	(1951) 0.34			
	(1949) 0.32	NPK_0		
	(1950) 0.33			
	(1951) 0.34			
	(1949) 0.18	N_0P_0K		
	(1950) 0.23			
	(1951) 0.20			
	(1949) 0.26	NP_0K		
	(1950) 0.27			
	(1951) 0.20			
	(1949) 0.19	N_0PK		
	(1950) 0.24			
	(1951) 0.33			
	(1949) 0.27	NPK		
	(1950) 0.30			
	(1951) 0.21			
Potassium	(1949) 1.35	$N_0P_0K_0$		
	(1950) 1.72			
	(1951) 0.81			
	(1949) 0.78	NP_0K_0		
	(1950) 0.96			
	(1951) 0.79			
	(1949) 1.33	N_0PK_0		
	(1950) 1.76			
	(1951) 1.94			
	(1949) 0.94	NPK_0		
	(1950) 0.89			
	(1951) 0.70			
	(1949) 2.49	N_0P_0K		
	(1950) 3.74			
	(1951) 3.49			

Table 1.93 Perennial Ryegrass with other grasses and clover (continued)

	July		Notes	Reference
Major Elements %DM (cont'd)				
Potassium	(1949) 2.14 (1950) 3.26 (1951) 3.59	NP_0K		463
	(1949) 2.89 (1950) 3.48 (1951) 4.19	N_0PK		
	(1949) 2.16 (1950) 2.91 (1951) 3.44	NPK		
Sodium	(1949) 1.03 (1950) 0.97 (1951) 1.08	$N_0P_0K_0$		
	(1949) 1.17 (1950) 1.05 (1951) 0.58	NP_0K_0		
	(1949) 0.65 (1950) 0.68 (1951) 0.69	N_0PK_0		
	(1949) 1.31 (1950) 1.33 (1951) 1.07	NPK_0		
	(1949) 0.35 (1950) 0.23 (1951) 0.11	N_0P_0K		
	(1949) 0.96 (1950) 0.62 (1951) 0.28	NP_0K		
	(1949) 0.34 (1950) 0.30 (1951) 0.19	N_0PK		
	(1949) 0.94 (1950) 0.73 (1951) 0.25	NPK		
Trace elements μg/g DM				
Manganese	(1949) 62 (1950) 54 (1951) 78	$N_0P_0K_0$		
	(1949) 70 (1950) 70 (1951) 54	NP_0K_0		
	(1949) 78 (1950) 78 (1951) 140	N_0PK_0		
	(1949) 54 (1950) 70 (1951) 54	NPK_0		
	(1949) 47 (1950) 54 (1951) 62	N_0P_0K		
	(1949) 47 (1950) 31 (1951) 54	NP_0K		
	(1949) 62 (1950) 62 (1951) 93	N_0PK		
	(1949) 54 (1950) 39 (1951) 39	NPK		

continued

Table 1.93 Perennial Ryegrass with other grasses and clover (continued)

	March	April	May	June	July	August	September	October	November	Notes	Reference
Mixed pasture of Perennial Ryegrass, Cocksfoot and *Agrostis* with about 20% White Clover										Cut when 10 cm high a : control plots b : magnesian limestone: 5 tonnes/ha	322
Major elements %DM Magnesium	a 0.14	0.14	0.13	0.16	0.20	0.19	0.19	–	– }(1957)	Magnesium sulphate had a small effect in the first year;	
	b 0.20	0.15	0.14	0.17	0.22	0.22	0.22	–	–		
	a 0.14	0.13	0.16	0.16	0.20	0.22	0.24	0.24	– }(1958)	carboniferous limestone reduced Mg in pasture.	
	b 0.18	0.17	0.18	0.19	0.26	0.30	0.32	0.31	–		
	a 0.17	0.17	0.16	0.14	0.18	0.19	–	0.20	0.16 }(1959)	See also ryegrass S22, cocksfoot S37, clover S123 and clover S100.	
	b 0.21	0.22	0.22	0.19	0.24	0.27	–	0.31	0.19		
	a –	0.19	0.17	0.18	0.21	0.20	–	0.20	– }(1960)		
	b –	0.26	0.25	0.24	0.31	0.31	–	0.30	–		
	a 0.20	0.19	0.18	0.16	0.16	0.21	0.24	0.22	– }(1961)		
	b 0.29	0.27	0.25	0.22	0.24	0.30	0.31	0.30	–		

Table 1.94 Italian Ryegrass, S22

	January			February			March			April			May			June		
Digestibility of DM%											82.5		76.7	73.8*	65.1	61.2	58.6 73.8*	
Digestibility of OM%											84.6		79.0	76.1*	66.8	62.6	60.0 75.5*	
Proximate constituents %DM Crude protein									(1958) (1959) (1960)	N_4 N_2 N_4 N_2 N_4 N_2 N_4 N_2	23.9* 20.1* 22.9* 16.9*	25.5 20.9		19.9* 16.4*	14.7 12.0 12.9* 9.8*	9.4 7.5 9.9 8.2		20.0* 15.6* 16.2* 13.0*
																		(1960)
												(1959)	23.8					(1957)
											(1962)	N_2 N_4	17.1 20.0				8.0 12.1	
								(1964)	N_1 N_2 N_3	29.6 35.7 35.6	24.9 37.6 41.6	16.8 26.7 34.7		12.0 18.6 23.9	9.5 14.4 18.4	8.0 10.9 14.9		
Carbohydrate fractions %DM Soluble carbohydrate												(1959)	19.5					(1957)
Total sugars Glucose +Fructose +Sucrose																		(1960)
Fructosan																		(1960)
Major elements %DM Potassium									2.059		2.870	2.422		2.204	1.642	1.279		1.232
Sodium									0.104		0.138	0.133		0.122	0.078	0.090		0.060
Magnesium							(1959) (1960)	0.13			0.14 0.13			0.14 0.20			0.13 0.18	

July			August			September			October			November			December			Notes	Reference
	65.7*		71.1*			69.4*												Digestibility by sheep *Regrowth	356
	68.6*		73.2*			72.4*													
15.5 14.2	22.7* 21.9*	16.4* 12.8*			19.7* 16.0*	10.2 9.7 9.1 8.6 15.6* 12.5*		25.7* 22.8*		21.0* 19.1*	23.6 18.2 18.4 18.1							N_4 : Nitrochalk 500 kg/ha N_2 : Nitrochalk 250 kg/ha *Regrowth	301, 300
N_0 N_1 N_2 N_3 N_4	21.2 35.1 36.4 34.2 34.9	16.8 25.2 32.2 22.6 31.8	14.2 17.7 26.4 17.9 24.7	12.9 14.5 22.6 16.8 23.2														N_0 : No extra N N_1 : 62 kg N/ha } as ammonium N_2 : 125 kg N/ha } sulphate N_3 : 62 kg N/ha } as sodium N_4 : 125 kg N/ha } nitrate	392
18.8									19.0 (1958)	13.8									350
		9.3 12.1			11.6 19.0				17.2 22.7									N_2 : Nitrochalk 250 kg/ha N_4 : Nitrochalk 500 kg/ha	270
																		N_1 : 27.5 kg N/ha N_2 : 82 kg N/ha N_3 : 137 kg N/ha } as Nitrochalk one week before first sample The means given are for 8 replicates in 3 years of N content of herbage harvested 1,2,3,4,5 or 6 weeks after application of N fertiliser Harvesting dates between 2 April and 5 June	552
15.1									16.1 (1958)	16.2									350
N_0 N_1 N_2 N_3 N_4	7.7 6.2 6.6 6.0 7.7	9.0 9.2 8.0 9.6 8.3	9.0 12.7 10.3 11.1 10.0	11.4 11.3 10.9 10.3 9.5														See crude protein, same reference Total soluble carbohydrate, the sum of sugars + fructosan	392
N_0 N_1 N_2 N_3 N_4	4.1 0.8 0.6 0.6 1.2	10.0 3.4 1.9 6.0 2.0	11.0 5.9 2.2 5.9 2.8	16.0 9.6 5.5 8.6 5.9															
																			254
	0.17 0.22			0.19 0.22			0.22 0.22						0.20					Cut when about 10 cm high. A dressing of magnesian limestone, 5 tonnes/ha, raised Mg content by 10% in 1959 and by 16% in 1960. Magnesium sulphate had little effect and carboniferous limestone reduced Mg in herbage.	322

Table 1.95 Italian Ryegrass, named varieties other than S22

	January			February			March			April			May			June		
New Zealand H_1 *Digestibility of* DM%										77.4 73.4		78.1	80.5*	72.0	68.6 65.4	70.6*		
										74.0		78.6		71.3	67.6 65.2			
Digestibility of OM%										80.7 81.6		80.3	82.9*	74.0	70.1 67.3	72.1*		
Proximate constituents %DM Crude protein, New Zealand H_1									(1959) (1960)	(1958) N_4 N_2 N_4 N_2	N_4 N_2 N_4 N_2 25.3* 19.0* 22.2* 16.6*	28.7 21.5		20.2* 15.3*	15.0 11.5 13.6* 9.4*	10.3 7.4 9.4 7.9		17.8* 13.8* 14.3* 11.6*
New Zealand									(1959) (1960)	(1958) N_4 N_2 N_4 N_2	N_4 N_2 N_4 N_2 27.1* 20.8* 21.8* 17.4*	29.2 20.9		21.2* 14.9*	16.8 11.9 10.5* 9.0*	10.5 7.5 10.9 7.7		17.8* 12.7* 15.2* 10.4*
											(1962)	N_2 N_4	17.8 21.2				8.4 10.9	
Ayrshire									(1959) (1960)	(1958) N_4 N_2 N_4 N_2	N_4 N_2 N_4 N_2 22.3* 19.5* 23.4* 17.9*	25.2 21.0		19.2* 17.3*	13.1 11.8 12.5* 10.0*	9.6 8.2 9.2 7.5		18.6* 14.5* 16.5* 12.4*
Danish									(1959) (1960)	(1958) N_4 N_2 N_4 N_2	N_4 N_2 N_4 N_2 21.3* 18.4* 21.7* 17.8*	25.1 19.9		18.5* 14.4*	13.8 11.8 11.5* 9.7*	9.4 7.4 9.3 7.8		17.9* 14.6* 16.8* 12.4*
Mommersteeg									(1959)	(1958) N_4 N_2	N_4 N_2 N_4 N_2 28.6* 25.8*	27.2 25.8			18.7 14.4 15.2* 11.2*	10.8 8.5 12.6 8.5		17.6* 14.0*
											(1962)	N_2 N_4	19.4 21.8				8.1 11.6	
Sceempter									(1959)	(1958) N_4 N_2	N_4 N_2 N_4 N_2 26.7* 24.3*	27.6 24.1			19.0 12.8 11.0* 11.0*	11.0 7.4 12.6 8.0		17.7* 17.1*
											(1962)	N_2 N_4	18.1 20.5				8.5 10.3	
Imperial											(1962)	N_2 N_4	17.4 22 6				8.7 10.9	
Melle											(1962)	N_2 N_4	17.2 21.1				8.1 12.0	
C.I.V.											(1962)	N_2 N_4	19.0 22.3				8.8 10.3	
Tetrone											(1962)	N_2 N_4	22.3 24.3				9 3 11 5	
E.F. 486											(1962)	N_2 N_4	19.5 21.9				8.4 10 8	
Hinderupgaard											(1962)	N_2 N_4	18.1 21.6				8.4 11.0	
Roskilde											(1962)	N_2 N_4	18.6 22.9				8.4 10.4	

July			August			September			October			November			December			Notes	Reference
68.2*					68.5*													*In vivo* with sheep *Regrowth	356
(1) (2) (3)	55.8 55 9 56.5																	*In vivo* (1) Barn dried hay. (2) Cold silage. (3) Hot silage	267
70.4*					71.6*													*In vivo* with sheep *Regrowth	356
16.1 14.0	23.1* 18.7*	15.3* 12.1*			17.8* 13.0*	10.6 9.2 9.7 8.5 12.4* 10.4*		25.5* 20.7*		20.5* 14.3*	21.6 17.7 20.3 17.9							N_4 : Nitrochalk 500 kg/ha N_2 : Nitrochalk 250 kg/ha *Regrowth	301, 300
15.3 13.4	21.2* 18.9*	17.5* 12.2*			19.0* 13.8*	10.1 8.8 10.3 8.7 14.3* 10.5*		26.0* 20.6*		20.6* 17.6*	21.6 18.7 21.7 18.4							N_4 : Nitrochalk 500 kg/ha N_2 : Nitrochalk 250 kg/ha *Regrowth	301, 300
		8.7 12.8			11.4 17.9				17.1 25.9										270
13.8 12.8	22.6* 18.4*	16.7* 12.6*			19.7* 15.0*	8.8 9.2 8.9 8.4 16.1* 11.7*		25.9* 21.3*		21.9* 18.3*	19.7 18.5 20.0 18.7								301, 300
15.3 12.8	22.0* 18.7*	16.4* 11.4*			20.3* 14.6*	9.2 9.0 9.9 8.2 14.9* 12.0*		25.1* 21.1*		20.5* 17.5*	20.4 17.7 19.7 18.0								
16.5 13.8		17.1* 13.1*				9.7 8.1 9.7 8.7 14.7* 11.4*				21.4* 17.0*	21.5 19.0 20.7 19.1								
		9.4 12.8			13.1 18.3				16.6 23.3										270
16.2 13.2		18.0* 13.4*				9.8 8.6 10.9 8.1 14.3* 11.2*				22.9* 18.9*	21.9 19.0 22.1 19.3								301, 300
		9.2 13.6			12.7 18.8				16.4 22.8										270
		8.9 12.8			12.3 14.4				16.6 23.8										
		8.9 13.7			10.7 16.6				15.7 22.8										
		9.2 12.3			11.7 18.3				17.1 23.3										
		8.8 12.8			11.9 17.9				16.4 22.1										
		9.2 12.8			12.3 16.8				16.5 25.1										
		8.9 13.2			12.4 19.7				16.2 25.6										
		8.9 12.5			11.3 17.7				16.4 22.6										

continued

Table 1.95 Italian Ryegrass, named varieties other than S22 (continued)

	January			February			March			April			May			June		
Irish										(1958)	N_4 N_2 N_4 N_2	26.9 21.6			13.6 11.4	9.9 7.7 9.5 7.5		
									(1959) (1960)	N_4 N_2 N_4 N_2	25.6* 20.5* 22.3* 17.8*			19.0* 15.5*	13.3* 9.3*			18.7* 14.9* 17.3* 13.4*
											(1962)	N_2 N_4	18.4 22.9				8.8 12.0	
Westernwolth																		19.8
										(1958)	N_4 N_2 N_4 N_2	27.3 24.1			18.3 13.7	11.6 9.2 11.9 8.8		
									(1959)	N_4 N_2	27.8* 22.8*				15.3* 13.0*			19.9* 17.5*
Carbohydrate fractions %DM **Irish**: Soluble carbohydrate																		
Westernwolth: Water-soluble carbohydrate																	(1955)	14.5
Lignin																		4.1

July			August			September			October			November			December			Notes	Reference
15.7 13.4	23.0* 20.1*	16.1* 12.5*			20.1* 15.1*	9.7 8.7 9.6 8.6 15.0* 12.8*		26.8* 22.1*		22.2* 18.7*	18.9 17.9 19.3 18.5							N_4 : Nitrochalk 500 kg/ha N_2 : Nitrochalk 250 kg/ha *Regrowth	301, 300
		10.0 12.0			12.1 16.9				17.3 25.3									N_2 : Nitrochalk 250 kg/ha N_4 : Nitrochalk 500 kg/ha	270
		12.0																	350
15.3	(1)9.4		(2)	4.3														Given 500 kg/ha of mixed fertiliser (1) Probably flowering (2) Seed shedding	511
16.4 14.1		17.9* 15.8*				11.0 11.4 11.1 10.8 16.5* 14.5*				22.7* 20.3*	22.9 20.1 22.6 20.7							N_4 : Nitrochalk 500 kg/ha N_2 : Nitrochalk 250 kg/ha *Regrowth	301, 300
		17.0																	350
12.2	(1)18.2		(2)	4.4														(1) Probably flowering (2) Seed shedding	511
7.0	(1) 7.9		(2)	10.6															

Table 1.96 Italian Ryegrass, Fenland study

	Year		June	June	July	July	August	September	September	October	November	Notes	Reference
Proximate constituents %DM													
Crude protein	(1949)	H	22.2		10.8	11.8	10.1		12.4	14.3		Effect of water level below surface on growth. H: high (36–41cm) M: medium (51–67cm) L: low (91–104cm) Means of 4 replicate plots	194
		M	31.6		19.8	20.2	13.2		15.9	19.8			
		L	31.2		25.2	26.6	20.4		20.4	24.5			
	(1949)	H				9.4			11.7*			Cut for hay in July *Regrowth	
		M				17.6			15.4*				
		L				20.4			19.0*				
	(1950)	H	20.7	17.7		10.7	12.0	14.0			13.7	Effect of water level below surface on growth. H: high (42–50cm) M: medium (53–65cm) L: low (69–90cm) Each value the mean of two, after kale and after celery crop	388
		M	24.3	21.9		15.7	16.3	18.4			17.7		
		L	24.7	23.2		18.2	18.6	19.5			19.0		
Ether extract	(1950)	H	3.8	3.8		2.6	3.2	3.4			3.4		
		M	4.1	4.2		3.6	3.6	4.2			3.9		
		L	3.8	4.0		3.8	3.6	4.3			3.6		
Nitrogen-free extract	(1950)	H	45.2	44.7		51.6	49.2	49.7			53.6		
		M	39.5	39.6		44.7	45.1	44.8			52.0		
		L	39.4	39.1		42.4	43.7	43.3			51.9		
Crude fibre	(1949)	H	18.1		16.8	21.0	23.1		24.8	21.9		See crude protein. same reference	194
		M	18.6		19.6	22.3	24.3		24.8	20.9			
		L	16.0		17.6	20.9	22.2		22.9	19.6			
	(1949)	H				25.2			29.6*				
		M				25.4			29.1*				
		L				24.3			26.3*				
	(1950)	H	19.8	22.7		26.9	26.4	22.4			19.2	See crude protein. same reference	388
		M	20.4	22.5		26.1	25.3	21.9			16.9		
		L	20.4	22.0		25.2	24.4	22.0			16.1		
Ash	(1949)	H	11.8		8.0	9.6	10.2		11.7		11.2	See crude protein. same reference	194
		M	14.3		11.0	11.6	10.6		12.0		11.8		
		L	15.9		11.0	11.6	10.2		11.3		11.7		
	(1949)	H				8.9			9.5*				
		M				11.0			9.1*				
		L				11.1			8.6*				
	(1950)	H	10.6	11.1		8.2	9.2	10.5			10.2	See crude protein, same reference	388
		M	11.7	11.8		9.9	9.8	10.8			9.6		
		L	11.8	11.7		10.4	9.7	10.8			9.4		
Major elements %DM													
Calcium	(1949)	H				0.77			1.17*			See crude protein, same reference	194
		M				0.78			1.00*				
		L				0.81			1.17*				
Phosphorus	(1949)	H				0.35			0.38*				
		M				0.35			0.44*				
		L				0.26			0.40*				
Magnesium	(1949)	H				0.13			0.16*				
		M				0.25			0.22*				
		L				0.26			0.22*				
Potassium	(1949)	H				2.35			1.59*				
		M				3.11			2.09*				
		L				3.41			1.81*				
Chlorine	(1949)	H				1.82			0.87*				
		M				2.45			1.36*				
		L				1.95			1.21*				

Table 1.97 Italian Ryegrass, unnamed varieties

	May	July	September	October	November	December	Notes	Reference
Proximate constituents %DM Crude protein					(1950) N_0(Aa)18.0 N_2(Aa)20.2 N_4(Aa)21.4		N_0 : No extra nitrogen N_2 : Nitrochalk 250 kg/ha N_4 : Nitrochalk 500 kg/ha A : N applied in autumn S : N applied in spring Plot experiment in which all plots were cut in summer. The (a) plots were cut, the (b) plots were not cut in the autumn	3
(1951)	N_0(Aa)13.5 N_0(Sa) 13.7 N_0(Ab)14.4 N_2(Aa)14.6 N_2(Sa) 14.8 N_2(Ab)13.4 N_4(Aa)13.9 N_4(Sa) 16.2 N_4(Ab)13.9		N_0(Aa)12.4 N_0(Sa) 13.2 N_0(Ab)13.0 N_2(Aa)13.0 N_2(Sa) 11.4 N_2(Ab)13.7 N_4(Aa)11.6 N_4(Sa) 14.0 N_4(Ab)12.4	N_0(Aa)16.0 N_2(Aa)19.5 N_4(Aa)22.5				
(1952)	N_0(Aa)10.2 N_0(Sa) 12.8 N_0(Ab)13.0 N_2(Aa)11.2 N_2(Sa) 12.8 N_2(Ab)12.4 N_4(Aa)12.0 N_4(Sa) 13.4 N_4(Ab)11.2	N_0(Aa) 7.8 N_0(Sa) 8.2 N_0(Ab) 8.0 N_2(Aa) 7.8 N_2(Sa) 7.7 N_2(Ab) 7.7 N_4(Aa) 7.4 N_4(Sa) 6.8 N_4(Ab) 7.5						
Carbohydrate fractions %DM							Cut for silage. For proximate constituents and metabolisable energy see table 1.226	351
Glucose	2.2							
Fructose	1.9							
Sucrose	8.4							
Other oligosaccharides	1.2							
Fructosan	5.9							
Cellulose	18.3							
Xylan	7.4							
Araban	3.2							
Galactan	0.6							
Total hemicelluloses	11.2							
Lignin	4.6							

	March	April	May	June	July	August	September	October	Notes	Reference
Vitamins μg/g DM Carotene	460	210	180	80	120	150	350*	340*	*Regrowth	362

Table 1.98 Italian Ryegrass with other grasses and clover

	April			May			June			August			September			October			Notes	Reference
Italian Ryegrass (Irish strain), White Clover, 10% perennial ryegrass																				
Proximate constituents %DM																				
Crude protein	18.5		16.6		11.7	15.6		9.1	11.2		8.9	9.1	12.1	11.5			13.8		Rotational grazing for bullocks. Pre-grazing samples.	341
		12.0		12.7		10.1	9.6		10.9			9.2	13.6	13.3			12.5	17.6	Caged areas cut at end of grazing periods	
Ether extract	6.0		5.3		3.1	4.3		3.1	3.4		3.6	3.2	4.4	3.9			4.3		Rotational grazing for bullocks Pre-grazing samples	
		4.2		3.7		3.2	3.0		3.8			3.2	4.0	4.8			4.2	5.4	Caged areas cut at end of grazing periods	
Nitrogen-free extract	44.1		47.9		54.1	45.7		51.5	51.3		54.8	50.8	46.6	44.8			41.5		Rotational grazing for bullocks Pre-grazing samples	
		54.5		56.1		54.0	52.4		51.0			50.1	46.8	42.0			44.0	37.7	Caged areas cut at end of grazing periods.	
Crude fibre	20.9		21.0		22.6	25.4		27.6	26.1		26.2	28.0	26.7	29.1			28.7		Rotational grazing for bullocks Pre-grazing samples	
		21.7		19.3		25.1	28.0		26.5			28.4	25.1	26.5			26.4	25.9	Caged areas cut at end of grazing periods	
Ash	10.5		9.2		8.5	9.0		8.7	8.0		6.5	8.9	10.2	10.8			11.8		Rotational grazing for bullocks Pre-grazing samples	
		7.6		8.2		7.6	7.0		7.8			9.1	10.5	13.4			12.9	13.4	Caged areas cut at end of grazing periods	
Italian Ryegrass, Perennial Ryegrass, Broad Red and late-flowering red clovers																				
Crude protein %DM (0)		23.3	20.0	15.6	14.3	11.0	10.5												(0) Freshly cut (1) Wilted in field for 48 hr. (2) Wilted under cover for 48 hr. (3) Wilted in scrim bag for 48 hr. (4) Wilted in field for 4 days (5) Wilted in field for 6-7 days	210
(1)		23.0	17.8	16.3	15.0	11.7	10.6													
(2)			19.0	16.8		11.1														
(3)			19.6	16.0		11.5														
(4)							11.0													
(5)							11.4													
Major elements %DM																				
Calcium (0)		1.0	1.1	0.9	0.9	0.9	1.0													
(1)		1.0	1.1	0.9	1.0	1.0	1.0													
(5)							0.5													
Phosphorus (0)		0.5	0.5	0.3	0.3	0.3	0.3													
(1)		0.5	0.4	0.3	0.3	0.3	0.3													
(5)							0.3													
Potassium (0)		3.5	3.0	3.1	2.6	2.0	1.7													
(1)		3.0	3.3	3.1	2.6	1.4	1.4													
(5)							1.6													
Chlorine (0)		1.5	1.2	1.1	1.1	0.9	0.9													
(1)		1.3	1.1	1.0	1.0	0.9	1.0													
(5)							1.1													
Vitamins μg/g DM																				
Carotene (0)		450	387	323	233	217	212													
(1)		284	226	273	201	184	159													
(2)			254	293		175														
(3)			257	284		175														
(4)							126													
(5)							51													

continued

Table 1.98 Italian Ryegrass with other grasses and clover (continued)

	January			February			March			April			May			June		
Italian ryegrass and broad red clover																		
Proximate constituents %DM																		
Crude protein										D			9.8				11.3	
										A	30.9		16.2			25.4		
										B	29.2		16.8				21.7	
										C	30.7		15.1				10.8	
										E	18.8			12.4			12.9	
										F		11.6			11.6			
										G	17.3				9.6			
										H	19.2				8.2			
										I	28.7		16.8			24.6		
Major elements %DM										D			0.86				0.83	
Calcium										A	1.21		0.80			0.75		
										I	0.68		0.61			0.75		
Phosphorus										D			0.28				0.32	
										A	0.59		0.32			0.34		
										I	0.49		0.35			0.43		
Vitamins μg/g DM										D			149				272	
Carotene										A	376		326			434		
										B	409		353				315	
										C	364		233				261	
										E	368			251			309	
										F		229			201			
										G	343				191			
										H	335				177			
										I	375		200			309		
Italian ryegrass and red clover																		
Proximate constituents %DM																		
Crude protein													1T	16.3				
													1G	17.5				
													2T	12.5				
													2G	14.3				
													3T	11.3				
													3G	9.6				
															(1)	20.9	18.7	
															(2)		13.7	
Ether extract													1T	4.5				
													1G	4.3				
													2T	3.5				
													2G	3.6				
													3T	3.5				
													3G	2.7				
Nitrogen-free extract													1T	47.2				
													1G	46.6				
													2T	51.7				
													2G	49.6				
													3T	52.0				
													3G	55.4				
Crude fibre													1T	23.3				
													1G	22.8				
													2T	25.6				
													2G	25.6				
													3T	26.3				
													3G	26.5				
Ash													1T	8.8				
													1G	8.8				
													2T	6.7				
													2G	6.9				
													3T	7.0				
													3G	5.9				

July			August			September			October			November			December			Notes	Reference
13.5			19.7			21.6												Plot experiments. For technique see Reference 284. The experiments simulated grass production for drying. All cuts were made before flowering. There is no statement about primary growth and regrowth. Fertiliser treatment was as follows, with Nitrochalk as source of N: D: No extra N A: 2250 kg/ha as 750 in March, May and July B: 1500 kg/ha as 750 in March and May C: 750 kg/ha in March E: 2000 kg/ha as 250 in March and 250 after each cut F: 875 kg/ha as 125 in March and 125 after each cut G: 1500 kg/ha as 250, 750 and 500 in March, May and July H: 750 kg/ha as 250 in March, May and July I: 1250 kg/ha Nitrochalk and 1250 kg/ha of a complete fertiliser in 3 equal lots in March, May and July	285
18.3		19.9			22.7				26.3		24.9								
17.3			15.5			15.9				20.3									
12.8			17.4			20.1				21.0									
16.7		20.4			19.3			25.1			27.1								
12.6		15.9			15.2			21.9			23.9								
18.6		16.8			23.5				27.3										
12.0		18.0			20.0				20.1										
16.7		18.7		21.8					22.6										
0.87			1.62			2.05													
0.79		0.82			1.02				1.06		0.88								
0.74		0.73		0.84					0.79										
0.34			0.34			0.38													
0.38		0.41			0.32				0.39		0.38								
0.34		0.39		0.44					-0.42										
296			467			418													
459		464			494				521		422								
335			420			412				431									
361			466							493									
438		460			430			508			487								
296		365			343			500			452								
426		432			462				601										
288		366			436														
438		462		485					412										
																		1T: Farm 1, tripod cured 1G: Farm 1, ground cured 2T: Farm 2, tripod cured 2G: Farm 2, ground cured 3T: Farm 3, tripod cured 3G: Farm 3, ground cured	106
																		(1) Flower heads formed (2) Flower heads emergent	187
																		See same reference above	106

Table 1.99 Ryegrass, unnamed

	May	June	July	August	September	October	Notes	Reference
Major elements %DM Magnesium	(1953-1957)	(1) 0.11 (2) 0.12 (3) 0.12 (4) 0.12 (5) 0.10					Cut for hay. Effect of soil type. (1) Old red sandstone; 13 samples, range 0.08–0.14 (2) Granite; 18 samples, range 0.10–0.20 (3) Basic; 12 samples, range 0.08–0.19 (4) Slate; 16 samples, range 0.06–0.17 (5) Sands and gravels; 12 samples, range 0.08–0.14	433
		(1960) 0.13		0.14*		0.21*	Effect of cutting and applying K. Means averaged over all treatments. * Regrowth	
(1)	0.11	0.17*		0.18*		0.22*	Effect of regular cutting (1) Area 1	
(2)	0.10	0.13*	0.18*	0.20*	0.21*	0.23*	(2) Area 2	
Trace elements μg/g DM Barium		(1953) (1) 10 (1953) (2) 12 (1954) (1) 11 (1954) (2) 13					Date of harvesting not given except for 1955 (1) Well drained soil (2) Poorly drained soil	359
Chromium		(1953) (1) 0.16 (1953) (2) 0.12 (1954) (1) 0.17 (1954) (2) 0.16					(0) No treatment (3) + Ca (4) + Mo + Cu + Zn (5) + Mo + Cu + Zn + Ca (6) + Co (7) + Cu	
Cobalt		(1953) (1) 0.18 (1953) (2) 1.50 (1954) (1) 0.11 (1954) (2) 0.64 1955) (0) 0.06 (6) 0.56 (0) 0.05 (6) 0.73					Ca = 7500 kg/ha $CaCO_3$ Mo = 1.1 kg/ha $Na_2MoO_4.2H_2O$ Zn = 22 kg/ha $ZnSO_4.7H_2O$ Cu = 22 kg/ha $CuSO_4 . 5H_2O$ Co = 1.6 kg/ha $CoSO_4 . 7H_2O$	
Copper		(1953) (1) 4.0 (1953) (2) 3.4 (1954) (1) 3.2 (1954) (2) 2.6 (1952) (0) 2.9 (3) 2.9 (4) 3.2 (5) 3.1 (1953) (0) 3.1 (3) 3.0 (4) 3.4 (5) 3.1 (1955) (0) 2.4 (7) 3.2 (0) 2.7 (7) 3.3						
Iron		(1953) (1) 34 (1953) (2) 47 (1954) (1) 28 (1954) (2) 44						

Table 1.99 Ryegrass, unnamed (continued)

	Sample	May	June	July	August	September	October	Notes	Reference
Trace elements μg/g DM (cont'd) Manganese	(1953) (1)		88						
	(1953) (2)		116						
	(1954) (1)		62						
	(1954) (2)		92						
	(1952) (0)		76						
	(1952) (3)		49						
	(1952) (4)		81						
	(1952) (5)		52						
	(1953) (0)		76						
	(1953) (3)		54						
	(1953) (4)		97						
	(1953) (5)		65						
Molybdenum	(1953) (1)		0.7						
	(1953) (2)		1.2						
	(1954) (1)		1.3						
	(1954) (2)		1.0						
	(1952) (0)		0.9						
	(1952) (3)		1.2						
	(1952) (4)		1.5						
	(1952) (5)		1.9						
	(1953) (0)		0.8						
	(1953) (3)		1.1						
	(1953) (4)		2.0						
	(1953) (5)		3.4						
Nickel	(1953) (1)		1.0						
	(1953) (2)		3.4						
	(1954) (1)		0.9						
	(1954) (2)		2.1						
Strontium	(1953) (1)		12						
	(1953) (2)		11						
	(1954) (1)		12						
	(1954) (2)		10						
Titanium	(1953) (1)		2.8						
	(1953) (2)		2.0						
	(1954) (1)		1.8						
	(1954) (2)		2.3						
Vanadium	(1953) (1)		0.08						
	(1953) (2)		0.05						
	(1954) (1)		0.06						
	(1954) (2)		0.05						
Zinc	(1953) (1)		27						
	(1953) (2)		29						
	(1954) (1)		25						
	(1954) (2)		21						
	(1952) (0)		24						
	(1952) (3)		19						
	(1952) (4)		24						
	(1952) (5)		23						
	(1953) (0)		29						
	(1953) (3)		28						
	(1953) (4)		29						
	(1953) (5)		26						
Vitamins μg/g DM Thiamin					2.4			Preflowering	493
Riboflavin					13.1				
Nicotinic acid					38.4				
Pantothenic acid					17.4				
Biotin					0.38				
Pyridoxine					12.9				

Table 1.100 Ryegrass, unnamed, with other grasses

		April			May			June			July			August			September			Notes	Reference
Proximate constituents %DM																					
Crude protein		26.0	23.6	17.4	15.4	13.8	10.7	9.2	8.1	7.4										Ryegrass, unnamed 25%; cocksfoot 40 to 45%; timothy 20%. Samples taken in April and up to May 20 were mainly leafy, preflowering; samples at end of May, cocksfoot and ryegrass were in full flower; in June, cocksfoot and ryegrass were late flowering: timothy in full flower.	221
Ether extract		3.8	3.7	3.0	3.2	2.9	2.3	2.1	2.1	1.9											
Nitrogen-free extract		40.4	44.6	51.2	53.4	52.6	53.0	51.1	52.1	52.9											
Crude fibre		17.7	17.8	17.9	19.7	22.5	26.2	30.2	30.8	31.4											
Ash		12.1	10.3	10.5	8.3	8.2	7.8	7.4	6.9	6.4											
Major elements %DM																					
Calcium		0.7	0.7	0.5	0.5	0.4	0.4	0.4	0.4	0.4											
Phosphorus		0.4	0.4	0.4	0.4	0.4	0.3	0.3	0.3	0.2											
Magnesium		0.2	0.2	0.2	0.2	0.2	0.1	0.1	0.1	0.1											
Potassium		2.5	2.5	2.4	2.5	2.3	2.1	1.9	1.9	1.6											
		May			June			July			August			September			October				
Crude protein	N_1			13.3		13.0						14.3						17.7		Leafy ryegrass, *Poa* sp and *Agrostis* sp. Fertilisers: N as Nitrochalk N_1 = 38.5 kg/ha N_2 = 96 kg/ha N_3 = 115 kg/ha applied in one or divided dressings N_4 = 190 kg/ha N_5 = 228 kg/ha applied in one or divided dressings N_6 = 343 kg/ha N_7 = N_6 + 143 kg P_2O_5/ha and 184kg K_2O/ha For technique see Ref. 284. The experiments simulated grass production for drying. All cuts were made preflowering. There is no statement about primary growth or regrowth.	285
	N_2			14.6			15.4			14.1						18.9					
	N_3		19.3	12.4		16.7	14.8			16.7	14.9					19.6		18.3			
	N_4			14.8		13.5				17.9				17.0				24.3			
	N_5		20.8	14.2		25.7 21.4			18.2 15.7				19.2	15.6				21.4	20.5		
	N_6		18.5			24.5			15.8			23.9					24.6				
	N_7	19.8				23.1		16.3				21.1					20.8				
Vitamins µg/g DM																					
Carotene	N_1			228		280						333						463			
	N_2			301			332			509						502					
	N_3		333	299		374	311			526	553					433		437			
	N_4			231		265				412				509				470			
	N_5		372	307		480 433			474 398				525	562				556	413		
	N_6		416			483			445			548					608				
	N_7	387				414		275				504					483				

Table 1.101 Ryegrass, unnamed, with other grasses and clover

	July			August			September			October			November			December			Notes	Reference
Ryegrass, Cocksfoot, Timothy and clover *Proximate constituents* %DM Crude protein		16.2			18.2			20.0			22.3			24.3					Dates refer to months of "putting up" of plots for winter use. All sampled 26 Feb. of the next year.	155
	21.7	19.7	22.3 16.1	20.6	19.4	25.9	21.8	20.3	19.8 18.0										Mean of 2 samples cut at the grazing stage. Mean of 1–3 samples cut at preflowering stage. Mean of 3 samples cut at early flowering stage.	45
						15.1													Mean value for 5 cuts between June and October from caged areas in sheep feeding.	486
Ether extract		3.7			4.2			4.5			4.7			4.8						155
	3.1	3.3	3.4 2.9	3.6	3.5	4.0	4.3	4.1	3.4 3.6											45
						2.8														486
Nitrogen-free extract		40.0			39.0			40.2			39.2			39.2						155
	45.2	46.6	42.9 48.8	46.3	44.3	39.6	41.5	44.7	42.5 44.3											45
						49.7														486
Crude fibre		26.8			25.2			22.7			22.1			20.9						155
	20.2	22.0	21.9 24.1	21.2	23.6	21.0	22.7	21.6	20.7 24.6											45
						23.0														486
Ash		13.4			13.4			12.6			11.7			10.8						155
	9.9	8.3	9.4 8.1	8.4	9.2	9.4	9.7	9.3	13.6 9.6											45
						9.4														486
Carbohydrate fractions %DM Total sugars						6.6														486
Cellulose						32.5														
Lignin						4.5														
Major elements %DM Calcium		0.45			0.42			0.47			0.51			0.53						155
						0.55														486
Phosphorus		0.28			0.31			0.35			0.37			0.37						155
						0.27														486
Magnesium						0.20														
Potassium						3.03														
Sodium						0.10														
Chlorine						0.97														
Trace elements μg/g DM Cobalt						0.09														
Copper						8.1														
Iron						330														
Manganese						225														
Vitamins μg/g DM Carotene						273														

Table 1.101 Ryegrass, unnamed, with other grasses and clover (continued)

	January			February			March			April			May			June		
Ryegrass, Meadow grass and White Clover																		
Proximate constituents %DM																		
Crude protein											21.7	19.9	13.3	10.6	9.5	9.9		11.6
												16.3	13.2	10.0	10.0	9.6		9.6
Ether extract											7.2	4.9	5.0	3.8	3.6	3.2		3.9
												4.4	4.7	3.5	3·5	3.3		3.0
Nitrogen-free extract											44.4	48.5	53.6	59.9	53.8	53.9		49.9
												48.1	54.8	59.9	55.9	51.3		34.9
Crude fibre											17.1	18.0	19.9	18.2	25.3	25.0		25.5
												21.1	19.4	19.7	23.0	27.8		43.8
Ash											9.6	8.7	8.2	7.5	7.8	8.0		9.0
												10.1	7.9	6.9	7.6	8.0		8.7
Ryegrass, Meadow grass, Yorkshire Fog, and White Clover																		
Proximate constituents %DM																		
Crude protein												(1947)	17.3	15.3	12.0 13.4		10.6	11.9
												(1947)		15.9	13.0 11.0		9.5	8.6
											(1948)	16.8 18.1	14.8		14.2	12.4	16.0	13.2
											(1948)	17.0	15.2	14.0	13.3		11.7	
Ether extract												(1947)	5.2	4.1	4.9 4.2		3.5	3.4
												(1947)		4.6	4.2 4.3		3.6	5.2
											(1948)	4.9 4.6	4.6		4.3	3.5	4.0	4.2
											(1948)	4.9	5.1	3.9	3.6		3.8	

July			August			September			October			November			December			Notes	Reference
15.8	17.1	13.9	17.5 16.2	12.5	17.4 12.7	19.6 14.3	16.6 13.3 13.3											Area 1 Means of 10–12 samples 2 from 3 areas used for 3 rotational grazing of bullocks	342
10.5	14.9	16.6 12.3	16.1	17.8	15.1	18.9	18.0 19.5 16.9					22.5 22.3 16.7						Area 1 Caged Means of 12 samples 2 from 3 caged areas 3 grazed as above.	
4.9	5.7	4.2	5.7 5.3	3.9	5.5 5.1	5.7 4.4	4.2 4.4 3.9											Area 1 2 3	
3.5	4.1	5.6 4.3	4.9	5.0	4.8	5.6	5.3 5.4 5.7					6.1 6.5 4.8						Area 1 Caged 2 3	
45.9	42.8	42.4	40.6 42.8	46.8	42.7 45.6	35.5 44.0	36.8 36.7 43.0											Area 1 2 3	
48.7	44.4	41.3 44.5	41.0	44.0	43.6	34.6	37.5 37.8 40.0					35.3 36.3 42.9						Area 1 Caged 2 3	
22.4	23.1	27.6	23.1 23.9	25.0	22.6 25.1	25.8 24.9	25.7 26.4 23.4											Area 1 2 3	
27.5	25.6	25.0 27.5	24.7	22.1	24.6	26.7	25.7 24.5 25.0					21.8 21.1 23.0						Area 1 Caged 2 3	
11.0	11.3	11.9	13.1 11.8	11.8	11.8 11.5	13.4 12.4	16.7 19.2 16.4											Area 1 2 3	
9.8	11.0	11.5 11.4	13.3	11.0	11.9	14.2	13.5 12.8 12.4					14.3 13.8 12.6						Area 1 Caged 2 3	
13.1	12.8		13.7	15.4	13.1 15.7	13.8 15.1	17.8 14.3 15.4				14.2 13.7 14.6							Plot 1 Pre-grazing samples from 2 3 areas used in rotational 3 grazing for bullocks.	343
8.5	16.7	12.8		14.1	14.5	16.3	19.3 17.1 17.8				18.3 17.1 15.2			20.9 18.4 20.1				Plot 1 Samples from areas 2 caged during grazing. 3	
	11.8		11.4		14.9	16.8 16.7												Plot 1 Pre-grazing as above 2 3	
11.5		10.7									17.5 17.2 13.4							Plot 1 Caged areas as above. 2 3	
5.0	3.9		5.0	5.2	5.3 5.8	4.6 5.4	5.1 4.2 4.8				4.7 4.6 4.8							Plot 1 2 3	
3.6	4.2	4.2		5.4	5.7	5.7	5.1 5.2 5.4				5.9 4.9 4.7			5.1 4.3 5.1				Plot 1 Caged 2 3	
	3.6		4.0		4.7	5.2 5.4												Plot 1 2 3	
4.0		4.1									6.0 5.5 4.4							Plot 1 Caged 2 3	

continued

Table 1.101 Ryegrass, unnamed, with other grasses and clover (continued)

	January			February			March			April			May			June		
Proximate constituents (cont'd) Nitrogen-free extract												(1947)	49.9	50.6	47.0 52.8		48.2	48.4
												(1947)		51.8	52.2 49.9		47.6	46.1
											(1948)	47.4 47.4	51.3		48.1	49.6	44.0	48.4
											(1948)	38.3	50.1	49.4	49.6		46.4	
Crude fibre												(1947)	17.7	21.1	26.7 20.7		29.2	28.0
												(1947)		18.7	21.2 26.4		31.2	31.3
											(1948)	20.8 19.9	21.0		23.5	24.6	26.8	24.1
											(1948)	29.1	21.4	23.5	24.8		29.0	
Ash												(1947)	9.9	8.9	9.4 8.9		8.5	8.3
												(1947)		9.0	9.4 8.4		8.1	8.8
											(1948)	10.1 10.0	8.3		9.9	9.9	9.2	10.1
											(1948)	10.7	8.2	9.2	8.7		9.1	
Ryegrass, Sweet Vernal, Yorkshire Fog, other grasses and clover. *Proximate constituents* %DM Crude protein												14.7		11.5	12.7	11.8		10.5
Ether extract												3.5		2.6	3.0	3.1		2.8
Nitrogen-free extract												55.9		56.8	56.6	53.2		52.0
Crude fibre												16.4		18.8	20.4	24.1		27.5
Ash												9.5		10.2	7.3	7.8		7.2
Major elements %DM Calcium												0.85		0.87	0.69	0.58		0.59
Phosphorus												0.38		0.34	0.35	0.38		0.37
Vitamins μg/g DM Carotene												382		240	226	165		137

July			August			September			October			November			December			Notes	Reference
44.9				44.9	45.6		37.9				44.3							Plot 1	343
	45.9				45.9	43.7	44.7				43.7							2	
			43.4			45.5	43.0				44.5							3	
	43.8				45.7		39.5				39.7			37.1				Plot 1 Caged	
		44.9				42.1	41.4				42.7			40.0				2	
48.5				47.4			39.2				43.0			38.0				3	
			46.9			40.7												Plot 1	
	49.3					41.8												2	
					44.8													3	
											39.0							Plot 1 Caged	
											38.8							2	
47.5		47.9									47.6							3	
26.7				24.9	25.6		28.2				25.4							Plot 1	
	28.0				23.3	27.9	26.4				27.0							2	
			28.8			25.0	26.9				25.5							3	
	25.2				24.5		25.2				24.0			24.9				Plot 1 Caged	
		27.3				26.0	26.1				23.7			25.1				2	
31.7				24.2			27.2				26.1			24.9				3	
			27.6			25.8												Plot 1	
	26.0					24.7												2	
					24.6													3	
											25.3							Plot 1 Caged	
											26.3							2	
27.7		25.8									26.9							3	
10.3				9.6	10.4		11.0				11.4							Plot 1	
	9.4				9.3	10.0	10.4				11.0							2	
			9.1			9.0	9.9				10.6							3	
	10.1				9.6		10.9				12.1			12.0				Plot 1 Caged	
		10.8				9.9	10.2				11.6			12.2				2	
7.7				8.9			10.4				11.0			11.9				3	
			10.1			11.5												Plot 1	
	9.3					11.4												2	
					11.0													3	
											12.1							Plot 1 Caged	
											12.2							2	
9.3		11.5									7.7							3	
9.4	8.5																	Mean of 5 replicates; composition: Ryegrass 27%, Sweet Vernal 13%, Yorkshire Fog 10%, others 33%, Clover 9%	361
2.7	2.0																		
50.8	50.0																		
29.2	31.5																		
7.8	8.1																		
0.56	0.52																		
0.31	0.31																		
102	121																		

Table 1.102 Ryegrass, unnamed, with clover

	January	February	May	June	September	October	Notes	Reference
Ryegrass and White Clover							Digestibility by sheep	429
***Digestibility of* DM%**			(1949) 80.8, 80.1	(1949) 73.4, 64.2		(1949) 68.5		
			(1950) 75.1, 69.7	(1950) 71.8, 65.9				
	(1952) 72.5		(1951) 78.3		(1951) 69.4			
***Digestibility of* OM%**			(1949) 83.9, 82.7	(1949) 75.9, 65.6		(1949) 73.6		
			(1950) 78.2, 73.8	(1950) 74.7, 67.9				
	(1952) 76.4		(1951) 81.5		(1951) 72.1			
Ryegrass and Red Clover	January	February	May	June	September	October		
***Crude protein* %DM**			8.7	9.5			Cut at heads emergent stage for hay experiment	456
Ryegrass and Red Clover	May	June	July	August	September	October		
***Proximate constituents* %DM**							Herbage contained ryegrass 20%, clover 80%; aftermath, with clover mature and stemmy. Date of cutting not stated (August assumed).	208
Crude protein				17.4				
Ether extract				4.4				
Nitrogen-free extract				45.7				
Crude fibre				25.2				
Ash				7.3				
***Major elements* %DM**							For composition of silages see table 1.134 same reference	
Calcium				1.74				
Phosphorus				0.30				
Potassium				1.33				
Chlorine				0.90				

continued

Table 1.102 Ryegrass, unnamed, with clover (continued)

	January			February			March			April			May			June		
Digestibility of DM%											(1949)	(S2) (S3)	78.0 78.6		(1949)	(L) (S1) (S2)	71.9 73.3 75.1	(1948)
(1952)	(L)	78.0				(1952)	(L)	79.0	(1950)	(L)	73.7	(1952)	(L)	78.4				
														(1953)	(L) (H)	76.2 75.9		(1949)
Digestibility of OM%													(S2) (S3)	81.2 81.6		(L) (S1) (S2)	74.5 76.0 77.7	
Proximate constituents %DM Crude protein														11.0 12.2		6.3 6.7		
Ether extract														3.6 3.8		2.9 5.0		
Nitrogen-free extract														54.6 52.7		57.3 55.6		
Crude fibre														23.0 23.3		27.2 26.6		
Ash														7.8 8.0		6.2 6.1		

July			August			September			October			November			December			Notes	Reference
(L) (S)	67.3 69.3 	 (1950) (1952)	 (L) (L)	 75.6 79.4	 (1949) (1952)	 (L) (L)	 75.4 80.0	 (1951)	 (L)	 79.0								*In vivo* L: Lambs S: Sheep S1: 1-year-old sheep S2: 2-year-old sheep S3: 3-year-old sheep	427
(L) (H)	64.3 64.2			(1950)	(L) (H)	72.1 69.4												*In vivo* with sheep Effect of level of feeding (L) approx. 80% of (H)	428
																		In vivo see DM, same reference	427
																		May and June samples from different centres. Upper row, tripod cured; lower row, ground cured. For hay table 1.126, same reference	106

Table 1.103 Meadow Fescue S53

	January			February			March			April			May			June		
Proximate constituents %DM Crude protein		(1951)	G F	23.5 13.6														
(1952) (1953)	W G F	13.6 19.4 11.9	12.6		12.6	13.9 20.4 11.8												
										(1952)	W L S	(1951) 24.8 25.2 15.4		15.0	18.1 24.1 12.6	11.8	18.5 18.9 9.9	16.3
			(1956) (1957)	W G F W G F	16.6 26.2 15.4	18.8 25.8 14.1												
Carbohydrate fractions %DM Total soluble sugar										(1951) (1952)	W L S	10.3 5.7 12.4		26.3	8.5 5.2 11.5	20.3	14.0 13.6 23.4	11.9
Hexoses										(1951) (1952)	W L S	1.9 1.2 2.6		3.0	2.0 1.4 3.0	3.9	1.8 1.8 2.2	1.8
Sucrose										(1951) (1952)	W L S	5.3 3.8 3.9		10.3	3.8 3.1 3.9	7.4	6.2 6.2 5.5	3.7
Fructosan										(1951) (1952)	W L S	2.9 0.6 5.1		11.6	2.4 0.5 4.0	7.7	5.2 4.8 13.9	6.9
Trace elements µg/g DM Barium																		H 1.9 L 3.9 S 5.3
Boron																		H 8 L 10 S 5
Chromium																		H 0.14 L 0.32 S 0.15
Cobalt																		H 0.03 L 0.03 S 0.03
Copper																		H 4.6 L 4.9 S 3.5
Iron																		H 40 L 109 S 19
Lead																		H 0.5 L 1.6 S 0.7
Manganese																		H 16.0 L 29.0 S 40.0
Molybdenum																		H 0.23 L 0.60 S 0.20
Nickel																		H 2.6 L 0.8 S 0.7
Strontium																		H 1.7 L 14.0 S 5.1
Tin																		H 0.14 L 0.21 S 0.18
Titanium																		H 1.3 L 6.0 S 1.1
Vanadium																		H 0.06 L 0.22 S 0.05
Zinc																		H 37 L 16 S 14
Vitamins µg/g DM Tocopherol (total)										300	302	265	243		137	100	90	

July			August			September			October			November			December			Notes	References
																		Sown in 5cm drills, autumn 1948 G: Green F: Frost burnt	296
								(1951)	W G F		13.5 14.6 11.6				13.9 16.7 11.3			W: Whole plant G: Green F: Frost burnt	
26.1 26.7 15.4	25.7	24.4 25.6 26.0 16.9		25.0	26.4 23.1 23.7 13.9			19.2 19.2 12.0	21.6									W: Whole plant L: Leaf S: Stem There is a small error in the September figures. Leaf: stem ratio 33:1	509
														(1954)	W G F	15.2 16.6 8.1		W: Whole plant G: Green F: Frost burnt	337
(1954-55) (1955-56) (1956-57)	15.2 13.2 14.8																	Means of all cuts and plots	240
6.6 6.5 8.0	4.3	5.0 3.5 3.5 4.4		8.2	8.8 5.8 5.5 9.9			9.3 9.0 16.7	7.2									W: Whole plant L: Leaf S: Stem Leaf: stem ratio in June 23:1	509
0.9 0.8 2.1	1.1	1.6 0.9 0.9 1.3		1.6	2.2 1.1 1.1 1.7			1.1 1.1 1.8	1.2										
5.0 5.0 4.0	2.0	2.0 2.3 2.3 1.9		5.6	5.5 3.5 3.5 3.6			5.6 5.6 5.2	3.2										
0.4 0.4 1.5	1.1	1.1 0.2 0.2 1.0		0.6	0.7 0.8 0.6 4.0			1.9 1.7 8.5	2.3										
																		H: Head L: Leaf S: Stem Early mature stage, Heads well formed Dates not given	236
	184 – 243 121 – 150			23		26												Sown in summer 1950 Average of cuts at 20-25cm high, presumably regrowth The same, dried	100

Table 1.104 Meadow Fescue S215

	January			February			March			April			May			June		
Digestibility of DM%													78.7	73.2	68.0	62.5	62.7	
													78.7	73.2	68.0	59.7 75.2*	62.7	73.1*
Digestibility of OM %													81.2	74.7	68.9	60.9 77.5*	63.4	75.0*
Crude protein % DM		(1951)	{G F	24.8 11.6														
					(1951) (1952) (1953) (1954)			18.4 15.2 14.2			25.8 23.4 22.0			18.2 22.0 17.0				
													18.1	15.6	11.9	8.8 16.9*	9.4	18.8*
Carbohydrate fractions % DM Water soluble carbohydrate																		
Major elements % DM Calcium														0.59				
Phosphorus														0.26				
Magnesium														0.14				
Potassium														2.70				
Sodium														0.09				
Potassium									2.36		2.82	2.68		2.32	1.83	1.73		1.54
Sodium									0.06		0.07	0.04		0.04	0.03	0.03		0.04
Trace elements μg/g DM Iodine														0.22				

Table 1.105 Meadow Fescue, named varieties

	January			February			March			April			May			June		
Proximate constituents %DM Crude protein S53 and S215			(1950)	G F G F	22.4 11.8 22.2 13.5													
11 named varieties													19.5	20.7		20.1 9.2} 7.5}		
Crude fibre													17.9	22.4		18.0 34.2} 32.4}		
Carbohydrate fractions %DM Soluble carbohydrate													21.2	11.5		19.5 12.5} 18.0}		

July			August			September			October			November			December			Notes	Reference
																		In vivo with sheep (2 usually or 3)	267
													W G B	61 76 46				*In vitro* W: Whole sample G: Green fraction B: Burnt fraction	355
					71.5*													*In vivo* with sheep * Regrowth	356
					74.4*														
																		Sown in 5cm drills autumn 1948 G: Green F: Frost burnt	296
										17.7 14.6 11.8						16.5 13.0 10.9			249
													W	10.5				W: Whole sample	355
					16.9*													* Regrowth	356
													W G B	12 19 4				W: Whole sample G: Green fraction B: Burnt fraction	
																		Mean of 6 successive cuts of primary growth from April to June	542
																			254
										0.22*								Mean values for 7 stages of growth between 10/4 and 22/6 and 4 stages of regrowth (*) between 4/9 and 28/11	542

July			August			September			October			November			December			Notes	References
																		G: green F: Frost burnt Sown in 5cm drills, autumn 1948 Sown broadcast autumn 1948	296
19.7*} 11.8*}			21.9															For the individual varieties only yields and digestibilities of dry matter are given.	166
24.7*} 28.2*}			24.8															* Regrowth	
6.6*} 14.9*}			5.7			10.7			12.2									Cuts in June, hay, and July, 1st aftermath (bracketed) are from different centres.	

		May			June			August			September			Notes	Reference
Proximate constituents	% DM														
Crude protein	(1952)	U	15.8											U: Unchopped	376
		C	14.7											C: Chopped	
		L	13.7											L: Lacerated	
		UW	13.9											UW: Unchopped, wilted 24 hr.	
		CW	13.6											CW: Chopped, wilted 24 hr.	
	(1953)		U	12.4										For the composition of the silages see table 1.135 same reference	
			C	13.1											
			C	11.7											
			L	11.9											
			C	10.4											
				A 14.9						B 19.5				A: Trial 2 B: Trial 3 – Autumn cut C: Trial 6 All unchopped, dates not given.	381
				C 17.8											
Ether extract	(1952)	U	2.9												376
		C	2.7												
		L	2.7												
		UW	2.6												
		CW	2.4												
	(1953)		U	2.3											
			C	2.5											
			C	2.4											
			L	2.5											
			C	2.0											
				A 3.0						B 3.2					381
				C 3.2											
Nitrogen-free extract	(1952)	U	41.4												376
		C	42.0												
		L	43.2												
		UW	41.3												
		CW	42.6												
	(1953)		U	47.4											
			C	47.1											
			C	48.5											
			L	47.6											
			C	50.0											
				A 46.0						B 41.5					381
				C 40.3											
Crude fibre	(1952)	U	30.3												376
		C	31.4												
		L	31.4												
		UW	33.3												
		CW	32.4												
	(1953)		U	30.3											
			C	29.4											
			C	30.0											
			L	30.4											
			C	30.5											
				A 27.9						B 25.0					381
				C 28.0											
Ash	(1952)	U	9.6												376
		C	9.2												
		L	9.0												
		UW	8.9												
		CW	9.0												
	(1953)		U	7.6											
			C	7.9											
			C	7.4											
			L	7.6											
			C	7.1											
				A 8.2						B 10.8					381
				C 10.7											

continued

Table 1.106 Meadow Fescue, unnamed (continued)

	January			February			March			April			May			June		
Major elements %DM Calcium												0.7		0.6			0.4	
																		H 0.1 L 0.9 S 0.4
Phosphorus												0.3		0.3			0.2	
																		H 0.5 L 0.3 S 0.3
Magnesium												0.3		0.3			0.3	
																		H 0.5 L 0.3 S 0.3
Potassium												2.6		2.5			2.0	
Sodium												0.21		0.16			0.18	
Chlorine												0.8		0.7			0.6	
Trace elements μg/g DM Cobalt												0.20		0.18			0.17	
Copper												5.6		23.0			6.9	
														3.6 5.4				
Iron												560		240			210	
Manganese												43.3		30.3			22.3	
														13.0 14.0				
Vitamins μg/g DM Carotene									631			314			312			142
Tocopherols														184 to 243				
Total														121 to 150				
α – % of total														100				
B Vitamins mg/g DM Thiamin Riboflavin Nicotinic acid Pantothenic acid Pyridoxine Biotin																		

July			August			September			October			November			December			Notes	References
	0.3			0.3														Dates uncertain. Apr/May, flowering. June, mid flowering. July, late flowering. August, mature	490
																		Early mature stage, heads well formed. No dates given. H: Head L: Leaf S: Stem	236
	0.2			0.2														As above, same reference	490
																		As above, same reference	236
	0.2			0.2														As above, same reference	490
																		As above, same reference	236
	1.6			1.5														As above, same reference	490
	0.10			0.07															
	0.5			0.6															
	0.12			0.14														As above, same reference	490
	5.5			6.6															
																		Untreated. Sprayed with one application of $MnSO_4$ 33kg/ha and $CuSO_4$ 22kg/ha Date uncertain.	441
	127			114														As above, same reference	490
	26.9			23.1														As above, same reference	490
																		As above, same reference	441
		141		186		481*				483*								April/August: cut close to ground. July: flowering completely. September/October: regrowth (*)	362
																		Fresh Dried Date uncertain, cut at 20-25 cm height	100
					3.8 13.4 43.6 5.1 15.0 0.16													Preflowering	493

Table 1.107 Meadow Fescue with White Clover

		CP %DM	Major elements % DM						Notes	Reference
			Ca	P	Mg	K	Na			
Meadow Fescue S53 with White Clover S100	N_0	20.1								288
	N_1	18.9								
	N_2	19.6								
	W_0	17.9	0.9	0.4	0.3	2.2	0.21	(1958)	W_0: control; no water, liquid manure or dry fertiliser	126
	W_1	17.6	0.9	0.4	0.3	2.3	0.19		W_1: water; low rate	
	W_2	17.8	0.9	0.4	0.3	2.4	0.20		W_2: water; high rate	
	M_1	16.6	0.7	0.4	0.2	2.9	0.12		M_1, M_2: Liquid manure: low: N/acre = dry fertiliser level; high: N/acre = dry fertiliser level	
	M_2	15.9	0.6	0.4	0.2	3.1	0.11			
	M_3	17.0	0.8	0.4	0.2	2.8	0.13		M_3, M_4: Liquid manure: low + 500 kg/ha/year superphosphate; high + 500 kg/ha/year superphosphate	
	M_4	15.8	0.6	0.4	0.2	3.1	0.07			
	F_1	16.8	0.8	0.4	0.3	2.7	0.14		F_1: Dry fertiliser: low: 625 kg Nitrochalk + 187 kg muriate of potash + 500 kg superphosphate/ha/year	
	F_2	16.3	0.6	0.4	0.3	2.9	0.09		F_2: Dry fertiliser: high: 1250 kg Nitrochalk + 312 kg muriate of potash + 500 kg superphosphate/ha/year	
	W_0	18.2	1.0	0.4	0.3	2.2	0.24	(1959)	Lay out in 4 replicate plots. Cut 5 times/year at 15–20cm height. Clover usually 20–30% ground cover.	
	W_1	19.2	1.2	0.4	0.3	2.2	0.31			
	W_2	18.9	1.0	0.4	0.3	2.9	0.19			
	M_1	19.0	1.1	0.4	0.2	1.8	0.29			
	M_2	17.7	1.1	0.4	0.3	1.6	0.33			
	M_3	20.0	1.2	0.4	0.3	1.4	0.35			
	M_4	17.2	1.0	0.4	0.3	3.0	0.17			
	F_1	17.8	1.1	0.4	0.3	2.5	0.27			
	F_2	16.6	1.2	0.4	0.3	1.5	0.35			
	W_0	13.8	0.9	0.4	0.4	1.2	0.26	(1960)		
	W_1	14.5	0.9	0.4	0.4	1.2	0.25			
	W_2	14.2	0.9	0.4	0.4	1.1	0.25			
	M_1	16.8	0.8	0.3	0.2	2.4	0.24			
	M_2	16.7	0.7	0.3	0.2	2.9	0.17			
	M_3	16.7	0.8	0.4	0.3	2.1	0.28			
	M_4	16.6	0.7	0.3	0.2	2.9	0.20			
	F_1	15.2	0.8	0.4	0.3	1.6	0.23			
	F_2	15.6	0.7	0.3	0.3	2.0	0.22			
Meadow Fescue S53 with White Clover S100, S184		14.7						(1947)	Means of several cuts taken during the grazing season, mainly between 16 August and 10 September	295
		19.6						(1948)		
		12.2						(1949)		

Table 1.108 Meadow Fescue with clover, unnamed

	Proximate constituents % DM					Notes	Reference
	CP	EE	NFE	CF	Ash		
Meadow Fescue and clover, unnamed	19.5	2.7	47.8	20.1	9.9	Silage cut: date not given	379
	16.5	2.8	45.8	26.0	8.9	'Later growth stage'	
						Ley for silage. For composition of silage see table 1.135, same reference	

Table 1.109 Meadow Fescue and Lucerne

	January			February			March			April			May			June		
Proximate constituents %DM Crude protein Meadow fescue S215 and Lucerne A.F.1.																		
Meadow fescue S215 and Lucerne Du Puits																		
Meadow fescue, unnamed and lucerne, unnamed																		
																	(1954) (1955)	11.0 10.3
Ether extract																		
																	(1954) (1955)	2.0 2.2
Nitrogen-free extract																		
																	(1954) (1955)	46.2 47.8
Crude fibre																		
																	(1954) (1955)	33.2 32.5
Ash																		
																	(1954) (1955)	7.6 7.2
Carbohydrate fractions % DM Total sugars																	(1954)	10.4
Fructosan																	(1954)	0.9
Vitamins Vitamin D IU per g (air-dry material)																(1954)	0.32 (1955)	(0.28 – 0.04 0.04

July			August			September			October			November			December			Notes	References
										18.6								Sown in rows 26/6/1956 4 replicates. Means only. Lucerne – 13% of DM.	159
										18.5								Sown in rows 26/6/1956 4 replicates. Means only. Lucerne – 33% of DM	
		14.8 14.7																Leys used in silage trials. No date given For the composition of the silage see table 1.135 same reference	379
																		Cut for hay. 3 replicates Means only. Grass: lucerne 50:50	378
		2.6 2.5																As above, same reference	379
																		As above, same reference	378
		52.0 52.3																As above, same reference	379
																		As above, same reference	378
		22.7 22.9																As above, same reference	379
																		As above, same reference	378
		7.9 7.6																As above, same reference	379
																		As above, same reference	378
0.37) (0.02 – 0.05) (0.02 – 0.06)																		Light petroleum extract of saponified material Mean values and true fiducial limits at P = 0.95 for groups of 8 rats.	277

Table 1.110 Tall Fescue S170

	January			February			March			April			May			June		
Digestibility of DM%																		
									(1960) (1961) (1960)	77.4	77.0 77.7	74.2 73.2	70.0 68.1	67.8*	64.2 61.2			
										W LB LS SI	83 84 81 86	80 81 75 81	78 81 73 82	73 79 65 72	68 77 66 64	66 76 63 63		
Digestibility of OM%									(1960) (1961) (1960)	81.9	79.5 80.7	76.1 75.4	71.0 70.0	70.0*	65.7 62.5			
Proximate constituents %DM Crude protein						(1953)	16.0											
	(1956) (1957)	W {W G F			15.8	20.1 22.6 11.5												
									(1960) (1961) (1960)	21.3	18.8 18.1	15.0 13.8	13.1 13.1	20.0*	10.0 8.1			
Carbohydrate fractions %DM Water soluble carbohydrate																		
Major elements %DM Magnesium											0.13	0.13	0.12	0.13	0.15	0.13	0.14	
Potassium									2.05		3.05	2.77		2.63	2.05	1.88		1.66
Sodium									0.10		0.15	0.11		0.10	0.09	0.07		0.06
Calcium														0.49				
Phosphorus														0.24				
Magnesium														0.15				
Potassium														3.08				
Sodium														0.15				
Trace elements μg/g DM Iodine														0.2				

July			August			September			October			November			December			Notes	References
													W G B	66 75 48				*In vitro* W: Whole sample G: Green fraction B: Burnt fraction	355
						65.2*												*In vivo* with sheep * Regrowth	356
																		In vitro W: Whole plant LB: Leaf blade LS: Leaf sheath SI: Stem and inflorescence	465
						68.8*												*In vivo* with sheep * Regrowth	356
																			296
															(1954)	W G F	13.9 17.1 7.2	W: Whole plant G: Green F: Frost burnt	337
														9.8				Whole sample 35% winter burnt foliage	355
						18.8*												* Regrowth	356
													W G B	15 20 5				As Digestibility, same reference	355
																			21
																			254
																		Means of 6 successive cuts of primary growth from April to June	542
									0.2*									Mean values for 7 stages of growth between 10/4 and 22/6 and 4 stages of regrowth (*) between 4/9 and 28/11.	

Table 1.111 Tall Fescue, unnamed and mixed

		January			February			March			April			May			June		
Tall fescue, unnamed																			
***Proximate constituents* %DM**																			
Crude protein												13.4				9.6	5.0		
Ether extract												1.9				1.8	1.0		
Nitrogen-free extract												52.2				54.0	53.6		
Crude fibre												22.7				26.4	34.2		
Ash												9.8				8.2	6.2		
Carbohydrate fractions % DM																			
Total sugars													(1960)	F D	11.8 11.9				
							(1961)	F D	13.7 13.1										
Fructosan													(1960)	F D	3.3 4.3				
							(1961)	F D	3.5 3.8										
Cellulose												33.5				38.8	39.9		
Lignin												4.5				3.8	6.4		
Major elements % DM																			
Calcium													0.5					0.2 0.3	
Phosphorus													0.3					0.2 0.3	
Magnesium													0.4					0.2 0.4	
Potassium																		1.4	
Sodium													0.16					0.22 0.10	
Chlorine													0.6					0.5 0.6	
Trace elements μg/g DM																			
Cobalt															0.18			0.09	
Copper															5.4			2.2	
Iron															447			137	
Manganese															32.5			23.0	
Vitamins μg/g DM																			
B Vitamins																			
Thiamin																			
Riboflavin																			
Nicotinic acid																			
Pantothenic acid																			
Pyridoxine																			
Biotin																			
Tall fescue, Alta, Meadow Fescue and White Clover S100																			
***Proximate constituents* %DM**																			
Crude protein	W	14.3	15.7	14.5	15.9	16.2	14.5												
	G		23.7	20.7	25.1	23.2	20.6												
	F		14.1	10.7	11.7	10.7	11.0												

July			August			September			October			November			December			Notes	References
3.8																		April, May: preflowering June: full flower July: seeding	9
0.8																			
57.0																			
32.1																			
6.3																			
																		F: Fresh D: Oven dried for 4 hr at 80°	
																			313
35.8																			9
7.5																			
	0.3 0.2			0.2 0.3														April: preflowering June: midflowering July: late flowering August: mature	490
	0.2 0.2			0.2 0.2															
	0.2 0.3			0.1 0.3															
	1.8			1.7															
	0.17 0.09			0.09 0.07															
	0.5 0.4			0.4 0.3															
	0.09			0.08														May: preflowering	
	3.4			5.5															
	206			342															
	17.1			19.7															
						4.04 13.7 44.9 7.7 7.0 0.14													493
																		Sampled before winter grazing with sheep. W: Whole plant G: Green F: Frost burnt	337

Table 1.112 Fine-leaved Red Fescue

	May	June	July	August	September	October	Notes	Reference
Proximate constituents %DM Crude protein			†				W : whole plant L : leaf S : stem M : grown in mixture P : grown in pure plots D : grown in drills Mixture with cocksfoot, timothy and meadow foxtail †Mean of 4 cuts taken on 5/5, 4/6, 5/7 and 28/9 Hay cut taken 19/6 *Regrowth	214
		WM 6.3	10.2	8.7*				
		WP 5.3	9.3	7.8*				
		WD 6.6	11.0	9.0*				
		LM 9.0	10.8	8.9*				
		LP 6.8	10.3	8.4*				
		LD 10.3	11.4	9.6*				
		SM 5.3	7.6	7.6*				
		SP 4.5	8.1	6.8*				
		SD 6.1	10.3	8.1*				
							F_0 : no fertiliser F_1 : P + K F_2 : P + K + N F_3 : P + K + N_2 P = 250 kg superphosphate/ha K = 500 kg Kainit/ha N = ammonium sulphate + sodium nitrate or Nitrochalk = 110 kg N/ha N_2 = ammonium sulphate =110 kg N/ha	215
	F_0 9.7						Grown at 750 feet above sea level	
	F_1 10.5							
	F_2 10.2							
			(1931) F_0 † 13.3				Grown at 300 feet above sea level †Mean of 4 cuts taken on 19/5, 19/6, 21/7, 28/9 ††Mean of 3 cuts taken on 13/6, 15/8, 18/10	
			(1931) F_3 16.6	††				
			(1932) F_0	12.3				
			(1932) F_3	15.4				
			(1931) F_0 † 15.2				Grown at 850 feet above sea level †As above ††As above	
			(1931) F_3 15.8	††				
			(1932) F_0	10.8				
			(1932) F_3	11.6				
Major elements %DM Calcium			†				As above, same reference	214
		WM 0.4	0.5	0.5*				
		WP 0.3	0.6	0.6*				
		WD 0.3	0.5	0.5*				
		LM 0.5	0.5	0.5*				
		LP 0.6	0.6	0.6*				
		LD 0.7	0.6	0.6*				
		SM 0.3	0.3	0.3*				
		SP 0.2	0.3	0.4*				
		SD 0.2	0.3	0.3*				
	F_0 0.3						As above, same reference	215
	F_1 0.3							
	F_2 0.4							
			(1931) F_0 † 0.5					
			(1931) F_3 0.6	††				
			(1932) F_0	0.5				
			(1932) F_3	0.5				
			(1931) F_0 † 0.4					
			(1931) F_3 0.5	††				
			(1932) F_0	0.4				
			(1932) F_3	0.6				

R

Table 1.112 Fine-leaved Red Fescue (continued)

	May	June	July	August	September	October	Notes	Reference
Major elements %DM (cont'd) Calcium		N_0 0.13 N_1 0.13 N_2 0.16 N_3 0.15 N_4 0.16 N_5 0.17					N_0 : no fertiliser, control plot N_1 : 167 kg ammonium sulphate/ha N_2 : 335 kg basic slag/ha N_3 : 110 kg muriate of potash/ha annually N_4 : $N_1 + N_2 + N_3$ N_5 : farmyard manure. 20 tonnes/ha/annum	489
	0.5	0.4	0.4	0.3			Date uncertain May, preflowering, June midflowering; July, late flowering; August, mature	490
Phosphorus		WM 0.2 WP 0.2 WD 0.2 LM 0.2 LP 0.3 LD 0.3 SM 0.2 SP 0.2 SD 0.2	† 0.3 0.3 0.3 0.3 0.3 0.3 0.3 0.3 0.4	0.2* 0.3* 0.3* 0.2* 0.3* 0.3* 0.3* 0.3* 0.3*			As above same reference	214
	F_0 0.2 F_1 0.2 F_2 0.2						As above same reference	215
			(1931) F_0 † 0.3 F_3 0.3					
			(1932) F_0 F_3	†† 0.3 0.3				
			(1931) F_0 † 0.2 F_3 0.3					
			(1932) F_0 F_3	†† 0.2 0.3				
		N_0 0.12 N_1 0.14 N_2 0.18 N_3 0.14 N_4 0.21 N_5 0.19					As above, same reference	489
	0.3	0.3	0.2	0.2			As above, same reference	490

continued

Table 1.112 Fine-leaved Red Fescue (continued)

	May			June			July			August			September			October			Notes	Reference
Major elements %DM (cont'd) Magnesium				N_0 0.06 N_1 0.10 N_2 0.07 N_3 0.09 N_4 0.07 N_5 0.09															As above, same reference	489
		0.2			0.2			0.2			0.1								As above, same reference	490
Potassium				WM WP WD LM LP LD SM SP SD	2.3 1.9 2.4 3.0 3.0 2.6 1.8 1.9 2.0		† 2.6 2.5 2.9 2.7 2.5 2.9 2.5 2.6 3.0					2.5* 2.6* 2.5* 2.7* 2.6* 2.5* 2.6* 2.5* 2.9*							As above, same reference	214
	F_0 F_1 F_2	1.8 2.0 1.9					(1931) (1932) (1931) (1932)	F_0 F_3 F_0 F_3 F_0 F_3 F_0 F_3	† 1.7 1.6 † 1.4 1.4		†† 1.4 1.7 †† 1.4 1.4								As above, same reference	215
				N_0 1.20 N_1 1.11 N_2 1.29 N_3 1.43 N_4 1.38 N_5 1.39															As above, same reference	489
		2.2			1.9			1.8			1.4								As above, same reference	490
Sodium				N_0 0.48 N_1 0.37 N_2 0.42 N_3 0.35 N_4 0.38 N_5 0.45															As above, same reference	489
		0.04			0.07			0.03			0.02								As above, same reference	490
Chlorine				WM WP WD LM LP LD SM SP SD	0.5 0.5 0.6 0.8 0.7 0.8 0.6 0.6 0.6		† 0.6 0.6 0.7 0.6 0.4 0.7 0.5 0.6 0.8					0.6* 0.4* 0.9* 0.5* 0.6* 1.0* 0.6* 0.6* 0.6*							As above, same reference	214

Table 1.112 Fine-leaved Red Fescue (continued)

	May	June	July	August	September	October	Notes	Reference
Major elements %DM (cont'd) Chlorine (cont'd)	F_0 0.3 F_1 0.4 F_2 0.4		(1931) F_0 † 0.7 F_3 0.7 (1932) F_0 F_3 (1931) F_0 † 0.4 F_3 0.5 (1932) F_0 F_3	†† 0.5 0.6 †† 0.4 0.7			As above, same reference	215
		N_0 0.4 N_1 0.4 N_2 0.4 N_3 0.5 N_4 0.6 N_5 0.5					As above, same reference	489
	0.5	0.4	0.5	0.4			As above, same reference	490
Sulphur, inorganic		N_0 0.08 N_1 0.06 N_2 0.06 N_3 0.07 N_4 0.09 N_5 0.08					As above, same reference	489
Sulphur, organic		N_0 0.01 N_1 0.04 N_2 0.04 N_3 0.05 N_4 0.03 N_5 0.04						
Trace elements µg/g DM Cobalt		N_0 0.23 N_1 0.24 N_2 0.18 N_3 0.21 N_4 0.16 N_5 0.17					As above, same reference	489
	0.20	0.19	0.17	0.21			As above, same reference	490
Copper		N_0 3.6 N_1 3.5 N_2 3.6 N_3 3.8 N_4 4.5 N_5 5.0					As above, same reference	489
	11.1	14.5	6.7	8.7			As above, same reference	490
Iron		N_0 105 N_1 241 N_2 115 N_3 104 N_4 57 N_5 181					As above, same reference	489
	327	241	228	302			As above, same reference	490

Table 1.113 Red Fescue, named varieties

	February	April	June	July	August	September	Notes	References
Red Fescue S59								
Proximate constituents %DM								
Crude protein	(1956) W 18.1						W : whole plant	337
	G 23.3						G : green	
	F 11.0						F : frost burnt	
	(1957) W 17.9							
	G 27.3							
	F 13.8							
Ash		10.2	8.7	9.6		8.6		
Carbohydrate fractions %DM							April preflowering	141
Soluble carbohydrate		10.2	9.7	7.3		10.5		
Total carbohydrate		32.2	52.1	48.4		46.5		
Lignin		6.7	6.2	6.4		5.8		
Red Fescue var. *genuina*							Cut at early flowering stage, height 30–36cm.	12
Proximate constituents %DM								
Crude protein			6.4					
Amino acids % crude protein								
Cystine			0.55					
Lysine			2.49					
Methionine, DL form			1.27					
L form			1.24					
Tryptophan			1.67					
Tyrosine			3.40					
Vitamins							Preflowering	493
B vitamins mg/g DM								
Thiamin						2.8		
Riboflavin						12.8		
Nicotinic acid						42.9		
Pantothenic acid						12.3		
Pyridoxine						12.2		
Biotin						0.12		
Red Fescue (Chewings fescue)								
Major elements % DM								
Calcium			0.24	0.25	0.21		Dates uncertain	490
Phosphorus			0.22	0.19	0.17		June: mid-flowering	
Magnesium			0.23	0.22	0.17		July: full maturity	
Potassium			1.39	1.77	1.70		August: in decline	
Sodium			0.22	0.17	0.09			
Chlorine			0.53	0.52	0.42			
Trace elements μg/g DM								
Cobalt			0.13	0.22	0.13			
Copper			3.9	7.8	7.2			
Iron			170	328	361			
Manganese			13.9	18.3	20.6			
Vitamins							Preflowering	493
B Vitamins mg/g DM								
Thiamin						4.4		
Riboflavin						17.5		
Nicotinic acid						34.9		
Pantothenic acid						15.5		
Pyridoxine						14.8		
Biotin						0.24		

Table 1.114 Timothy S48

	January			February			March			April			May			June		
Digestibility of DM%											(1960) (1961) (1960)	79.1	76.9	75.0	73.6 73.3	67.9	65.1 61.0	66.6 64.8
														74.1			73.8	
										W LB LS SI	85 85 (88) 83 (12)	W LB LS S Dd I 84 84 (83) -83 (17)	84 83 86 83 84 (74) 80 (23) (3)	78 83 81 80 82 (59) 74 (24) 84 (17)	76 79 70 85 78 81 (55) 70 (25) 81 (20)	71 76 66 69 68 73 77 (41) 70 (24) 71 (36)	66 72 (31) 61 (22) 64 (47)	67 77 59 65 65 77 63 75 (23) 56 (20) 61 (57)
Digestibility of OM%											(1960) (1961) (1960)	83.3	80.1	77.3	75.4 75.3	69.1	66.8 62.9	67.3 65.9
Proximate constituents %DM Crude protein																		
				(1947) (1948) (1949) (1950)	25.5 13.9 18.0			26.4 16.3 19.3			20.1 17.7 20.4							
	(1950) (1951)	14.1	17.3															
	G G B B	19.5 20.6 14.8 10.2	23.0 23.1 11.8 13.0															
											(1951) (1952)	W W L S		18.1 18.5 11.2	22.6 24.5 16.1		8.4 19.1 24.0 11.9	
																	(1954-55) 16.0 (1955-56) 13.8 (1956-57) 14.6	
												16.4			8.6			
											18.4					12.7		16.7
																	H L S	11.9 15.6 5.6
											(1960) (1961) (1960)	17.5	15.0	10.6	15.6 10.6	11.0	8.8 8.1	6.9 6 9
Ash Water-soluble ash											8.7 5.6					7.4 5.1		
Ash											(1960) (1961) (1960)	10.8	9.5	8.8	8.2 8.6	6.5	5.6 7.7	4.9 6.0

July			August			September			October			November			December			Notes	Reference
		69.3*			67.9*													*In vivo* with sheep *Regrowth	356
													W G F	59 75 45				*In vitro* W : Whole sample G : Green F : Frost burnt	355
																		In vivo with sheep	423
63 74 58 58 62 68																		*In vitro* W : Whole plant LB : Leaf blade LS : Leaf sheath S : Stem Dd : Dead I : Inflorescence SI : Stem and inflorescence Proportions in brackets underneath	465
		74.2*			71.9*													*Regrowth	356
		14.8 18.1																No date but late in season	209
																17.8 14.5 18.9		Sown 23/8/46 Dead leaf : Dec. 20% Feb. 30% Cut to simulate grazing	205
											16.2							Oct. cut of pasture rested since July G : Green 5cm drills / Broadcast B : Burnt 5cm drills / Broadcast Sown autumn	296
	16.9 21.8 23.8 14.6	25.2 26.6 15.5			14.6 25.2 25.6 14.7		21.9	21.1 21.5 12.9										W : Whole plant L : Leaf S : Stem	509
																		Sown April 1954 Means of all cuts and plots for each season	240
	4.6						2.4											Third year of growth. July, probably flowering; Sept., seed shedding stage	511
			17.4															Leafy, preflowering at start; July, seeding	141
																		H : Head L : Leaf S : Stem Early mature stage, heads well formed. Date not given	236
		15.6*			17.5*													*Regrowth	356
														10.3				Whole plant 52% Winter burnt foliage	355
		8.1 5.7				8.6 6.0													141
		8.9*			9.0*													*Regrowth	356

continued

Table 1.114 Timothy S48 (continued)

	January			February			March			April			May			June		
Carbohydrate fractions %DM Total carbohydrate											47.0					57.6		
Water-soluble carbohydrate												19.9			19.7			
											18.9					14.4		
										N_0 N_1 N_2 N_3	11.0 4.8 5.8 7.4			15.8 3.4 2.7 3.4		11.5 7.2 5.9 4.7		
Total sugars											(1951) (1952)	L S W		8.4 14.8 12.7	9.4 8.4 9.8		16.3 8.8 14.3 11.0	
Hexoses											(1951) (1952)	L S W		1.3 2.5 1.8	2.3 2.9 2.5		1.8 1.2 1.8 1.4	
Sucrose											(1951) (1952)	L S W		4.4 2.5 4.8	6.7 3.9 6.4		3.9 6.3 3.4 5.1	
Fructosan											(1951) (1952)	L S W		2.3 8.7 5.7	0.1 1.3 0.7		9.3 0.9 8.0 3.8	
Lignin												3.6			4.6			
											6.6					5.7		
Major elements %DM Calcium																	H L S	0.3 0.9 0.2
									(1964) (1965)	0.4	0.3	0.5 0.4		0.7 0.4	0.5 0.4	0.4 0.3		0.4 0.3
Phosphorus																	H L S	0.2 0.1 0.1
									(1964) (1965)	0.3	0.5	0.2 0.5		0.3 0.3	0.4 0.4	0.3 0.3		0.2 0.2
Magnesium																	H L S	0.1 0.3 0.2
									(1964) (1965)	0.1	0.2	0.1 0.2		0.1 0.1	0.1 0.1	0.1 0.1		0 1 0.1
Potassium																	H L S	1.4 1.7 1.6
									(1964) (1965)	2.8	3.6	3.8 3.5		2.6 2.1	2.9 3.0	2.8 2.5		2.6 2.2
											2.6	2.4		2.4	1.9	1.9		2.2
Sodium										0.06	0.08	0.10 0.12		0.16 0.04	0.09 0.06	0.05 0.06		0.06 0.05
											0 06	0.04		0.03	0.03	0.02		0.02

July			August			September			October			November			December			Notes	Reference
		52.8				49.8													141
	14.2						11.4											Third year of growth July, probably flowering; Sept., seed shedding stage	511
		12.9				13.5													141
													W G B	11 20 3				W : Whole sample G : Green B : Burnt	355
19.9 20.3 10.4 11.2			7.1 2.4 2.7 6.4			11.1 6.2 7.1 7.2		8.0 6.3 8.5 7.9										N_0 : no fertiliser N_1 : sodium nitrate N_2 : ammonium sulphate N_3 : urea Fertiliser to supply 88 kg N/ha 1 month before first cut and after each cut except the last	314
	6.7 4.8 8.2 5.5	7.9 9.7 8.2			14.3 6.2 7.9 6.3		5.6	8.6 17.2 8.9										L : Leaf S : Stem W : Whole plant Cut at height of 20-25 cm Leaf: stem ratio (Hexoses) in August 26:1; September 25:1 Leaf: stem ratio in May (Sucrose) 4·3:1 Figures as given	509
	1.0 1.1 2.2 1.3	1.6 2.6 1.7			1.7 1.3 2.4 1.3		0.8	1.1 1.4 0.9											
	3.9 3.6 3.4 3.6	6.0 4.3 5.8			5.9 4.5 3.2 4.4		3.9	6.2 4.5 6.1											
	1.4 0 2.2 0.5	0 2.3 0.3			5.8 0.2 1.9 0.3		0.6	0.6 10.0 1.0											
	9.2						11.9											See above, same reference	511
		6.1				4.6													509
																		H : Head L : Leaf S : Stem Early mature stage, heads well formed. Date not given.	236
																			542
																		As Ca. same reference	236
																			542
																		As Ca, same reference	236
																			542
																		As Ca, same reference	236
																			542
																			254
																			542
																			254

continued

Table 1.114 Timothy S48 (continued)

	January			February			March			April			May			June		
Trace elements μg/g DM Barium																	H L S	2.7 4 9 4.1
Boron																	H L S	15 17 2
Chromium																	H L S	0.21 0.29 0.15
Cobalt																	H L S	0.03 0 02 0.02
Copper																	H L S	6.5 4.6 3.2
Iodine												N_0 N_1	0.8* 0.3*	0.31				0.6* 0.2*
Iron																	H L S	50 42 11
Lead																	H L S	0.7 1.3 0.4
Nickel																	H L S	2.8 0.7 0.6
Manganese																	H L S	42 38 28
Molybdenum																	H L S	0.3 0.6 0.2
Strontium																	H L S	2.4 7.4 3 0
Tin																	H L S	0.1 0.2 0.1
Titanium																	H L S	2 6 3.7 1.8
Vanadium																	H L S	0 09 0.16 0.02
Zinc																	H L S	36 19 17
Vitamins μg/g DM Carotene																		
α-Tocopherol											217	222	195			191	163	

July			August			September			October			November			December			Notes	Reference
																		As Ca, same reference	236
								0.4* 0.2*		0.31*								Mean values for 7 stages of growth between 10/4 and 22/6 and 4 stages of regrowth between 4/9 and 28/11 N_0 : no fertiliser N_1 : monthly applications of 500 kg ammonium sulphate/ha *Regrowth	542
																		As Ca. same reference	236
		310																No date but late in season	209
72 184 - 100 -		249 172			48		09											Average of cuts at 20–25cm high, presumably regrowth The same dried	100

Table 1.115 Timothy. Scotch strain

	April			May			June			July			August			September			Notes	Reference
Proximate constituents %DM Crude protein					6.9			11.5			8.1			10.0*			17.5		May, young. June, flowering. July, old. Aug., aftermath*. Sept., grazed. Dates uncertain	460
							W S L H	22.6	20.2 15.1 22.4	20.3 13.6 21.8	13.2 6.8 17.5 14.8	8.9 4.5 15.9 13.3			5.5 2.9 10.8 12.5		6.0 2.8 12.8 13.0	4.6 2.4 8.8	W:whole plant S: stem L: leaf H: head	515
					W S L	18.6 14.1 21.6	14.2 11.4 17.2	9.9 7.5 16.9	6.7 4.6 13.0	6.8 3.4 11.6	5.5 3 4 11.1	4.2 1.9 10.7		3.9 2.2 8.7		2.5 2.3 6.1		1.9 1.8 5.9	12/6 to 4/7 – preflowering 11/7 to 30/7–midflowering 20/8 to 30/9–late flowering First year	
Ether extract					W S L	3.6 2.8 3.9	3.7 2.5 4.7	3.0 2.3 4.3	2.2 1.7 3.4	1.8 1.5 3.1	1.6 1.3 3.2	1.4 1.2 2.9		1.0 0.6 2.7		1.0 0.8 2.1		0.7 0.5 2.1	20/5 to 16/6–preflowering 23/6 to 21/7–midflowering 11/8 to 25/9–late flowering Third year	
Nitrogen-free extract					W S L	47.7 48.7 46.3	48.6 48.4 47.8	50.0 51.5 44.5	53.4 54.9 48.2	54.1 57.6 48.3	55.2 57.9 46.1	56.8 59.4 47.3		54.8 57.1 47.5		57.4 58.3 49.0		51.8 52.8 46.5		
Crude fibre					W S L	21.8 24.7 20.6	26.3 29.7 23.8	29.8 31.7 25.5	31.5 33.3 27.0	31.2 31.9 27.9	32.1 32.4 30.6	32.3 32.7 30.2		35.1 35.7 31.8		34.7 34.9 33.8		39.6 40.2 35.9		
Ash					W S L	8.3 9.7 7.6	7.2 8.0 6.5	7.3 7.0 8.8	6.2 5.5 8.4	6.1 5.6 9.1	5.6 5.0 9.0	5.3 4.8 8.9		5.2 4.4 9.3		4.4 3.7 9.0		6.0 4.7 9.6		
Major elements %DM Calcium					S L	0.1 0.4	0.2 0.5	0.1 0.5	0.1	0.1 0.6	0.1	0.1								
(1964) (1965)	0.4	 0.3	0.6 0.4		0.6 0.4	0.4 0.3	0.4 0.2		0.3 0.2											542
Phosphorus					S L	0.9 0.7	0.4 0.6	0.4 0.5	0.4	0.4 0.3	0.3	0.3							See above, third year	515
(1964) (1965)	0.4	 0.5	0.4 0.4		0.3 0.4	0.3 0.3	0.2 0.2		0.2 0.2											542
Magnesium (1964) (1965)	0.1	 0.2	0.1 0.2		0.1 0.1	0.1 0.1	0.1 0.1		0.1 0.1											
Potassium (1964) (1965)	2.8	 3.6	4.0 2.9		2.7 2.6	2.8 2.4	2.3 1.8		2.0 1.6											
Sodium (1964) (1965)	0 05	 0.10	0.09 0.06		0.11 0.07	0.11 0.07	0.05 0.06		0.08 0.07											
Vitamins μg/g DM Carotene					223		 LS HF	201 160 112			112			197*			261		As Crude protein, same reference LS: leaf and stem HF:head or flower	460
							W L S H	408	332 432 96	312 484 110	253 336 84 112	170 312 46 103			138 277 56 64		117 202 35 95	141 201 43	W: whole plant L: leaf S: stem H: head 1st year growth, as above	515
					W L S	274 372 99	156 341 61	118 310 58	73 215 32	66 205 35	88 203 48	44 120 23		42 64 34		33 33 14		9 14 12	W: whole plant L: leaf S: stem 3rd year growth, as above	

Table 1.116 Timothy, named varieties

	January			February			March			August			September			October			Notes	Reference
Timothy S50 Crude protein %DM	(1950) (1951)	 13.7	17.6															14.9	Oct. cut of pasture rested since July 1950	296
	(1950)	G	21.2																G : green B : burnt 5cm drills	
		G	23.0																Broadcast	
		B	12.8																5cm drills	
		B	12.7																Broadcast	
(1951)	G	18.7																	5cm drills; pasture rested since July	
	G	19.9																	Broadcast	
	B	11.0																	5cm drills; pasture rested since July	
	B	11.8																	Broadcast	
	May			June			July			August			September			October				
Timothy: 19 named varieties *Proximate constituents* %DM Crude protein	17.8	20.1		21.4 8.2 5.5			14.3* 11.5*			22.3									For the individual varieties only yields and digestibilities of dry matter are given.	166
Crude fibre	18.3	22.9		18.8 31.7} 34.7}			29.5* 30.1*			24.1									*Regrowth	
Carbohydrate fractions %DM Soluble carbohydrate	22.9	8.1		13.1 14.5} 16.0}			8.3* 11.4*			4.3			8.1			10.0			Cuts in June, hay, and July, 1st aftermath, (bracketed) are from different centres.	
	May			June			July			August			September			October				
Timothy S51 *Crude protein* %DM									16.6 17.7										No date but late in season	209
Vitamins μg/g DM Carotene									320											
	January			February			March			October			November			December				
Timothy S48 and S50 Crude protein %DM	(1952) (1953)	14.7	13.3		13.3	15.5					(1951)	14.7				14.9			Sown April 1950, 2.5cm drill experimental plots.	296

Timothy 1.117 Timothy, unnamed

	January			February			March			April			May			June		
Proximate constituents %DM Crude protein																WM WP WD LM LP LD SM SP SD	7.3 6.9 8.2 8.7 7.9 10.2 5.1 5.5 7.1	
								11.6 13.2										
												13.5						
												10.6						
																	8.8	
Amino acids % crude protein Cystine Histidine Lysine Methionine, DL form L form Tryptophan Tyrosine																	0.48 2.06 4.16 1.72 1.69 2.10 2.30	
Crude fibre												22.4						
												25.2						
Ash																		
Carbohydrate fractions %DM Soluble Carbohydrate								44.6 44.8										
Total sugars						(1960)							F D	9.8 8.5				
						(1961)	F D	12.2 11.9										
Fructosan						(1960)							F D	4.5 4.9				
						(1961)	F D	4.0 4.7										

July			August			September			October			November			December			Notes	Reference
† 10.5 11.1 11.6 11.4 10.9 13.4 8.6 9.8 10.1					8.0* 8.8* 8.1* 8.8* 9.4* 10.6* 5.7* 5.9* 6.0*													W : Whole plant L : Leaf S : Stem M : Grown in mixture P : Grown in pure plots D : Grown in pure drills †Average of 4 cuts taken 5/5, 4/6, 5/7 and 28/9 Hay cut taken 19/6 *Regrowth	214
					A B	{N_0 {N_1 {N_0 {N_1		14.6 18.0 13.8 14.8										N_0 : No fertiliser N_1 : Superphosphate 250 kg/ha + Kainit 500 kg/ha A:Grown at 300 feet above sea level B:Grown at 750 feet above sea level	215
																		Dates uncertain but probably 14/1 to 15/5 High percentage burnt	255
																		Leafy stage, no date given	14
																		Preflowering stage. no date given	494
																		Cut at early flowering stage, height 20-22 cm.	12
																		Leafy stage no date given	14
																		Preflowering stage, no date given	494
					A B	{N_0 {N_1 {N_0 {N_1		7.1 7.5 6.8 6.7										As Crude protein, same reference	215
																		As Crude protein, same reference	255
																		F : Fresh D : Dried in unitherm oven for 4 hr. at 80°C	313

continued

Table 1.117 Timothy, unnamed (continued)

	January			February			March			April			May			June		
Major elements %DM																		
Calcium																WM	0.4	
																WP	0.4	
																WD	0.4	
																LM	0.6	
																LP	0.6	
																LD	0.6	
																SM	0.2	
																SP	0.2	
																SD	0.3	
								0.4 0.4										
														0.5			0.4	
												0.54						
Phosphorus																WM	0.2	
																WP	0.3	
																WD	0.3	
																LM	0.3	
																LP	0.2	
																LD	0.2	
																SM	0.3	
																SP	0.3	
																SD	0.3	
								0.3 0.3										
														0.3			0.2	
												0.32						
Magnesium														0.3			0.3	
Potassium																WM	2.9	
																WP	2.7	
																WD	2.4	
																LM	2.5	
																LP	2.5	
																LD	2.2	
																SM	2.7	
																SP	2.9	
																SD	3.1	
														1.9			2.3	
Sodium														0.22			0.29	
Chlorine																WM	0.8	
																WP	0.9	
																WD	0.8	
																LM	0.7	
																LP	0.6	
																LD	0.4	
																SM	1.0	
																SP	1.1	
																SD	1.2	
														0.4			0.5	

July			August			September			October			November			December			Notes	Reference
† 0.5 0.5 0.5 0.7 0.7 0.7 0.3 0.3 0.3					0.6* 0.5* 0.4* 0.7* 0.7* 0.5* 0.6* 0.4* 0.3*													As Crude protein, same reference	214
					A B	N_0 N_1 N_0 N_1		0.7 0.7 0.4 0.6										As Crude protein, same reference	215
																		As Crude protein, same reference	255
	0.3			0.4														Dates uncertain. May, young leaf; June, late leaf; July, midflowering; August, mature	490
																		Leafy stage, no dates given	14
† 0.3 0.3 0.3 0.3 0.3 0.3 0.3 0.4 0.3					0.2* 0.2* 0.2* 0.2* 0.2* 0.2* 0.2* 0.2* 0.2*													As Crude protein, same reference	214
					A B	N_0 N_1 N_0 N_1		0.3 0.3 0.2 0.3										As Crude protein, same reference	215
																		As Crude protein, same reference	255
	0.2			0.2														As Calcium, same reference	490
																		Leafy stage, no dates given As Crude protein, same reference	14
	0.2			0.2														As Calcium, same reference	490
† 2.6 2.6 2.7 2.6 2.5 2.8 2.9 2.7 3.1					2.1* 1.7* 1.8* 2.1* 2.1* 2.3* 2.1* 2.1* 2.4*													As Crude protein, same reference	214
					A B	N_0 N_1 N_0 N_1		1.4 1.5 1.8 2.0										As Crude protein, same reference	215
	2.2			2.3														As Crude protein, same reference	490
	0.24			0.13															
† 0.7 0.9 0.8 0.7 0.6 0.6 0.8 0.8 0.8					0.6* 0.7* 0.9* 0.6* 0.7* 0.5* 0.6* 0.6* 0.6*													As Crude protein, same reference	214
					A B	N_0 N_1 N_0 N_1		0.7 0.9 0.7 0.7										As Crude protein, same reference	215
	0.7			0.7														As Calcium, same reference	490

continued

Table 1.117 Timothy, unnamed (continued)

	January			February			March			April			May			June		
Trace elements µg/g DM Cobalt														0.1			0.2	
																N_0 N_1 N_0 N_1	0.03 0.25 0.03 0.40	
Copper														8.2			6.8	
																N_0 N_1 N_0 N_1	2.7 3.4 2.7 3.4	
Iron														448			292	
												128						
Manganese														35.3			34.0	
Vitamins µg/g DM Carotene								582			338				275			128
Tocopherols Total tocopherol																	184–249 100–172	
a-tocopherol % of total																	100	
B vitamins mg/g DM Thiamin Riboflavin Nicotinic acid Pantothenic acid Biotin Pyridoxine																		

July			August			September			October			November			December			Notes	Reference
	0.1			0.2														As Calcium, same reference	490
																		N_0 : no fertiliser N_1 : Co $SO_4.7H_2O$. 1.65 kg/ha in April 1954 and 1955. Sampled in June. Replicate plots	359
	7.3			7.6														As Calcium, same reference	490
																		N_0 : no fertiliser N_1 : Cu $SO_4.5H_2O$. 22kg/ha in April 1954 and 1955. Sampled in June. Replicate plots	359
	349			408														As calcium, same reference	490
																		Preflowering stage, no date given	494
	27.0			25.3														As Calcium, same reference	490
		110		121			408*			447*								Cut close to ground in April and Aug. July, flowering completed. August, leaf only. *Regrowth	362
																		(18.4–24.9) Fresh (10.0–17.2) Dried Dates uncertain. Grass 20 to 25 cm high	100
				3.9 12.6 37.9 9.1 0.17 11.7														Late flowering experimental plots	493

Table 1.118 Timothy with other grasses and/or clover

	April			May			June			July			Notes	Reference
Timothy S48 and Meadow Fescue S215, S53 *Digestibility of* DM%					76.3									265
Digestibility of OM%					78.4									

	May			June			July			August			Notes	Reference
Timothy S48 and White Clover S100 Crude protein %DM									N_1 17.4 N_2 18.1 N_0 18.1				N_1 Nitrochalk 250 kg/ha after each cut N_2 Nitrochalk 250 kg/ha after first and last cuts only N_0 No nitrogen fertiliser Four-five cuts per year (1951-53) Weighted means only. No replicates. 26-59% clover; 13-27% other grasses	288
								(1957) (1958) (1959)	N_0 20.0 N_1 19.2 N_2 19.8 N_0 16.7 N_1 16.6 N_2 17.5 N_0 14.6 N_1 15.1 N_2 16.0				N_0 No N N_1 Nitrochalk 1250 kg/ha given in spring and after each cut except the last N_2 Same amount given after second, third and fourth cuts only 5 cuts per year. 3 replicate plots.	432
Timothy S48 and White Clover S100, S184 Crude protein %DM					(1947) A 13.5 (1948) B 21.5 (1949) C 15.5								A: Grazed from 21/5 to 3/9 B: Grazed from 8/4 to 15/9 C: Grazed from 21/4 to 19/7 Mean of several cuts during growing season	295
Timothy Scotia and White Clover S100 Crude protein %DM									N_1 20.4 N_2 20.6 N_0 22.0				See above, same reference	288
Timothy Scots and White Clover S100 Crude protein %DM									N_1 17.9 N_2 20.2 N_0 19.1					

	March			April			May			June			Notes	Reference
Timothy and White Clover *Digestibility of* DM%			(1949)	80.7		(1951)	78.6 76.1			76.0			Digestibility trials with 2 or 3 sheep These values are not stated to apply to cuts of the same crop	429
Digestibility of OM%			(1949)	84.3		(1951)	81.2 79.6			78.5				
Timothy and White Clover (dried) *Digestibility of* DM%						(1950)	77.8							
Digestibility of OM%						(1950)	80.6							

	May			June			July			August			Notes	Reference
Timothy, Meadow Fescue and Clover Crude protein %DM				O F O F O F	22.4 19.9 20.0 21.1 19.7 22.4								O : ordinary grazing F : close-folding Over 3 periods of grazing on plots between April and August. No exact dates given.	414
				(1958) (1959)	17.6 21.1 20.6 20.3 21.0 19.4 17.8								First and second year leys. Four grazing periods between May 1st and September, 1958; and three between April 21st and September, 1959.	291
Timothy, Meadow Fescue, Cocksfoot and Clover Crude protein %DM				O F O F O F O F	16.9 17.3 19.2 20.8 19.3 20.0 15.9 16.3								O : ordinary grazing F : close-folding Second year sward. Pre-grazing samples. No exact dates given.	413

Table 1.119 Timothy and Lucerne

	May			June			July			August			Notes	Reference
Proximate constituents %DM Crude protein				14.9 15.5 14.1 15.3					18.3 18.9 18.5 17.9			24.2	The 4 values represent parts of the crop ensiled in different ways. For composition of the silages see table 1.136, same reference June, first cut (final value wilted 24 hr.) July, second cut (final value wilted 24-48 hr.) Aug., third cut, bulked sample Low proportion of timothy in second and third cuts.	375
	(1952) (1953)	U C L U UW CW LW	11.3 11.5 11.4 16.0 16.4 14.9 15.1										For composition of the silages see table 1.136, same reference. C : Chopped L : Lacerated U : Unchopped UW: Unchopped, wilted 24 hr. CW: Chopped, wilted 24 hr. LW: Lacerated, wilted 24 hr. More lucerne than timothy	376
			1 {21.6 21.6 4 {18.1 17.5 17.4									5 {20.4 20.8	For composition of the silages see table 1.136, same reference Trial 1 Slightly wilted in field. Chopped before analysis. Dates not given Presumably green, leafy. Probably bulked sample. Dates not given. Trial 4 Presumably green/leafy Third value, chopped. Trial 5 Autumn cut of crop. Dates not given. More lucerne than timothy.	381
Ether extract				2.4 2.4 2.4 2.5					2.3 2.3 2.1 2.3			2.4	As Crude protein, same reference	375
	(1952) (1953)	U C L U UW CW LW	2.1 2.0 2.0 2.1 2.3 2.2 1.9											376
			1 {2.6 2.6 4 {2.8 2.6 2.8									5 {2.9 3.0		381
Nitrogen-free extract				48.0 47.4 48.1 47.7					48.5 46.9 48.4 47.3			39.4		375
	(1952) (1953)	U C L U UW CW LW	45.4 44.7 45.0 45.2 44.7 45.7 46.4											376
			1 {40.8 40.8 4 {44.3 45.3 44.7									5 {41.7 40.8		381

continued

Table 1.119 Timothy and Lucerne (continued)

	May			June			July			August			Notes	Reference
Proximate constituents (cont'd)														
Crude fibre				25.8 25.4 26.4 25.8					22.2 22.7 22.1 24.1			22.5		375
	(1952) (1953)	U C L U UW CW LW	33.7 34 4 34.2 28.6 28.6 28 9 28.8											376
			1 {25.4 25.4 4 {26.1 26.5 26.5									5 {25.6 25.3		381
Ash				8.9 9.3 9.0 8.7					8.7 9.2 8.9 8.4			11.5		375
	(1952) (1953)	U C L U UW CW LW	7.5 7.4 7.4 8 1 8.0 8.3 7.8											376
			1 {9.6 9.6 4 {8.7 8.1 8.6									5 {9.4 10.1		381
Vitamins μg/g DM														
β-carotene				209					222 304					375

Table 1.120 Pasture grass, unnamed

		January			February			March			April			May			June		
Proximate constituents %DM																			
Crude protein	(1)											27.1			26.4			22.6	
	(2)											21.1			21.1			19.0	
	(3)														21.7			18.1	
	(1933)																	10.6	
	(1934)														22.0	16.6	13.6		
Ether extract	(1933)																	4.3	
	(1934)														6.4	5.1	3.7		
Nitrogen-free extract	(1933)																	49.5	
	(1934)														37.7	42.3	43.3		
Crude fibre	(1933)																	29.6	
	(1934)														24.2	26.9	30.8		
Ash	(1933)																	6.0	
	(1934)														9.7	9.1	8.6		
Major elements %DM																			
Calcium	(1)											0.66			0.62			0.55	
	(2)											0.49			0.63			0.61	
	(3)														0.66			0.54	
	(1933)																	0.51	
	(1934)														0.49	0.49	0.49		
Phosphorus	(1)											0.40			0.37			0.34	
	(2)											0.36			0.34			0.28	
	(3)														0.32			0.25	
	(1933)																	0.23	
	(1934)														0.39	0.42	0.32		
Magnesium	(1)											0.20			0.21			0.20	
	(2)											0.16			0.20			0.19	
	(3)														0.19			0.22	
Potassium	(1)											2.78			2.89			2.93	
	(2)											3.35			2.99			2.84	
	(3)														2.61			2.34	
	(1933)																	2.25	
	(1934)														3.49		3.10		
Sodium	(1)											0.33			0.37			0.36	
	(2)											0.21			0.24			0.15	
	(3)														0.36			0.40	
Chlorine	(1)											1.12			1.30			1.42	
	(2)											1.08			1.09			1.40	
	(3)														1.30			1.31	
	(1933)																	1.12	
	(1934)														1.26		1.33		
Total sulphur	(1928)												0.64						0.73
	(1928)													0.74					0.78
	(1929)												0.53		0.62		0.72		
																		0.47	
																		0.42	
Sulphur, inorganic	(1928)												0.29						0.38
	(1928)													0.28					0.38
	(1929)												0.25		0.26		0.31		
																		0.22	
																		0.24	
Sulphur, total	(1)											0.46			0.46			0.46	
	(2)											0.33			0.38			0.34	
	(3)														0.38			0.42	
Trace elements µg/g DM																			
Iron	(1)											610			920			720	
	(2)											180			710			560	
	(3)														610			730	
Manganese	(1)											7.6			11.7			10.0	
	(2)											6.4			14.2			17.0	
	(3)														11.4			11.5	
Vitamins µg/g DM																			
Thiamin																			
Vitamin D iu/g DM																			
Light petroleum extract	(1952)														0.08 (0.06–0.10)			0.14 (0.11–0.18)	
	(1953)											0.07 (0.05–0.09)			0.30 (0.22–0.47)			0.14 (0.11–0.17)	
	(1954)												0.15 (0.12–0.18)						0.03 (approx.)
Light petroleum extract of saponified material	(1953)																		
	(1954)												0.29 (0.23–0.41)						0.03 (approx.)

July	August	September	October	November	December	Notes	Reference
23.3 22.6 15.6	24.2 26.5 17.6	27.2 27.4 16.4	26.4			Intensively managed pasture sampled at grazing stage each month. (1) Marton, N. Yorks; (2) Romsey, Herts; (3) Bracknell, Berks.	223
9.5 15.3							208
3.0 4.3							
54.3 42.7							
28.7 28.6							
4.5 9.1							
0.56 0.66 0.54	0.55 0.61 0.62	0.51 0.58 0.49	0.55			As for Crude protein	223
0.50 0.74							208
0.33 0.32 0.22	0.36 0.38 0.22	0.37 0.42 0.19	0.37			As for Crude protein	223
0.17 0.26							208
0.20 0.22 0.24	0.22 0.25 0.28	0.22 0.25 0.26	0.20			As for Crude protein	223
3.00 3.29 2.32	3.25 3.70 2.02	3.13 3.65 2.55	3.12			As for Crude protein	223
							208
0.33 0.14 0.20	0.29 0.20 0.22	0.52 0.33 0.26	0.23			As for Crude protein	223
1.46 1.34 1.33	1.50 1.57 1.38	1.55 1.59 1.53	1.54			As for Crude protein	223
							208
0.73 0.75 0.66	0.62 0.71 0.63	0.87 0.58				Cut over approximately 4 weeks centred at the dates shown. Weekly 2–weekly Monthly	198
						Early cut hay	
0.38 0.39 0.40	0.37 0.34 0.32	0.39 0.38				As for Total sulphur	
0.51 0.35 0.36	0.46 0.34 0.36	0.45 0.36 0.33	0.48			As for Crude protein	223
1110 400 380	630 230 380	900 340 800	1410			As for Crude protein	223
9.0 13.2 11.1	10.8 12.0 12.8	13.2 9.5 7.8	14.4				
	5.4					Estimation by thiochrome method. Mean of 5 samples, June-October	405
0.18 (0.15–0.23) 0.10 (0.08–0.13)	0.12 (0.10–0.14) 0.09 (0.06–0.12) 0.20 (0.15–0.40)	0.06 (approx.) 0.23 (0.16–0.39) 0.14 (0.10–0.19)	0.12 (0.09–0.14) 0.31 (approx.) 0.04 (0.02–0.07)	0.05 (approx.) 0.74 (approx.) 0.09 (approx.)		Mean and range (true fiducial limits at P = 0.95) for 6–12 rats. As for 1952	277
	0.04 (0.03–0.05)	0.04 (0.02–0.05) 0.06 (0.05–0.08)	0.14 (0.11–0.16)	0.34 (approx.) 0.03 (approx)		Mean and range (true fiducial limits at P = 0.95) for 6–12 rats.	

Table 1.121 Grass, unnamed. Effect of fertilisers (a) Pasture

		May		June		July		August		September			October	Notes	Reference
Proximate constituents %DM															
Crude protein	F_0	10.3	8.3	9.9	8.7	8.5	8.2	7.6	7.8	7.4	7.5	7.2	7.0	Pasture grass. Samples are of 14 days' growth	476
	F_1	17.0	15.8	15.5	13.2	11.6	12.1	11.8	13.3	13.3	14.3	13.5	8.0	Fertilisers as follows:	
	F_2	16.6	15.5	15.1	11.4	10 8	12.0	10.3	14.4	12.6	11.8	11.1	7.5	F_0 : None	
	F_3	15.7	14.4	14.5	12.5	11.2	14.6	12.4	12.8	11.7	11.5	10.0	7.0	F_1 : Basic slag, 625 kg/ha every 3rd year.	
	F_4	17.7	15.4	14.9	12.3	11.5	12.7	13.5	13.4	13.0	11.5	10.8	9.6	F_2 : Basic slag and 110 kg muriate of potash/ha every 3rd year.	
Ether extract	F_0	1.9	1.1	1.4	1.4	1.1	1.2	1.1	1.4	1.0	1.0	1.3	0.7	F_3 : Basic slag and 2500 kg burnt lime/ha every 3rd year.	
	F_1	2.7	1.5	2.2	1.7	0.8	1.2	1.2	2.0	1.4	1.6	1.3	0.9	F_4 Basic slag and 125 kg sodium nitrate/ha every 3rd year.	
	F_2	2.4	1.4	1.6	1.3	1.2	1.0	1.0	1.4	1.2	1.2	1.1	0.8		
	F_3	2.7	1.7	1.6	1.4	1.4	1.6	1.5	2.0	1.0	2.1	1.5	0.6		
	F_4	2.8	1.7	1.4	1.1	1.3	1.4	1.9	1.6	1.8	1.4	0.8	0.7		
Nitrogen-free extract	F_0	62.2	64.4	62.1	62.7	62.4	66.2	66.4	65.0	64.0	65.6	62.1	65.7		
	F_1	52.6	53.0	53.5	57.0	59.6	56.5	59.5	56.8	55.7	55.8	58.4	68.6		
	F_2	55.2	52.3	54.5	59.4	59.5	60.3	62.9	58.0	59.2	60.6	61.1	67.4		
	F_3	56.6	50.8	54.0	57.0	55.8	55.5	59.4	57.4	61.8	61.4	64.8	67.4		
	F_4	53.2	50.5	54.6	58.0	57.9	58.5	57.6	55.6	59.5	59.1	64.7	66.3		
Crude fibre	F_0	21.3	21.9	21.7	22.7	23.5	20.3	20.3	22.1	21.7	19.9	19.6	19.4		
	F_1	19.7	22.2	21.4	21.6	22.5	23.4	20.1	20.4	21.7	20.1	18.3	13.6		
	F_2	18.4	23.5	20.9	21.9	22.5	20.2	19.0	18.6	18.0	17.7	19.0	15.0		
	F_3	19.9	24.9	21.0	20.9	24.2	20.4	18.0	18.4	16.1	14.9	12.2	13.1		
	F_4	19.0	25.6	21.9	22.4	23.5	21.5	20.3	21.9	17.6	20.3	15.4	15.0		
Ash	F_0	4.3	4.3	4.9	4.5	4.5	4.1	4.6	3.7	5.9	6.0	9.8	7.2		
	F_1	8.0	7.5	7.4	6.5	5.5	6.8	7.4	7.5	7.9	8.2	8.5	8.9		
	F_2	7.4	7.3	7.9	6.0	6.0	6.5	6.8	7.6	9.0	8.7	7.7	9.3		
	F_3	5.1	8.2	8.9	8.2	7.4	7.9	8.7	9.4	9.4	10.1	11.5	11.9		
	F_4	7.3	6.8	7.2	6.2	5.8	5.9	6.7	7.5	8.1	7.7	8.3	8.4		
Major elements %DM															
Calcium	F_0	0.5	0.4	0.4	0.5	0.6	0.6	0.6	0.3	0.5	0.5	0.5	0.4		
	F_1	0.9	0.7	0.5	0.6	0.4	0.6	0.6	0.5	0.5	0.5	0.5	0.6		
	F_2	0.3	0.3	0.6	0.6	0.6	0.7	0.7	0.8	1.0	0.5	0.6	0.7		
	F_3	0.4	0.8	1.2	1.4	1.6	1.2	1.3	0.9	0.7	1.1	0.7	1.0		
	F_4	0.4	0.4	0.4	0.6	0.5	0.6	0.4	0.6	0.4	0.5	0.7	0.5		
Phosphorus	F_0	0.2	0.2	0.1	0.2	0.2	0.2	0.2	0.2	0.2	0.2	0.2	0.1		
	F_1	0.5	0.5	0.5	0.4	0.3	0.4	0.4	0.5	0.4	0.5	0.5	0.4		
	F_2	0.5	0.5	0.5	0.4	0.4	0.4	0.4	0.4	0.5	0.5	0.5	0.4		
	F_3	0.4	0.5	0.6	0.4	0.4	0.5	0.5	0.6	0.5	0.5	0.4	0.5		
	F_4	0.5	0.6	0.7	0.7	0.5	0.5	0.4	0.4	0.5	0.5	0.5	0.7		

continued

Table 1.121 Grass, unnamed. Effect of fertilisers (b) Pasture, some clover

		April	May	June	July	August	September	October	Notes	Reference
Proximate constituents %DM										
Crude protein	F_0	19.3	18.9	17.0	16.8	19.4	18.1	22.2	Pasture grass; under 10% clover and 10% weeds.	360
	F_1	18.7	19.1	16.9	17.5	19.1	19.0	22.4		
	F_2	17.7	19.1	17.0	17.1	19.9	19.5	23.1		
	F_3	19.8	20.7	17.6	17.6	19.5	19.8	23.0	Application of fertiliser as follows, in kg or tonnes/ha in March (initial) and monthly:	
	F_4	22.3	24.2	20.8	19.5	21.0	22.3	25.7		
	F_5	22.3	24.2	21.0	20.5	21.5	23.6	26.5		
Ether extract	F_0	2.8	3.6	3.4	3.3	4.2	3.4	4.2	F_0 : None	
	F_1	3 0	3.2	3.5	3.4	4.2	3.4	3.9	F_1 : Potassium sulphate 250 and 125 kg	
	F_2	2.3	3.3	3.4	3.4	4.4	3.6	4.1	F_2 : Calcium carbonate 10 tonnes	
	F_3	2.9	3.5	3.7	3.6	4.5	3.9	4.2	F_3 : Superphosphate 750 and 500 kg	
	F_4	2.8	3.0	3.4	3.5	4.7	3.7	3.9	F_4 : Sodium nitrate 164 and 164 kg	
	F_5	3.0	3.6	3.8	3.8	4.4	3.7	4.1	F_5 : Ammonium sulphate 125 and 125 kg	
Nitrogen-free extract	F_0	52.3	50.1	50.9	50.7	45.9	49.8	44.9		
	F_1	51.3	49.6	49.8	49.0	45.2	47.9	44.8		
	F_2	53.6	48.5	51.0	49.5	45.1	48.2	44 8		
	F_3	49.2	47.1	48.0	47.5	43.7	46.5	43.9		
	F_4	49.4	46.1	48.5	47.4	43.5	46.2	43.4		
	F_5	49.9	45.2	46 4	45.2	42.5	44.4	41.5		
Crude fibre	F_0	18.0	20.0	20.8	20.6	20.7	20.0	19.6		
	F_1	18.0	20.1	21.0	21.3	21.6	20.4	19.5		
	F_2	16.4	20.1	20.4	21.7	21.3	19.9	18.8		
	F_3	18.5	19.9	20.8	21.7	21.5	20.0	18.9		
	F_4	18.0	19.7	19.4	21.6	22.1	20.0	18.9		
	F_5	17.0	19.2	20.3	21.9	22.2	19.5	18.4		
Ash	F_0	7.6	7.4	7.9	8.6	9.8	8.7	9.1		
	F_1	9.0	8.0	8.8	8.8	9.9	9.3	9.4		
	F_2	10.0	9.0	8.2	8.3	9.3	8.8	9.2		
	F_3	9.6	8.8	9.9	9.6	10.8	9.8	10.0		
	F_4	7.5	7.0	7.9	8.0	8.7	7.8	8.1		
	F_5	7.8	7.8	8.5	8.6	9.4	8.8	9.5		
Vitamins μg/g DM									The increases in carotene content with potassium sulphate were not significant. The increases with nitrogen, except for April and sodium nitrate in May, were highly significant.	
Carotene	F_0	245	327	382	370	461	445	610		
	F_1	260	351	397	405	500	459	645		
	F_2	256	340	370	369	495	452	586		
	F_3	249	315	437	409	445	475	603		
	F_4	315	363	507	483	592	564	758		
	F_5	318	402	502	513	632	603	713		

continued

Table 1.121 Grass unnamed. Effect of fertilisers (c) Plot experiments

		May			June			July			August			September			October			Notes	Reference
Proximate constituents %DM																					
Crude protein	(1932) F_0				19.6			17.8		19.0	19.4		21.3			19.9		19.5		Pasture grass. Means for 5 replicate plots. F_0: No fertiliser. F_1: 1930: lime 5000 kg/ha; 1931: Nitrochalk 625 kg/ha; 1932: Nitrochalk 375 kg/ha as 125-kg dressings at intervals.	477
	F_1				19.9			17.8		16.5	19.8		20.3			21.0		19.7			
	(1933) F_0	22.3		24.2	22.2		21.0		21.7	21.9		20.4	20.2		19.7		22.6	21.1		Same areas: 7 replicate plots; means given. F_0: No fertiliser. F_1: Nitrochalk 125 kg/ha twice. Results corrected for possible soil contamination by Woodman formula, which assumes a normal silica content of 1.72% for clean grass. The figures for nitrogen-free extract are as given in the original.	482
	F_1	20.9		23.3	21.4		22.9		21.1	21.6		20.7	18.8		20.6		24.6	23.9			
Ether extract	(1932) F_0				3.8			3.2		2.8	3.2		2.6			3.2		0.9			477
	F_1				3.7			3.8		2.6	4.1		3.1			3.5		0.9			
	(1933) F_0	3.2		3.0	3.1		2.5		2.7	2.5		2.9	2.4		1.5		3.1	2.6			482
	F_1	3.1		3.1	4.0		2.7		2.8	2.4		2.8	2.6		1.7		3.2	2.0			
Nitrogen-free extract	(1932) F_0				49.7			49.9		49.9	49.8		47.7			45.8		50.6			477
	F_1				51.0			50.0		51.1	48.5		48.7			45.6		51.6			
	(1933) F_0	44.1		40.1	42.7		48.9		45.5	45.7		44.7	45.6		45.5		38.4	40.6			482
	F_1	46.0		42.2	46.0		45.3		46.5	45.5		45.5	49.4		46.8		36.1	41.8			
Crude fibre	(1932) F_0				20.4			22.1		22.3	20.4		19.1			19.0		20.2			477
	F_1				19.8			21.7		24.2	20.9		18.9			19.5		18.7			
	(1933) F_0	21.6		22.1	21.3		19.0		20.9	20.3		21.8	20.5		21.6		22.2	20.9			482
	F_1	20.8		22.5	21.2		20.2		21.8	20.9		21.6	20.3		20.2		22.2	19.2			
Ash	(1932) F_0				6.5			7.0		6.1	7.2		9.3			12.2		9.0			477
	F_1				5.6			6.7		5.6	6.7		9.1			10.5		9.1			
	(1933) F_0	10.1		10.5	9.1		8.6		9.1	9.6		10.4	11.2		11.7		15.2	14.8			482
	F_1	9.2		8.9	6.6		9.0		7.9	9.3		9.5	9.0		10.7		13.5	13.1			
Major elements %DM																					
Calcium	(1932) F_0				0.7			0.9		1.1	1.6		1.4								477
	F_1				0.9			1.1		1.1	1.4		1.6								
	(1933) F_0	0.9		0.9	1.0		0.9		0.9	0.8		0.9	1.0		0.9		1.2	0.9			482
	F_1	0.8		0.8	1.0		0.8		0.9	0.9		1.0	0.9		0.9		1.1	0.9			
Phosphorus	(1932) F_0				0.5			0.5		0.4	0.5		0.6			0.6		0.6			477
	F_1				0.4			0.4		0.4	0.5		0.5			0.5		0.7			
	(1933) F_0	0.6		0.7	0.6		0.6		0.5	0.6		0.5	0.5		0.5		0.6	0.5			482
	F_1	0.5		0.6	0.5		0.5		0.4	0.5		0.5	0.5		0.5		0.5	0.5			

Table 1.122 Pasture grass and clover. Effect of fertilisers

			January	February	March		April			May			June		
Proximate constituents %DM															
Crude protein	(1929)	A							26.9	26.3		22.9	18.0	20.3	21.3
		B							24.6	24.4		22.7	18.6	20.0	21.4
		C							25.9	22.6		22.2	17.3	21.2	21.9
		D								23.6			17.8		
	(1930)	A			25.1			22.9			18.7			19.7	
		B				24.6			22.4			17.3		20.8	
		C					24.1		21.5			19.0			20.0
		D					22.9			21.1			19.2		
Ether extract	(1929)	A							7.3	6.2		5.6	4.4	4.0	5.6
		B							6.9	7.2		5.3	4.5	4.2	4.2
		C							7.4	6.1		5.3	4.4	5.2	4.3
		D								5.7			3.9		
	(1930)	A			5.2			5.1			5.2			5.5	
		B				5.4			4.9			4.7		5.9	
		C					5.2		4.5			4.2			5.4
		D					5.3			5.0			5.0		
Nitrogen-free extract	(1929)	A							41.2	41.6		44.3	46.8	45.7	47.1
		B							43.5	43.7		45.4	45.3	45.4	49.0
		C							41.5	46.2		46.2	47.0	45.9	48.8
		D								46.3			47.9		
	(1930)	A			42.2			40.7			41.4			43.3	
		B				44.6			42.4			44.1		43.3	
		C					44.7		42.8			41.9			43.8
		D					44.5			43.4			38.7		
Crude fibre	(1929)	A							12.9	14.7		17.0	20.5	19.3	15.3
		B							12.8	14.0		16.6	21.9	19.4	15.2
		C							13.2	14.3		16.0	21.6	16.8	15.4
		D								14.1			21.1		
	(1930)	A			17.6			19.8			23.2			20.8	
		B				16.0			18.6			22.5		19.7	
		C					16.3		19.6			23.7			20.3
		D					17.0			20.1			24.7		
Ash	(1929)	A							11.7	11.2		10.2	10.3	10.7	10.7
		B							12.2	10.7		10.0	9.7	11.0	10.2
		C							12.0	10.8		10.3	9.7	10.9	9.6
		D								10.3			9.3		
	(1930)	A			9.9			11.5			11.5			10.7	
		B				9.4			11.7			11.4		10.3	
		C					9.7		11.6			11.2			10.5
		D					10.3			10.4			12.4		
Major elements %DM															
Calcium	(1929)	A							1.3	1.1		1.1	1.1	1.3	1.3
		B							1.5	1.2		1.2	1.2	1.5	1.5
		C							1.4	1.1		1.3	1.1	1.5	1.4
		D								1.0			1.1		
	(1930)	A			0.9			0.8			0.8			0.9	
		B				1.1			1.0			0.9		1.1	
		C					1.0		0.9			0.9			1.1
		D					0.9			0.8			0.9		
Phosphorus	(1929)	A							0.45	0.42		0.41	0.34	0.36	0.32
		B							0.40	0.38		0.40	0.32	0.35	0.33
		C							0.42	0.39		0.40	0.34	0.36	0.34
		D								0.37			0.33		
	(1930)	A			0.41			0.48			0.45			0.41	
		B				0.39			0.45			0.49		0.43	
		C					0.42		0.45			0.46			0.43
		D					0.41			0.48			0.46		
Potassium	(1929)	A							2.7	2.8		3.1	2.7	2.7	2.4
		B							2.3	2.5		2.5	2.6	2.4	2.1
		C							2.5	2.7		2.8	2.9	2.7	2.2
		D								2.4			2.5		
	(1930)	A			3.0			3.3			3.2			2.9	
		B				2.6			3.1			3.1		2.9	
		C					2.8		3.1			3.2			2.7
		D					2.7			3.4			3.0		
Sodium	(1929)	A							0.20	0.32		0.23	0.18	0.30	0.38
		B							0.17	0.31		0.24	0.23	0.30	0.36
		C							0.20	0.29		0.27	0.13	0.28	0.30
		D								0.38			0.27		
	(1930)	A			0.18			0.21			0.25			0.39	
		B				0.16			0.19			0.15		0.36	
		C					0.16		0.20			0.15			0.32
		D					0.14			0.17			0.17		
Chlorine	(1929)	A							0.4	0.4		0.5	0.5	0.5	0.5
		B							0.4	0.4		0.4	0.4	0.5	0.5
		C							0.4	0.4		0.5	0.4	0.4	0.4
		D								0.4			0.4		
	(1930)	A			0.4			0.5			0.5			0.4	
		B				0.4			0.4			0.5		0.4	
		C					0.4		0.5			0.5			0.5
		D					0.4			0.5			0.6		

July			August			September			October			November			December			Notes	Reference
18.2	22.0	21.6		23.4	21.1		21.8	20.4	22.3	28.2	27.8							Pasture grass and clover: Grasses include: ryegrass, cocksfoot, timothy, red fescue, and creeping bent. Manurial treatment, in kg/ha: Sub-plots A: 625 kg ground lime + 625 kg superphosphate + 250 kg potassium sulphate + ammonium sulphate. Sub-plots B: 625 kg ground lime. Sub-plots C: 625 kg ground lime + 375 kg superphosphate + 250 kg potassium sulphate. Sub-plots D: No treatment.	587
19.1	21.3	21.1		22.6	20.4	20.6	20.2	21.1	19.4	22.5	23.5								
18.1	22.3	20.7		21.6	18.2	21.2	18.8	16.6	19.6	22.2	23.3								
17.6		19.4			17.8			18.6			21.3								
20.0			19.8			21.1			23.7										
	20.6			21.7		22.3			23.6										
		20.3		21.9			21.8			22.8									
19.7		20.4			19.2			19.6			21.8								
5.0	4.8	5.6		6.4	5.8		6.1	4.3	4.6	6.1	6.0								
5.3	4.1	4.9		6.0	5.7	6.4	5.2	4.2	4.8	5.8	6.1								
4.2	5.0	4.9		6.0	5.0	5.9	5.0	5.1	4.4	5.3	6.5								
5.9		6.0			5.7			4.3			4.8								
5.3			5.6			5.5			6.1										
	5.4			5.6		5.6			5.5										
		5.0		5.2			5.7			5.3									
5.4		5.3			5.8			5.1			6.0								
51.5	46.1	45.6		44.1	47.2		45.3	44.7	44.8	38.9	38.8								
49.6	47.5	47.4		44.9	47.0	47.0	47.0	43.0	48.6	45.1	40.6								
51.1	45.9	46.5		46.2	49.3	46.6	45.9	47.5	48.6	45.0	40.2								
51.9		47.1			46.7			41.9			42.8								
44.8			43.8			43.5			42.2										
	44.5			43.2		43.9			43.6										
		45.7		43.0			43.4			44.1									
43.6		44.9			45.0			44.9			44.0								
15.2	16.9	16.9		15.8	15.5		16.5	16.8	15.3	15.5	13.2								
15.6	17.0	16.4		15.7	15.4	16.2	17.0	16.6	16.2	15.8	14.6								
15.9	16.1	17.0		15.5	16.0	16.3	16.6	17.4	14.9	15.1	14.8								
15.6		16.6			15.7			16.0			14.0								
19.9			19.9			19.3			17.1										
	19.9			19.2		18.3			16.5										
		19.3		19.0			18.9			16.8									
20.7		19.8			19.3			20.2			16.6								
10.1	10.2	10.3		10.3	10.4		10.3	13.8	13.0	11.3	14.2								
10.4	10.1	10.2		10.8	11.5	9.8	10.6	15.1	11.0	10.8	15.2								
10.7	10.7	10.9		10.7	11.5	10.0	13.7	13.4	12.5	12.4	15.2								
9.0		10.9			14.1			19.2			17.1								
10.0			10.9			10.6			10.9										
	9.6			10.3		9.9			10.8										
		9.7		10.9			10.2			11.0									
10.6		9.6			10.7			10.2			11.6								
1.1	1.3	1.2		1.2	1.0		1.0	1.2	1.1	1.0	1.1								
1.3	1.3	1.5		1.5	1.2	1.2	1.2	1.4	1.2	1.1	1.2								
1.2	1.4	1.3		1.4	1.1	1.2	1.1	1.2	1.3	1.2	1.2								
1.1		1.1			1.0			1.2			1.0								
0.9			0.9			0.9			0.9										
	1.1			1.1		1.1			1.1										
		1.0		1.1			1.0			1.0									
1.0		0.9			1.1			0.9			1.0								
0.29	0.30	0.29		0.31	0.26		0.25	0.22	0.29	0.36	0.40								
0.31	0.31	0.27		0.32	0.26	0.26	0.24	0.23	0.29	0.34	0.36								
0.31	0.32	0.30		0.30	0.25	0.27	0.24	0.24	0.28	0.35	0.38								
0.28		0.27			0.24			0.19			0.34								
0.38			0.47			0.51			0.46										
	0.38			0.47		0.50			0.48										
		0.40		0.47			0.52			0.51									
0.44		0.41			0.46			0.53			0.50								
2.0	2.3	2.0		2.6	2.3		2.2	1.9	1.8	2.3	2.6								
1.9	2.0	1.7		2.3	2.0	2.0	1.9	1.5	1.6	1.7	1.9								
2.0	2.2	1.8		2.3	2.0	2.2	1.9	1.6	1.6	1.8	2.2								
1.8		1.6			1.8			1.5			1.8								
2.3			2.9			2.4			2.5										
	2.2			2.8		2.4			2.5										
		2.2		2.7			2.5			2.7									
3.0		2.3			2.7			2.4			2.4								
0.30	0.32	0.32		0.25	0.32		0.29	0.22	0.19	0.20	0.12								
0.36	0.33	0.32		0.25	0.28	0.33	0.27	0.24	0.19	0.20	0.13								
0.27	0.36	0.34		0.31	0.32	0.31	0.20	0.21	0.17	0.12	0.15								
0.33		0.26			0.24			0.21			0.12								
0.42			0.23			0.33			0.18										
	0.43			0.21		0.34			0.18										
		0.41		0.28			0.25			0.23									
0.38		0.30			0.25			0.23			0.10								
0.5	0.5	0.5		0.5	0.5		0.5	0.5	0.4	0.4	0.4								
0.5	0 4	0.4		0.5	0.5	0.5	0.5	0.5	0.4	0.4	0.4								
0.4	0.4	0.4		0.5	0.5	0.5	0.5	0.5	0.4	0.4	0.4								
0.4		0.4			0.5			0.4			0.4								
0.5			0.5			0.5			0.4										
	0.4			0.5		0.5			0.5										
		0.4		0.5			0.5			0.5									
0.5		0.4			0.5			0.5			0.5								

Table 1.123 Grass, unnamed, with or without clover

Crop studies. Cut for silage	Year	Treatment	May	June	August	September	Notes	Reference
Proximate constituents %DM								
Crude protein	(1953)	TO		12.9			Grass and clover Herbage cut at long leafy stage in 1953 and 1954, cut earlier at young leafy stage (exact date not given) in 1955 TO: no treatment TW: wilted for 17½, 20 or 22½ hr in 1953: mean of 6 samples MO: mown, no treatment: 1 sample MW: mown, wilted for 8 or 27 hr in 1954: 2 samples BO: bruised, no other treatment: 1 sample BW: bruised, wilted for 4 or 8 hr in 1954: 2 samples AO: no additive: 1 sample AM: 11.4 kg molasses added /1000 kg herbage: 1 sample AA: 5.5 kg MBS (sodium metabisulphite) added/ 1000 kg herbage: 1 sample AOW: wilted 18½ hr, no additive: 1 sample AMW: wilted 18½ hr, molasses added: 1 sample AAW: wilted 18½ hr, MBS added: 1 sample *Ether extract was assumed to be 2 per cent of DM and nitrogen-free extract was calculated by difference with that assumption For the composition of the silages see table 1.138, same reference	383
		TW		12.6				
	(1954)	MO		12.3				
		MW		10.7				
		BO		11.9				
		BW		11.9				
	(1955)	AO	15.1					
		AM	14.7					
		AA	14.9					
		AOW	11.8					
		AMW	13.0					
		AAW	11.8					
Nitrogen-free extract*	(1953)	TO		48.9				
		TW		48.6				
	(1954)	MO		48.4				
		MW		50.5				
		BO		48.5				
		BW		47.8				
	(1955)	AO	54.1					
		AM	55.7					
		AA	54.0					
		AOW	56.1					
		AMW	54.3					
		AAW	56.3					
Crude fibre	(1953)	TO		27.7				
		TW		28.6				
	(1954)	MO		29.5				
		MW		29.5				
		BO		28.1				
		BW		29.0				
	(1955)	AO	19.7					
		AM	18.9					
		AA	20.0					
		AOW	22.2					
		AMW	22.0					
		AAW	21.6					
Ash	(1953)	TO		8.5				
		TW		8.2				
	(1954)	MO		7.8				
		MW		7.3				
		BO		9.5				
		BW		9.3				
	(1955)	AO	9.1					
		AM	8.7					
		AA	9.1					
		AOW	7.9					
		AMW	8.7					
		AAW	8.3					

Mixed grasses	Year	May	May	June	June	July	August	Notes	Reference
Proximate constituents %DM									
Crude protein	(1934)	19.2	16.7	14.1	13.0			Grasses mixed. Unnamed	578
	(1935)		19.0	19.5					
Ether extract	(1934)	7.9	7.0	6.1	5.4				
	(1935)		3.9	4.6					
Nitrogen-free extract	(1934)	44.9	48.2	50.1	50.9				
	(1935)		44.7	44.9					
Crude fibre	(1934)	19.0	19.4	21.2	21.6				
	(1935)		20.3	21.1					
Ash	(1934)	9.0	8.7	8.5	9.1				
	(1935)		12.1	9.9					

continued

Table 1.123 Grass, unnamed, with or without clover (continued)

Aftermath. Effect of fertilisers	September	October	November	December	Notes	Reference
Proximate constituents %DM Crude protein		(1929) F_0 9.2 (1930) F_0 11.4 F_1 12.0 F_2 11.4			Grass and clover, aftermath 1929, cut 2/10. No extra fertiliser 1930, cut 3/10 F_0 : No extra fertiliser F_1 : Ground chalk 5000 kg/ha, superphosphate 625 kg/ha potassium sulphate + 250 kg/ha F_2 : Ground Chalk 5000 kg/ha, superphosphate 625 kg/ha + potassium sulphate 250 kg/ha + ammonium sulphate 250 kg/ha	585
Ether extract		(1929) F_0 6.1 (1930) F_0 3.7 F_1 3.6 F_2 3.6				
Nitrogen-free extract		(1929) F_0 51.8 (1930) F_0 48.4 F_1 46.8 F_2 46.6				
Crude fibre		(1929) F_0 21.3 (1930) F_0 25.6 F_1 26.7 F_2 27.7				
Ash		(1929) F_0 11.6 (1930) F_0 10.9 F_1 10.9 F_2 10.7				
Major elements %DM Calcium		(1929) F_0 1.5 (1930) F_0 1.0 F_1 1.1 F_2 1.0				
Phosphorus		(1929) F_0 0.3 (1930) F_0 0.4 F_1 0.4 F_2 0.4				

Ley. Effect of fertilisers	July	August	September	October	Notes	Reference
Proximate constituents %DM Crude protein	Area 1 F_0 16.9 F_1 18.6 F_2 18.4 F_3 17.8 Area 2 F_0 17.9 F_1 18.3 F_2 17.7 F_3 19.1				Ley grass and clover. Mean values of monthly cuts April to October inclusive F_0 : No extra fertiliser F_1 : Nitrochalk 812 kg/ha as 8 monthly dressings. F_2 : superphosphate 500 kg/ha in March. F_3 : Kainit 500 kg/ha in February.	219
Major elements %DM Calcium	Area 1 F_0 0.9 F_1 0.8 F_2 1.0 F_3 0.9 Area 2 F_0 1.0 F_1 1.1 F_2 1.2 F_3 1.1					
Phosphorus	Area 1 F_0 0.3 F_1 0.4 F_2 0.4 F_3 0.5 Area 2 F_0 0.4 F_1 0.4 F_2 0.5 F_3 0.4					
Potassium	Area 1 F_0 3.4 F_1 3.6 F_2 3.7 F_3 3.7 Area 2 F_0 1.8 F_1 1.5 F_2 1.4 F_3 2.1					
Chlorine	Area 1 F_0 0.6 F_1 0.5 F_2 0.5 F_3 0.6 Area 2 F_0 0.8 F_1 0.8 F_2 0.7 F_3 1.3					

continued

Table 1.123 Grass, unnamed, with or without clover (continued)

Pasture. Effect of fertilisers	August		September	October	November	Notes	Reference
Major elements %DM						Pasture grass, unnamed Cut 6/8. Sampled 5/10. F_0 : No fertiliser F_1 : High-protein concentrates* 330 kg/ha annually F_2 : Basic slag 625 kg/ha every 3rd year. F_3 : Superphosphate 937 kg/ha every 3rd year. F_4 : Basic slag as F_2 + muriate of potash 110 kg/ha every 3rd year. F_5 : Basic slag as F_2 + lime 2500 kg/ha every 3rd year. F_6 : Basic slag as F_2 + nitrate of soda 125 kg/ha every 3rd year. *Described as "fed to sheep", Here applied as fertiliser.	
Calcium	F_0	0.38					478
	F_1	0.50					
	F_2	0.58					
	F_3	0.56					
	F_4	0.59					
	F_5	0.79					
	F_6	0.59					
Phosphorus	F_0	0.22					
	F_1	0.31					
	F_2	0.48					
	F_3	0.49					
	F_4	0.46					
	F_5	0.47					
	F_6	0.50					
Magnesium	F_0	0.13					
	F_1	0.23					
	F_2	0.22					
	F_3	0.21					
	F_4	0.21					
	F_5	0.25					
	F_6	0.25					
Potassium	F_0	1.69					
	F_1	2.95					
	F_2	3.28					
	F_3	2.73					
	F_4	3.00					
	F_5	2.79					
	F_6	2.93					
Sodium	F_0	0.05					
	F_1	0.11					
	F_2	0.16					
	F_3	0.15					
	F_4	0.11					
	F_5	0.31					
	F_6	0.15					
Chlorine	F_0	0.27					
	F_1	0.59					
	F_2	0.58					
	F_3	0.57					
	F_4	0.60					
	F_5	0.52					
	F_6	0.52					
Total sulphur	F_0	0.36					
	F_1	0.53					
	F_2	0.52					
	F_3	0.44					
	F_4	0.42					
	F_5	0.57					
	F_6	0.48					
Inorganic sulphur	F_0	0.24					
	F_1	0.30					
	F_2	0.32					
	F_3	0.31					
	F_4	0.28					
	F_5	0.39					
	F_6	0.31					
Trace elements µg/g DM							
Cobalt	F_0	0.088					
	F_1	0.081					
	F_2	0.071					
	F_3	0.079					
	F_4	0.062					
	F_5	0.204					
	F_6	0.174					
Copper	F_0	12.2					
	F_1	6.48					
	F_2	5.58					
	F_3	5.33					
	F_4	6.50					
	F_5	6.00					
	F_6	5.90					
Iron	F_0	346					
	F_1	222					
	F_2	150					
	F_3	133					
	F_4	120					
	F_5	187					
	F_6	136					
Manganese	F_0	595				The reduction of Mn content by F_5 treatment is mentioned in the text.	
	F_1	573					
	F_2	105					
	F_3	507					
	F_4	279					
	F_5	38					
	F_6	287					

Table 1.123 Grass, unnamed, with or withour clover (continued)

Meadow grass. Effect of fertilisers	Year		June Air-dried	June Oven-dried	July	August Air-dried	September	Notes	Reference
Vitamins μg/g DM									
Thiamin	(1946)	F_0				1.97		Meadow grass Cut 3/8/1946 and 27/6/1947. Fertiliser treatment/ha as follows: F_0 : None F_1 : 20 tonnes dung, 41 kg ammonium sulphate, 55 kg sodium nitrate, 165 kg basic slag. 55 kg muriate of potash F_2 : 20 tonnes dung F_3 : 165 kg sulphate of ammonia F_4 : 330 kg basic slag F_5 : 110 kg muriate of potash F_6 : 165 kg ammonium sulphate + 330 kg basic slag F_7 : 165 kg ammonium sulphate + 110 kg muriate of potash F_8 : 330 kg basic slag + 110 kg muriate of potash F_9 : 165 kg ammonium sulphate + 330 kg basic slag + 110 kg muriate of potash F_{10}: 82.5 kg ammonium sulphate + 110 kg sodium nitrate + 330 kg basic slag + 110 kg muriate of potash	493
		F_1				1.52			
		F_2				1.53			
		F_3				1.65			
		F_4				2.34			
		F_5				2.46			
		F_6				2.03			
		F_7				2.45			
		F_8				1.88			
		F_9				2.77			
		F_{10}				1.91			
	(1947)	F_0	3.83	1.70					
		F_1	3.30	2.45					
		F_2	3.39	1.82					
		F_3	3.78	2.50					
		F_4	3.75	1.54					
		F_5	3.75	1.65					
		F_6	3.38	2.09					
		F_7	2.55	1.64					
		F_8	2.51	2.30					
		F_9	3.44	1.80					
		F_{10}	2.79	2.23					
Riboflavin	(1946)	F_0				7.8			
		F_1				6.4			
		F_2				8.3			
		F_3				7.6			
		F_4				10.2			
		F_5				8.4			
		F_6				7.1			
		F_7				9.3			
		F_8				9.7			
		F_9				10.0			
		F_{10}				9.9			
	(1947)	F_0	11.6						
		F_1	8.9						
		F_2	11.1						
		F_3	13.4						
		F_4	12.1						
		F_5	11.3						
		F_6	12.9						
		F_7	11.8						
		F_8	12.8						
		F_9	12.1						
		F_{10}	13.2						
Pantothenic acid	(1946)	F_0				8.6			
		F_1				7.8			
		F_2				8.6			
		F_3				11.6			
		F_4				9.1			
		F_5				7.1			
		F_6				8.8			
		F_7				8.4			
		F_8				9.2			
		F_9				8.2			
		F_{10}				7.3			
	(1946)	F_0	15.4	8.6					
		F_1	14.6	7.5					
		F_2	14.0	7.5					
		F_3	14.1	9.3					
		F_4	17.7	7.7					
		F_5	14.8	8.1					
		F_6	17.7	10.8					
		F_7	16.6	8.1					
		F_8	14.3	8.8					
		F_9	13.8	8.9					
		F_{10}	17.9	9.9					
Nicotinic acid	(1946)	F_0				33.9			
		F_1				22.0			
		F_2				27.7			
		F_3				27.7			
		F_4				27.3			
		F_5				27.9			
		F_6				25.5			
		F_7				36.1			
		F_8				34.3			
		F_9				38.1			
		F_{10}				25.9			
	(1947)	F_0	43.5						
		F_1	30.1						
		F_2	33.2						
		F_3	31.6						
		F_4	36.6						
		F_5	32.8						
		F_6	44.7						
		F_7	29.7						
		F_8	33.1						
		F_9	32.6						
		F_{10}	34.9						

continued

Table 1.123 Grass, unnamed, with or without clover (continued)

Meadow grass. Effect of fertilisers (continued)	Year	Treatment	June Air-dried	June Oven-dried	July	August Air-dried	September	Notes	Reference
Vitamins (cont'd) μg/g DM									
Pyridoxine	(1946)	F_0				8.8			493
		F_1				7.9			
		F_2				7.6			
		F_3				8.7			
		F_4				8.7			
		F_5				8.9			
		F_6				9.0			
		F_7				9.8			
		F_8				9.0			
		F_9				9.3			
		F_{10}				8.8			
	(1947)	F_0	11.0						
		F_1	10.3						
		F_2	11.3						
		F_3	11.5						
		F_4	10.1						
		F_5	11.4						
		F_6	12.4						
		F_7	11.1						
		F_8	13.5						
		F_9	12.8						
		F_{10}	12.3						
Biotin	(1946)	F_0				0.179			
		F_1				0.151			
		F_2				0.150			
		F_3				0.190			
		F_4				0.222			
		F_5				0.174			
		F_6				0.204			
		F_7				0.207			
		F_8				0.156			
		F_9				0.146			
		F_{10}				0.150			
	(1947)	F_0	0.179						
		F_1	0.195						
		F_2	0.202						
		F_3	0.166						
		F_4	0.183						
		F_5	0.192						
		F_6	0.219						
		F_7	0.143						
		F_8	0.196						
		F_9	0.194						
		F_{10}	0.224						
Tryptophan %DM	(1946)	F_0				0.161		Tryptophan, as precursor of nicotinic acid is included here.	
		F_1				0.117			
		F_2				0.134			
		F_3				0.112			
		F_4				0.112			
		F_5				0.103			
		F_6				0.101			
		F_7				0.123			
		F_8				0.122			
		F_9				0.114			
		F_{10}				0.102			

Table 1.124 Grass or grass and clover, artificially dried

	May			June			July			August			Notes	Reference
Cocksfoot, unnamed														
Proximate constituents %DM														
Crude protein							13.5						Grass cut at advanced leaf or early flowering stage, height 25–30 cm. Commercial samples.	12
Amino acids % of crude protein														
Cystine							0.37							
Lysine							4.50							
Methionine: DL form							1.59							
L form							1.42							
Tryptophan							1.60							
Tyrosine							3.95							
Italian Ryegrass, unnamed														
Proximate constituents %DM														
Crude protein							12.8							
Amino acids % of crude protein														
Cystine							0.39							
Lysine							3.90							
Methionine: DL form							1.65							
L form							1.75							
Tryptophan							1.59							
Tyrosine							3.25							
Ryegrass and Red Clover, unnamed														
Proximate constituents %DM														
Crude protein				13.1	12.8	18.6			16.9	15.4	17.7		Not wilted	210
				13.2	13.2	17.5			17.1	15.2	16.8		Wilted for 12–72 h	
Vitamins µg/g DM														
Carotene				192	180	231			188	233	213		Not wilted	
				190	173	224			150	130	195		Wilted for 12–72 h	
Grass, unnamed														
Proximate constituents %DM														
Crude protein				14.0	14.0	13.3		17.2			18.1	15.4	Not wilted	209
				13.9	13.5	12.6		17.0			17.6	14.9	Wilted for 12 h	
Vitamins µg/g DM														
Carotene				280	300	180		250			260	210	Not wilted	
				250	260	180		210			230	200	Wilted for 12 h	
Grass, unnamed														
Proximate constituents %DM														
Crude protein								13.9 (12.7–16.0)					Mean and range for grass dried at 19 Ministry of Agriculture Grass Drying Centres. Cut between May and October.	334
								14.5 (8.0 – 21.8)					Means and ranges for 26 samples from commercial grass-drying plants. Correlation between CP and riboflavin: r= 0.811 (P <0.001)	101
Vitamins µg/g DM														
Riboflavin								15.3 (5.0 –24.4)						
Carotene								611					4 samples, as packed.	332
								441 (375 – 510)					Mean and range of 36 samples stored under different conditions	
Digestibility of DM%								76					Mean for 4 sheep	149
Proximate constituents %DM														
Crude protein								12						
Vitamins µg/g DM														
Carotene								193						
Caroteniods, total								290						

Table 1.125 Cocksfoot and Lucerne: hay

	June		Notes	Reference
Proximate constituents %DM				
Crude protein	14.5		Lucerne predominant: mean values (no date)	35
	12.7		Cocksfoot predominant: mean values (no date)	36
	13.7		Lucerne predominant: mean values (no date)	37
	14.5		Lucerne predominant: mean values (no date)	33
	15.1		Sown under barley 1942: cut 19/6/1943	154
		15.6 15.4	Lucerne predominant: cut 22/June, baled 28/June, swath cured. Lucerne predominant: cut 22/June, baled 11/July, tripod cured.	378
Ether extract	1.3			35
	1.6			36
	1.6			37
	1.7			33
		1.8 1.8		378
Nitrogen-free extract	40.7			35
	41.8			36
	39.2			37
	39.1			33
		42.0 40.5		378
Crude fibre	34.3			35
	36.3			36
	36.1			37
	35.6			33
	28.4			154
		32.6 34.5		378
Ash	9.2			35
	7.6			36
	9.4			37
	9.1			33
	7.0			154
		8.0 7.8		378
Carbohydrate fractions %DM				
Cellulose	37.1			37
	35.7			33
Cellulosans	8.8			37
	8.7			33
Hexosan cellulose	27.0			33
Pentosans not in cellulose	9.0			37
	9.4			33
Lignin	7.6			33
Vitamins iu/g DM				
Vitamin D	1 0.70 (0.58 – 0.84) 2 0.42 (0.35 – 0.51)		Mean values and true Fiducial Limits at P = 0.95 for groups of 8 rats. 1. Swath cured 2. tripod cured	277

Table 1.126 Ryegrass and Ryegrass mixtures: hay

Ryegrass	June			Notes	Reference
Proximate constituents %DM Crude protein	7.3			Mature when cut	34
	7.5 (5.4 – 10.4) 7.5 (5.8 – 9.8)			Cut early June: Tripod cured Ground cured	106
Ether extract	1.8				37
	2.3 (1.7 – 3.0) 2.4 (1.6 – 3.4)			Tripod cured Ground cured	106
Nitrogen-free extract	54.9				34
	50.2 (43.1 – 56.2) 49.5 (44.8 – 55.7)			Tripod cured Ground cured	106
Crude fibre	29.9				34
	32.9 (30.3 – 36.4) 33.4 (29.2 – 36.7)			Tripod cured Ground cured	106
Ash	6.1				34
	7.1 (5.7 – 9.2) 7.2 (6.7 – 8.9)			Tripod cured Ground cured	106
Carbohydrate fractions %DM Cellulose	46.0				34
Ryegrass and Clover *Digestibility of* DM%	a b c	68.8 67.3 65.0		Hay cut 16 June. Each value mean for 4 cows with rumen fistulae. Hay ration: a 4.5 kg; b 6.8 kg; c *ad lib*	115
Digestibility of OM%	a b c	70.4 69.4 67.4			
Proximate constituents %DM Crude protein	10.6			Hay cut 6 June	210
	11.6 (7.9 – 15.9) 11.1 (7.5 – 16.2)			Cut early June: 6 1 acre swards Tripod dried Ground dried For metabolisable energy see table 1.229, same reference	106
		8.7		As for Digestibility	115
Ether extract	2.6 (2.2 – 3.4) 2.6 (2.1 3.6)			As for Crude protein	106
		1.8		As for Digestibility	115
Nitrogen-free extract	45.6 (33.5 – 51.2) 48.0 (39.7 – 53.3)			As for Crude protein	106
		52.1		As for Digestibility	115
Crude fibre	32.1 (28.2 – 37.6) 31.0 (27.6 – 33.4)			As for Crude protein	106
		31.1		As for Digestibility	115
Ash	8.1 (7.4 – 9.6) 7.3 (6.7 – 8.9)			As for Crude protein	106
		6.3		As for Digestibility	115
Vitamins μg/g DM Carotene	49			As for Crude protein	210

continued

Table 1.126 Ryegrass and Ryegrass mixtures: hay (continued)

Ryegrass, Cocksfoot, Timothy and Clover		July			Notes	Reference
Digestibility of OM %		54.6			Ryegrass predominant. Hay cut 9/July, baled 2/Aug. Digestibility by sheep and composition of hay as baled. For parallel plots with increasing amounts of herbs see next entry, same reference	487
Proximate constituents %DM						
Crude protein		6.3				
Digestibility %		33.3				
Ether extract		1.2				
Nitrogen-free extract		49.2				
Crude fibre		37.6				
Digestibility %		57.5				
Ash		5.7				
Carbohydrate fractions %DM						
Cellulose		48.2				
Digestibility %		63.9				
Total sugars		3.0				
Digestibility %		98.2				
Lignin		9.1				
Digestibility %		6.7				
Major elements %DM						
Calcium		0.39				
Phosphorus		0.19				
Magnesium		0.14				
Sodium		0.22				
Potassium		1.82				
Chloride		0.58				
Trace elements μg/g DM						
Cobalt		0.11				
Copper		4.9				
Iron		204				
Manganese		261				
Ryegrass, Cocksfoot, Timothy, Clover and Herbs						
Digestibility of OM %	M_1	53.4			M_1 : 50:50 herbs:grass and clover.	
	M_2	53.3			M_2 : 35:65 herbs:grass and clover.	
	M_3	51.0			M_3 : 20:80 herbs:grass and clover	
Proximate constituents %DM					Sown 5/1950; cut 9/7/1950; baled 2/8/1950. Herbs: chicory, plantains, yarrow, sainfoin, burnet. For basic plot without herbs see preceding entry, same reference	
Crude protein	M_1	6.3				
	M_2	6.4				
	M_3	6.4				
Digestibility %	M_1	25.7				
	M_2	29.8				
	M_3	34.1				
Ether extract	M_1	1.3				
	M_2	1.2				
	M_3	1.3				
Nitrogen-free extract	M_1	50.0				
	M_2	49.6				
	M_3	48.2				
Crude fibre	M_1	36.3				
	M_2	36.4				
	M_3	37.8				
Digestibility %	M_1	55.0				
	M_2	53.8				
	M_3	51.3				
Ash	M_1	6.1				
	M_2	6.4				
	M_3	6.3				

Table 1.126 Ryegrass and Ryegrass mixtures: hay (continued)

Ryegrass, Cocksfoot, Timothy, Clover and Herbs (cont'd)		July			Notes	Reference
Carbohydrate fractions %DM						
Cellulose	M_1 M_2 M_3	46.8 48.1 48.2				487
Digestibility %	M_1 M_2 M_3	63.0 62.4 59.1				
Total sugars	M_1 M_2 M_3	3.6 3.3 1.3				
Digestibility %	M_1 M_2 M_3	99.9 100 95.1				
Lignin	M_1 M_2 M_3	9.2 8.9 9.2				
Digestibility %	M_1 M_2 M_3	4.0 0.05 2.1				
Major elements %DM						
Calcium	M_1 M_2 M_3	0.47 0.50 0.43				
Phosphorus	M_1 M_2 M_3	0.22 0.21 0.20				
Magnesium	M_1 M_2 M_3	0.17 0.16 0.18				
Sodium	M_1 M_2 M_3	0.30 0.26 0.17				
Potassium	M_1 M_2 M_3	1.91 2.14 1.85				
Chloride	M_1 M_2 M_3	0.59 0.63 0.54				
Trace elements µg/g DM						
Cobalt	M_1 M_2 M_3	0.05 0.06 0.06				
Copper	M_1 M_2 M_3	5.5 6.8 3.8				
Iron	M_1 M_2 M_3	226 182 180				
Manganese	M_1 M_2 M_3	269 244 238				

Table 1.127 Fescue mixtures: hay

Fine-leaved Fescue and bent			Notes	Reference
Proximate constituents %DM				
Crude protein	F_1	15.6	Fescue, 60-65%, bent 20-30%. Samples from open hill plots given fertilisers since 1930. F_1: Superphosphate 500 kg/ha F_2: Superphosphate 500, Kainit 500 or muriate of potash 154 and ammonium sulphate 125 kg/ha F_3: Limestone 5 tonnes/ha F_4: as F_2 and F_3 together No date specified	151
	F_2	17.0		
	F_3	15.8		
	F_4	15.3		
Crude fibre	F_1	25.8		
	F_2	24.2		
	F_3	25.6		
	F_4	28.0		
Major elements %DM				
Calcium	F_1	0.41		
	F_2	0.39		
	F_3	0.60		
	F_4	0.66		
Phosphorus	F_1	0.33		
	F_2	0.35		
	F_3	0.26		
	F_4	0.36		
Fine-leaved Fescue and *Molinia*			Open hill plots, *Molinia caerulea* association. Fescue 10-20%, *Molinia* 70% Fertilisers as above	
Proximate constituents %DM				
Crude protein	F_1	12.7		
	F_2	12.8		
	F_3	14.0		
	F_4	12.0		
Crude fibre	F_1	28.0		
	F_2	31.2		
	F_3	30.0		
	F_4	31.5		
Major elements %DM				
Calcium	F_1	0.40		
	F_2	0.29		
	F_3	0.64		
	F_4	0.63		
Phosphorus	F_1	0.33		
	F_2	0.34		
	F_3	0.26		
	F_4	0.34		

Table 1.128 Seeds grass: hay

		Range	Notes	Reference
Proximate constituents %DM				
Crude protein	a)	9.4 – 15.6	Ranges only for 33 samples of seeds hay from 16 farms, taken from the swath(a) and winter rick (b)	344
	b)	8.9 – 16.6		
Crude fibre	a)	25.3 – 35.0		
	b)	26.7 – 35.5		
Ash	a)	5.9 – 9.3		
	b)	7.0 – 10.3		
Major elements %DM				
Calcium	a)	0.65 – 2.02		
	b)	0.72 – 1.79		
Phosphorus	a)	0.20 – 0.29		
	b)	0.19 – 0.37		
Potassium	a)	0.84 – 2.48		
	b)	1 07 – 2.26		
		June		
Proximate constituents %DM				
Crude protein		13.2	Mean value from main cut hay and aftermath, 3 farms	344
Crude fibre		30.2		
Total ash		8.7		
Major elements %DM				
Calcium		0.94		
Phosphorus		0.25		
Potassium		1.67		
Proximate constituents %DM				
Ash		5.98 (3.46 – 8.50)	Mean and range of 75 samples. No dates or details	20
Major elements %DM				
Calcium		0.65 (0.21 – 1.31)		
Phosphorus		0.22 (0.09 – 0.33)		
Magnesium		0.19 (0.13 – 0.35)	As for ash but 43 samples only.	
Chloride		0.66 (0.39 – 1.11)		
Trace elements μg/g DM				
Cobalt		0.27	Mean for small number of samples. Four (4) of 6 samples from Old Red Sandstone, where pining cured by cobalt, had less than 0.1 μg/g. Highest value 0.35 for a single sample of seeds hay.	
Copper		6.4 (2.9 – 10.8)	As for ash but 55 samples only.	
Iron		115 (5.3 – 194)	As for ash but 47 samples only.	
Manganese		100 (13 – 297)	As for ash but 55 samples only.	

continued

Table 1.128 Seeds grass: hay (continued)

	Range	Notes	Reference
Proximate constituents %DM			
Crude protein	6.1 (4.1 – 9 4)	Mean and range of 62 samples collected over 3 years. Cut late June–early July	458
Ether extract	1.2 (0.7 – 1.6)		
Nitrogen-free extract	56.6 (49.7 – 62.1)		
Crude fibre	30.2 (26.7 – 34.6)		
Ash	5.9 (3.8 – 7.9)		
Major elements %DM			
Calcium	0.45 (0.29 – 0.86)		
Phosphorus	0.21 (0.13 – C.30)		
Chloride	0.42 (0.2 – 0.6)	Mean of 22 samples from 1 year.	

Table 1.129 Meadow grass and meadow grass and clover: hay

Meadow grass		July	Notes	Reference
Proximate constituents				
Crude protein	F_0	8.4	Old hay plots: cut 19th July, sampled 20th July. Means of 5 samples. F_0: no fertilisers F_1: Dung 20,000 kg/ha F_2: Ammonium sulphate 165 kg/ha F_3: Basic slag 330 kg/ha alternating each year with superphosphate F_4: Muriate of potash 110 kg/ha F_5: Ammonium sulphate 165 kg/ha + basic slag 330 kg/ha F_6: Ammonium sulphate 165 kg/ha + muriate of potash 110 kg/ha F_7: Basic slag 330 kg/ha + muriate of potash 110 kg/ha F_8: Ammonium sulphate 165 kg/ha + basic slag 330 kg/ha + muriate of potash 110 kg/ha	197
	F_1	7.9		
	F_2	8.0		
	F_3	7.9		
	F_4	8.1		
	F_5	8.0		
	F_6	7.6		
	F_7	9.5		
	F_8	8.0		
		10.7 (7.6 – 14.1) 10.9 (7.9 – 14.0)	Means and ranges: 70 samples from 60 farms. Sampled at swath stage. 75 samples from 60 farms. Sampled from winter rick.	344
		9.6 (6.1 – 12.9)	Sampled from farm stacks during winter and spring.66 samples.	19
Ether extract	F_0	1.5	As for Crude protein	197
	F_1	1.4		
	F_2	1.7		
	F_3	1.9		
	F_4	1.5		
	F_5	1.7		
	F_6	2.2		
	F_7	2.0		
	F_8	2.2		
Nitrogen-free extract	F_0	52.4		
	F_1	53.7		
	F_2	52.5		
	F_3	52.6		
	F_4	53.1		
	F_5	54.2		
	F_6	54.1		
	F_7	50.6		
	F_8	52.7		
Crude fibre	F_0	32.3		
	F_1	31.8		
	F_2	33.3		
	F_3	31.6		
	F_4	32.1		
	F_5	30.6		
	F_6	32.9		
	F_7	31.1		
	F_8	31.4		
		30.4 (25.6 – 35.2) 30.6 (25.0 – 36.8)	As for Crude protein, same reference.	344
		30.8 (24.1 – 36.9)	As for Crude protein, same reference.	19

continued

Table 1.129 Meadow grass and meadow grass and clover: hay (continued)

Meadow grass (continued)	July	Notes	Reference
Proximate constituents (cont'd)			
Ash	F_0 5.5 F_1 5.2 F_2 4.5 F_3 6.0 F_4 5.2 F_5 5.5 F_6 3.2 F_7 6.8 F_8 5.7	These values for ash have been computed on the assumption that the published values for NFE are correct	197
	7.5 (4.8 – 11.0) 8.8 (6.4 – 11.3)	As for Crude protein, same reference.	344
	6.5 (4.7 – 8.8)	As for Crude protein, same reference.	19
Major elements %DM			
Calcium	F_0 0.28 F_1 0.31 F_2 0.19 F_3 0.46 F_4 0.25 F_5 0.38 F_6 0.10 F_7 0.57 F_8 0.32	As for Crude protein, same reference	197
	0.66 (0.27 – 1.11) 0.76 (0.24 – 1.20)	As for Crude protein, same reference	344
	0.63 (0.34 – 1.40)	As for Crude protein, reference 19	20
Phosphorus	F_0 0.18 F_1 0.22 F_2 0.18 F_3 0.28 F_4 0.17 F_5 0.27 F_6 0.12 F_7 0.30 F_8 0.27	As for Crude protein, same reference	197
	0.22 (0.13 – 0.31) 0.22 (<0.15 – >0.28)	As for Crude protein, same reference	344
	0.22 (0.13 – 0.31)	As for Crude protein, reference 19	20
Magnesium	0.18 (0.07 – 0.28)	As for Crude protein, reference 19, only 33 samples.	20
Potassium	1.42 (0.77 – 2.21) 1.48 (0.77 – 2.31)	As for Crude protein, same reference	344
Chloride	0.68 (0.40 – 1.60)	As for Crude protein, reference 19	20
Trace elements μg/g DM			
Cobalt	0.27 (0.1 – 0.35)	As for Crude protein, reference 19 only 24 samples.	20
Copper	7.4 (4.0 – 13.6)	As for Crude protein, reference 19 only 45 samples.	
Iron	135 (51 – 294)	As for Crude protein, reference 19 only 45 samples.	
Manganese	122 (20 – 506)	As for Crude protein, reference 19 only 44 samples.	
Vitamins μg/g DM			
Carotene	8 (4 – 11)	Mean and range of 5 samples.	148

Table 1.129 Meadow grass and meadow grass and clover: hay (continued)

Meadow grass (continued)	July		Notes	Reference
Vitamins μg/g DM (cont'd)				
Thiamin	2,64 (1.87 – 3.15)		Means and ranges from 11 long-term manurial treatment plots. Cut 2/7/1947	493
Riboflavin	11.9 (7.8 – 14.3)			
Pantothenic acid	13.3 (9.9 – 16.4)			
Nicotinic acid	34.8 (29.7 – 44.7)			
Pyridoxine	11.6 (10.1 – 13.5)			
Biotin	0.190 (0.143 – 0.224)			
Miscellaneous				
Xanthophyll mg/100 g	3.5 (2.6 – 4.2)		Means and ranges of 5 samples.	148
Chlorophyll-a	60 (55 – 65)			
Chlorophyll-b	38 (34 – 42)			
Phaeophytin-a	8 (6 – 12)			
Phaeophytin-b	1.6 (1 – 3)			
Total tetrapyrroles	108 (96 – 119)			
Meadow grass and clover	**June**			
Proximate constituents %DM				
Crude protein (1929)	F_0	11.3	F_0: No manure F_1: Ground chalk 5000 kg/ha + superphosphate 625 kg/ha + sulphate of potash 250 kg/ha F_2: as above + ammonium sulphate 250 kg/ha	585
(1930)	F_0	10.7		
	F_1	11.3		
	F_2	10.3		
Ether extract (1929)	F_0	4.6		
(1930)	F_0	3.1		
	F_1	3.3		
	F_2	3.0		
Nitrogen-free extract (1929)	F_0	51.0		
(1930)	F_0	47.2		
	F_1	47.5		
	F_2	46.1		
Crude fibre (1929)	F_0	24.2		
(1930)	F_0	30.3		
	F_1	28.6		
	F_2	31.8		
Ash (1929)	F_0	8.9		
(1930)	F_0	8.7		
	F_1	9.3		
	F_2	8.8		
Major elements %DM				
Calcium (1929)	F_0	1.13		
(1930)	F_0	0.96		
	F_1	1.08		
	F_2	0.75		
Phosphorus (1929)	F_0	0.28		
(1930)	F_0	0.33		
	F_1	0.34		
	F_2	0.31		

Table 1.130 Grass and grass and clover: hay

	July			Notes	Reference
Proximate constituents %DM					
Crude protein (1)	9.5			(1) As stacked	208
(2)	10.2			(2) As used	
Ether extract (1)	2.3				
(2)	3.0				
Nitrogen-free extract (1)	54.4				
(2)	52.2				
Crude fibre (1)	28.2				
(2)	28.6				
Ash (1)	5.6				
(2)	6.0				
Major elements %DM					
Calcium (1)	0.55				
(2)	0.60				
Phosphorus (1)	0.16				
(2)	0.17				
Unspecified grass					
Proximate constituents %DM					
Crude protein	7.6 (4.7 – 11.4)			Survey for quality: means and ranges for 33 samples from farms, as used. Eire, 1940–45	449
Ether extract	1.5 (1.0 – 2.2)				
Nitrogen-free extract	50.8 (44.1 – 58.7)				
Crude fibre	33.6 (27.6 – 39.7)				
Ash	6.5 (5.0 – 7.9)				
Grass and Clover	August				
Proximate constituents %DM					
Crude protein (a)	F_0	9.4		Meadow grass with White and Red clover. Cut 7 Aug., stacked 16 Sept., Sampled 4 Dec.	491
	F_1	7.7		F_0: Untreated	
	F_2	8.4		F_1: 20 tonnes FYM/0.13 ha plot.	
(b)	F_0	9.4		F_2: 330 kg basic slag/0.2 ha plot.	
	F_1	8.1			
	F_2	8.0			
Ether extract (a)	F_0	1.6		(a) On stacking 16 Sept.	
	F_1	1.3		(b) From stack 4 Dec.	
	F_2	1.6			
(b)	F_0	1.5			
	F_1	1.4			
	F_2	1.9			
Nitrogen-free extract (a)	F_0	49.3			
	F_1	49.1			
	F_2	50.2			
(b)	F_0	50.5			
	F_1	49.2			
	F_2	49.0			
Crude fibre (a)	F_0	32.3			
	F_1	33.9			
	F_2	32.0			
(b)	F_0	31.0			
	F_1	33.7			
	F_2	33.1			
Ash (a)	F_0	7.4			
	F_1	8.0			
	F_2	7.8			
(b)	F_0	7.6			
	F_1	7.6			
	F_2	8.0			

Table 1.130 Grass and grass and clover: hay (continued)

Grass and Clover (continued)		August	Notes	Reference
Major elements %DM				
Calcium (a)	F_0	0.49		491
	F_1	0.40		
	F_2	0.66		
(b)	F_0	0.58		
	F_1	0.40		
	F_2	0.63		
Phosphorus (a)	F_0	0.17		
	F_1	0.26		
	F_2	0.27		
(b)	F_0	0.18		
	F_1	0.27		
	F_2	0.28		
		June		
Digestibility of OM %		61.8 (61.1 – 63.1)	Mean and range for F_0, F_{10}, F_{11} and F_{12}	480
Proximate constituents %DM				
Crude protein	F_0	9.0	Botanical composition included Ryegrass, Cocksfoot, Fescues: not more than 7% clover in ground cover. Cut 30 June. Fertilisers/ha.	
	F_1	6.7	F_0 : None	
	F_2	6.6	F_1 : FYM 20 tonnes + ammonium sulphate 41 kg + sodium nitrate 55 kg + basic slag 165 kg + muriate of potash 55 kg	
	F_3	6.5	F_2 : FYM 1 year and artificials as in F_1 on alternate years	
	F_4	6.0	F_3 : FYM only in alternate years	
	F_5	8.1	F_4 : FYM every 4th year and artificials as in F_1 on other 3 years	
	F_6	8.4	F_5 : ammonium sulphate 165 kg	
	F_7	7.7	F_6 : muriate of potash 110 kg	
	F_8	7.9	F_7 : sulphate of ammonia 165 kg + basic slag 330 kg	
	F_9	6.8	F_8 : ammonium sulphate 165 kg + muriate of potash	
	F_{10}	6.3	F_9 : basic slag 330 kg + muriate of potash 110 kg	
	F_{11}	7.4	F_{10}: FYM each year	
	F_{12}	7.0	F_{11}: basic slag 330 kg	
Ether extract	F_0	1.5	F_{12}: ammonium sulphate 165 kg + basic slag 330 kg + muriate of potash 110 kg	
	F_1	1.5		
	F_2	1.7		
	F_3	1.5		
	F_4	1.5		
	F_5	1.6		
	F_6	1.8		
	F_7	1.4		
	F_8	1.6		
	F_9	1.6		
	F_{10}	1.5		
	F_{11}	1.6		
	F_{12}	1.5		
Nitrogen-free extract	F_0	50.8		
	F_1	48.6		
	F_2	49.7		
	F_3	48.4		
	F_4	52.0		
	F_5	50.9		
	F_6	51.0		
	F_7	49.9		
	F_8	53.4		
	F_9	49.8		
	F_{10}	47.5		
	F_{11}	50.0		
	F_{12}	51.3		
Crude fibre	F_0	27.7		
	F_1	29.7		
	F_2	29.9		
	F_3	30.3		
	F_4	29.6		
	F_5	27.8		
	F_6	25.9		
	F_7	29.2		
	F_8	27.5		
	F_9	30.3		
	F_{10}	31.6		
	F_{11}	29.2		
	F_{12}	29.2		

continued

Table 1.130 Grass and grass and clover: hay (continued)

Grass and Clover (continued)	June		Notes	Reference
Proximate constituents (cont'd)				
Ash	F_0	11.0		
	F_1	13.5		
	F_2	12.1		480
	F_3	13.3		
	F_4	10.9		
	F_5	11.6		
	F_6	12.9		
	F_7	11.8		
	F_8	9.6		
	F_9	11.5		
	F_{10}	13.1		
	F_{11}	11.8		
	F_{12}	11.0		
Major elements %DM				
Calcium	F_0	0.97		
	F_1	0.46		
	F_2	0.41		
	F_3	0.38		
	F_4	0.37		
	F_5	0.21		
	F_6	0.66		
	F_7	0.65		
	F_8	0.24		
	F_9	0.56		
	F_{10}	0.44		
	F_{11}	0.62		
	F_{12}	0.51		
Phosphorus	F_0	0.14		
	F_1	0.28		
	F_2	0.24		
	F_3	0.24		
	F_4	0.23		
	F_5	0.13		
	F_6	0.13		
	F_7	0.25		
	F_8	0.15		
	F_9	0.24		
	F_{10}	0.24		
	F_{11}	0.25		
	F_{12}	0.24		
Magnesium	F_0	0.14		
	F_1	0.07		
	F_2	0.10		
	F_3	0.14		
	F_4	0.13		
	F_5	0.07		
	F_6	0.12		
	F_7	0.06		
	F_8	0.06		
	F_9	0.06		
	F_{10}	0.07		
	F_{11}	0.13		
	F_{12}	0.10		
Potassium	F_0	0.81	As for Crude protein in reference 480	481
	F_1	1.60		
	F_2	1.34		
	F_3	0.96		
	F_4	1.10		
	F_5	1.02		
	F_6	1.08		
	F_7	0.51		
	F_8	0.84		
	F_9	0.91		
	F_{10}	1.48		
	F_{11}	0.61		
	F_{12}	0.93		

Table 1.130 Grass and grass and clover: hay (continued)

Grass and Clover (continued)		June		Notes	Reference
Major elements %DM (cont'd)					
Sodium		F_0	0.10		
		F_1	0.10		481
		F_2	0.18		
		F_3	0.11		
		F_4	0.11		
		F_5	0.07		
		F_6	0.07		
		F_7	0.21		
		F_8	0.03		
		F_9	0.15		
		F_{10}	0.15		
		F_{11}	0.12		
		F_{12}	0.13		
Chloride		F_0	0.54		
		F_1	0.73		
		F_2	0.55		
		F_3	0.58		
		F_4	0.66		
		F_5	0.48		
		F_6	0.56		
		F_7	0.44		
		F_8	0.41		
		F_9	0.61		
		F_{10}	0.48		
		F_{11}	0.43		
		F_{12}	0.68		
Trace elements µg/g DM					
Cobalt		F_0	0.19		
		F_1	0.17		
		F_2	0.16		
		F_3	0.17		
		F_4	0.16		
		F_5	0.14		
		F_6	0.20		
		F_7	0.15		
		F_8	0.16		
		F_9	0.18		
		F_{10}	0.17		
		F_{11}	0.14		
		F_{12}	0.18		
Copper		F_0	6.0		
		F_1	9.6		
		F_2	9.2		
		F_3	9.4		
		F_4	5.4		
		F_5	6.4		
		F_6	7.6		
		F_7	5.8		
		F_8	6.6		
		F_9	7.4		
		F_{10}	8.8		
		F_{11}	9.2		
		F_{12}	8.2		
Iron		F_0	83		
		F_1	92		
		F_2	125		
		F_3	80		
		F_4	69		
		F_5	240		
		F_6	109		
		F_7	94		
		F_8	92		
		F_9	60		
		F_{10}	118		
		F_{11}	186		
		F_{12}	84		

continued

Table 1.130 Grass and grass and clover: hay (continued)

Grass and Clover (continued)	June		Notes	Reference
Trace elements μg/g DM (cont'd)				
Manganese	F_0	362		481
	F_1	86		
	F_2	168		
	F_3	129		
	F_4	169		
	F_5	314		
	F_6	396		
	F_7	282		
	F_8	344		
	F_9	135		
	F_{10}	101		
	F_{11}	213		
	F_{12}	368		
Zinc	F_0	37		
	F_1	32		
	F_2	36		
	F_3	25		
	F_4	31		
	F_5	37		
	F_6	43		
	F_7	40		
	F_8	31		
	F_9	31		
	F_{10}	29		
	F_{11}	45		
	F_{12}	35		

Table 1.131 Uncultivated grasses: hay

	March	September	November	December	Notes	Reference
Bent grass hay: *Agrostis* spp.						
Proximate constituents %DM						
Crude protein		9.4		9.2	16 Sept.: analysed on stacking, 40 days after cutting on 7 Aug.	491
Ether extract		1.6		1.8	4 Dec.: sampled after 79 days in stack	
Nitrogen-free extract		52.0		51.0	One plot received ammonium sulphate 165 kg/ha yearly since 1897.	
Crude fibre		30.8		31.6		
Ash		6.2		6.4		
Major elements %DM						
Calcium		0.4		0.4		
Phosphorus		0.2		0.2		
Poa pratensis						
Proximate constituents %DM						
Crude protein	7.3				Means of 5 samples	20
Ash	5.02					
Major elements %DM						
Calcium	0.72					
Phosphorus	0.16					
Magnesium	0.11					
Chlorine	0.62					
Trace elements μg/g DM						
Copper	3.6					
Iron	79					
Manganese	24					

Table 1.132 Alderman's hays Reference 5, (For proximate constituents, digested nutrients and metabolisable energy see table 1.230)

Cocksfoot/Meadow Fescue/clover hay: Code 632
Cut 1/6/65. Trace of clover. Fairly leafy, cut after light spring grazing.

	May			June			July			August		
Carbohydrate fractions %DM Soluble carbohydrates				7.8								
Pentosans				19.1								
Hexosans				32.2								
Lignin				8.3								
Major elements %DM Calcium				0.4								
Phosphorus				0.2								
Magnesium				0.1								
Potassium				1.8								
Sodium				0.19								
Chlorine				0.9								
Sulphur Sulphate sulphur				0.3								
Trace elements μg/g DM Cobalt				0.09								
Copper				5.4								
Iodine				0.4								
Manganese				153								
Molybdenum				0.2								
Zinc				75								

Perennial Ryegrass/Cocksfoot/Timothy/clover hay: Code 723
Cut 6/65. Timothy fairly stemmy.

	May			June			July			August		
Carbohydrate fractions %DM Soluble carbohydrates					11.1							
Pentosans					18.8							
Hexosans					34.6							
Lignin					7.9							
Major elements %DM Calcium					0.3							
Phosphorus					0.2							
Magnesium					0.1							
Potassium					1.9							
Sodium					0.01							
Chlorine					0.5							
Sulphur Sulphate sulphur					0.3							
Trace elements μg/g DM Cobalt					0.1							
Copper					4.6							
Iodine					0.2							
Manganese					49							
Molybdenum					0.3							
Zinc					27							

continued

Table 1.132 Alderman's hays Reference 5 (continued)
Perennial Ryegrass/Cocksfoot/Timothy/Meadow Fescue/White Clover hay: Code 680, 700
Both cuts 31/5/65.

	May			June			July			August		
Carbohydrate fractions %DM												
Soluble carbohydrates			7.0									
			5.6									
Pentosans			16.5									
			16.9									
Hexosans			29.9									
			30.4									
Lignin			5.2									
			5.7									
Major elements %DM												
Calcium			0.3									
			0.4									
Phosphorus			0.3									
			0.2									
Magnesium			0.1									
			0.1									
Potassium			2.9									
			2.9									
Sodium			0.14									
			0.16									
Chlorine			0.8									
			1.0									
Sulphur												
Sulphate sulphur			0.5									
			0.5									
Trace elements μg/g DM												
Cobalt			0.2									
			0.2									
Copper			7.6									
			7.3									
Iodine			0.2									
			0.2									
Manganese			99									
			96									
Molybdenum			0.2									
			0.3									
Zinc			29									
			34									

Table 1.132 Alderman's hays Reference 5 (continued)
Perennial Ryegrass/Timothy/White Clover S100 hay: Code 6056, 6057, 6058
(1) Cut 27/6/65. Ley undersown in 1960; no N; moderately leafy; baled on 30/6/65 and 1/7/65.
(2) Cut 27/6/65. Ley undersown in 1960; 20 units N on 25/2/65; fairly leafy; baled on 30/6/65 and 1/7/65.
(3) Cut 11/6/65. Ley undersown in 1960; 40 units N on 25/2/65; fairly leafy; baled on 17/6/65 and 21/6/65.

1 Unit N/acre ≡ 1/100 cwt ≡ 1.25 kg N/ha.

	May			June			July		August		
Carbohydrate fractions %DM Soluble carbohydrates				(3)	8.3	(1) 15.4 (2) 14.5					
Pentosans				(3)	19.5	(1) 18.7 (2) 18.3					
Hexosans				(3)	30.9	(1)28.3 (2) 28.5					
Lignin				(3)	6.6	(1) 6.4 (2) 6.5					
Major elements %DM Calcium				(3)	0.4	(1) 0.4 (2) 0.4					
Phosphorus				(3)	0.2	(1) 0.2 (2) 0.2					
Magnesium				(3)	0.1	(1) 0.1 (2) 0.1					
Potassium				(3)	2.2	(1) 2.1 (2) 2.1					
Sodium				(3)	0.17	(1) 0.06 (2) 0.11					
Chlorine				(3)	0.5	(1) 0.7 (2) 0.5					
Sulphur Sulphate sulphur				(3)	0.5	(1) 0.5 (2) 0.5					
Trace elements μg/g Cobalt				(3)	0.2	(1) 0.07 (2) 0.4					
Copper				(3)	6.0	(1) 4.1 (2) 5.1					
Iodine				(3)	0.1	(1) 0.1 (2) 0.2					
Manganese				(3)	30	(1) 19 (2) 43					
Molybdenum				(3)	2.0	(1) 1.2 (2) 2.0					
Zinc				(3)	27	(1) 25 (2) 27					

continued

Table 1.132 Alderman's hays Reference 5 (continued)
Perennial Ryegrass/White Clover hay: Code 605, 604, 603
Ley in its eighth year. Cut 22/6/64, baled 26/6/64 at approx. 25% moisture and barn dried.
N_0, no N; N_1, 80 units N; N_2, 160 units N/acre

1 Unit N = 1/100 cwt/acre ≡ 1.25 kg/ha.

	May			June			July			August		
Carbohydrate fractions %DM Soluble carbohydrates					N_0 N_1 N_2	14.7 8.8 5.5						
Pentosans					N_0 N_1 N_2	16.7 16.7 17.4						
Hexosans					N_0 N_1 N_2	27.9 28.8 28.1						
Lignin					N_0 N_1 N_2	5.8 6.3 6.7						
Major elements %DM Calcium					N_0 N_1 N_2	0.4 0.3 0.3						
Phosphorus					N_0 N_1 N_2	0.3 0.2 0.2						
Magnesium					N_0 N_1 N_2	0.1 0.1 0.2						
Potassium					N_0 N_1 N_2	2.2 2.4 2.2						
Sodium					N_0 N_1 N_2	0.08 0.07 0.15						
Chlorine					N_0 N_1 N_2	0.7 0.7 0.6						
Sulphur Sulphate sulphur					N_0 N_1 N_2	0.7 0.7 0.6						
Trace elements μg/g DM Cobalt					N_0 N_1 N_2	0.07 0.08 0.05						
Copper					N_0 N_1 N_2	5.1 5.0 6.3						
Iodine					N_0 N_1 N_2	0.08 0.09 0.09						
Manganese					N_0 N_1 N_2	33 44 44						
Molybdenum					N_0 N_1 N_2	1.0 1.2 0.9						
Zinc					N_0 N_1 N_2	29 29 32						

Table 1.132 Alderman's hays Reference 5 (continued)

Italian Ryegrass S22/Kersey White Clover hay: Code 739
Cut 23/6/64. Ley undersown in 1962; 70 units N on 10/2/64; made quickly and well in good weather.
1 unit N/acre ≡ 1/100 cwt ≡ 1.25 kg/ha.

	May			June			July			August		
Carbohydrate fractions %DM Soluble carbohydrates						21.8						
Pentosans						16.1						
Hexosans						26.5						
Lignin						6.5						
Major elements %DM Calcium						0.4						
Phosphorus						0.1						
Magnesium						0.09						
Potassium						1.7						
Sodium						0.04						
Chlorine						0.6						
Sulphur Sulphate sulphur						0.6						
Trace elements μg/g DM Cobalt						0.1						
Copper						4.0						
Iodine						0.1						
Manganese						23						
Molybdenum						0.7						
Zinc						27						

Ryegrass/Cocksfoot/clover hay: Code 719
Cut 15/6/65.

	May			June			July			August		
Carbohydrate fractions %DM Soluble carbohydrates					5.3							
Pentosans					18.2							
Hexosans					34.7							
Lignin					8.7							
Major elements %DM Calcium					0.4							
Phosphorus					0.2							
Magnesium					0.1							
Potassium					2.5							
Sodium					0.12							
Chlorine					0.7							
Sulphur Sulphate sulphur					0.6							
Trace elements μg/g DM Cobalt					0.1							
Copper					6.3							
Iodine					0.2							
Manganese					88							
Molybdenum					0.3							
Zinc					32							

continued

Table 1.132 Alderman's hays Reference 5 (continued)
Ryegrass/Timothy/Cocksfoot/White Clover hay: Code 658
Cut 15/6/63. Fairly leafy, badly weathered.

	May			June			July			August		
Carbohydrate fractions %DM Soluble carbohydrates					2.7							
Pentosans					18.5							
Hexosans					34.0							
Lignin					9.1							
Major elements %DM Calcium					0.6							
Phosphorus					0.3							
Magnesium					0.2							
Potassium					2.1							
Sodium					0.28							
Chlorine					0.9							
Sulphur Sulphate sulphur					0.6							
Trace elements μg/g DM Cobalt					0.09							
Copper					7.1							
Iodine					0.2							
Manganese					46							
Molybdenum					0.4							
Zinc					30							

Ryegrass/Timothy/Cocksfoot hay: Code 630
Cut 6/63. Mainly Ryegrass with some Timothy and Cocksfoot. No data for major elements or trace elements.

	May			June			July			August		
Carbohydrate fractions %DM Soluble carbohydrates					11.3							
Pentosans					18.6							
Hexosans					30.8							
Lignin					7.2							

Table 1.132 Alderman's hays Reference 5 (continued)
Ryegrass/Timothy/Meadow Fescue hay: Code 606/65A, 659
(1) Cut 7/64, medium quality.
(2) Cut 10/7/64, fairly stemmy, badly made.

	May			June			July			August		
Carbohydrate fractions %DM Soluble carbohydrates							(1) (2)	11.7 5.4				
Pentosans							(1) (2)	17.6 18.5				
Hexosans							(1) (2)	32.0 34.0				
Lignin							(1) (2)	7.3 8.3				
Major elements %DM Calcium							(1) (2)	0.3 0.4				
Phosphorus							(1) (2)	0.2 0.2				
Magnesium							(1) (2)	0.1 0.2				
Potassium							(1) (2)	2.0 2.2				
Sodium							(1) (2)	0.18 0.12				
Chlorine							(1) (2)	0.8 1.5				
Sulphur Sulphate sulphur							(1) (2)	0.5 0.4				
Trace elements µg/g DM Cobalt							(1) (2)	0.1 0.4				
Copper							(1) (2)	5.2 7.6				
Iodine							(1) (2)	0.1 0.2				
Manganese							(1) (2)	49 110				
Molybdenum							(1) (2)	0.2 0.2				
Zinc							(1) (2)	34 27				

continued

Table 1.132 Alderman's hays Reference 5 (continued)
Ryegrass/Fescue/clover hay: Code 557
Cut 1965, no date. Trace of clover. Moderately stemmy, poor hay, purchased.

	May			June			July			August		
Carbohydrate fractions %DM												
Soluble carbohydrates				6.5								
Pentosans				15.9								
Hexosans				29.5								
Lignin				8.7								
Major elements %DM												
Calcium				0.8								
Phosphorus				0.2								
Magnesium				0.1								
Potassium				1.5								
Sodium				0.07								
Chlorine				0.7								
Sulphur												
Sulphate sulphur				0.4								
Trace elements μg/g DM												
Cobalt				0.3								
Copper				7.0								
Iodine				0.4								
Manganese				50								
Molybdenum				1.3								
Zinc				34								

Ryegrass hay: Code 592
Cut 1965, no date. Moderately stemmy, poor hay purchased.

	May			June			July			August		
Carbohydrate fractions %DM												
Soluble carbohydrates				11.1								
Pentosans				19.4								
Hexosans				32.9								
Lignin				7.2								
Major elements %DM												
Calcium				0.3								
Phosphorus				0.2								
Magnesium				0.2								
Potassium				2.1								
Sodium				0.03								
Chlorine				0.9								
Sulphur												
Sulphate sulphur				0.5								
Trace elements μg/g DM												
Cobalt				0.2								
Copper				5.1								
Iodine				0.3								
Manganese				173								
Molybdenum				0.3								
Zinc				34								

Table 1.132 Alderman's hays Reference 5 (continued)
Ryegrass/White Clover hay: Code 6053
Cut 6/64. In its third year, slightly stemmy.

	May			June			July			August		
Carbohydrate fractions %DM Soluble carbohydrates					11.7							
Pentosans					16.9							
Hexosans					28.4							
Lignin					6.8							
Major elements %DM Calcium					0.4							
Phosphorus					0.3							
Magnesium					0.1							
Potassium					2.7							
Sodium					0.11							
Chlorine					1.5							
Sulphur Sulphate sulphur					0.4							
Trace elements μg/g DM Cobalt					0.1							
Copper					5.4							
Iodine					0.1							
Manganese					96							
Molybdenum					0.3							
Zinc					25							

Timothy/Ryegrass/White Clover hay: Code 545/66
Cut 6/65. Fairly stemmy; third year of ley.

	May			June			July			August		
Carbohydrate fractions %DM Soluble carbohydrates					9.5							
Pentosans					17.8							
Hexosans					31.1							
Lignin					7.1							
Major elements %DM Calcium					0.5							
Phosphorus					0.3							
Magnesium					0.2							
Potassium					2.7							
Sodium					0.10							
Chlorine					1.5							
Sulphur Sulphate sulphur					0.4							
Trace elements μg/g DM Cobalt					0.2							
Copper					6.0							
Iodine					0.1							
Manganese					30							
Molybdenum					2.0							
Zinc					27							

continued

Table 1.132 Alderman's hays Reference 5 (continued)
Fine Fescue/*Poa*/Ryegrass hay: Code 549
Cut 8/8/65. No rain on swath.

	May			June			July			August		
Carbohydrate fractions %DM Soluble carbohydrates										16.1		
Pentosans										17.5		
Hexosans										28.1		
Lignin										6.0		
Major elements %DM Calcium										0.3		
Phosphorus										0.3		
Magnesium										0.1		
Potassium										2.1		
Sodium										0.07		
Chlorine										0.9		
Sulphur Sulphate sulphur										0.6		
Trace elements μg/g DM Cobalt										0.07		
Copper										6.8		
Iodine										0.3		
Manganese										90		
Molybdenum										2.3		
Zinc										34		

Timothy/Cocksfoot/clover hay: Code 654
1964. No date. Slightly stemmy; made after spring grazing. Only small amount of clover present

	May			June			July			August		
Carbohydrate fractions %DM Soluble carbohydrates					10.6							
Pentosans					18.0							
Hexosans					33.1							
Lignin					7.5							
Major elements %DM Calcium					0.5							
Phosphorus					0.2							
Magnesium					0.1							
Potassium					2.0							
Sodium					0.08							
Chlorine					0.9							
Sulphur Sulphate sulphur					0.4							
Trace elements μg/g DM Cobalt					0.1							
Copper					6.1							
Iodine					0.2							
Manganese					30							
Molybdenum					1.9							
Zinc					22							

Table 1.132 Alderman's hays Reference 5 (continued)
Timothy/Meadow Fescue/clover hay: Code 798, 797, 765, 6000, 6001, 550
(1) Cut 24/6/1964; 5 days in field but only 0.03 inch of rain
(2) Cut 24/6/1964, no rain on swath
(3) Cut with flail mower, 29/7/1964; 7 days in swath with 0.34 inch of rain
(4) Cut 30/6/1965; slightly stemmy, 2 days in field with no rain
(5) Cut 30/6/1965; moderately stemmy, 6 days in field with 0.02 inch of rain
(6) Cut 11/8/1965, no rain on swath.

	May			June			July			August		
Carbohydrate fractions %DM Soluble carbohydrates					(1) (2) (4) (5)	3.3 5.7 12.2 9.3		(3)	6.0	(6)	14.8	
Pentosans					(1) (2) (4) (5)	19.7 17.8 18.7 19.6		(3)	18.3	(6)	18.3	
Hexosans					(1) (2) (4) (5)	35.5 32.4 30.7 31.4		(3)	34.9	(6)	29.5	
Lignin					(1) (2) (4) (5)	5.5 6.6 6.5 6.5		(3)	7.9	(6)	6.1	
Major elements %DM Calcium					(1) (2) (4) (5)	0.3 – 0.4 0.4		(3)	0.3	(6)	0.3	
Phosphorus					(1) (2) (4) (5)	0.3 – 0.3 0.3		(3)	0.2	(6)	0.3	
Magnesium					(1) (2) (4) (5)	0.1 – 0.1 0.1		(3)	0.1	(6)	0.1	
Potassium					(1) (2) (4) (5)	2.6 – 2.4 2.4		(3)	1.4	(6)	2.1	
Sodium					(1) (2) (4) (5)	0.03 – 0.03 0.05		(3)	0.07	(6)	0.07	
Chlorine					(1) (2) (4) (5)	0.8 – 0.8 0.8		(3)	0.6	(6)	0.9	
Sulphate sulphur					(1) (2) (4) (5)	0.6 – 0.7 0.5		(3)	0.3	(6)	0.6	

continued

Table 1.132 Alderman's hays Reference 5 (continued)
Timothy/Meadow Fescue/clover hay: (continued)

	May			June			July			August		
Trace elements µg/g DM Cobalt					(1) (2) (4) (5)	0.08 – 0.1 0.2		(3)	0.5	(6)	0.07	
Copper					(1) (2) (4) (5)	7.7 – 5.8 6.2		(3)	11.0	(6)	6.8	
Iodine					(1) (2) (4) (5)	0.3 – 0.3 0.2		(3)	1.0	(6)	0.3	
Manganese					(1) (2) (4) (5)	153 – 58 84		(3)	68	(6)	90	
Molybdenum					(1) (2) (4) (5)	1.1 – 1.2 1.6		(3)	1.5	(6)	2.3	
Zinc					(1) (2) (4) (5)	47 – 27 29		(3)	59	(6)	34	

Timothy/Meadow Fescue/Ryegrass/White Clover hay: Code 631
Cut 6/64. Well made (Calf hay).

	May			June			July			August		
Carbohydrate fractions % DM Soluble carbohydrates					14.4							
Pentosans					17.1							
Hexosans					32.0							
Lignin					6.9							
Major elements %DM Calcium					0.5							
Phosphorus					0.2							
Magnesium					0.1							
Potassium					2.4							
Sodium					0.12							
Chlorine					1.5							
Sulphur Sulphate sulphur					0.4							
Trace elements µg/g DM Cobalt					0.08							
Copper					4.5							
Iodine					0.1							
Manganese					84							
Molybdenum					0.2							
Zinc					26							

Table 1.132 Alderman's hays Reference 5 (continued)
Timothy/Meadow Fescue/Cocksfoot/White Clover S100 hay: Code 591, 593, 592
(1) Cut 25/5/64, weather bright, warm and windy during making (well made). Sown 14/3/62; 60 units N on 12/2/64.
(2) Cut 16/6/64, weather mostly cloudy with occasional showers and light breeze. Sown 14/3/62, 60 units N on 12/2/64.
(3) Cut 29/6/64, well made, weather bright, warm and windy. Sown 14/3/62, 60 units N on 12/2/64.
1 unit N/acre ≡ 1/100 cwt ≡ 1.25 kg N/ha.

	May			June			July			August		
Carbohydrate fractions %DM Soluble carbohydrates		(1)	5.5	(2)	5.0	(3) 7.8						
Pentosans		(1)	16.4	(2)	19.4	(3) 18.8						
Hexosans		(1)	27.5	(2)	32.0	(3) 32.0						
Lignin		(1)	5.8	(2)	6.6	(3) 7.5						
Major elements %DM Calcium		(1)	0.4	(2)	0.4	(3) 0.3						
Phosphorus		(1)	0.2	(2)	0.2	(3) 0.2						
Magnesium		(1)	0.1	(2)	0.1	(3) 0.1						
Potassium		(1)	2.8	(2)	2.1	(3) 2.0						
Sodium		(1)	0.03	(2)	0.02	(3) 0.02						
Chlorine		(1)	0.4	(2)	0.4	(3) 0.7						
Sulphur Sulphate sulphur		(1)	0.6	(2)	0.4	(3) 0.5						
Trace elements μg/gDM Cobalt		(1)	0.07	(2)	0.09	(3) 0.06						
Copper		(1)	5.6	(2)	5.1	(3) 3.8						
Iodine		(1)	0.1	(2)	0.1	(3) 0.1						
Manganese		(1)	60	(2)	28	(3) 55						
Molybdenum		(1)	0.5	(2)	0.5	(3) 0.4						
Zinc		(1)	29	(2)	28	(3) 28						

Timothy/Meadow Fescue/White Clover hay: Code 702, 715
(1) Cut 8/6/64, leafy, dried in field (0.38 inch of rain) and then brought in for barn drying on 12/6/64.
(2) Cut 5/65.

	May			June			July			August		
Carbohydrate fractions %DM Soluble carbohydrates	(2)	7.0		(1)	5.2							
Pentosans	(2)	14.4		(1)	16.5							
Hexosans	(2)	30.2		(1)	31.7							
Lignin	(2)	5.2		(1)	7.2							
Major elements %DM Calcium	(2)	0.3		(1)	0.5							
Phosphorus	(2)	0.3		(1)	0.3							
Magnesium	(2)	0.1		(1)	0.2							
Potassium	(2)	3.3		(1)	1.9							
Sodium	(2)	0.03		(1)	0.08							
Chlorine	(2)	0.7		(1)	0.5							
Sulphur Sulphate sulphur	(2)	0.5		(1)	0.8							
Trace elements μg/g DM Cobalt	(2)	0.2		(1)	0.3							
Copper	(2)	8.1		(1)	11.5							
Iodine	(2)	0.2		(1)	0.4							
Manganese	(2)	110		(1)	88							
Molybdenum	(2)	0.5		(1)	2.0							
Zinc	(2)	46		(1)	46							

continued

Table 1.132 Alderman's hays Reference 5 (continued)
Timothy/Meadow Fescue hay: Code 743
Cut 16/6/64. Timothy moderately leafy. Hay provided by local farmer, well made.

	May			June			July			August		
Carbohydrate fractions %DM Soluble carbohydrates					8.5							
Pentosans					17.1							
Hexosans					31.3							
Lignin					7.7							
Major elements %DM Calcium					0.4							
Phosphorus					0.3							
Magnesium					0.1							
Potassium					2.8							
Sodium					0.02							
Chlorine					1.4							
Sulphur Sulphate sulphur					0.3							
Trace elements μg/g DM Cobalt					0.07							
Copper					6.6							
Iodine					0.1							
Manganese					22							
Molybdenum					0.5							
Zinc					27							

Table 1.132 Alderman's hays Reference 5 (continued)
Timothy/Meadow Fescue/S 100 white clover hay: Code 727, 731, 735, 502
(1) Cut 8/6/64. Ley undersown in 1962; 60 units N on 14/2/64. Well made but baled slightly green and heated in stack.
(2) Cut 8/6/64. Ley undersown in 1962; 60 units N on 14/2/64.. Weathered in swath by rain.
(3) Cut 24/6/64. Ley undersown in 1962; 60 units N on 14/2/64. Well made.
(4) Cut 19/6/65. 60 units N on 7/2/65. Well made, moderately stemmy.
1 unit N/acre ≡ 1/100 cwt ≡ 1.25 kg N/ha.

	May			June			July			August		
Carbohydrate fractions %DM Soluble carbohydrates				(1) 5.0 (2) 2.5	(4) 9.4	(3) 11.2						
Pentosans				(1) 18.8 (2) 21.3	(4) 19.4	(3) 18.0						
Hexosans				(1) 33.0 (2) 38.2	(4) 31.7	(3) 33.4						
Lignin				(1) 6.7 (2) 7.7	(4) 6.3	(3) 7.1						
Major elements %DM Calcium				(1) 0.4 (2) 0.4	(4) 0.4	(3) 0.4						
Phosphorus				(1) 0.2 (2) 0.1	(4) 0.2	(3) 0.2						
Magnesium				(1) 0.1 (2) 0.1	(4) 0.1	(3) 0.1						
Potassium				(1) 2.5 (2) 1.5	(4) 2.2	(3) 2.0						
Sodium				(1) 0.01 (2) 0.01	(4) 0.01	(3) 0.01						
Chlorine				(1) 0.5 (2) 0.3	(4) 0.5	(3) 0.4						
Sulphur Sulphate sulphur				(1) 0.4 (2) 0.2	(4) 0.4	(3) 0.3						
Trace elements µg/g DM Cobalt				(1) 0.1 (2) 0.1	(4) 0.07	(3) 0.06						
Copper				(1) 5.8 (2) 5.4	(4) 4.4	(3) 4.6						
Iodine				(1) 0.1 (2) 0.2	(4) 0.1	(3) 0.1						
Manganese				(1) 24 (2) 44	(4) 26	(3) 14						
Molybdenum				(1) 0.6 (2) 0.5	(4) 0.5	(3) 0.8						
Zinc				(1) 25 (2) 22	(4) 22	(3) 18						

continued

Table 1.132 Alderman's hays Reference 5 (continued)

Timothy hay: Code 747, 633
(1) Cut 7/64. Very mature Timothy hay bought from local dealer, stemmy and coarse leaved; well made.
(2) Cut 2/6/65. Mainly Timothy, cut after light spring grazing.

	May			June			July			August		
Carbohydrate fractions %DM Soluble carbohydrates			(2)	3.8			(1)	6.5				
Pentosans			(2)	21.6			(1)	18.8				
Hexosans			(2)	34.7			(1)	32.4				
Lignin			(2)	9.3			(1)	9.4				
Major elements %DM Calcium			(2)	0.3			(1)	0.3				
Phosphorus			(2)	0.3			(1)	0.2				
Magnesium			(2)	0.09			(1)	0.1				
Potassium			(2)	2.3			(1)	1.9				
Sodium			(2)	0.01			(1)	0.06				
Chlorine			(2)	1.3			(1)	1.0				
Sulphur Sulphate sulphur			(2)	0.2			(1)	0.3				
Trace elements μg/g DM Cobalt			(2)	0.05			(1)	0.1				
Copper			(2)	5.5			(1)	5.9				
Iodine			(2)	0.2			(1)	0.2				
Manganese			(2)	74			(1)	80				
Molybdenum			(2)	0.2			(1)	0.4				
Zinc			(2)	30			(1)	47				

Table 1.132 Alderman's hays Reference 5 (continued)
Meadow Grass/Clover hay: Code 815
Cut 7/65. Meadow grass with a little clover, fairly leafy, rather yellow in colour, purchased from local farmer.

	May			June			July			August		
Carbohydrate fractions %DM Soluble carbohydrates								15.1				
Pentosans								14.8				
Hexosans								28.4				
Lignin								7.3				
Major elements %DM Calcium								0.6				
Phosphorus								0.2				
Magnesium								0.1				
Potassium								1.6				
Sodium								0.28				
Chlorine								1.2				
Sulphur Sulphate sulphur								0.5				
Trace elements μg/g DM Cobalt								0.2				
Copper								5.6				
Iodine								0.2				
Manganese								136				
Molybdenum								0.3				
Zinc								20				

Bent/Fine Fescue/*Poa*/Ryegrass hay: Code 764/65
Cut with flail mower, 5/8/64. 9 days in swath with 1.55 inch of rain.

	May			June			July			August		
Carbohydrate fractions %DM Soluble carbohydrates										3.8		
Pentosans										19.9		
Hexosans										33.5		
Lignin										9.8		
Major elements %DM Calcium										0.3		
Phosphorus										0.2		
Magnesium										0.1		
Potassium										1.4		
Sodium										0.06		
Chlorine										0.7		
Sulphur Sulphate sulphur										0.3		
Trace elements μg/g DM Cobalt										0.3		
Copper										8.9		
Iodine										0.6		
Manganese										65		
Molybdenum										1.8		
Zinc										50		

continued

Table 1.132 Alderman's hays Reference 5 (continued)
Hay Unnamed: Code 676, 635, 803, 802
(1) No date given. Poor quality hay obtained from farmer
(2) No date given. Poor quality hay obtained from farmer
(3) Cut 6/1965. Average meadow hay, purchased from local farmer
(4) Cut 7/1965. Poor meadow hay, purchased from local farmer; dark brown in colour and stemmy. Reference 5

	May			June			July			August		
Carbohydrate fractions %DM Soluble carbohydrates				(3)	15.4		(4)	7.6	(1) (2)	5.9 4.3		
Pentosans				(3)	18.9		(4)	18.1	(1) (2)	19.7 20.8		
Hexosans				(3)	27.2		(4)	33.1	(1) (2)	34.4 34.7		
Lignin				(3)	6.1		(4)	7.9	(1) (2)	7.3 8.0		
Major elements %DM Calcium				(3)	0.4		(4)	0.3	(1) (2)	0.3 0.5		
Phosphorus				(3)	0.2		(4)	0.2	(1) (2)	0.2 0.2		
Magnesium				(3)	0.1		(4)	0.2	(1) (2)	0.1 0.2		
Potassium				(3)	1.1		(4)	1.3	(1) (2)	2.2 1.4		
Sodium				(3)	0.47		(4)	0.13	(1) (2)	0.07 0.36		
Chlorine				(3)	0.8		(4)	0.5	(1) (2)	0.9 0.9		
Sulphur Sulphate sulphur				(3)	0.5		(4)	0.2	(1) (2)	0.5 0.8		
Trace elements μg/g DM Cobalt				(3)	0.1		(4)	0.5	(1) (2)	0.1 0.1		
Copper				(3)	6.0		(4)	8.9	(1) (2)	6.6 7.6		
Iodine				(3)	0.2		(4)	0.5	(1) (2)	0.2 0.2		
Manganese				(3)	22		(4)	310	(1) (2)	50 82		
Molybdenum				(3)	0.8		(4)	0.9	(1) (2)	0.4 1.3		
Zinc				(3)	18		(4)	54	(1) (2)	27 36		

Table 1.133 Cocksfoot and Cocksfoot mixtures: Silage

		Notes	Reference
Cocksfoot, unnamed *Amines* mg/kg DM		Estimated chromatographically	346
Histamine	A 326 B 774 C 844 D 711	A = lacerated, inoculated with lactobacilli B = lacerated C = untreated D = inoculated with lactobacilli	
Tyramine	A 998 B 1874 C 3084 D 2541		
pH	A 4.20 B 4.50 C 4.85 D 4.90		
Vitamins μg/g DM α - tocopherol	301–310	Silage of Cocksfoot S143 cut at 23 cm height	100
Cocksfoot and Timothy *Proximate constituents* %DM		Herbage mature aftermath ensiled with AIV acids (2N HCL and H_2SO_4) 90 litres/100kg. Ensiled 15-16 September in a wooden pit silo. Opened 22 November. Means of 4 samples	401
Crude protein	12.8		
Ether extract	4.0		
Nitrogen-free extract	45.9		
Crude fibre	29.3		
Ash	8.0		
Major elements %DM		For composition of the crop before ensiling see table 1.86, same reference	
Calcium	0.60		
Phosphorus	0.20		
pH	3.6		
Cocksfoot, Meadow fescue and Yorkshire Fog *Proximate constituents* %DM		Survey of silage quality	386
Crude protein	9.6	1 centre	
Organic acids %DM			
Lactic	1.79		
Acetic	0.18		
Butyric	0.03		
pH	5.0		
Cocksfoot, Timothy and Meadow fescue *Proximate constituents* %DM			
Crude protein	(1955) 11.0	1 centre	
Organic acids %DM			
Lactic	2.48		
Acetic	0.31		
Butyric	0.52		
pH	4.5		
Cocksfoot and Clover *Proximate constituents* %DM			
Crude protein	(1955) 11.5		
Organic acids %DM			
Lactic	2.78		
Acetic	0.33		
Butyric	0.13		
pH	4.1		

continued

Table 1.133 Cocksfoot and Cocksfoot mixtures: Silage (continued)

		Notes	Reference
Cocksfoot and Clover ley			
Volatile bases %DM	C 0.6 L 1.2	Crude protein in fresh herbage low, 10%	376
Organic acids %DM		Herbage either chopped (C) or lacerated (L). Concrete silos	
Lactic	C 7.3 L 5.6		
Acetic	C 1.8 L 1.8		
Butyric	C 0.1 L 0.8		
Propionic	C 0.1 L 0.2		
Higher acids as valeric	C 0.1 L 0.2		
Cocksfoot, Ryegrass and Clover			
Proximate constituents %DM			386
Crude protein	(1955) 11.6 (1957) 11.0 (9.3–12.8)	Mean of 2 samples from 2 centres Mean and range of 6 samples from 4 centres	
Organic acids %DM			
Lactic	(1955) 2.51 (1957) 2.15 (1.04–2.86)		
Acetic	(1955) 0.22 (1957) 0.61 (0.07–1.42)		
Butyric	(1955) 0.12 (1957) 0.05 (Nil–0.21)		
pH	(1955) 4.3 (1957) 4.2 (4.0–4.5)		
Cocksfoot, Ryegrass, Clover and Chicory			
Proximate constituents %DM			
Crude protein	(1957) 10.9	Mean of 2 samples from 1 centre	
Organic acids %DM			
Lactic	2.11		
Acetic	0.50		
Butyric	0.66		
pH	4.9		

Table 1.134 Ryegrass and Ryegrass mixtures: Silage

		Notes	Reference
Perennial Ryegrass and a little Clover			
Carbohydrate fractions %DM of the fresh herbage		Crop cut at leafy stage: 6th May	589
		Artificially inoculated with mixed strains of lactobacilli.	
Glucose	E1 8.3 E2 3.0		
Fructose	E1 6.2 E2 10.8	E1 : Ensiled 8 days E2 : Ensiled 8 months	
Sucrose	E1 0.3 E2 nil		
Fructosan	E1 1.8 E2 nil	For composition of the crop before ensiling see table 1.89, same reference.	
Galactose	E1 1.0 E2 0.7		
Arabinose	E1 nil E2 1.2		
Xylose	E1 nil E2 1.3		
Oligosaccharides	E1 present E2 absent		
pH	E1 3.8 E2 3.5		
Ryegrass, Italian			
Proximate constituents %DM		Survey of silage quality	386
Crude protein	(1955) 13.8	1 centre	
Organic acids %DM			
Lactic	0.70		
Acetic	0.47		
Butyric	1.35		
pH	5.2		
Ryegrass, Perennial and Italian			
Proximate constituents %DM			
Crude protein	(1955) 14.5	1 centre	
Organic acids %DM			
Lactic	2.26		
Acetic	0.87		
Butyric	0.63		
pH	4.7		
Ryegrass, unnamed			
Proximate constituents %DM			
Crude protein	(1955) 9.0	1 centre	
Organic acids %DM			
Lactic	3.98		
Acetic	0.30		
Butyric	0.49		
pH	4.3		
Proximate constituents %DM			
Crude protein	(1957) 10.2	Mean of 2 samples from 2 centres	
Organic acids %DM			
Lactic	2.25		
Acetic	0.47		
Butyric	0.20		
pH	4.35		

continued

Table 1.134 Ryegrass and Ryegrass mixtures: Silage (continued)

		Notes	Reference
Ryegrass and Timothy			
Proximate constituents %DM			386
Crude protein	(1955) 14.1 (1957) 4.5	1 centre 1 centre	
Organic acids %DM			
Lactic	(1955) 2.04 (1957) 1.24		
Acetic	(1955) 0.14 (1957) 0.37		
Butyric	(1955) 0.32 (1957) 0.01		
pH	(1955) 4.9 (1957) 4.7		
Ryegrass, Timothy and Meadow Fescue			
Proximate constituents %DM			
Crude protein	(1955) 9.45	1 centre	
Organic acids %DM			
Lactic	0.61		
Acetic	0.52		
Butyric	Nil		
pH	4.2		
Ryegrass, Timothy, Fescue and Clover:			
Proximate constituents %DM			
Crude protein	(1957) 10.1	1 centre	
Organic acids %DM			
Lactic	0.60		
Acetic	0.08		
Butyric	0.08		
pH	7.1		
Ryegrass and Cocksfoot:			
Proximate constituents %DM			
Crude protein	(1955) 12.6	Mean of samples from 4 centres	
Organic acids %DM			
Lactic	1.89		
Acetic	0.33		
Butyric	0.63		
pH	4.8		
Ryegrass, Cocksfoot and Yorkshire Fog			
Proximate constituents %DM			
Crude protein	(1957) 13.9	1 centre	
Organic acids %DM			
Lactic	2.22		
Acetic	0.51		
Butyric	0.41		
pH	4.5		
Ryegrass and Yorkshire Fog			
Proximate constituents %DM			
Crude protein	(1955) 8.8 (1957) 10.3	1 centre 1 centre	
Organic acids %DM			
Lactic	(1955) 3.10 (1957) 1.36		
Acetic	(1955) 0.29 (1957) 0.53		
Butyric	(1955) 0.14 (1957) 0.58		
pH	(1955) 4.0 (1957) 4.2		

Table 1.134 Ryegrass and Ryegrass mixtures: Silage (continued)

		Notes	Reference
Ryegrass and Clover *Proximate constituents* %DM			
Crude protein	(1955) 11.5 (1957) 12.3 (8.4–19.0)	Mean of 2 samples from 2 centres Mean and range of 12 samples from 6 centres	386
Organic acids %DM			
Lactic	(1955) 0.55 (1957) 2.09 (0.22–4.32)		
Acetic	(1955) 0.24 (1957) 0.64 (Nil–2.33)		
Butyric	(1955) 1.20 (1957) 0.31 (Nil–1.54)		
pH	(1955) 5.1 (1957) 4.4 (3.6–5.4)		
Ryegrass, Yorkshire Fog and Clover *Proximate constituents* %DM			
Crude protein	(1955) 11.0	3 samples from 3 centres	
Organic acids %DM			
Lactic	3.06		
Acetic	0.42		
Butyric	0.09		
pH	4.7		
Ryegrass, Meadow Fescue and Clover *Proximate constituents* %DM			
Crude protein	(1955) 11.7	1 centre	
Organic acids %DM			
Lactic	1.05		
Acetic	0.57		
Butyric	0.06		
pH	4.1		
Ryegrass, *Agrostis* and Clover *Proximate constituents* %DM			
Crude protein	(1957) 13.8	1 centre	
Organic acids %DM			
Lactic	5.58		
Acetic	0.88		
Butyric	0.06		
pH	4.1		
Ryegrass, *Agrostis, Cynosurus* and Clover *Proximate constituents* %DM			
Crude protein	(1957) 13.0	1 centre	
Organic acids %DM			
Lactic	2.30		
Acetic	0.37		
Butyric	0.12		
pH	4.1		
Ryegrass and Lucerne *Proximate constituents* %DM			
Crude protein	(1957) 12.2	1 centre	
Organic acids %DM			
Lactic	3.15		
Acetic	0.30		
Butyric	0.10		
pH	4.5		

continued

Table 1.134 Ryegrass and Ryegrass mixtures: Silage (continued)

		Notes	Reference
Ryegrass, *Poa*, Clover and Lucerne			
Proximate constituents %DM			386
Crude protein	(1957) 8.8		
Organic acids %DM			
Lactic	3.66		
Acetic	0.05		
Butyric	0.46		
pH	4.3		
Ryegrass and Red Clover			
Proximate constituents %DM		Ryegrass 20%; Red Clover 80%. Aftermath, clover mature, stemmy. Pit: no additive Tower: acid, A.I.V. fluid.	208
Crude protein	A 25.1		
	B 16.4		
	C 19.7		
	D 16.8	A Pit Edge	
Ether extract	A 1.8	B Pit Centre	
	B 5.8	C Tower Edge	
	C 2.9	D Tower Centre	
	D 4.4		
Nitrogen-free extract	A 28.5		
	B 33.8		
	C 37.1		
	D 41.3		
Crude fibre	A 30.6		
	B 34.8		
	C 30.0		
	D 29.1		
Ash	A 14.0		
	B 9.2		
	C 10.3		
	D 8.4		
Major elements %DM			
Calcium	A 2.48		
	B 1.97		
	C 2.29		
	D 1.54		
Phosphorus	A 0.47		
	B 0.27		
	C 0.40		
	D 0.22		
Potassium	1.52		
Chloride	0.85		
Ryegrass mixed and Clover, Red and White			
Amines mg/kg DM		Estimated chromatographically	346
Histamine	255		
Tyramine	532		
pH	4.65		

Table 1.134 Ryegrass and Ryegrass mixtures: Silage (continued)

		Notes	Reference
Ryegrass, Cocksfoot, Timothy and Clover			
Proximate constituents %DM		Analysis of sward before cutting; per cent:	185
Crude protein	16.0 (12.4–18.8)	Italian Ryegrass 6	
		Perennial Ryegrass 49	
Ether extract	5.2 (3.5–6.6)	Cocksfoot 12	
		Timothy 9	
		Red Clover 21	
Nitrogen-free extract	33.6 (26.7–39.9)	*Poa* spp 2½	
		Other species ½	
Crude fibre	35.5 (28.2–39.5)	Cut 22–30/5. Concrete clamp silo. Mean and range of 40 samples.	
Ash	9.7 (8.0–13.9)		
pH	4.55 (3.95–5.60)		

Table 1.135 Meadow Fescue and mixtures: Silage

		Notes	Reference
Meadow Fescue: *Proximate constituents* %DM Crude protein	(1952) U 20.3 C 19.9 L 17.6 UW 17.1 CW 17.1 (1953) U 17.1 C 16.5 C 13.6 L 14.3 CM 12.6 CM 12.7	Tower silos U = unchopped C = chopped L = lacerated UW = unchopped, wilted 24 hr. CW = chopped, wilted 24 hr. CM = chopped, molasses (1/50 kg) added For the composition of the crop before ensiling sec table 1.106, same reference	376
Ether extract	(1952) U 4.8 C 3.2 L 3.2 UW 2.2 CW 2.3 (1953) U 3.2 C 3.0 C 2.9 L 3.0 CM 2.7 CM 3.0		
Nitrogen-free extract	(1952) U 26.8 C 31.2 L 30.8 UW 25.5 CW 36.4 (1953) U 33.2 C 35.7 C 38.6 L 36.0 CM 43.0 CM 40.2		
Crude fibre	(1952) U 38.8 C 36.1 L 38.9 UW 34.2 CW 34.2 (1953) U 37.6 C 36.0 C 36.5 L 38.7 CM 33.1 CM 35.8		
Ash	(1952) U 9.3 C 9.6 L 9.5 UW 11.0 CW 10.0 (1953) U 8.9 C 8.8 C 8.4 L 8.0 CM 8.6 CM 8.3		

Table 1.135 Meadow Fescue and mixtures: Silage (continued)

Meadow Fescue (continued)				Notes	Reference
Organic acids %DM				In 1952 the silage was sampled at 3 levels in the silo for organic acid determination; in 1953 eight samples taken at equal intervals down the silo were bulked for analysis.	376
Lactic	(1952)	U	0.6		
		C	1.2		
		L	0.3		
		UW	0.8		
		CW	4.5		
	(1953)	U	0.2		
		C	2.6		
		C	4.3		
		L	0.3		
		CM	5.8		
		CM	6.1		
Acetic	(1952)	U	3.4		
		C	2.6		
		L	5.0		
		UW	0.9		
		CW	1.8		
	(1953)	U	2.0		
		C	2.9		
		C	3.9		
		L	3.0		
		CM	3.1		
		CM	4.7		
Butyric	(1952)	U	5.0		
		C	3.1		
		L	1.8		
		UW	0.8		
		CW	0.5		
	(1953)	U	4.6		
		C	2.4		
		C	0.2		
		L	2.4		
		CM	0.4		
		CM	0.3		
Propionic	(1952)	U	1.5		
		C	0.8		
		L	1.0		
		UW	0.2		
		CW	0.1		
	(1953)	U	0.9		
		C	0.9		
		C	0.6		
		L	0.7		
		CM	0.3		
		CM	0.6		
Higher acids as valeric	(1952)	U	3.4		
		C	2.1		
		L	0.8		
		UW	0.1		
		CW	0.1		
	(1953)	U	2.3		
		C	1.3		
		C	0.1		
		L	0.8		
		CM	0.1		
		CM	0.1		
pH	(1952)	U	5.1		
		C	5.0		
		L	5.2		
		UW	5.2		
		CW	4.7		
	(1953)	U	5.4		
		C	4.7		
		C	4.3		
		L	4.9		
		CM	4.3		
		CM	4.2		

continued

Table 1.135 Meadow Fescue and mixtures: Silage (continued)

		Notes	Reference
Meadow Fescue: small scale trials *Proximate constituents* %DM			381
Crude protein	A0 16.9 A1 17.9 A2 * A3 16.5	Trial 2 Herbage unchopped. fresh A0 No additive A1 1.8 kg sodium metabisulphite/ 1000 kg herbage A2 3.6 kg " " A3 5.5 kg " " * No value for CP is given From the other data it would be 20.5	
Ether extract	A0 3.3 A1 3.1 A2 3.2 A3 2.9		
Nitrogen-free extract	A0 35.9 A1 36.2 A2 35.2 A3 37.4		
Crude fibre	A0 35.3 A1 34.9 A2 32.5 A3 34.9		
Ash	A0 8.6 A1 7.9 A2 8.6 A3 8.3		
pH	A0 4.5 A1 4.4 A2 4.5 A3 4.5		
Proximate constituents %DM		Trial 3 Autumn cut Herbage unchopped A0 Unwilted, no additive A1 Unwilted, 4 kg sodium metabisulphite (MBS)/1000 kg herbage A2 Unwilted, 7.2 kg MBS/1000 kg A3 Wilted, 5 kg MBS/1000 kg	
Crude protein	A0 21.1 A1 21.0 A2 20.8 A3 19.8		
Ether extract	A0 4.4 A1 4.2 A2 3.8 A3 3.7		
Nitrogen-free extract	A0 33.7 A1 34.9 A2 37.9 A3 40.2		
Crude fibre	A0 30.0 A1 25.9 A2 22.9 A3 23.6		
Ash	A0 10.8 A1 14.0 A2 14.6 A3 12.7		
Volatile bases %DM	A0 3.38 A1 2.79 A2 2.33 A3 1.26		
Organic acids %DM Lactic	A0 3.28 A1 5.79 A2 3.98 A3 0.67		
Volatile acids as acetic	A0 7.38 A1 6.37 A2 5.23 A3 3.99		

Table 1.135 Meadow Fescue and mixtures: Silage (continued)

Meadow Fescue: small scale trials (continued)		Notes	Reference
pH	A0 4.8 A1 4.8 A2 4.9 A3 5.4	Trial 3 (cont'd)	381
Proximate constituents %DM		Trial 6 Herbage unchopped, unwilted A0 No additive A1 Sovilon (halogenated acetate of glycol) 165 g/1000 kg herbage For the composition of the crop before ensiling see table 1.106 same reference	
Crude protein	A0 18.1 A1 17.6		
Ether extract	A0 4.4 A1 4.2		
Nitrogen-free extract	A0 32.4 A1 35.4		
Crude fibre	A0 33.4 A1 31.4		
Ash	A0 11.7 A1 11.4		
Volatile bases %DM	A0 3.32 A1 2.49		
Organic acids %DM			
Lactic	A0 3.68 A1 2.75		
Volatile acids as acetic	A0 8.63 A1 8.05		
pH	A0 4.8 A1 4.7		
Meadow Fescue and Timothy: *Proximate constituents* %DM			
Crude protein	T0 15.4 T1 15.4 T2 16.5	Herbage chopped before filling. Low temperature, maximum 71–78°F.	380
Nitrogen-free extract	T0 50.5 T1 52.1 T2 50.4	T0: no treatment T1: molasses 9 kg/1000 kg herbage T2: wilted, no additive	
Amino acids (total) %DM	T0 3.5 T1 4.7 T2 3.2		
Volatile bases %DM	T0 1.2 T1 1.6 T2 1.4		
Organic acids %DM			
Lactic	T0 12.4 T1 11.2 T2 3.1		
Acetic	T0 2.0 T1 2.9 T2 1.3		
Butyric	T0 0.1 T1 0.3 T2 0.1		

continued

Y

Table 1.135 Meadow Fescue and mixtures: Silage (continued)

		Notes	Reference
Meadow Fescue and clover unspecified *Proximate constituents* %DM		Herbage cut for silage (no date) in (1) small scale trial, (2) large scale trial. T0: no treatment T1: 11.4 kg molasses/ 1000 kg herbage T2: 3.6 kg metabisulphite/ 1000 kg herbage.	379
Crude protein	(1) T0 } T1 } 19.5 T2 }		
	(2) T0 16.1 T2 16.9		
Ether extract	(1) T0 } T1 } 2.7 T2 }		
	(2) T0 2.7 T2 2.8	For composition of crop before ensiling see table 1.108 same reference	
Nitrogen-free extract	(1) T0 } T1 } 47.8 T2 }		
	(2) T0 46.6 T2 45.2		
Crude fibre	(1) T0 } T1 } 20.1 T2 }		
	(2) T0 26.2 T2 25.8		
Ash	(1) T0 } T1 } 9.9 T2 }		
	(2) T0 8.4 T2 9.3		

Table 1.135 Meadow Fescue and mixtures: Silage (continued)

		Notes	Reference
Meadow Fescue and Clover: *Proximate constituents* %DM		Herbage unwilted (1) small scale trial. (2) large scale trial.	379
Crude protein	(1) T0 19.9 T1 21.6 T2 24.2 (2) T0 20.1 T2 18.7	T0: no treatment T1: 11.4 kg molasses/1000 kg herbage T2: 3.6 kg sodium metabisulphite powder/ 1000 kg herbage.	
Ether extract	(1) T0 4.8 T1 4.8 T2 4.1		
Nitrogen-free extract	(1) T0 35.9 T1 38.4 T2 37.6	For composition of crop before ensiling see table 1.108, same reference	
Crude fibre	(1) T0 28.1 T1 24.2 T2 22.5		
Ash	(1) T0 11.3 T1 11.0 T2 11.6		
Volatile bases %DM	(1) T0 3.8 T1 2.4 T2 2.6 (2) T0 7.7 T2 3.9		
Organic acids %DM			
Lactic	(1) T0 4.0 T1 12.2 T2 8.3 (2) T0 0.2 T2 4.7		
Acetic	(2) T0 2.8 T2 4.2		
Butyric	(2) T0 6.5 T2 2.9		
Propionic	(2) T0 1.1 T2 0.9		
Volatile acids as acetic	(1) T0 11.9 T1 6.8 T2 6.4		
pH	(1) T0 4.8 T1 4.2 T2 4.5 (2) T0 5.8 T2 5.1		

Table 1.136 Timothy mixture: Silage

		Notes	Reference
Timothy, Meadow Fescue and Clover:			
Proximate constituents %DM		Survey of silage quality	386
Crude protein	(1955) 17.9 (1957) 14.8	1 centre Mean of 2 samples from 1 centre	
Organic acids %DM			
Lactic	(1955) 0.68 (1957) 3.56		
Acetic	(1955) 0.29 (1957) 0.52		
Butyric	(1955) 1.88 (1957) 0.23		
pH	(1955) 5.4 (1957) 4.0		

Table 1.137 Lucerne mixtures: silage

		Notes	Reference
Lucerne and Cocksfoot *Proximate constituents* %DM			
Crude protein	W1 19.3 W2 20.8 W3 20.0	Cut late May. Tower silos. Wilted as follows: W1: 0 W2: 4½ hours W3: 8 hours	374
	T0 19.0 T1 19.0 T2 18.3	Low temperature – max. 65–75°F.. Herbage chopped before filling. T0: no treatment T1: 4 kg sodium metabisulphite/ 1000 kg herbage T2: wilted only	380
Nitrogen-free extract	T0 43.8 T1 44.1 T2 43.1		
Amino acids %DM	W1 6.1 W2 4.7 W3 4.8		374
	T0 6.3 T1 5.0 T2 4.8		380
Volatile bases %DM	W1 8.2 W2 1.2 W3 1.1		374
	T0 9.1 T1 2.3 T2 3.4		380
Organic acids %DM			374
Lactic	W1 0.1 W2 1.3 W3 1.2		
	T0 0.2 T1 4.3 T2 5.7		380
Acetic	T0 3.0 T1 0.8 T2 1.2		
Butyric	T0 6.0 T1 0.1 T2 1.2		
Total volatile acid %DM	W1 10.0 W2 1.5 W3 1.7		374
pH	W1 5.3 W2 5.1 W3 5.0		

continued

Table 1.137 Lucerne mixtures: silage (continued)

		Notes	Reference
Lucerne and Meadow Fescue ley *Proximate constituents* %DM		Small scale trial.	379
Crude protein	T0 16.8 T1 16.7 T2 17.7 W0 17.5 W1 16.3	T0: herbage untreated T1: with 10 kg molasses T2: 4 kg metabisulphite per 1000 kg of herbage	
Ether extract	T0 4.1 T1 3.1 T2 4.6 W0 3.7 W1 3.3	W0: wilted W1: wilted with 4.5 kg metabisulphite/per 1000 kg herbage	
Nitrogen-free extract	T0 37.6 T1 44.0 T2 39.5 W0 41.4 W1 43.9	For composition of crop before ensiling see table 1.109, same reference	
Crude fibre	T0 30.1 T1 26.9 T2 27.3 W0 28.0 W1 26.0		
Ash	T0 11.4 T1 9.3 T2 10.9 W0 9.4 W1 10.5		
Volatile bases %DM	T0 4.5 T1 2.8 T2 2.3 W0 2.0 W1 1.4		
Organic acids %DM			
Lactic acid	T0 4.5 T1 8.4 T2 4.5 W0 4.9 W1 2.2		
Volatile acid as acetic	T0 9.9 T1 6.4 T2 5.8 W0 4.3 W1 3.3		
pH	T0 5.0 T1 4.5 T2 4.8 W0 4.7 W1 5.1		

Table 1.137 Lucerne mixtures: silage (continued)

		Notes	Reference
Lucerne and Timothy: *Proximate constituents* %DM		Tower silos C1 = 1st cut, 6-9 June C2 = 2nd cut, 24-27 June C3 = 3rd cut, 31 August - 3rd September	375
Crude protein	C1 16.5 C2 19.6 C3 27.1		
Ether extract	C1 3.3 C2 3.2 C3 3.0	Cuts 1 and 3 had 7.3 kg glycollic acid, cut 2 had 10 kg formic acid/ 1000 kg unchopped herbage	
Nitrogen-free extract	C1 42.4 C2 44.0 C3 32.8		
Crude fibre	C1 28.7 C2 23.8 C3 25.5	The Timothy content was appreciably lower for the 2nd and 3rd cuts than for the 1st.	
Ash	C1 9.1 C2 9.4 C3 11.6	Means of 8 samples taken down the silo.	
Organic acids %DM			
Lactic	C1 4.5 C2 2.5 C3 1.9	For the composition of the crop before ensiling see table 1.119, same reference	
Acetic	C1 3.1 C2 3.2 C3 6.2		
Butyric	C1 0.73 C2 0.13 C3 0.13		
Propionic	C1 0.17 C2 0.07 C3 0.70		
Higher acids as valeric	C1 0.38 C2 0.11 C3 0.07		
pH	C1 4.23 C2 4.14 C3 4.96		
Proximate constituents %DM		Tower silos C1 = 1st cut, 6-9 June C2 = 2nd cut, 24-27 June C3 = 3rd cut, 31 August - 3 September	
Crude protein	C1 16.6 C2 19.4 C3 28.2		
Ether extract	C1 3.5 C2 3.4 C3 3.3	27 kg molasses added/1000 kg green unchopped herbage	
Nitrogen-free extract	C1 39.9 C2 42.8 C3 29.3	The Timothy content was appreciably lower for the 2nd and 3rd cuts than for the 1st.	
Crude fibre	C1 30.6 C2 24.6 C3 26.5	Means of 8 samples taken down the silo.	
Ash	C1 9.4 C2 9.8 C3 12.7	For composition of the crop before ensiling see table 1.119 same reference	

continued

Table 1.137 Lucerne mixtures: silage (continued)

		Notes	Reference
Lucerne and Timothy (cont'd) *Organic acids* %DM			
Lactic	C1 5.4 C2 6.3 C3 0.2		
Acetic	C1 2.9 C2 3.3 C3 7.4		
Butyric	C1 1.77 C2 0.21 C3 1.60		
Propionic	C1 0.28 C2 0.13 C3 1.63		
Higher acids as valeric	C1 0.82 C2 0.10 C3 0.68		
pH	C1 4.35 C2 4.40 C3 5.40		
Proximate constituents %DM		Tower silos 1 kg barley meal added/ 20 kg chopped herbage C1 = 1st cut, 6-9 June C2 = 2nd cut, 24-27 July C3 = 3rd cut, 31 August - 3 September The Timothy content was appreciably lower for the 2nd and 3rd cuts than for the 1st Means of 8 samples taken down the silo For the composition of the crop before ensiling see table 1.119, same reference	375
Crude protein	C1 16.8 C2 18.8 C3 26.3		
Ether extract	C1 3.1 C2 3.3 C3 3.3		
Nitrogen-free extract	C1 42.7 C2 45.0 C3 34.7		
Crude fibre	C1 28.2 C2 23.8 C3 24.2		
Ash	C1 9.2 C2 9.1 C3 11.5		
Organic acids %DM			
Lactic	C1 7.8 C2 5.6 C3 0.4		
Acetic	C1 5.3 C2 2.6 C3 11.7		
Butyric	C1 0.15 C2 0.02 C3 0.19		
Propionic	C1 0.31 C2 0.02 C3 0.94		
Higher acids as valeric	C1 0.11 C2 0.05 C3 0.26		
pH	C1 4.36 C2 4.39 C3 5.10		

Table 1.137 Lucerne mixtures: silage (continued)

		Notes	Reference
Lucerne and Timothy (cont'd) *Proximate constituents %DM*		Tower silos C1 = 1st cut, 6 –9 June	375
Crude protein	C1 17.5 C2 21.5 C3 24.6	C2 = 2nd cut, 24 – 27 June C3 = 3rd cut, 31 August to 3 September	
Ether extract	C1 2.4 C2 2.5 C3 2.7	All cuts were left in the field for periods varying between 24 and 48 hr. before ensiling unchopped	
Nitrogen-free extract	C1 41.5 C2 41.3 C3 37.1		
Crude fibre	C1 28.5 C2 25.2 C3 23.7	The Timothy content was appreciably lower for the 2nd and 3rd cuts than for the 1st	
Ash	C1 10.1 C2 9.5 C3 11.9	Means of 8 samples taken down the silo	
Organic acids %DM		For the composition of the crop before ensiling see table 1.119, same reference	
Lactic	C1 2.2 C2 0.3 C3 2.1		
Acetic	C1 1.3 C2 1.5 C3 2.1		
Butyric	C1 0.10 C2 0.02 C3 0.02		
Propionic	C1 0.03 C2 0.02 C3 0.10		
Higher acids as valeric	C1 0.12 C2 0.12 C3 0.03		
pH	C1 4.86 C2 4.88 C3 5.40		
Lucerne and Timothy, unnamed *Proximate constituents % DM*			
Crude protein	(1952) U 13.5 C 15.1 L 12.9 (1953) U 19.4 UW 19.8 CW 18.8 LW 17.1	Tower silos. Cut, 1952, 21 May to 3 June; 1953, 26 May to 5 June U = unchopped C = chopped L = lacerated W = wilted for 24 hr.	376
Ether extract	(1952) U 2.6 C 3.1 L 3.2 (1953) U 2.5 UW 2.2 CW 2.3 LW 2.9	For the composition of the crop before ensiling see table 1.119, same reference	
Nitrogen-free extract	(1952) U 36.4 C 34.9 L 36.7 (1953) U 35.0 UW 37.5 CW 36.4 LW 35.1		

continued

Table 1.137 Lucerne mixtures: silage (continued)

			Notes	Reference
Lucerne and Timothy, unnamed (continued)				376
Proximate constituents (cont'd)				
Crude fibre	(1952) U	39.0		
	C	38.3		
	L	38.7		
	(1953) U	33.9		
	UW	31.0		
	CW	33.1		
	LW	36.1		
Ash	(1952) U	8.5		
	C	8.6		
	L	8.5		
	(1953) U	9.2		
	UW	9.5		
	CW	9.4		
	LW	8.8		
Organic acids %DM				
Lactic	(1952) U	0.1		
	C	0.1		
	L	0.5		
	(1953) U	1.0		
	UW	0.6		
	CW	1.5		
	LW	0.5		
Acetic	(1952) U	1.2		
	C	1.8		
	L	3.8		
	(1953) U	2.0		
	UW	1.6		
	CW	1.7		
	LW	2.3		
Butyric	(1952) U	2.2		
	C	0.1		
	L	0.2		
	(1953) U	5.4		
	UW	1.2		
	CW	2.8		
	LW	3.3		
Propionic	(1952) U	0.5		
	C	0.4		
	L	0.4		
	(1953) U	1.2		
	UW	0.3		
	CW	0.9		
	LW	0.6		
Higher acids as valeric	(1952) U	1.2		
	C	0.1		
	L	0.1		
	(1953) U	2.0		
	UW	0.1		
	CW	0.6		
	LW	1.1		
pH	(1952) U	5.0		
	C	5.6		
	L	4.7		
	(1953) U	5.3		
	UW	5.1		
	CW	5.5		
	LW	5.3		

Table 1.137 Lucerne mixtures: silage (continued)

			Notes	Reference
Lucerne and Timothy				
Proximate constituents %DM			Trial 1: Crop wilted, chopped	381
Crude protein	A0	22.0	A0: no additive	
	A1	21.1	A1: 3.6 kg sodium metabisulphite/ 1000 kg herbage	
Ether extract	A0	3.6		
	A1	3.2		
Nitrogen-free extract	A0	39.1		
	A1	40.5		
Crude fibre	A0	26.2		
	A1	25.5		
Ash	A0	9.1		
	A1	9.7		
Volatile bases %DM	A0	3.54		
	A1	2.11		
Organic acids %DM				
Lactic	A0	5.15		
	A1	5.39		
Volatile acids as acetic	A0	2.57		
	A1	1.40		
pH	A0	4.8		
	A1	4.6		
Volatile bases %DM	A0	9.57	Trial 4:	
	A1	6.55	A0: unchopped, no additive	
	A2	5.84	A1: unchopped, 5.5kg MBS/1000 kg herbage	
Organic acids %DM			A2: chopped, 3.6 kg MBS/1000 kg herbage	
Lactic	A0	0.03		
	A1	0.13		
	A2	2.02		
Volatile acids as acetic	A0	9.91		
	A1	6.24		
	A2	9.09		
Volatile bases %DM	A0	10.57	Trial 5: All herbage unchopped	
	A1	8.86		
Organic acids %DM			A0: no additive	
Lactic	A0	0.35	A1: 3.6 kg MBS/1000 kg herbage	
	A1	1.17		
Acetic	A0	3.56		
	A1	5.43	For the composition of the crop before ensiling see table 1.119, same reference	
Butyric	A0	9.08		
	A1	7.19		
Propionic	A0	2.70		
	A1	2.49		
Higher acids as valeric	A0	1.55		
	A1	1.04		
Volatile acids as acetic	A0	12.1		
	A1	13.0		
pH	A0	5.6		
	A1	5.4		

Table 1.138 Miscellaneous fodders: Silage

				Notes	Reference
Yorkshire Fog and Crested Dogstail					386
Proximate constituents %DM				Survey of silage quality 1 Centre	
Crude protein	(1957)		10.3		
Organic acids %DM					
Lactic			1.00		
Acetic			0.30		
Butyric			Nil		
pH			6.0		
Grasses, unspecified					
Proximate constituents %DM				Wooden Tower Silos.	208
				A0 = no additive	
				A1 = A.I.V. Acids	
Crude protein	(1933)	A0	11.9	A2 = 45.4 kg molasses in 212 litres water, 5.3 litres/50 kg grass, sprayed.	
	(1933)	A1	11.7		
	(1933)	A2	12.3		
	(1933)	A3	12.9	A3 = Acid as in A1 + 1 kg molasses/200 kg herbage	
	(1934)	A2 RI	12.4		
	(1934)	A2 I	14.6	A4 = Whey 5.3 litres/50 kg cut grass.	
	(1934)	A2 FM	15.3		
	(1934)	A4 ME	14.0		
	(1934)	A4 MC	10.6	1933: Herbage cut 16 June at fairly leafy stage.	
Ether extract	(1933)	A0	3.6		
	(1933)	A1	3.3	1934:	
	(1933)	A2	3.8	RI Rain soaked immature grass, cut 18 May	
	(1933)	A3	3.0	I Immature grass, cut 30 May	
	(1934)	A2 RI	10.4	FM Fairly mature grass, cut 19 July	
	(1934)	A2 I	8.4		
	(1934)	A2 FM	5.3	ME Mature grass: edge sample, cut 7 June	
	(1934)	A4 ME	3.7		
	(1934)	A4 MC	5.8	MC Mature grass: centre sample, cut 7 June	
Nitrogen-free extract	(1933)	A0	45.4		
	(1933)	A1	47.1		
	(1933)	A2	47.1		
	(1933)	A3	46.0		
	(1934)	A2 RI	30.4		
	(1934)	A2 I	35.7	For composition of crop before ensiling see table 1.120, same reference	
	(1934)	A2 FM	37.3		
	(1934)	A4 ME	41.1		
	(1934)	A4 MC	41.5		
	(1933)	A0	32.2		
	(1933)	A1	30.7		
	(1933)	A2	28.7		
	(1933)	A3	30.1		
	(1934)	A2 RI	36.0		
	(1934)	A2 I	30.0		
	(1934)	A2 FM	33.0		
	(1934)	A4 ME	31.7		
	(1934)	A4 MC	32.3		
Ash	(1933)	A0	6.9		
	(1933)	A1	7.2		
	(1933)	A2	8.1		
	(1933)	A3	8.0		
	(1934)	A2 RI	10.8		
	(1934)	A2 I	11.3		
	(1934)	A2 FM	9.1		
	(1934)	A4 ME	9.5		
	(1934)	A4 MC	9.8		

Table 1.138 Miscellaneous fodders: Silage (continued)

		Notes	Reference
Grasses, unspecified (cont'd) *Major elements* %DM (cont'd) Calcium	(1933) A0 0.83 A1 0.65 A2 0.76 A3 0.60		208
	(1934) A2 RI 0.57 A2 I 0.74 A2 FM 0.94 A4 ME 0.48 A4 MC 0.81		
Phosphorus	(1933) A0 0.29 A1 0.22 A2 0.22 A3 0.25		
	(1934) A2 RI 0.20 A2 I 0.30 A2 FM 0.21 A4 ME 0.27 A4 MC 0.27		
Potassium	(1933) A0 – A1 2.02 A2 2.75 A3 2.29		
	(1934) A2 RI 2.87 A2 I – A2 FM – A4 ME – A4 MC 2.95		
Chloride	(1933) A1 1.38 A2 1.39 A3 1.72		
	(1934) A2 1.07 A4 1.34		
Meadow Grass, unspecified, old pasture			
Proximate constituents %DM		Survey of silage quality	386
Crude protein	(1955) 10.1 (1957) 11.2	Mean of 3 samples from 3 centres 1 centre	
Organic acids %DM			
Lactic	(1955) 0.79 (1957) 3.91	Mean of 2 samples from 2 centres 1 centre	
Acetic	(1955) 0.72 (1957) 0.20	Mean of 2 samples from 2 centres 1 centre	
Butyric	(1955) 0.89 (1957) 0.09	Mean of 2 samples from 2 centres 1 centre	
pH	(1955) 5.3 (1957) 4.1	Mean of 3 samples from 3 centres 1 centre	
Grass, unspecified 1st and 2nd year leys:			
Proximate constituents %DM			
Crude protein	(1955) 13.3 (10.8–19.6) (1957) 16.3 (10.6–23.1)	Mean and range of 4 samples from 3 centres Mean and range of 4 samples from 2 centres	
Organic acids %DM			
Lactic	(1955) 2.55 (0.46–4.77) (1957) 2.82 (1.47–3.47)		

continued

Table 1.138 Miscellaneous fodders: Silage (continued)

		Notes	Reference
Grass, unspecified (cont'd) *Organic acids* %DM (cont'd)			386
Acetic	(1955) 0.27 (nil–0.54) (1957) 0.68 (0.32–1.48)		
Butyric	(1955) 0.78 (0.01–1.92) (1957) 0.64 (0.12–2.03		
pH	(1955) 4.5 (3.9–5.3) (1957) 4.4 (3.8–5.6)		
Pasture grass, old pasture			
Proximate constituents %DM			
Crude protein	(1955) 13.0 (10.1–17.0) (1957) 12.9	Mean and range of 9 samples from 7 centres 1 centre	
Organic acids %DM			
Lactic	(1955) 1.77 (0.30–2.93) (1957) 1.78		
Acetic	(1955) 0.55 (0.22–1.63) (1957) 0.41		
Butyric	(1955) 0.61 (0.12–1.24) (1957) 0.88		
pH	(1955) 4.9 (4.3–5.3) (1957) 4.5		
Grass, unspecified (mixed)			
Proximate constituents %DM			
Crude protein	(1955) 13.8 (1957) 13.0 (8.7–24.4)	Mean of 3 samples from 3 centres Mean and range of 17 samples from 6 centres.	
Organic acids %DM			
Lactic	(1955) 0.80 (1957) 1.67 (0.23–3.24)		
Acetic	(1955) 0.58 (1957) 0.51 (0.10–1.93)		
Butyric	(1955) 0.26 (1957) 0.30 (nil–1.24)		
pH	(1955) 5.3 (1957) 4.7 (4.0–5.8)		
Grass and Clover			
Proximate constituents %DM			
Crude protein	(1955) 12.7 (1957) 11.5	Mean of 3 samples from 3 centres Mean of 3 samples from 3 centres	
Organic acids %DM			
Lactic	(1955) 2.74 (1957) 3.51		
Acetic	(1955) 0.52 (1957) 0.68		
Butyric	(1955) 0.49 (1957) 0.16		
pH	(1955) 4.7 (1957) 4.3		

Table 1.138 Miscellaneous fodders. Silage (continued)

			Notes	Reference
Grass and Lucerne				
Proximate constituents %DM				
Crude Protein	(1955)	14.7	1 centre	386
Organic acids %DM				
Lactic		1.71		
Acetic		0.60		
Butyric		0.67		
pH		4.3		
Grass, cereals and legumes				
Proximate constituents %DM				
Crude protein	(1955)	12.3	1 centre	
Organic acids %DM				
Lactic		2.24		
Acetic		0.32		
Butyric		0.32		
pH		4.2		
Oats, Peas and Vetches				
Proximate constituents %DM				
Crude protein	(1955)	11.9	1 centre	
Organic acids %DM				
Lactic		0.77		
Acetic		0.51		
Butyric		0.79		
pH		5.6		
Oats and Ryegrass				
Proximate constituents %DM				
Crude protein	(1955)	9.5	1 centre	
Organic acids %DM				
Lactic		1.85		
Acetic		1.43		
Butyric		Nil		
pH		4.5		

continued

Table 1.138 Miscellaneous fodders: Silage (continued)

		Notes	Reference
Grass, unspecified		Two sets of 13 silages, 1–4 samples from each. No additive	349
Proximate constituents %DM			
Crude protein	14.6 (9.7–23.7)		
	14.8 (10.1–23.2)		184
Organic acids mg/100g fresh silage			
Lactic acid			
Total	1332 (105–2049)		349
Volatile	118 (8–247)		
Total	999 (13–2021)		184
Volatile	86* (0–205)	*or 112 allowing for 3 samples with none	
Acetic acid			
Total	355 (210–576)		349
Volatile	314 (168–546)		
Total	432 (288–683)		184
Volatile	406 (274–641)		
Butyric acid			
Total	193* (0–698)	*or 228 allowing for 2 samples with none	349
Volatile	180* (0–668)	*or 213 allowing for 2 samples with none	
Total	221* (0–633)	*or 359 allowing for 5 samples with none	184
Volatile	209* (0–607)	*or 340 allowing for 5 samples with none	
pH	4.4 (3.7–5.2)		349
	4.5 (3.7–5.4)		184

Table 1.138 Miscellaneous fodders: Silage (continued)

		Notes	Reference
Grass, unspecified			
Proximate constituents %DM		A study of the effect of temperature in the silo.	
Crude protein	A 13.7 (12.1–16.0) B 15.3 (11.0–17.8)	A: Mean of top samples from 3 silos, unchopped herbage B: Mean of top samples from 5 silos, chopped herbage	380
Nitrogen-free extract	A 46.8 (46.5–47.4) B 48.4 (45.3–50.3)		
Amino Acids %DM	U 3.1 (1.7–3.9) C 4.5 (1.3–6.4)	U: Mean of 11 samples from 3 silos, unchopped herbage C: Mean of 21 samples from 5 silos, chopped herbage	
Volatile bases %DM	U 6.0 (4.2–8.3) C 3.6 (1.5–10.0)		
Organic acids %DM		Between lactic acid and maximum temperature there was a negative correlation	
Lactic	U 0.47 (0.1–1.1) C 6.1 (0.1–16.4)		
Acetic	U 2.4 (1.6–3.5) C 3.0 (1.4–5.1)		
Butyric	U 4.2 (3.6–5.2) C 2.2 (0.1–6.6)		
Grass, unspecified			
Amines mg/kg DM		Pit silage	346
Histamine	Trace	Amines estimated by chromatography	
Tyramine	90		
pH	3.80		
Grassland herbage			
Proximate constituents %DM		Silages collected for study of protein break-down. No information about crops.	535
Crude protein	O 14.6 M 15.4 A 14.8	O: Weighted means of 38 samples in 4 groups by pH; no additive.	
Volatile bases %DM	O 2.43 M 1.45 A 1.10	M: Weighted means of 38 samples in 4 pH groups with molasses, 2.2 to 4.4 kg/100 kg fresh grass.	
Amino Acids %DM	O 2.77 M 3.77 A 2.54	A: Weighted means of 143 samples in 5 pH groups with AIV fluid as recommended by Virtanen.	

continued

Table 1.138 Miscellaneous fodders: Silage (continued)

		Notes	Reference
Grass, unspecified and Clover			
Proximate constituents %DM		Herbage semi-mature, stemmy Silage as follows:	384
Crude protein	S1 12.2 S2 12.5 S3 10.9 S4 13.5	S1 : cut 13/6–18/6, wilted 48 hr. Walled Trench silo. S2 : cut 4/7–12/7, wilted 18 hr. Walled Trench silo. S3 : cut 6/6–28/6, wilted 70 hr. Sunk Trench silo. S4 : cut 27/6–3/7, wilted 23 hr. Sunk Trench silo. Means of 3 samples.	
Organic acids %DM Lactic	S1 3.19 S2 2.39 S3 4.46 S4 6.65		
Acetic	S1 0.52 S2 0.26 S3 0.23 S4 0.34		
Butyric	S1 0.13 S2 0.14 S3 0.01 S4 0.21		
pH	S1 4.2 S2 4.8 S3 4.2 S4 4.3		
Grass and Clover, unspecified:			
Proximate constituents %DM Crude protein	(1953) TO 13.3 TW 13.7 (13.0–15.1) (1954) MO 12.3 MW 12.1 BO 12.3 BW 12.3 (1955) AO 16.2 AM 16.6 AA 17.6 AOW 14.6 AMW 13.6 AAW 14.6	Herbage cut beginning of June at long, leafy stage in 1953 and 1954. Young leafy in 1955. Small tower silos. TO: no treatment, 2 samples TW: wilted for 17½, 20 or 22½ h in 1953, mean and range of 6 samples. MO: Cut with mower, no treatment, 1 sample	383
Nitrogen-free extract*	(1953) TO 41.7 TW 42.1 (41.2–42.9) (1954) MO 44.0 MW 44.3 BO 41.1 BW 42.7 (1955) AO 47.9 AM 48.0 AA 45.8 AOW 49.7 AMW 50.2 AAW 50.1	MW: Cut with mower, wilted for 8 or 27h in 1954, 2 samples BO: bruised by cutting with Shearmow forage harvester, no other treatment, 1954, 1 sample. BW: bruised, wilted for 4 or 8 h, 1954, 2 samples AO: no additive, 1955, 1 sample	
Crude fibre	(1953) TO 32.9 TW 31.9 (30.8–32.8) (1954) MO 32.6 MW 31.9 BO 31.9 BW 30.4 (1955) AO 22.9 AM 22.1 AA 21.6 AOW 22.9 AMW 23.7 AAW 22.0	AM: 11.4 kg molasses/1000 kg herbage, 1 sample AA: 5.5 kg herbage, MBS (sodium metabisulphite)/1000 kg 1 sample AOW: wilted 18½ h, no additive, 1 sample AMW: wilted 18½ h, molasses, 1 sample AAW: wilted 18½ h, MBS, 1 sample	

Table 1.138 Miscellaneous fodders: Silage (continued)

		Notes	Reference
Grass and Clover, unspecified: *Proximate constituents* %DM Ash (cont'd)	(1953) TO 9.1 TW 9.3 (9.0–9.9) (1954) MO 8.1 MW 8.7 BO 11.7 BW 11.6 (1955) AO 10.0 AM 10.3 AA 12.0 AOW 9.8 AMW 9.5 AAW10.3	*Ether extract was assumed to be 3% of DM and nitrogen-free extract was calculated by difference with that assumption For the composition of the crop before ensiling, see table 1.123, same reference.	383
Organic acids %DM			
Lactic	(1953) TO 7.81 TW 5.55 (5.15–6.07) (1954) MO 8.47 MW 4.39 BO 7.69 BW 7.49 (1955) AO 13.61 AM 11.96 AA 7.21 AOW 9.13 AMW 7.97 AAW 5.79		
Acetic	(1953) TO 2.42 TW 0.82 (0.50–1.27) (1954) MO 3.27 MW 1.05 BO 3.29 BW 3.15 (1955) AO 1.70 AM 1.81 AA 0.40 AOW 1.07 AMW 0.90 AAW 0.70		
Butyric	(1953) TO 0.18 TW 0.11 (0.05–0.21) (1954) MO 0.00 MW Trace BO 0.00 BW 0.00 (1955) AO 0.00 AM 0.20 AA 0.15 AOW 0.12 AMW 0.11 AAW 0.81		
pH	(1953) TO 4.0 TW 4.2 (4.0–4.4) (1954) MO 3.8 MW 4.3 BO 3.9 BW 3.9 (1955) AO 4.1 AM 3.9 AA 4.6 AOW 4.0 AMW 4.1 AAW 5.0		

continued

Table 1.138 Miscellaneous fodders. Silage (continued)

		Notes	Reference
Maize *Proximate constituents %DM*			
Crude protein Ether Extract Nitrogen-free extract Crude fibre Ash	7.9 2.5 60.5 24.3 4.8	Kingscrost KF1. Reading 1947. Cut at dough stage	129
Crude protein Ether extract Nitrogen-free extract Crude fibre Ash pH	(1) 11.2 2.4 51.0 28.5 6.9 3.5–3.7	Wisconsin 275. Reading 1948. Cut at dough stage (1) Opened March 1949 (2) Opened January 1950	
Crude protein Ether extract Nitrogen-free extract Crude fibre Ash pH	(2) 9.5 2.1 58.1 24.6 5.7 3.7		
Crude protein Ether extract Nitrogen-free extract Crude fibre Ash pH	9.5 2.2 55.5 25.8 7.0 5.5	Var. unspecified Newport, Shropshire, 1959	373
Crude protein Ether extract Nitrogen free extract Crude fibre Ash	9.0 2.4 51.8 28.7 8.1	Wisconsin 275 with Marrow-stem. Kale. Maize cut 7/10/48. Silo opened January 1949	129
Cereals and Legumes *Proximate constituents %DM*			535
Crude protein	T0 13.4 T1 19.8 T2 13.9	Mainly oats and vetches, some peas and beans.	
Amino acids (total %DM)	T0 3.31 T1 2.54 T2 3.01	T0 = no treatment: 14 samples T1 = 1–2kg molasses per 100kg fresh herbage: 4 samples. T2= Mineral acid AIV process: 6 samples	
Volatile bases %DM	T0 1.50 T1 3.71 T2 1.11		
pH	T0 4.31 T1 4.73 T2 4.14	CP calculated as N from amino acids and volatile bases x 6.25.	

For Sugar beet tops silage and Oats, peas, beans and tares silage see table 1.186

Table 1.139 Grass and Clover Silage. Fatty acids and pH

Reference 46

pH range	Fatty acids mg/100 g DM									Total volatile fatty acids %DM	Notes
	C_1	C_2	C_3	n–C_4	n–C_5	*iso*–C_5	C_6	C_7	C_8		
3.61–3.80	30	2073	314	247	9	0	0	0	0	2.67	A study of fatty acid variation with pH.
3.81–4.00	49	2502	292	381	8	0	0	0	0	3.23	42 samples altogether.
4.01–4.20	0	1330	304	376	0	0	0	0	0	2.01	Estimated chromatographically after steam distillation.
4.21–4.40	16	1835	285	303	141	0	77	0	0	2.66	
4.41–4.60	41	4058	663	1309	232	0	595	0	0	6.90	
4.61–4.80	97	3850	645	2012	218	0	578	0	0	7.40	
4.81–5.00	0	2030	224	535	84	0	770	108	358	4.11	
5.01–5.20	59	5290	853	2084	464	0	563	0	0	9.31	
5.41–5.60	234	5201	730	1485	716	490	1439	0	0	10.29	
6.21–6.40	0	330	107	314	0	114	35	0	0	0.90	
6.41–6.60	89	1610	346	377	99	0	0	0	0	2.52	
6.81–7.00	0	266	14	16	0	0	0	0	0	0.30	
7.01–7.20	0	333	20	28	0	0	0	0	0	0.38	

Table 1.140 Lucerne: Provence

	January			February			March			April			May			June		
Digestibility of DM%																	1C 62.8	
												1A 73.3				1C 63.4	1F 57.9	
Digestibility of OM%																	1C 65.5	
												1A 75.8				1C 69.2	1F 60.2	
Proximate constituents %DM Crude protein												1A 35.0	1A 33.4	1A 31.1	1A 23.1	1C 22.8	1C 20.4	1F 16.2
															1A 23.0	1B 21.5		1C 20.3
																	(a) (b)	1E 17.6 17.4
													1A 27.9		2A 32.6			3A 26.7
												1A 25.3				1C 20.7	1F 17.1	
											W {	1A 24.4			2A 29.6	1C 19.1 1F 18.0		3A 25.6 2C 24.3
											L {	1A 30.1			2A 34.8	1C 27.5 1F 23.5		3A 30.7 2C 29.7
											S {	1A 17.0			2A 19.1	1C 12.5 1F 9.4		3A 14.1 2C 15.2
																W L S	16.4 24.0 9.9	
											(1951) {	N_0 N_0 N_1 N_1 N_2 N_2 N_0 N_1 N_2	(Aa) (Ab) (Aa) (Ab) (Aa) (Ab) (Sa) (Sa) (Sa)	27.2 25.6 26.8 26.8 27.0 27.2 26.8 28.2 27.1				
											(1952) {	N_0 N_0 N_1 N_1 N_2 N_2 N_0 N_1 N_2	(Aa) (Ab) (Aa) (Ab) (Aa) (Ab) (Sa) (Sa) (Sa)	24.5 23.9 25.2 25.8 24.5 26.3 25.2 25.2 25.7				N_0 N_0 N_1 N_1 N_2 N_2 N_0 N_1 N_2
												25.7	24.7		20.0			15.2
																	(1949) {	
									(1950) {	(a) (b) (a) (b) (a) (b) (a) (b)	18.6 18.6 19.4 19.4 18.7 18.4 19.8 19.3				(a) 15.3 (b) 14.4 (a) 15.4 (b) 14.8 (a) 13.9 (b) 15.6 (a) 16.2 (b) 15.7			

July			August			September			October			November			December			Notes	Reference
1F 55.9			2F 57.7															In the data from references 580 and 581 there are distinquished: primary growth (1), 4 regrowths after cutting (2-5), and 7 stages of development at cutting: prebud (A), early bud (B), bud (C), half flower (D), early flower (E), flower (F), and late flower (G). There are two sets of metabolism data as shown under Digestibility. The first is for lucerne cut at Howe Hill in 1932 at three dates and stages: 1C (mid June), 1F (early July) and 2F (early August). The second is for 6 dates in 1933, the first at Howe Hill at stage 1A, the rest at Willingham at stages and dates: 1C (early June), 2C (early July), 3C (mid August), 1F (mid June) and 2F (late July). The same notation has been applied to the Crude protein data. The corresponding data for other proximate constituents and major elements should be matched against Crude protein.	580
2C 63.4		2F 61.2		3C 60.6															581
1F 60.4			2F 62.0																580
2C 65.6		2F 64.4		3C 64.8															581
1F 17.4	2B 23.5		2F 19.3		2F 21.1 3B 27.1		3F 23.3												580
			2D 21.3			3E 24.4													
					2G 17.3														
	4A 29.6			5A 23.8														Digestibilities are for 2 sheep For metabolisable energy see table 1.232 (a) Seed inoculated "old culture" (b) Seed inoculated "acid culture"	581
2C 22.4		2F 19.3		3C 21.4		3F 21.5													
	4A 26.1	3C 20.6 2F 19.5		3F 19.9															
	4A 30.6	3C 28.1 2F 26.3																W : whole plant L : leaf S : stem	556
	4A 13.3	3C 10.6 2F 10.3																	
W L S	17.4 27.5 10.9																	W : whole plant L : leaf S : stem Planted September, 1951. Sampled 1952 (upper group) and 1953 (lower group)	170
(Aa) (Ab) (Aa) (Ab) (Aa) (Ab) (Sa) (Sa) (Sa)	17.2 17.6 17.6 17.8 18.2 17.7 17.2 17.4 16.7					N_0 N_0 N_1 N_1 N_2 N_2 N_0 N_1 N_2	(Aa) (Ab) (Aa) (Ab) (Aa) (Ab) (Sa) (Sa) (Sa)	20.7 20.8 20.4 21.5 19.1 20.0 19.9 20.4 21.7	N_0 (Aa) 21.0 N_1 (Aa) 21.0 N_2 (Aa) 23.0		1950	N_0 N_1 N_2	(Aa) (Aa) (Aa)	24.0 26.8 27.2				N_0 : no fertiliser N_1 : Nitrochalk 250 kg/ha in autumn N_2 : Nitrochalk 500 kg/ha in autumn A : N applied in autumn S : N applied in spring Plot experiments in which all plots were cut in summer. The (a) plots were and the (b) plots were not cut in the previous autumn	3
16.5*	17.1*	19.3*		19.9*	20.4*		21.6*				22.0*							First cut: primary growth Second cut, Third cut, Fourth cut: *Regrowth	161
(a) (b) (a) (b) (a) (b) (a) (b)	17.1 18.2 16.3 15.8 18.3 18.7 14.9 15.0						(a) (b) (a) (b) (a) (b) (a) (b)	17.0 17.2 15.9 15.6 17.3 17.6 15.6 15.8										(a) Cut previous autumn (b) Grazed previous autumn 1949 : Third growth 1950 : Primary, second and third growth	42

continued

Table 1.140 Lucerne: Provence (continued)

	January			February			March			April			May			June		
Proximate constituents %DM (cont'd)																		
Crude protein (cont'd)														(1948)	N_0	18.4		
															N_1	18.3		
															N_2	18.7		
															N_3	18.8		
														(1949)	N_0	17.7		
															N_1	17.8		
															N_2	17.6		
															N_3	17.8		
														(1950)	N_0	18.7		
															N_1	18.9		
															N_2	19.1		
															N_3	20.1		
														(1951)	N_0	16.1		
															N_1	16.9		
															N_2	17.3		
															N_3	17.4		
															(1948)		15.6	
															(1949)		14.4	
															(1950)	16.2		
												26.2		22.6		20.4		15.7
																(1956)	W	15.8
																	L	22.3
																	S	10.0
																(1957)	W	14.3
																	L	21.2
																	S	8.6
																	(1958)	W
																		L
																		S
														First year	(1)	17.0		
															(2)	16.4		
															(3)	16.9		
															(4)	17.9		
															(5)	18.1		
															(6)	19.1		
															(7)	19.2		
															(8)	19.3		
														Second year	(1)	16.2		
															(2)	15.4		
															(3)	15.6		
															(4)	15.8		
															(5)	15.5		
															(6)	17.4		
															(7)	17.2		
															(8)	17.6		
														Third year	(1)	17.4		
															(2)	18.1		
															(3)	18.1		
															(4)	18.9		
															(5)	18.8		
															(6)	18.9		
															(7)	20.2		
															(8)	19.6		
															(1955)	(C3)	17.3	
																(G4)	17.9	
																(G5)	17.1	
																(G3C)	16.8	
																(G4C)	18.5	
																(CG3)	18.6	
																(CG4)	17.4	
															(1956)	(C3)	17.9	
																(G4)	17.6	
																(G5)	17.2	
																(G6)	17.6	
																(G7)	17.6	
																(G3C)	17.4	
																(G4C)	17.9	
																(G5C)	18.4	
																(CG3)	17.2	
																(CG4)	17.7	
																(CG5)	18.4	

July			August			September			October			November			December			Notes	Reference
		19.1 19.1 18.9 19.0 17.4 17.8 17.8 18.1 18.1 17.1 17.7 17.9						20.1 19.9 19.7 19.9 14.8 15.1 15.2 15.3		14.8 15.0 14.7 14.8								N_0 : No fertiliser N_1 : Cuts taken after dressing in February, June and August; 3 cuts N_2 : Cuts taken after dressing in February and June, February and August, and June and August; 3 cuts N_3 : Cut after dressing in February, June and August; 1 cut Fertiliser: ammonium sulphate, 125 kg/ha	438
		16.0			15.6			18.8	19.0									Grazed by sheep in August Cut each time at late bud/ early flowering stage	439
18.5*	18.3*	17.7*		19.9*	19.8*		18.7*				22.6*							First cut: primary growth Second cut, Third cut, Fourth cut } *Regrowth	164
15.1 22.3 10.5																		W : whole plant L : leaf S : stem Sown July 1955 in close drills	590
	15.9 19.5 18.6			19.0														1 cut 1956 3 cuts (May, July, August 1957) 3 cuts (June, July, September 1958) 3 cuts (June, July, August 1959) Sown in rows 22/5/1956. Weighted means for all cuts in each year 1957-59	132
			19.6 19.6 19.1 18.9 18.9 18.4 17.7 18.9 19.3 20.4 19.4 19.6 20.3 19.2 19.4 18.2 20.1 20.0 20.8 21.3 19.7 19.3 19.9 20.3								16.6 16.9 17.4 15.6 16.7 16.7 17.1 17.1 15.9 16.2 16.5 15.6 16.7 17.8 15.6 19.4 13.2 15.9 15.8 16.6 17.3 16.8 18.0 16.0							Plants cm apart (1) 2.5 (2) 5.1 (3) 7.6 (4) 15.2 (5) 22.9 (6) 30.5 (7) 61.0 (8) 91.4 Population density and yield experiment	308
	(C3) 19.1 (G4) 19.1 (G5) 19.2 (G6) 20.3 (G3C) 19.4 (G4C) 19.1 (G5C) 19.8 (CG3) 19.9 (CG4) 18.9 (CG5) 19.8	(C3) 19.2 (G4) 19.9 (G5) 20.3 (G6) 20.3 (G7) 20.6 (G3C) 19.4 (G4C) 19.9 (G5C) 20.1 (CG3) 19.8 (CG4) 19.4 (CG5) 21.1								(C3) 16.2 (G4) 17.0 (G5) 18.3 (G6) 18.1 (G3C) 15.8 (G4C) 16.7 (G5C) 17.4 (CG3) 17.7 (CG4) 17.7 (C3) 18.8 (G4) 19.2 (G5) 19.8 (G6) 20.1 (G7) 20.9 (G3C) 18.9 (G4C) 19.7 (G5C) 19.8 (CG3) 20.4 (CG4) 19.3 (CG5) 20.1								Authors' code: C = Cut G = Grazed: number gives number of times cut or grazed	541
	(1959) 18.0 (1960) 19.7																	3 cuts each year. Means only Sown spring 1957	158

continued

Table 1.140 Lucerne: Provence (continued)

	January			February			March			April			May			June		
Proximate constituents %DM (cont'd)																		
Ether extract												3.3	3.2	2.9	2.4	3.1	2.6	3.0
															2.4	3.9		3.2
																	(a) (b)	2.1 2.0
													3.4		3.1			2.4
												2.4						
															2.2		1.6	
											W	3.8			3.7	2.6 2.5		4.0 3.0
											L	3.4			3.2	3.0 3.0		4.1 4.2
											S	1.2			1.1	1.2 1.0		1.5 1.7
															(1952)	W L S	2.2 3.3 1.2	(1953)
												3.0	2.8		2.7			2.3
												3.7		3.5		2.9		2.0
Nitrogen-free-extract												35.7	38.6	37.5	41.1	39.5	42.3	41.1
															39.7	37.0		39.1
																	(a) (b)	38.4 38.9
													38.7		35.4			38.9
												38.1						
															40.7		41.3	
											W	41.4			33.9	39.3 39.1		37.4 35.5
											L	41.2			37.6	43.5 46.2		41.6 40.8
											S	41.2			32.1	36.0 38.2		35.9 33.6
															(1952)	W L S	40.3 44.6 36.6	(1953)
												44.1	44.1		42.0			38.7
												46.4		43.6		40.2		40.9

July			August			September			October			November			December			Notes	Reference
2.1	2.9		2.3		2.6 3.3		2.8											See Crude protein, same reference	580
			2.8			2.3													
					1.9														
	3.0			3.0														See Crude protein, same reference	581
2.6		2.5		2.3		2.4													
	4.1 4.5 1.8	3.5 3.0 5.0 4.0 1.5 1.6		2.7															556
W L S	2.4 3.9 1.4																	See Crude protein, same reference	170
2.5*	2.3*	3.0*		2.9*	3.0*		2.7*				3.9*							First cut, primary growth Second cut, Third cut, Fourth cut } *Regrowth	161
2.8*	2.6*	2.7*		2.4*	2.4*		2.7*							3.0*				First cut, primary growth Second cut, Third cut, Fourth cut } *Regrowth	164
39.3	38.9		39.0		37.2 34.8		35.8											See Crude protein, same reference	580
			36.0			35.0													
					37.5														
	39.4			43.1														See Crude protein, same reference	581
39.3		39.9		40.3		42.1													
	38.4 40.4 36.6	38.2 37.5 41.8 43.3 34.6 35.2		38.2															556
W L S	38.0 40.6 36.3																	See Crude protein, same reference	170
38.4*	38.5*	39.3*		40.0*	41.1*		40.3*				39.6*							See Crude protein, same reference	161
37.8*	39.2*	39.7*		38.3*	40.5*		42.6*				40.4*							See Crude protein, same reference	164

continued

Table 1.140 Lucerne: Provence (continued)

	January			February			March			April			May			June		
Proximate constituents (cont'd) Crude fibre												13.7	14.4	17.5	24.3	25.4	23.9	29.2
															26.2	29.7		28.5
																	(a) (b)	33.3 32.3
													20.7		17.5			22.3
												22.1						
															28.0		29.9	
											W	18.7			16.5	28.1 29.6		20.7 24.4
											L S	13.0 30.8			11.6 32.3	12.3 13.4 42.5 44.4		11.2 12.2 37.0 39.1
															(1952)	W L S	30.6 11.9 46.7	(1953)
												15.3	17.5		24.3			33.9
												13.6		20.3		27.8		33.2
																(1956) (1957)	W L S W L S (1958)	27.1 10.9 41.8 28.3 11.6 42.3 W L S
Ash												12.3	10.4	10.8	9.1	9.2	10.8	10.5
															8.7	7.9		8.9
																	(a) (b)	8.7 9.1
													9.3		11.4			9.7
												12.1						
															8.4		10.1	
											W	11.7			16.3 D	10.9 10.8 10.8		12.3 12.8
											L S	12.3 9.8			12.8 15.4	13.7 13.9 7.8 7.0		12.4 13.1 11.5 10.4
															(1952)	W L S	10.5 16.2 5.6	(1953)
												11.9	10.9		11.0			9.9
												10.1		10.0		8.7		8.2

July			August			September			October			November			December			Notes	Reference
29.7	24.7		28.5		29.8 34.8		35.8											See Crude protein, same reference *Regrowth	580
			30.8			29.1													
					35.6*														
	17.8			18.4														See Crude protein, same reference	581
26.1		28.8	25.4			22.3													
	18.9	25.8 29.2		27.5															
	11.8 36.7	11.6 13.2 44.2 45.6																	556
W L S	33.4 13.3 46.3																	See Crude protein, same reference	170
33.4*	32.8*	27.7*		27.0*	25.5*		24.1*				22.3*							See Crude protein, same reference	161
31.9*	30.9*	31.4*		30.8*	29.0*		26.6*				23.4*							See Crude protein, same reference	164
32.1 12.3 44.7																		See Crude protein, same reference	590
11.5	10.0		10.9		9.1 9.5		11.2											See Crude protein, same reference	580
			9.1			9.2													
					7.7														
	10.2			11.7														See Crude protein, same reference	581
9.6		9.5	10.6			11.7													
	12.5	11.9 10.8		11.7															
	12.7 11.6	13.5 13.2 9.1 7.3																	556
W L S	8.8 14.7 5.1																	See Crude protein, same reference	170
9.2*	9.3*	10.7*		10.2*	10.0*		11.3*				12.2*							See Crude protein, same reference	161
9.0*	9.0*	8.5*		8.6*	8.3*		9.4*				10.4*							See Crude protein, same reference	164

continued

Table 1.140 Lucerne: Provence (continued)

	January			February			March			April			May			June		
Major elements %DM Calcium												2.2	2.2	2.4	2.4	2.5	3.1	3.4
															1.8	2.0		2.3
																	(a) (b)	2.0 2.0
													2.3		2.1			2.6
												2.1						
															2.5		2.9	
											W	1.8			2.3	2.3 2.5		2.4 2.4
											L	3.0			2.3	3.8 4.0		2.9 3.3
											S	1.2			1.4	1.4 1.3		1.6 1.4
Phosphorus												0.6	0.7	0.6	0.4	0.4	0.4	0.3
															0.4	0.3		0.3
																	(a) (b)	0.2 0.2
													0.4		0.5			0.3
												0.4						
															0.3		0.2	
											W	0.4			0.5	0.3 0.3		0.3 0.4
											L	0.4			0.5	0.3 0.3		0.4 0.4
											S	0.3			0.4	0.2 0.2		0.3 0.3
Potassium												2.2	2.0	1.8	1.1	1.1	1.0	0.7
															1.6	1.1		1.1
																	(a) (b)	1.6 1.8
											L	2.2			3.0	1.1 1.8		2.1 2.0
											S	3.3			5.7	1.2 1.9		3.7 3.3
Sodium												0.19	0.18	0.09	0.14	0.17	0.21	0.30
															0.12	0.30		0.25
																	(a) (b)	0.25 0.27
											L	0.30			0.01	0.18 0.05		0.35 0.21
											S	0.21			0.02	0.57 0.03		0.30 0.14

July			August			September			October			November			December			Notes	Reference
3.5	2.6		3.2		2.5 2.1*		2.9											See Crude protein, same reference	580
			2.3			2.0													
					1.4														
	2.6			3.3														See Crude protein, same reference	581
2.4		2.6		2.9		3.3													
	2.5	2.6 2.4		2.6														See Crude protein, same reference	581
	2.9 1.6	3.4 3.5 1.2 1.1																	556
0.3	0.4		0.3		0.3 0.5		0.3											See Crude protein, same reference	580
			0.3			0.3													
					0.2*														
	0 4			0.3														See Crude protein, same reference	581
0.3		0.3		0.3		0.2													
	0.4	0.3 0 3		0.3														See Crude protein, same reference	581
	0.4 0.3	0.4 0.3 0.3 0.3																	556
0.8	1.5		0.9		0.9 1.7		2.6											See Crude protein, same reference	580
			1.2			1.6													
					2.4														
	2.3 3.7	2.4 1.6 3.0 2.2																See Crude protein, same reference	556
0.31	0.33		0.32		0.36 0.24		0.14											See Crude protein, same reference	580
			0.22			0.24													
					0.04														
	0.05 0.08	0.10 0.16 0.16 0.01																See Crude protein, same reference	556

continued

Table 1.140 Lucerne: Provence (continued)

	January			February			March			April			May			June		
Major elements (cont'd) Chlorine												0.3	0.3	0.3	0.2	0.2	0.2	0.2
															0.2	0.1		0.1
																	(a) (b)	0.3 0.3
													0.1		0.2			0.1
												0.2				0.1	0.2	
																	0.2	
											W	{ 0.2			0.2 D	0.2 0.2 0.2		0.3 0.3
											L	{ 0.4			0.3	0.4 0.5		0.4 0.4
											S	{ 0.5			0.8	0.5 0.4		0.8 0.7
Sulphur, Total												0.5	0.5	0.6	0.6	0.7	0.4	0.6
Inorganic												0.2	0.2	0.2	0.2	0.2	0.2	0.1
Total Inorganic															0.5 0.1	0.5 0.1		0.4 0.1
Total																	(a) (b)	0.5 0.5
Inorganic																	(a) (b)	0.08 0.06

July			August			September			October			November			December			Notes	Reference
0.3	0.3		0.3		0.6 0.3		0.4											See Crude protein, same reference	580
			0.2			0.2													
					0.3														
	0.2			0.1														See Crude protein, same reference	581
0.1		0.1		0.1		0.2													
	0.3	0.2 0.3		0.3														See Crude protein, same reference	581
	0.5 0.8	0.4 0.6 0.4 0.4																	556
0.4 0.2	0.9 0.2		0.5 0.2		0.9 0.9 0.2 0.3		1.0 0.2											See Crude protein, same reference *Regrowth	580
			0.8 0.2			0.9 0.2													
					0.8* 0.09														

2A

Table 1.141 Lucerne: Du Puits

	January			February			March			April			May			June		
Proximate constituents %DM																		
Crude protein														(1952)	W L S	16.8 24.8 9.8		
																(1953)	W L S	16.9 26.1 11.0
												25.7	23.9	20.9	17.5		19.4*	17.0*
													(1954)	W L S				
													(1955)	L S F Sd L S	32.2 22.3			31.1 10.7
														(1955)	20.8		30.6	24.4
														(1956)	21.7			23.8
															(1956)	W L S	16.8 26.1 10.2	
															(1957)	W L S	16.1 24.0 9.0	
															(1958)	W L S	18.4 27.8 10.7	
																	L	24.8
											(1957)	26.4				24.8		
									(1957)	30.0	26.9		28.1	19.4		15.6		14.4
Ether extract														(1952)	W L S	2.7 4.2 1.5		
														(1953)	W L S			2.0 3.5 1.0
												3.5	3.5	3.2	2.6		3.0*	3.0*
														(1955)	2.4		3.6	3.1
Nitrogen-free extract														(1952)	W L S	43.4 48.3 39.1		
														(1953)	W L S			42.3 45.7 40.0
												40.4	39.2	38.6	42.0		41.2*	40.6*
														(1955)	43.1		35.9	36.3

July			August			September			October			November			Dec.	January	Notes	Reference
																	W: Whole plant L: Leaf S: Stem Planted September 1951. Sampled 1952, (upper line) and 1953 (lower line). Early variety, based on time of flowering	170
16.9*	18.3* 22.6*	19.8*	19.0* 23.8*	21.4*		20.5*	21.8*										First cut, primary growth Second cut } Third cut } *Regrowth Fourth cut }	161
29.2 8.1	22.3	21.2 22.0 6.9 24.5		19.9	21.5 19.6 6.3 25.9	30.4 9.1 20.3* 6.4*		26.5 9.4 18.6 5.5	16.8 5.2 24.7* 8.0*		25.5 10.4		22.6 9.3	3.6		(1956) 3.0	W: Whole plant L: Leaf S: Stem F: Flower Sd: Seed 1954. 1st year growth: 1955, 2nd year growth: Sept., aftermath with flowers and seeds; Oct., aftermath vegetative * Regrowth	279
17.7 20.8 21.6 24.8 19.4	17.1 19.0 20.3																First cut: mean of 4 cuts, early bud Second cut stage Third cut } Fourth cut } mean of 4 cuts, early flowering Fifth cut } First cut: mean of 4 cuts, early bud Second cut stage Third cut } mean of 4 cuts, Fourth cut } early flowering	164
																	W: Whole plant L: Leaf S: Stem	590
			L 28.8 L 28.2 (26.8–29.5) S 12.9 (10.4–14.7)														First and second cuts, leaf only Weighted means, leaf and stem separately, of 9 cuts at same stage of development as, and including, the first two.	156
	20.1 20.2 18.8			19.5													1 cut, 1956, year of sowing 3 cuts, 1957, 1958 and 1959 Weighted means	132
		21.0				20.3											Sown in 1954 and grazed and mown in 1956	141
	(1959) (1960)	17.3 19.3															3 cuts each year. Means only Sown Spring 1957	158
																	All first cuts from 15-20 cm height to 90 cm in full flower	553
																	See Crude protein, same reference	170
2.8*	2.8* 3.1*	2.9*	3.1* 3.1*	2.8*		3.3*	3.4*										First cut, primary growth Second cut } Third cut } *Regrowth Fourth cut }	161
2.0 2.5 1.9 2.7																	First cut: mean of 4 cuts, early bud Second cut stage Third cut } Fourth cut } mean of 4 cuts, early flowering Fifth cut }	164
																	See Crude protein, same reference	170
41.4*	45.0* 41.3*	44.5*	44.5* 39.8*	40.3*		43.5*	43.0*										First cut, primary growth Second cut Third cut *Regrowth Fourth cut	161
37.1 39.3 39.5 42.1																	First cut: mean of 4 cuts, early bud stage, Second cut Third cut: } Fourth cut: } mean of 4 cuts, all early flowering Fifth cut: }	164

continued

Table 1.141 Lucerne: Du Puits (continued)

	January			February			March			April			May			June		
Proximate constituents (cont'd) Crude fibre														(1952)	W L S	28.4 10.7 43.8 (1953)	W L S	30.9 12.5 42.9
												18.5	22.2	27.0	29.1		26.3*	30.0*
														(1955) (1956)	24.1 24.3		17.3	24.9 24.7
															(1956) (1957) (1958)	W L S W L S W L S	30.5 12.0 43.8 27.9 12.0 42.2 30.5 12.4 45.4	
Ash														(1952)	W L S	8.7 12.0 5.8 (1953)	W L S	7.9 12.2 5.1
												11.9	11.2	10.3	8.8		10.1*	9.4*
														(1955)	9.6		12.6	11.3
										(1957)	9.7					8.6		
Amino acids N% of total N before hydrolysis																		
Alanine										5.3	6.0		5.4	5.2		5.1		5.0
γ-Aminobutyric acid										0.8	0.8		0.8	0.7		0.9		0.7
Arginine										9.8	9.5		9.3	8.4		8.6		8.3
Aspartic acid										7.7	8.2		7.1	6.6		7.4		8.0
Cystine										0.8	0.8		0.8	0.8		0.8		0.9
Glutamic acid										5.1	5.2		5.3	5.2		4.9		4.7
Glycine										4.9	5.1		5.4	5.6		5.3		5.2
Histidine										3.1	3.1		3.0	3.0		3.3		3.1
Isoleucine										2.5	2.6		2.7	2.7		2.7		2.5
Leucine										4.4	4.4		4.5	4.7		4.6		4.1
Lysine										5.6	5.4		4.8	4.7		5.6		4.7
Methionine										0.6	0.4		0.4	0.6		0.6		0.5
Phenylalanine										2.0	2.2		2.4	2.3		2.3		2.2
Proline										3.1	3.4		3.3	3.6		3.5		3.3
Serine										3.1	3.2		3.3	3.5		3.3		3.2
Threonine										2.9	2.9		2.9	2.8		2.9		3.0
Tryptophan										1.3	1.2		1.4	1.3		1.4		1.4
Tyrosine										1.5	1.6		1.6	1.5		1.6		1.6
Valine										3.6	3.7		4.1	3.8		3.9		3.6
Carbohydrate fractions %DM Water-soluble carbohydrate										(1957)	11.1					11.6		
Glucose													(1955)	L S F Sd	0.6 1.8			(1954) 0.6 2.6
Fructose													(1955)	L S F Sd	0.6 1.7			(1954) 0.6 2.0

July			August			September			October			November			Dec.	January	Notes	Reference
																	See Crude protein, same reference	170
29.4*	23.9* 21.9*	22.6*	23.2* 22.0*	25.7*		21.5*	19.5*										First cut, primary growth Second cut } Third cut } *Regrowth Fourth cut }	161
33.6 26.9 26.1 17.1 31.3	33.8 31.3 26.1																First cut: mean of 4 cuts, early bud Second cut stage Third cut } Fourth cut } mean of 4 cuts, Fifth cut } early flowering First cut: mean of 4 cuts, early bud Second cut stage Third cut } mean of 4 cuts, Fourth cut } all early flowering	164
																		590
																	See Crude protein, same reference	170
9.5*	10.0* 11.1*	10.2*	10.2* 11.3*	9.8*		11.2*	12.3*										First cut, primary growth Second cut } Third cut } *Regrowth Fourth cut }	161
9.6 10.5 10.9 13.3																	First cut: mean of 4 cuts, early bud Second cut: stage Third cut } Fourth cut } mean of 4 cuts, Fifth cut } early flowering	164
		8.9				8.6												141
																	N recovered as: Amino acids: 65.8% of total Ammonia: 12.6% Humin 7.0% Amino acid values are means of duplicate hydrolysates except for cystine and tryptophan which represent single analyses of special hydrolysates. For amino acids of a 'cytoplasmic protein preparation' from the 2nd, 4th and 5th samples, see table 1.67 Concentrates.	553
		6.7				7.8												141
{ W L S 0.3 2.0	0.3	0.5 1.7 1.9 3.1		0.5 (1955)	0.6 0.2 0.5 1.1 { L S	1.1 1.2 0.1* 0.5*		1.4 1.2 0.2 0.2		0.1 0.2 0.5* 0.5*	0.3 1.1		0.7 0.5	0.1		(1956) trace	See Crude protein, same reference	279
{ W L S 0.3 2.0	0.3	0.5 0.5 1.6 3.5		0.6 (1955)	0.8 0.2 0.7 1.4 { L S	0.4 1.3 0.1* 0.6*		0.6 1.7 0.1 0.2		0.1 0.3 0.7* 0.6*	0.4 1.9		1.2 0.6	0.1		(1956) trace		

continued

Table 1.141 Lucerne: Du Puits (continued)

	January			February			March			April			May			June		
Carbohydrate fractions (cont'd)																		
Sucrose																		(1954)
													(1955)	L	1.4			1.2
														S	1.7			2.1
														F				
														Sd				
Oligosaccharides																		(1954)
													(1955)	L	0.1			0.1
														S	0.01			0.4
														F				
														Sd				
Glucosan																		(1954)
													(1955)	L	4.8			11.2
														S	0.9			0.8
														F				
														Sd				
Cellulose																		(1954)
													(1955)	L	11.8			5.8
														S	20.8			26.0
														F				
														Sd				
Xylan																		(1954)
													(1955)	L	3.6			1.5
														S	6.4			8.9
														F				
														Sd				
Araban																		(1954)
													(1955)	L	4.4			3.5
														S	3.2			2.5
														F				
														Sd				
Galactan																		(1954)
													(1955)	L	3.7			0.9
														S	1.3			1.3
														F				
														Sd				
Mannan																		(1954)
													(1955)	L	0.6			1.1
														S	0.7			1.8
														F				
														Sd				
Uronic anhydride													(1955)	L	12.2			10.8
														S	9.3			8.8
														F				
														Sd				

July			August			September			October			November			Dec.	January	Notes	Reference
W	0.01	1.2		2.0	1.9													
L						1.3		2.0			2.4		3.2					
S						3.1		5.6			5.9		6.9					
0.8		0.4			0.6			1.6		1.2						(1956) trace		
3.2		3.4			1.3			2.0		1.9			0.1	0.1				
		1.7																
					1.9													
				(1955)	L	1.8*				2.2*								
					S	2.7*				5.6*								
W	0.3	0.8		0.8	0.6													
L						0.8		0.9			0.4		1.2					
S						0.4		0.4			0.6		1.0					
0.2		0.4			0.5			0.2		0.4								
0.4		0.2			0.2			0.2		0.2				0.1				
		0.2																
					0.6													
				(1955)	L	0.3*				0.1*								
					S	2.5*				0.3*								
W	9.5	9.2		2.2	1.5													
L						1.9		1.9			1.5		0.8					
S						1.4		0.7			1.5		0.6					
9.3		2.6			4.0			1.7		1.3						(1956) 0.5	As for Crude protein	
2.1		5.7			0.8			0.3		0.4				1.5				
		2.8																
					1.2													
				(1955)	L	2.5*				1.7*								
					S	1.4*				0.6*								
W	7.6	10.4		14.3	16.1													
L						6.5		5.9			5.2		6.9					
S						30.1		23.7			26.2		32.8					
5.9		14.9			12.0			7.1		9.5						(1956) 43.1		
25.2		33.0			32.9			33.7		31.6				26.3				
		8.1																
					13.2													
				(1955)	L	7.7*				6.0*								
					S	33.1*				28.2*								
W	1.9	2.7		2.9	2.6													
L						3.2		1.2			3.5		3.2					
S						10.7		6.2			8.8		8.5					
3.5		4.1			9.5			2.5		3.2						(1956) 8.5		
8.2		8.9			11.1			6.6		11.7				15.3				
		4.4																
					9.0													
				(1955)	L	1.2*				1.4*								
					S	1.0*				9.5*								
W	2.3	2.9		1.3	2.4													
L						2.2		2.2			1.5		2.5					
S						2.1		2.3			2.6		2.8					
1.8		2.0			1.7			2.2		1.9						(1956) 1.4		
3.1		2.1			2.9			1.8		2.3				2.7				
		4.9																
					3.4													
				(1955)	L	1.7*				2.2*								
					S	3.7*				1.9*								
W	1.3	1.2		0.7	1.2													
L						0.9		1.0			1.1		1.3					
S						0.3		1.3			2.2		1.1					
1.0		0.4			1.5			0.9		0.7						(1956) 0.9		
1.6		0.8			1.5			0.6		1.0				1.8				
		2.1																
					1.4													
				(1955)	L	1.0*				0.7*								
					S	1.5*				1.6*								
W	0.7	1.0		1.5	0.8													
L						0.6		0.3			0.7		0.6					
S						1.6		0.6			2.4		1.9					
0.4		0.3			0.5			0.3		0.9						(1956) 2.5		
1.6		2.2			1.7			2.4		3.0				3.7				
		1.6																
					1.5													
				(1955)	L	0.7*				0.4*								
					S	1.7*				1.2*								
		11.7			10.5					9.9								
		6.6			7.7					7.4				5.2				
		11.5																
					28.4													
				(1955)	L	11.9*												
					S	9.8*												

continued

Table 1.141 Lucerne: Du Puits (continued)

	January			February			March			April			May			June		
Carbohydrate fractions (cont'd) Lignin													(1955)	L S F Sd	3.5 6.4			(1954) 7.6 10.6
										(1957)	5.4					6.7		

July			August			September			October			November			Dec.	January	Notes	Reference
{ W	5.2	3.7		8.4	9.3													
L						6.0		.7.5			6.9		7.6					
S						13.7		12.6			18.2		18.7					
4.8		6.1			7.3			6.0		5.2						(1956) 19.6		
11.1		16.9			14.3			15.6		15.6				15.4				
		5.5																
					8.9													
				(1955)	{ L	6.7*				7.6*								
					S	14.6*				12.2*								
		8.2				7.7												141

Table 1.142 Lucerne: Hungarian

	May			June			July			August			Notes	Reference
Proximate constituents %DM														
Crude protein	(1)			21.6					17.5			20.7	1) 9 cm apart	580
	(2)		26.4	22.7		15.7	23.5	20.6	16.3			19.8	2) 18 cm apart	
	(3)			22.1					16.8			21.1	3) 27 cm apart	
	(4)			22.4					17.3			21.2	4) 36 cm apart	
Ether extract	(1)			3.0					2.1			4.0	1-year-old crop. No fertiliser	
	(2)		2.9	3.1		2.9	3.5	2.6	2.3			2.9		
	(3)			3.0					2.5			4.0	May/June – first growth	
	(4)			2.8					2.4			3.5	July – second growth. August – third growth	
Nitrogen-free extract	(1)			38.1					42.7			38.3		
	(2)		36.5	36.7		41.7	39.2	39.7	40.9			38.8		
	(3)			37.2					40.4			38.5		
	(4)			35.8					40.4			39.4		
Crude fibre	(1)			26.8					27.4			24.7		
	(2)		22.2	26.2		31.4	21.7	25.6	30.4			26.7		
	(3)			26.8					30.2			24.4		
	(4)			26.8					29.5			23.4		
Ash	(1)			10.5					10.3			12.3		
	(2)		12.0	11.3		8.3	12.1	11.5	10.1			11.8		
	(3)			10.9					10.1			12.0		
	(4)			12.2					10.4			12.5		
Major elements %DM														
Calcium	(1)			1.6					2.3			2.8		
	(2)		1.7	1.7		1.8	2.1	2.1	2.0			2.5		
	(3)			1.6					2.1			2.7		
	(4)			1.7					2.1			2.6		
Phosphorus	(1)			0.4					0.3			0.3		
	(2)		0.5	0.4		0.3	0.4	0.3	0.3			0.3		
	(3)			0.4					0.3			0.3		
	(4)			0.4					0.3			0.3		
Potassium	(1)			3.2					2.3			2.4		
	(2)		3.8	3.2		1.9	3.3	3.2	1.8			2.2		
	(3)			3.2					2.3			2.4		
	(4)			3.2					2.5			2.8		
Sodium	(1)			0.21					0.13			0.05		
	(2)		0.02	0.20		0.16	0.24	0.19	0.12			0.10		
	(3)			0.20					0.06			0.05		
	(4)			0.08					0.06			0.08		
Chlorine	(1)			0.16					0.10			0.07		
	(2)		0.16	0.15		0.06	0.05	0.15	0.08			0.10		
	(3)			0.19					0.10			0.10		
	(4)			0.15					0.09			0.14		
Sulphur														
Total sulphur	(1)			0.60					0.72			0.70		
	(2)		0.72	0.54		0.55	0.96	0.77	0.75			0.73		
	(3)			0.59					0.50			0.57		
	(4)			0.60					0.67			0.73		
Inorganic sulphur	(1)			0.11					0.18			0.24		
	(2)		0.16	0.12		0.05	0.26	0.26	0.16			0.19		
	(3)			0.12					0.18			0.21		
	(4)			0.12					0.17			0.22		

Table 1.143 Lucerne: Hungarian White Seal

	April	April	April	May	May	May	June	June	June	July	July	July	August	August	August	October	October	October	Notes	Reference
Proximate constituents %DM Crude protein	(1950)	27.7				19.1	20.9 15.8					23.4 20.4			27.0 29.6 25.0				Sown 17/5/1949 Samples taken from 3 areas in 1950 and 1951	152
					(1951)	18.4 19.3 20.0							24.6	24.3 24.6		24.0 23.6 21.3				
											(1952) 16.4								Mean of 2 cuts taken on 25-29/5 and 29/7 or 13-14/8/1952	157
Major elements %DM Calcium											(1950 & 1951)	2.0 2.1 2.5							Means for all cuts taken in 1950 and 1951 from 3 areas	152
											(1952)	1.7							Mean of 2 cuts taken on 25-29/5 and 29/7 or 13-14/8/1952	157
Phosphorus											(1950 & 1951)	0.3 0.3 0.2							Means for all cuts taken in 1950 and 1951 from 3 areas	152
											(1952)	0.3							Mean of 2 cuts taken on 25-29/5 and 29/7 or 13-14/8/1952	157
Vitamins µg/g DM β-carotene	(1950)	334				270	291												Samples taken from 3 areas	152
					(1951)	250 256 270														

Table 1.144 Lucerne: Other named varieties. Proximate constituents

	May			June			July			August			Notes	Reference
Argentine Saladina *Proximate constituents* %DM Crude protein							W L S	15.4 24.2 8.1					W: Whole plant L: Leaf S: Stem Planted September 1951; Sampled 1952 Extra late strain, based on time of flowering	170
Ether extract							W L S	2.0 3.1 1.1						
Nitrogen-free extract							W L S	42.0 46.3 38.1						
Crude fibre							W L S	31.7 12.6 47.7						
Ash							W L S	8.9 13.8 5.0						
Chartrainvillers *Proximate constituents* %DM Crude protein			W W L L S S	15.5 23.1 8.8 		 17.3 26.9 11.3							W: Whole plant L: Leaf S: Stem Planted September 1951. No fertiliser. Means only Sampled 1952 (upper line) and 1953 (lower line) Early strain based on time of flowering	
			 (1956) (1957) (1958) 	W L S W L S W L S	16.6 25.1 9.8 15.7 23.4 9.0 17.6 26.9 10.2								Sown July 1955 in close drills	590
					L	25.1		L 28.3 (27.3–29.1) S 13.1 (11.3–14.8)			L	29.5	First and second cuts, leaf only Weighted means, leaf and stem separately, of 9 cuts at same stage of development as, and including, the first two.	156
								18.9 20.2 18.7			19.0		One cut, 1956, year of sowing 3 cuts, 1957, 1958 and 1959 Weighted means	132
Ether extract			W W L L S S	2.4 3.6 1.4 		 2.2 3.7 1.2							See Crude protein, same reference	170
Nitrogen-free extract			W W L L S S	46.5 51.6 42.0 		 37.6 42.3 34.7								
Crude fibre			W W L L S S	27.5 10.4 42.5 		 34.9 13.4 48.2								
			 (1956) (1957) (1958) 	W L S W L S W L S	29.7 12.5 43.3 29.0 12.4 43.3 31.0 11.4 46.6									590
Ash			W W L L S S	8.1 11.3 5.3 		 8.0 13.7 4.6							See Crude protein, same reference	170

Table 1.144 Lucerne: Other named varieties. Proximate constituents (continued)

	May			June			July			August			Notes	Reference
Eynsford														
Proximate constituents %DM														
Crude protein			W	16.4									W: Whole plant	170
			W			16.2							L: Leaf	
			L	24.6									S: Stem	
			L			25.1							Planted September 1951.	
			S	9.4									Sampled 1952 (upper line) and	
			S			10.5							1953 (lower line)	
													Early strain, based on time of flowering	
			(1956)	W	16.7								Sown July 1955 in close drills	590
				L	25.3									
				S	10.2									
			(1957)	W	16.0									
				L	23.9									
				S	9.1									
			(1958)	W	18.5									
				L	27.7									
				S	11.1									
											19.0		One cut in 1956, year of sowing	132
								19.0					3 cuts taken in May, July, Aug. 1957	
								20.7					3 cuts taken in June, July, Sept. 1958	
								18.3					3 cuts taken in June, July, Aug. 1959	
													Weighted means for all cuts	
Ether extract			W	2.7									See Crude protein, same reference	170
			W			2.3								
			L	4.0										
			L			3.8								
			S	1.6										
			S			1.3								
Nitrogen-free extract			W	43.9										
			W			40.8								
			L	48.2										
			L			45.5								
			S	40.0										
			S			37.7								
Crude fibre			W	28.4										
			W			32.9								
			L	10.9										
			L			12.7								
			S	43.5										
			S			45.9								
			(1956)	W	29.7								See Crude protein, same reference	590
				L	11.6									
				S	43.3									
			(1957)	W	28.0									
				L	12.1									
				S	41.8									
			(1958)	W	30.4									
				L	11.7									
				S	45.4									
Ash			W	8.6									See Crude protein, same reference	170
			W			7.8								
			L	12.3										
			L			12.9								
			S	5.5										
			S			4.6								
Eynsford No. 1													W: Whole plant	170
Proximate constituents %DM													L: Leaf	
Crude protein		W	19.9										S: Stem	
		W			18.8								Planted September 1951.	
		L	27.8										Sampled 1952 (upper line)	
		L			27.0								and 1953 (lower line)	
		S	10.2											
		S			11.6								Semi-late strain, based on time of flowering	
Ether extract		W	3.2											
		W			2.5									
		L	4.5											
		L			4.0									
		S	1.5											
		S			1.2									
Nitrogen-free extract		W	40.7											
		W			39.9									
		L	42.4											
		L			44.8									
		S	38.8											
		S			35.6									

continued

Table 1.144 Lucerne: Other named varieties. Proximate constituents (continued)

Eynsford (continued)	June	July	August	September	Notes	Reference
Proximate constituents (cont'd) Crude fibre	W 25.6 L 11.2 S 43.3	W 30.6 L 12.5 S 46.5				279
Ash	W 10.6 L 14.1 S 6.2	W 8.2 L 11.7 S 5.1				
Grimm *Proximate constituents* %DM Crude protein		W 18.6 W 17.0 L 26.7 L 27.1 S 9.0 S 10.1			W: Whole plant L: Leaf S: Stem Planted September 1951. Sampled 1952 (upper line) and 1953 (lower line) Late strain, based on time of flowering	170
	(1956) W 15.9 L 23.8 S 10.0 (1957) W 16.4 L 24.0 S 10.3	(1958) W 15.6 L 22.5 S 10.0				590
		L 25.3 L 28.2 (26.0–29.2) S 14.9 (12.0–16.6)		L 26.7	First and second cuts of 2, leaf only. (Value for second cut computed from value for first and mean of 2) Weighted means, leaf and stem separately, of 2 cuts taken at same stage of development (above), 3 cuts taken at fixed dates, 31/5, 14/7, 1/9; and 4 cuts, also at fixed dates, 21/5, 23/6, 10/8 and 29/9.	156
		21.3 22.6 20.3	19.0		1 cut, 1956, year of sowing 3 cuts, 1957, 1958 and 1959 Weighted means	132
Ether extract		W 2.4 W 2.2 L 3.5 L 3.6 S 1.1 S 1.2			See Crude protein, same reference	170
Nitrogen-free extract		W 40.8 W 39.1 L 43.3 L 43.1 S 37.7 S 36.4				
Crude fibre		W 28.7 W 34.0 L 13.7 L 13.6 S 46.6 S 47.9				
	(1956) W 30.2 L 12.8 S 43.3 (1957) W 30.4 L 14.0 S 43.7	(1958) W 30.7 L 13.0 S 45.0				590
Ash		W 9.5 W 7.7 L 12.8 L 12.6 S 5.6 S 4.4			See Crude protein, same reference	170

Table 1.144 Lucerne: Other named varieties. Proximate constituents (continued)

	June			July			August			September			Notes	Reference
Hunter River ***Proximate constituents*** **%DM** **Crude protein**		W L S	15.1 22.8 7.2										W: Whole plant L: Leaf S: Stem Planted September 1951; Sampled 1952 Semi-late variety, based on time of flowering	170
Ether extract		W L S	2.0 3.0 1.0											
Nitrogen-free extract		W L S	43.8 47.6 40.0											
Crude fibre		W L S	30.1 12.9 47.6											
Ash		W L S	9.0 13.7 4.2											
N. Z. Marlborough ***Proximate constituents*** **%DM** **Crude protein**	W L S	16.4 24.6 9.5											W: Whole plant L: Leaf S: Stem Planted September 1951; Sampled 1952 Mid-season strain, based on time of flowering	170
Ether extract	W L S	2.4 3.6 1.3												
Nitrogen-free extract	W L S	40.9 43.6 38.8												
Crude fibre	W L S	30.0 12.3 44.9												
Ash	W L S	10.3 15.9 5.5												
N. Z. Strain B ***Proximate constituents*** **%DM** **Crude protein**	W W L L S S	16.2 24.4 9.0			16.9 27.1 10.4								W: Whole plant L: Leaf S: Stem Planted September 1951. Mid-season strain, based on time of flowering. Sampled 1952 (upper line) and 1953 (lower line)	170
	(1956) (1957) (1958)	W L S W L S W L S	16.9 24.2 10.7 15.3 23.2 9.9	15.7 23.2 11.3									Sown July, 1955 in close drills	590
					19.0 22.1 19.0			19.4					One cut in 1956, year of sowing 3 cuts taken in May, July, Aug. 1957 3 cuts taken in June, July, Sept. 1958 3 cuts taken in June, July, Aug. 1959 Weighted means	132
Ether extract	W W L L S S	2.1 3.3 1.0			2.3 4.1 1.2								See Crude protein, same reference	170
Nitrogen-free extract	W W L L S S	39.8 44.0 36.2			38.7 41.6 36.8									

continued

Table 1.144 Lucerne: Other named varieties. Proximate constituents (continued)

	June			July			August			September			Notes	Reference
Proximate constituents %DM														
Crude fibre	W	32.1												
	W				33.9									
	L	13.3												
	L				14.0									
	S	48.7												
	S				46.6									
	(1956)	W	28.9										See Crude protein, same reference	590
		L	12.0											
		S	43.1											
	(1957)	W	32.1											
		L	12.7											
		S	45.3											
	(1958)	W		33.1										
		L		12.8										
		S		45.1										
Ash	W	9.8											See Crude protein, same reference	170
	W				8.2									
	L	15.0												
	L				13.2									
	S	5.1												
	S				5.0									
Rhizoma														
Proximate constituents %DM														
Crude protein				W	17.2								W: Whole plant L: Leaf S: Stem Late strain, based on time of flowering. Planted September 1951. Sampled in 1952 (upper line) and 1953 (lower line)	170
				W	18.0									
				L	26.4									
				L	27.8									
				S	8.0									
				S	10.8									
		L	26.0			L 30.2 (28.1–31.3) S 15.6 (12.2–17.7)			L	30.2			First and second cuts of 2, leaf only. (Value for second cut computed from value for first and mean of 2) Weighted means, leaf and stem separately, of 2 cuts taken at same stage of development (above), 3 cuts taken at fixed dates, 31/5, 14/7, 1/9; and 4 cuts, also at fixed dates, 21/5, 23/6, 10/8 and 29/9.	156
								18.7					One cut in 1956, year of sowing 3 cuts taken in May, July, Aug. 1957 3 cuts taken in June, July, Sept. 1958 3 cuts taken in June, July, Aug. 1959 Weighted means	132
					20.5									
					21.5									
					20.2									
Ether extract				W	2.6								See Crude protein, same reference	170
				W	2.3									
				L	4.0									
				L	4.0									
				S	1.1									
				S	1.1									
Nitrogen-free extract				W	42.0									
				W	37.8									
				L	44.7									
				L	42.9									
				S	39.1									
				S	34.1									
Crude fibre				W	30.2									
				W	34.7									
				L	13.1									
				L	14.2									
				S	47.5									
				S	49.6									
Ash				W	8.0									
				W	7.2									
				L	11.8									
				L	11.1									
				S	4.3									
				S	4.4									

continued

Table 1.144 Lucerne: Other named varieties. Proximate constituents (continued)

		June			July			August			September			Notes	Reference
W268														W: Whole plant L: Leaf S: Stem Planted September 1951; Sampled 1952 (upper line) and 1953 (lower line) Early strain, based on time of flowering	170
Proximate constituents %DM	W	15.5													
Crude protein	W			17.4											
	L	23.2													
	L			26.1											
	S	8.7													
	S			11.5											
Ether extract	W	2.5													
	W			2.5											
	L	3.6													
	L			4.1											
	S	1.5													
	S			1.5											
Nitrogen-free extract	W	45.8													
	W			37.8											
	L	51.9													
	L			41.4											
	S	40.5													
	S			35.3											
Crude fibre	W	28.1													
	W			33.5											
	L	10.0													
	L			13.6											
	S	44.0													
	S			46.9											
Ash	W	8.1													
	W			8.8											
	L	11.3													
	L			14.8											
	S	5.3													
	S			4.8											

Table 1.145 Lucerne: Other named varieties. Crude protein, Crude fibre

	June			July			August			September			Notes	Reference
FD 100													Sown July 1955, in close drills W: Whole plant L: Leaf S: Stem	590
Proximate constituents %DM														
Crude protein (1956)	W	16.7												
	L	25.3												
	S	9.8												
(1957)	W	15.6												
	L	23.1												
	S	9.0												
(1958)	W	17.3												
	L	25.2												
	S	10.2												
Crude fibre (1956)	W	29.7												
	L	12.3												
	S	43.8												
(1957)	W	29.5												
	L	12.3												
	S	44.6												
(1958)	W	30.0												
	L	11.0												
	S	47.1												
Gamma													Sown July 1955 in close drills W: Whole plant L: Leaf S: Stem	590
Proximate constituents %DM														
Crude protein (1956)	W	17.5												
	L	26.0												
	S	10.5												
(1957)	W	16.4												
	L	24.7												
	S	9.8												
(1958)	W	19.3												
	L	27.5												
	S	11.7												
Crude fibre (1956)	W	28.6												
	L	11.4												
	S	42.8												
(1957)	W	29.3												
	L	12.6												
	S	42.6												
(1958)	W	28.8												
	L	11.5												
	S	44.7												
Marais													Sown July 1955 in close drills About 20% of plants flowering W: Whole plant L: Leaf S: Stem	590
Proximate constituents %DM														
Crude protein	(1956)	W	15.6											
		L	24.3											
		S	9.1											
	(1957)	W	14.5											
		L	23.2											
		S	9.1											
	(1958)	W		16.8										
		L		26.3										
		S		10.8										
				(1956)				20.9					One cut	132
				(1957)	18.6								3 cuts – May, July, Aug.	
				(1958)	20.7								3 cuts – June, July, Sept.	
				(1959)	18.0								3 cuts – June, July, Aug. Sown in rows 22/5/1956 Weighted means for all cuts in each year 1957–59	
Crude fibre	(1956)	W	30.2										Sown July 1955 in close drills About 20% of plants flowering W: Whole plant L: Leaf S: Stem	590
		L	11.8											
		S	44.1											
	(1957)	W	32.6											
		L	12.9											
		S	44.9											
	(1958)	W		34.8										
		L		13.6										
		S		46.8										
Poitou													W: Whole plant L: Leaf S: Stem	590
Proximate constituents %DM														
Crude protein	(1956)	W	15.9											
		L	23.7											
		S	9.9											
	(1957)	W	15.1											
		L	23.1											
		S	9.6											
	(1958)	W		16.1										
		L		26.7										
		S		10.9										
Crude fibre	(1956)	W	29.7											
		L	11.7											
		S	43.7											
	(1957)	W	31.3											
		L	12.3											
		S	44.5											
	(1958)	W		34.7										
		L		13.3										
		S		45.2										

Table 1.146 Lucerne: Other named varieties. Crude protein, %DM

	June			July			August			September			October			November			Notes	Reference
AFI (Grimm x Hungarian)					(1959) 19.5 (1960) 20.8														3 cuts each year Means only	158
Alfa					(1959) 18.3 (1960) 19.2															
Arizona	14.0	15.9 14.4				15.0			14.3			18.1	17.1						1948: First and second crops 1949: First and third crops 1950: First, second and third crops Plots cut or grazed 3 times each year at late bud or early flower stage.	439
Buffalo (1948) (1949) (1950)	15.2	16.3 14.6				15.4			14.6			17.4 17.8							See above, Arizona	
Cardinal (1959) (1960)						17.3 19.0													3 cuts each year Means only Sown Spring 1957	158
Cossack (1948) (1949) (1950)	15.6	16.4 13.8				15.7			16.0			18.8	15.3						See above, Arizona	439
Flandria					19.4 20.0 18.7			19.2											1 cut 1956 3 cuts (May, July, August 1957) 3 cuts (June, July, September 1958) 3 cuts (June, July, August 1959) Sown in rows 22/5/1956 Weighted means for all cuts in each year 1957–1959	132
G.P.R. 1					(1959) 17.6 (1960) 19.3														See above, Cardinal	158
G.P.R. 2					(1959) 18.1 (1960) 19.6															
Ladak (1948) (1949) (1950)	15.6	16.1 15.3				17.3			14.7			18.9	19.1						See above, Arizona	439
Omega					(1959) 18.7 (1960) 20.0														See above, Cardinal	158
Orchesienne					19.0 21.3 19.0			18.3											See above, Flandria	132

continued

Table 1.146 Lucerne: Other named varieties. Crude Protein, %DM (continued)

	June			July			August			September			October			November			Notes	Reference
Poitou 1		L	25.2		L S L S L S	30.0 15.9 28.0 13.0 27.0 12.5		L	28.8										L: Leaf S: Stem First and second cuts of 2, leaf only. (Value for second cut computed from value for first and mean of 2) Weighted means, leaf and stem separately, of 2 cuts taken at same stage of development (above), 3 cuts taken at fixed dates, 31/5, 14/7, 1/9; and 4 cuts, also at fixed dates, 21/5, 23/6, 10/8 and 29/9	156
Poitou 4			L	27.1	L S L S L S	29.1 15.3 27.5 12.8 26.4 12.7				L	25.7									
Provence 2			L	26.6	L S L S L S	28.3 14.5 26.2 12.3 25.6 12.7				L	24.6									
Ranger (1948) (1949) (1950)	 15.5	17.0 14.5				 15.6			15.9			 18.9	 19.0						See above, Arizona	439
Vendee		L	24.0		L S L S L S	29.1 15.5 28.0 13.1 26.2 11.6			L	28.4									See above, Poitou 1	156
Vernal					(1959) (1960)	18.9 20.1													See above, Cardinal	158

Table 1.147 Lucerne: unnamed

	January			February			March			April			May			June		
Digestibility of DM%												(1950)	68.6					
Digestibility of OM%												(1950)	73.8					
Proximate constituents %DM Crude protein											35.9		29.5		23.1	21.2		
														32.7				
(1951) (1952)		9.4				9.5												
(1951) (1952)		9.4				9.5												
(1952 + 1953)		8.0																
		(1953)	7.8		7.8													
						(1953)	16.4											
														23.6				
Ether extract											1.6		2.0		2.0	1.3		
Nitrogen-free extract											34.3		40.1		38.8	38.2		
Crude fibre											14.7		18.8		26.4	29.6		
														19.2				
														20.9				
Ash											13.5		9.6		9.7	9.7		
Proximate constituents %DM Crude protein																21.2		
Amino acids % of crude protein Cystine Lysine Methionine DL form L form Tryptophan Tyrosine																 0.35 2.10 0.97 0.91 1.44 2.27		
Carbohydrate fractions %DM Cellulose											17.7		18.0		24.7	29.9		
Lignin											3.6		4.1		6.7	8.0		

July			August			September			October			November			December			Notes	Reference
62.0		(1952)	64.5 63.5															Digestibility trials with 2 or 3 sheep	429
65.0		(1952)	65.9 66.4															These values are not stated to apply to cuts of the same crop	
18.9																		Air dried 7–10 days; from experimental plots	9
																		Preflowering. Nursery plots	13
											16.1 25.3 11.7 8.4				9.9 9.9	 8.0		Young } Green foliage Old } } Burnt' foliage } Sown with Cocksfoot } Sown with Timothy and Meadow Fescue Sown with Tall Fescue A study of winter grazing	296
		21.0																	133
													16.3 24.8 8.3	13.5 23.4 7.9		12.2 22.9 8.3	11.5 20.0 8.2	Whole plant (green + frostburnt) Leaf Stem	337
																		Date uncertain. Probably leafy preflowering.	420
		15.9 15.8 14.6 16.3 16.3																Date not given. Data from different centres	1
1.4																		See Crude protein, same reference	9
42.2																			
27.8																			
																		See Crude protein, same reference	13
																		See Crude protein, same reference	420
9.7																		See Crude protein, same reference	9
																		Cut pre-flowering, height 41–45 cm.	12
25.9																		See Crude protein, same reference	9
35.7																		No date	35
7.2																		See Crude protein, same reference	9
7.6																		No date	35

continued

Table 1.147 Lucerne: Unnamed (continued)

	January			February			March			April			May			June		
Major elements %DM Calcium														1.7				
																		(a) (b)
											2.0			1.9			2.1	
Phosphorus														0.6				
											0.7			0.5			0.3	
Magnesium																		
											0.6			0.6			0.6	
Potassium																		
											3.9			3.3			2.3	
Sodium											0.04			0.05			0.07	
Chlorine											0.7			0.6			0.4	
Sulphur (total)																		
Trace elements μg/g DM Barium																		
Boron																		
Cobalt																		
											0.3			0.1			0.2	
Copper																		
											12.1			11.5			8.2	
Iron														856.5				
											558			229			217	
Lead																		
Manganese																		
											60.2			29.6			30.8	
Molybdenum																		
Nickel																		
Strontium																		
Tin																		
Titanium																		
Vanadium																		
Zinc																		
Vitamins μg/g DM α-tocopherol								90										
								35			60			85			105	
B Vitamins Thiamin Riboflavin Nicotinic acid Pantothenic acid Pyridoxine Biotin																		

July			August			September			October			November			December			Notes	Reference
																		Preflowering. Nursery plots	13
1.69 1.63		(1.62 – 1.86) (1.60 – 1.68)																Dates not given. (a) Chemical methods. Means and range of 4 centres. (b) Spectrochemical methods Mean and range of 4 centres	1
	2.2			2.6														Dates uncertain. April/May, pre-flowering. June, midflowering. July, late flowering. August, mature.	490
																		See Calcium, same reference	13
		0.3																	133
0.26		(0.25 – 0.26)																Dates not given. Data from 5 different centres	1
	0.3			0.2														See Calcium, same reference	490
0.15		(0.14 – 0.18)																Dates not given. Data from 6 different centres	1
	0.5			0.5														See Calcium, same reference	490
		1.7																	133
2.14		(2.11 – 2.26)																Dates not given. Data from 5 different centres	1
	2.0			1.5														See Calcium, same reference	490
	0.10			0.10															
	0.5			0.4															
0.20 0.17																		Dates not given. Data from 2 different centres	1
		8.8																Date not given. Data from 5 different centres	1
35		(28 – 41)																	1
		S 31 C 33																S : Spectrographic method. C : Colorimetric method.	450
								45.1 61.6											542
		0.25																Date not given	1
	0.1			0.1														See Calcium, same reference	490
18.6		(16.6 – 20.3)																Date not given. Data from 4 different centres	1
	7.0			6.8														See Calcium, same reference	490
																		Date uncertain. Probably leafy, preflowering	1
341		(302 – 365)																Date not given. Data from 6 different centres	1
	234			219														See Calcium, same reference	490
		<1																Date not given	1
38		(35 – 43)																Date not given. Data from 4 different centres	1
	29.4			34.5														See Calcium, same reference	490
0.27		(0.19 – 0.48)																Date not given. Data from 4 different centres	1
								0.84 1.06											542
		1.18																Date not given	1
		37																	
		<1																	
		40																	
		0.38																	
		20																	
	200			260															87
	115			120			230			160			240						86
						2.6 15.1 40.4 43.4 11.6 0.5												Early flowering Oven-dried herbage	493

Table 1.148 Lucerne Mixtures

Lucerne, White Seal and White Clover S100		April			May			June			July			August			October			Notes	Reference
Proximate constituents %DM Crude protein																				Sown 17/5/1949. 10 to 20% clover except in area C where 1.5% only	152
	1950		29.0					22.8					23.8			26.8				Area A	
							20.5						19.7			28.8				Area B	
								18.7								24.4				Area C	
	1951						17.7								21.2		23.4			Area A	
							19.0								22.5		22.5			Area B	
							23.4							23.1			21.3			Area C	
Major elements %DM Calcium	%DM 1950 and 1951										1.8									Area A	
													1.8							Area B	
													2.4							Area C } Means only	
Phosphorus	1950 and 1951										0.3									Area A	
													0.3							Area B	
													0.2							Area C } Means only	
Vitamins μg/g DM β-carotene	1950							315												Area A	
											251									Area B	
												266								Area C } Means only	
	1951												249							Area A	
													256							Area B	
													263							Area C } Means only	

Table 1.148 Lucerne Mixtures (continued)

Lucerne, White Seal and White Clover S100		April			May			June			July			August			October			Notes	Reference
Proximate constituents %DM Crude protein																				Sown 17/5/1949. 10 to 20% clover except in area C where 1.5% only	152
	1950		29.0					22.8					23.8			26.8				Area A	
							20.5						19.7			28.8				Area B	
								18.7								24.4				Area C	
	1951						17.7								21.2		23.4			Area A	
							19.0								22.5		22.5			Area B	
							23.4							23.1			21.3			Area C	
Major elements %DM Calcium	%DM 1950 and 1951										1.8									Area A } Means only	
														1.8						Area B } Means only	
														2.4						Area C } Means only	
Phosphorus	1950 and 1951										0.3									Area A } Means only	
														0.3						Area B } Means only	
														0.2						Area C } Means only	
Vitamins μg/g DM β-carotene	1950							315												Area A } Means only	
											251									Area B } Means only	
												266								Area C } Means only	
	1951													249						Area A } Means only	
														256						Area B } Means only	
														263						Area C } Means only	
Lucerne and Timothy		May			June			July			August			September			October				
Proximate constituents %DM																				Crop grown for silage. GF = cuts which had glycollic or formic acid added before ensiling M = cuts which had molasses added before ensiling B = cuts which had barley meal added before ensiling W = cuts which were left to wilt before ensiling The Timothy content was appreciably lower for the 2nd (June 24–27) and 3rd (August 31 –September 3) cuts than for the 1st (June 6–9) For composition of the silages see table 1.137, same reference	375
Crude protein				GF	14.9		18.3														
				M	15.5		18.9								24.4						
				B	14.1		18.5														
				W	15.3		17.9														
Ether extract				GF	2.4		2.3														
				M	2.4		2.3								2.4						
				B	2.4		2.1														
				W	2.5		2.3														
Nitrogen-free extract				GF	48.0		48.5														
				M	47.4		46.9								39.4						
				B	48.1		48.4														
				W	47.7		47.3														
Crude fibre				GF	25.8		22.2														
				M	25.4		22.7								22.5						
				B	26.4		22.1														
				W	25.8		24.1														
Ash				GF	8.9		8.7														
				M	9.3		9.2								11.5						
				B	9.0		8.9														
				W	8.7		8.4														

continued

Table 1.148 Lucerne mixtures (continued)

Lucerne and Timothy	May/June	Notes	Reference
Proximate constituents %DM		Trial 1, 1952	376
Crude protein	(1952) U 11.3	Crop cut for silage and used:	
	C 11.5	Unchopped (U), Chopped (C)	
	L 11.4	or Lacerated (L)	
	(1953) U 16.0	Trial 3, 1953	
	UW 16.4	Crop: Unchopped (U),	
	CW 14.9	Unchopped wilted (UW),	
	LW 15.1	Chopped wilted (CW) or	
Ether extract	(1952) U 2.1	Lacerated wilted (LW)	
	C 2.0		
	L 2.0		
	(1953) U 2.1		
	UW 2.3	For composition of silages see	
	CW 2.2	table 1.137, same reference	
	LW 1.9		
Nitrogen-free extract	(1952) U 45.4		
	C 44.7		
	L 45.0		
	(1953) U 45.2		
	UW 44.7		
	CW 45.7		
	LW 46.4		
Crude fibre	(1952) U 33.7		
	C 34.4		
	L 34.2		
	(1953) U 28.6		
	UW 28.6		
	CW 28.9		
	LW 28.8		
Ash	(1952) U 7.5		
	C 7.4		
	L 7.4		
	(1953) U 8.1		
	UW 8.0		
	CW 8.3		
	LW 7.8		
	June	Cut for silage (no date)	381
Proximate constituents %DM			
Crude protein	21.6	Trial 1	
Ether extract	2.6		
Nitrogen-free extract	40.8		
Crude fibre	25.4		
Ash	9.6		
Crude protein	18.1	Trial 4, A0	
	17.5	Trial 4, A1	
	17.4	Trial 4, A2	
Ether extract	2.8	A0	
	2.9	A1	
	3.0	A2	
Nitrogen-free extract	44.3	A0	
	45.0	A1	
	44.5	A2	
Crude fibre	26.1	A0	
	26.5	A1	
	26.5	A2	
Ash	8.7	A0	
	8.1	A1	
	8.6	A2	
Crude protein	20.4	Trial 5, A0	
	20.8	Trial 5, A1	
Ether extract	2.9	A0	
	3.0	A1	
Nitrogen-free extract	41.7	A0	
	40.8	A1	
Crude fibre	25.6	A0	
	25.3	A1	
Ash	9.4	A0	
	10.1	A1	
		For additives and composition of the silages see table 1.137, same reference	

continued

Table 1.148 Lucerne Mixtures (continued)

Lucerne and Meadow Fescue ley	June	Notes	Reference
Proximate constituents %DM			
Crude protein		Crop cut (no date) for silage, small scale trial. Treatments as follows	379
	T_0, T_1, T_2: 14.8	T_0: untreated T_1: with 10 kg molasses T_2: 4 kg metabisulphite/1000 kg of herbage	
	W_0, W_1: 14.7		
Ether extract	T_0, T_1, T_2: 2.6	W_0: wilted W_1: wilted with 4.5 kg metabisulphite/1000 kg	
	W_0, W_1: 2.5		
Nitrogen free extract	T_0, T_1, T_2: 52.0		
	W_0, W_1: 52.3	For composition of silages see table 1.137, same reference	
Crude fibre	T_0, T_1, T_2: 22.7		
	W_0, W_1: 22.9		
Ash	T_0, T_1, T_2: 7.9		
	W_0, W_1: 7.6		

Table 1.149 Lucerne, artifically dried

	Crude protein %DM	Total Tocopherol µg/g DM	Notes	Reference
Du Puits		264	α-form, >95% of total	100
Grimm		280		
Ontaria variegated		276		
Provence		218		
Lucerne, unnamed				
Sample I	(a) 24.0 (b) 25.3		(a) Freeze dried, milled (b) Freeze dried, milled with residue in mill collected and returned	27
	(c) 25.0 (d) 25.3 (e) 24.7		(c) Freeze dried, unmilled (d) Oven-dried, milled (e) Oven-dried, unmilled (f) Oven-dried, milled with residue in mill collected and returned	
Sample II	(a) 18.3 (b) 19.4 (d) 19.2 (f) 19.6			

Table 1.150 Lucerne (Provence) and Lucerne mixture: hay

	May	June	July	August	Notes	Reference
Provence						
Digestibility of DM%	(1933) 60.2, 61.0		50.3, 55.1	(1932) 57.7, 59.0	With sheep 1-year-old crop	554
Digestibility of OM%	(1933) 62.0, 62.9		51.8, 56.4	(1932) 60.4, 61.8		
Proximate constituents %DM						
Crude protein	(1933) 21.1		16.1	(1932) 22.5		
Ether extract	(1933) 1.1		1.3	(1932) 1.3		
Nitrogen-free extract	(1933) 35.3		34.4	(1932) 36.4		
Crude fibre	(1933) 31.1		38.1	(1932) 30.2		
Ash	(1933) 11.4		10.1	(1932) 9.6		
Major elements %DM						
Calcium	(1933) 2.1		2.0	(1932) 2.3		
Phosphorus	(1933) 0.4		0.2	(1932) 0.3		
Chlorine	(1933) 0.2		0.3	(1932) 0.2		
Lucerne and Cocksfoot:						
Proximate constituents %DM						
Crude protein			(a) 13.4 (b) 14.3 (c) 15.6		(a) Tripoded and stacked (b) Piked and stacked (c) "Fenced" and baled Methods compared	416
Vitamins µg/g DM						
Carotene			(a) 29.0 (b) 28.1 (c) 89 3			

Table 1.151 Lucerne, unnamed: hay

Digestibility of DM%	Proximate constituents %DM CP	EE	NFE	CF	Ash	Notes	Reference
(a) 59.3 (57.5 – 62.4) (b) 60.3 (58.3 – 63.5)	13.8	1.8	36.9	38.3	9.2	Digestibility: means and ranges for 6 cows (a) water unrestricted (b) water restricted Proximate constituents for bulked samples.	38
	17.0	0.9	42.2	31.1	8.8	Used in milk production trials Ca: 1.3%DM P: 0.3%DM	56
	17.5	2.6	37.1	31.8	11.0	Commercial samples	569
	17.5	2.5	36.7	32.2	11.1	Commercial samples	572
	13.3	2.0	40.5	35.9	8.3		26
	12.8			36.7		Stemmy: used for wintering store cattle	402
	11.1			33.8		Tripoded: used for wintering store cattle	403
	14.6	1.8	42.6	31.2	9.8		23
(a)	15.1	1.6	44.1	29.7	9.5	(a) Means of 4 samples	
						Mean values for samples taken weekly	24
(a)	10.4	1.5	44.4	36.3	7.4	(a) Hay as fed	
(b)	20.9	1.9	44.9	19.8	12.5	(b) Leaf residue	
	14.5	1.7	39.1	35.6	9.1	Used in a cattle feeding experiment	33
						Weekly samples bulked for analysis at end of season	99
(a)	10.3	1.1	42.3	38.6	7.7	(a) 1954–55	
(b)	13.7	1.4	40.5	34.5	9.9	(b) 1955–56	

Table 1.152 Wild White Clover

		April	May	June	July	August	September	October	Notes	Reference
English Wild White *Proximate constituents* %DM Crude protein	(1928)	31.2	28.5		25.5	28.3		29.1 22.4*	Sown in 1927. Given basic slag 750 kg/ha in February, 1928 *Regrowth	550
	(1929)		22.4	23.2		22.3, 20.5*	28.2			
Crude fibre	(1928)	10.6	17.2		22.4	19.2		18.9 19.0*		
	(1929)		16.9	19.3		25.5, 24.1*	21.8			
Major elements %DM Calcium	(1928)	1.9	2.6		2.2	1.8		2.0 2.0*		
	(1929)		2.6	2.5		1.5, 1.7*	1.9			
Phosphorus	(1928)	0.2	0.1		0.4	0.2		0.4 0.4*		
	(1929)		0.3	0.4		0.4, 0.3*	0.5			
Kentish Wild White *Proximate constituents* %DM Crude protein					27.1				No date. Probably leafy. preflowering. Oven-dried and ground	420
Crude fibre					13.8					
Trace elements μg/g DM Iron					718					
N.Z. Wild White *Proximate constituents* %DM Crude protein	(1928)				28.0 28.8				Basic slag 750 kg/ha in February 1928 As above + 93 kg/ha Nitrochalk after each cut. Sown in 1927. Cut 4-5 times each year. Means only	550
	(1929)				24.6 23.8					
Crude fibre	(1928)				17.7 18.1					
	(1929)				23.5 23.0					
Major elements %DM Calcium	(1928)				2.2 2.2					
	(1929)				2.3 2.1					
Phosphorus	(1928)				0.2 0.3					
	(1929)				0.4 0.3					

Table 1.152 Wild White Clover (continued)

	May			June			July			August			September			October			Notes	Reference
Wild White, unnamed *Proximate constituents* %DM Crude protein												23.8							Preflowering. Oven-dried and ground.	13
				F_0 F_1 F_2	19.6 19.8 19.3									22.6* 21.5* 21.9*					F_0: no fertiliser F_1: Ammonium sulphate 187 kg/ha F_2: Ammonium sulphate 375 kg/ha Date uncertain. *Regrowth.	519, 520
Crude fibre												20.9							See Crude protein, same reference	13
Major elements %DM Calcium												1.4								13
				F_0 F_1 F_2	1.6 1.5 1.5									1.7* 1.7* 1.7*						519, 520
	(1957)	F_0 F_1 F_2 F_3 F_4 F_5	2.03 1.91 2.03 1.89 2.01 2.04				1.70 1.64 1.70 1.67 1.70 1.71			1.85 1.84 1.90 1.83 1.89 1.79			1.83 1.86 1.82 1.87 1.78 1.76						Clover hand separated from a sward with perennial ryegrass, cocksfoot and timothy. F_0: no fertiliser F_1: NaCl 500 kg/ha F_2: $MgSO_4$ 250 kg/ha F_3: Ammonium sulphate 375 kg/ha F_4: Superphosphate 375 kg/ha F_5: Muriate of potash 250 kg/ha Cut at silage stage each time, first cut in late May, last cut in early September. Dates of second and third cuts not given; assumed to be early July and early August. Fertilisers applied in spring each year, except N, applied after each cut. Clover died out on N-treated plots after first year.	274
	(1958)	F_0 F_1 F_2 F_3 F_4 F_5	1.18 1.14 1.16 — 1.30 1.13				1.13 1.09 1.09 — 1.13 1.10			1.18 1.20 1.15 — 1.15 1.16			1.10 1.00 1.00 — 1.03 1.03							
	(1959)	F_0 F_1 F_2 F_3 F_4 F_5	0.99 0.94 0.98 — 1.03 1.01				0.78 0.68 0.76 — 0.83 0.79			0.81 0.75 0.78 — 0.84 0.73			0.73 0.67 0.70 — 0.75 0.74							
Phosphorus									F_0 F_1	0.2 0.4									F_0: no fertiliser F_1: Superphosphate 750 kg/ha Moor pastures. Aphosphorosis investigation. No date given	145
												0.4							See Crude protein, same reference	13
				F_0 F_1 F_2	0.2 0.2 0.3									0.3* 0.3* 0.3*						519 520

continued

Table 1.152 Wild White Clover (continued)

	Year	Treatment	May	June	July	August	September	October	Notes	Reference
Major elements %DM (cont'd) Magnesium	(1957)	F_0	0.221		0.229	0.259	0.201		See Calcium, same reference †Computed from initial value and 'response' to muriate of potash.	274
		F_1	0.214		0.217	0.262	0.206			
		F_2	0.238		0.229	0.265	0.212			
		F_3	0.200		0.243	0.278	0.223			
		F_4	0.219		0.224	0.253	0.205			
		F_5	0.207		0.210	0.235	0.197			
	(1958)	F_0	0.239		0.222	0.214	0.142			
		F_1	0.224		0.210	0.223	0.121			
		F_2	0.280		0.220	0.226	0.158			
		F_3	—		—	—	—			
		F_4	0.255		0.230	0.198	0.136			
		F_5	0.230		0.175	0.189	0.118			
	(1959)	F_0	0.201		0.195	0.190	0.204			
		F_1	0.186		0.166	0.185	0.177			
		F_2	0.223		0.212	0.216	0.220			
		F_3	—		—	—	—			
		F_4	0.204		0.191	0.191	0.206			
		F_5	0.186		0.173	0.156	0.182			
Potassium	(1957)	F_0	1.85		1.65	1.86	1.55			
		F_1	1.91		1.67	1.91	1.56			
		F_2	1.78		1.76	1.94	1.66			
		F_3	1.93		1.60	1.64	1.35			
		F_4	1.80		1.60	1.87	1.55			
		F_5	2.38		2.09	2.41	2.02			
	(1958)	F_0	1.94		1.76	1.62	1.66			
		F_1	2.02		1.77	1.59	1.61			
		F_2	1.96		1.76	1.50	1.73			
		F_3	—		—	—	—			
		F_4	1.83		1.59	1.59	1.51			
		F_5	2.99		2.66	2.43	2.55			
	(1959)	F_0	1.77		1.64	2.02	1.72			
		F_1	1.87		1.85	2.28	2.06			
		F_2	1.81		1.60	1.79	1.77			
		F_3	—		—	—	—			
		F_4	1.79		1.60	2.06	1.61			
		F_5	2.92		2.70	3.44	3.03			
Sodium	(1957)	F_0	0.273		0.265	0.231	0.248			
		F_1	0.469		0.373	0.343	0.336			
		F_2	0.261		0.238	0.218	0.221			
		F_3	0.338		0.331	0.342	0.336			
		F_4	0.277		0.267	0.255	0.238			
		F_5	0.244		0.183	0.137	0.134			
	(1958)	F_0	0.209		0.247	0.273	0.292			
		F_1	0.314		0.320	0.364	0.427			
		F_2	0.162		0.262	0.267	0.292			
		F_3	—		—	—	—			
		F_4	0.159		0.227	0.294	0.317			
		F_5	0.052		0.130	0.092	0.014			
	(1959)	F_0	0.279		0.328	0.343	0.506			
		F_1	0.381		0.398	0.416	0.706			
		F_2	0.258		0.291	0.321	0.506			
		F_3	—		—	—	—			
		F_4	0.294		0.366	0.343	0.536			
		F_5	0.092		0.046	0.000†	0.126			

Table 1.152 Wild White Clover (continued)

Trace elements µg/g DM	Year		May	June	July	August	September	October	Notes	Reference
Copper	(1957)	F_0	10.5		9.5	19.0	16.5		Clover from a sward with ryegrass, cocksfoot and timothy F_0: no fertiliser F_1: superphosphate, 375 kg/ha F_2: muriate of potash, 250 kg/ha F_3: ammonium sulphate, 375 kg/ha F_4: ammonium sulphate + superphosphate F_5: ammonium sulphate + muriate of potash Cut at silage stage each time, first cut in late May, last cut in mid-September Fertilisers applied in spring each year, except N applied after each cut. Clover died out on N-treated plots after first year.	275
		F_1	8.5		9.0	13.5	18.5			
		F_2	12.5		12.5	12.5	15.0			
		F_3	10.0		13.0	18.5	19.0			
		F_4	11.0		13.0	18.5	19.0			
		F_5	11.0		11.5	14.5	19.0			
	(1958)	F_0	13.5		15.0	13.0	11.0			
		F_1	12.5		11.5	14.5	10.5			
		F_2	10.0		10.5	13.5	17.5			
	(1959)	F_0	10.5		11.5	12.0	11.5			
		F_1	10.5		9.0	11.5	13.0			
		F_2	9.5		10.4	12.0	10.8			
Iron	(1957)	F_0	250		160	283	405			
		F_1	222		128	438	288			
		F_2	145		148	178	280			
		F_3	120		175	305	345			
		F_4	175		150	253	450			
		F_5	275		128	270	300			
	(1958)	F_0	158		165	330	305			
		F_1	195		230	375	350			
		F_2	135		260	415	265			
	(1959)	F_0	100		125	100	141			
		F_1	125		123	85	181			
		F_2	100		90	75	130			
Manganese	(1957)	F_0	20		20	29	29			
		F_1	34		31	32	31			
		F_2	25		29	34	31			
		F_3	47		39	44	29			
		F_4	41		41	42	45			
		F_5	31		32	34	31			
	(1958)	F_0	18		23	31	32			
		F_1	27		32	34	29			
		F_2	29		31	31	29			
	(1959)	F_0	28		31	38	34			
		F_1	29		31	33	38			
		F_2	31		34	38	32			
Molybdenum	(1957)	F_0	3.0		2.8	3.6	4.0			
		F_1	2.8		2.5	3.2	3.0			
		F_2	1.8		2.4	2.6	2.8			
		F_3	1.7		1.8	2.2	2.5			
		F_4	2.8		1.9	2.7	2.6			
		F_5	2.5		2.4	3.3	3.0			
	(1958)	F_0	2.8		2.6	3.6	2.3			
		F_1	2.1		1.6	2.7	1.8			
		F_2	1.8		1.5	1.3	2.0			
	(1959)	F_0	2.0		1.9	1.9	1.8			
		F_1	1.5		1.4	1.5	1.3			
		F_2	1.0		0.9	0.7	0.6			

continued

Table 1.152 Wild White Clover (continued)

	March	April	May	June	July	August	September	October	Notes	Reference
Vitamins μg/g DM										
Carotene	384	530	465	398	274	311	669*	561*	Cut close to ground in April and August. *Regrowth Estimated by Moon's method.	362
Tocopherol										
Total						202			No date given	100
a-tocopherol % of total						>95				
Total						(a) 126 – 130			(a) Artificially dried.	
a-tocopherol % of total						>95				

Table 1.153 White clover

	April	May	June	July	August	September	October	Notes	Reference
White S100 *Proximate constituents* %DM Crude protein				F 22.5 1a 18.7 1b 17.3 2a 22.5 2b 21.2 2c 22.8				Tests with clover S100 grown on waste ash from electricity generating stations, 1 and 2, and on a farm F. The samples of leaf analysed from the farm were of 2-year old clover; from the stations 1-year old. F: farm 1a: ash with 2 inch layer of fine soil 1b: ash alone 2a: ash with 250-375 kg of an NPK fertiliser, not further defined, /ha 2b: ash alone 2c: ash with sewage	431
	26.4		25.6	26.0		27.5			141
		27.4		23.3	25.3		28.3	Various fertilisers (unspecified) gave little variation.	140
					N_0 28.1 N_1 27.3 N_0 18.5 N_1 16.0			N_0: no fertiliser, N_1: 110 kg N/ha } Leaf+petiole N_0, N_1 } Stolon+petiole base	
Ash	9.8		9.2	10.2		9.8			141
Carbohydrate fractions %DM Water-soluble carbohydrate	12.0		13.8	7.7		10.7			
Lignin	5.6		6.5	5.6		4.4			
Major elements %DM Calcium				F 2.1 1a 2.4 1b 2.6 2a 2.5 2b 2.2 2c 2.0				See Crude protein, same reference	431
Phosphorus				F 0.4 1a 0.3 1b 0.4 2a 0.4 2b 0.5 2c 0.4					
	May	June	July	August	September	October	November		
Magnesium (1959)	0.17	0.19	0.19	0.20	0.18		0.17	Cut when about 10 cm high. A dressing of magnesian limestone, 5 tonnes/ha, raised Mg content by 11% in 1959 and by 34% in 1960. Magnesium sulphate had little effect and carboniferous limestone reduced Mg in herbage	322
Magnesium (1960)	0.18	0.20	0.19	0.21	0.21				

continued

Table 1.153 White clover (continued)

	April	May	June	July	August	September	October	Notes	Reference
Major elements %DM (cont'd) Potassium				F 2.8 1a 1.4 1b 1.3 2a 2.1 2b 1.5 2c 1.4				See Crude protein, same reference	431
Vitamins μg/g DM Carotene				F 230 1a 240 1b 250 2a 190 2b 190 2c 230					

Early White	March	April	May	June	July	August	September	October	Notes	Reference
Vitamins μg/g DM Carotene	499	506	351	347	345	273	548*	670*	Cut close to ground in April and August. *Regrowth	426

	May	June	July	August	September	October	Notes	Reference
Dutch White *Proximate constituents* %DM Crude protein			(1928) 27.2, 26.4 (1929) 21.7, 21.5	20.0*		23.5*	Basic slag 750 kg/ha in Feb. 1928. As above + Nitrochalk 93 kg/ha after each cut. Sown in 1927. Cut 4-5 times each year Means only for primary growth. *Regrowth	550
Crude fibre			(1928) 19.7, 19.1 (1929) 23.4, 24.5	28.3*		23.0*		
Major elements %DM Calcium			(1928) 2.1, 1.9 (1929) 2.1, 2.0	1.4*		2.2*		
Phosphorus			(1928) 0.3, 0.3 (1929) 0.4, 0.4	0.4*		0.4*		

Table 1.153 White clover (continued)

	May	June	July	August	September	October	Notes	Reference
N.Z. White *Proximate constituents* %DM Crude protein				(1929) 22.0*		(1928) 23.2*	*Regrowth Sown in 1927. Given 750 kg basic slag /ha in February, 1928	550
Crude fibre				(1929) 25.5*		(1928) 22.1*		
Major elements %DM Calcium				(1929) 1.3*		(1928) 2.4*		
Phosphorus				(1929) 0.3*		(1928) 0.4*		
N.Z. 'Ordinary White *Proximate constituents* %DM Crude protein				(1929) 20.3*		(1928) 23.6*	*Regrowth Sown in 1927. Given 750 kg basic slag /ha in February, 1928	
Crude fibre				(1929) 24.3*		(1928) 22.2*		
Major elements %DM Calcium				(1929) 1.2*		(1928) 2.1*		
Phosphorus				(1929) 0.4*		(1928) 0.5*		
N.Z. Stubble White *Proximate constituents* %DM Crude protein			(1928) {27.1, 25.5 (1929) {23.0, 22.7	20.7*		24.1*	Basic slag 750 kg/ha in Feb. 1928. As above + Nitrochalk 93 kg/ha after each cut. Sown in 1927. Cut 4-5 times each year Means only for primary growth. *Regrowth	
Crude fibre			(1928) {17.8, 18.3 (1929) {23.4, 22.6	26.9*		20.2*		
Major elements %DM Calcium			(1928) {2.2, 2.1 (1929) {2.2, 2.1	1.4*		2.2*		
Phosphorus			(1928) {0.3, 0.3 (1929) {0.4, 0.3	0.4*		0.4*		
N.Z. Large Leaved *Proximate constituents* %DM Crude protein				(1929) 22.5*		(1928) 23.2*	*Regrowth Sown in 1927. Given 750 kg basic slag /ha in February, 1928	
Crude fibre				(1929) 24.2*		(1928) 21.5*		
Major elements %DM Calcium				(1929) 1.4*		(1928) 2.2*		
Phosphorus				(1929) 0.4*		(1928) 0.5*		

continued

Table 1.153 White clover (continued)

		May		June		July		August	September	October	Notes	Reference
Unnamed White *Crude protein* %DM		25.0		18.4		16.5		24.4†	21.6		Date uncertain. May, young; June, flowering or heading; July, old; August, aftermath† September, being grazed or recently been grazed.	460
	N_0				22.8						N_0: No N	519, 520
	N_1				22.1						N_1: ammonium sulphate 187 kg/ha	
	N_2				21.9						N_2: ammonium sulphate 375 kg/ha	
(1955)	N_0	23.2	22.0	24.7	25.3	21.0		21.7	22.1	25.0	N_0: no fertiliser	464
	N_1	23.6	27.2	24.0	26.2	23.3		22.6	25.0	28.1	N_1: Nitrochalk 750 kg/ha	
	N_2	23.1	24.8	25.4	25.4	23.5		22.2	26.7		N_2: Nitrochalk 1500 kg/ha	
(1956)	N_0	25.7	25.7			25.2		26.2	30.0	31.9	N_1* : annually since 1951	
	N_1							24.6	31.5			
	N_1*		23.5									
	N_2	31.9	31.9							26.4		
Major elements %DM Calcium	N_0				1.7						See Crude protein, same reference	519, 520
	N_1				1.8							
	N_2				1.9							
Phosphorus	N_0				0.2							
	N_1				0.2							
	N_2				0.2							
Trace elements μg/g DM Copper	N_0						3.5				N_0: no fertiliser	441
	N_1						5.1				N_1: copper sulphate 22 kg/ha and manganese sulphate 33 kg/ha	
Manganese	N_0						28				No date given. Grown on machair land DM basis assumed	
	N_1						30					
Vitamins μg/g DM Carotene	W	506		331		196		315†	405		W: whole plant	460
	L+S			216							L+S: leaf + stem (at flowering)	
	HF			95							HF: head or flower	
											See Crude protein, same reference	

	March	April	May	June	July	August	September	October	November	Notes	Reference
a-tocopherol	105	90	115	120	120	125	135	180	210	Closely mown crop	86

Table 1.154 Red Clover

	May			June			July			August			September			October			Notes	Reference
Digestibility of DM% EARLY VARIETIES																				
Essex D.C.			70.8						61.6						63.1				Means of 2 samples. Cambridge 1961 Averages and ranges (all samples) for the 3 cuts of early varieties were: Cut 1: 69.6 (67.0–71.0) Cut 2: 63.6 (61.2–67.3) Cut 3: 63.8 (61.4–65.0)	168
Sussex (Berry's)			68.7						63.1						63.7					
Pearse's			71.0						61.2						64.4					
Cotswold D.C.			70.8						62.4						62.1					
Dorset Marl			70.6						62.1						64.4					
Sussex (Drewitt's)			70.4						64.0						64.5					
S151			67.0						67.2						65.0					
Cotswold S.C.			68.5						67.3						64.5					
Vale of Clwyd			68.6						63.3						62.9					
LATE VARIETIES S151					66.1							61.6							Cambridge 1961 Average, and ranges for the 2 cuts of late varieties were: Cut 1: 68.1 (66.1–72.4) Cut 2: 62.5 (61.0–65.4)	
Cotswold S.C.					67.5							63.0								
S123					67.4							62.6								
Altaswede					66.5							64.0								
Essex S.C.					66.8							61.6								
Montgomery N.Z.					69.2							61.2								
Cornish Marl					68.1							62.0								
Ulva					72.4							65.4								
Rea 28					67.5							62.4								
S123 no mark					69.2							61.0								

	April			May			June			July			August			October			Notes	Reference
S123, Late Red ***Proximate constituents*** **%DM** **Crude protein**					(1950)	24.0			23.4 13.7			21.9			25.5 26.7 25.2					152
									(1951)	16.0 16.5 16.9						21.5 21.6 22.5				
					F_3 F_4 F_5	34.1 33.6 32.6													See Phosphorus, same reference for fertilisers	169
Major elements **%DM** **Calcium**					(1950)	1.8			1.5			2.0								152
Phosphorus					(1950)	0.3			0.3			0.3								
(1956)	0.3	0.4	0.3	0.3	0.3	0.3			0.3			0.2				(1957)	0.3 0.3		Leaf without petiole. 1957, late spring, date uncertain. F_0 : No fertiliser F_1 : Sulphate of potash 375 kg/ha F_2 : Superphosphate 750 kg/ha F_3 : $F_1 + F_2$ F_4 : $F_1 + F_2$ + Nitrochalk 375 kg/ha F_5 : $F_1 + F_2$ + Nitrochalk 750 kg/ha F_6 : $F_1 + F_2$ + magnesium sulphate 750 kg/ha F_7 : basic slag 2.5 tonnes/ha F_8 : $F_1 + F_2$ + borax 25 kg/ha F_9 : $F_1 + F_2$ + borax 50 kg/ha F_{10} : $F_1 + F_2$ + borax 74 kg/ha	169
	(1956)	F_0 F_1 F_2 F_3 F_4 F_5	0.3 0.3 0.4 0.4 0.4 0.5			0.3 0.3 0.3 0.3 0.3 0.3														
				(1957)	F_0 F_1 F_2 F_3 F_7	0.2 0.3 0.3 0.3 0.3														

continued

Table 1.154 Red Clover (continued)

S123 Late Red (continued)	April			May			June			July			August			October			Notes	Reference
Major elements (cont'd) Magnesium (1956)	0.2	0.2	0.2	0.2	0.3												0.3			
																(1957)	0.3			
				(1956)	F_0	0.3														
					F_3	0.3														
					F_6	0.3														
					(1957)	F_0 0.2														
						F_3 0.2														
						F_6 0.3														
	May			June			July			August			September			November				
	(1959)	0.20			0.24			0.25			0.30			0.24			0.20		Cut when about 10 cm high. A dressing of magnesian limestone, 5 tonnes/ha raised Mg content by 19% of the mean in 1959 and by 27% in 1960. Magnesium sulphate had little effect and carboniferous limestone reduced Mg in herbage.	322
	(1960)	0.26			0.29			0.29			0.30			0.29						
	April			May			June			July			August			October				
Potassium (1956)	1.8	1.7	1.7	1.6	1.4	1.1			1.1			1.1					2.1		See Phosphorus, same reference	169
																(1957)	2.1			
	(1956)	F_0	1.7			1.1														
		F_1	1.9			1.3														
		F_2	1.8			1.0														
		F_3	1.9			1.2														
		F_4	2.0			1.3														
		F_5	2.5			1.5														
				(1957)	F_0	2.0														
					F_1	2.3														
					F_2	1.9														
					F_3	2.2														
					F_7	2.0														
Trace elements μg/g DM Boron (1956)	25	26	31	30	30	30			43			49					30			
																(1957)	32			
	(1956)	F_0	32			31														
		F_3	29			28														
		F_8	39			33														
				(1957)	F_0	32														
					F_3	28														
					F_8	31														
					F_9	33														
					F_{10}	40														
Vitamins μg/g DM β-carotene				(1950)	A B C	372			294			232							A : 3 cuts, B : 2 cuts, C : 2 cuts — in first year, means only	152
								(1951)	A	237										
									B	232										
									C	242										

Table 1.154 Red Clover (continued)

	March			April			May			June			July			October			Notes	Reference
Cotswold Single cut *Proximate constituents* %DM Crude protein								F_1 F_1 F_1	32.8 31.9 29.7										Leaf without petiole 1957, late spring: date uncertain	169
Major elements %DM Phosphorus				(1956)	F_0 F_1 F_2 F_3 F_4 F_5	0.3 0.3 0.4 0.4 0.4 0.4			0.2 0.2 0.3 0.3 0.3 0.3										F_0: No fertiliser F_1: Sulphate of potash 375 kg/ha F_2: Superphosphate 750 kg/ha F_3: $F_1 + F_2$ F_4: $F_1 + F_2$ + Nitrochalk 375 kg/ha F_5: $F_1 + F_2$ + Nitrochalk 750 kg/ha F_6: $F_1 + F_2$ + magnesium sulphate 750 kg/ha F_7: basic slag 2.5 tonnes/ha F_8: $F_1 + F_2$ + borax 25 kg/ha F_9: $F_1 + F_2$ + borax 50 kg/ha F_{10}: $F_1 + F_2$ + borax 74 kg/ha	
								(1957)	F_0 0.2 F_7 0.3 F_1 0.3 F_2 0.3 F_3 0.3											
			(1956)	0.3	0.4	0.3	0.3	0.3	0.3			0.3			0.2	(1957)	0.4 0.4			
Magnesium							(1956)	F_0 F_3 F_6	0.3 0.3 0.3											
								(1957)	F_0 0.2 F_3 0.2 F_6 0.2											
			(1956)	0.2	0.2	0.2	0.2	0.3								(1957)	0.3 0.3			
Potassium				(1956)	F_0 F_1 F_2 F_3 F_4 F_5	1.8 1.8 1.6 1.8 1.9 1.3			1.1 1.3 0.9 1.1 1.1 1.2											
								(1957)	F_0 1.7 F_7 2.0 F_1 2.0 F_2 1.7 F_3 2.0											
			(1956)	1.7	1.7	1.8	1.6	1.3	1.1			1.0			1.0	(1957)	2.2 2.3			
Trace elements μg/g DM Boron				(1956)	F_0 F_3 F_8	32 31 38			30 28 30											
								(1957)	F_0 34 F_3 29 F_8 32 F_9 37 F_{10} 47											
			(1956)	27	29	31	32	29	31			47			54	(1957)	31 36			

continued

Table 1.154 Red Clover (continued)

	March			April			May			June			August			October			Notes	Reference
Essex double cut *Proximate constituents* %DM Crude protein								F_3 F_4 F_5	33.6 33.0 32.6										Leaf without petiole 1957, late spring, date uncertain 1958, sown under barley, after lucerne on land deficient in phosphate.	169
Major elements %DM Phosphorus			(1956)	0.4	0.4	0.3	0.3	0.3	0.3							(1957)	0.4 0.4		F_0: No fertiliser F_1: Sulphate of potash 375 kg/ha F_2: Superphosphate 750 kg/ha F_3: F_1 + F_2 F_4: F_1 + F_2 + Nitrochalk 375 kg/ha F_5: F_1 + F_2 + Nitrochalk 750 kg/ha F_6: F_1 + F_2 + magnesium sulphate 750 kg/ha F_7: basic slag 2.5 tonnes/ha F_8: F_1 + F_2 + borax 25 kg/ha F_9: F_1 + F_2 + borax 50 kg/ha F_{10}: F_1 + F_2 + borax 74 kg/ha *Regrowth	
				(1956)	F_0 F_1 F_2 F_3 F_4 F_5	0.3 0.3 0.4 0.4 0.4 0.4			0.3 0.3 0.3 0.3 0.3 0.3											
								(1957)	F_0 0.2 F_1 0.3 F_2 0.3 F_3 0.3 F_4 0.3 F_7 0.3											
					(1958)	F_0 F_1 F_2 F_3 F_4	0.2 0.2 0.4 0.3							0.2* 0.2* 0.3* 0.3* 0.3*						
Magnesium			(1956)	0.2	0.2	0.2	0.2	0.2								(1957)	0.3 0.3			
						(1956)	F_0 F_3 F_6	0.3 0.3 0.3												
Potassium			(1956)	1.8	1.6	1.7	1.5	1.3	1.2							(1957)	2.3 2.3			
				(1956)	F_0 F_1 F_2 F_3 F_4 F_5	1.7 1.7 1.7 1.7 1.8 2.2			1.1 1.5 1.1 1.3 1.5 1.6											
								(1957)	F_0 2.1 F_1 2.4 F_2 2.0 F_3 2.4 F_4 2.2 F_7 2.2											
					(1958)	F_0 F_1 F_2 F_3 F_4	1.7 2.0 1.4 2.0							1.7* 2.1* 1.3* 1.7* 1.7*						
Trace elements μg/g DM Boron			(1956)	25	29	32	33	30	33							(1957)	30 33			
				(1956)	F_0 F_3 F_8	32 30 37			35 30 34											

Table 1.154 Red Clover (continued)

	June	July	August	September	Notes	Reference
Montgomery Red *Proximate constituents* %DM Crude protein		(1931) N_0L 22.0 N_1L 22.7 N_0U 17.9 N_1U 19.1			N_0: no fertiliser N_1: manured L : Lowland 50–400 feet above sea level U : Upland 750–900 feet above sea level "Manured" = the residual effects of fertilisers applied in 1927	215
Ash		(1931) N_0L 10.4 N_1L 9.7 N_0U 9.4 N_1U 9.1				
Major elements %DM Calcium		(1931) N_0L 2.2 N_1L 2.4 N_0U 1.7 N_1U 1.8				
Phosphorus		(1931) N_0L 0.3 N_1L 0.3 N_0U 0.2 N_1U 0.3				
Potassium		(1931) N_0L 1.8 N_1L 1.9 N_0U 1.6 N_1U 1.8				
Chlorine		(1931) N_0L 0.5 N_1L 0.5 N_0U 0.4 N_1U 0.5				
Montgomery Red, Late-flowering *Proximate constituents* %DM Crude protein		H 23.1 L+P 28.1 S 10.0			H : Head L+P : Leaf + petiole S : Stem Date uncertain, flowering stage. Given 500 kg superphosphate, 250 kg muriate of potash and 250 kg calcium ammonium nitrate per ha. *As given. The major elements below total 4.4%. From the ratios to ash of the corresponding totals for Head and Stem the ash of Leaf + petiole might be 10.3%.	236
Ash		H 7.3 L+P 0.3* S 6.5				
Major elements %DM Calcium		H 1.1 L+P 2.1 S 1.1				
Phosphorus		H 0.4 L+P 0.3 S 0.2				
Magnesium		H 0.3 L+P 0.3 S 0.2				
Potassium		H 2.1 L+P 1.7 S 1.7				
Trace elements μg/g DM Barium		H 3.1 L+P 6.4 S 13.0				
Boron		H 26 L+P 26 S 20				
Chromium		H 0.1 L+P 0.2 S 0.2				
Cobalt		H 0.10 L+P 0.11 S 0.07				
Copper		H 11 L+P 17 S 14				
Iron		H 64 L+P 134 S 24				
Lead		H 0.4 L+P 1.1 S 0.7				
Manganese		H 60 L+P 136 S 19				

continued

Table 1.154 Red Clover (continued)

Montgomery Red Late (cont'd)	May	June	July		August	September	October	Notes	Reference
Trace elements (cont'd) Molybdenum			H L+P S	0.3 0.3 0.5					236
Nickel			H L+P S	4.6 2.7 1.5					
Strontium			H L+P S	15 37 37					
Tin			H L+P S	0.08 0.13 0.12					
Titanium			H L+P S	1.7 3.0 1.4					
Vanadium			H L+P S	0.17 0.27 0.08					
Zinc			H L+P S	40 42 12					

	March	April	May	June	July	August	September	October	Notes	Reference
Broad Red *Proximate constituents* %DM Crude Protein					22.6				No date Oven dried and ground	420
Crude fibre					17.7					
Trace elements µg/g DM Iron					755					
Vitamins µg/g DM Carotene	594	461	487	228	356	402	425*	591*	Cut close to ground in April and August *Regrowth	362

Table 1.154 Red Clover (continued)

	August			September			October			November			Notes	Reference
Double Cut *Proximate constituents* %DM Crude protein					33.2 (31.9 – 34.6)								Means and ranges for 50 leaf samples taken from 9 commercial farms, autumn 1956 Leaf without petiole	169
Major elements %DM Phosphorus					0.4									
Magnesium					0.25 (0.2 – 0.3)									
Potassium					2.2 (1.9 – 2.4)									
Trace elements μg/g DM Boron					36 (32 – 42)									
	May			June	July	Aug.	September			October				
Single Cut *Proximate constituents* %DM Crude protein								31.4 (29.3 – 34.1)					Mean and range of 50 leaf samples taken from 10 commercial farms, autumn 1956.	
Major elements %DM Phosphorus	0.43		0.3					0.37					Means of 50 leaf samples taken from 9, 6 and 10 commercial farms, respectively in 1956, 1955 and 1956	
Magnesium			0.3					0.25					Means of 50 leaf samples taken from 6 and 10 commercial farms, respectively in 1955 and 1956	
Potassium	2.2		1.4					2.2					As Phosphorus, above	
Trace elements μg/g DM Boron	28		36					35					As Phosphorus, above	
	June			July			August			September				
Red, Late-flowering *Proximate constituents* %DM Crude protein									29.9				Preflowering. Oven dried and ground	13
Crude fibre									20.7					
Major elements %DM Calcium									1.7					
Phosphorus									0.4					

continued

Table 1.154 Red Clover (continued)

	May	June	July	August	Notes	Reference
Red, unnamed						
Major elements **%DM**						
Phosphorus			N_0 0.1		N_0: no fertiliser N_1: Superphosphate 750 kg/ha Moor pastures. Aphosphorosis investigation. No date	145
			N_1 0.2			
Magnesium		(O) 0.39 (0.26 – 0.50)			Means and ranges for samples grown in soil of the group: Old Red Sandstone (O) 13; Granitic (G) 18; Basic (B) 12; Slate (S) 16; and Sands and Gravels (SG) 12.	433
		(G) 0.33 (0.19 – 0.49)				
		(B) 0.42 (0.26 – 0.67)				
		(S) 0.37 (0.16 – 0.49)				
		(SG) 0.31 (0.17 – 0.44)				
Trace elements **μg/g DM**						
Cobalt		(1954) N_0 0.04			Fertilisers applied: N_0: none N_1: Ca N_2: Mo, Cu, Zn N_3: Mo, Cu, Zn, Ca N_4: Cu only N_5: Co only Mo: $Na_2MoO_4.2H_2O$, 1.1 kg/ha Cu: $CuSO_4.5H_2O$, 22 kg/ha Zn: $ZnSO_4.7H_2O$, 22 kg/ha Ca: $CaCO_3$, 7500kg/ha (applied later than the others) Co: $CoSO_4.7H_2O$, 1.7 kg/ha	359
		(1954) N_5 0.87				
		(1955) N_0 0.06				
		(1955) N_5 1.10				
Copper		(1952) N_0 4.7				
		(1952) N_1 6.8				
		(1952) N_2 8.8				
		(1952) N_3 7.7				
		(1953) N_0 3.0				
		(1953) N_1 4.3				
		(1953) N_2 7.0				
		(1953) N_3 6.2				
		(1954) N_0 1.4				
		(1954) N_4 7.8				
		(1955) N_0 1.3				
		(1955) N_4 8.4				
Manganese		(1952) N_0 49				
		(1952) N_1 45				
		(1952) N_2 43				
		(1952) N_3 39				
		(1953) N_0 36				
		(1953) N_1 35				
		(1953) N_2 46				
		(1953) N_3 26				
Molybdenum		(1952) N_0 2.0				
		(1952) N_1 9.3				
		(1952) N_2 11.4				
		(1952) N_3 16.7				
		(1953) N_0 0.7				
		(1953) N_1 3.4				
		(1953) N_2 18.6				
		(1953) N_3 35.6				
Zinc		(1952) N_0 33				
		(1952) N_1 27				
		(1952) N_2 47				
		(1952) N_3 33				
		(1953) N_0 33				
		(1953) N_1 29				
		(1953) N_2 75				
		(1953) N_3 44				

Table 1.155 Red Clover. Trace elements: effect of soil drainage

Trace elements μg/g DM	Well drained		Poorly drained		Notes	Reference
	1953	1954	1953	1954		
Barium	10	12	14	24	No date given	359
Chromium	0.14	0.13	0.15	0.13		
Cobalt	0.16	0.17	1.4	1.2		
Copper	7.9	8.9	10.3	10.2		
Iron	69	79	78	77		
Manganese	39	33	51	46		
Molybdenum	1.0	2.0	3.1	2.4		
Nickel	2.0	1.0	5.9	3.0		
Strontium	57	39	48	68		
Titanium	4.0	1.8	1.3	2.3		
Vanadium	0.08	0.05	0.08	0.04		
Zinc	38	35	38	37		

Table 1.156 Clover Mixtures, named varieties

	May			June			July			August			September			October			Notes	Rerference
White S100 and Kentish Wild White *Proximate constituents* %DM Crude protein		N_0 N_1 N_2	24.7 24.0 23.7																N_0: no fertiliser N_1: Ammonium sulphate 187 kg/ha N_2: Ammonium sulphate 375 kg/ha	519, 520
Major elements %DM Calcium		N_0 N_1 N_2	1.7 1.7 1.8																	
Phosphorus		N_0 N_1 N_2	0.3 0.3 0.3																	
Montgomery Red, Dutch White and Wild White *Proximate constituents* %DM Crude protein									N_0 N_1 N_2	20.2 20.3 21.0										
Major elements %DM Calcium									N_0 N_1 N_2	1.5 1.5 1.7										
Phosphorus									N_0 N_1 N_2	0.2 0.3 0.3										
Red S151, S123 and White S100 *Proximate constituents* Crude protein					N_0 N_1 N_2	20.2 21.6 21.4								21.1 19.8 19.7						
Major elements %DM Calcium					N_0 N_1 N_2	1.7 1.3 1.3								2.1 2.1 2.0						
Phosphorus					N_0 N_1 N_2	0.3 0.2 0.3								0.3 0.3 0.3						

Table 1.157 Clover Mixtures, unnamed

	April			May			June			July			August			September			Notes	Reference
Major elements %DM																				
Calcium			(1956)	2.03		1.82						1.85			2.19				Sown in 1955. Trials with ammonium sulphate 375 kg/ha, superphosphate 375 kg/ha, muriate of potash 250 kg/ha, sodium chloride, 500 kg/ha, and in 1957 only, additional magnesium sulphate 250 kg/ha and ammonium sulphate 375kg/ha after each cut. Means for untreated plots. Responses to fertilisers given only as increases or decreases	273
			(1957)				2.03				1.70			1.85				1.83		
Phosphorus			(1956)	0.32		0.22						0.14			0.16					
			(1957)				0.16				0.19			0.19				0.22		
Magnesium			(1956)	0.26		0.23						0.23			0.22					
			(1957)				0.22				0.23			0.26				0.21		
Potassium			(1956)	3.12		2.85						1.90			1.86					
			(1957)				1.85				1.65			1.86				1.55		
Sodium			(1956)	0.10		0.06						0.09			0.09					
			(1957)				0.27				0.26			0.23				0.25		
Trace elements %DM																				
Copper							N_0	21									19		N_0: no fertiliser	551
							N_1	21									18		N_1: + N	
							N_2	22									17		N_2: + P	
							N_3	22									18		N_3: + NP	
							N_4	24									11		N_4: + K	
							N_5	18									16		N_5: + NK	
							N_6	23									22		N_6: + PK	
							N_7	21									15		N_7: + NPK	
							N_8	18									15		N_8: + farmyard manure	
							N_9	19									20		N_9: + farmyard manure + NPK	
Manganese							N_0	82									130		N = 19 kg/ha as Nitrochalk	
							N_1	78									163		P = 62 kg/ha as superphosphate	
							N_2	87									130		K = 125 kg/ha as potassium sulphate	
							N_3	78									98		Farmyard manure: 37.5 tonnes/ha	
							N_4	78									78		All + hydrated lime 3750 kg/ha	
							N_5	82									65		Two cuts; dates assumed	
							N_6	69									98			
							N_7	96									78			
							N_8	87									98			
							N_9	82									98			
Molybdenum							N_0	0.7									1.6			
							N_1	0.7									2.0			
							N_2	0.9									2.5			
							N_3	1.0									2.1			
							N_4	0.7									1.9			
							N_5	0.6									1.7			
							N_6	0.6									1.4			
							N_7	0.8									1.7			
							N_8	0.4									1.1			
							N_9	0.3									1.0			
Zinc							N_0	80									72			
							N_1	72									72			
							N_2	66									66			
							N_3	82									76			
							N_4	56									64			
							N_5	74									66			
							N_6	66									61			
							N_7	60									61			
							N_8	72									69			
							N_9	56									53			

Table 1.158 Alsike Clover

	May			June			July			August			Notes	Reference
Proximate constituents %DM														
Crude protein								24.5					No date given. Oven dried and ground	420
Crude fibre								16.4						
Proximate constituents %DM														
Crude protein					20.3								Cut at pre-flowering stage, height 25-30 cm.	12
Amino acids % of crude protein														
Cystine					0.35									
Histidine					2.00									
Lysine					3.86									
Methionine, DL form					0.97									
L form					0.92									
Tryptophan					1.81									
Tyrosine					2.48									
Major elements %DM														
Calcium		1.8		2.0				2.0			2.1		May and June, preflowering; July, late flowering; August, mature	490
Phosphorus		0.4		0.4				0.3			0.2			
Magnesium		0.6		0.6				0.7			0.7			
Potassium		2.5		2.4				1.9			1.8			
Sodium		0.03		0.04				0.04			0.05			
Chlorine		0.5		0.5				0.4			0.3			
Trace elements µg/g DM														
Cobalt		0.2		0.2				0.2			0.1			
Copper		12.4		8.8				8.9			12.0			
Iron		491		296				298			205			
Manganese		60.7		52.3				62.2			59.9			
Vitamins µg/g DM														
Thiamin												2.99	Meadow plots, Air dried	493
Riboflavin												12.1		
Nicotinic acid												27.7		
Pantothenic acid												22.1		
Pyridoxine												14.1		
Biotin												0.39		

Table 1.159 Clover Hay

Variety	Proximate constituents %DM		Major elements %DM		Notes	Reference
	Crude protein	Crude fibre	Ca	P		
English White	22.3 21.2	21.9 15.7	2.0 1.7	0.3 0.2	Sown 1927. Given 750 kg basic slag/ha in February, 1928 First samples cut 12/7/1928 Second samples cut 10/6/1929	550
Dutch White	17.5 18.1	25.4 18.0	1.7 1.5	0.3 0.3		
N.Z. White	21.0 19.9	23.4 15.7	2.1 1.9	0.3 0.3		
N.Z. Large-leaved White	19.7 21.0	23.6 15.8	2.1 1.8	0.3 0.3		
N.Z. Ordinary White	19.2 18.9	23.4 14.8	1.9 1.9	0.3 0.2		
N.Z. Stubble White	18.9 19.3	24.4 17.1	2.0 2.2	0.3 0.3		

Table 1.160 Sainfoin

		April			May			June			July			August			September			Notes	Reference
Proximate constituents %DM																					
Crude protein	W					23.7	18.8		14.5	19.4	19.6		18.0		21.3 20.1	22.2			23.5	W: Whole plant L: Leaf S: Stem There were 3 cuts of primary, second and third growth, shown in separate lines, each at 3 stages of growth, preflowering, early flower and full flower, shown at consecutive dates; and a cut taken at full flower after a cut at preflowering (fourth line) H: hay cut at full flower Sodium and chlorine computed from Chloride (as NaCl).	28
	L					28.2	24.7		22.2	23.2	24.6		24.8		27.9 26.6	28.3			26.1		
	S					19.1	13.9		11.7	16.0	16.2		15.2		16.0 15.4	15.5			14.9		
	H								14.1												
Ether extract	W					3.1	2.7		2.3	3.0	2.5		2.6		2.4 2.4	2.7			2.7		
	L					4.7	3.4		3.6	3.5	3.0		3.0		2.9 3.1	3.1			3.0		
	S					2.4	2.0		1.8	2.3	2.1		2.0		1.8 1.6	1.9			2.0		
	H								1.4												
Nitrogen-free extract	W					51.5	51.6		49.7	51.5	50.3		46.8		47.5 46.6	46.1			49.8		
	L					47.8	52.7		53.2	53.4	51.3		49.8		48.4 49.4	48.4			52.5		
	S					53.8	53.3		46.8	50.3	49.7		45.0		43.8 45.7	43.7			46.2		
	H								45.3												
Crude fibre	W					14.5	20.9		27.1	18.3	20.2		25.0		21.9 23.6	22.4			16.8		
	L					11.9	12.1		12.7	11.7	11.6		12.6		12.2 12.4	12.1			10.7		
	S					17.7	25.7		35.5	24.7	26.1		31.8		32.8 31.8	33.7			30.8		
	H								31.8												

continued

Table 1.160 Sainfoin (continued)

	April			May			June			July			August			September			Notes	Reference
Proximate constituents (cont'd) Ash				W	7.2	6.0		6.4	7.8	7.4		7.6		6.9 7.3	6.6			7.2		
				L	7.4	7.1		8.3	8.2	9.5		9.8		8.6 8.5	8.1			7.7		
				S	7.0	5.1		4.2	6.7	5.9		6.0		5.6 5.5	5.2			6.1		
				H				7.4												
Crude protein								20.9											Cut at early flowering, height 41-45 cm.	12
Amino acids % of crude protein																				
Cystine								0.35												
Lysine								3.55												
Methionine DL form								1.04												
L form								0.96												
Tryptophan								1.37												
Tyrosine								3.32												
***Major elements* %DM**																				
Calcium			1.3		1.1		1.0		0.9		0.9								Pre-flowering April, May; Mid-flowering early June; Late flowering late June; Mature July.	490
				W	1.32	1.12		1.29	1.79	1.84		1.70		1.47 1.63	1.50			1.52	See Crude protein, same reference	28
				L	1.71	1.78		2.24	2.19	2.67		2.70		2.14 2.24	2.09			1.79		
				S	0.86	0.69		0.76	1.02	1.16		1.04		0.93 1.01	0.94			1.08		
				H				1.45												
Phosphorus			0.6		0.6		0.3		0.3		0.2								See Calcium, same reference	490
				W	0.35	0.30		0.19	0.23	0.30		0.28		0.22 0.28	0.22			0.28	See Crude protein, same reference	28
				L	0.33	0.30		0.18	0.22	0.28		0.28		0.23 0.28	0.23			0.29		
				S	0.31	0.27		0.17	0.24	0.31		0.28		0.21 0.28	0.21			0.25		
				H				0.27												

Table 1.160 Sainfoin (continued)

	April			May			June			July			August			September			Notes	Reference
Major elements %DM (cont'd)			0.8		0.8		0.7		1.1		0.5								See Calcium, same reference	490
Magnesium				W	0.26	0.22		0.20	0.30	0.31		0.28		0.25 0.27	0.27			0.29	See Crude protein, same reference	28
				L	0.22	0.24		0.26	0.28	0.36		0.34		0.28 0.30	0.27			0.28		
				S	0.28	0.22		0.17	0.23	0.27		0.22		0.21 0.21	0.22			0.29		
				H				0.28												
Potassium			3.7		3.3		2.3		1.9		1.6								See Calcium, same reference	490
				W	1.76	1.42		1.22	1.25	1.20		1.38		1.38 1.42	1.18			1.48	See Crude protein, same reference	28
				L	1.34	1.56		1.00	0.97	1.01		1.13		1.19 1.22	1.15			1.38		
				S	2.21	1.11		1.00	1.24	1.28		1.55		1.40 1.33	1.20			1.40		
				H				1.68												
Sodium			0.02		0.04		0.04		0.05		0.05								See Calcium, same reference	490
				W	0.22	0.15		0.12	0.33	0.33		0.38		0.29 0.27	0.28			0.36	See Crude protein, same reference	28
				L	0.12	0.11		0.18	0.29	0.37		0.43		0.28 0.27	0.28			0.34		
				S	0.28	0.20		0.17	0.26	0.27		0.32		0.27 0.26	0.26			0.37		
				H				0.21												
Chlorine			0.2		0.2		0.1		0.1		0.1								See Calcium, same reference	490
				W	0.33	0.23		0.19	0.51	0.50		0.59		0.45 0.41	0.42			0.52	See Crude protein, same reference	28
				L	0.19	0.17		0.27	0.44	0.57		0.67		0.42 0.42	0.44			0.52		
				S	0.42	0.30		0.25	0.41	0.42		0.49		0.41 0.39	0.41			0.57		
				H				0.33												

continued

Table 1.160 Sanfoin (continued)

	April			May			June			July			August			September			Notes	Reference
Trace elements μg/g DM																				
Cobalt			0.2		0.2		0.1		0.1		0.1								See Calcium, same reference	490
Copper			10.4		10.0		7.2		7.8		5.0									
Iron			360		265		200		199		131									
Manganese			70.3		50.7		40.9		30.1		27.5									
Vitamins μg/g DM																				
β - Carotene				W	252	93		91	236	216		256		197 228	224			236	See Crude protein, same reference	28
				L	393	192		213	302	326		290		335 379	289			304		
				S	87	21		38	174	142		238		101 124	162			102		
Thiamin															2.5				Oven dried. Preflowering stage.	493
Riboflavin															16.3					
Nicotinic acid															47.3					
Pantothenic acid															15.4					
Pyridoxine															10.7					
Biotin															0.47					

Table 1.161 Oat: whole plant as fodder

Variety			Proximate constituents %DM CP	EE	NFE	CF	Ash	Notes	Reference
Ayr Line Potato	1952	W (2)	5.0	1.6	59.0	29.8	4.6	Plant harvested at three stages of growth: (2) Grain, milky; straw green (3) Grain, early cheesy; straw half ripe (4) Ripe	389
	1953		5.8	1.4	55.7	33.5	3.6		
	1954		4.5	1.3	58.2	31.9	4.1		
	1952	W (3)	4.6	2.6	62.1	26.7	4.0		
	1953		5.6	2.3	58.8	29.7	3.6		
	1954		4.4	2.3	59.2	30.5	3.6		
	1952	W (4)	4.4	2.5	64.7	24.5	3.9	W : whole plant L : leaf S : stem I : inflorescence	
	1953		6.7	3.0	57.9	28.8	3.6		
	1954		5.0	3.1	58.2	29.7	4.0		
	1953	L (2)	10.1	4.7	52.6	26.1	6.5		
	1954		8.9	4.3	50.2	27.8	8.8		
	1953	L (3)	6.2	3.4	50.4	34.0	6.0		
	1954		5.2	3 4	49.4	34.7	7.3		
	1953	L (4)	5.1	2.4	50.2	37.5	4.8		
	1954		4.3	2.9	50.9	35.6	6.3		
	1953	S (2)	3.6	0 7	57.3	34.6	3.8		
	1954		2.5	0.7	54.5	38.6	3.7		
	1953	S (3)	2.7	0.9	56.4	35.9	4.1		
	1954		1.9	0.6	54.8	38.9	3.8		
	1953	S (4)	2.1	0.6	51.2	41.8	4.3		
	1954		1.9	0.6	51.0	42.5	4.0		
	1953	I (2)	7.9	2.3	60.7	25.6	3.5		
	1954		6.6	1.5	64.4	23.7	3.8		
	1953	I (3)	11.0	4.3	62.0	19.3	3.4		
	1954		8.0	5.6	63.5	18.8	4.1		
	1953	I (4)	11.1	4.7	64.5	16.1	3.6		
	1954		9.1	6.2	64.8	15.8	4.1		
Ceirch-du-Bach			13.4	5.2	56.2	16.8	8.4	1 cut — Cut at height of 30cm.	211
			13.3	4.3	52.5	21.6	8.3	3 cuts — Artificially dried.	
Haidd Garw			14.3	5.9	50.0	21.8	8.0	1 cut	
			14.5	5.0	47.0	24.5	9.0	3 cuts	
Golden Rain			20.6	6.9	44.6	18.7	9.2	1 cut	
			18.4	5.9	45.0	21.9	8.8	3 cuts	
S84	1950	(1)	18.2					Cut at 5 stages: (1) leafy (2) early milky (3) early cheesy (4) nearly ripe (5) ripe	276
	1951		19.9						
	1949	(2)	6.9						
	1950		8.8						
	1951		7.4						
	1949	(3)	6.4						
	1950		8 7						
	1951		7.2						
	1949	(4)	6.2						
	1950		8.2						
	1951		7.0						
	1949	(5)	6.2						
	1950		8.3						
	1951		7.0						
S147	1950-53	(1)	27.4					Grazed by sheep (1) at height of 5-7.5cm; (2) at height of 15-18cm.	150
		(2)	19.5						

continued

Table 1.161 Oat: whole plant as fodder (continued)

Variety		Major elements %DM			Trace elements		Vitamins	Notes	Reference
		Ca	P	K	Mn μg/g DM		Carotene μg/g DM		
Ayr Line Potato									
1953	W	0.24	0.14					Ripe stage	389
1954	W	0.24	0.10						
1953	L	1.00	0.18						
1954	L	0.99	0.17						
1953	S	0.15	0.08						
1954	S	0.16	0.06						
1953	I	0.20	0.17						
1954	I	0.17	0.16						
Ceirch-du-Bach		0.58 0.51	0.22 0.26	2.66 2.37				See Crude protein, same reference	211
Haidd Garw		0.59 0.54	0.28 0.30	3.00 3.11					
Golden Rain		0.58 0.52	0.30 0.35	3.57 2.69					
Triumph		0.35	1.92		170			Extra P applied. Means of 12 cuts	182
Early Miller							126	Cut at height of 18cm.	
							96	Cut at height of 38cm.	
							10	Later cut.	
							149	Cut at height of 18cm. } Ammonium sulphate 125kg/ha	
							147	Cut at height of 38cm. } Ammonium sulphate 125kg/ha	
							48	Later cut } Ammonium sulphate 125kg/ha	

Table 1.162 Oat hay

Variety	Year	Digestibility of OM%					Notes	Reference
Ayr Line Potato								
	(1953) (2)	49.2					Digestibility by sheep.	389
	(1953) (5)	51.3						
	(1954) (0)	47.2					Cut in 1953 at two stages. (2) as described for fresh crop (Table 1.161) and (5) late cheesy, stored 1-3 months in loft. Cut in 1954 at three stages (0) flowering, (2) as before, (5) as before. Stored 6-9 months in loft.	
	(1954) (2)	47.0						
	(1954) (5)	47.5						
		Proximate constituents %DM						
		CP	EE	NFE	CF	Ash		
	(1953) (2)	6.9	1.5	47.9	38.2	5.5		
	(1953) (5)	7.4	2.7	49.4	35.5	5.0		
	(1954) (0)	5.0	2.4	53.0	34.2	5.4		
	(1954) (2)	5.1	2.2	49.2	38.1	5.4		
	(1954) (5)	6.2	1.8	47.3	39.7	5.0		
	(1952)	5.6	2.2	52.2	34.4	5.6	For fresh crop at stages (2), (3) and (4) see Table 1.161, Proximate constituents, same reference.	
	(1953) (2)	6.7	1.5	50.6	36.6	4.6		
	(1954)	4.7	1.6	50.8	38.0	4.9		
	(1952)	4.2	2.5	61.2	27.6	4.5		
	(1953) (3)	6.4	2.8	57.1	29.7	4.0		
	(1954)	4.6	2.4	55.3	33.3	4.4		
	(1952)	4.3	2.8	62.1	26.7	4.1		
	(1953) (4)	6.8	2.8	58.7	27.9	3.8		
	(1954)	4.8	2.8	56.9	31.1	4.4		
	(1957)	6.6	2.2	52.4	33.6	5.2	Grown in Scottish uplands	167
	(1958)	5.3	2.0	58.5	29.6	4.6		
	(1959)	5.9	3.1	56.6	29.3	5.1		
	(1960)	5.7	1.8	58.8	29.8	3.9		
	(1961)	6.5	1.7	53.1	34.6	4.1		
Blenda							West of England and Wales	163
	(1955)	6.4	2.7	53.3	32.6	5.0	One lowland centre	
	(1955)	7.8	2.0	49.2	34.9	6.1	Four upland centres	
	(1956)	7.8	1.7	50.1	35.0	5.4	Two lowland centres	
	(1956)	8.2	1.5	45.5	37.9	6.9	Four upland centres	
	(1957)	7.4	2.8	51.4	32.6	5.8	Two lowland centres	
	(1957)	9.6	2.4	49.6	32.1	6.3	Three upland centres	
	(1957)	6.9	3.0	54.6	30.9	4.6	Scotland. South West Uplands	167
	(1958)	5.1	2.0	59.4	29.1	4.4		
	(1959)	5.8	3.8	60.2	25.6	4.6		
	(1960)	6.2	2.9	59.6	27.6	3.7		
	(1961)	6.5	2.6	55.9	31.2	3.8		
Maldwyn	(1955)	6.8	2.5	51.1	34.1	5.5	As Blenda, same reference	163
	(1955)	7.2	2.3	49.3	35.4	5.8		
	(1956)	7.7	1.6	48.7	36.2	5.8		
	(1956)	9.1	1.6	45.8	36.2	7.3		
	(1957)	7.2	2.6	50.2	34.0	6.0		
	(1957)	9.8	2.0	46.8	34.6	6.8		
	(1957)	6.7	2.5	52.9	32.9	5.0	Scotland. South West Uplands	167
	(1958)	5.0	2.0	58.4	30.5	4.1		
	(1959)	5.8	3.0	57.9	28.3	5.0		
	(1960)	5.8	2.4	60.0	28.0	3.8		
	(1961)	6.5	2.0	55.0	32.6	3.9		
Radnorshire Sprig								
	(1955)	6.7	2.6	51.1	34.2	5.4	As Blenda, same reference	163
	(1955)	7.3	1.9	49.1	35.7	6.0		
	(1956)	7.6	1.6	49.8	35.6	5.4		
	(1956)	8.8	1.5	45.3	36.2	8.2		
	(1957)	7.7	2.0	50.2	34.1	6.0		
	(1957)	10.2	1.8	46.5	34.7	6.8		

Table 1.163 Oat straw and straw pulp

Variety		Digestibility % DM	Digestibility % OM				Notes	Reference
Unnamed	(1937)	75.8	78.1				Straw pulp. Digestibility by sheep	224
	(1938)	68.0	69.8					
	(1939)	57.2	61.4					
	(1940)	60.8	63.2					
Unnamed		44.0	40.9				Straw. Digestibility by cows. For proximate constituents and metabolisable energy see Table 1.233	116
		39.3	40.8					
		Proximate constituents %DM						
		CP	EE	NFE	CF	Ash		
Ceirch Llwyd		2.7	2.3	44.9	45.9	4.2	Straw as used for feeding: Mid Wales	18
Ceirch Llwyd Cwta		3.0	2.2	45.9	44.0	4.9		
Grey Winter hybrid		2.7	2.0	44.9	45.0	5.4		
Golden Rain		2.1	2.0	42.2	47.4	6.3		
Marvellous		2.5	2.0	46.6	42.2	6.7		
Scotch Potato		3.1	2.5	47.7	42.5	4.2		
Record		2.3	2.1	43.5	45.1	7.0		
Victory		2.4	1.6	44.8	44.5	6.7		
Victory x Red Algerian		3.4	2.5	47.0	41.7	5.4		
Unnamed		3.1	1.6	45.6	40.5	9.2	Mean of 2 samples	54
Unnamed		2.2	1.6	46.7	41.9	7.6		136
Eagle	(1952) N_0	2.9	1.9	47.5	39.8	7.9	Binder ripe. N_0: no extra fertiliser; N_1: Nitrochalk, 250 kg/ha. Average time between binder and combine cuts: 16 days in 1952; 19 days in 1953; 18 days in 1954.	162
	N_1	3.5	1.8	47.2	39.3	8.2		
	N_0	2.7	1.9	46.5	41.9	7.0	Combine ripe	
	N_1	3.5	1.7	46.3	41.3	7.2		
	(1953) N_0	3.8	1.6	45.1	41.4	8.1	Binder ripe	
	N_1	4.4	1.5	44.3	41.9	7.9		
	N_0	3.5	1.3	45.9	42.7	6.6	Combine ripe	
	N_1	4.1	1.2	44.3	44.0	6.4		
	(1954) N_0	4.6	1.2	46.7	39.9	7.6	Binder ripe	
	N_1	5.4	1.1	45.3	40.8	7.4		
	N_0	4.6	1.0	46.4	41.2	6.8	Combine ripe	
	N_1	5.1	1.0	45.6	42.1	6.2		
Onward	(1952) N_0	2.6	1.6	49.3	38.9	7.6	Binder ripe	
	N_1	3.2	1.6	48.8	38.8	7.6		
	N_0	2.7	1.7	47.5	41.9	6.2	Combine ripe	
	N_1	3.6	1.7	47.3	41.2	6.2		
	(1953) N_0	3.7	1.6	46.9	40.8	7.0	Binder ripe	
	N_1	4.4	1.5	45.7	41.3	7.1		
	N_0	3.4	1.2	46.9	43.0	5.5	Combine ripe	
	N_1	4.3	1.3	45.5	43.4	5.5		
	(1954) N_0	4.6	1.1	47.2	40.7	6.4	Binder ripe	
	N_1	5.3	1.0	46.7	40.5	6.5		
	N_0	4.5	0.9	47.1	42.1	5.4	Combine ripe	
	N_1	5.2	0.8	46.0	42.7	5.3		
Milford	(1952) N_0	3.3	1.8	48.7	38.5	7.7	Binder ripe	
	N_1	4.0	1.8	47.7	38.8	7.7		
	N_0	2.9	1.6	46.1	42.4	7.0	Combine ripe	
	N_1	4.2	1.6	46.1	41.1	7.0		
	(1953) N_0	3.9	1.7	46.2	40.6	7.6	Binder ripe	
	N_1	4.7	1.5	44.6	41.9	7.3		
	N_0	3.7	1.3	45.6	43.0	6.4	Combine ripe	
	N_1	4.5	1.2	44.1	44.0	6.2		
	(1954) N_0	5.0	1.2	46.3	40.4	7.1	Binder ripe	
	N_1	5.6	1.1	45.8	40.5	7.0		
	N_0	4.7	1.0	45.9	42.4	6.0	Combine ripe	
	N_1	5.3	0.8	44.9	43.3	5.7		

continued

Table 1.163 Oat straw and straw pulp (continued)

		Proximate constituents %DM					Notes	Reference
		CP	EE	NFE	CF	Ash		
Ceirch-du-Bach	(1)	2.6	1.8	44.3	44.5	6.8	(1) Normal maturity	211
	(2)	3.3	2.7	45.4	42.1	6.5	(2) Stubble, mature after crop cut at 30 cm.	
Golden Rain	(1)	3.2	2.0	49.0	39.8	6.0	} England	
	(2)	3.0	1.9	46.0	43.5	5.6		
Haidd Garw	(1)	2.4	1.8	44.7	45.5	5.6		
	(2)	2.6	1.8	46.3	42.5	6.8		
Yielder	N_0	2.1					N_0 : No extra nitrogen	195
	N_1	3.1					N_1 : ammonium sulphate 125 kg/ha	
	N_2	3.0					N_2 : ammonium sulphate 250 kg/ha	
	N_3	6.8					N_3 : ammonium sulphate 375 kg/ha	
Minor	N_0	2.2						
	N_1	3.0						
	N_2	2.2						
	N_3	3.4						
		Major elements %DM						
		Ca	P	K	Cl			
Ceirch Llwyd		0.29	0.06	0.8	0.7		Straw as used for feeding	18
Ceirch Llwyd Cwta		0.29	0.10	0.9	0.9		Mid Wales	
Grey Winter hybrid		0.34	0.07	0.9	0.9			
Golden Rain		0.29	0.13	1.4	0.9			
Marvellous		0.33	0.09	1.2	0.7			
Scotch Potato		0.31	0.07	1.0	0.8			
Record		0.29	0.11	1.2	0.8			
Victory		0.34	0.08	1.2	1.1			
Victory x Red Algerian		0.34	0.07	0.9	0.9			
Unnamed		0.6	0.4				Mean of 2 samples	54
Ceirch-du-Bach	(1)	0.39	0.10	2.4			See Proximate constituents, same reference	211
	(2)	0.34	0.18	2.6				
Golden Rain	(1)	0.30	0.08	2.0				
	(2)	0.37	0.10	2.1				
Haidd Garw	(1)	0.32	0.08	2.2				
	(2)	0.30	0.16	2.1				
		Vitamin Thiamin µg/g DM		Date of cutting				
Marvellous		1.3		Late July				280
		0.8		Mid August				
		0.5		Late August				
		0.1		Mid September				
Star		0.9		Late July				
		0.4		Mid August				
		0.4		Late August				
		0.4		Early September				
		0.1		Mid September				

Table 1.164 Wheat: whole plant as fodder

		Notes	Reference
Major elements %DM			
Calcium	F_0 0.37 P 0.37	Variety: Progress F_0 : No extra fertiliser P : Extra phosphate	182
Phosphorus	F_0 0.22 P 0.24		
Potassium	F_0 2.44 P 2.43		
Sulphur	F_0 140 P 170		
Trace elements µg/g DM			
Barium	7, <4 6, 5, 5 15	A study of deficiencies in Romney Marsh pastures. Top line: 2 samples. Alkaline soil, Lydd and Broomhill areas. Variety Atle. Second line: 3 samples. Alkaline soil, Ivychurch area. Variety Capelle Desprez. Third line: 1 sample. Acid soil, Old Romney. Variety Capelle Desprez.	444
Chromium	0.4, 0.3 0.3, 0.5, 0.4 0.4		
Cobalt	<0.04, <0.03 <0.03, 0.05, 0.04 0.07		
Copper	7.8, 4.6 6.0, 8.2, 8.3 6.7		
Iron	114, 82 82, 148, 123 83		
Lead	1.6, 3.1 1.1, 1.5, 0.9 1.3		
Manganese	22,19 14, 11, 18 74		
Molybdenum	0.57, 0.83 0.55, 0.52, 0.52 0.18		
Nickel	0.60, 0.30 0.27, 0.39, 0.40 0.44		
Strontium	11, 8 17, 14, 16 8		
Tin	<1, <1 <1, <1, <1 <1		
Titanium	6.8, 5 3.3, 17, 13 5		
Vanadium	0.26, 0.3 0.16, 0.6, 0.4 0.2		
Zinc	38, 51 23, 50, 35 43		

	Total	a - % of total		
Vitamins µg/g DM Tocopherol	92.6 87.3	81.6 95.5	Variety: Atle 12-36 cm high Green ear, 100 cm high	248
	34.0 62.9	81.0 } 94.1 }	Unripe ear Plant without ear	
	15.5 37.5	37.1 } 77.1 }	Almost ripe ear Plant without ear	

Table 1.165 Wheat straw and straw pulp

		Notes	Reference
Digestibility of DM%	65.1	Pulp. Digestibility by sheep, 23 trials over 3 years.	224
	66.8 73.6 69.3	Digestibility by bullock " " sheep " " pig	570
Digestibility of OM%	70.3	As for DM, same reference	224
	– 76.5 71.5	For metabolisable energy see Table 1.233	570
Proximate constituents %DM Crude protein	1.7 2.7 3.4 2.4 3.0 (2.8 – 3.2) 2.1	Straw Victor. Average of 3 years Hybrid 46. Average of 2 years Unnamed Unnamed. Mean and range for same sample analysed at 6 centres Unnamed	329 527 263 133 1 136
Carbohydrate fractions %DM Cellulose	43.5 80.5	Straw. Fibre (Norman's method) less lignin Pulp	136 33
Hexosan cellulose Cellulosans Pentosans not in cellulose	66.0 14.5 4.6		
Lignin	9.9 1.8	Straw. By H_2SO_4 method corrected for ash and protein	136 33
Major elements %DM Calcium	0.24 0.16	Straw Stubble } Variety Victor	329
	L 0.18 M 0.06 H 0.07	Pot cultures. Variety Atle: low K, L; medium K, M; high K, H	522
	F_0 0.06 F_e 0.06	4 varieties with no lime: F_0 With extra lime: F_e	181
	0.32 (0.25 – 0.36)	Mean and range for same sample, unnamed, analysed at 9 centres	1
	0.24 (0.18 – 0.31)	Mean and range of 9 treatments: unnamed Broadbalk. 1851–1892; 1901–1921	134
Phosphorus	0.06 0.04	See Calcium, same reference	329
	L 0.31 M 0.26 H 0.35	See Calcium, same reference	522
	F_0 0.06 F_e 0.07	See Calcium, same reference	181
	0.06 0.05	2 samples, unnamed	133
	0.12 0.10	2 samples, unnamed	499
	0.07 (0.06 – 0.08)	Mean and range for same sample, unnamed, analysed at 6 centres	1
Magnesium	L 0.07 M 0.06 H 0.05	See Calcium, same reference	522
	0.08 (0.06 – 0.09)	Mean and range for same sample, unnamed, analysed at 9 centres	1
	0.06 (0.05 – 0.07)	See Calcium, same reference	134

continued

Table 1.165 Wheat straw and straw pulp (continued)

		Notes	Reference
Major elements %DM (cont'd) Potassium	0.56 0.93	See Calcium, same reference	329
	L 1.99 M 2.32 H 2.66	See Calcium, same reference	522
	F_0 0.59 F_e 0.46	See Calcium, same reference	181
	0.85 0.83	2 samples, unnamed	133
	0.87 (0.51 – 1.18)	See Calcium, same reference	134
	0.96 0.67	2 samples, unnamed	499
	1.02 (0.90 – 1.12)	Mean and range for same sample, unnamed, analysed at 9 centres	1
Sodium	L 0.11 M 0.09 H 0.15	See Calcium, same reference	522
	F_0 0.06 F_e 0.06	See Calcium, same reference	181
	0.0022 (0.0005 – 0.0054)	See Calcium, same reference	134
Sulphur, total	L 0.13 M 0.14 H 0.24	See Calcium, same reference	522
	0.21 (0.17 – 0.23)	Mean and range for same sample, unnamed, analysed at 5 centres	1
Trace elements μg/g DM Boron	4.9 4.0	Spectrographic method Colorimetric method	450
	12 4 – 20	Mean and range for same sample analysed at 5 centres	1
Copper	9 5 – 16	Mean and range of 10 samples from manurial trials	551
	4.2 3.0 – 5.5	Mean and range for the same sample analysed at 5 centres	1
Iron	103 78 – 111	Mean and range for same sample analysed at 8 centres	1
Manganese	70 52 – 87	Mean and range of 10 samples from manurial trials	551
	F_0 92 F_e 68	See Calcium, same reference	181
	48 44 – 51	Mean and range for same sample analysed at 6 centres	1
Molybdenum	0.14 0.09 – 0.18	Mean and range of 10 samples from manurial trials	551
	0.22 0.16 – 0.31	Mean and range for the same sample analysed at 5 centres	1
Zinc	39 21 – 54	Mean and range of 10 samples from manurial trials	551
	18 13 – 23	Mean and range for the same sample analysed at 4 centres	1
Barium Cobalt Lead Nickel Strontium Tin Titanium Vanadium	42 0.08 1.3 0.39 13 0.5 5.9 0.25	Analysed at 1 centre only, unnamed.	1

Table 1.166 Barley: whole plant as fodder

			Notes		Reference
Proximate constituents %DM					
Crude protein	1945	1946	Variety Plumage Archer		284
			Cut		
	N_0 19.7	N_0 16.4	1	N_0 : no extra nitrogen	
	16.0	16.0	2	N_1 : 437 kg/ha in 1, 2 or	
	12.7	16.5	3	3 dressings.	
	16.0	19.4	4		
	14.0	14.5	5		
	N_1 26.9	N_1 21.0	1		
	22.7	18.9	2		
	17.9	19.8	3		
	16.6	20.1	4		
	13.7	15.9	5		
			Variety B8(8)		459
	N_0 13.8		Preflowering	N_0 : no extra	
	7.5		Ears appearing	nitrogen	
	4.6		Grain full		
	N_1 15.6		Preflowering	N_1 : ammonium	
	8.3		Ears appearing	sulphate	
	5.4		Grain full	125 kg/ha	
Vitamins µg/g DM	Crude	β-carotene			
Carotene	N_0 93	72			
	57	45			
	12	9			
	N_1 130	104			
	50	43			
	15	12			
Tocopherol	Total	α-% of total	Height in cm: days after planting		
	117	94.5	12-13: 24	Whole plant	248
	194	100	19-20: 38		
	67	100	90: 74		
	24	100			
	35	100	: 89	Unripe ears only	
	88	80.6	: 93	Green ears only	
	139	84.9	90-100:106	Plant *less* ears	
	50	33.9	:106	Unripe ears only	
	210	48.5	90-100:122	Plant *less* ears	
	45	22.3	:122	Ripe ear only	

Table 1.167 Barley straw and straw pulp

		Notes	Reference
Proximate constituents **%DM** **Crude protein**	3.8	Straw Rothamsted Survey 1852–1937. All varieties	446
	2.9	Unnamed variety	136
	N_0 3.2 N_1 2.7	Variety B8(8) E. Scotland N_0 : no extra nitrogen N_1 : ammonium sulphate 125 kg/ha	459
	3.8 3.7	Unnamed variety W. Midland	133
	4.9 4.7	Unnamed variety Reading	499
Ether extract	2.0	Unnamed variety	136
Nitrogen-free extract	45.0		
Crude fibre	42.5		
Ash	7.6		
Carbohydrate fractions %DM Cellulose	44.2	Fibre by Norman's method less lignin. Lignin by sulphuric acid method corrected for ash and protein	
Lignin	8.2		
Major elements %DM Calcium	F_1 0.31 F_2 0.12 F_3 0.09	Spratt Archer F_1 : low potassium F_2 : medium potassium F_3 : high potassium	522
	F_0 0.22 F_1 0.26	Freja	181
	F_0 0.21 F_1 0.25	Kenia F_0 : no extra calcium F_1 : CaO 3750 kg/ha	
	F_0 0.22 F_1 0.25	Ymer	
	F_0 0.20 F_1 0.23	No. 5	
Phosphorus	0.06	All varieties. See Crude protein	446
	F_1 0.38 F_2 0.31 F_3 0.31	Spratt Archer. See Calcium	522
	F_0 0.06 F_1 0.09	Freja	181
	F_0 0.06 F_1 0.09	Kenia	
	F_0 0.06 F_1 0.08	Ymer	
	F_0 0.06 F_1 0.07	No. 5	
	0.06 0.08	Unnamed variety	133
	0.14	Unnamed variety	499
Magnesium	F_1 0.08 F_2 0.08 F_3 0.05	Spratt Archer. See Calcium	522

Table 1.167 Barley straw and straw pulp (continued)

		Notes	Reference
Major elements %DM (cont'd) Potassium	0.83	All varieties. See Crude protein	446
	F_1 1.33 F_2 1.68 F_3 2.98	Spratt Archer. See Calcium	522
	F_0 0.25 F_1 0.34	Freja	181
	F_0 0.35 F_1 0.41	Kenia	
	F_0 0.37 F_1 0.39	Ymer	
	F_0 0.33 F_1 0.34	No. 5	
	0.46 0.50	Unnamed variety	133
	1.09	Unnamed variety	499
Sodium	F_1 0.35 F_2 0.11 F_3 0.08	Spratt Archer, See Calcium	522
	F_0 0.09 F_1 0.09	Freja	181
	F_0 0.09 F_1 0.10	Kenia	
	F_0 0.10 F_1 0.11	Ymer	
	F_0 0.09 F_1 0.09	No. 5	
Sulphur (total)	F_1 0.32 F_2 0.21 F_3 0.36	Spratt Archer. See Calcium	522
Trace elements μg/g DM Copper	8 5 – 12	Mean and range for 10 samples, different fertilisers	551
Manganese	F_0 240 F_1 200	Freja	181
	F_0 300 F_1 180	Kenia	
	F_0 250 F_1 190	Ymer	
	F_0 270 F_1 180	No. 5	
	70 52 – 96	Unnamed variety Mean and range for 10 samples, different fertilisers.	551
Molybdenum	0.175 0.09 – 0.24		
Zinc	44 30 – 78		
Digestibility of DM %	61.3	Straw pulp. Digestibility by sheep: 4 trials	224
Digestibility of OM %	65.1		

Table 1.168 Maize: whole plant and parts as fodder

Variety and place*		Proximate constituents %DM					Notes	Reference
		CP	EE	NFE	CF	Ash		
WHOLE PLANT								
Kingscrost KF1								
Reading	(1947)	6.5	1.9	67.8	19.5	4.3	No N, thinned	128
	”	6.2	1.9	67.4	19.6	4.9	No N, unthinned	
	”	8.1	2.1	66.9	18.4	4.5	+ N, thinned	
	”	7.8	1.7	67.3	18.7	4.5	+ N, unthinned	
							+ N plots received 1 dressing of sulphate of ammonia 750 kg/ha	
Wisconsin Hybrid 275								
Reading	(1948)	10.4	1.4	61.3	20.7	6.2	No N, thinned	
	”	10.1	1.4	61.2	21.0	6.3	No N, unthinned	
	”	11.1	1.4	61.0	20.2	6.3	+ N, thinned	
	”	11.4	1.5	60.1	20.6	6.4	+ N, unthinned	
							+ N plots received 1 dressing of sulphate of ammonia 750 kg/ha	
Reading	(1949)	7.6	2.2	66.7	19.2	4.3	No N, thinned	
	”	7.2	1.9	66.6	20.2	4.1	No N, unthinned	
	”	8.0	2.0	65.6	20.3	4.1	+ N, thinned	
	”	7.4	2.1	66.6	19.9	4.0	+ N, unthinned	
							+ N plots received 2 dressings each of sulphate of ammonia 375 kg/ha	
Kingscrost KF1							For silage see Table 1.138	129
Reading	(1947)	6.5	1.7	67.1	20.1	4.6		
Wisconsin Hybrid 275								
Reading	(1948)(1)	10.8	1.5	60.1	21.1	6.5	(1) Ensiled 15-17 September	
	(1948)(2)	9.9	1.6	63.2	20.0	5.3	(2) Ensiled 7-8 October	
	(1949)	7.9	1.9	65.1	20.1	5.0		
							Maize grown for silage, Sprowston. 1949. Days required to reach maturity for silage: cobs cheesy	74
Wisconsin 240		7.8					130	
Wisconsin 255		8.8					137	
Wisconsin 275		8.4					140	
Wisconsin 355		8.6					138	
Canbred 150		9.6					130	
De Kalb 43		8.3					140	
Funks G.35		8.9					138	
							Data for whole plant calculated from ear and stover (plant less ear), proportional data	
Grand Roux Basque								108, 109 and unpublished details
Cambridge	(1954)	8.4	0.7	57.9	26.3	6.7		
Jaune Gros								
Kent	(1952)	8.3		63.1	16.9		$5/m^2$ Plot I	
	”	8.0		62.5	18.8		” Plot II	
	”	7.8		62.7	17.8		$10/m^2$ Plot I	
	”	7.2		61.6	20.2		” Part II	
Devon								
Starcross	(1951)	7.5		59.2	23.5			
Kingsbridge	(1952)	8.8		64.1	20.8			
Starcross	(1952)	6.7		68.8	19.4			
Starcross	(1953)	8.1		65.5	20.8			
Nodak 301								
Cambridge	(1954)	8.1	0.8	59.3	26.5	5.3		
Wisconsin 416								
Kent	(1952)	7.5		61.1	20.4		$5/m^2$ Plot I	
	”	7.7		61.6	19.9		” Plot II	
	”	6.9		61.3	21.0		$10/m^2$ Plot I	
	”	7.3		60.5	20.5		” Plot II	
Cambridge	(1953)	7.7	1.2	63.2	22.2	5.7		
	(1954)	7.9	1.0	60.4	24.7	6.0		
	(1955)	8.5	–	–	21.3	–		
Devon								
Starcross	(1951)	6.1		60.7	24.4			
Kingsbridge	(1952)	9.6		62.0	21.4			
Starcross	(1952)	6.7		66.5	22.6			
Starcross	(1953)	8.2		63.3	22.4			

*Data for Wytham (Oxon), found to correspond to averages for Kent, have been omitted.

Table 1.168 Maize: whole plant and parts as fodder (continued)

Variety and place		Proximate constituents %DM					Notes	Reference
		CP	EE	NFE	CF	Ash		
Wisconsin 355								
Cambridge	(1953)	7.7	1.4	63.4	21.8	5.7		108, 109 and unpublished details (as above)
Devon	(1951)	6.4		57.7	26.1			
Wisconsin 275A								
Kent	(1953)	8.6		59.5	21.7		$5/m^2$ Plot I	
	"	7.4		60.2	23.5		" Plot II	
	"	7.6		59.6	23.3		$10/m^2$ Plot I	
	"	6.5		59.9	22.5		" Plot II	
Devon								
Starcross	(1951)	6.3		61.1	25.0			
Kingsbridge	(1952)	10.1		60.1	23.0			
Starcross	(1952)	6.1		68.1	21.5			
Starcross	(1953)	8.8		60.3	25.2			
Cambridge	(1953)	8.4	1.4	64.7	20.6	4.9		
Wisconsin 464								
Devon								
Starcross	(1951)	8.3		60.9	23.6			
Kingsbridge	(1952)	10.0		60.2	22.0			
Starcross	(1952)	9.3		62.9	23.0			
Starcross	(1953)	7.5		63.3	23.5			
Wisconsin 341A								
Devon								
Kingsbridge	(1952)	10.2		59.5	22.9			
Starcross	(1952)	6.2		66.5	23.0			
Starcross	(1953)	8.1		60.4	24.7			
Kent	(1952)	8.1		58.5	21.1		$5/m^2$ Plot I	
	"	7.7		60.0	23.0		" Plot II	
	"	7.9		58.7	20.9		$10/m^2$ Plot I	
	"	8.2		57.4	21.9		" Plot II	
	(1953)	6.8		58.2	24.9		$5/m^2$ Plot I	
	"	7.6		57.2	25.2		" Plot II	
	"	6.6		56.8	25.8		$10/m^2$ Plot I	
	"	6.7		56.6	23.8		" Plot II	
Cambridge	(1953)	7.1	1.2	61.4	24.3	6.0		
	(1955)	8.4	–	–	21.7	–		
White Horsetooth								
Devon								
Starcross	(1951)	6.1		52 8	32.0			
Kingsbridge	(1952)	14 8		49.3	26.8			
Starcross	(1952)	6.3		64.0	25.2			
Starcross	(1953)	9.2		59.0	24.7			
Cambridge	(1952)	6.5	1.0	58.2	26.5	7.8		
	(1953)	7.9	1.3	58.3	25.5	7.0		
	(1954)	9.5	1.2	52.4	28.3	8.6		
	(1955)	9.9	–	–	24.2	–		
Kent	(1952)	6.6		55.4	24.0		$5/m^2$ Plot I	
	"	7.2		56.8	22.8		" Plot II	
	"	7.4		52.1	25.2		$10/m^2$ Plot I	
	"	7.4		54.6	25.6		" Plot II	
	(1953)	6.6		55.5	26.2		$5/m^2$ Plot I	
	"	8.1		53.2	25.6		" Plot II	
	"	7.8		54.6	24.4		$10/m^2$ Plot I	
	"	7.7		52.1	27.4		" Plot II	
Warwick 696								
Cambridge	(1954)	8.4	0.8	56.4	26.8	7.6		

continued

Table 1.168 Maize: whole plant and parts as fodder (continued)

Variety		Proximate constituents %DM CP	EE	NFE	CF	Ash	Notes	Reference
FRESH COBS Unnamed		12.0	5.8	76.5	2.3	3.4	Scotland, 1935	506
Kingscrost KF1	N_0 N_1	7.6 9.2	2.6 2.5	73.8 74.4	13.8 11.7	2.2 2.2	Reading, 1947 N_0 : no extra nitrogen N_1 : sulphate of ammonia, 500 kg/ha	128
Wisconsin 275	N_0 N_1	10.0 11.4	1.1 1.2	67.1 66.0	17.6 17.1	4.2 4.3	Reading, 1948 N_0 : no extra nitrogen N_1 : sulphate of ammonia, 750 kg/ha	
	N_0 N_1	8.2 8.4	2.6 2.7	72.8 73.0	14.0 13.7	2.4 2.2	Reading, 1949 N_0 : no extra nitrogen N_1 : sulphate of ammonia, 375 kg/ha x 2	
FRESH LEAF Kingscrost KF1	N_0 N_1	7.6 9.5	2.2 2.5	52.8 50.9	27.7 27.8	9.7 9.3	Reading, 1947 N_0 : no extra nitrogen N_1 : sulphate of ammonia, 375 kg/ha x 2	
Wisconsin 275	N_0 N_1	18.8 19.2	3.6 3.5	44.1 43.8	21.3 21.2	12.2 12.3	Reading, 1948 N_0 : no extra nitrogen N_1 : sulphate of ammonia, 375 kg/ha x 2	
	N_0 N_1	12.6 12.5	3.4 3.3	50.0 49.8	23.6 23.8	10.4 10.6	Reading, 1949 N_0 : no extra nitrogen N_1 : sulphate of ammonia, 375 kg/ha x 2	
FRESH STEM Kingscrost KF1	N_0 N_1	3.3 4.8	0.5 0.6	66.8 64.8	24.1 24.4	5.3 5.4	Reading, 1948 N_0 : no extra nitrogen N_1 : sulphate of ammonia, 750 kg/ha	
Wisconsin 275	N_0 N_1	6.8 7.8	0.8 1.0	61.7 61.0	24.4 23.8	6.3 6.4	Reading, 1948 N_0 : no extra nitrogen N_1 : sulphate of ammonia, 750 kg/ha	
	N_0 N_1	3.6 4.2	0.7 0.6	64.6 62.0	26.9 29.0	4.2 4.2	Reading, 1949 N_0 : no extra nitrogen N_1 : sulphate of ammonia, 375 kg/ha x 2	

Table 1.168 Maize: whole plant and parts as fodder (continued)

Variety		Proximate constituents %DM CP	EE	NFE	CF	Ash	Notes	Reference
FRESH EAR								
Jaune Gros	(1952)	8.6	3.5	73.6	12.0	2.3	$5/m^2$ Plot I	108, 109 and unpublished details
Kent	”	8.0	3.1	72.9	13.8	2.2	” Plot II	
	”	8.3	3.7	74.6	11.0	2.4	$10/m^2$ Plot I	
	”	7.5	3.1	73.0	14.0	2.4	Plot II	
	Means	8.1	3.4	73.5	12.7	2.3		
Devon								
Starcross	(1951)	8.7	3.8	73.1	11.8	2.6		
Kingsbridge	(1952)	8.8	1.6	69.6	16.8	3.2		
Starcross	(1952)	7.3	3.2	77.9	9.5	2.1		
”	(1953)	8.9	1.7	72.8	14.0	2.6		
	Means	8.4	2.6	73.4	13.0	2.6		
White Horsetooth								
Devon								
Starcross	(1951)	11.4	1.8	66.7	15.8	4.3		
Kingsbridge	(1952)	13.8	0.9	57.9	22.0	5.4		
Starcross	(1952)	10.2	0.9	69.7	14.6	4.6		
”	(1953)	13.2	0.8	71.0	10.1	4.9		
	Means	12.2	1.1	66.3	15.6	4.8		
Wisconsin 275A								
Devon								
Starcross	(1951)	8.2	2.5	74.8	11.7	2.8		
Kingsbridge	(1952)	10.0	1.3	65.9	19.4	3.4		
Starcross	(1952)	7.4	1.2	75.1	13.6	2.7		
”	(1953)	9.9	1.6	72.9	13.1	2.5		
	Means	8.9	1.6	72.2	14.4	2.9		
Kent	(1953)	9.0	1.9	72.5	14.2	2.4	$5/m^2$ Plot I	
	”	8.6	1.7	69.7	17.8	2.2	” Plot II	
	”	9.8	1.4	69.7	16.7	2.4	$10/m^2$ Plot I	
	”	8.6	2.2	71.9	14.9	2.4	” Plot II	
	Means	9.0	1.8	70.9	15.9	2.4		
Wisconsin 341A								
Cambridge	(1952)	10.7	1.3	70.2	14.1	3.7		
Devon								
Kingsbridge	(1952)	10.9	0.9	67.0	16.8	4.4		
Starcross	(1952)	8.2	1.2	71.6	16.0	3.0		
”	(1953)	11.9	1.2	67.9	15.4	3.6		
	Means	10.3	1.1	68.8	16.1	3.7		
Kent	(1952)	10.0	2.6	69.7	13.9	3.8	$5/m^2$ Plot I	
	”	8.8	2.7	73.1	12.6	2.8	” Plot II	
	”	9.6	3.0	73.9	10.6	2.9	$10/m^2$ Plot I	
	”	10.2	2.8	71.5	12.3	3.2	” Plot II	
	(1953)	8.8	1.7	68.8	18.5	2.2	$5/m^2$ Plot I	
	”	9.0	2.6	67.1	18.9	2.4	” Plot II	
	”	9.2	3.1	67.4	17.8	2.5	$10/m^2$ Plot I	
	”	9.3	3.5	69.3	15.3	2.6	” Plot II	
$5/m^2$	Means	9.2	2.4	69.6	16.0	2.8		
$10/m^2$	Means	9.6	3.1	70.5	14.0	2.8		
Wisconsin 355								
Devon								
Starcross	(1951)	8.1	3.0	73.6	12.4	2.9		
Wisconsin 416								
Cambridge	(1952)	10.5	1.3	70.4	14.8	3.0		
Devon								
Starcross	(1951)	9.3	2.8	70.5	13.2	4.2		
Kingsbridge	(1952)	10.6	1.0	66.6	18.4	3.4		
Starcross	(1952)	8.9	0.9	73.1	14.2	2.9		
”	(1953)	11.8	1.2	70.0	13.7	3.3		
	Means	10.2	1.5	70.0	14.9	3.4		
Kent	(1952)	8.6	2.5	72.7	13.5	2.7	$5/m^2$ Plot I	
	”	9.0	2.4	72.3	13.5	2.8	” Plot II	
	”	8.6	2.2	72.2	14.3	2.7	$10/m^2$ Plot I	
	”	9.0	2.4	72.9	13.1	2.6	” Plot II	
	Means	8.8	2.4	72.5	13.6	2.7		

continued

Table 1.168 Maize: whole plant and parts as fodder (continued)

Variety and place		Proximate constituents %DM					Notes	Reference
		CP	EE	NFE	CF	Ash		
FRESH EAR (cont'd)								
Wisconsin 464								
Devon								
Starcross	(1951)	10.8	1.7	64.8	19.4	3.3		108, 109 and unpublished details
Kingsbridge	(1952)	10.7	1.0	65.5	17.3	5.5		
Starcross	(1952)	15.4	1.7	63.2	17.0	2.7		
"	(1953)	9.4	1.2	69.5	17.0	2.9		
	Means	11.6	1.4	65.7	17.7	3.6		
Cambridge	(1952)	9.1	1.7	72.1	14.3	2.8		

Variety and place		Major elements %DM			Notes
		Ca	P	K	
Jaune Gros	(1952)	0.06	0.34	0.74	5/m^2 Plot I
Kent	"	0.07	0.30	0.54	" Plot II
	"	0.07	0.34	0.69	10/m^2 Plot I
	"	0.08	0.27	0.85	" Plot II
Wisconsin 275A					
Kent	(1953)	0.04	0.30	0.82	5/m^2 Plot I
	"	<0.04	0.30	0.75	" Plot II
	"	0.06	0.30	0.83	10/m^2 Plot I
	"	<0.04	0.30	0.87	" Plot II
Wisconsin 341A					
Kent	(1952)	0.11	0.30	0.90	5/m^2 Plot I
	"	0.08	0.30	0.75	" Plot II
	"	0.09	0.30	0.90	10/m^2 Plot I
	"	0.12	0.27	1.01	" Plot II
	(1953)	<0.04	0.27	0.75	5/m^2 Plot I
	"	<0.04	0.27	0.72	" Plot II
	"	<0.04	0.27	0.73	10/m^2 Plot I
	"	<0.04	0.28	0.85	" Plot II
Wisconsin 416					
Kent	(1952)	0.06	0.27	0.78	5/m^2 Plot I
	"	0.07	0.28	0.80	" Plot II
	"	0.09	0.27	0.70	10/m^2 Plot I
	"	0.08	0.27	0.84	" Plot II

Variety and place		Proximate constituents %DM					Notes
		CP	EE	NFE	CF	Ash	
FRESH STOVER							
Jaune Gros							
Devon							
Starcross	(1951)	6.5	0.8	47.1	34.0	11.6	
Kingsbridge	(1952)	8.8	1.0	59.6	24.0	6.6	
Starcross	(1952)	6.1	1.4	60.3	28.6	3.6	
"	(1953)	7.5	0.6	60.0	26.0	5.9	
	Means	7.2	1.0	56.7	28.2	6.9	
Kent	(1952)	7.9	1.8	57.8	24.0	8.5	5/m^2 Plot I
	"	7.9	1.4	54.6	27.0	9.1	" Plot II
	"	7.1	1.7	56.4	26.8	8.0	10/m^2 Plot I
	"	6.9	1.6	59.5	25.2	6.8	" Plot II
	Means	7.4	1.6	57.1	25.8	8.1	

Table 1.168 Maize: whole plant and parts as fodder (continued)

Variety and place		Proximate constituents %DM					Notes	Reference
		CP	EE	NFE	CF	Ash		
FRESH STOVER (cont'd)								
White Horsetooth								
Devon								
Starcross	(1951)	5.8	1.2	52.0	32.5	8.5		108, 109 and unpublished details
Kingsbridge	(1952)	14.8	1.2	49.2	26.9	7.9		
Starcross	(1952)	6.0	1.3	63.6	26.0	3.1		
"	(1953)	9.0	1.1	58.6	25.2	6.1		
	Means	8.9	1.2	55.8	27.7	6.4		
Kent	(1952)	6.6	1.8	60.7	24.0	6.9	5/m² Plot I	
	"	7.2	1.6	62.1	22.8	6.3	" Plot II	
	"	7.4	1.8	58.5	25.2	7.1	10/m² Plot I	
	"	7.4	1.5	59.2	25.6	6.3	" Plot II	
	Means	7.2	1.7	60.1	24.4	6.6		
Wisconsin 275A								
Devon								
Starcross	(1951)	4.9	1.1	51.5	34.5	8.0		
Kingsbridge	(1952)	10.3	1.0	57.0	24.9	6.8		
Starcross	(1952)	5.1	1.8	62.5	27.8	2.8		
"	(1953)	8.1	1.0	53.1	32.1	5.7		
	Means	7.1	1.2	56.0	29.8	5.8		
Kent	(1953)	8.3	1.2	56.9	26.8	6.8	5/m² Plot I	
	"	6.8	1.6	59.7	26.7	5.2	" Plot II	
	"	6.2	1.3	58.7	27.7	6.1	10/m² Plot I	
	"	5.5	1.2	60.5	26.2	6.6	" Plot II	
	Means	6.7	1.3	59.0	26.8	6.2		
Wisconsin 341A								
Cambridge	(1952)	5.4	0.9	58.8	26.0	8.9		
Devon								
Kingsbridge	(1952)	10.0	0.9	57.3	24.6	7.2		
Starcross	(1952)	5.2	1.7	61.8	28.6	2.7		
"	(1953)	7.1	1.0	58.4	27.1	6.4		
	Means	7.4	1.2	59.2	26.8	5.4		
Kent	(1952)	6.6	1.8	57.0	27.2	7.4	5/m² Plot I	
	"	6.6	1.8	54.5	29.8	7.3	" Plot II	
	"	6.8	1.8	56.2	27.6	7.6	10/m² Plot I	
	"	7.1	1.8	56.8	27.0	7.3	" Plot II	
	Means	6.8	1.8	56.1	27.9	7.4		
	(1953)	5.4	0.8	57.7	29.4	6.7	5/m² Plot I	
	"	5.3	0.6	57.9	29.8	6.4	" Plot II	
	"	6.7	0.8	56.6	29.1	6.8	10/m² Plot I	
	"	5.5	0.6	58.4	27.8	7.7	" Plot II	
	Means	5.7	0.7	57.7	29.0	6.9		
Wisconsin 355								
Devon								
Starcross	(1951)	5.3	1.4	47.8	34.6	10.9		
Wisconsin 416								
Devon								
Starcross	(1951)	4.4	1.1	55.5	30.5	8.5		
Kingsbridge	(1952)	9.8	0.8	58.3	23.9	7.2		
Starcross	(1952)	5.2	1.7	61.8	28.6	2.7		
"	(1953)	6.9	0.9	60.8	25.7	5.7		
	Means	6.6	1.1	59.1	27.2	6.0		
Cambridge	(1952)	7.0	1.1	58.3	26.3	7.3		
Kent	(1952)	6.3	1.7	56.4	28.0	7.6	5/m² Plot I	
	"	6.5	1.6	59.4	25.6	6.9	" Plot II	
	"	5.9	1.6	61.6	24.8	6.1	10/m² Plot I	
	"	6.4	1.7	60.9	24.5	6.5	" Plot II	
	Means	6.3	1.7	59.6	25.7	6.8		

continued

Table 1.168 Maize: whole plant and parts as fodder (continued)

Variety and place		Proximate constituents %DM					Notes	Reference
		CP	EE	NFE	CF	Ash		
FRESH STOVER (cont'd)								
Wisconsin 464								
Devon								
Starcross	(1951)	6.6	1.0	58.2	26.4	7.8		108, 109 and unpublished details
Kingsbridge	(1952)	9.8	0.8	58.3	23.9	7.2		
Starcross	(1952)	5 4	1.9	62.6	26.8	3.3		
"	(1953)	6.9	0.9	61.2	25.7	5.3		
	Means	7.2	1.1	60.1	25.7	5.9		
Cambridge	(1952)	5.8	1.1	58.3	26.7	8.1		
		Major elements %DM						
		Ca	P	K				
Jaune Gros	(1952)	0.79	0.20	1.40			$5/m^2$ Plot I	
Kent	"	0.69	0.23	0.65			" Plot II	
	"	0.94	0.15	1.05			$10/m^2$ Plot I	
	"	0.87	0.20	0.99			" Plot II	
White Horsetooth								
Kent	(1952)	0.71	0.23	1.00			$5/m^2$ Plot I	
	"	0.87	0.20	1.25			" Plot II	
	"	1.29	0.23	1.34			$10/m^2$ Plot I	
	"	0.62	0.20	0.97			" Plot II	
Wisconsin 275A								
Kent	(1953)	0.62	0.18	1.21			$5/m^2$ Plot I	
	"	0 50	0 18	1 14			" Plot II	
	"	0.61	0.17	1.09			$10/m^2$ Plot I	
	"	0 48	0.18	1.58			" Plot II	
Wisconsin 341A								
Kent	(1952)	0.70	0.19	1.01			$5/m^2$ Plot I	
	"	0.70	0.17	0.64			" Plot II	
	"	0.75	0.18	1.03			$10/m^2$ Plot I	
	"	0.87	0.17	0.72			" Plot II	
	(1953)	0.66	0.20	1.36			$5/m^2$ Plot I	
	"	0.67	0.18	1.24			" Plot II	
	"	0.64	0.25	1.17			$10/m^2$ Plot I	
	"	0.67	0 23	1.56			" Plot II	
Wisconsin 416								
Kent	(1952)	0.75	0.16	1.08			$5/m^2$ Plot I	
	"	0.71	0.15	0.95			" Plot II	
	"	0.70	0.22	0.60			$10/m^2$ Plot I	
	"	1.09	0.17	0.75			" Plot II	

Table 1.169 Rye: whole plant as fodder, and straw. Reference 328

Variety		Years	Crude protein %DM	No. Trials	Crude fibre %DM	No. Trials	Notes
Petkus Spring	A	1957–1961	20.3 (15.5–25.9)	13	17.6 (15.7–19.1)	13	Twelve (12) varieties were tested. Four (4) centres took part: Cambridge, Seale Hayne, Sparsholt and Trawscoed, but not all varieties were tested at all centres. [It is not possible from the table to say how many centres tested any particular variety].
	B	”	16.1 (10.8–22.2)	6	21.2 (19.4–23.4)	4	All the plots were cut between mid-April and mid-May: some had an earlier cut also, between late March and early April. Those cuts are given in the Table as
	C	1959–1961	13.6 (18.5, 15.5, 6.7)	4	25.0 (20.9, 29.0)	2	A: early; B: later following early; and C: later cut only.
Petkus Tetraploid	A	1957–1959	21.4 (22.8, 21.8, 19.5)	9	15.4 (17.7, 16.8, 11.8)	9	Means and ranges are given here for 4 or more yearly averages, single year averages for less than 4.
	B	”	16.1 (22.4, 14.1, 11.7)	3	21.2 (22.7, 21.2, 19.6)	3	[There are many cuts for which analyses are not given.]
	C	1959	20.2	1	20.6	1	
Gartons Large-grained	A	1957–1959	20.0 (23.5, 20.5, 16.0)	8	16.2 (19.2, 17.3, 12.2)	8	
	B	1957, 1958	12.1 (12.9, 11.4)	2	22.0 (23.4, 20.7)	2	
Bernburg Fodder	A	1958–1963	21.5 (16.5–25.9)	14	16.7 (14.7–18.4)	14	
	B	”	16.8 (12.4–21.2)	10	24.0 (21.2–25.9)	8	
	C	1959–1963	14.7 (10.6–19.2)	9	23.7 (20.5–29.0)	8	
C.R.D.	A	1958–1960	22.6 (26.2, 22.1, 19.5)	9	16.6 (17.3, 17.2, 15.4)	9	
	B	”	18.5 (21.1, 21.1, 13.2)	4	23.4 (21.1, 25.7)	2	
	C	1959, 1960	16.9 (17.9, 15.9)	3	22.2 (20.6, 23.8)	2	
Lovaszpatonai	A	1960–1963	21.3 (15.1–27.0)	8	17.8 (17.0–18.6)	8	
	B	”	17.2 (13.8–21.0)	8	23.9 (23.0, 23.0, 25.6)	6	
	C	”	13.4 (10.2–16.8)	8	27.1 (26.0–32.0)	7	
Feniks	A	1960	26.6	3	17.1	3	
	B	”	21.8	2	—	—	
	C	”	18.2	2	18.9	1	
Ovari	A	1960, 1961	22.4 (25.3, 19.6)	5	16.6 (16.4, 16.9)	5	
	B	”	16.3 (18.9, 13.8)	4	24.8	2	
	C	”	12.7 (15.1, 10.3)	4	24.5 (20.2, 28.8)	3	
Cd. 80	A	1962, 1963	18.8 (22.6, 15.1)	3	17.1 (15.8, 18.5)	3	
	B	”	18.1 (21.9, 14.2)	4	22.4 (22.2, 22.6)	4	
	C	”	13.2 (16.3, 10.1)	4	27.8 (26.6, 29.0)	4	
Petkus Normal straw	A	1957–1962	21.0 (17.0–26.6)	15	15.9 (11.9–18.1)	15	
	B	”	16.0 (11.5–21.8)	9	22.0 (19.2–24.3)	7	
	C	1959–1962	15.3 (12.0–20.1)	7	24.9 (27.2, 26.0, 21.6)	5	
Petkus Short straw	A	1957–1959	20.6 (22.6, 19.7, 19.5)	9	14.9 (16.6, 16.3, 11.8)	9	
	B	”	15.9 (22.7, 14.0, 11.1)	3	20.9 (23.2, 20.0, 19.4)	3	
	C	1959	20.4	1	21.1	1	

Table 1.170 Grass straw

Variety	Proximate constituents %DM					Notes	Reference
	CP	EE	NFE	CF	Ash		
Cocksfoot	3.7	2.9	45.1	42.5	5.8		153
Perennial Ryegrass	4.2	2.0	52.1	36.1	5.6		
Red Fescue	2.6	1.6	55.6	34.8	5.4		
Timothy	3.2	2.1	53.9	37.1	3.7		
		Major elements %DM					
		Ca	P	K			
Cocksfoot		0.3	0.1	1.6			
Perennial Ryegrass		0.3	0.2	1.6			
Red Fescue		0.2	0.1	1.5			
Timothy		0.2	0.1	1.3			

Table 1.171 Fibrous residues from protein extracts

Variety	Digestibility % DM	Digestibility % OM	Proximate constituents %DM CP	Fibre	Ash	Notes	Reference
Italian Ryegrass	69.2 (68.0–70.4)	71.6	19.6*			Fibrous residue after extraction of protein from fresh herbage. 3 sheep.	429
Not stated	63.5	66.5	18.5*			*Organic matter basis	
	60.3	64.1	17.8*				
Italian Ryegrass and Red Clover	63.5	66.5	15.0			Roller pressed } 3 sheep	426
	60.3	64.1	14.4			Screw pressed } 3 sheep	
Grass unnamed	60.3	63.2	13.6	48.5*	8.0	7 sheep. *Ad lib* feeding	430
	62.3	65.1				7 sheep ⅔ of high level *Normal acid fibre	
Lucerne	58.1 (55.2–59.8)	60.7 (57.4–62.9)	15.2* (13.5–17.1)			2 or 3 sheep *Organic matter basis	429
	68.6	73.8	21.5*			3 sheep	
	59.6	62.9	19.4			Roller pressed once	426
	55.2	57.4	13.8			Roller pressed twice }	
	59.8	62.9	18.8			Roller pressed twice }	
	59.8	61.6	16.9			Roller pressed three times	
	55.9	58.8	15.6			Roller pressed four times 3 sheep	
Lucerne and grass	57.8	60.3	13.8			Roller pressed twice 3 sheep	

Table 1.172 Kale, Marrowstem, named varieties

Variety		Proximate constituents %DM						Reference
		Crude protein		EE	NFE	CF	Ash	
		Cut 1	Cut 2					
Medium Choumellier	W	19.0	18.0	2.4	51.7	15.5	11.4	315
	L	24.6	24.6	4.0	47.2	10.6	13.6	
	S	14.6	15.4	1.2	55.2	19.4	9.6	
Giant Choumellier	W	16.9	18.6	1.8	53.6	16.4	11.3	
	L	23.2	25.9	3.4	49.3	10.4	13.7	
	S	13.3	15.9	0.8	56.3	19.7	9.9	
Green Stem	W	17.0	16.4	2.3	51.4	16.9	12.4	
	L	22.9	24.7	3.6	47.2	10.6	15.7	
	S	13.9	14.2	1.6	53.6	20.1	10.8	
Purple Stem	W	20.4	20.5	2.6	50.5	14.7	11.8	
	L	25.1	26.9	3.9	46.6	10.0	14.4	
	S	15.2	17.2	1.2	55.3	19.8	8.5	

Notes:

Cut 1 : 26 November – 12 December, 1952

Cut 2 : (Crude protein only): 20 January – 2 February, 1953

W : Whole plan
L : Leaf
S : Stem

Variety		Major elements %DM							
		Ca	P	Mg	K	Na	Cl	S	
								Total	Sulphate
Medium Choumellier	W	1.07	0.40	0.19	2.42	0.62	1.31	0.83	0.54
	L	1.79	0.42	0.19	1.95	0.61	1.59	1.09	0.70
	S	0.53	0.38	0.19	2.79	0.63	1.11	0.63	0.42
Giant Choumellier	W	1.16	0.36	0.20	2.17	0.85	1.39	0.77	0.54
	L	2.05	0.38	0.22	1.56	0.78	1.71	1.23	0.88
	S	0.66	0.35	0.19	2.52	0.88	1.21	0.51	0.35
Green Stem	W	1.25	0.37	0.21	2.27	0.92	1.55	0.80	0.57
	L	2.36	0.37	0.24	1.54	0.82	2.01	1.26	0.95
	S	0.69	0.36	0.20	2.75	0.99	1.33	0.57	0.38
Purple Stem	W	1.36	0.39	0.21	2.21	0.59	1.39	0.75	0.38
	L	2.06	0.42	0.23	1.87	0.67	1.74	1.02	0.55
	S	0.57	0.37	0.18	2.55	0.51	1.05	0.55	0.28

Variety		Trace elements μg/g DM								Reference
		B	Cu	Fe	Pb	Mn	Mo	Zn		317
Medium Choumellier	W	28.2				25.9	0.6			
	L	35.2				46.0	1.1			
	S	22.5				9.2	0.3			
Giant Choumellier	W	27.5				28.7	0.5			
	L	35.7				63.0	1.0			
	S	23.0				9.8	0.3			
Green Stem	W	28.9				26.1	0.5			
	L	35.5				55.0	1.0			
	S	25.2				9.6	0.3			
Purple Stem	W	28.2	2.7	36.2	Trace	36.8	0.7	25.7		
	L	33.5	3.3	61.0	0.4	62.0	0.9	19.0		
	S	23.2	2.3	19.4	Trace	12.5	0.3	32.5		
Cannell's	W		2.6	32.2	0.5			22.6		
	L		3.3	61.8	0.8			18.0		
	S		2.4	20.7	0.4			24.5		
Makanta Groena	W		2.5	35.8	Trace			22.5		
	L		3.0	62.5	0.6			18.6		
	S		2.2	21.7	Trace			25.0		
White French	W		2.4	35.0	0.4			19.2		
	L		3.0	58.6	0.4			18.0		
	S		2.0	20.5	0.4			20.0		

Table 1.173 Kale, Marrowstem, unnamed

	September			October			November			December			January			February			Notes	Reference
Digestibility of DM%											81.7 82.9								Digestibility by sheep Digestibility by rabbit	539
		U	78.5		S	78.4							U	78.1	S	76.4			U : unthinned. S : singled Digestibility by sheep. For proximate constituents and metabolisable energy see table 1.234	577
Digestibility of OM%											85.0 82.7								See above, same reference	539
				85.2															Digestibility of OM% for 4 samples without dates were: 85.7, 82.3, 80.8 and 82.1. For proximate constituents and metabolisable energy see table 1.234	538
		U	79.8		S	80.8							U	80.1	S	78.3			See above, same reference	577
Proximate constituents %DM Crude protein				SL SR SM	13.9 7.7 12.0														U : unthinned S : singled W : whole plant L : leaf R : stem rind M : stem marrow	577
											27.3								Overwintering studies with hill sheep. No date	255
				TW UW	17.9 10.7		TW UW TL UL TS US	16.3 8.2 12.7 13.6 7.3 6.9	TLl TLp TSR TSM	12.8 6.2 9.2 13.0									T : thinned U : unthinned W : whole plant L : leaf Ll : leaf lamina Lp : leaf petiole SR : stem rind SM : stem marrow	216
										T U	20.0 19.6								T : thinned U : unthinned Means for monthly harvests from 12/10/1942 to 14/1/1943	217
									(1943) (1944)	12.7 14.3 20.7 23.8										225
		9.0																		129
											11.0								Mean of 3 samples used in cattle feeding trials. No date	112
											13.1 11.3 14.8 14.8								4 samples cut for silage	238

Table 1.173 Kale, Marrowstem, unnamed (continued)

	September	October	November	December	January	February	Notes	Reference
Crude protein (cont'd)			(1948) L { N_0 13.4 N_1 16.9	16.5 20.8			N_0 : no fertiliser N_1 : ammonium sulphate 750 kg/ha L : leaf S : stem	127
			(1949) L { N_0 29.5 N_1 33.5	30.4 32.4				
			(1950) L { N_0 15.0 N_1 17.6	17.3 19.4				
			(1951) L { N_0 16.9 N_1 22.7	18.1 23.1				
			(1948) S { N_0 4.5 N_1 8.1	5.0 8.7			Planted late April/early May. Means for first cut in November and for 3 to 4 cuts from November to February. The outstandingly high CP and other values in 1949 and 1949–50 are mentioned in the text, but no explanation is offered.	
			(1949) S { N_0 18.1 N_1 22.0	19.2 23.2				
			(1950) S { N_0 5.5 N_1 9.1	7.0 9.8				
			(1951) S { N_0 6.5 N_1 13.0	7.0 12.2				
				(1) 16.7	(2) 13.0		(1) No date. Used in feeding trials with cows and heifers (2) Same crop slightly more mature	111
				L(a) 25.0 L(b) 22.8 L(c) 24.1 T(a) 20.2 T(b) 15.9 Tr(c) 16.4 Tx(c) 17.6 Tm(c) 15.5 B(a) 14.2 B(b) 10.5 Br(c) 12.1 Bx(c) 8.5 Bm(c) 11.6			(a) Means for 19/11, 1/12, 9/12. 1958 (b) Means for 19/11, 15/12, 6/1, 1959-60 (c) No date, 1961-62 L : Leaf T : Stem top Tr : Stem top, rind Tx : Stem top, xylem Tm : Stem top, marrow B : Stem bottom Br : Stem bottom, rind Bx : Stem bottom, xylem Bm : Stem bottom, marrow	321
Ether extract		SL 3.3 SR 0.8 SM 1.3					See Crude protein, same reference	577
		TW 5.0 UW 3.5	TW 3.7 UW 2.4 TL 3.7 UL 3.7 TS 1.9 US 1.5	TLl 4.6 TLp 1.9 TSR 0.5 TSM 0.8			See Crude protein, same reference	216
				T 3.9 U 3.6			See Crude protein, same reference	217
				(1943) { 1.9 1.9 (1944) { 1.7 2.1			See Crude protein, same reference	225
	1.9							129
				2.0			See Crude protein, same reference	112
				1.6 1.5 2.0 1.8			4 samples cut for silage	238

continued

Table 1.173 Kale, Marrowstem, unnamed (continued)

	September	October	November	December	January	February	Notes	Reference
Proximate constituents %DM Ether extract (cont'd) (cont'd)			(1948)L N_0 3.6 (1948)L N_1 3.6 (1949)L N_0 3.4 (1949)L N_1 3.2 (1950)L N_0 3.3 (1950)L N_1 3.5 (1951)L N_0 3.5 (1951)L N_1 3.7 (1948)S N_0 1.2 (1948)S N_1 1.1 (1949)S N_0 1.3 (1949)S N_1 1.3 (1950)S N_0 1.1 (1950)S N_1 1.0 (1951)S N_0 1.0 (1951)S N_1 0.9				See Crude protein, same reference	127
				(1) 1.7	(2) 1.4		See Crude protein, same reference	111
				L(a) 2.3 L(b) 2.0 L(c) 3.4 T(a) 0.8 T(b) 0.9 Tr(c) 1.8 Tx(c) 1.5 Tm(c) 1.1 B(a) 0.6 B(b) 0.8 Br(c) 1.9 Bx(c) 1.5 Bm(c) 1.0			See Crude protein, same reference	321
Nitrogen-free extract		SL 55.5 SR 55.4 SM 58.4					See Crude protein, same reference	577
				43.7			See Crude protein, same reference	255
		TW 45.5 UW 50.6	TW 49.8 UW 60.5 TL 55.0 UL 47.3 TS 61.6 US 57.9 TL1 TLp TSR TSM	55.1 60.9 48.7 57.5			See Crude protein, same reference	216
				T 43.1 U 46.1			See Crude protein, same reference	217
			(1943) (1944)	55.1 53.8 47.2 43.9			See Crude protein, same reference	225
	56.6							129
				59.0			See Crude protein, same reference	112
				57.9 52.7 54.8 55.6			4 samples cut for silage	238

Table 1.173 Kale, Marrowstem, unnamed (continued)

	September	October	November	December	January	February	Notes	Reference
Nitrogen-free extract (cont'd)			(1948)L { N_0 60.7 N_1 54.9 (1949)L { N_0 40.9 N_1 38.8 (1950)L { N_0 57.6 N_1 53.1 (1951)L { N_0 55.7 N_1 49.5 (1948)S { N_0 64.2 N_1 59.3 (1949)S { N_0 42.2 N_1 40.3 (1950)S { N_0 63.3 N_1 58.5 (1951)S { N_0 65.3 N_1 59.2				See Crude protein, same reference	127
				(1) 52.5	(2) 57.3		See Crude protein, same reference	111
				L(a) 45.0 L(b) 45.5 L(c) 48.5 T(a) 47.0 T(b) 52.0 Tr(c) 56.7 Tx(c) 52.5 Tm(c) 59.2 B(a) 41.9 B(b) 45.5 Br(c) 60.7 Bx(c) 42.1 Bm(c) 61.5			See Crude protein, same reference	321
Crude fibre		SL 12.4 SR 29.1 SM 13.1					See Crude protein, same reference	577
		TW 16.1 UW 23.4	TW 18.1 UW 21.3 TL 19.2 UL 23.6 TS 20.5 US 25.1 TL1 17.7 TLp 22.3 TSR 33.7 TSM 16.5				See Crude protein, same reference	216
				T 21.6 U 20.8			See Crude protein, same reference	217
			(1943) { 19.0 19.1 (1944) { 20.2 19.2				See Crude protein, same reference	225
	22.2							129
				18.8			See Crude protein, same reference	112
				16.2 23.0 14.6 13.9			4 samples cut for silage	238

continued

Table 1.173 Kale, Marrowstem, unnamed (continued)

	September	October	November	December	January	February	Notes	Reference
Proximate constituents %DM Crude fibre (cont'd) (cont'd)			(1948)L {N_0 11.4, N_1 11.1 (1949)L {N_0 10.0, N_1 9.9 (1950)L {N_0 10.4, N_1 11.1 (1951)L {N_0 9.7, N_1 9.7 (1948)S {N_0 23.6, N_1 23.0 (1949)S {N_0 22.6, N_1 21.3 (1950)S {N_0 22.1, N_1 21.3 (1951)S {N_0 18.5, N_1 17.0				See Crude protein, same reference	127
				(1) 15.8	(2) 16.7		See Crude protein, same reference	111
				L(a) 12.4 L(b) 11.9 L(c) 12.2 T(a) 19.4 T(b) 17.9 Tr(c) 14.2 Tx(c) 19.0 Tm(c) 11.0 B(a) 31.3 B(b) 31.3 Br(c) 14.2 Bx(c) 42.0 Bm(c) 12.6			See Crude protein, same reference	321
Ash		SL 14.9 SR 7.0 SM 15.2					See Crude protein, same reference	577
		TW 15.5 UW 11.8.	TW 12.1 UW 7.6 TL 9.4 UL 11.8 TS 8.7 US 8.6	TLl 9.8 TLp 8.7 TSR 7.9 TSM 12.2			See Crude protein, same reference	216
				T 11.4 U 9.9			See Crude protein. same reference	217
				(1943) {11.3, 10.9 (1944) {10.2, 11.0			See Crude protein, same reference	225
	10.3							129
				9.2			See Crude protein, same reference	112
				11.2 11.5 13.8 13.9			4 samples cut for silage	238

Table 1.173 Kale, Marrowstem, unnamed (continued)

	September	October	November	December	January	February	Notes	Reference
Ash (cont'd)			(1948) L N_0 10.9 (1948) L N_1 13.5 (1949) L N_0 16.2 (1949) L N_1 14.6 (1950) L N_0 13.7 (1950) L N_1 14.7 (1951) L N_0 14.2 (1951) L N_1 14.4 (1948) S N_0 6.5 (1948) S N_1 8.5 (1949) S N_0 15.8 (1949) S N_1 15.1 (1950) S N_0 8.0 (1950) S N_1 10.1 (1951) S N_0 8.7 (1951) S N_1 9.9				See Crude protein, same reference	127
				(1) 13.3	(2) 11.6		See Crude protein, same reference	111
				L(a) 15.3 L(b) 17.8 L(c) 11.8 T(a) 12.6 T(b) 13.3 Tr(c) 10.9 Tx(c) 9.4 Tm(c) 13.2 B(a) 12.0 B(b) 11.9 Br(c) 11.1 Bx(c) 5.9 Bm(c) 13.3			See Crude protein, same reference	321
Major elements %DM Calcium	UW 2.7	SW 2.3 SL 3.6 SR 0.9 SM 1.0			UW 1.8 SW 1.8		See Crude protein, same reference	577
				1.6			See Crude protein, same reference	255
		TW 2.9 UW 2.5	TW 2.6 UW 1.7 TL 2.5 UL 2.9 TS 0.9 US 1.2	TL1 2.8 TLp 1.8 TSR 0.8 TSM 0.7			See Crude protein, same reference	216
				T 1.7 U 1.9			See Crude protein, same reference	217
				L(a) 2.8 L(b) 3.2 L(c) 3.1 T(a) 1.1 T(b) 1.0 Tr(c) 1.2 Tx(c) 0.6 Tm(c) 0.6 B(a) 1.1 B(b) 1.0 Br(c) 1.3 Bx(c) 0.4 Bm(c) 0.7			See Crude protein, same reference	321

continued

Table 1.173 Kale, Marrowstem, unnamed (continued)

	September			October			November			December			January			February			Notes	Reference
Major elements %DM (cont'd) Phosphorus		UW	0.3	 SL SR SM	SW 0.3 0.3 0.5	0.3							UW	0.4	SW	0.4			See Crude protein, same reference	577
											0.7								See Crude protein, same reference	255
				TW UW	0.5 0.3		TW UW TL UL TS US	0.5 0.3 0.3 0.3 0.5 0.3	TLl TLp TSR TSM	0.3 0.3 0.3 0.6									See Crude protein, same reference	216
										T U	0.5 0.4								See Crude protein, same reference	217
										L(a) L(b) L(c) T(a) T(b) Tr(c) Tx(c) Tm(c) B(a) B(b) Br(c) Bx(c) Bm(c)	0.4 0.4 0.4 0.4 0.5 0.3 0.4 0.5 0.3 0.4 0.3 0.2 0.4								See Crude protein, same reference	321
Magnesium		UW	0.3	 SL SR SM	SW 0.2 0.2 0.3	0.3							UW	0.2	SW	0.4			See Crude protein, same reference	577
										L(a) L(b) L(c) T(a) T(b) Tr(c) Tx(c) Tm(c) B(a) B(b) Br(c) Bx(c) Bm(c)	0.1 0.1 0.2 0.1 0.1 0.2 0.2 0.2 0.1 0.1 0.2 0.1 0.3								See Crude protein, same reference	321
Potassium		UW	3.3	 SL SR SM	SW 2.0 2.1 5.1	3.2							UW	3.2	SW	3.6			See Crude protein, same reference	577
										L(a) L(b) L(c) T(a) T(b) Tr(c) Tx(c) Tm(c) B(a) B(b) Br(c) Bx(c) Bm(c)	2.5 3.5 2.4 3.9 4.9 3.8 3.2 5.8 3.5 4.1 3.7 2.0 5.1								See Crude protein. same reference	321

Table 1.173 Kale, Marrowstem, unnamed (continued)

	September			October			November			December			January			February			Notes	Reference
Sodium		UW	0.4		SW	0.3							UW	0.2	SW	0.2			See Crude protein, same reference	577
				SL	0.2															
				SR	0.3															
				SM	1.5															
										L(a)	0.3								See Crude protein, same reference	321
										L(b)	0.5									
										L(c)	0.2									
										T(a)	0.4									
										T(b)	0.6									
										Tr(c)	0.2									
										Tx(c)	0.3									
										Tm(c)	0.3									
										B(a)	0.6									
										B(b)	0.7									
										Br(c)	0.4									
										Bx(c)	0.3									
										Bm(c)	0.5									
Chlorine		UW	1.5		SW	1.8							UW	1.3	SW	1.4			See Crude protein, same reference	577
				SL	1.9															
				SR	0.7															
				SM	2.0															
										L(a)	0.9								See Crude protein. same reference	321
										L(b)	1.6									
										L(c)	1.6									
										T(a)	0.8									
										T(b)	1.1									
										Tr(c)	1.6									
										Tx(c)	1.0									
										Tm(c)	1.6									
										B(a)	0.6									
										B(b)	1.0									
										Br(c)	1.4									
										Bx(c)	0.4									
										Bm(c)	0.6									
Sulphur																				
Total sulphur		UW	1.4		SW	1.2							UW	1.1	SW	1.0			See Crude protein, same reference	577
				SL	1.5															
				SR	0.7															
				SM	1.1															
Inorganic sulphur		UW	0.8		SW	0.7							UW	0.5	SW	0.5				
				SL	0.9															
				SR	0.3															
				SM	0.6															
Total sulphur										L(a)	1.2								See Crude protein, same reference	321
										L(b)	1.2									
										L(c)	1.3									
										T(a)	0.8									
										T(b)	0.7									
										Tr(c)	1.0									
										Tx(c)	0.8									
										Tm(c)	0.9									
										B(a)	0.5									
										B(b)	0.4									
										Br(c)	0.8									
										Bx(c)	0.5									
										Bm(c)	0.8									

continued

Table 1.173 Kale, Marrowstem, unnamed (continued)

	September	October	November	December	January	February	Notes	Reference
Major elements %DM (cont'd)								
Sulphate sulphur				L(a) 0.8				321
				L(b) 0.8				
				L(c) 0.5				
				T(a) 0.4				
				T(b) 0.4				
				Tr(c) 0.5				
				Tx(c) 0.3				
				Tm(c) 0.3				
				B(a) 0.3				
				B(b) 0.3				
				Br(c) 0.3				
				Bx(c) 0.1				
				Bm(c) 0.3				
Vitamins μg/g DM								
Tocopherols								
Total tocopherol				L 787			L : Leaf S : Stem	100
α-tocopherol % of total				L >95				
Total tocopherol				S 9–15				
α-tocopherol % of total				S >95				

Table 1.174 Kale, Thousand-headed, named varieties

Variety		Proximate Constituents %DM						Notes		Reference
		Crude protein		EE	NFE	CF	Ash			
		Cut 1	Cut 2					Cut 1: 26 November – 12 December, 1952		315
Cannell's	W	16.9	20.8	2.8	53.3	16.7	10.3	Cut: 2: (Crude protein only): 20 January – 2 February, 1953		
	L	19.8	24.4	3.8	52.8	11.3	12.3			
	S	11.9	15.8	1.2	54.2	25.8	6.9			
Canson	W	17.2	18.8	2.8	55.1	15.9	9.0			
	L	20.2	22.5	3.7	54.9	10.9	10.3			
	S	11.6	13.9	1.1	55.3	25.4	6.6			
New Zealand	W	16.6	18.0	3.0	52.8	16.7	10.9	W: Whole plant		
	L	19.8	22.5	4.0	51.6	11.5	13.1	L: Leaf		
	S	11.5	13.2	1.4	55.1	24.8	7.2	S: Stem		
		Major elements %DM								
		Ca	P	Mg	K	Na	Cl	S		
								Total	Sulphate	
Cannell's	W	1.16	0.40	0.19	2.32	0.34	1.23	0.74	0.42	
	L	1.59	0.41	0.20	2.41	0.36	1.49	0.91	0.52	
	S	0.47	0.38	0.19	2.15	0.33	0.78	0.45	0.24	
Canson	W	0.98	0.40	0.18	2.18	0.20	0.97	0.62	0.33	
	L	1.27	0.42	0.19	2.24	0.22	1.14	0.73	0.37	
	S	0.43	0.37	0.17	2.07	0.16	0.76	0.43	0.24	
New Zealand	W	1.40	0.38	0.21	2.30	0.29	1.31	0.84	0.55	
	L	1.97	0.39	0.22	2.34	0.30	1.63	1.10	0.74	
	S	0.49	0.38	0.18	2.22	0.27	0.83	0.46	0.28	
		Trace elements µg/g DM								
		B	Cu	Fe	Pb	Mn	Mo	Zn		317
Cannell's	W	30.9	3.0	29.4	Trace	40.3	0.6	17.3		
	L	35.0	4.1	41.5	0.8	57.0	0.9	16.0		
	S	23.7	2.0	17.1	Trace	10.8	0.1	18.6		
Canson	W	27.6	2.7	31.8	0.5	44.5	0.6	17.5		
	L	30.7	3.0	41.6	0.6	64.0	0.8	15.6		
	S	21.2	2.3	17.8	0.4	9.3	0.2	20.6		
New Zealand	W	31.2				35.0	0.6			
	L	36.7				50.0	0.8			
	S	22.0				10.6	0.3			

Table 1.175 Kale, Thousand-headed, mixture, named varieties

	Part	November	December	January	February	Notes	Reference
Proximate constituents %DM							
Crude protein	L		23.4	25.8	26.0	L: Leaf TS: Top Stem BHS: Bottom half stem Varieties: Cannell's Canson Chou de Cholet Chou branchu du Poitou Chou fourrager de la Sarthe	320
	TS		19.4	20.2	20.4		
	BHS		13.8	13.5	13.3		
Ether extract	L		2.2	2.5	3.5		
	TS		0.9	0.9	1.0		
	BHS		0.7	0.6	0.8		
Nitrogen-free extract	L		46.0	46.9	46.7		
	TS		44.9	46.6	46.1		
	BHS		36.4	38.9	37.1		
Crude fibre	L		12.8	11.0	11.2		
	TS		23.8	21.9	22.4		
	BHS		39.1	37.5	40.6		
Ash	L		15.6	13.8	12.6		
	TS		11.0	10.4	10.1		
	BHS		10.0	9.5	8.2		
Major elements %DM							
Calcium	L		2.3	1.8	1.7		
	TS		1.0	0.9	0.9		
	BHS		1.0	0.9	0.8		
Phosphorus	L		0.4	0.5	0.4		
	TS		0.4	0.4	0.4		
	BHS		0.3	0.3	0.3		
Magnesium	L		0.2	0.2	0.2		
	TS		0.2	0.2	0.2		
	BHS		0.1	0.1	0.1		
Potassium	L		3.1	3.1	3.1		
	TS		3.4	3.2	3.3		
	BHS		2.7	2.5	2.4		
Sodium	L		0.2	0.2	0.2		
	TS		0.3	0.3	0.3		
	BHS		0.5	0.5	0.4		
Chlorine	L		0.9	0.8	0.8		
	TS		0.6	0.6	0.6		
	BHS		0.5	0.4	0.4		
Sulphur							
Total sulphur	L		1.0	1.1	1.0		
	TS		0.7	0.7	0.7		
	BHS		0.4	0.4	0.4		
Sulphate sulphur	L		0.5	0.5	0.5		
	TS		0.4	0.3	0.3		
	BHS		0.3	0.2	0.2		

Table 1.176 Kale, Thousand-headed, unnamed

	October			November			December			January			February			March			Notes	Reference
Digestibility of DM%					76.8 79.1										74.5 77.1				Digestibility by 2 sheep	577
Digestibility of OM%					79.4 81.6										76.4 79.4				For proximate constituents and metabolisable energy see table 1.234	
Proximate constituents %DM Crude protein								10.1											Winter feeding of dairy cows No date	55
		30.6			21.6														Harvested unthinned	216
						21.2													Cut monthly 12/10/1942 to 14/1/1943 Mean values only given	217
			(1948)	N_o N_1	5.6 8.0				6.0 10.0										N_o: no fertiliser N_1: ammonium sulphate 750 kg/ha	127
			(1949)	N_o N_1	20.0 20.8				21.0 22.5										Means for first harvest in November and for 3 to 4 harvests from November to February (Crude protein only)	
			(1950)	N_o N_1	6.7 8.4				8.7 9.9										Stem only	
			(1951)	N_o N_1	7.2 10.9				7.8 11.0										The outstandingly high CP values in 1949 and 1949-50 are mentioned in the text, but no explanation is offered.	
											L Tr Tx Tm Br Bx Bm	22.1 19.4 14.5 11.5 15.3 7.7 9.5							L Leaf Tr Top stem, rind Tx Top stem, xylem Tm Top stem, marrow Br Bottom half stem, rind Bx Bottom half stem, xylem Bm Bottom half stem, marrow	320
Ether extract								1.0											See Crude protein, same reference	55
		6.5			4.0														See Crude protein, same reference	216
						3.5													See Crude protein, same reference	217
			(1948) (1949) (1950) (1951)	N_o / N_1 N_o / N_1 N_o / N_1 N_o / N_1	1.3 / 1.3 1.2 / 1.3 1.1 / 1.0 1.2 / 1.1														See Crude protein, same reference	127
											L Tr Tx Tm Br Bx Bm	4.0 1.7 1.0 0.7 1.6 1.1 0.6							See Crude protein, same reference	320

continued

Table 1.176 Kale, Thousand-headed, unnamed (continued)

	October			November			December			January			February			March			Notes	Reference
Proximate constituents (cont'd) Nitrogen-free extract								62.0											See Crude protein, same reference	55
		33.1			46.5														See Crude protein, same reference	216
						45.8													See Crude protein, same reference	217
			(1948) (1949) (1950) (1951)	{No {N1 {No {N1 {No {N1 {No {N1	58.9 55.3 39.7 36.3 58.8 57.2 61.8 60.3														See Crude protein, same reference	127
											L Tr Tx Tm Br Bx Bm	51.1 57.6 48.7 74.5 57.7 41.7 69.3							See Crude protein, same reference	320
Crude fibre								19.8											See Crude protein, same reference	55
		13.7			16.3														See Crude protein, same reference	216
						19.2													See Crude protein, same reference	217
			(1948) (1949) (1950) (1951)	{No {N1 {No {N1 {No {N1 {No {N1	28.4 28.8 26.5 29.3 26.7 26.3 23.3 21.0														See Crude protein, same reference	127
											L Tr Tx Tm Br Bx Bm	12.2 14.2 31.6 6.1 17.6 46.0 11.0							See Crude protein, same reference	320
Ash								7.1											See Crude protein, same reference	55
		16.1			11.6														See Crude protein, same reference	216
						10.3													See Crude protein, same reference	217
			(1948) (1949) (1950) (1951)	{No {N1 {No {N1 {No {N1 {No {N1	5.8 6.6 12.6 12.3 6.7 7.1 6.5 6.7														See Crude protein, same reference	127
											L Tr Tx Tm Br Bx Bm	10.6 7.1 4.2 7.2 7.8 3.5 9.6							See Crude protein, same reference	320

Table 1.176 Kale, Thousand-headed, unnamed (continued)

	October			November			December			January			February			March			Notes	Reference
Major elements %DM Calcium					1.8										1.7				See Digestibility, same reference	577
		3.8			3.1														See Crude protein, same reference	216
						1.5													See Crude protein, same reference	217
											L Tr Tx Tm Br Bx Bm	2.1 1.3 0.5 0.6 1.7 0.4 1.1							See Crude protein, same reference	320
Phosphorus					0.3										0.4				See Digestibility, same reference	577
		0.5			0.5														See Crude protein, same reference	216
						0.4													See Crude protein, same reference	217
											L Tr Tx Tm Br Bx Bm	0.3 0.3 0.2 0.2 0.2 0.1 0.2							See Crude protein, same reference	320
Magnesium					0.3										0.3				See Digestibility, same reference	577
											L Tr Tx Tm Br Bx Bm	0.09 0.09 0.07 0.10 0.11 0.09 0.14							See Crude protein, same reference	320
Potassium					2.2										3.2				See Digestibility, same reference	577
											L Tr Tx Tm Br Bx Bm	2.3 2.0 1.4 2.7 1.7 1.1 3.3							See Crude protein, same reference	320
Sodium					0.16										0.10				See Digestibility, same reference	577
											L Tr Tx Tm Br Bx Bm	0.16 0.13 0.11 0.26 0.27 0.19 0.49							See Crude protein, same reference	320

continued

Table 1.176 Kale, Thousand-headed, unnamed (continued)

	October	November	December	January	February	March	Notes	Reference
Major elements (cont'd) Chlorine		0.9			1.1		See Digestibility, same reference	577
				L 1.0 Tr 0.4 Tx 0.1 Tm 0.8 Br 0.3 Bx 0.1 Bm 0.9			See Crude protein, same reference	320
Sulphur Total sulphur Inorganic sulphur		0.9 0.5			1.0 0.5		See Digestibility, same reference	577
Total sulphur				L 1.3 Tr 0.9 Tx 0.6 Tm 0.7 Br 0.8 Bx 0.4 Bm 0.7			See Crude protein, same reference	320
Sulphate sulphur				L 0.6 Tr 0.4 Tx 0.2 Tm 0.3 Br 0.3 Bx 0.1 Bm 0.4				
Trace elements μg/g DM Iron		65					See Digestibility, same reference	577
Vitamins μg/g DM Carotene			W 114 L 192 S 15				W: Whole plant L: Leaf S: Stem Winter feeding of dairy cows. No date	55
Tocopherols Total α-tocopherol % of total			L 371 S 12 – 16 L > 95 S > 95				L: Leaf S: Stem No date	100

Table 1.177 Kale, Digestibility *in vitro*, variety trials

	October			November			December			January			February			March			April			Notes	Reference
PB1 Hybrid ***Digestibility of* DM%**			81.4					79.4			76.1											Stem only Cambridge 1960-61	168
						85.7 90.3 78.1		79.5 88.3 78.7		78.6 86.7 78.7												Leaf Stem top Stem bottom } Trawscoed 1960-61	
											73.5				78.3				77.5			Stem only Cambridge 1961	
											72.2		72.8			76.2						Stem only Sutton Bonington 1961	
											81.8*											* No date Stem only Wye 1961-62	
											71.7* 79.0*											Stem only Whole plant Sutton Bonington 1961-62	
											78.8*											Stem only Cambridge1961-62	
											78.9*											Stem only Trawscoed1961-62	
											80.7* 79.4*											Whole plant, Cut 1 Whole plant, Cut 3 } Seale-Hayne 1961-62	
										79.6			70.8									Stem only Cambridge 1962	
***Proximate constituents* %DM** **Crude protein**			13.3					16.1			16.9											Stem only Cambridge 1960-61	
						23.5 19.3 17.3		21.8 21.5 18.3		24.6 23.5 22.3												Leaf Stem top Stem bottom } Trawscoed 1960-61	
											16.8				15.4				14.5			Stem only Cambridge 1961	
											14.6		11.2			9.4						Stem only Sutton Bonington 1961	
Crude fibre			19.0					16.8			18.3											Stem only Cambridge 1960-61	
						12.1 10.8 18.7		12.9 13.3 19.7		12.9 11.9 17.2												Leaf Stem top Stem bottom } Trawscoed 1960-61	
											17.2				19.3				21.4			Stem only Cambridge 1961	
											20.1		20.4			20.8						Stem only Sutton Bonington 1961	
											16.1*											* No date Stem only Wye 1961-62	
											16.7*											*No date Stem only Cambridge1961-62	

continued

Table 1.177 Kale, Digestibility *in vitro*, variety trials (continued)

	October	November	December	January	February	March	April	Notes		Reference
Canson *Digestibility of* DM%				69.6	72.2		67.3	Stem only	Cambridge 1961	
				62.6	70.0	72.2		Stem only	Sutton Bonington 1961	
				78.4*				* No date Stem only	Wye 1961-62	
				73.5* 73.3*				Stem only Cut 1 Whole plant Cut 2	Sutton Bonington 1961-62	
				72.6*				Stem only	Cambridge 1961-62	
				71.6*				Stem only	Trawscoed 1961-62	
				79.2* 75.2*				Whole plant: Cut 1 Whole plant: Cut 3	Seale-Hayne 1961-62	
				72.5	65.0			Stem only	Cambridge 1962	
Proximate constituents %DM Crude protein				16.9	18.0		13.2	Stem only	Cambridge 1961	
				11.9	13.0	9.7		Stem only	Sutton Bonington 1961	
Crude fibre				24.5	23.6		29.4	Stem only	Cambridge 1961	
				30.4	26.6	26.9		Stem only	Sutton Bonington 1961	
				20.0*				*No date	Wye 1961-62	
				31.2*				*No date	Cambridge 1961-62	
Blattstamkohl *Digestibility of* %DM				77.7	71.9		66.7	Stem only	Cambridge 1961	
				64.4	72.3	69.4		Stem only	Sutton Bonington 1961	
Proximate constituents %DM Crude protein				17.0	16.0		13.6	Stem only	Cambridge 1961	
				12.3	13.7	13.6		Stem only	Sutton Bonington 1961	
Crude fibre %DM				19.2	29.7		30.9	Stem only	Cambridge 1961	
				27.0	25.6	26.0		Stem only	Sutton Bonington 1961	

Table 1.177 Kale, Digestibility *in vitro*, variety trials (continued)

	October	November	December	January	February	March	April	Notes	Reference
MARROWSTEM unnamed *Digestibility of* DM%	(2) 69.0 (12) 70.4		73.1 69.6	74.4 67.7				Seedsmen's stock, samples (2) and (12) Stem only Cambridge 1960-61	168
		82.4 86.0 62.8	78.9 81.8 62.3	76.8 85.2 58.2				Leaf only Stem, top Stem, bottom } Trawscoed 1960-61	
				65.3	63.0		67.6	Stem only Cambridge 1961	
				64.6	66.7	49.9		Stem only Sutton Bonington 1961	
				71.0*				* No date Stem only Wye 1961-62	
				64.3* 72.5*				Stem only Whole plant } Sutton Bonington 1961-62	
				64.7*				Stem only Cambridge 1961-62	
				66.7*				Stem only Trawscoed 1961-62	
				77.4* 75.6*				Whole plant, Cut 1 Whole plant, Cut 3 } Seale-Hayne 1961-62	
				72.0				Stem only Cambridge 1962	
Proximate constituents %DM Crude protein	10.8		15.0	12.0				Stem only Cambridge 1960-61	
		26.3 23.2 16.6	24.7 21.9 15.9	26.8 23.7 15.1				Leaf only Stem, top Stem, bottom } Trawscoed 1960-61	
				12.6	15.9		13.2	Stem only Cambridge 1961	
				10.6	9.4	8.8		Stem only Sutton Bonington 1961	
Crude fibre %DM	26.8		24.1	23.9				Stem only Cambridge 1960-61	
		13.1 17.2 34.4	12.0 17.2 31.8	12.5 15.3 30.8				Leaf only Stem, top Stem, bottom } Trawscoed 1960-61	
				23.0	26.2		27.5	Stem only Cambridge 1961	
				26.1	30.1	28.3		Stem only Sutton Bonington 1961	
				25.5* 23.6*				* No date Wye 1961-62 * No date Cambridge 1961-62	
MARROWSTEM unnamed *Digestibility of* %DM				(13) 71.9* (70.4 – 76.0)				* No date Means and ranges of seedsmen's stocks. No. of samples in brackets before mean. Stem only Wye 1961-62	

continued

Table 1.177 Kale, Digestibility *in vitro*, variety trials (continued)

	October	November	December	January	February	March	April	Notes	Reference
MARROWSTEM unnamed *Digestibility of* DM% (cont'd)				(13) 66.4* (61.8 – 70.5) 70.9* (64.6 – 74.1)				Stem only Whole plant } Sutton Bonington 1961-62	
				(13) 67.3* (64.7 – 69.0)				Stem only Cambridge 1961-62	
				(13) 67.7* (64.5 – 71.4)				Stem only Trawscoed 1961-62	
				(13) 76.0* (73.7 – 78.2) 73.0* (70.6 – 75.6)				Whole plant Cut 1 Whole plant Cut 3 } Seale-Hayne 1961-62	
Proximate constituents %DM Crude fibre				(13) 24.8* (21.5 – 26.6)				Stem only Wye 1961-62	
THOUSAND-HEADED unnamed *Digestibility of* DM%	63.9		65.6	60.2				Seedsmen's stock, sample no. 21 Stem only Cambridge 1960-61	
		83.0 75.8 54.8	81.6 70.5 57.1	77.7 66.9 50.9				Leaf only Stem, top Stem, bottom } Trawscoed 1960-61	
				66.5	59.5		53.0	Stem only Cambridge 1961	
				60.7	63.3	59.8		Stem only Sutton Bonington 1961	
				63.2	53.2			Stem only Cambridge 1962	
				68.8* 47.0*				Stem, top Stem, bottom } Trawscoed 1962 * No date	
Proximate constituents %DM Crude protein	12.5		13.5	14.6				Stem only Cambridge 1960-61	
		24.9 17.9 13.0	23.3 17.6 13.7	26.0 19.0 15.4				Leaf only Stem, top Stem, bottom } Trawscoed 1960-61	
				14.2	14.9		10.0	Stem only Cambridge 1961	
				8.6	9.4	7.5		Stem only Sutton Bonington 1961	
Proximate constituents %DM Crude fibre	31.6		30.4.	33.0				Stem only Cambridge 1960-61	
		12.8 26.5 41.9	13.4 27.2 35.8	18.1 26.4 38.0				Leaf only Stem, top Stem, bottom } Trawscoed 1960-61	
				28.3	33.2		38.4	Stem only Cambridge 1961	
				32.6	33.0	35.4		Stem only Sutton Bonington 1961	
				28.9*				* No date Cambridge 1961-62	
THOUSAND-HEADED unnamed *Digestibility of* DM%				(12) 60.6 (54.3 – 68.7)	52.8 (46.8 – 59.8)			* No date Means and ranges of seedsmen's stocks. No. of samples in brackets before mean. Stem only Cambridge 1961-62	
				(12) 71.7* (68.8 – 74.8) 48.5* (45.4 – 54 4)				Stem, top Stem, bottom } Trawscoed 1961-62	
Proximate constituents %DM Crude fibre				(12) 29.7* (24.1 – 34.9)				Cambridge 1961-62	

Table 1.178 Kale, unnamed

	Proximate constituents %DM					Notes	Reference
	CP	EE	NFE	CF	Ash		
	18.0					Mean value for 3 farms	217
	15.4						80
	12.8	2.2	59.9	14.6	10.5		47
	L 17.1 S 14.1					L: Leaf S: Stem	133
	10.3	5.2	55.0	19.0	10.5	Used in studies with dairy cows Means of weekly samples for 5 months	287
	Major elements %DM						
	Ca	P	Mg	K	Na		
	1.5	0.3					217
	L 2.5 S 0.9	0.25 0.32	0.13 0.16	2.4 3.4	0.32 0.30	Means of samples over 3 years	272
	1.7						287
	L S	0.3 0.3		2.3 3.0			133
	Trace elements μg/g DM						
	Boron	Copper	Manganese	Molybdenum	Zinc		
	28						450
	F_0 F_1 F_2 F_3 F_4 F_5 F_6 F_7 F_8 F_9	5 6 4 5 3 4 3 5 3 7	39 44 31 26 26 31 17 44 26 44	0.6 0.8 0.4 0.3 0.6 0.8 0.4 0.4 0.5 0.3	37 46 34 41 28 33 33 38 32 30	F_0: Control F_1: N, Nitrochalk 125 kg/ha F_2: P, superphosphate 62 kg/ha F_3: N + P, as before F_4: K, potassium sulphate 125 kg/ha F_5: N + K F_6: P + K F_7: N + P + K F_8: farmyard manure 37.5 tonnes/ha F_9: N + P + K + farmyard manure All given hydrated lime 3750 kg/ha	551

Table 1.179 Kale, silage

Sample	Proximate constituents %DM					Notes	Reference
	CP	EE	NFE	CF	Ash		
Marrowstem kale	12.3	3.1	45.2	23.6	15.8	With AIV solution	538
(1944) (1945)	13.2 15.1 20.8 19.9	2.8 3.3 3.5 3.4				Concrete silos Cut late November/December; opened January/February following year	225
T E E	12.0 14.4 14.4	2.0 2.9 3.0	52.0 50.1 47.7	22.8 19.1 19.4	11.2 13.5 15.5	T: Tower silo (15 tons) E: Small experimental silo (10 cwt)	238
Kale, unnamed	14.0 15.9 17.5					Values for 3 farms	217
A_0 A_1 A_2	13.6 14.1 15.3	2.3 2.7 2.9	55.0 55.2 55.5	16.5 15.0 13.4	12.6 13.0 12.9	A_0: No additive A_1: 1 kg molasses/110 kg kale A_2: 43 g sulphur dioxide/97 kg kale	47
	Major elements %DM						
	Ca	P					
	1.4 1.1 1.5	0.2 0.2 0.2				Values for 3 farms	217

Table 1.180 Rape and Rape-Type Kale

Variety		Proximate constituents %DM							Notes	Reference
		Crude protein			EE	NFE	CF	Ash		
		Cut 1	Cut 2	Cut 3						
RAPE										
Broad-Leaved Essex	W	21.8	25.8		3.2	52.3	12.7	10.0	Since more than half the information on Rape-type kale is recorded with that on Rape (References 316, 317) all information is so presented here, including that reported with other Kales (Reference 315). Where there were 2 cuts, each at 2 centres, the dates were: (1) 24 November and 2 December, 1952; (2) 30 January and 2 February, 1953. Where there were 3 cuts at 1 centre, the dates were: 2 December 1952, 30 January and 11 March 1953. Copper, iron, lead and zinc were estimated in pooled samples from 2 centres in the winter 1953/54: boron, manganese and molybdenum in samples from 2 of the 3 that gave samples in the winter 1952/53 for proximate constituents and major elements. W : whole plant; L : leaf; S : stem	316
	L	25.2	31.3		4.1	49.3	10.2	11.2		
	S	16.7	19.8		1.9	57.1	16.4	7.9		
NZ Broad-Leaved Essex	W	19.0	23.7		2.9	53.7	15.1	9.3		
	L	23.0	31.0		3.9	50.7	11.3	11.1		
	S	13.1	17.3		1.5	58.2	20.6	6.6		
Giant, Club Root Resistant	W	21.7	25.6		3.2	50.1	14.8	10.2		
	L	27.0	34.7		3.8	46.3	10.8	12.1		
	S	13.3	15.8		2.1	56.4	21.1	7.1		
Giant	W	22.3	28.8		3.0	49.8	14.5	10.4		
	L	26.0	35.0		3.8	47.5	10.5	12.2		
	S	15.8	22.9		1.5	54.1	21.8	6.8		
NZ Giant	W	18.5			3.0	53.3	15.3	9.9		
	L	24.9			3.7	48.6	10.6	12.2		
	S	11.2			1.9	59.4	20.4	7.1		
	W	20.5	24.4	21.6						
	L	24.2	31.1	29.1						
	S	15.7	16.8	16.1						
RAPE-TYPE KALE										
Hungry Gap	W	22.7	26.0		2.8	51.7	13.1	9.7		315
	L	25.8	31.8		3.6	48.6	10.5	11.5		
	S	17.6	20.5		1.6	56.7	17.3	6.8		
	W	23.3	27.8		2.9	51.5	12.1	10.2		316
	L	25.8	33.9		3.6	48.7	10.1	11.8		
	S	18.5	22.2		1.5	56.9	16.1	7.0		
	W	23.6	25.7	25.6						
	L	25.7	33.6	36.2						
	S	19.6	19.7	21.4						
Rape-Kale	W	20.5	23.6		2.7	53.0	13.9	9.9		315
	L	25.6	29.7		3.3	48.6	10.4	12.1		
	S	15.6	18.9		1.9	56.8	19.0	6.7		
	W	21.5	24.9		2.8	52.4	13.0	10.3		316
	L	24.8	31.1		3.2	49.7	10.2	12.1		
	S	15.5	19.8		2.1	57.6	17.9	6.9		
	W	21.2	21.9	23.8						
	L	23.5	29.4	33.5						
	S	15.8	16.5	19.3						

Table 1.180 Rape and Rape-Type Kale (continued)

		Major elements %DM								Notes	Reference
		Ca	P	Mg	K	Na	Cl	S			
								Total	Sulphate		
RAPE											315
Broad-Leaved Essex	W	0.94	0.44	0.19	2.40	0.22	1.15	0.74	0.35		
	L	1.23	0.49	0.21	2.74	0.18	1.44	0.76	0.35		
	S	0.46	0.36	0.18	1.87	0.29	0.69	0.73	0.37		
NZ Broad-Leaved Essex	W	0.63	0.39	0.19	2.32	0.25	1.05	0.69	0.36		
	L	0.81	0.44	0.20	2.76	0.21	1.34	0.71	0.37		
	S	0.37	0.32	0.17	1.68	0.31	0.60	0.65	0.35		
Giant, Club Root Resistant	W	0.56	0.44	0.19	2.67	0.20	1.16	0.74	0.39		
	L	0.70	0.49	0.20	3.14	0.13	1.43	0.75	0.36		
	S	0.34	0.36	0.18	1.93	0.30	0.74	0.73	0.42		
Giant	W	1.22	0.43	0.22	2.68	0.24	1.36	0.69	0.33		
	L	1.54	0.48	0.24	3.02	0.19	1.74	0.74	0.34		
	S	0.54	0.35	0.19	2.00	0.33	0.59	0.62	0.31		
NZ Giant	W	1.16	0.38	0.22	2.31	0.27	1.22	0.62	0.35		
	L	1.60	0.45	0.25	2.63	0.18	1.60	0.74	0.43		
	S	0.54	0.31	0.18	1.95	0.36	0.78	0.52	0.28		
RAPE-TYPE KALE	W	0.90	0.47	0.21	2.54	0.18	1.21	0.73	0.38		
Hungry Gap	L	1.13	0.50	0.22	2.86	0.13	1.64	0.79	0.42		
	S	0.46	0.42	0.19	2.01	0.25	0.47	0.64	0.33		
	W	1.06	0.45	0.21	2.69	0.19	1.34	0.77	0.39		316
	L	1.34	0.47	0.22	3.00	0.16	1.77	0.82	0.43		
	S	0.52	0.41	0.19	2.09	0.26	0.48	0.69	0.34		
Rape-Kale	W	0.90	0.45	0.22	2.55	0.24	1.26	0.79	0.40		315
	L	1.18	0.50	0.24	2.99	0.20	1.73	0.91	0.50		
	S	0.46	0.38	0.19	1.93	0.29	0.48	0.66	0.29		
	W	1.10	0.43	0.22	2.71	0.22	1.39	0.77	0.41		316
	L	1.39	0.47	0.24	3.07	0.20	1.84	0.85	0.48		
	S	0.52	0.37	0.18	2.12	0.24	0.49	0.65	0.29		

		Trace elements μg/g DM							Notes	Reference
		B	Cu	Fe	Pb	Mn	Mo	Zn		
RAPE										317
Broad-leaved Essex	W	31.7	4.8	103	Trace	41.8	0.4	22.3		
	L	37.0	5.9	161	0.4	57.0	0.5	19.0		
	S	24.0	4.0	56	Trace	19.1	0.2	25.0		
NZ Broad-Leaved Essex	W	29.5				40.3	0.4			
	L	34.5				54.0	0.6			
	S	22.0				20.5	0.2			
Giant. Club Root Resistant	W	25.3				42.7	0.5			
	L	31.0				58.0	0.7			
	S	16.5				18.6	0.2			
Giant	W	30.6	3.6	95	Trace	50.9	0.4	20.9		
	L	35.0	5.3	162	0.4	70.0	0.6	18.6		
	S	22.5	2.7	51	Trace	18.1	0.2	22.5		
NZ Giant	W	30.0				45.6	0.4			
	L	35.7				64.0	0.6			
	S	23.2				17.6	0.2			
RAPE-TYPE KALE	W	31.2	4.2	117	0.6	44.3	0.5	24.9		
	L	35.5	6.0	190	0.8	56.0	0.7	20.2		
Hungry Gap	S	22.7	2.8	56	0.4	20.3	0.2	28.6		
Rape-Kale	W	33.5	4.9	85	Trace	42.5	0.4	26.8		
	L	37.0	6.7	150	0.6	55.0	0.6	21.4		
	S	27.7	3.6	41	Trace	19.0	0.1	30.0		

Table 1.181 Rape. Digestibility and crude fibre content

Variety	Digestible DM%	CF %DM	Notes	Reference
Early Giant	(1) 73.3 (2) 74.2 (3) 80.2	(2) 18.8	(1) Whole plant, Sutton Bonington (2) Whole plant, Trawscoed (3) Stem only, Cockle Park A study at 3 centres of the relation between the digestibility of DM *in vitro* of Rape and its crude fibre content. Digestible DM% and the correlation coefficients are shown for each centre. Values for crude fibre are given only for (2) Trawscoed. There were 12 varieties. For 9 of them 1 value only from each centre is given, each value the mean for 4 to 22 samples. Of the remaining 3, there are 8 values for Giant, 2 for Broad-Leaved Essex and 4 for Dwarf Essex at each centre. Only the general means are given here, the total number of samples, in the same order, being 113, 21 and 12. The weighted mean for digestible DM in Giant is 73.4, the general mean 74.0. Correlation coefficients for crude fibre content and digestible dry matter % are given as: (1) −0.881; (2) −0.899; (3) −0.945.	168
Giant	(1) 75.1 (2) 74.0 (3) 81.2	(2) 19.2		
Giant Broad-Leaved	(1) 71.8 (2) 72.3 (3) 77.8	(2) 21.4		
Giant Essex	(1) 78.2 (2) 72.9 (3) 77.8	(2) 20.7		
Essex	(1) 74.7 (2) 74.5 (3) 81.8	(2) 18.5		
Broad-Leaved Essex	(1) 77.3 (2) 75.0 (3) 82.9	(2) 18.1		
Dwarf Essex	(1) 77.5 (2) 74.7 (3) 85.6	(2) 16.2		
Late Dwarf	(1) 77.8 (2) 76.4 (3) 83.6	(2) 17.3		
Leafy	(1) 78.1 (2) 75.6 (3) 82.6	(2) 17.9		
Clubroot resistant	(1) 76.7 (2) 73.7 (3) 82.6	(2) 19.1		
Radish	(1) 63.0 (2) 72.8 (3) 78.4	(2) 24.4		
Rapeseed	(1) 74.3 (2) 74.3 (3) 80.6	(2) 19.0		

Table 1.182 Cattle Cabbage

Variety		CP	EE	NFE	CF	Ash	Notes	Reference
		Proximate constituents %DM					N_0: No extra nitrogen; N_1: Ammonium sulphate, 750 kg/ha	
Flatpoll type	N_0	14.0	2.1	62.8	10.7	10.4	First harvest, November 1948-49	127
	N_1	19.8	1.6	57.8	10.5	10.3		
	N_0	17.5					Means of 4 harvests, November, December, January, February	
	N_1	23.1						
Giant ox-drumhead	N_0	28.3	1.8	–	–	14.3	First harvest, November 1949-50	
	N_1	30.7	1.9	40.7	12.8	13.9		
	N_0	31.2					Means of 2 harvests, November, December	
	N_1	33.3						
Purple drumhead	N_0	15.1	2.8	62.5	9.5	10.1	First harvest, November 1950-51	
	N_1	17.4	2.5	60.1	9.4	10.6		
	N_0	17.9					Means of 3 harvests, November, January, February	
	N_1	19.5						
	N_0	15.3	2.5	61.5	9.8	10.9	First harvest, November 1951-52	
	N_1	19.1	2.4	58.3	9.7	10.5		
	N_0	15.6					Means of 3 harvests, November, December, January	
	N_1	20.6						
Purple flat-poll		18.7 (17.8 – 20.5)	1.7 (1.6 – 1.8)	53.3 (51.1 – 55.8)	13.3 (12.1 – 14.6)	13.0 (12.1 – 14.0)	Means and ranges for 5 harvests, 2 in November, 2 in December, 1 in January. Part of a complex manurial trial, 10 treatments, 5 harvests, for 7 years	319

Variety	Ca	P	Mg	K	Na	Cl	S total	S sulphate	Notes
	Major elements %DM								
Purple flat-poll	2.23 (2.05–2.58)	0.26 (0.05–0.29)	0.13 0.14	2.22 (2.05–2.40)	0.53 (0.50–0.57)	0.41 (0.37–0.48)	0.90 (0.89–0.92)	0.56 (0.52–0.60)	Means and ranges for 5 harvests, as above, except for Mg, harvests of Nov. and January only

Table 1.183 Mangold, Fodder Beet and Sugar Beet

Variety		CP	EE	NFE	CF	Ash		Notes	Reference
		Proximate constituents %DM							
MANGOLD									
Yellow Globe		10.0	0.4	75.2	5.3	9.1	Root 1947	Harvested Nov. 6- 14, 1947	130
	N_0	7.0	0.3	76.6	5.8	10.3	Root 1948	N_0: No fertiliser; N_1: superphosphate 375 kg/ha in 1948 and 1949 plus muriate of potash 125 kg/ha in 1949	
	N_1	11.0	0.4	70.8	6.0	11.8			
	N_0	14.5	2.6	48.8	9.9	24.2	Tops 1948		
	N_1	15.8	2.5	49.0	10.2	22.5			
	N_0	20.2	0.6	53.1	6.9	19.2	Root 1949	Harvested Nov. 5–11, 1948; Nov. 8, 1949	
	N_1	21.5	0.6	50.3	7.2	20.4			
	N_0	21.2					Tops 1949		
	N_1	21.6							
	N_0	6.2	0.3	76.4	5.5	11.6	Root 1950	N_2: superphosphate 500 kg/ha and muriate of potash 250 kg/ha in 1950; superphosphate 375 kg/ha in 1951	131
	N_2	11.8	0.4	69.9	6.1	11.8			
	N_0	12.9					Tops 1950		
	N_2	17.7							
	N_0	7.6	0.4	76.1	5.5	10.4	Root 1951	Harvested Nov. 2, 3 and 4, 1950 and Oct. 29, 1951	
	N_2	13.4	0.4	68.4	6.2	11.6			
	N_0	15.2					Tops 1951		
	N_2	20.1							
		18.6	0.8	42.0	8.6	30.0			34
Kirches Ideal		9.8	0.3	76.0	5.0	8.9		See also: Proximate constituents and ME for Mangold, unnamed, table 1.235, Reference 538	130
Mammoth Long Red		10.4	0.5	75.3	5.4	8.4			
Sugar Mangold		9.7	0.5	76.9	5.4	7.5			

continued

Table 1.183 Mangold, Fodder Beet and Sugar Beet (continued)

MANGOLD (cont'd)		Carbohydrate fractions %DM				
		Sugar		Cellulose		
Yellow Globe		9.1	Root 1947		See Proximate constituents, same reference	130
	N_0	4.3	Root 1948			
	N_1	4.3				
	N_0	3.8	Root 1949			
	N_1	3.7				
	N_0	6.0	Root 1950		See Proximate constituents, same reference	131
	N_2	6.2				
	N_0	7.2	Root 1951			
	N_2	6.1				
				10.3		34
Kirches Ideal		10.0				130
Mammoth Long Red		9.8				
Sugar Mangold		10.4				

FODDER BEET		Proximate constituents %DM							
		CP	EE	NFE	CF	Ash			
Barres Øtofte X	N_0	5.1	0.3	81.9	5.2	7.5	Root 1950	See Yellow Globe Mangold, same reference	131
	N_2	9.0	0.3	77.7	5.3	7.7			
	N_0	13.3					Tops 1950		
	N_2	15.8							
	N_0	7.8	0.3	79.0	5.5	7.4	Root 1951		
	N_2	9.7	0.3	76.4	5.7	7.9			
	N_0	17.9					Tops 1951		
	N_2	20.2							
Øtofte		6.4	0.3	85.2	3.8	4.3	Root 1947		130
	N_0	4.9	0.2	82.9	4.8	7.2	Root 1948		
	N_1	7.2	0.3	79.7	4.7	8.1			
	N_0	13.4	2.8	50.7	10.9	22.2	Tops 1948		
	N_1	16.4	2.7	54.0	9.7	17.2			
	N_0	10.9	0.3	73.6	5.2	10.0	Root 1949		
	N_1	11.4	0.2	73.5	5.2	9.7			
	N_0	19.2					Tops 1949		
	N_1	20.9							
Red Øtofte X	N_0	4.4	0.2	84.7	4.6	6.1	Root 1950		131
	N_2	8.1	0.3	80.9	4.5	6.2			
	N_0	12.8					Tops 1950		
	N_2	14.5							
	N_0	6.2	0.3	82.7	4.7	6.1	Root 1951		
	N_2	8.9	0.3	79.1	4.7	7.0			
	N_0	14.3					Tops 1951		
	N_2	19.5							
Pajbjerg Rex IX		6.7	0.3	84.5	4.1	4.4	Root 1947		130
	N_0	4.9	0.3	82.3	4.9	7.6	Root 1948		
	N_1	7.7	0.2	80.1	4.8	7.2			
	N_0	14.2	3.1	54.0	9.6	19.1	Tops 1948		
	N_1	17.2	2.5	52.7	9.7	17.9			
	N_0	12.3	0.4	72.0	5.2	10.1	Root 1949		
	N_1	12.5	0.3	69.8	5.6	11.8			
	N_0	20.1					Tops 1949		
	N_1	21.5							
Pajbjerg Rex X	N_0	4.3	0.2	84.7	4.6	6.2	Root 1950		131
	N_2	7.2	0.3	81.0	4.6	6.9			
	N_0	13.0					Tops 1950		
	N_2	15.6							
	N_0	6.1	0.3	80.9	4.8	7.9	Root 1951		
	N_2	8.6	0.3	78.6	5.0	7.5			
	N_0	15.1					Tops 1951		
	N_2	19.9							

Table 1.183 Mangold, Fodder Beet and Sugar Beet (continued)

Variety		Proximate constituents %DM						Notes	Reference
		CP	EE	NFE	CF	Ash			
Pajbjerg Rex X (cont'd)		8.4 (7.8 – 9.2)	0.1 (0.1 – 0.2)	77.4 (75.2 – 79.2)	5.9 (4.9 – 8.0)	8.2 (6.9 – 9.6)	Root	From Clamp, October to April	125
		26.0 (22.1 – 29.9)					Tops		
		Carbohydrate fractions %DM							
		Sugar							
Barres Øtofte X	N_0	10.0	Root 1950					See Yellow Globe Mangold, same reference	131
	N_2	9.2							
	N_0	10.2	Root 1951						
	N_2	9.7							
Øtofte		17.0	Root 1947						130
	N_0	15.7	Root 1948						
	N_1	14.7							
	N_0	11.2	Root 1949						
	N_1	11.6							
Red Øtofte X	N_0	11.2	Root 1950						131
	N_2	11.6							
	N_0	13.2	Root 1951						
	N_2	11.9							
Pajbjerg Rex IX		15.7	Root 1947						130
	N_0	14.8	Root 1948						
	N_1	14.4							
	N_0	9.8	Root 1949						
	N_1	9.9							
Pajbjerg Rex X	N_0	15.9	Root 1950						131
	N_2	14.9							
	N_0	13.0	Root 1951						
	N_2	12.2							
		Ash	Ca	P					
SUGAR BEET		7.09 (5.59 – 8.15)	0.58 (0.42 – 0.84)	0.07 (0.06 – 0.07)				Pulp, molassed. Means and ranges of 244 samples from 18 factories over 2 seasons	193

Table 1.184 Turnip and Swede

TURNIP		Proximate constituents %DM					Notes	Reference
		CP	EE	NFE	CF	Ash		
Swede turnip		8.0	0.6	76.9	8.8	5.7	See also: Proximate constituents and ME for Swede turnip, table 1.235, Reference 336	400

		Major elements %DM						Notes	Reference
		Ca	P	Mg	K	Na			271
Turnip, unnamed	No	0.45	0.20	0.08	2.35	0.14	Root	Means of 4 plot experiments, 2 in 1955 and 2 in 1956. Plot size was 0.01 acre, 4 m^2, and fertilisers were given at the rates below: No: no fertiliser; N1: salt, 562 kg/ha; N2: muriate of potash, 187 kg/ha; N3: superphosphate, 562 kg/ha; N4: sulphate of ammonia, 250 kg/ha	
	N1	0.45	0.23	0.08	2.56	0.20	Root		
	N2	0.44	0.13	0.08	2.54	0.11	Root		
	N3	0.47	0.24	0.08	2.34	0.15	Root		
	N4	0.43	0.22	0.08	2.35	0.15	Root		
	No	1.96	0.17	0.17	2.73	0.28	Top		
	N1	1.89	0.18	0.17	2.79	0.44	Top		
	N2	1.89	0.19	0.16	3.09	0.22	Top		
	N3	1.90	0.20	0.18	2.84	0.34	Top		
	N4	2.00	0.18	0.17	2.67	0.30	Top		

SWEDE, unnamed				Notes	Reference
Proximate constituents %DM					400
Crude protein			8.0		
Ether extract			0.6		
Nitrogen free extract			76.9		
Crude fibre			8.8		
Ash			5.7		
Crude protein	L	1960	8.5	L: large roots S: small roots For metabolisable energy see table 1.235	336
	L	1961	12.0		
	S	1960	7.1		
	S	1961	12.3		
Ether extract	L	1960	0.6		
	L	1961	0.7		
	S	1960	0.6		
	S	1961	0.6		
Nitrogen-free extract	L	1960	73.5		
	L	1961	69.8		
	S	1960	74.2		
	S	1961	69.1		
Crude fibre	L	1960	11.2		
	L	1961	10.7		
	S	1960	12.7		
	S	1961	11.4		
Ash	L	1960	6.3		
	L	1961	6.9		
	S	1960	5.4		
	S	1961	6.7		

Table 1.185 Potato

Variety or State	Proximate constituents %DM		Notes	Reference
	CP	Ash		
Gladstone, raw	2.9 (2.6 – 3.5)	1.0 (0.7 – 1.1)	Mean and range for 7 trials with increasing proportion of K in NPK fertiliser on K-deficient chalk soil. Increasing K depressed N in tuber.	32
Majestic, raw	2.4 (2.1 – 2.8)	1.0 (0.9 – 1.1)	Mean and range for 9 samples grown on sand or warp. Vitamin C in mg/g: 0.12 (0.11 – 0.13)	
			See also: Proximate constituents and ME in raw and processed potato, see table 1.235. Reference 261 and 538	

Variety or State	Organic acids %DM and pH				Notes	Reference
Unnamed, silage	Acetic	Butyric	Lactic	pH	Experimental silos P: pulped; C: chopped; S: steamed a: with added commercial preparation of phosphoric acid (Phossilan) For proximate constituents and ME see table 1.235, Reference 538	385
P+a	3.1	2.6	1.3	4.9		
P	4.8	1.4	9.4	4.3		
P	2.7	2.3	1.0	5.0		
C	1.5	1.4	4.2	4.6		
S	3.6	0.3	9.3	4.0		
S	2.8	1.3	9.2	4.4		

	Potato peeled, boiled	Potato protein	Notes	Reference
Amino acids g/16gN			Protein heat-coagulated, separated by centrifuge *Extrapolated to zero time of hydrolysis	292
Alanine	2.4	4.7		
γ-Aminobutyric acid	1.3	0		
Arginine	4.6	*5.5		
Aspartic acid	22.7	13.0		
Cystine	1.0	1.6		
Glutamic acid	23.8	11.3		
Glycine	1.9	4.9		
Histidine	1.8	2.4		
Isoleucine	2.4	6.8		
Leucine	3.7	11.1		
Lysine	4.8	8.3		
Methionine	1.3	*2.8		
Phenylalanine	3.4	6.2		
Proline	2.1	5.1		
Serine	2.8	*5.8		
Threonine	3.2	*5.7		
Tryptophan	1.1	1.8		
Tyrosine	3.3	6.1		
Valine	4.1	8.0		

Table 1.186 Miscellaneous fodders

Type and State	Proximate constituents %DM					Notes	Reference
	CP	EE	NFE	CF	Ash		
Lupin, silage	19.0	7.6	25.0	35.9	12.5		393
Lupin, Whole	20.1					Crop grown in green manuring experiments with kale	353
Top	22.4						
Root	11.8						
Beans, Peas and Tares	21.9						457
Pea pods	14.8	2.0	53.4	23.4	6.4	Fresh pods from canning factory	564
	12.8	3.7	46.6	30.4	6.5	Silage. Digestibility by sheep of DM% 68.2; of OM% 72.3	
Pea haulms and pods	14.8	6.1	36.5	25.3	17.3	Silage, molasses added. Digestibility by sheep of DM% 58.3; of OM% 65.8	

	Carbohydrate fractions %DM		Major elements %DM			Notes
	Invert sugar	Sucrose	Ca	P	Cl	
Pea pods	13.5	8.5	1.4	0.2	0.1	Fresh pods from canning factory
			1.3	0.2	0.1	Silage
Pea haulms and pods			1.6	0.3	0.2	Silage. molasses added

Table 1.187 Horsetail: *Equisetum*

	May	June	July	August	Notes	Reference
Corn horsetail *(E. arvense)*						
Proximate constituents %DM						
Crude Protein				8.2	1 Sample; mature	220
Ether extract				1.9	Sampled 1952	
Crude fibre				16.4		
Nitrogen-free extract				46.8		
Ash				26.7		
Major elements %DM						
Calcium				3.68		
Phosphorus				0.16		
Magnesium				0.41		
Potassium				1.22		
Sodium				0.23		
Chloride				1.24		
Trace elements μg/g DM						
Cobalt				0.27		
Copper				10.8		
Iron				478		
Manganese				84		
Great water horsetail *(E. maximum)*						
Ash %DM				19.7	Sampled 1955	
Marsh horsetail *(E. palustre)*						
Ash %DM				18.2		

Table 1.188 Bracken: *Pteris aquilina (Pteridium aquilinum)*

For 10 studies of Bracken, fresh and as hay and silage, proximate constituents, digested components and metabolisable energy see table 1.237

	January			February			March			April			May			June		
Proximate constituents %DM Crude protein															P(M) P(E) L(M) L(E)	21.4 25.2 	17.7 16.1 21.3 20.4	13.3 14.8 16.6 17.1
															L(G)		17.4	
															S(M) S(E)	15.0 	9.3 10.2	4.8 5.5
															S(G)		8.7	
													 L(P)(0)	L(P) (1) 26.94	20.69 (1)20.75	(2) (2)	16.4 17.1	 (3)
	R(P)(0) 8.44									(0)	7.12			(1)	8.50	(2)	7.1	
Ether extract															P(M) P(E) L(M) L(E)	2.1 2.5 	2.3 1.8 2.8 2.4	1.9 1.7 2.3 2.0
															L(G)		2.8	
															S(M) S(E)	1.4 	0.9 0.9	0.8 0.5
															S(G)		0.9	
Nitrogen-free extract															L(G) S(G)		54.0 33.9	
Crude fibre															L(G) S(G)		18.3 49.4	
Ash															P(M) P(E) L(M) L(E)	7.9 8.4 	5.6 6.5 5.7 6.5	5.6 6.4 6.0 6.6
															L(G)		7.5	
															S(M) S(E)	7.1 	5.2 6.3	4.4 5.6
															S(G)		7.2	
	R(P)(0) 6.6									(0)	6.5		 L(P)(0) 	L(P) (1) 11.2 (1)	8.4 (1)8.5 6.7	(2) (2) (2)	7.0 7.7 8.0	 (3)

July			August			September			October			November			December			Notes	Reference
	10.3 11.1 12.5 13.2		10.6 8.9 12.9 10.9	8.6 7.4 10.7 8.9	8.2 8.2 10.3 9.9		6.8 6.1 8.4 7.3	6.3 5.5 7.9 6.4		5.1 3.5 6.5 4.1	3.9 3.1 5.1 3.7	3.9 3.0 5.0 3.8						P : Whole plant L : Leaf S : Stem R : Rhizome Area: (M) Berkshire, Maidenhead (E) Berkshire, Easthampstead (G) Selkirk, Galashiels (A1)Aberdeenshire (good) (A2)Aberdeenshire (poor) (P) Perthshire Rhizome: branches (l) leaders (s) secondaries (t) tertiaries Leaf (frond) : stage of growth (0) Pinnae not unfolded (1) 1 pair unfolded (2) 2 pairs unfolded (3) 7 pairs unfolded (4) Fully expanded, many tips brown (5) 1 pair pinnae brown (6) Many pinnae brown (7) Almost completely brown (8) Completely brown	227
16.0		15.6	14.1		12.5		8.1	8.7	6.3										365
	2.8 3.6		2.9 2.3	2.2 2.0	2.3 1.9		2.0 1.8	2.0 1.8		2.0 1.5	1.2 1.6	1.7 1.3							227
5.6		2.6	2.0		1.3		1.0		2.4										365
(3) 14.1	14.8	(4)12.1 (4)	(4)13.2 11.7	(5)	12.6	 (6)	(7) 9.2	8.8 (7)	 3.8	(8)	6.3								304
(3)	5.6	(4)6.8	(4)6.6	(5)	6.7		(7)	6.0		(8)	5.1					(8)	7.0*		
				Fr Pi	10.7 14.9			R(A1) R(A2)	(l) (s) (t) (l) (s) (t)	6.5 6.4 5.9 5.7 5.4 5.5									
	2.0 1.3 2.4 1.4		0.9 1.1 1.1 1.3	1.0 1.0 1.2 1.2	1.3 0.9 1.6 1.1		1.0 0.6 1.2 0.7	1.5 1.2 1.8 1.4		1.6 1.5 2.1 1.8	1.2 1.0 1.6 1.3	1.5 1.7 1.9 2.2						*The stages of growth refer to the fronds of the plants from which the rhizomes came. Fronds (Fr) and Pinnules (Pi) analysed separately: means for 8 areas (10 including sub-areas) in Scotland Some values for NFE have been adjusted on the assumption that the other values, as published, are correct.	227
1.4		2.8	1.5		1.3		1.3	2.2	1.1										365
	0.8 0.8		0.4 0.3	0.4 0.3	0.5 0.3		0.2 0.2	0.5 0.5		0.5 0.6	0.2 0.2	0.7 0.6							227
1.2		0.6	0.7		0.5		0.6		0.8										365
59.3 33.0		56.9 31.6	50.4 32.4		54.7 34.5		54.9 31.6	50.5	54.7 31.3										
17.2 54.7		16.5 60.0	26.7 59.3		23.9 57.7		29.6 60.0	33.2	32.8 59.7										
	5.4 5.8 5.8 6.1		6.1 6.4 6.3 6.9	5.8 6.4 6.2 6.9	6.6 8.2 7.2 8.9		6.0 6.5 6.2 6.9	6.3 6.9 6.6 7.5		6.1 7.0 6.8 7.7	4.7 5.3 5.0 5.9	4.4 4.7 5.0 5.5							227
6.1		8.2	7.3		7.6		6.1	5.4	5.2										365
	3.9 4.9		5.4 4.8	4.6 4.8	5.2 5.8		5.2 4.9	5.3 5.0		4.4 5.0	3.9 3.7	3.2 2.8							227
5.5		5.2	5.7		6.0		6.8		5.8										365
(3) 5.6 (3)	6.2 7.0	(4)6.0 (4) (4)7.2	(4)6.5 5.2 (4)5.7	(5) (5)	6.3 6.2	 (6)	(7) (7)	6.8 (7) 5.9	 4.5	(8) (8)	5.8 6.4					 (8)	 6.2		304
								R(A1) R(A2)	(l) (s) (t) (l) (s) (t)	4.6 4.2 3.9 6.6 4.4 2.9									

continued

Table 1.188 Bracken: (continued)

	January			February			March			April			May			June		
Carbohydrate fractions %DM																		
Total sugars															P(M)	5.47	7.39	8.84
															P(E)		5.85	6.91
															L(M)	4.67	7.50	8.96
															L(E)		5.08	7.08
															S(M)	6.81	7.13	8.54
															S(E)		6.88	6.25
Reducing sugars															P(M)	4.94	6.84	8.36
															P(E)		5.26	5.40
															L(M)	3.83	6.88	8.50
															L(E)		4.33	5.25
															S(M)	6.81	6.75	8.00
															S(E)		6.50	6.00
Starch															P(M)	3.02		
															L(M)	3.15		
															S(M)	2.81		
True cellulose															P(M)	20.44	23.47	25.77
															P(E)		24.08	23.77
															L(M)	15.23	18.07	21.79
															L(E)		17.04	20.28
															S(M)	29.20	36.12	36.07
															S(E)		33.48	37.53
Pentosan in cellulose															P(M)	3.31	4.28	5.41
															P(E)		4.39	4.15
															L(M)	2.22	2.92	4.04
															L(E)		2.27	3.29
															E(M)	5.14	7.48	8.95
															S(E)		7.22	7.52
Furfural															P(M)	2.21	2.04	1.76
															P(E)		2.08	2.44
															L(M)	1.81	1.71	1.51
															L(E)		1.67	2.05
															S(M)	2.89	2.81	2.39
															S(E)		2.62	3.96
Lignin															P(M)	19.87	20.10	26.45
															P(E)		22.48	28.21
															L(M)	21.72	20.35	27.51
															L(E)		23.75	28.56
															S(M)	16.75	19.53	23.72
															S(E)		20.79	26.84
Undetermined															P(M)	17.29	17.19	10.97
															P(E)		16.81	11.71
															L(M)	18.23	19.58	11.20
															L(E)		20.77	13.09
															S(M)	15.70	11.56	10.37
															S(E)		11.51	6.30

July			August			September			October			November			December			Notes	Reference
	10.17		10.18	10.00	7.26		6.81	8.60		4.37	2.70	0.88							227
	10.09		9.70	7.03	7.49		7.78	8.37		2.75	3.83	1.33							
	10.00		9.75	10.00	5.83		5.67	7.72		2.80	3.62	0.54							
	10.50		9.50	6.50	6.75		6.92	7.81		1.70	3.71	0.50							
	10.75		11.63	10.00	11.33		10.38	10.94		8.08	0.60	1.50							
	8.63		10.38	8.92	10.17		10.88	10.50		5.92	4.17	3.09							
	8.94		9.53	5.85	6.28		6.54	8.40		3.80	2.53	0.79							
	9.09		8.25	4.98	5.89		6.93	7.05		2.52	3.76	1.33							
	8.63		9.13	5.00	4.70		5.44	7.45		2.19	3.38	0.52							
	9.25		7.81	4.19	4.75		5.94	6.15		1.48	3.63	0.50							
	10.00		10.88	8.50	10.80		10.00	10.94		7.60	0.60	1.27							
	8.50		9.73	7.75	10.00		10.50	10.50		5.63	4.13	3.09							
												2.31							
												2.50							
												1.88							
	25.68		23.23	26.30	27.55		26.29	26.19		27.73	30.99	33.91							
	25.05		25.41	26.68	27.53		25.12	27.99		29.00	30.33	34.60							
	21.81		20.32	22.85	23.84		23.57	23.15		23.84	27.58	29.34							
	21.69		22.39	23.71	25.01		22.59	25.48		26.18	27.40	29.96							
	38.72		32.95	37.04	38.10		34.86	34.34		36.89	38.79	42.20							
	37.07		35.43	37.18	36.58		34.19	37.60		37.47	38.62	44.52							
	5.19		4.26	4.85	5.20		4.24	4.93		3.80	5.03	6.22							
	5.15		5.15	5.05	5.20		4.02	5.13		4.50	5.28	6.16							
	4.29		3.51	3.95	4.31		3.73	3.98		2.80	4.31	5.31							
	4.06		4.26	4.09	4.49		3.20	4.46		3.75	4.55	5.26							
	8.20		6.75	7.64	7.75		5.84	7.48		6.15	6.68	7.86							
	9.06		8.12	8.45	7.77		6.95	7.67		6.77	7.35	8.08							
	2.06		2.29	2.44	2.18		3.08	2.62		2.97	3.64	3.11							
	2.00		2.29	2.52	2.50		2.61	2.63		3.08	3.89	2.88							
	1.87		2.05	2.16	1.89		2.93	2.32		2.65	3.06	2.50							
	1.89		1.95	2.24	2.15		2.29	2.38		2.59	3.34	2.02							
	2.70		3.11	3.30	2.99		3.57	3.44		3.73	5.00	4.23							
	2.41		3.42	3.51	3.76		3.78	3.59		4.55	5.46	4.72							
	26.27		27.05	26.72	26.83		28.29	26.65		29.79	33.80	33.99							
	26.30		26.60	27.42	27.36		29.35	27.83		30.52	34.67	34.84							
	27.24		27.71	27.68	27.60		29.62	27.63		31.28	35.80	36.23							
	26.83		26.88	27.92	27.90		30.67	28.39		31.79	36.89	37.32							
	23.00		24.85	23.73	24.65		24.12	24.03		26.29	29.24	29.92							
	24.42		25.67	25.65	25.40		24.59	25.70		26.69	28.37	29.55							
	12.93		15.39	14.21	14.85		17.52	17.00		18.50	14.15	12.18							
	13.19		14.53	16.49	12.61		17.96	14.14		18.20	12.52	10.81							
	14.08		16.39	15.22	17.50		18.70	18.89		21.25	14.01	14.20							
	14.31		15.99	18.53	13.82		19.41	16.21		20.38	13.22	13.40							
	9.09		12.03	11.06	7.28		13.75	11.90		12.00	14.38	8.53							
	9.13		9.64	9.22	8.26		12.76	7.71		11.67	10.53	5.27							

continued

Table 1.188 Bracken (continued)

	January			February			March			April			May			June		
Major elements %DM																		
Calcium															P(M)	0.19	0.29	0.34
																P(E)	0.24	0.34
															L(M)	0.22	0.36	0.43
																L(E)	0.33	0.39
															L(G)	(1947)	0.29	
															S(M)	0.12	0.11	0.12
																S(E)	0.13	0.15
															S(G)	(1947)	0.18	
														L(P)(1)	0.11	(2)	0.15	
													L(P)(0)	0.15(1)	0.16	(2)	0.20	(3)
	R(P)(0)	0.16								(0)	0.19			(1)	0.28	(2)	0.37	
Phosphorus															P(M)	0.44	0.27	0.21
																P(E)	0.31	0.26
															L(M)	0.54	0.30	0.25
																L(E)	0.39	0.29
															L(G)	(1947)	0.39	
															S(M)	0.27	0.18	0.09
																S(E)	0.22	0.14
															S(G)	(1947)	0.20	
														L(P)(1)	0.41	(2)	0.13	
													L(P)(0)	0.64	(1)0.50	(2)	0.35	(3)
	R(P)(0)	0.07								(0)	0.06			(1)	0.06	(2)	0.05	
Potassium															P(M)	3.14	2.14	1.71
																P(E)	2.67	2.17
															L(M)	2.97	2.33	1.61
																L(E)	2.49	2.15
															L(G)	(1947)	2.32	
															S(M)	3.24	2.07	1.74
																S(E)	2.90	2.29
															S(G)	(1947)	2.90	
														L(P)(1)	3.64	(2)	2.81	
													L(P)(0)	4.65	(1)3.72	(2)	2.88	(3)
	R(P)(0)	1.33								(0)	1.24			(1)	1.40	(2)	1.24	
Sodium															P(M)	0.27	0.14	0.07
																P(E)	0.19	0.06
															L(M)	0.29	0.13	0.08
																L(E)	0.19	0.06
															S(M)	0.24	0.18	0.06
																S(E)	0.19	0.07
															L(P)	(2)	0.08	
													L(P)(0)	0.06	(1)0.06	(2)	0.13	(3)
	R(P)(0)	0.24								(0)	0.20			(1)	0.22	(2)	0.25	

July			August			September			October			November			December			Notes	Reference
	0.36 0.36 0.42 0.43		0.44 0.56 0.52 0.68	0.44 0.56 0.54 0.66	0.51 0.65 0.63 0.77		0.54 0.65 0.63 0.77	0.55 0.69 0.66 0.81		0.64 0.86 0.83 1.08	0.57 0.76 0.73 0.95	0.56 0.68 0.76 0.90							227
0.29		0.46	0 54		0.54		0.60	0.77	0.60										365
	0.14 0.14		0.15 0.15	0.14 0.20	0.16 0.20		0.23 0.20	0.25 0.23		0.19 0.23	0.21 0.20	0.19 0.20							227
0.16		0.20	0.22		0.29		0.29		0.21										365
(3)	0.19	(4)0.28	(4)0.29	(5)	0.35		(7)	0.41		(8)	0.40								304
0.25		(4)	0.37			(6)	0.51	(7)	0.50										
				Fr	0.22														
				Pi	0.29														
(3)	0.35	(4)0.26	(4)0.18	(5)	0.17		(7)	0.18		(5)	0.15					(8)	0.18		
								R(A1)	(l)	0.30									
									(s)	0.26									
									(t)	0.16									
								R(A2)	(l)	0.20									
									(s)	0.09									
									(t)	0.04									
	0.16 0.17 0.18 0.19		0.13 0.13 0.16 0.15	0.15 0.11 0.19 0.13	0.13 0.11 0.17 0.13		0.13 0.09 0.16 0.10	0.10 0.11 0.12 0.13		0.08 0.05 0.10 0.06	0.06 0.04 0.08 0.04	0.06 0.03 0.07 0.04							227
0.32		0.36	0.35		0.20		0.16	0.17	0.08										365
	0.07 0.10		0.06 0.07	0.05 0.05	0.04 0.04		0.04 0.06	0.03 0.03		0.03 0.02	0.01 0.02	0.02 0.01							227
0.12		0.10	0.06		0.07		0.05		0.09										365
(3)	0.19	(4)0.15	(4)0.12	(5)	0.10		(7)	0.09		(8)	0.10								304
0.23		(4)	0.14			(6)	0.10	(7)	0.10										
				Fr	0.14														
				Pi	0.20														
(3)	0.05	(4)0.05	(4)0.05	(5)	0.06		(7)	0.06		(8)	0.06					(8)	0.06		
								R(A1)	(l)	0.035									
									(s)	0.037									
									(t)	0.034									
								R(A2)	(l)	0.037									
									(s)	0.038									
									(t)	0.028									
	1.79 2.04 1.50 2.04		1.85 1.80 2.05 1.81	1.88 1.72 1.93 1.70	1.95 1.88 2.06 1.79		1.91 1.73 2.18 1.68	2.03 1.88 2.32 1.83		1.39 1.19 1.61 1.08	0.65 0.46 0.93 0.21	0.61 0.38 1.13 0.26							227
1.83		2.04	1.32		1.56		0.98	0.70	0.61										365
	1.88 2.02		1.78 1.78	1.86 1.79	1.92 2.26		1.82 1.93	1.93 2.12		1.30 1.53	0.52 1.14	0.33 0.65							227
2.34		1.35	2.02		1.83		1.87		0.85										365
(3)	2.57	(4)2.48	(4)2.69	(5)	2.15		(7)	2.00		(8)	1.25								304
2.23		(4)	1.64			(6)	1.34	(7)	0.71										
				Fr	1.83														
				Pi	1.94														
(3)	1.05	(4)1.29	(4)1.24	(5)	1.30		(7)	1.08		(8)	1.05					(8)	1.20		
								R(A1)	(l)	1.33									
									(s)	1.25									
									(t)	0.94									
								R(A2)	(l)	1.47									
									(s)	1.12									
									(t)	0.81									
	0.08 0.18 0.07 0.19 0.10 0.15		0.23 0.24 0.24 0.25 0.17 0.20	0.10 0.15 0.09 0.16 0.13 0.12	0.10 0.20 0.10 0.23 0.10 0.10		0.18 0.18 0.15 0.18 0.24 0.20	0.11 0.15 0.08 0.14 0.19 0.17		0.16 0.24 0.14 0.24 0.21 0.23	0.04 0.05 0.03 0.04 0.09 0.10	0.04 0.03 0.02 0.015 0.07 0.04							227
(3)	0.08	(4)0.07	(4)0.14	(5)	0.15		(7)	0.14		(8)	0.13								304
0.10		(4)	0.20			(6)	0.19	(7)	0.10										
				Fr	0.13														
				Pi	0.11														
(3)	0.24	(4)0.28	(4)0.26	(5)	0.24		(7)	0.25		(8)	0.23					(8)	0.25		
								R(A1)	(l)	0.34									
									(s)	0.21									
									(t)	0.20									
								R(A2)	(l)	0.17									
									(s)	0.24									
									(t)	0.11									

continued

Table 1.188 Bracken (continued)

	January	February	March	April	May	June	
Major elements %DM (cont'd)							
Chloride						L(G) (1947) 0.57 S(G) (1947) 0.69	
					L(P)(1) 0.19 L(P)(0) 0.26 (1)0.21	(2) 0.15 (2) 0.18	(3)
	R(P)(0) 0.15			(0) 0.21	(1) 0.29	(2) 0.36	
Trace elements μg/g DM							
Barium				R(P)(0) 124	L(P)(0) 27 (1)21	L(P)(2) 22 (2) 27	(3)
Chromium				R(P)(0) 2.0	L(P)(0) 1.1 (1)1.2	L(P)(2) 1.7 (2) 3.4*	(3)
Cobalt				R(P)(0) 0.80	L(P)(0) 0.32 (1)0.46	L(P)(2) 0.30 (2) 0.53	(3)
Copper				R(P)(0)10.0	L(P)(0)23.5 (1)16.0	L(P)(2)13.0 (2)20.5	(3)
Iron				R(P)(0) 790	L(P)(0) 141 (1)158	L(P)(2) 336 (2) 146	(3)
Lead				R(P)(0) 2.9	L(P)(0) 5.1 (1)6.7	L(P)(2) 8.8 (2) 7.7	(3)
Manganese				R(P)(0) 132	L(P)(1) 101 L(P)(0)65.5 (1)90 6	(2) 150 (2)83.6	(3)
Molybdenum				R(P)(0) 0 43	L(P)(0) 0.84 (1)0.22	L(P)(2) 0.13 (2) 0.22	(3)
Nickel				R(P)(0) 4.60	L(P)(0) 20* (1)23*	L(P)(2) 11* (2) 20*	(3)
Silver				R(P)(0) 0.17	L(P)(0) 0.2 (1)0 4	L(P)(2) 0.08 (2) 0.1	(3)
Strontium				R(P)(0) 12	L(P)(0) 3.8 (1)3.4	L(P)(2) 6.8 (2) 5.9	(3)
Tin				R(P)(0) 2.8	L(P)(0) 7.2 (1)6.2	L(P)(2) 3.1 (2) 2.3	(3)
Titanium				R(P)(0) 111	L(P)(0) 16 (1)12	L(P)(2) 19 (2) 6.1	(3)
Vanadium				R(P)(0) 2.20	L(P)(0) 0.43 (1)low	L(P)(2) 0.38 (2) 0.25	(3)
Zinc				R(P)(0) 49	L(P)(0) 98 (1)105	L(P)(2) 67 (2) 51	(3)

July			August			September			October			November			December			Notes	Reference
0.51 0.63		0.71 0.76	0.76 0.97		0.89 0.97		0.55 0.66	0.32	0.41 0.40										365
(3) 0.17	0.14	(4)0.11 (4)	(4)0.11 0.17	(5)	0.10	(6)	(7) 0.16	0.09 (7)	0.15	(8)	0.09								304
(3)	0.34	(4)0.28	(4)0.18	Fr Pi (5)	0.16 0.23 0.20		(7)	0.21 R(A1) R(A2)	(l) (s) (t) (l) (s) (t)	(8) 0.28 0.21 0.16 0.20 0.09 0.04	0.18					(8)	0.24		
(3) 20 (3)	35 154	(4)38 (4)	(4)60 36	(5) Fr Pi	68 27 27	(6)	(7) 34	86 (7)	37	(8)	99							*Unexplained abnormal values	304
(3) 0.4 (3)	1.1 1.7	(4)0.9 (4)	(4)0.7 0.4	(5) Fr Pi	0.7 0.5 0.6	(6)	(7) 0.6	0.7 (7)	0.6	(8)	1.5								
(3) 0.31 (3)	0.22 1.07	(4)0.29 (4)	(4)0.24 0.24	(5) Fr Pi	0.38 0.52 0.57	(6)	(7) 0.28	0.37 (7)	0.29	(8)	0.65								
(3) 11.0 (3)	11.0 10.5	(4)8.5 (4)	(4)7.5 7.5	(5) Fr Pi	8.6 9.9 10.7	(6)	(7) 8.4	11.0 (7)	6.6	(8)	10.0								
(3) 64 (3)	213 838	(4)257 (4)	(4)167 116	(5) Fr Pi	208 206 175	(6)	(7) 121	327 (7)	188	(8)	530								
(3) 5.7 (3)	5.1 6.6	(4)5.3 (4)	(4)4.8 3.3	(5) Fr Pi	4.2 3.4 3.5	(6)	(7) 9.8	10.3 (7)	10.7	(8)	9.2								
(3) 122 (3)	207 121	(4)229 (4)	(4)319 122	(5) Fr Pi	350 243 319	(6)	(7) 216	474 (7)	209	(8)	370								
(3) 0.14 (3)	0.04 0.46	(4)0.16 (4)	(4)0.15 0.13	(5) Fr Pi	0.16 0.105 0.11	(6)	(7) 0.15	0.22 (7)	0.25	(8)	0.37								
(3) 4.0 (3)	3.06 6.40	(4)3.60 (4)	(4)1.34 0.63	(5) Fr Pi	2.16 3.33 4.38	(6)	(7) 1.43	2.24 (7)	1.57	(8)	2.64								
(3) 0.1 (3)	0.06 0.31	(4)0.10 (4)	(4)0.05 0.8	(5) Fr Pi	0.10 0.1 0.1	(6)	(7) 0.2	0.17 (7)	0.2	(8)	0.23								
(3) 5.5 (3)	8.2 17	(4)10 (4)	(4)15 10	(5) Fr Pi	22 12 12	(6)	(7) 12	23 (7)	15	(8)	20								
(3) 1.5 (3)	1.7 3.9	(4)1.7 (4)	(4)0.9 0.8	(5) Fr Pi	1.4 1.8 1.0	(6)	(7) 1.3	1.5 (7)	1.8	(8)	2.0								
(3) 3.9 (3)	11 97	(4)15 (4)	(4)22 6.6	(5) Fr Pi	34 6.6 8.5	(6)	(7) 6.9	34 (7)	9.9	(8)	66								
(3) 0.15 (3)	0.26 2.10	(4)0.61 (4)	(4)0.47 0.30	(5) Fr Pi	0.53 0.17 0.19	·(6)	(7) 0.49	0.47 (7)	0.63	(8)	1.75								
(3) 37 (3)	45 64	(4)46 (4)	(4)39 47	(5) Fr Pi	73 43 48	(6)	(7) 61	72 (7)	83	(8)	110								

Table 1.189 Bent grass: *Agrostis*

For Studies of *Agrostis* and other uncultivated plants, proximate constituents, digested components and metabolisable energy see table 1.237

	March			May			June			July			September			November			Notes	Reference
Heath Bent grass: *A. setacea* *Proximate constituents* %DM Crude protein	5.4					8.1			10.5			11.4		9.4		10.2			Moorland	189
Ether extract	2.1					2.1			3.0			3.0		3.1		2.3				
Nitrogen-free extract	50.6					52.4			50.2			50.7		47.8		48.4				
Crude fibre	35.3					31.5			29.3			27.4		34.8		32.8				
Ash	6.6					5.9			7.0			7.5		4.9		6.3				
Major elements %DM Calcium	0.06					0.09			0.1			0.1		0.1		0.08				
Phosphorus	0.06					0.08			0.09			0.1		0.1		0.1				
Potassium	0.5					0.8			1.2			1.4		1.3		1.0				
Sodium	0.07					0.08			0.08			0.13		0.11		0.11				
New Zealand Bent grass: *A. tenuis* *Proximate constituents* %DM Crude protein					18.0		15.3					17.0	15.1						Leafy, preflowering at start. Sown in 1954, grazed and mown in 1956 and sampled in 1957	141
Ash					11.4		8.4					8.9	8.4							
Carbohydrate fractions %DM Water soluble carbohydrate					17.2		18.0					12.5	18.7							
Lignin					5.4		5.4					6.0	4.8							
	March			April			May			June			July			August				
Bent grass, unnamed *Proximate constituents* %DM Crude protein											14.6									218
									12.4										Probably mean values for 3 upland and 3 lowland centres, altitude 50–900 feet above sea level	215
																8.9			One plot received ammonium sulphate 165 kg/ha yearly since 1897	491
Ether extract																2.2				
Nitrogen-free extract																53.0				
Crude fibre											25.5									218
																30.0			Fertiliser – see above	491
Ash											6.7									218
																5.9			Fertiliser – see above	491
Major elements %DM Calcium											0.3									218
									0.3										See above	215
																0.4			Fertiliser – see above	491
Phosphorus											0.2									218
									0.2										See above	215
																0.2			Fertiliser – see above	491
Potassium											2.2									218
Chlorine											0.9									

Table 1.190 Meadow Foxtail: *Alopecurus pratensis,* and mixture

Meadow Foxtail	May	June	July	August	Notes	Reference
Proximate constituents %DM						
Crude protein	WM	9.8	† 12.7	14.6*	W: whole plant L: leaf S: stem M: Grown in mixture with cocksfoot, timothy and fine-leaved fescue P: grown in pure plots D: grown in drills †: Mean of 4 cuts taken on 5 May, 4 June, 5 July and 28 September Hay cut taken on 19 June * Regrowth	214
	WP	8.7	12.2	9.3*		
	WD	11.5	12.9	13.0*		
	LM	10.2	12.1	15.8*		
	LP	11.1	13.3	10.5*		
	LD	14.0	14.6	13.3*		
	SM	7.0	9.6	10.1*		
	SP	6.4	9.1	6.9*		
	SD	8.3	9.8	8.9*		
Major elements %DM						
Calcium	WM	0.5	† 0.6	0.6*		
	WP	0.4	0.6	0.6*		
	WD	0.3	0.6	0.4*		
	LM	0.6	0.6	0.7*		
	LP	0.5	0.6	0.7*		
	LD	0.4	0.6	0.5*		
	SM	0.2	0.4	0.3*		
	SP	0.3	0.3	0.3*		
	SD	0.2	0.3	0.3*		
Phosphorus	WM	0.2	† 0.4	0.2*		
	WP	0.2	0.4	0.3*		
	WD	0.2	0.3	0.2*		
	LM	0.2	0.3	0.2*		
	LP	0.3	0.3	0.3*		
	LD	0.3	0.3	0.2*		
	SM	0.2	0.4	0.2*		
	SP	0.3	0.4	0.3*		
	SD	0.2	0.3	0.3*		
Potassium	WM	2.8	† 2.8	2.4*		
	WP	2.6	2.9	2.2*		
	WD	2.6	3.0	2.5*		
	LM	2.3	2.5	2.3*		
	LP	2.5	2.8	2.3*		
	LD	2.4	2.9	1.9*		
	SM	3.0	3.2	2.3*		
	SP	2.9	3.3	2.1*		
	SD	3.3	3.3	2.7*		
Chlorine	WM	0.8	† 0.8	0.6*		
	WP	0.9	0.8	0.7*		
	WD	0.8	0.8	0.9*		
	LM	0.7	0.8	0.5*		
	LP	0.9	0.7	0.6*		
	LD	0.6	0.7	0.6*		
	SM	1.1	0.8	0.8*		
	SP	1.1	0.9	1.0*		
	SD	1.0	0.7	0.9*		

Meadow foxtail and Tall Fescue	Late February	March	April	May	Notes	Reference
Proximate constituents %DM						
Crude protein	28.4 (27.8–28.8)				Rotationally grazed pasture. Sampled 28 February. Means and ranges of 4 samples.	207
Ether extract	6.4 (5.9–7.4)					
Nitrogen-free extract	39.0 (37.8–40.3)					
Crude fibre	15.6 (15.2–16.1)					
Ash	10.6 (10.2–11.2)					
Major elements %DM						
Calcium	0.55 (0.54–0.56)					
Phosphorus	0.345(0.34–0.35)					

Table 1.191 Crested Dogstail: *Cynosurus cristatus*

For proximate constituents, digested components and metabolisable energy see table 1.237

		June	July	August	September	Notes	Reference
Proximate constituents %DM							
Crude protein		8.0				Cut at early flowering stage, height 22–25 cm.	12
Amino acids % of crude protein							
Cystine		0.45					
Histidine		1.61					
Lysine		4.28					
Methionine: DL form		2.08					
L form		1.87					
Tryptophan		1.75					
Tyrosine		2.69					
Major elements % DM							
Calcium	F_0	0.15				F_0: no extra fertiliser	489
	F_1	0.19				F_1: 20 tonnes farmyard manure/ha/annum	
	F_2	0.14				F_2: 165 kg sulphate of ammonia/ha	
	F_3	0.20				F_3: 330 kg basic slag/ha	
	F_4	0.08				F_4: 110 kg muriate of potash /ha	
	F_5	0.17				F_5: Last 3 together	
		(1) Pre-flowering 0.46; mid-flowering 0.32; late-flowering 0.29; mature 0.25				(1) Intervals between cuts, after the first, 3, 6, 8 and 12 weeks	490
		Average value, 5 samples, no date, 0.47					20
Phosphorus	F_0	0.16					489
	F_1	0.29					
	F_2	0.27					
	F_3	0.23					
	F_4	0.14					
	F_5	0.24					
		(1) Pre-flowering 0.27; mid-flowering 0.24; late-flowering 0.22; mature 0.19					490
		Average value, 5 samples, no date, 0.15					20
Potassium	F_0	1.18					489
	F_1	1.97					
	F_2	1.10					
	F_3	1.42					
	F_4	1.80					
	F_5	1.31					
		(1) Pre-flowering 2.24; mid-flowering 1.92; late-flowering 1.72; mature 1.40					490
Magnesium	F_0	0.11					489
	F_1	0.13					
	F_2	0.10					
	F_3	0.12					
	F_4	0.08					
	F_5	0.09					
		(1) Pre-flowering 0.26; mid-flowering 0.29; late-flowering 0.29; mature 0.25					490
		Average value, 5 samples, no date, 0.10					20
Sodium	F_0	0.24					489
	F_1	0.21					
	F_2	0.30					
	F_3	0.23					
	F_4	0.19					
	F_5	0.30					
		(1) Pre-flowering 0.09; mid-flowering 0.08; late-flowering 0.08; mature 0.07					490
Chlorine	F_0	0.56					489
	F_1	0.76					
	F_2	0.56					
	F_3	0.66					
	F_4	0.81					
	F_5	0.86					
		(1) Pre-flowering 0.52; mid-flowering 0.55; late-flowering 0.64; mature 0.53					490
		Average value, 5 samples, no date, 0.64					20

Table 1.191 Crested Dogstail (continued)

	June			July			August			September			Notes	Reference
Trace elements μg/gDM														
Cobalt	F_0	0.23											See Calcium, same reference	489
	F_1	0.31												
	F_2	0.16												
	F_3	0.12												
	F_4	0.20												
	F_5	0.13												
	N_0	0.05											N_0: no fertiliser N_1: 1.65 kg $CoSO_4 \cdot 7H_2O$/ha in April 1954 and 1955. Replicate plots	359
	N_0	0.06												
	N_1	0.59												
	N_1	0.72												
	(1) Pre-flowering 0.17; mid-flowering 0.23; late flowering 0.17; mature 0.15													490
Copper	F_0	4.4											See Calcium, same reference	
	F_1	5.9												
	F_2	4.9												
	F_3	6.9												
	F_4	4.8												
	F_5	6.4												
	N_0	2.8											N_0: no fertiliser N_1: 22 kg $CuSO_4.5H_2O$/ha in April 1954 and 1955 Replicate plots	359
	N_0	2.9												
	N_1	3.7												
	N_1	3.8												
	(1) Pre-flowering 9.6; mid-flowering 6.8; late-flowering 8.8; mature 6.8													490
	Average value, 5 samples, no date, 3.7													20
Iron	F_0	61											See Calcium, same reference	489
	F_1	47												
	F_2	76												
	F_3	69												
	F_4	61												
	F_5	53												
	(1) Pre-flowering 332; mid-flowering 178; late-flowering 200; mature 163													490
	Average value, 5 samples, no date, 69													20
Manganese	F_0	132											See Calcium, same reference	489
	F_1	95												
	F_2	96												
	F_3	119												
	F_4	257												
	F_5	139												
	(1) Pre-flowering 40.5; mid-flowering 33.9; late-flowering 32.0; mature 24.9													490
	Average value, 5 samples, no date, 28													20
Vitamins μg/gDM														
Thiamin									1.98				Meadow plots Pre-flowering	493
Riboflavin									20.4					
Nicotinic acid									46.3					
Pantothenic acid									15.5					
Biotin									0.407					
Pyridoxine									14.7					

Table 1.192 Yorkshire Fog: *Holcus lanatus,* and mixtures

	January			February			March			April			May			June		
Yorkshire Fog																		
Proximate constituents %DM																		
Crude protein													19.7		16.0			16.5
(1)	13.6			12.7			13.8			13.8								
(2)	12.9			14.1			13.8			15.1								
										Mean of 5 samples, hand sorted, no date: 7.2								
Ether extract													3.2		3.1			3.0
(1)	2.8			2.2			3.2			2.4								
(2)	2.9			2.0			2.0			1.8								
Nitrogen-free extract													48.7		49.6			51.0
(1)	46.4			51.2			50.0			50.8								
(2)	47.9			50.7			50.8			52.0								
Crude fibre													19.6		23.2			21.6
(1)	31.3			29.2			28.4			28.2								
(2)	30.3			27.6			28.2			26.1								
Ash													8.9		8.1			7.9
(1)	5.9			4.7			4.7			4.8								
(2)	6.0			5.6			5.2			5.0								
										Mean of 5 samples, hand sorted, no date: 5.46								
Major elements %DM																		
Calcium													0.93		1.01			1.09
(1)	0.61			0.44			0.40			0.43								
(2)	0.49			0.34			0.39			0.55								
															F_0	0.25	0.23	
															F_1	0.25	0.21	
															F_2	0.21	0.18	
															F_3	0.23	0.21	
															F_4	0.24	0.19	
															F_5	0.30	0.25	
										Mean of 5 samples, hand sorted, no date: 0.45								
Phosphorus													0.46		0.45			0.43
(1)	0.40			0.37			0.37			0.33								
(2)	0.40			0.41			0.35			0.36								
															F_0	0.40	0.31	
															F_1	0.27	0.20	
															F_2	0.28	0.24	
															F_3	0.40	0.29	
															F_4	0.30	0.23	
															F_5	0.37	0.29	
										Mean of 5 samples, hand sorted, no date: 0.21								
Magnesium															F_0	0.16	0.11	
															F_1	0.14	0.14	
															F_2	0.18	0.14	
															F_3	0.11	0.13	
															F_4	0.15	0.12	
															F_5	0.14	0.13	
										Mean of 5 samples, hand sorted, no date: 0.05								
Potassium															F_0	2.81	1.90	
															F_1	2.28	1.63	
															F_2	2.36	1.78	
															F_3	2.45	1.63	
															F_4	2.64	1.84	
															F_5	2.30	1.68	
Sodium															F_0	0.12	0.13	
															F_1	0.13	0.18	
															F_2	0.23	0.14	
															F_3	0.20	0.20	
															F_4	0.12	0.22	
															F_5	0.17	0.23	
Chlorine															F_0	0.99	0.80	
															F_1	0.90	0.75	
															F_2	0.94	0.82	
															F_3	1.08	0.75	
															F_4	1.18	0.84	
															F_5	1.21	0.82	
										Mean of 5 samples, hand sorted, no date: 0.66								
Sulphur																		
Inorganic															F_0	0.11	0.08	
															F_1	0.07	0.07	
															F_2	0.08	0.06	
															F_3	0.07	0.10	
															F_4	0.07	0.10	
															F_5	0.07	0.12	
Organic															F_0	0.13	0.10	
															F_1	0.13	0.10	
															F_2	0.14	0.12	
															F_3	0.14	0.10	
															F_4	0.14	0.11	
															F_5	0.14	0.10	

July			August			September			October			November			December			Notes	Reference
		18.8			20.5		23.3		(1) (2)	13.9 13.3			12.4 14.3			11.3 13.2		(1) Ungrazed (2) Grazed from October	471
																			20
		3.3			4.0		3.6		(1) (2)	3.2 3.2			3.0 3.0			2.8 2.9		(1) Ungrazed (2) Grazed from October	471
		46.1			45.9		42.6		(1) (2)	49.6 50.5			49.5 48.5			49.2 47.2			
		22.7			20.7		20.9		(1) (2)	25.5 25.7			27.6 26.9			30.5 30.2			
		9.1			8.9		9.6		(1) (2)	7.8 7.3			7.5 7.3			6.2 6.5			
																			20
		1.14			1.02		0.98		(1) (2)	0.58 0.65			0.44 0.59			0.59 0.57		(1) Ungrazed (2) Grazed from October	471
																		F_0 :Farmyard manure 20 tonnes/ha p.a. F_1 :Control F_2 :165 kg sulphate of ammonia/ha F_3 :330 kg basic slag/ha F_4 :110 kg muriate of potash/ha F_5 :Last 3 together	489
																		Cut preflowering	20
		0.49			0.48		0.55		(1) (2)	0.39 0.37			0.33 0.38			0.37 0.35		(1) Ungrazed (2) Grazed from October	471
																		See Calcium, same reference	489
																			20
																		See Calcium, same reference	489
																			20
																		See Calcium, same reference	489
																			20
																		See Calcium, same reference	489

continued

Table 1.192 Yorkshire Fog (continued)

	January			February			March			April			May			June		
Trace elements **µg/g DM**																		
Cobalt															F_0	0.17	0.17	
															F_1	0.19	0.19	
															F_2	0.14	0.26	
															F_3	0.14	0.17	
															F_4	0.14	0.24	
															F_5	0.14	0.14	
										Mean of 5 samples, hand sorted, no date: 5.2								
Copper															F_0	11.5	12.4	
															F_1	7.9	8.9	
															F_2	6.9	8.4	
															F_3	8.0	8.7	
															F_4	6.1	8.9	
															F_5	8.5	9.4	
Iron															F_0	175	66	
															F_1	42	67	
															F_2	68	75	
															F_3	47	71	
															F_4	62	79	
															F_5	76	62	
										Mean of 5 samples, hand sorted, no date: 94								
Manganese															F_0	229	189	
															F_1	389	437	
															F_2	380	383	
															F_3	174	425	
															F_4	479	380	
															F_5	287	370	
										Mean of 5 samples, hand sorted, no date: 54								
Vitamins µg/g DM																		
Carotene									600		(1)	300			290			140
Yorkshire Fog and other grasses																		
Proximate constituents %DM																		
Crude protein												(1934)	19.7		16.0			16.5
(1935)	13.6			12.7			13.8			13.8								
(1935)	12.9			14.1			13.8			15.1								
Ether extract												(1934)	3.2		3.1			3.0
(1935)	2.8			2.3			3.2			2.5								
(1935)	2.9			2.0			2.0			1.8								
Nitrogen-free extract												(1934)	48.6		49.6			51.0
(1935)	46.4			51.1			49.9			50.7								
(1935)	47.9			50.7			50.8			52.0								
Crude fibre												(1934)	19.6		23.2			21.6
(1935)	31.3			29.2			28.4			28.2								
(1935)	30.3			27.6			28.2			26.1								
Ash												(1934)	8.9		8.1			7.9
(1935)	5.9			4.7			4.7			4.8								
(1935)	6.0			5.6			5.2			5.0								
Major elements %DM																		
Calcium												(1934)	0.93		1.01			1.09
(1935)	0.61			0.44			0.40			0.43								
(1935)	0.49			0.34			0.39			0.55								
Phosphorus												(1934)	0.46		0.45			0.43
(1935)	0.40			0.37			0.37			0.33								
(1935)	0.40			0.41			0.35			0.36								
Proximate constituents **%DM**																		
Crude protein															22.3	24.6	25.0	24.8
															22.2		20.7	23.5
																20.1		23.6
																	18.4	
															F_0		19.9	
															F_1		21.1	
															F_2		19.1	
															F_3		19.7	
Vitamins **µg/g DM**																		
Carotene															320	286	435	359
															423		430	376
																221		266
																	343	
															F_0		291	
															F_1		382	
															F_2		261	
															F_3		348	

July			August			September			October			November			December			Notes	Reference
																		See Calcium, same reference	489
																			20
																		See Calcium, same reference	489
																			20
																		See Calcium, same reference	489
																			20
	(2)	140	(3)	170		(4)	440		(4)	620								(1) Cut close to ground (2) Flowering completed (3) Leaf only. Cut close to ground for second time (4) Aftermath	362
		18.8			20.5		23.3											*Holcus* 70% cover; not more than 5% clover. Winter grazing.	471
									(1934)	13.9			12.4			11.3		1934. Plots rested.	
									(1934)	13.3			14.3			13.2		Last 3 months of 1934, early 1935. Ungrazed plots.	
		3.3			4.0		3.7												
									(1934)	3.2			3.1			2.8		Last 3 months of 1934, early 1935. Grazed plots.	
									(1934)	3.2			3.0			2.9			
		46.1			45.9		42.5											Plots rested.	
									(1934)	49.6			49.4			49.2		Ungrazed plots.	
									(1934)	50.5			48.5			47.2		Grazed plots.	
		22.7			20.7		20.9												
									(1934)	25.5			27.6			30.5			
									(1934)	25.7			26.9			30.2			
		9.1			8.9		9.6												
									(1934)	7.8			7.5			6.2			
									(1934)	7.3			7.3			6.5			
		1.14			1.02		0.98												
									(1934)	0.58			0.44			0.59			
									(1934)	0.65			0.59			0.57			
		0.49			0.48		0.55												
									(1934)	0.39			0.33			0.37			
									(1934)	0.37			0.38			0.35			
																		Old pasture. *Holcus* 70% cover. All plots cut to ground on May 13.	
22.9	23.0	21.9	23.5	22.6	21.1	25.0	24.6	23.5										1 week's growth	484
19.8		20.9	20.8	21.8		22.9	22.8	23.7										2 weeks' growth	
	20.1		20.4		20.5		21.1		23.5									3 weeks' growth	
19.2			19.5			19.9		22.2										4 weeks' growth	
18.2			14.8			20.1		24.1										F_0: no fertiliser	
20.5			19.9			21.8		25.4										F_1: sulphate of ammonia 125 kg/ha	
18.2			20.1			20.1		25.2										F_2: sulphate of iron 55 kg/ha	
16.7			17.9			21.0		21.5										F_3: carbonate of lime 7500 kg/ha	
497	406	474	421	435	613	570	530	725										1 week's growth	
451		479	394	413		660	492	836										2 weeks' growth	
	328		192		524		594		518									3 weeks' growth	
411			394			528		681										4 weeks' growth	
306			385			575		612										As for Crude protein	
514			532			712		762											
210			305			616		680											
360			530			584		589											

Table 1.193 Flying Bent: *Molinia caerulea*

	January			February			March			April			May			June		
Proximate constituents %DM																		
Crude protein											(1)	19.3		(1)	17.9			14.8
																	16.2	
Ether extract											(1)	1.9		(1)	2.0			1.8
Nitrogen-free extract											(1)	47.8		(1)	46.9			41.9
Crude fibre											(1)	24.7		(1)	26.0			32.1
																	32.00	
Ash											(1)	6.3		(1)	7.2			9.4
																	4.55	
Crude protein																		
Amino acids % of crude protein																		
Cystine																		
Lysine																		
Methionine, DL form																		
L form																		
Tryptophan																		
Tyrosine																		
Major elements %DM																		
Calcium																	0.16	
											(1)	0.124		(1)	0.128			0.119
						0.101							0.117					0.142
Phosphorus																	0.18	
											(1)	0.455		(1)	0.326			0.274
						0.068							0.404					0 216
Magnesium						0.051							0.178					0.141
Potassium																	1.31	
						0.13							2.40					2.31
Sodium						–							–					0.046
Chloride																	0.35	
						0.047							0.452					0 345
Total Sulphur						0.089							0.231					0.226

July			August			September			October			November			December			Notes	Reference
				14.4		10.7				7.6								(1) New growth only	473
																		For study of composition, digested nutrients and metabolisable energy see table 1.237	218
				2.3		1.6				1.9									
				48.1		51.3				52.3									473
				31.1		32.4				34.3									
																			218
				4.1		4.0				3.9									473
																			218
	14.5																	Cut flowering; height 30–36 cm.	12
	0.57 3.95 1.87 1.70 1.32 4.20																		
																			218
				0.197		0.229				0.164									473
	0.152			0.174			0.203												492
																			218
				0.206		0.155				0.087									473
	0.200			0.151			0.120												492
	0.138			0.158			0.192												
																			218
	1.29			1.07			0.88												492
	0.025			0.024			0.031												
																			218
	0.334			0.350			0.377												
	0.227			0.229			0.216												492

continued

Table 1.193 Flying Bent: (continued)

	February			May			June			July			August			September			Notes	Reference
Trace elements μg/g DM Barium				(1)	(1946)	8		4		(1)	(1947)	7 13	(1) (2) (3)	(1948) 12 8	9	15			(1) Natural moorland. Sample hand sorted (2) Wet moorland (3) Dry moorland	358
Cobalt			0.086	0.077					0.106		0.054			0.087			0.041			492
				(1)	(1946)	0.06	0.05			(1)	(1947)	0.07 0.05	(2) (3)	0.03 0.03		0.04				358
Copper			6.59	19.56					6.30		7.98			4.96			5.22			492
				(1)	(1946)	10.4	7.2			(1)	(1947)	6.3 6.0	(1) (2) (3)	(1948) 7.0 10.5	5.5	3.5				358
Iron			139	116					109		94			106			104			492
				(1)	(1946)	91	86			(1)	(1947)	95 80	(2) (3)	(1948) 81 128	58	57				358
Lead				(1)	(1946)	0.5		1.0		(1)	(1947)	1.4 1.6	(1) (2) (3)	(1948) 4.5 16.2	2.3	3.5				
Manganese			295	283					382		428			330			434			492
				(1)	(1946)	130		85		(1)	(1947)	50 170	(2) (3)	(1948) 64 140	75	90				358
Molybdenum				(1)	(1946)	0.16		0.24		(1)	(1947)	0.22 0.18	(2) (3)	(1948) 0.21 0.26	0.19	0.20				
Nickel				(1)	(1946)	0.49		0.83		(1)	(1947)	0.77 0.47	(2) (3)	(1948) 0.36 0.62	0.23	0.43				
Strontium				(1)	(1946)	3		4		(1)	(1947)	7 10	(1) (2) (3)	(1948) 8 5	8	11				
Tin				(1)	(1946)	0.1		0.2		(1)	(1947)	0.4 0.2	(1) (2) (3)	(1948) 1.7 0.4	0.4	0.2				
Titanium				(1)	(1946)	1.3		1.3		(1)	(1947)	1.7 2.6	(1) (2) (3)	(1948) 2.1 2.2	1.6	1.8				
Zinc				(1)	(1946)	44		40		(1)	(1947)	47 35	(2) (3)	(1948) 71 69	36	33				
Vitamins μg/g DM Thiamin										4.24									Preflowering	493
Riboflavin										14.0										
Nicotinic acid										42.8										
Pantothenic acid										4.6										
Biotin										0.273										
Pyridoxine										7.7										

Table 1.194 Mat grass or White Bent: *Nardus stricta*

For proximate constituents, digested components and metabolisable energy see table 1.237

	January			February			March			April			May			June		
Proximate constituents %DM																		
Crude protein																	10.47	
		(3)	5.5									(1) (2)	9.2 12.5			(1) (2)	11.3 16.0	
Ether extract		(3)	1.3									(1) (2)	1.2 1.6			(1) (2)	0.8 0.5	
Nitrogen-free extract		(3)	52.3									(1) (2)	50.6 52.2			(1) (2)	47.9 44.5	
Crude fibre																	32.10	
		(3)	33.0									(1) (2)	30.6 25.8			(1) (2)	32.1 32.1	
Ash																	4.50	
		(3)	7.9									(1) (2)	8.4 7.9			(1) (2)	7.9 6.9	
Major elements %DM																		
Calcium																	0.16	
		(3)	0.04									(1) (2)	0.08 0.08			(1) (2)	0.06 0.08	
							0.06					0.084					0.082	
Phosphorus																	0.13	
		(3)	0.07									(1) (2)	0.16 0.23			(1) (2)	0.17 0.24	
							0.113					0.242					0.211	
Magnesium							0.032					0.093					0.115	
Potassium																	1.32	
							0.17					1.52					1.82	
Sodium							0.040					0.082					0.074	
Chloride																	0.28	
							0.087					0.644					0.422	
Sulphur, total							0.096					0.238					0.191	
Trace elements µg/g DM																		
Cobalt						(1947)	0.097					0.080					0.092	
Copper						(1947)	3.55					7.74					4.48	
Iron						(1947)	111					141					75	
Manganese						(1947)	137					718					469	

July			August			September			October			November			December			Notes	Reference
																			218
(3)	9.8		(3)	7.2				(3)	6.3									(1) Dead + green (2) Green only (3) Dead or bleached	475
(3)	0.8		(3)	1.1				(3)	0.9										
(3)	50.8		(3)	50.8				(3)	50.1										
																			218
(3)	32.1		(3)	33.7				(3)	35.5										475
																			218
(3)	6.5		(3)	7.2				(3)	7.2									See Crude protein, same reference	475
																			218
(3)	0.07		(3)	0.08				(3)	0.07										475
		0.089		0.089			0.075												492
																			218
(3)	0.17		(3)	0.13				(3)	0.13										475
		0.176			0.096		0.103												492
		0.097			0.106		0.075												
																			218
		1.17			1.19		0.54												492
		0.036			0.024		0.028												
																			218
		0.401			0.338		0.241												492
		0.181			0.175		0.146												
		0.078			0.112		0.117												
		5.62			4.06		3.80												
		85			86		89												
		548			588		350												

Table 1.195 Uncultivated grasses, miscellaneous

	April			May			June			July			August			September			Notes	Reference
Tufted Hairgrass, Yorkshire Fog and Sweet Vernal: *Aira caespitosa, Holcus lanatus* and *Anthoxanthum odoratum*																				
Proximate constituents %DM																				
Crude protein						18.9		18.7	18.8		15.4		10.1						Old, unimproved sward, grazed	341
Ether extract						5.3		5.2	4.6		4.3		3.4							
Nitrogen-free extract						41.6		40.3	38.5		43.9		48.9							
Crude fibre						24.6		25.9	27.3		26.7		28.4							
Ash						9.6		9.9	10.8		9.7		9.2							
Sweet Vernal grass: *Anthoxanthum odoratum*																				
Trace elements μg/g DM																				
Barium					(1946)	17		11				33				20			Natural moorland grazing Sample hand sorted	358
					(1947)	16			25			36								
					(1948)				14						36					
Cobalt					(1946)	0.23		0.14				0.16				0.13				
					(1947)	0.24			0.19			0.15								
					(1948)				0.15						0.20					
Copper					(1946)	6.7		4.6				8.5				4.2				
					(1947)	5.0			5.5			5.2								
					(1948)				4.1						5.2					
Iron					(1946)	68		60				92				111				
					(1947)	56			66			56								
					(1948)				41						93					
Lead					(1946)	1.4		1.2				1.4				4.0				
					(1947)	1.1			2.2			1.6								
					(1948)				2.1						4.1					
Manganese					(1946)	400		450				700				200				
					(1947)	400			550			840								
					(1948)				210						340					
Molybdenum					(1946)	0.56		0.69				0.90				0.35				
					(1947)	0.78			1.47			0.48								
					(1948)				0.25						0.55					
Nickel					(1946)	1.26		0.71				1.72				0.65				
					(1947)	1.19			0.73			1.03								
					(1948)				0.83						1.08					
Strontium					(1946)	16		9				18				11				
					(1947)	10			20			22								
					(1948)				8						20					
Tin					(1946)	0.4		0.2				0.4				0.4				
					(1947)	0.2			0.5			0.4								
					(1948)				0.4						0.6					
Titanium					(1946)	3.5		2.8				1.3				1.6				
					(1947)	3.3			6.1			2.2								
					(1948)				2.6						3.0					
Zinc					(1946)	41		30				38				37				
					(1947)	36			–			30								
					(1948)				30						51					

Table 1.195 Uncultivated grasses, miscellaneous (continued)

	March	April	May	June	July	August	September	October	Notes	Reference
Tall Oat grass: *Arrhenatherum elatius* *Vitamins* μg/g DM Carotene	670	(1) 316	359*	218*	(1) 135*	(2) 106*	369*	322*	Samples from nursery plots *Regrowth (1) Flowering completed Cut close to ground (2) Leaf only. Cut for 2nd time	362

	February	March	April	May	June	July	Notes	Reference
Reed Canary grass: *Phalaris tuberosa* *Proximate constituents* %DM Crude protein	(1956) W 17.1 G 23.1 B 16.3 (1957) W 18.7 G 19.8 B 11.7						Used as winter grazing for sheep. Sampled before grazing in late February. W: whole plant G: green B: Frostburnt	337

	May	June	July	August	September	October	Notes	Reference
Rough-stalked Meadow grass: *Poa trivialis* *Vitamins* μg/g DM							Air-dried herbage Early flowering	493
Thiamin				1.6				
Riboflavin				13.6				
Nicotinic acid				37.5				
Pantothenic acid				11.6				
Pyridoxine				14.2				
Biotin				0.27				

	March	April	June	July	September	November	Notes	Reference
Heath grass: *Triodia procumbens* *Proximate constituents* %DM							The NFE values for May and June have been adjusted on the assumption that the other values, as published, are correct.	190
Crude protein	7.1	10.9	10.3	9.1	8.1	8.3		
Ether extract	2.4	2.4	2.6	2.9	2.6	3.2		
Nitrogen-free extract	52.7	58.7	59.1	50.6	51.8	52.2		
Crude fibre	32.5	22.2	22.7	32.7	33.3	31.4		
Ash	5.3	5.8	5.3	4.7	4.2	4.9		
Major elements %DM								
Calcium	0.10	0.11	0.10	0.08	0.07	0.11		
Phosphorus	0.09	0.15	0.14	0.13	0.09	0.10		

Table 1.196 Sedges

	January			February			March			April			May			June		
Carnation sedge: *Carex panicea*																		
Proximate constituents %DM																		
Crude protein	9.7									11.4					12.6			11.8
Ether extract	2.0									1.6					2.2			2.8
Nitrogen-free extract	55.3									59.0					59.3			56.7
Crude fibre	28.3									23.2					18.7			23.8
Ash	4.7									4.8					7.2			4.9
Major elements %DM																		
Calcium	0.16									0.14					0.11			0.11
Phosphorus	0.11									0.13					0.14			0.13
Trace elements μg/g DM																		
Cobalt													(1946)		0.24		0.33	
													(1946)		0.06		0.05	
													(1947)		0.40			0.53
													(1947)		0.08			0.05
													(1948)					0.17
													(1948)					0.04
Copper													(1946)		11.9		8.1	
													(1946)		11.5		7.4	
													(1947)		15.0			7.2
													(1947)		11.0			8.5
													(1948)					7.2
													(1948)					6.5
Iron													(1946)		85		62	
													(1946)		99		80	
													(1947)		107			81
													(1947)		110			88
													(1948)					44
													(1948)					47
Lead													(1946)		1.4		1.0	
													(1946)		1.4		2.2	
													(1947)		2.8			2.2
													(1947)		1.4			1.8
													(1948)					1.3
													(1948)					1.2
Manganese													(1946)		250		370	
													(1946)		350		110	
													(1947)		450			–
													(1947)		230			170
													(1948)					210
													(1948)					100
Molybdenum													(1946)		0.38		0.27	
													(1946)		0.48		0.60	
													(1947)		0.52			0.34
													(1947)		0.62			0.65
													(1948)					0.31
													(1948)					0.64
Nickel													(1946)		1.58		2.50	
													(1946)		0.68		0.61	
													(1947)		3.15			2.14
													(1947)		0.72			0.51
													(1948)					0.71
													(1948)					0.57
Tin													(1946)		0.4		0.1	
													(1946)		0.2		0.1	
													(1947)		0.9			0.2
													(1947)		0.1			0.2
													(1948)					0.4
													(1948/9)					0.2
Titanium													(1946)		1.7		1.1	
													(1946)		5.0		7.0	
													(1947)		4.8			1.4
													(1947)		10.0			4.2
													(1948)					2.4
													(1948)					1.8
Zinc													(1946)		42		37	
													(1946)		44		48	
													(1947)		66			30
													(1947)		51			53
													(1948)					24
													(1948)					58

July			August			September			October			November			December			Notes	Reference
	11.4						10.8					10.6						Sheep grazing study	191
	2.6						2.2					2.3							
	55.2						56.6					56.1							
	26.3						25.3					26.9							
	4.5						5.1					4.1							
	0.11						0.12					0.18							
	0.13						0.12					0.12							
		0.30 – 0.20 –			0.15 0.05	0.46 0.06 0.36 0.03												Natural moorland grazing One value at each date from each of two districts	358
		7.7 – 6.0 –			10.0 4.7	7.0 6.4 7.5 5.6													
		97 – 41 –			71 75	119 108 65 46													
		2.4 – 1.3 –			1.3 1.9	1.1 1.4 1.2 1.0													
		280 – – –			250 77	740 120 310 290													
		0.28 – 0.20 –			0.29 0.77	0.49 0.79 0.26 0.63													
		1.52 – 0.74 –			1.42 0.65	0.91 0.38 1.54 0.24													
		0.3 – 0.2 –			0.4 0.4	0.3 0.2 0.3 –													
		2.0 – 2.5 –			1.2 3.0	1.4 11.0 2.3 3.6													
		35 – 40 –			28 40	36 43 48 50													

continued

Table 1.196 Sedges (continued)

	March	April	June	July	August	September	Notes	Reference
Cotton grass, Draw moss: *Eriophorum vaginatum* *Proximate constituents* %DM Crude protein		(1) 10.9 (2) 11.8	(1) 10.3 (2) 9.3				(1) Whole Plant with flower heads in June (2) Long leaf basis (= scallion) only	467
Ether extract		(1) 1.1 (2) 1.3	(1) 2.2 (2) 2.1					
Nitrogen-free extract		(1) 55.2 (2) 53.6	(1) 54.1 (2) 53.0					
Crude fibre		(1) 29.6 (2) 28.6	(1) 30.1 (2) 32.0					
Ash		(1) 3.2 (2) 4.7	(1) 3.3 (2) 3.6					

	April	May	June	July	August	September	Notes	Reference
Proximate constituents %DM Crude protein			12.4				Cut flowering, height 25–30 cm.	12
Amino acids % of crude protein								
Cystine			0.33					
Lysine			3.79					
Methionine, DL form			1.58					
L form			1.61					
Tryptophan			1.50					
Tyrosine			2.71					

Table 1.196 Sedges (continued)

	March	April	June	July	August	September		
Major elements %DM Calcium	(1) (2)	0.099 0.007	(1) 0.007 (2) 0.011				See Crude protein, same reference	467
	0.146	0.146	0.181	0.155	0.172 0.178 0.146	0.138 0.157 0.076	Leaf and butt Leaf-butt Leaf	492
Phosphorus	(1) (2)	0.280 0.409	(1) 0.294 (2) 0.307					467
	0.144	0.378	0.263	0.264	0.224 0.445 0.215	0.159 0.312 0.194	Leaf and butt Leaf-butt Leaf	492
Magnesium	0.129	0.220	0.182	0.187	0.165 0.233 0.173	0.173 0.227 0.178		
Potassium	0.29	0.99	1.07	0.98	0.88 2.12 0.83	1.16 1.81 1.10		
Sodium	0.053	–	0.040	0.036	0.037 0.064 0.048	0.057 0.039 0.036		
Chloride	0.108	0.113	0.073	0.051	0.053 0.297 0.075	0.107 0.411 0.046		
Sulphur Total	0.094	0.167	0.240	0.169	0.180 0.207 0.201	0.172 0.208 0.181		
Trace elements μg/g DM Cobalt	0.080	0.079	0.058	0.063	0.033 0.260 0.067	0.082 0.143 0.082		
Copper	5.94	9.61	8.65	7.52	12.07 12.67 7.62	8.24 11.30 7.78		
Iron	108	103	205	96	176 164 124	143 280 123		
Manganese	224	296	177	180	149 212 141	139 283 151		
	April	**May**	**June**	**July**	**August**	**September**		
Vitamins μg/g DM Thiamin		3.01					Late-flowering	493
Riboflavin		12.4						
Nicotinic acid		44.0						
Pantothenic acid		7.4						
Biotin		0.23						
Pyridoxine		11.9						

continued

Table 1.196 Sedges (continued)

	February	May	June	July	August	September	Notes	Reference
Deer's grass: *Scirpus caespitosus*								
Proximate constituents %DM								
Crude protein			11.76	10.89				
			Average value for leaf, no date, 9.48					
Ether extract			Average value for leaf, no date, 3.84					218
Nitrogen-free extract			Average value for leaf, no date, 50.43					
Crude fibre			29.00	31.98				
			Average value for leaf, no date, 31.40					
Ash			4.94	4.82				
			Average value for leaf, no date, 4.85					
Major elements %DM								
Calcium			0.19	0.18				
			Average value for leaf, no date 0.18					218
	0.141	0.144	0.105	0.139	0.195	0.236	Sampled and sorted by hand on hill land	492
Phosphorus			0.27	0.20				
			Average value for leaf, no date, 0.19					218
	0.169	0.260	0.157	0.137	0.156	0.099		492
Magnesium	0.117	0.171	0.135	0.131	0.141	0.166		
Potassium		(1930)	2.81	1.77				218
			Average value for leaf, no date, 2.01					
	0.43	1.52	1.42	1.18	1.32	1.25		492
Sodium	–	–	0.034	0.038	0.027	0.040		
Chloride		(1930)	0.71	0.39				218
			Average value for leaf, no date, 0.46					
	0.084	0.281	0.312	0.257	0.276	0.321		492
Sulphur, total	0.108	0.154	0.212	0.240	0.249	0.266		

Table 1.196 Sedges (continued)

	February			May			June			July			August			September			Notes	Reference
Deer's grass (cont'd) *Trace elements* %DM Barium					(1946) (1946) (1947) (1948)	17 11		13 12	11 12			11 17	(1947)	15	13	25 30			Natural moorland Wet moorland	358
Cobalt					(1946) (1946) (1947) (1948)	0.10 0.09		0.62 0.06	0.04 0.03			0.04 0.04	(1947)	0.03	0.02	0.04 0.05			(Same sequence for each element from this reference)	
			0.087	0.115					0.058	0.084				0.103			0.039			492
Copper					(1946) (1946) (1947) (1948)	10.5 11.0		11.0 6.4	8.1 6.6			5.0 4.4	(1947)	8.0	2.8	3.1 4.6				358
			5.07	9.90					17.56	7.87				7.57			4.23			492
Iron					(1946) (1946) (1947) (1948)	100 91		233 103	81 51			65 102	(1947)	92	69	104 92				358
			113	125					99	105				105			110			492
Lead					(1946) (1946) (1947) (1948)	0.7 1.4		7.8 1.3	2.9 2.7			0.9 2.8	(1947)	3.5	1.9	3.2 3.7				358
Manganese					(1946) (1946) (1947) (1948)	200 260		450 200	250 160			85 130	(1947)	59	80	112 95				
			247	312					325	353				502			656			492
Molybdenum					(1946) (1946) (1947) (1948)	0.35 0.34		1.23 0.44	0.38 0.30			0.41 0.37	(1947)	0.28	0.33	0.55 0.49				358
Nickel					(1946) (1946) (1947) (1948)	1.06 1.43		2.90 1.08	0.72 0.56			0.79 0.30	(1947)	0.46	0.26	0.36 0.27				
Strontium					(1946) (1946) (1947) (1948)	6 5		6 5	5 4			6 6	(1947)	7	5	14 12				
Tin					(1946) (1946) (1947) (1948)	0.1 0.2		1.6 0.3	0.1 0.3			0.2 0.3	(1947)	0.2	0.3	0.3 0.3				
Titanium					(1946) (1946) (1947) (1948)	2.9 2.9		6.8 2.3	2.5 1.5			1.4 2.9	(1947)	2.6	1.8	2.0 3.3				
Zinc					(1946) (1946) (1947) (1948)	40 52		93 37	32 43			36 30	(1947)	77	22	34 47				

Table 1.197 Rushes

	January			February			March			April			May			June			
Heath Rush or Steel Bent: *Juncus squarrosus*																			
Proximate constituents %DM																			
Crude Protein																14.0			
										Average value for leaf, no date 8.75									
Ether extract																1.6			
										Average value for leaf, no date 3.28									
Nitrogen-free extract																56.8			
										Average value for leaf, no date 56.64									
Crude fibre																23.9			
										Average value for leaf, no date 27.51									
Ash																3.6			
										Average value for leaf, no date 3.82									
Proximate constituents %DM																			
Crude protein																10.6			
Amino acids % of Crude protein																			
Cystine																0.46			
Lysine																3.97			
Methionine, DL form																1.55			
L form																1.35			
Tryptophan																1.69			
Tyrosine																3.50			
Major elements %DM																			
Calcium																0.13			
							0.069					0.069					0.079		
										Average value for leaf, no date 0.19									
Phosphorus																0.22			
							0.200					0.182					0.212		
										Average value for leaf, no date 0.10									
Magnesium							0.128					0.143					0.172		
Potassium							1.23					1.81					2.07		
										Average value for leaf, no date 1.52									
Sodium																	0.156		
Chlorine							0.668					0.867					0.665		
										Average value for leaf, no date 0.46									
Sulphur, total							0.147					0.242					0.220		
Trace elements µg/g DM																			
Cobalt							0.081					0.072					0.089		
Copper							5.19					8.62					8.09		
Iron							140					146					129		
Manganese							106					165					270		
Vitamins µg/g DM																			
Thiamin																		2.26	
Riboflavin																		11.5	
Nicotinic acid																		42.9	
Pantothenic acid																		9.7	
Biotin																		0.321	
Pyridoxine																		12.9	
Common Rush: *Juncus communis*																			
Proximate constituents %DM																			
Crude protein															12.42				
Crude fibre															27.10				
Ash															7.25				
Ether extract										Mean value for the 2 samples, 27 May and 9 July: 3.56									
Major elements %DM																			
Calcium															0.26				
Phosphorus															0.25				
Potassium															2.76				
Chloride															1.10				
Woodrush: *Lazula campestris*																			
Proximate constituents %DM																			
Crude protein																			8.1
Ether extract																			2.4
Nitrogen-free extract																			58.5
Crude fibre																			28.4
Ash																			2.6
Major elements %DM																			
Calcium																			0.11
Phosphorus																			0.11
Potassium																			0.92
Chlorine																			0.21

July			August			September			October			November			December			Notes	Reference
F	9.0 8.5				8.4						7.8					7.8		Sample hand picked from heather moor F: flower heads only	468
																			218
F	1.5 2.1				1.4						1.4					1.3			468
																			218
F	57.6 61.6				57.5						56.9					56.7			468
																			218
F	28.3 24.0				29.8						30.5					31.8			468
																			218
F	3.6 3.8				2.9						3.4					2.4			468
																			218
																		Cut flowering, height 15–20 cm.	12
F	0.10 0.10				0.10						0.10					0.08		See Crude protein, same reference	468
		0.079		0.076			0.068												492
																			218
F	0.15 0.22				0.13						0.13					0.11			468
		0.183		0.126			0.126												492
																			218
		0.160		0.165			0.153												
		1.52		1.43			1.15												492
																			218
		0.154		0.141			0.125												492
		0.583		0.604			0.563												
																			218
		0.200		0.160			0.134												
		0.078		0.106			0.048												492
		6.49		12.76			7.15												
		107		116			155												
		280		168			162												
																		Early flowering	493
7.96 29.09 5.97																		Average values for leaf; no date	218
0.39 0.17 2.05 0.78																			

Table 1.198 Bog Asphodel: *Narthecium ossifragum*

	Year	Site	May	June		July	August		September	October	Notes	Reference
Proximate constituents %DM												
Crude protein										7.4	Whole plant analysed.	218
Ether extract										4.5		
Nitrogen-free extract										41.5		
Crude fibre										40.6		
Ash										6.0		
Major elements %DM												
Calcium										0.6		
Phosphorus										0.1		
Potassium										1.1		
Chlorine										0.9		
Trace elements μg/g DM												
Barium	(1946)	R	9	9		7			11		R : area Rhigolter, Sutherland.	358
		C	7	3		6			4		C : area Carbreck, Sutherland.	
	(1947)	R	15		17	10			14		G : area Glendye, Kincardineshire	
		C	5		4	16			4		Natural woodland grazing	
		G					14				All parts of herbage used	
	(1948)	R			18			5				
		C			3			2				
Cobalt	(1946)	R	0.35	0.21		0.16			0.35			
		C	0.25	0.24		0.10			0.07			
	(1947)	R	0.53		0.20	0.11			0.18			
		C	0.31		0.12	0.21			0.07			
		G					0.04					
	(1948)	R			0.14			0.17				
		C			0.09			0.09				
Copper	(1946)	R	14.0	8.4		6.2			4.4			
		C	16.0	11.0		8.1			5.4			
	(1947)	R	17.0		9.8	6.3			5.4			
		C	15.5		15.5	9.2			5.6			
		G					8.4					
	(1948)	R			8.2			5.4				
		C			12.0			7.0				
Iron	(1946)	R	76	69		80			96			
		C	86	102		80			73			
	(1947)	R	97		70	32			83			
		C	86		93	111			82			
		G					92					
	(1948)	R			52			91				
		C			75			73				
Lead	(1946)	R	1 3	4.2		1.3			3.0			
		C	1.5	1.3		1.0			4.3			
	(1947)	R	1.9		1.4	0:8			3.1			
		C	2.9		2.0	2.7			1 7			
		G					2.7					
	(1948)	R			1.7			2.4				
		C			1.6			2.1				
Manganese	(1946)	R	320	270		190			430			
		C	420	180		120			120			
	(1947)	R	430		460	310			530			
		C	190		130	250			110			
		G					180					
	(1948)	R			200			120				
		C			108			78				

Table 1.198 Bog Asphodel (continued)

Trace elements μg/g DM	Year		May	June			July			August			September			October			Notes	Reference
Molybdenum (cont'd)	(1946)	R	0.20		0.14				0.16				0.18							
		C	0.23		0.25				0.22				0.25							
	(1947)	R	0.18			0.13			0.09				0.17							
		C	0.17			0.19			0.46				0.25							
		G									0.20									
	(1948)	R				0.15						0.16								
		C				0.23						0.24								
Nickel	(1946)	R	2.20		1.33				1.25				0.55							
		C	2.48		1.84				1.39				0.40							
	(1947)	R	5.25			1.48			0.77				1.22							
		C	2.98			2.09			1.06				0.55							
		G									0.92									
	(1948)	R				1.81						2.44								
		C				1.82						1.19								
Strontium	(1946)	R	16		17				12				23							
		C	47		13				14				23							
	(1947)	R	22			24			14				25							
		C	18			22			19				23							
		G									39									
	(1948)	R				13						9								
		C				12						11								
Tin	(1946)	R	0.4		0.2				0.2				0.4							
		C	0.4						0.2				0.3							
	(1947)	R	1.9			0.5			0.2				0.7							
		C	0.5			0.2			0.4				0.3							
		G									0.3									
	(1948)	R				0.3						0.5								
		C				0.3						0.7								
Titanium	(1946)	R	1.8		1.2				4.0				2.2							
		C	2.4		7.2				4.1				4.3							
	(1947)	R	3.3			1.7			1.0				2.4							
		C	4.4			4.1			7.1				3.5							
		G									4.3									
	(1948)	R				0.7						2.2								
		C				1.7						4.5								
Zinc	(1946)	R	49		31				38				29							
		C	53		60				50				34							
	(1947)	R	72			61			22				41							
		C	62			53			33				40							
		G									67									
	(1948)	R				55						36								
		C				44						28								

Table 1.199 Uncultivated field legumes

	April			May			June			July			August			September			Notes	Reference
Nonsuch: *Medicago lupulina* *Proximate constituents* %DM Crude protein		29.1		24.2		19.9	19.4			18.0									Air-dried, 7–10 days.	9
									22.4										No date given. Probably leafy, preflowering. Oven-dried and ground.	420
Ether extract		2.5		2.2		2.0	1.2			2.2										9
Nitrogen-free extract		43.9		43.8		48.9	45.1			45.1										
Crude fibre		14.0		20.3		21.6	27.3			27.0										
									18.9			1 .9								420
Ash		10.5		9.5		7.6	7.0			7.7										9
Carbohydrate fractions %DM Cellulose		17.4		22.1		23.6	26.4			27.4										
Lignin		4.0		4.8		6.8	8.5			8.4										
Major elements %DM Calcium			1.8		1.9		1.7		1.6		1.3								Dates uncertain. April/May = Preflowering. Early June = mid-flowering. Late June = late flowering. July = mature	490
Phosphorus			0.5		0.3		0.3		0.3		0.3									
Magnesium			0.9		0.8		0.7		0.8		0.6									
Potassium			2.7		2.4		2.0		2.0		2.3									
Sodium			0.11		0.15		0.15		0.15		0.09									
Chlorine			0.7		0.6		0.5		0.5		0.5									
Trace elements μg/g DM Cobalt			0.21		0.16		0.29		0.15		0.20									
Copper			9.6		7.0		7.1		7.4		5.3									
Iron			442		375		279		445		373									
									556											420
Manganese			51.0		45.0		38.3		40.4		42.5									490
Birdsfoot Trefoil: *Lotus corniculatus* *Proximate constituents* %DM Crude protein						17.5													Flowering stage	220
						2.3														
Nitrogen-free extract						49.1														
Crude fibre						22.0														
Ash						9.1														
Major elements %DM Calcium						1.0														
Phosphorus												A 0.1 B 0.2							A : untreated. B : superphosphate 750 kg/ha No date given. Moor pasture. Aphosphorosis investigation.	145
						0.2														220
Magnesium						0.3														
Potassium						2.0														
Sodium						0.8														
Chlorine						0.7														
Vitamins μg/g DM Thiamin															3.5				Air-dried. Early flowering stage.	493
Riboflavin															16.0					
Nicotinic acid															34.5					
Pantothenic acid															20.2					
Pyridoxine															8.6					
Biotin															0.47					

Table 1.199 Uncultivated field legumes (continued)

	April			May			June			July			August			September			Notes	Reference
Spring Vetch																				
Crude protein %DM																				284
(1945)								N_0	29.4		28.0								N_0 : no fertiliser	
								N_1	31.6		29.1								N_1 : Nitrochalk 437 kg/ha applied in one dressing	
								N_2	29.6		28.6								N_2 : Nitrochalk 437 kg/ha applied in two dressings	
								N_3	29.0		27.1								N_3 : Nitrochalk 437 kg/ha applied in three dressings	
(1946)							N_0	27.9		23.4										
							N_1	28.9		26.3										
							N_2	30.0			29.3									
							N_3	27.8			28.2									
Tares																				
Proximate constituents %DM																				
Crude protein											23.1								Young plants grown for experiments in green manuring of soils.	146
Ether extract											2.2									
Carbohydrate fractions %DM																				
Cellulose											9.7									
Lignin											12.6									
Total furfuraldehyde											5.5									
Major elements %DM																				
Potassium											2.4									
Tufted Vetch																				
Proximate constituents %DM																				
Crude protein						36.3													Early flowering stage	220
Ether extract						1.9														
Nitrogen-free extract						31.6														
Crude fibre						21.0														
Ash						9.2														
Major elements %DM																				
Calcium						0.8														
Phosphorus						0.5														
Magnesium						0.2														
Potassium						3.0														
Sodium						0.14														
Chlorine						0.8														
Trace elements μg/g DM																				
Cobalt						0.2														
Copper						14.4														
Iron						273														
Manganese						58														
Vetch, unnamed																				
Crude protein %DM											19.1								Whole plant	353
											20.3								Tops	
											12.3								Roots	
																			Green manuring experiment Mean values for years 1936–39.	

Table 1.200 Gorse or whin: *Ulex europaeus*

	March	April	May	June	July	August	October	Notes	Reference
Proximate constituents % DM Crude protein					16.8			Leaf only. No date. (Components of DM as given Total 104.0%	218
	11.9	13.6	(1) 11.3 (2) 11.5	15.2	(1) 16.3 (2) 15.3	15.3	13.4	May: (1) in flower bud, (2) in flower. July: (1) tips with seed, (2) seed only.	310
Ether extract					4.7			Leaf only. No date.	218
	2.9	2.3	(1) 1.0 (2) 1.6	1.3	(1) 1.4 (2) 1.6	2.1	2.3	As Crude protein, same reference	310
Nitrogen-free extract					42.0			Leaf only. No date.	218
	44.1	44.4	(1) 55.9 (2) 58.0	53.0	(1) 48.4 (2) 45.3	35.7	31.3	As Crude protein, same reference.	310
Crude fibre					33.1			Leaf only. No date	218
	37.7	36.4	(1) 28.6 (2) 26.1	26.2	(1) 29.7 (2) 34.0	43.2	49.8	As Crude protein, same reference.	310
Ash					7.4			Leaf only. No date.	218
	3.4	3.3	(1) 3.2 (2) 2.8	4.3	(1) 4.2 (2) 3.8	3.7	3.2	As Crude protein, same reference.	310
Major elements %DM Calcium					0.5			Leaf only. No date.	218
	0.4	0.4	(1) 0.4 (2) 0.4	0.5	(1) 0.4 (2) 0.3	0.4	0.3	As Crude protein, same reference.	310
Phosphorus					0.3			Leaf only. No date.	218
	0.1	0.1	(1) 0.1 (2) 0.1	0.2	(1) 0.2 (2) 0.2	0.2	0.1	As Crude protein, same reference.	310
Magnesium	0.2	0.1	(1) 0.1 (2) 0.1	0.2	(1) 0.2 (2) 0.2	0.2	0.1		
Potassium					2.7			Leaf only. No date.	218
	0.7	0.6	(1) 0.6 (2) 0.6	0.9	(1) 0.8 (2) 0.9	1.0	0.9	As Crude protein, same reference.	310
Sodium	0.40	0.37	(1) 0.30 (2) 0.30	0.34	(1) 0.41 (2) 0.31	0.37	0.24		
Chlorine					0.5			Leaf only. No date.	218
	0.4	0.2	(1) 0.2 (2) 0.2	0.2	(1) 0.3 (2) 0.2	0.3	0.3	As Crude protein, same reference.	310
Trace elements μg/g DM Cobalt	–	0.03	(1) 0.07 (2) 0.06	0.07	(1) 0.07 (2) –	0.07	0.06		
Copper	6	4	(1) 4 (2) 3	3	(1) 5 (2) –	5	5		
Iron	250	191	(1) 157 (2) 138	226	(1) 211 (2) 552	182	108		
Manganese	58	41	(1) 65 (2) 61	74	(1) 100 (2) 68	63	51		
Zinc	42	42	(1) 33 (2) 49	52	(1) 45	49	46		

Table 1.201 Sorrel, Broad Dock: *Rumex*

		Notes	Reference
Sheep's sorrel: *Rumex acetosella* *Proximate constituents* %DM Crude protein	S 18.27 L 27.03	Sampled 16 May 1930 S : stem L : leaf	218
Crude fibre	S 21.27 L 17.40		
Ash	S 10.87 L 15.00		
Major elements %DM Calcium	S 0.42 L 0.96	As for Crude protein	218
	F_0 0.45 F_1 0.51 F_2 0.50 F_3 0.66 F_4 0.49 F_5 0.69	F_0 : control. Sampled 3–15 June. Pre-flowering. F_1 : farmyard manure 20 tonnes/ha annually F_2 : ammonium sulphate 165 kg/ha F_3 : basic slag 330 kg/ha F_4 : muriate of potash 110 kg/ha F_5 : last 3 together	489
Phosphorus	S 0.57 L 0.38	As for Crude protein	218
	F_0 0.40 F_1 0.60 F_2 0.48 F_3 0.61 F_4 0.51 F_5 0.60	As for Calcium	489
Magnesium	F_0 0.16 F_1 0.16 F_2 0.17 F_3 0.17 F_4 0.14 F_5 0.16	As for Calcium	489
Potassium	S 2.33 L 3.32	As for Crude protein	218
	F_0 3.55 F_1 4.01 F_2 3.06 F_3 3.42 F_4 4.24 F_5 3.77	As for Calcium	489
Sodium	F_0 0.25 F_1 0.20 F_2 0.31 F_3 0.24 F_4 0.18 F_5 0.23	As for Calcium	489
Chloride	S 1.03 L 3.26	As for Crude protein	218
	F_0 1.24 F_1 1.25 F_2 1.21 F_3 0.99 F_4 1.65 F_5 1.45	As for Calcium	489
Sulphur, inorganic	F_0 0.05 F_1 0.04 F_2 0.09 F_3 0.10 F_4 0.07 F_5 0.14	As for Calcium	489
Sulphur, organic	F_0 0.25 F_1 0.26 F_2 0.22 F_3 0.12 F_4 0.15 F_5 0.15	As for Calcium	489

continued

Table 1.201 Sorrel, Broad Dock: (continued)

		Notes	Reference
Sheep's sorrel (cont'd)			
Trace elements µg/g DM			
Cobalt	F_0 0.27 F_1 0.22 F_2 0.22 F_3 0.12 F_4 0.22 F_5 0.20	As for Calcium	489
Copper	F_0 14.2 F_1 11.0 F_2 11.7 F_3 5.4 F_4 8.5 F_5 14.2		
Iron	F_0 144 F_1 139 F_2 163 F_3 137 F_4 191 F_5 196		
Manganese	F_0 299 F_1 135 F_2 201 F_3 214 F_4 372 F_5 502		
Broad Dock: *Rumex obtusifolius*			
Proximate constituents %DM			
Crude protein	25.09	Sampled 28th May. Preflowering	220
Ether Extract	1.73		
Nitrogen-free extract	50.54		
Crude fibre	12.89		
Ash	9.75		
Major elements %DM			
Calcium	0.83		
Phosphorus	0.53		
Magnesium	0.26		
Potassium	3.20		
Sodium	0.33		
Chloride	1.23		
Trace elements µg/g DM			
Cobalt	0.15		
Copper	10.2		
Iron	234		
Manganese	38		

Table 1.202 Buttercup: *Ranunculus* spp.

			Notes	Reference
Proximate constituents %DM			*R. repens* in Ref. 218; *R. bulbosus* in Ref. 220	
Crude protein		16.2	Average value for leaf, no date	218
		12.0	Average value 5 samples	20
		25.1	Early flowering 28 March	220
Ether extract		5.8	Average value for leaf, no date	218
		2.3	As for Crude protein	220
Nitrogen-free extract		54.8		218
		41.6		220
Crude fibre		16.8		218
		15.3		220
Ash		6.4		218
		7.4		20
		15.7		220
Major elements %DM				
Calcium		0.94		218
	F_0	1.02	F_0: Control plot; no fertiliser	489
	F_1	1.06	F_1: farmyard manure 20 tonnes/ha/annum	
	F_2	0.99	F_2: sulphate of ammonia 165 kg/ha	
	F_3	1.06	F_3: basic slag 330 kg/ha	
	F_4	1.00	F_4: muriate of potash 110 kg/ha	
	F_5	1.04	F_5: Last 3 together	
		1.63	As for Crude protein	20
		1.08		220
Phosphorus		0.33		218
	F_0	0.26	As for Calcium	489
	F_1	0.41		
	F_2	0.32		
	F_3	0.39		
	F_4	0.28		
	F_5	0.41		
		0.30	As for Crude protein	20
		0.63		220
Magnesium	F_0	0.20	As for Calcium	489
	F_1	0.20		
	F_2	0.22		
	F_3	0.22		
	F_4	0.18		
	F_5	0.21		
		0.23	As for Crude protein	20
		0.22		220
Potassium		1.72	As for Crude protein	218
	F_0	2.02	As for Calcium	489
	F_1	2.54		
	F_2	2.01		
	F_3	2.13		
	F_4	2.59		
	F_5	2.47		
		5.61	As for Crude protein	220

continued

Table 1.202 Buttercup (continued)

			Notes	Reference
Major elements %DM (cont'd)				
Sodium	F_0	0.15	As for Calcium	
	F_1	0.19		
	F_2	0.23		489
	F_3	0.24		
	F_4	0.20		
	F_5	0.24		
		0.64	As for Crude protein	220
		0.88		218
Chlorine	F_0	0.20	As for Calcium	
	F_1	0.25		489
	F_2	0.24		
	F_3	0.33		
	F_4	0.45		
	F_5	0.56		
		0.43	As for Crude protein	20
		1.38		220
Sulphur, inorganic	F_0	0.06	As for Calcium	
	F_1	0.08		
	F_2	0.05		489
	F_3	0.05		
	F_4	0.07		
	F_5	0.06		
Sulphur, organic	F_0	0.15		
	F_1	0.15		
	F_2	0.17		
	F_3	0.18		
	F_4	0.12		
	F_5	0.17		
Trace elements µg/g DM				
Cobalt	F_0	0.26		
	F_1	0.25		
	F_2	0.26		
	F_3	0.22		
	F_4	0.24		
	F_5	0.25		
		0.14	As for Crude protein	220
Copper	F_0	13.5	As for calcium	
	F_1	17.5		
	F_2	16.0		489
	F_3	17.7		
	F_4	16.8		
	F_5	17.9		
		14.3	As for Crude protein	20
		15.0		220
Iron	F_0	170	As for Calcium	489
	F_1	56		
	F_2	140		
	F_3	92		
	F_4	125		
	F_5	117		
		125	As for Crude protein	20
		413		220
Manganese	F_0	134	As for Calcium	
	F_1	77		
	F_2	98		489
	F_3	82		
	F_4	149		
	F_5	122		
		31	As for Crude protein	20
		76		220

Table 1.203 Burnet: *Poterium sanguisorba*

	April			May			June			July			August			September			Notes	Reference
Proximate constituents %DM																				
Crude protein		17.4		15.5		9.2	7.1			6.1									1 and 2 preflowering, 3 full flower, 4 and 5 seeding	9
													18.6						Leafy, preflowering	15
													13.9						Preflowering	421
Ether Extract		2.5		2.7		1.8	1.9			1.3									As for Crude Protein	9
Nitrogen-free extract		65.2		61.6		61.9	60.8			62.5										
Crude fibre		7.6		12.6		21.3	24.5			24.2									As for Crude protein	9
													11.7						Leafy, preflowering	15
													11.99						Preflowering	421
Ash		7.3		7.6		5.8	5.7			5.9									As for Crude protein	9
Carbohydrate fractions %DM																				
Cellulose		11.2		14.5		21.6	25.0			23.0										
Lignin		3.4		3.2		5.4	12.2			13.1										
Major elements %DM																				
Calcium													1.32			1.20		1.38	1 preflowering, 2 midflowering, 3 late flowering	490
													1.3						Leafy, preflowering	15
Phosphorus													0.34			0.23		0.22	As for Calcium	490
													0.13						Leafy, preflowering	15
Magnesium													1.30			1.07		0.91	As for Calcium	
Potassium													2.0			1.75		1.70		490
Sodium													0.06			0.04		0.06		
Chlorine													0.15			0.13		0.16		
Trace elements μg/g DM																				
Cobalt													0.19			0.28		0.13		
Copper													10.0			6.3		6.7		
Iron													284			181		247		
													299						Preflowering	421
Manganese													47			23.2		19.7	As for Calcium	490
Vitamins μg/g DM																				
Thiamin															3.97				Early flowering	493
Riboflavin															11.9					
Nicotinic Acid															36.8					
Pantothenic Acid															11.7					
Biotin															0.207					
Pyridoxine															18.8					

Table 1.204 Heather: *Calluna vulgaris*

For studies of proximate composition, digested components and metabolisable energy see table 1.237

	Years since last burning	January			February			March			April			May			June		
Proximate constituents %DM																			
Crude protein																			
	–																	9.0	
	–																		
	–																		
	–																		
	2																		
	4																		
	6																		
	8																		
	>20																		
	4																		
	6																		
	8																		
	>20																		
	3		8.7																12.9
	5		7.6																8.6
	7		7.1																7.5
	3																	8.2	
	5																	7.9	
	7																	8.3	
	9																	7.4	
	4																		
	First																		
	1 & 2																		
	10–12																		
	First																		
	1 & 2																		
	12–14																		
	–																		
	–																		
	–						5.8		6.4	5.8	6.8	6.7							
	–																		
	–																		
	–			8.0															
	4				6.6									7.2					
	4 (Sept. sample)																		
	10 (Oct. sample)																		
	3																	7.0	
	7																	7.1	
	14																	7.2	
Ether extract	–																	7.8	
	2																		
	4																		
	6																		
	8																		
	>20																		
	4																		
	6																		
	8																		
	>20																		
	3		3.5																1.3
	5		3.9																2.9
	7		4.2																3.3
	3																	3.9	
	5																	4.4	
	7																	4.3	
	9																	4.0	
	–			2.6															
	4							3.5						3.4					
	4 (Sept. sample)																		
	10 (Oct. sample)																		
	3																	3.7	
	7																	3.0	
	14																	3.4	

July			August			September			October			November			December			Notes	Reference
																		When no sampling date is given the data are entered as mid-June. Here age of the heather is more important than date of sampling.	
																		Top shoots. Mid Wales	218
							8.54											Tips. Central Wales	
							17.69											Part unspecified. Central Wales	
							11.18											Part unspecified. Central Wales (autumn assumed as Sept.)	523
	11.9 8.6 7.7 7.0 6.9 7.7 7.2 6.5 6.4																	Newbiggin, Northumberland. Shoots, fresh, green, mainly tips. Shoots, green, slightly woody. (mid-summer taken as July)	466
		10.2 8.8 7.5					9.6 8.5 7.4			10.1 7.9 7.2								Newbiggin, Northumberland	469
																		Blanchland, Northumberland. Green leaves, tips and bloom.	474
	10.9 8.8 9.8 7.5						6.8 6.6			6.2 5.9								Blanchland, Northumberland: leafy shoots as eaten by sheep as eaten by sheep	470
	11.8																	Upland and semi-moor samples	145
					8.4													Mean of 2 samples Area 1	
				7.7 20.2						8.0								5 samples. Area 1 12 samples August 1 sample October. Area 2 1 sample, reseeded area within Area 2.	391
					8.9													2 samples. Area 3.	
																		Young shoots	11
							7.0		6.4									Fresh tips, shoots, flowers: some dry shoots	488
																		Mainly tips and shoots. No date.	461
																		As for Crude Protein	218
	3.1 4.3 4.6 5.1 5.2 3.9 3.9 4.2 4.1																		466
		2.4 3.8 3.5					5.1 4.5 4.2			4.1 4.3 4.3									469
																			474
					3.4 3.1													2 samples. Area 1 2 samples. Area 3	391
																			11
							3.7		3.8										488
																			461

continued

Table 1.204 Heather (continued)

	Years since last burning	January			February			March			April			May			June		
Proximate constituents (cont'd) Nitrogen-free extract	–																	46.8	
	2																		
	4																		
	6																		
	8																		
	>20																		
	4																		
	6																		
	8																		
	>20																		
	3		66.0																63.1
	5		67.3																65.4
	7		65.4																64.4
	3																	63.9	
	5																	64.8	
	7																	63.0	
	9																	64.1	
	–			57.7															
	4							64.4						65.1					
	4 (Sept. sample)																		
	10 (Oct. sample)																		
	3																	64.2	
	7																	64.6	
	14																	61.5	
Crude fibre	–																	28.9	
	–																		
	–																		
	–																		
	2																		
	4																		
	6																		
	8																		
	>20																		
	4																		
	6																		
	8																		
	>20																		
	3		17.7																17.9
	5		18.1																19.6
	7		20.1																20.9
	3																	20.5	
	5																	19.3	
	7																	21.1	
	9																	21.1	
	4																		
	First																		
	1 & 2																		
	10–12																		
	First																		
	1 & 2																		
	12–14																		
	–																		
	–						30.0		27.4	30.2	27.3	28.0							
	–																		
	–																		
	–			28.8															
	4				21.8									20.8					
	4 (Sept. sample)																		
	10 (Oct. sample)																		
	3																	21.3	
	7																	21.7	
	14																	24.4	

July			August			September			October			November			December			Notes	Reference
																			218
	58.3 62.1 62.8 64.4 63.7 62.2 63.8 64.2 61.7																		466
		63.7 63.5 64.1					59.8 62.8 63.8			63.6 64.0 64.5									469
																			474
					60.7 58.2													2 Samples. Area 1 2 Samples. Area 3	391
																			11
							65.2		64.6										488
																			461
																			218
							23.0 24.4 27.2												523
	22.2 21.7 21.5 20.4 21.1 23.0 21.5 22.1 24.8																		466
		19.4 20.6 21.1					21.2 21.1 20.8			18.0 20.2 20.1									469
																			474
	22.2 21.5 24.0 23.1						24.3 22.7			23.2 24.7									470
				23.7 18.7	24.3 25.5					29.9									391
																			11
							20.3		21.7										488
																			461

continued

Table 1.204 Heather (continued)

	Years since last burning	January			February			March			April			May			June		
Proximate constituents (cont'd) Ash	– – – –																	7.5	
	2 4 6 8 >20 4 6 8 >20																		
	3 5 7		4.1 3.1 3.2																4.8 3.5 3.9
	3 5 7 9																	3.5 3.6 3.3 3.4	
	4 First 1 & 2 10–12 First 1 & 2 12–14																		
	– –			2.9															
	4				3.6									3.5					
	4 (Sept. sample) 10 (Oct. sample)																		
	3 7 14																	3.8 3.5 3.4	
Amino acids g/16gN Alanine																		4.3 5.4	
Arginine																		4.1 4.7	
Aspartic acid																		14.7 9.2	
Cystine																		1.2 1.3	
Glutamic acid																		9.4 10.4	
Glycine																		3.8 5.3	
Histidine																		2.2 1.8	
Isoleucine																		3.2 4.0	
Leucine																		6.1 8.0	
Lysine																		4.6 5.4	
Methionine																		1.7 1.7	
Phenylalanine																		3.8 4.8	
Proline																		3.9 4.0	
Serine																		3.4 4.6	
Threonine																		3.6 4.4	
Tyrosine																		2.6 3.3	
Valine																		4.6 5.4	

July			August			September			October			November			December			Notes	Reference
							2.5 6.6 3.6												218 523
	4.5 3.3 3.7 3.1 3.1 3.2 3.6 3.0 3.0																		466
		4.3 3.3 3.8					4.3 3.1 3.8			4.2 3.6 3.9									469
																			474
	3.5 4.1 3.1 4.0						2.8 3.1			3.3 2.6									470
					3.2 4.3													Mean of 2 samples. Area 1 2 samples. Area 3	391
																			11
							3.8		3.5										488
																			461
																			369

continued

Table 1.204 Heather: (continued)

	Years since last burning	January			February			March			April			May			June		
Carbohydrate fractions %DM	First																		
Lignin	1 & 2																		
	First																		
	1 & 2																		
Major elements %DM																			
Calcium	–																	0.36	
	–																		
	–																		
	–																		
	2																		
	4																		
	6																		
	8																		
	>20																		
	4																		
	6																		
	8																		
	>20																		
	3		0.43																0.62
	5		0.61																0.48
	7		0.67																0.61
				0.16															
Phosphorus																		0.15	
	2																		
	4																		
	6																		
	8																		
	>20																		
	4																		
	6																		
	8																		
	>20																		
	3		0.13																0.17
	5		0.09																0.12
	7		0.10																0.10
	–																		
	–																		
	–																		
				0.11															
Magnesium	–			0.21															
Potassium																		0.86	
				0.44															
Sodium	–			0.06															
Chlorine																		0.07	
Major elements %DM																			
Calcium	2																	0.43	
	4																	0.48	
	6																	0.46	
	8																	0.43	
	10																	0.49	
Phosphorus	2																	0.16	
	4																	0.11	
	6																	0.22	
	8																	0.11	
	10																	0.14	
Magnesium	2																	0.22	
	4																	0.21	
	6																	0.27	
	8																	0.19	
	10																	0.24	

July			August			September			October			November			December			Notes	Reference
	22.5 22.6 22.0 23.2																		470
						0.53 0.15 0.39												Top shoots. Mid Wales Tips. Central Wales Part Unspecified. Central Wales Part Unspecified. Central Wales	218 523
	0.66 0.34 0.43 0.39 0.49 0.31 0.44 0.61 0.40																	Newbiggin, Northumberland Shoots, fresh, green, mainly tips. Shoots, green, slightly woody	466
		0.56 0.56 0.61					0.49 0.58 0.59			0.51 0.54 0.56								Newbiggin, Northumberland	469
																		Hill pasture	391
							0.10 0.04 0.09											As for Calcium	218 523
	0.21 0.10 0.11 0.10 0.10 0.08 0.10 0.09 0.09																		466
		0.18 0.12 0.12					0.13 0.11 0.10			0.14 0.10 0.10									469
	0.11 0.10 0.06																	Upland Semi-moor Moor	145
																		Hill, pasture	391
																		Hill pasture	391
																		Top shoots. Central Wales Hill pasture	218 391
																		Hill pasture	391
																		Top shoots. Central Wales	218
	0.47 0.48 0.46 0.46 0.48			0.50 0.47 0.48 0.45 0.49		F_0 F_1 F_2 F_3	0.44 0.62 0.42 0.47			0.59 0.60 0.53 0.50 0.48								Newbiggin, Northumberland, 1937. September values to show effect of fertilisers. Per ha: Lime, 12.5 tonnes ground limestone; P, 1500 kg North African phosphate; K, 375 kg sulphate of potash F_0: No fertiliser F_1: with lime F_2: with K and P F_3: with P	478
	0.16 0.11 0.17 0.13 0.12			0.13 0.09 0.14 0.11 0.10		F_0 F_1 F_2 F_3	0.16 0.16 0.17 0.18			0.15 0.11 0.14 0.11 0.10									
	0.27 0.28 0.28 0.24 0.26			0.24 0.24 0.26 0.21 0.25		F_0 F_1 F_2 F_3	0.24 0.20 0.24 0.26			0.24 0.24 0.29 0.23 0.26									

continued

Table 1.204 Heather (continued)

	Years since last burning	January			February			March			April			May			June		
Major elements %DM (cont'd) Potassium	2																	0.70	
	4																	0.62	
	6																	0.65	
	8																	0.66	
	10																	0.64	
Sodium	2																	0.063	
	4																	0.037	
	6																	0.047	
	8																	0.042	
	10																	0.043	
Chlorine	2																	0.21	
	4																	0.21	
	6																	0.22	
	8																	0.22	
	10																	0.21	
Sulphur, total	2																	0.24	
	4																	0.21	
	6																	0.25	
	8																	0.20	
	10																	0.18	
Sulphur, inorganic	2																	0.10	
	4																	0.08	
	6																	0.09	
	8																	0.09	
	10																	0.07	
Trace elements μg/g DM Cobalt	2																		
	4																		
	6																		
	8																		
	10																		
Copper	2																	14.2	
	4																	13.4	
	6																	14.3	
	8																	13.3	
	10																	16.9	
Iron	2																	304	
	4																	356	
	6																	302	
	8																	415	
	10																	371	
Manganese	2																	890	
	4																	890	
	6																	870	
	8																	850	
	10																	1110	

July	August	September		October	November	December	Notes	Reference
0.81	0.57			0.50				
0.59	0.53	F_0	0.49	0.50				478
		F_1	0.45					
		F_2	0.55					
		F_3	0.45					
0.56	0.53			0.49				
0.55	0.54			0.49				
0.63	0.60			0.50				
0.052	0.059			0.069				
0.051	0.060	F_0	0.062	0.083				
		F_1	0.055					
		F_2	0.061					
		F_3	0.054					
0.062	0.060			0.069				
0.059	0.064			0.073				
0.047	0.054			0.073				
0.20	0.19			0.15				
0.19	0.18	F_0	0.12	0.15				
		F_1	0.09					
		F_2	0.13					
		F_3	0.14					
0.19	0.16			0.14				
0.19	0.19			0.14				
0.19	0.17			0.15				
0.21	0.18			0.15				
0.20	0.16	F_0	0.15	0.14				
		F_1	0.15					
		F_2	0.17					
		F_3	0.18					
0.18	0.20			0.19				
0.17	0.15			0.17				
0.20	0.18			0.15				
0.08	0.07			0.06				
0.09	0.06	F_0	0.06	0.07				
		F_1	0.04					
		F_2	0.06					
		F_3	0.06					
0.07	0.06			0.07				
0.07	0.06			0.09				
0.08	0.07			0.06				
	0.22							
	0.17	F_0	0.14					
		F_1	0.12					
		F_2	0.09					
		F_3	0.13					478
	0.11							
0.17								
0.13								
16.8	10.8			9.8				
15.7	11.5	F_0	12.5	14.8				
		F_1	11.1					
		F_2	13.1					
		F_3	11.4					
13.4	11.8			14.2				
14.2	12.2			13.0				
14.1	12.6			14.1				
284	271			239				
242	403	F_0	211	225				
		F_1	249					
		F_2	171					
		F_3	250					
281	213			278				
282	223			255				
406	286			257				
860	730			760				
850	800	F_0	760	810				
		F_1	270					
		F_2	760					
		F_3	800					
830	740			740				
870	800			750				
870	880			760				

continued

Table 1.204 Heather (continued)

	Age of shoots, years	January			February			March			April			May			June		
Trace elements μg/g DM (cont'd)	C																		
Barium														(1946)		8		4	
														(1947)		11			7
														(1948)					12
Cobalt														(1946)		0.13		0.21	
														(1947)		0.12			0.10
														(1948)					0.09
Copper														(1946)		6.5		8.0	
														(1947)					11.0
														(1948)					6.5
Iron														(1946)		160		242	
														(1947)		183			137
														(1948)					105
Lead														(1946)		5.8		8.5	
														(1947)		8.1			4.5
														(1948)					4.0
Manganese														(1946)		400		170	
														(1947)		400			550
														(1948)					470
Molybdenum														(1946)		0.19		0.20	
														(1947)		0.28			0.17
														(1948)					0.24
Nickel														(1946)		0.64		(2.60)	
														(1947)		0.87			0.89
														(1948)					0.85
Strontium														(1946)		18		7	
														(1947)		15			7
														(1948)					15
Tin														(1946)		0.5		0.3	
														(1947)		0.6			1.1
														(1948)					0.4
Titanium														(1946)		27		38	
														(1947)		34			18
														(1948)					16
Zinc														(1946)		28		26	
														(1947)		20			29
														(1948)					30
Barium																			
Cobalt																			
Copper																			
Iron																			
Lead																			
Manganese																			
Molybdenum																			
Nickel																			
Strontium																			
Tin																			
Titanium																			
Zinc																			

July			August			September			October			November			December			Notes	Reference
		5 9			10	6 10												C : Current year's growth, leaf and stem	
		0.12 0.09			0.10	0.13 0.09													358
		8.1 8.3			6.6	8.0 8.5												Carbreck, Durness, Sutherland Moor	
		145 129			138	158 105													
		6.4 5.7			7.7	7.8 3.4													
		420 650			320	430 220													
		0.15 0.17			0.19	0.18 0.13													
		0.80 0.64			0.66	0.83 0.59												Value (2.60) in June questionable.	
		9 10			12	10 9													
		0.4 0.4			1.1	0.6 0.3													
		24 21			22	28 12													
		28 27			22	29 22													
				46 0.11 11.0 117 7.0 287 0.26 0.62 8 <0.5 18 37														Kincardine Dry Moor, 1947	

2L

Table 1.205 Heath: *Erica* spp.

	July	August	September	October	Notes	Reference
Cross-leaved heath: *E. tetralix*						
Proximate constituents %DM						
Crude protein		9.7			Average value for leaf only No date.	218
Ether extract		6.7				
Nitrogen-free extract		43.6				
Crude fibre		36.0				
Ash		4.0				
Major elements %DM						
Calcium		0.6				
Phosphorus		0.1				
Potassium		1.2				
Chlorine		0.7				
Trace elements μg/g DM						
Barium		(1) 88 (2) 97 (3) 34			(1) Dry moor sample (2) Dry moor sample, recently burnt. (3) Wet moor. Natural moorland grazing	358
Cobalt		(1) 0.2 (2) 0.2 (3) 0.2				
Copper		(1) 10.5 (2) 10.5 (3) 14.0				
Iron		(1) 151 (2) 217 (3) 214				
Lead		(1) 11.7 (2) 9.2 (3) 20.0				
Manganese		(1) 200 (2) 265 (3) 200				
Molybdenum		(1) 0.2 (2) 0.4 (3) 0.2				
Nickel		(1) 1.1 (2) 1.2 (3) 1.5				
Strontium		(1) 14 (2) 13 (3) 28				
Tin		(1) 0.5 (2) 0.7 (3) 1.0				
Titanium		(1) 15 (2) 30 (3) 18				
Zinc		(1) 38 (2) 35 (3) 48				

Table 1.205 Heath: (continued)

	July			August			September			October			Notes	Reference
Fine-leaved Heath (Bell heather): *E. cinerea* *Trace elements* μg/g DM														
Barium				(1) (2)	64 28								(1) dry moor sample (2) dry moor sample, recently burnt. Natural moorland grazing	358
Cobalt				(1) (2)	0.2 0.1									
Copper				(1) (2)	12.8 7.8									
Iron				(1) (2)	165 116									
Lead				(1) (2)	12.3 11.7									
Manganese				(1) (2)	165 115									
Molybdenum				(1) (2)	0.7 0.2									
Nickel				(1) (2)	1.7 1.5									
Strontium				(1) (2)	20 14									
Tin				(1) (2)	0.5 0.6									
Titanium				(1) (2)	18 11									
Zinc				(1) (2)	47 41									

Table 1.206 Bilberry or Blaeberry: *Vaccinium myrtillus*

	January			February			March			April			May			June		
Proximate constituents %DM																		
Crude protein																	15.7	
(1937)														18.6				13.5
(1938)		7.2																
Ether extract																	7.2	
(1937)														2.3				3.2
(1938)		1.8																
Nitrogen-free extract																	56.4	
(1937)														64.6				63.6
(1938)		49.8																
Crude fibre																	16.1	
(1937)														9.9				16.0
(1938)		38.8																
Ash																	4.6	
(1937)														4.6				3.7
(1938)		2.4																
Major elements %DM																		
Calcium																	0.54	
(1937) (1937)														0.71				0.76
(1938)		0.54																
																		0.699 0.662
Phosphorus																	0.34	
(1937) (1937)														0.24				0.23
(1938)		0.15																
																		0.236 0.187
Magnesium																		0.240 0.209
Potassium																	1.63	
																		0.79 0.92
Sodium																		0.049 0.065
Chloride																	0.14	
																		0.375 0.223
Sulphur,total																		0.287 0.204
Trace elements μg/g DM																		
Cobalt																		0.088 0.068
Copper																		13.08 14.25
Iron																		143 153
Manganese																		946 1648

July			August			September			October			November			December			Notes	Reference
																		Topshoots	218
		10.1		11.6 (1937)	9.7													Leaves Berries Stems + buds milled together	472
																		Topshoots	218
		4.1		3.4 (1937)	7.2													Leaves Berries Stems + buds milled together	472
																		Topshoots	218
		62.7		62.3 (1937)	67.4													Leaves Berries Stems + buds milled together	472
																		Topshoots	218
		17.2		17.7 (1937)	12.8													Leaves Berries Stems + buds milled together	472
																		Topshoots	218
		5.9		5.0 (1937)	2.9													Leaves Berries Stems + buds milled together	472
																		Topshoots	218
		1.31		0.95	0.25													Leaves Berries Stems + buds milled together	472
	0.883 0.742			0.979 0.601			1.06 0.696											Leaf Green stem	492
																		Topshoots	218
		0.18		0.22	0.22													Leaves Berries Stems + buds milled together	472
	0.219 0.195			0.255 0.218			0.204 0.160											Leaf Green stem	492
	0.315 0.211			0.424 0.187			0.383 0.176											Leaf Green stem	492
																		Topshoots	218
	1.40 0.45			0.68 0.34			0.84 0.41											Leaf Green stem	492
	0.050 0.057			0.063 0.053			0.026 0.057											Leaf Green stem	492
																		Topshoots	218
	0.214 0.189			0.187 0.051			0.133 0.047											Leaf Green stem	492
	0.289 0.188			0.180 0.122			0.181 0.103											Leaf Green stem	492
	0.088 0.087			0.115 0.086			0.116 0.106											Leaf Green stem	492
	13.90 14.60			14.40 –			12.91 17.74												
	188 137			136 124			193 126												
	1335 1624			2639 1868			2593 2028												

Table 1.207 Comfrey: *Symphytum* spp.

		April	May	June	July	August	September	Notes	Reference
Russian comfrey: *Symphytum peregrinum*									
Proximate constituents %DM									
Crude protein	(1957)	28.1	24.4	21.8	26.5		25.2	7 Varieties Plots sown 6/4/1956 Replacements 2/4/1957 Basal compound fertiliser in spring. Nitrochalk after June and August cuts.	548
	(1958)		23.9	20.6		17.4	21.7		
Ether extract	(1957)	2.4	2.7	2.2	2.9		1.5		
	(1958)		2.2	2.4		1.6	2.2		
Nitrogen-free extract	(1957)	41.3	42.2	44.7	38.9		38.7		
	(1958)		47.5	47.6		49.3	43.7		
Crude fibre	(1957)	9.8	13.7	11.8	12.5		12.7		
	(1958)		12.1	14.6		14.2	14.0		
Ash	(1957)	18.4	17.0	19.5	19.2		21.9		
	(1958)		14.3	14.8		17.5	18.4		
Major elements %DM									
Calcium	(1957)	1.33	1.93	1.29	1.89		2.22		
	(1958)		1.23	1.67		1.69	1.68		
Phosphorus	(1957)	0.55	0.31	0.50	0.44		0.46		
	(1958)		0.63	0.55		0.60	0.63		
Magnesium	(1957)	0.19	0.19	0.21	0.26		0.32		
Potassium	(1957)	6.60	4.93	6.17	6.85		6.50		
	(1958)		5.79	5.84		6.67	7.24		
Chloride	(1957)	1.08	0.94	1.08	0.91		1.06		
Prickly comfrey: *Symphytum asperrimum*									
Proximate constituents %DM									
Crude protein		25.5						Artificially dried. Leafy as cut	283
Ether extract		3.0							
Nitrogen-free extract		43.0							
Crude fibre		12.2							
Ash		16.3							
Major elements %DM									
Calcium		1.13							
Phosphorus		0.52							
Vitamins μg/g DM									
Carotene		449							

Table 1.208 Yellow rattle: *Rhinanthus crista-galli.*

	June	Notes	Reference
Proximate constituents %DM			
Crude protein	13.4	Mean of 5 samples	20
Ash	9.95	As for Crude protein	
Major elements %DM		Sampled 3-15th June.	489
Calcium	F_0 1.21 F_1 1.34 F_2 1.31 F_3 1.63 F_4 1.13 F_5 1.46	Pre-flowering stage. F_0 : control F_1 : farmyard manure 20 tonnes/ha/annum. F_2 : ammonium sulphate 165 kg/ha F_3 : basic slag 330 kg/ha F_4 : muriate of potash 110 kg/ha F_5 : last 3 together	
	1.36	As for Crude protein	20
Phosphorus	F_0 0.63 F_1 0.93 F_2 0.69 F_3 0.82 F_4 0.42 F_5 0.81	As for Calcium	489
	0.38	As for Crude protein	20
Magnesium	F_0 0.40 F_1 0.44 F_2 0.42 F_3 0.49 F_4 0.30 F_5 0.40	As for Calcium	489
	0.22	As for Crude protein	20
Potassium	F_0 2.82 F_1 3.10 F_2 2.76 F_3 2.70 F_4 3.53 F_5 3.09	As for Calcium	489
Sodium	F_0 0.33 F_1 0.15 F_2 0.32 F_3 0.36 F_4 0.34 F_5 0.27	As for Calcium	489
Chloride	F_0 1.30 F_1 1.40 F_2 1.21 F_3 1.37 F_4 1.14 F_5 1.41	As for Calcium	489
	0.88	As for Crude protein	20
Sulphur, inorganic	F_0 0.35 F_1 0.48 F_2 0.37 F_3 0.37 F_4 0.33 F_5 0.39	As for Calcium	489
Sulphur, organic	F_0 0.17 F_1 0.18 F_2 0.17 F_3 0.14 F_4 0.15 F_5 0.13	As for Calcium	489

continued

Table 1.208 Yellow rattle: (continued)

	June	Notes	Reference
Trace elements μg/g DM			
Copper	F_0 14.2 F_1 17.9 F_2 12.4 F_3 10.9 F_4 12.1 F_5 10.3	As for Calcium	489
	12.1	As for Crude protein	20
Iron	F_0 128 F_1 101 F_2 158 F_3 160 F_4 141 F_5 125	As for Calcium	489
	148	As for Crude protein	20
Manganese	F_0 413 F_1 345 F_2 479 F_3 301 F_4 897 F_5 513	As for Calcium	489
	95	As for Crude protein	20

Table 1.209 Plantain and mixture: *Plantago* spp.

	May	June	July	August	September	October	Notes	Reference
Narrow-leaved Plantain: *Plantago lanceolata*								
Proximate constituents %DM								
Crude protein			19.6	12.5			July: leafy, pre-flowering September: mature F_0 : no extra fertiliser F_1 : Farmyard manure 20 tonnes/ha/annum F_2 : Sulphate of ammonia 165 kg/ha F_3 : Basic slag 330 kg/ha F_4 : Muriate of potash 110 kg/ha F_5 : Last 3 together	15 421
			Average value for leaf, no date, 20.2					218
Ether extract			Average value for leaf, no date, 4.4					
Nitrogen-free extract			Average value for leaf, no date, 48.3					
Crude fibre			12.7	12.6				15 421
			Average value for leaf, no date, 14.8					218
Ash			Average value for leaf, no date, 12.2					
Crude protein	8.5						Cut at early flowering, height 25–30 cm; mid May assumed	12
Amino acids % of Crude protein								
Cystine	0.56							
Histidine	1.80							
Lysine	3.47							
Methionine, DL form	1.32							
L form	1.50							
Tryptophan	1.55							
Tyrosine	3.80							
Major elements %DM								
Calcium			2.34					15
			Average value for leaf, no date, 1.61					218
		F_0 1.59 F_1 1.73 F_2 1.53* F_3 1.62 F_4 1.34 F_5 1.62*					*"Calculated probable values". not explained.	489
			1.87	1.71	1.66 1.39			490
Phosphorus			0.13					15
			Average value for leaf, no date, 0.47					218
		F_0 0.21 F_1 0.35 F_2 0.28* F_3 0.41 F_4 0.24 F_5 0.35*						489
			0.37	0.34	0.33 0.28			490
Magnesium		F_0 0.09 F_1 0.10 F_2 0.10* F_3 0.09 F_4 0.08 F_5 0.08*						489
			0.67	0.63	0.64 0.51			490
Potassium			Average value for leaf, no date, 2.80					218
		F_0 1.29 F_1 1.60 F_2 1.27* F_3 1.28 F_4 2.07 F_5 1.53*						489
			2.57	2.95	2.84 2.60			490

continued

Table 1.209 Plantain (continued)

	May			June			July			August			September			October			Notes	Reference
Narrow-leaved Plantain (cont'd) *Major elements* %DM (cont'd) Sodium				F_0	0.77															489
				F_1	0.86															
				F_2	0.72*															
				F_3	0.76															
				F_4	0.36															
				F_5	0.71*															
									0.29		0.32		0.29	0.26						490
Chlorine							Average value for leaf, no date, 2.27													218
				F_0	0.40															489
				F_1	0.43															
				F_2	0.37*															
				F_3	0.45															
				F_4	0.74															
				F_5	0.75*															
									0.67		0.62		0.56	0.62						490
Sulphur, inorganic				F_0	0.30															489
				F_1	0.31															
				F_2	0.27*															
				F_3	0.26															
				F_4	0.27															
				F_5	0.29*															
Sulphur, organic				F_0	0.16															
				F_1	0.15															
				F_2	0.16*															
				F_3	0.13															
				F_4	0.14															
				F_5	0.13*															
Trace elements μg/g DM Cobalt				F_0	0.29															489
				F_1	0.22															
				F_2	0.26*															
				F_3	0.22															
				F_4	0.25															
				F_5	0.21*															
									0.31		0.12		0.19	0.18						490
Copper				F_0	16.0															489
				F_1	16.1															
				F_2	15.2*															
				F_3	15.7															
				F_4	16.1															
				F_5	16.2*															
									15.6		8.7		10.8	6.9						490
Iron										442										421
				F_0	156															489
				F_1	178															
				F_2	174*															
				F_3	122															
				F_4	138															
				F_5	139*															
									625		389		471	474						490
Manganese				F_0	121															489
				F_1	75															
				F_2	81*															
				F_3	50															
				F_4	105															
				F_5	119*															
									60		28		30	23						490
Vitamins μg/g DM																				493
Thiamin											4.07									
Riboflavin											10.5									
Nicotinic acid											38.4									
Pantothenic acid											5.6									
Biotin											0.25									
Pyridoxine											13.2									

Table 1.209 Plantain (continued)

	May			June			July			August			Notes	Reference
Broad-leaved Plantain: *Plantago major*														
Proximate constituents %DM														
Crude protein			18.6										Sampled preflowering	
Ether extract			2.6											
Nitrogen-free extract			50.0											220
Crude fibre			11.6											
Ash			17.2											
Major elements %DM														
Calcium			2.4											
Phosphorus			0.4											
Magnesium			0.2											
Potassium			3.8											
Sodium			0.25											
Chlorine			1.9											
Trace elements μg/g DM														
Cobalt			0.1											
Copper			13.7											
Iron			560											
Manganese			38											
Plantain, Yarrow, Chicory, Burnet and Sainfoin														
Proximate constituents %DM														
Crude protein								16.9					Sown early in July, 1948. No fertiliser dressing Mean values for 5 cuts (September, 1948 and May–August, 1949)	485
Ether extract								2.4						
Nitrogen-free extract								50.7						
Crude fibre								15.2						
Ash								14.8						
Carbohydrate fractions %DM														
Cellulose								24.1						
Lignin								7.0						
Major elements %DM														
Calcium								2.1						
Phosphorus								0.4						
Magnesium								0.3						
Potassium								3.6						
Sodium								0.1						
Chlorine								0.2						

Table 1.210 Yarrow: *Achillea millefolium*

	April			May			June			July			Notes	Reference
Proximate constituents %DM														
Crude protein		21.4		16.9		12.7	8.1			12.8			Stage of growth, in order of date: 1 and 2, preflowering; 3, full flower; 4 and 5, seeding	9
				Average value for leaf, no date: 19.9										218
				Mean of 5 samples, hand sorted, no date: 16.1										20
Ether extract		2.6		1.6		2.0	1.5			1.5			As for Crude protein	9
				Average value for leaf, no date: 4.4										218
Nitrogen-free extract		48.9		51.1		55.1	54.4			46.7			As for Crude protein	9
				Average value for leaf, no date: 40.8										218
Crude fibre		13.0		16.4		18.6	23.6			28.9			As for Crude protein	9
				Average value for leaf, no date: 24.3										218
Ash		14.1		14.0		11.6	12.4			10.1			As for Crude protein	9
				Average value for leaf, no date: 10.6										218
Crude protein %DM					8.1								Cut at early flowering, height 10–30cm, mid-May assumed	12
Amino acids % of Crude protein														
Cystine					0.60									
Histidine					1.86									
Lysine					4.30									
Methionine, DL form					1.40									
L form					1.59									
Tryptophan					1.91									
Tyrosine					3.30									
Carbohydrate fractions %DM														
Cellulose		16.2		19.4		21.2	28.3			28.8			As for Crude protein	9
Lignin (true)		3.3		4.3		5.0	8.6			6.9				
Major elements %DM														
Calcium			1.42		1.14		1.24						Dates assumed to correspond to stage of growth: preflowering, midflowering, mature.	490
				Average value for leaf, no date: 0.77										218
				Mean of 5 samples, hand sorted, no date: 1.23										20
Phosphorus			0.48		0.37		0.33						As for Calcium	490
				Average value for leaf, no date: 0.40										218
				Mean of 5 samples, hand sorted, no date: 0.32										20
Magnesium			0.65		0.56		0.53						As for Calcium	490
				Mean of 5 samples, hand sorted, no date: 0.28										20
Potassium			4.2		3.5		3.3						As for Calcium	490
				Average value for leaf, no date: 2.33										218
Sodium			0.05		0.02		0.05						As for Calcium	490
Chlorine			0.61		0.52		0.45							
				Average value for leaf, no date: 0.92										218
				Mean of 5 samples, hand sorted, no date: 1.23										20
Trace elements µg/g DM														
Cobalt			0.18		0.19		0.14						As for Calcium	490
Copper			12.6		9.3		9.3							
				Mean of 5 samples, hand sorted, no date: 11.2										20
Iron			333		304		250						As for Calcium	490
				Mean of 5 samples, hand sorted, no date: 191										20
Manganese			67.1		43.2		32.0						As for Calcium	490
				Mean of 5 samples, hand sorted, no date: 87										20
	May			June			July			August				
Vitamins µg/g DM														
Thiamin												4.25	Early flowering	493
Riboflavin												11.9		
Nicotinic acid												43.2		
Pantothenic acid												5.1		
Biotin												0.2		
Pyridoxine												12.5		

Table 1.211 Chicory: *Cichorium intybus*

	May			June			July			August			Notes	Reference
Proximate constituents %DM														
Crude protein								21.3					Leafy, preflowering	15
									17.5					421
Crude fibre								12.7					Leafy, preflowering	15
									11.7					421
Crude protein		10.9											Cut preflowering, height 51–66 cm, mid-May assumed	12
Amino acids % of Crude protein														
Cystine		0.39												
Histidine		1.49												
Lysine		3.24												
Methionine: DL form		0.84												
L form		0.97												
Tryptophan		1.51												
Tyrosine		2.52												
Major elements %DM														
Calcium								1.1		1.6	1.6		Date assumed to correspond to stage of growth, preflowering to mature.	490
								1.7					Leafy preflowering	15
Phosphorus								0.35		0.52	0.46		As for Calcium	490
								0.16						15
Magnesium								0.71		0.64	0.59		As for Calcium	490
Potassium								5.50		3.65	4.55			
Sodium								0.25		0.23	0.33			
Chlorine								1.08		1.00	0.69			
Trace elements μg/g DM														
Cobalt								0.19		0.23	0.19		As for Calcium	490
Copper								15.4		12.2	9.9			
Iron								482		536	388			
									621					421
Manganese								75		56	42		As for Calcium	490
Vitamins μg/g DM														
Thiamin											3.11			493
Riboflavin											13.6			
Nicotinic acid											57.5			
Pantothenic acid											18.8			
Biotin											0.24			
Pyridoxine											13.4			

Table 1.212 Pasture herbs: General composition

Proximate Constituents % DM		CP	EE	NFE	CF	Ash		Notes	Reference
Bellis perennis: Daisy	L	16.7	3.6	52.4	12.8	14.5		L leaf: S stem	218
	S	14.3	5.1	52.6	18.7	9.3			
Capsella bursa-pastoris: Shepherd's purse		27.3	2.0	32.2	26.7	11.8		Mid-flowering	220
Carduus arvensis: Creeping thistle		29.6	2.4	34.2	15.5	18.3		Preflowering	
Carduus heterophyllus: Melancholy thistle	a	17.5	3.1	51.7	14.6	13.1		a. mid-September 1935	404
	b	12.8	3.5	54.3	16.8	12.6		b. Early November 1936	
Carduus lanceolatus: Spear thistle		17.4	2.5	41.5	24.9	13.7		Early flowering	220
Centaurea nigra: Knapweed		20.2	6.7	42.2	19.0	11.9		Leaf only	218
Chrysanthemum leucanthemum: Ox-eye daisy		10.8	5.7	47.1	27.9	8.5		Leaf only	
Epilobium angustifolium: Rosebay willowherb		17.6	2.0	59.4	14.2	6.8		Preflowering	220
Gallium aparine: Cleavers		22.2	2.0	37.3	20.9	17.6		Preflowering	
Gallium cruciata: Crosswort (Mugwort)		15.6	2.8	47.0	21.4	13.2		Early flowering	
Gallium verum: Lady's bedstraw		16.5	5.0	50.3	22.1	6.1		Leaf only	218
Heracleum spondylium: Cowparsnip (Hogweed)		20.1	4.7	45.2	15.5	14.5		Leaf only	
Hypochoeris radicata: Cat's ear		19.6	4.6	40.1	21.1	14.6		Leaf only	
Lamium album: White deadnettle		11.5	2.2	56.1	18.7	11.5		Flowering	220
Leontodon sp.: Hawkbit		18.6	3.8	44.2	18.8	14.6		Leaf only	218
Prunella vulgaris: Selfheal		10.3	4.2	53.5	18.6	13.4		Leaf only	
Scabiosa succisa: Devil's bit scabious		12.9	5.6	49.1	21.4	11.0		Leaf only	
Senecio jacobaea: Ragwort		17.0	5.2	44.3	21.5	12.0		Leaf only	
		13.6	1.5	50.8	12.5	21.6		Preflowering	220
Sonchus oleraceus: Sowthistle		20.0	–	–	16.4	12.1		Leaf only: 26.5.31	218
		11.7	–	–	23.3	9.3		Leaf only: 27.7.31	
		17.8	5.5	40.9	20.8	15.0		Leaf only: no date	
Spiraea ulmaria: Meadowsweet		16.9	1.5	53.4	21.3	6.9		Preflowering	220
Stellaria holostea: Stitchwort		17.3	3.1	43.2	24.0	12.4		Early flowering	
Stellaria media: Chickweed		13.5	2.7	53.5	17.7	12.6		Early flowering	
Taraxacum officinale: Dandelion		19.4	5.2	42.8	17.1	15.5		Leaf only	218
		13.7	5.0	53.2	11.3	16.8		Flowering	220
Tussilago farfara: Coltsfoot		15.3	2.4	53.4	13.7	15.2		Post-flowering	
Urtica dioica: Nettle		22.4	3.5	21.9	32.7	19.5		Leaf only	218
		27.4	2.2	39.7	13.7	17.0		Preflowering	220
Veronica officinalis: Speedwell		12.6	5.8	50.4	22.9	8.3		Leaf only	218
Viola lutea: Wild pansy		11.8	4.3	48.6	24.0	11.3		Leaf only	

Major elements % DM		Ca	P	Mg	K	Na	Cl	Notes	Reference
Bellis perennis: Daisy	L	1.3	0.4	–	2.6	–	2.8	L leaf: S stem	218
	S	0.9	0.5	–	2.5	–	1.1		
Capsella bursa-pastoris: Shepherd's purse		2.0	0.6	0.2	2.8	0.37	0.6	mid-flowering	220
Carduus arvensis: Creeping thistle		2.1	0.5	0.2	4.3	0.27	3.3	Preflowering	
Carduus heterophyllus: Melancholy thistle	a	3.3	0.4	–	–	–	–	a. mid-September 1935	404
	b	3.1	0.4					b. Early November 1936	
Carduus lanceolatus: Spear thistle		1.8	0.4	0.2	3.5	0.35	1.7	Early flowering	220
Centaurea nigra: Knapweed		1.1	0.5		3.4		2.2	Leaf only	218
Chrysanthemum leucanthemum: Ox-eye daisy		0.8	0.3		1.7		0.5	Leaf only	
Epilobium angustifolium: Rosebay willowherb		0.8	0.4	0.4	2.2	0.25	0.6	Preflowering	220
Gallium aparine: Cleavers		1.5	0.7	0.1	5.2	0.39	1.0	Preflowering	
Gallium cruciata: Crosswort (Mugwort)		1.7	0.3	0.2	3.6	0.26	1.2	Early flowering	
Gallium verum: Lady's bedstraw		0.4	0.3		2.2		0.3	Leaf only	218
Heracleum spondylium: Cow parsnip (Hogweed)		1.2	0.4		4.2		0.8	Leaf only	
Hypochoeris radicata: Cat's ear		1.0	0.4		2.4		3.5	Leaf only	
Lamium album: White deadnettle		0.8	0.3	0.2	4.1	0.23	0.6	Flowering	220
Leontodon sp.: Hawkbit		1.7	0.4		1.5		1.9	Leaf only	218

Table 1.212 Pasture herbs (continued)

Major elements % DM (cont'd)	Ca	P	Mg	K	Na	Cl	Notes	Reference
Prunella vulgaris: Selfheal	0.9	0.3		2.9		0.7	Leaf only	218
Scabiosa succisa: Devil's bit scabious	0.9	0.3		2.3		0.5	Leaf only	
Senecio jacobaea: Ragwort	1.0	0.2		2.6		1.7	Leaf only	
	2.2	0.2	0.1	2.7	0.81	1.6	Preflowering	220
Sonchus oleraceus: Sowthistle	1.4	0.5		3.2		2.3	Leaf only: 26.5.31	218
	1.5	0.4		3.0		1.9	Leaf only: 27.7.31	
	1.5	0.5		2.7		1.7	Leaf only: no date	
Spiraea ulmaria: Meadowsweet	0.5	0.3	0.3	2.2	0.22	0.6	Preflowering	220
Stellaria holostea: Stitchwort	1.0	0.4	0.3	4.2	0.23	1.3	Early flowering	
Stellaria media: Chickweed	0.8	0.6	0.2	4.3	1.22	0.7	Early flowering	
Taraxacum officinale: Dandelion	1.3	0.4		3.4		2.0	Leaf Only	218
	2.8	0.3	0.5	3.6	0.64	2.2	Flowering	220
Tussilago farfara: Coltsfoot	1.9	0.3	0.3	4.1	0.26	1.7	Post-flowering	
Urtica dioica: Nettle	2.0	0.8		4.0		1.0	Leaf only	218
	2.8	0.7	0.3	4.1	0.23	0.6	Preflowering	220
Veronica officinalis: Speedwell	1.0	0.4		2.1		0.3	Leaf only	218
Viola lutea: Wild pansy	0.9	0.4		3.2		0.8	Leaf only	
Trace elements μg/g DM	**Co**	**Cu**	**Fe**	**Mn**				
Capsella bursa-pastoris: Shepherd's purse	0.1	10.7	182	34			Mid-flowering	220
Carduus arvensis: Creeping thistle	0.06	29.2	468	59			Preflowering	
Carduus lanceolatus: Spear thistle	0.2	24.2	297	36			Early flowering	
Epilobium angustifolium: Rosebay willowherb	0.2	11.3	136	331			Preflowering	
Gallium aparine: Cleavers	0.1	11.1	316	66			Preflowering	
Gallium cruciata: Crosswort (Mugwort)	0.1	14.5	482	79			Early flowering	
Lamium album: White dead nettle	0.08	13.2	341	34			Flowering	
Senecio jacobaea: Ragwort	0.16	15.6	594	52			Preflowering	
Spiraea ulmaria: Meadowsweet	0.08	9.5	155	305			Preflowering	
Stellaria holostea: Stitchwort	0.12	10.4	353	344			Early flowering	
Stellaria media: Chickweed	0.12	10.9	345	168			Early flowering	
Taraxacum officinale: Dandelion	0.13	18.2	619	48			Flowering	
Tussilago farfara: Coltsfoot	0.07	13.3	215	25			Post-flowering	
Urtica dioica: Nettle	0.13	22.1	606	81			Preflowering	
							Some of the Mn values are high. It is suggested that soil reaction affects Mn.	

For Carotene See Table 1.213 and for Tocopherols Table 1.214

Table 1.213 Pasture weeds or herbs: Carotene, μg/g DM

	May	June	July	August	September	October	Notes	Reference
Urtica dioica: Nettle	468	247	243	336	504	506	Whole plant	362
Ranunculus sp: Buttercup	417	312	388	639	570	383	Random samples from a pasture plot	
Achillea millefolium: Yarrow	466	332	308	444	345	381		
Veronica chamaedrys: Germander speedwell	344	349	287	398	432	329		
Plantago lanceolata: Ribwort or Narrow-leaved plantain	405	431	324	436	486	352		
Carduus arvensis: Creeping thistle	366	323	318	394	247	317		
Senecio jacobaea: Ragwort	357	214	382	469	555	418		

Table 1.214 Leaves of trees and herbs: *a*-Tocopherol, μg/g DM

	February	March	April	May	June	July	October	November	Notes	Reference
Achillea millefolium: Yarrow						180			In bloom	87
						110			Young regrowth	
						750			Yellow, senescent	
Betula alba: Birch				100	270		980			
Chenopodium album: Fat hen						25			Height 5 cm	
						70			Height 7 cm	
						120			Height 10 and 35 cm	
Ligustrum vulgare: Privet	1270		120							
Populus italica: Poplar			170			290	350			
Sambucus nigra: Elder		270						1300		
Taraxacum officinale: Dandelion				150			200			
Ulex europaeus: Gorse or whin				85					Young leaves	
				120					Old leaves	
Urtica dioica: Nettle		190	180			680	740	770		

Table 1.215 Seaweed: Phaeophyceae

	June	Notes	Reference
Ascophyllum nodosum MEAL			
Digestibility of DM%	48.3	Digestibility by sheep	451
Digestibility of OM%	45.8		
Proximate constituents %DM			
Crude protein	11.6		451
	8.3		66
	8.8		62
	9.1		64
	8.4		352
	9.0		63
	11.0		
	6.9		112
	10.2		262
	9.1		
	7.6		70
Ether extract	1.7		451
	3.3		62
	2.3		64
	3.3		112
	4.3		65
	1.9		262
	2.3		
	2.6		70
Nitrogen-free extract	56.1		451
	62.9		62
	57.1		64
	60.9		112
	60.9		262
	57.1		
	56.9		70
Crude fibre	5.0		451
	3.1		62
	4.0		64
	5.5		112
	3.5		262
	4.0		
	8.4		70
Ash	25.6		451
	17.3		66
	21.9		62
	27.5		64
	27.2		68
	19.5		
	20.4		
	19.6		63
	27.5		352
	22.3		63
	23.4		112
	23.5		262
	27.5		
	24.5		70
Carbohydrate fractions %DM			
Mannitol	9.0		66
	7.2		352
	8.8		63
	8.7		
Laminarin	4.3		66
	4.4		352
	6.2		63
	2.6		

		Notes	Reference
Carbohydrate fractions %DM (cont'd)			
Cellulose	4.0		63
	2.4		
	1.9		
	1.7		
	1.6		
	2.8		
	2.5		
Fucose as $C_6H_{12}O_5$	9.0		66
L-fucose	7.3		63
	6.9		
Alginic acid	24.4		66
	24.3		63
	23.3		
Major elements %DM			
Calcium	1.22		62
	1.58		64
	1.80		112
	1.26		262
	1.58		
	2.16		70
Phosphorus	0.12		62
	0.15		64
	0.11		112
	0.12		262
	0.15		
	0.09		70
Magnesium	0.82		70
Potassium	2.26		
Sodium	1.50		112
	2.90		70
Chlorine	5.98		62
	2.32		112
	5.56		262
	5.98		
	1.89		70
Sulphur, total	2.32		
Trace elements μg/g DM (except I)			
Barium	50		68
	13		
	18		
	27		70
Chromium	0.7		68
	1.0		
	1.9		
	2.9		70
Cobalt	0.41		68
	0.73		
	0.73		
	1.43		70
Copper	4		68
	4		
	12		
	61		70
Iodine %DM	0.05		

Table 1.215 Seaweed: Phaeophyceae (continued)

		Notes	Reference
A. nodosum (cont'd)			
Trace elements µg/g DM (cont'd)			
Iron	168		68
	283		
	1150		
	1132		70
Lead	6		68
	4		
	4		
	<4		70
Lithium	4		68
	–		
	–		
Manganese	50		68
	27		
	36		
	45		70
Molybdenum	0.69		68
	0.29		
	0.89		
	1.25		70
Nickel	1.5		68
	4.4		
	3.7		
	3.35		70
Rubidium	80		68
	–		
	–		
Silver	0.3		68
	0.2		
	0.1		
	< 0.3		70
Strontium	2600		68
	>700		
	570		
	560		70
Tin	1.0		68
	0.7		
	1.1		
	< 5		70
Titanium	9		68
	26		
	28		
	114		70
Vanadium	1.9		68
	1.5		
	2.8		
	5.9		70
Zinc	103		68
	60		
	116		
	110		70

		Notes	Reference
Vitamins µg/g DM			
Carotene	24		398
	25		
Carotenoids	38		398
	30		
β-Carotene	35, 68		70
Fucoxanthin	90		
D_3	0.01		
E	156		
Thiamin	1.4-5.4		
Riboflavin	7.5		
Niacin, nicotinic acid	12.3		
Pantothenic acid	0.2		
Folic acid	0.07		
B_{12}	0.004		
	0.08		
K	14.2		
SILAGE			
Proximate constituents %DM			
Crude protein	11.6		63
	9.4		
	9.3		
	9.3		
	9.7		
	10.0		
Ash	21.7		
	20.2		
	19.7		
	19.8		
	19.8		
	19.6		
Carbohydrate fractions %DM			
Mannitol	5.7		
	10.1		
	10.1		
	9.8		
	9.2		
	8.4		
Laminarin	2.8		
	4.5		
	4.3		
	4.2		
	3.9		
	4.5		
L-fucose	8.0		
	7.4		
Alginic acid	20.6		
	22.3		
	24.7		
	24.2		

continued

Table 1.215 Seaweed: Phaeophyceae (continued)

		Notes	Reference
Fucus serratus			
Proximate constituents %DM			
Crude protein	9.7 (6.4-14.9)		61
Ether extract	3.0		65
Ash	24.0 (19.6-30.6)		61
Carbohydrate fractions %DM			
Alginic acid	19.6 (17.0-22.1)		61
Mannitol	11.3 (6.3-16.5)		
Laminarin	4.6 (2.1-10.2)		
Sterol %DM			
Fucosterol	0.20		65
Trace elements %DM			
Iodine	0.08 (0.05-0.10)		61
MEAL			
Proximate constituents %DM			
Ash	27.4 21.8 18.0		68
Trace elements µg/gDM			
Barium	80 22 16		
Chromium	2.6 0.7 0.7		
Cobalt	0.47 0.84 0.63		
Copper	11 5 6		
Iron	717 375 320		
Lead	21 10 4		
Lithium	6		
Manganese	800 155 120		
Molybdenum	0.40 0.65 0.21		

		Notes	Reference
Trace elements µg/g DM (cont'd)			
Nickel	1.6 3.2 4.5		
Rubidium	170		
Silver	0.6 0.3 0.2		
Strontium	>2800 > 700 520		
Tin	1.2 1.3 0.5		
Titanium	9 20 7		
Vanadium	3.3 2.0 0.6		
Zinc	79 70 63		
Fucus spiralis			
Proximate constituents %DM			
Crude protein	8.3 (4.0-14.0)		61
Ether extract	4.8		65
Ash	21.2 (15.2-22.9)		61
Carbohydrate fractions %DM			
Mannitol	9.7 (5.6-12.0)		61
Laminarin	3.7 (1.5-6.9)		
Alginic acid	15.0 (13.0-16.6)		
Sterol %DM			
Fucosterol	0.20		65
Trace elements %DM			
Iodine	0.03 (0.02-0.04)		61
MEAL			
Proximate constituents %DM			
Ash	24.3 22.8		68
Trace elements µg/g DM			
Barium	19 64		
Chromium	0.9 3.7		

Table 1.215 Seaweed: Phaeophyceae (continued)

		Notes	Reference
F. spiralis (cont'd)			
Trace elements μg/g DM (cont'd)			
Cobalt	1.39 2.00		68
Copper	6 31		
Iron	638 3380		
Lead	5 5		
Manganese	104 121		
Molybdenum	0.29 1.32		
Nickel	6.0 9.3		
Silver	0.2 0.4		
Strontium	>700 420		
Tin	1.8		
Titanium	27 308		
Vanadium	1.9 11.9		
Zinc	62		
Fucus vesiculosus			
Proximate constituents %DM			
Crude protein	7.9 (5.9-10.9)		61
Ether extract	1.8 3.8		65
Ash	18.3 (13.8-20.7)		61
Carbohydrate fractions %DM			
Alginic acid	15.5 (13.8-17.2)		
Mannitol	12.1 (8.3-16.1)		
Laminarin	3.5 (2.0-4.9)		
Sterol %DM			
Fucosterol	0.25 0.26		65
Trace elements %DM			
Iodine	0.05 (0.03-0.06)		61

		Notes	Reference
MEAL			
Digestibility of DM%	37.5 38.7 28.9	Digestibility by sheep Sample dates. spring and autumn, 1942 autumn, 1944	451
Digestibility of OM%	29.3 31.0 31.4		
Proximate constituents %DM			
Crude protein	8.4	For digested components and metabolisable energy see table 1.238	
Ether extract	1.6		
Nitrogen-free extract	61.5		
Crude fibre	10.5		
Ash	18.0		
Proximate constituents %DM			
Crude protein	13.3 11.5 7.5 7.4		451 66
Ether extract	0.8 0.1		451
Nitrogen-free extract	44.5 44.8		
Crude fibre	11.6 11.9		
Ash	29 8 31.7 26.0 20.5 24.0 19.5 26.0		 66 68 67
Carbohydrate fractions %DM			
Fucose as $C_6H_{12}O_5$	10.1 9 7 9.2		66 67
Alginic acid	15.0 15.2 15.0		66 67
Mannitol	10.6 13.7 9.0		66 67
Laminarin	2.6 4.8 2.6		66 67
Trace elements μg/gDM			
Barium	44 22		68
Chromium	1.8 1.5		
Cobalt	0.65 0.91		
Copper	7 10		

continued

Table 1.215 Seaweed: Phaeophyceae (continued)

		Notes	Reference
F. vesiculosus (cont'd)			
Trace elements µg/g DM (cont'd)			
Iron	221		68
	730		
Lead	2		
	7		
Manganese	116		
	102		
Molybdenum	0.34		
	0.34		
Nickel	3.8		
	5.9		
Silver	0.2		
	0.2		
Strontium	>700		
	730		
Tin	0.5		
	1.1		
Titanium	28		
	27		
Vanadium	1.9		
	1.7		
Zinc	60		
	105		
Vitamins µg/g DM			
Carotene	35		398
	58		
Carotenoids, total	55		
	107		
Laminaria cloustoni			
Proximate constituents %DM			
Crude protein			
Frond	10-16		588
Stipe	8-11		
Lamina	10.5		69
Stipe	8.4		
Ether extract			
Frond	0.6		65
Stipe	0.6		
Ash			
Frond	15-43		588
	13-37		245
Stipe	35-43		588
	30-38		245
Lamina	43		69
Stipe	43		
Carbohydrate fractions %DM			
Mannitol			
Frond	5-27		588
	6-23		245

		Notes	Reference
L. cloustoni (cont'd)			
Carbohydrate fractions %DM (cont'd)			
Mannitol			
Stipe	5-10		588
	5-9		245
Lamina	4.0		69
Stipe	5.3		
Alginic acid			
Frond	11-24		588
	8-19		245
Stipe	21-27		588
	19-23		245
Laminarin			
Frond	0-36		588
	1-32		245
Stipe	0		588
	Nil		245
Cellulose			
Frond	4-5		588
Stipe	8-10		
Fucoidin (as L-glucose)			
Frond	4		
Stipe	3		
Sterol %DM			
Fucosterol			
Frond	0.07		65
Stipe	0.12		
MEAL			
Proximate constituents %DM			
Crude protein			
Whole	12.0		112
	14.1		262
	11.8		63
Frond	10.6		66
	7.4		62
	7.6		64
	11.9		63
	15.4		
	16.4		
	—		
	7.0		352
	7.4		262
	12.6		70
Stipe	8.9		352
	8.4		70
	11.7		63
	10.9		262
Ether extract			
Whole	0.7		112
	1.3		262
Frond	0.5		62
	1.7		64
	0.5		262
	0.4		70
Stipe	0.2		
	0.5		262
Nitrogen-free extract			
Whole	52.8		112
	37.9		262
Frond	66.4		62
	58.9		64
	66.3		262
	60.2		70

Table 1.215 Seaweed: Phaeophyceae (continued)

		Notes	Reference
L. cloustoni (cont'd)			
Proximate constituents %DM			
Nitrogen-free extract (cont'd)			
Stipe	44.2		
	43.4		262
Crude fibre			
Whole	8.7		112
	5.4		262
Frond	5.6		62
	4.0		64
	5.7		262
	5.0		70
Stipe	9.6		
	10.3		262
Ash			
Whole	25.9		112
	41.3		262
	34.8		63
Frond	20.3		* 66
	20.1		62
	27.8		64
	32.4		68
	22.4		
	32.2		
	28.0		63
	31.5		
	37.3		
	37.0		
	20.1		352
	20.3		* 67
	20.1		262
	21.8		70
Stipe	34.9		352
	37.6		70
	38.8		63
	34.9		262
Amino acids % of Crude protein			
Frond			
Alanine	5.1		139
Arginine	12.3		
Aspartic acid. Frond (dried)	7.9		
Cystine	+		
Glutamic acid	3.4		
Glycine	8.5		
Histidine	0		
Leucine + *iso*leucine	5.1		
Lysine	4.6		
Phenylalanine	1.1		
Proline	3.9		
Serine	3.2		
Threonine	3.1		
Tryptophan	+		

		Notes	Reference
Amino acids % of Crude protein (cont'd)			
Tyrosine	0.9		
Valine + Methionine	4.4		
Carbohydrate fractions %DM			
Mannitol			
Whole	13.9		63
Frond	15.4		66
	11.0		63
	9.4		
	5.0		
	9.1		
	15.5		352
	15.4		67
Stipe	4.1		352
	6.2		63
Alginic acid			
Whole	14.2		63
Frond	10.0		66
	16.7		63
	14.7		
	15.8		
	–		
	10.0		67
Stipe	19.3		63
Laminarin			
Whole	13.4		63
Frond	29.2		66
	9.2		63
	6.0		
	Nil		
	Nil		
	27.1		352
	29.2		67
Stipe	Nil		352
	Nil		63
Cellulose			
Frond	5.4		67
L-fucose			
Whole	3.5		63
Fucose as $C_6H_{12}O_5$			
Frond	3.14		66
Frond	3.14		67

*The data given in 66 and 67 are identical

		Notes	Reference
Major elements %DM			
Calcium			
Whole	1.88		112
	1.21		262
Frond	1.64		62
	2.46		64
	1.64		262
	1.04		70
Stipe	1.80		
	1.78		262
Phosphorus			
Whole	0.26		112
	0.36		262
Frond	0.17		62
	0.24		64
	0.17		262
	0.28		70

continued

Table 1.215 Seaweed: Phaeophyceae (continued)

		Notes	Reference
L. cloustoni (cont'd)			
Trace elements μg/g DM			
Vanadium (cont'd)			
Stipe	2.5		70
Zinc			
Frond	117		68
	136		
	76		
	170		70
Stipe	59		
Vitamins μg/g DM			
Carotene			
Frond	14		398
Stipe	0.1		
Carotenoids total			
Frond	19		
Stipe	0.5		
Non-carotene carotenoids			
Frond	5		
Stipe	0.4		
β-Carotene			
Frond	85		70
Vitamin E			
Frond	29.9		
Riboflavin			
Frond	2.4		
Nicotinic acid			
Frond	19.4		
SILAGE			
Proximate constituents %DM			
Crude protein			
Whole	10.8		63
Ash			
Whole	33.3		
Carbohydrate fractions %DM			
Mannitol			
Whole	8.7		
Alginic acid			
Whole	16.2		
Laminarin			
Whole	2.7		
L-fucose			
Whole	4.0		
Laminaria digitata			
Proximate constituents %DM			
Crude protein			
Frond	5-1		588
Stipe	6-1		
Ether extract			
Frond	1.1		65
	0.6		
Stipe	0.9		
	0.6		

		Notes	Reference
L. digitata (cont'd)			
Proximate constituents %DM (cont'd)			
Ash			
Frond	18-44		588
Stipe	35-46		
Carbohydrate fractions %DM			
Mannitol			
Frond	3-27		
Stipe	4-13		
Alginic acid			
Frond	17-27		
Stipe	27-33		
Laminarin			
Frond	0-25		
Stipe	0		
Cellulose			
Frond	4-5		
Stipe	6-7		
Fucoidin (as L-fucose)			
Frond	3		
Stipe	2		
Sterol %DM			
Fucosterol			
Frond	0.12		65
	0.10		
Stipe	0.14		
	0.13		
Digestibility of DM%		Digestibility by pig	
MEAL	69.7	For digested	
	65.2	components	454
Hydrolysed meal	63.8	and	
	60.1	metabolisable	
Frond	60.8	energy see	451
	66.6	table 1.238	
	52.4	Digestibility by sheep	
Digestibility of OM%		For digested	
FROND	55.6	components	
	65.7	and	
	52.1	metabolisable energy see table 1.238	
Proximate constituents %DM			
MEAL			
Crude protein	7.8		454
Ether extract	0.4		
Nitrogen-free extract	68.5		
Crude fibre	6.9		
Ash	16.4		
Hydrolysed meal			
Crude protein	7.6		
Ether extract	0.4		
Nitrogen-free extract	66.5		
Crude fibre	7.0		
Ash	18.5		
Frond			
Crude protein	9.9		451
Ether extract	1.0		
Nitrogen-free extract	42.4		
Crude fibre	12.7		
Ash	34.0		

Table 1.215 Seaweed: Phaeophyceae (continued)

		Notes	Reference
L. digitata (cont'd)			
Proximate constituents %DM (cont'd)			
Crude protein	12.5		451
	8.2		
	11.5		63
Ether extract	0.4		451
	0.1		
Nitrogen-free extract	47.0		
	48.3		
Crude fibre	8.8		
	9.0		
Ash	31.3		
	34.4		
	26.2		
	29.7		68
	41.5		
	31.8		
	40.8		63
Stipe			
Ash	39.7		68
	–		
	38.9		
	37.8		
Carbohydrate fractions %DM			
Mannitol			
Frond	10.5		63
Laminarin			
Frond	Nil		
Major elements %DM			
Calcium	2.0		454
Chlorine	2.6		
Trace elements μg/g DM		Date of collection	
Barium			
Frond	120 } Area 1	12 Jan.	68
	13	26 May	
	17	27 June	
	18 Area 2	27 June	
Stipe	60	The note refers to frond and stipe for each element	
	15		
	28		
	20		
Chromium			
Frond	1.1		
	0.4		
	1.1		
	1.8		
Stipe	1.4		
	0.4		
	1.8		
	2.9		
Cobalt			
Frond	1.46		
	0.29		
	0.31		
	0.22		
Stipe	0.43		
	0.92		
	0.46		
	0.62		
Copper			
Frond	20		
	<3		
	5		
	6		
Trace elements μg/g DM			
Copper (cont'd)			
Stipe	16		
	5		
	11		
	21		
Iron			
Frond	350		
	138		
	400		
	410		
Stipe	–		
	293		
	1260		
	1570		
Lead			
Frond	13		
	2		
	7		
	4		
Stipe	–		
	6		
	16		
	7		
Lithium			
Frond	8		
	–		
	–		
	–		
Stipe	4		
	–		
	–		
	–		
Manganese			
Frond	80		
	9		
	<30		
	<30		
Stipe	100		
	10		
	<30		
	<30		
Molybdenum			
Frond	0.32		
	0.19		
	0.16		
	0.17		
Stipe	0.15		
	0.10		
	0.28		
	0.34		
Nickel			
Frond	8.2		
	2.4		
	1.8		
	2.1		
Stipe	3.2		
	3.9		
	3.7		
	5.7		
Rubidium			
Frond	240		
	–		
	–		
	–		
Stipe	240		
	–		
	–		
	–		
Silver			
Frond	0.4		
	0.3		
	0.1		
	0.0		

continued

Table 1.215 Seaweed: Phaeophyceae (continued)

		Notes	Reference
L. digitata (cont'd)			
Trace elements μg/g DM			
Silver (cont'd)			
Stipe	0.2		68
	0.4		
	0.4		
	0.0		
Strontium			
Frond	4000		
	>700		
	690		
	950		
Stipe	4000		
	>700		
	1150		
	1120		
Tin			
Frond	0.7		
	1.0		
	0.6		
	0.2		
Stipe	–		
	1.7		
	2.8		
	0.2		
Titanium			
Frond	20		
	5		
	4		
	4		
Stipe	16		
	2		
	8		
	5		
Vandium			
Frond	2.0		
	0.6		
	0.5		
	0.7		
Stipe	2.2		
	0.3		
	0.7		
	1.2		
Zinc			
Frond	99		
	64		
	59		
	71		
Stipe	–		
	62		
	92		
	85		
Vitamins μg/g DM			
Carotene			
Frond	10		398
Stipe	1.4		
Carotenoids, total			
Frond	16		
Stipe	3.5		
Laminaria saccharina			
Proximate constituents %DM			
Crude protein			
Frond	5-15		588
Stipe	6-14		
Ether extract			
Frond	1.2		65
	1.2		

		Notes	Reference
Proximate constituents %DM			
Ether extract (cont'd)			
Stipe	0.9		
	0.9		
Ash			
Frond	18-46		588
Stipe	31-44		
Carbohydrate fractions %DM			
Mannitol			
Frond	4-26		588
Stipe	6-13		
Alginic acid			
Frond	11-21		
Stipe	20-28		
Laminarin			
Frond	0-27		
Stipe	0		
Cellulose			
Frond	4-5		
Stipe	7-8		
Fucoidin (as L-glucose)			
Frond	3		
Stipe	2		
Sterol %DM			
Fucosterol			
Frond	0.09		65
	0.09		
Stipe	0.11		
	0.13		
MEAL			
Proximate constituents %DM			
Crude protein			
Whole	4.2		262
Frond	12.1		63
Ether extract			
Whole	0.3		262
Nitrogen-free extract			
Whole	72.5		
Crude fibre			
Whole	4.8		
Ash			
Whole	18.2		
Frond	35.9		63
Carbohydrate fractions %DM			
Mannitol			
Frond	14.8		
Laminarin			
Frond	Nil		
Major elements %DM			
Whole			
Calcium	1.48		262
Phosphorus	0.13		
Chlorine	5.95		

Table 1.215 Seaweed: Phaeophyceae (continued)

		Notes	Reference
Laminaria stenophylla			
MEAL			
Digestibility of DM%			
Frond	48.3	Digestibility by sheep	451
Digestibility of OM%			
Frond	46.5	For digested components and metabolisable energy see table 1.238	
Proximate constituents %DM			
Frond			
Crude protein	10.6		
Ether extract	2.5		
Nitrogen-free extract	57.6		
Crude fibre	10.2		
Ash	19.1		
Pelvetia canaliculata			
Proximate constituents %DM			
Crude Protein	7.8		61
	(4.5-12.1)		
Ether extract	8.13		65
Ash	19.4		61
	(15.2-22.9)		
Carbohydrate fractions %DM			
Alginic acid	15.3		
	(13.3-18.6)		
Mannitol	9.3		
	(5.8-12.3)		
Laminarin	3.5		
	(1.8-5.2)		
Sterol %DM			
Fucosterol	0.28		65
Trace elements %DM			
Iodine	0.04		61
	(0.03-0.06)		
MEAL			
Proximate constituents %DM			
Crude protein	9.7		66
Ash	24.1		
	24.6		68
	21.6		
	23.3		
Carbohydrate fractions %DM			
Mannitol	8.5		66
Laminarin	2.5		
Fucose as $C_6H_{12}O_5$	11.2		
Trace elements μg/g DM			
Barium	70		68
	20		
	34		

		Notes	Reference
P. canaliculata (cont'd)			
Trace elements μg/g DM (cont'd)			
Chromium	0.6		
	1.2		
	1.2		
Cobalt	0.37		
	0.72		
	1.30		
Copper	5		
	5		
	16		
Iron	195		
	565		
	2040		
Lead	4		
	5		
	13		
Lithium	6		
Manganese	70		
	22		
	51		
Molybdenum	0.34		
	0.35		
	0.55		
Nickel	1.9		
	3.7		
	4.8		
Rubidium	100		
Silver	0.2		
	0.2		
	0.3		
Strontium	>2400		
	> 700		
	720		
Titanium	11		
	38		
	60		
Vanadium	1.2		
	2.6		
	3.2		
Zinc	40		
	47		
	90		
Vitamins μg/g DM			
Carotene	44		398
	38		
Carotenoids	73		
	97		
Saccorhiza bulbosa			
Proximate constituents %DM			
Crude protein			
Frond	7.5		59
	5.9		
	5.9		
	4.9		
	6.2		
Stipe	5.6		
	3.5		
	4.8		
	5.4		
	5.6		
Ash			
Frond	47.1		
	41.7		
	32.1		
	38.9		
	37.6		
Stipe	51.0		
	55.0		
	42.3		
	31.5		
	35.1		

continued

Table 1.215 Seaweed: Phaeophyceae (continued)

		Notes	Reference
L. cloustoni (cont'd)			
Major elements %DM			
Phosphorus (cont'd)			
Stipe	0.25		
	0.27		262
Magnesium			
Frond	0.58		70
Stipe	0.73		
Potassium			
Frond	5.25		70
Stipe	8.15		
Sodium			
Whole	4.69		112
Frond	2.88		70
Stipe	1.35		
Chlorine			
Whole	7.22		112
	13.82		262
Frond	6.58		62}
	–		64}
	6.58		262
	5.92		70
Stipe	12.48		
	10.04		262
Sulphur, total			
Frond	1.02		70
Stipe	0.85		
Trace elements µg/gDM (except I)			
Barium			
Frond	60		68
	65		
	31		
	28		70
Stipe	43		
Chromium			
Frond	1.2		68
	1.2		
	1.5		
	1.4		70
Stipe	1.3		
Cobalt			
Frond	0.56		68
	0.35		
	0.25		
	0.37		70
Stipe	0.48		
Copper			
Frond	–		68
	15		
	14		
	4.6		70
Stipe	5		
Iodine %DM			
Frond	0.50	as	
Stipe	0.33	given	
Iron			
Frond	283		68
	226		
	159		
	437		70
Stipe	446		
Trace elements µg/g DM (cont'd)			
Lead			
Frond	10		68
	26		
	12		
	7.9		70
Stipe	5.4		
Lithium			
Frond	6		
	4		
Manganese			
Frond	30		68
	30		
	10		
	<20		70
Stipe	47		
Molybdenum			
Frond	0.50		68
	0.24		
	0.14		
	0.25		70
Stipe	0.29		
Nickel			
Frond	2.0		68
	1.5		
	1.6		
	1.28		70
Stipe	2.88		
Rubidium			
Frond	250		68
	130		
Silver			
Frond	0.7		68
	0.5		
	0.3		
	0.2		70
Stipe	0.9		
Strontium			
Frond	3000		68
	2200		
	>700		
	650		70
Stipe	2500		
Tin			
Frond	0.7		68
	1.0		
	1.4		
	< 5		70
Stipe	< 5		
Titanium			
Frond	19		68
	20		
	10		
	18		70
Stipe	26		
Vanadium			
Frond	1.3		68
	1.0		
	0.9		
	1.0		70

Table 1.215 Seaweed: Phaeophyceae (continued)

		Notes	Reference
S. bulbosa (cont'd)			
Carbohydrate Fractions %DM			
Mannitol			
Frond	4.5		59
	22.0		
	28.2		
	21.7		
	16.0		
Stipe	5.8		
	11.9		
	23.1		
	21.5		
	19.5		
Alginic acid			
Frond	14.8		
	11.8		
	12.9		
	14.0		
	15.7		
Stipe	15.8		
	14.5		
	13.1		
	14.6		
	15.6		
Laminarin			
Frond	<1.0		
	<1.0		
	<1.0		
	<1.0		
	<1.0		
Stipe	<1.0		
	<1.0		
	<1.0		
	<1.0		
	<1.0		
Trace elements %DM			
Iodine			
Frond	0.12		
	0.06		
	0.10		
	0.05		
	0.16		
Stipe	0.12		
	0.09		
	0.10		
	0.13		
	0.11		

Table 1.216 Seaweed: Chlorophyceae

	June	Notes	Reference
Cladophora rupestris			
MEAL			
Proximate constituents %DM			
Crude protein	30.5	Collected in January	70
Ether extract	0.48		
Nitrogen-free extract	23.1		
Crude fibre	16.6		
Ash	29.3		
Major elements %DM			
Calcium	1.52		
Phosphorus	0.27		
Magnesium	0.73		
Potassium	3.28		
Sodium	2.50		
Chloride	6.34		
Sulphur, total	1.54		
C. Rupestris (cont'd)			
Trace elements μg/g DM (Except I)			
Barium	40		
Chromium	6.4		
Cobalt	16.2		
Copper	31		
Iodine %DM	0.11		
Iron	4400		
Lead	38		
Manganese	1260		
Molybdenum	2.44		
Nickel	20.0		
Silver	<0.7		
Strontium	112		
Tin	<5		
Titanium	550		
Vanadium	24		
Zinc	92		
Vitamins μg/g DM			
Carotene	8.0	Collected 3rd March	398
Carotenoids	196	Spectrophotometric estimation	
Thiamin	1.9	Collected in January	70
Riboflavin	5.9		
Nicotinic acid	26.2		
Enteromorpha intestinalis			
FRESH PROTEIN CONCENTRATE			
Amino acids % of protein N			
Alanine	5 to 10		139
Arginine	8.5		
Aspartic acid	5.2		
Cystine	+		
Glutamic acid	3.7		
Glycine	4.4		
Histidine	0.8		
Leucine + *iso*leucine	7.5		
Lysine	3.7		
Phenylalanine	2.1		
Proline	3.2		
Serine	1.8		
Threonine	2.0		
Tryptophan	+		
Tyrosine	1.2		
Valine + Methionine	6.8		
MEAL			
Vitamins μg/g DM			
Carotene	98		398
Carotenoids, total	399		

Table 1.217 Seaweed: Rhodophyceae

	June	Notes	Reference
Ahnfeltia sp MEAL			
Proximate constituents %DM			
Crude protein	25.4	Sampled 4th November	445
Ash	25.6		
Carbohydrate fractions %DM			
Cellulose	8.8		
Reducing sugars	41.4		
Uronic anhydride	3.4		
Major elements %DM			
Sulphur, total	0.53		
Ceramium rubrum MEAL			
Proximate constituents %DM			
Crude protein	19.4	Sampled 10th June	
Ash	27.5		
Carbohydrate fractions %DM			
Cellulose	4.7		
Reducing sugars	36.1		
Uronic anhydride	5.8		
Major elements %DM			
Sulphur, total	1.77		
Chondrus crispus MEAL			
Proximate constituents %DM			
Crude protein	19.4	Sampled 22nd February	
Ash	20.8		
Carbohydrate fractions %DM			
Cellulose	2.0		
Reducing sugars	39.3		
Uronic anhydride	3.2		
Major elements %DM			
Sulphur, total	5.58		
Corallina officinalis MEAL			
Proximate constituents %DM			
Crude protein	9.4	Sampled 6th November	
Ash	50.0		
Carbohydrate fractions %DM			
Cellulose	5.1		
Reducing sugars	10.8		
Uronic anhydride	4.5		
Major elements %DM			
Sulphur, total	0.17		
Dilsea edulis MEAL			
Proximate constituents %DM			
Crude protein	19.4	Sampled 3rd November	
Ash	19.8		
Carbohydrate fractions %DM			
Cellulose	3.1		
Reducing sugars	45.4		
Uronic anhydride	2.1		
Major elements %DM			
Sulphur, total	2.24		

		Notes	Reference
Endocladia muricata MEAL			
Proximate constituents %DM			
Crude protein	21.2	Sampled August	
Ash	18.1		
Carbohydrate fractions %DM			
Cellulose	1.0		
Reducing sugars	31.3		
Uronic anhydride	2.4		
Major elements %DM			
Sulphur, total	2.61		
Furcellaria fastigiata MEAL			
Proximate constituents %DM			
Crude protein	17.5	Sampled 15th December	
Ash	22.7		
Carbohydrate fractions %DM			
Cellulose	5.7		
Reducing sugars	37.2		
Uronic anhydride	2.6		
Major elements %DM			
Sulphur, total	3.61		
Gelidium cartilageam MEAL			
Proximate constituents %DM			
Crude protein	20.6	S. Africa 26/1	
Ash	22.6		
Carbohydrate fractions %DM			
Cellulose	9.0		
Reducing sugars	33.9		
Uronic anhydride	2.4		
Major elements %DM			
Sulphur, total	0.80		
Gelidium coulteri MEAL			
Proximate constituents %DM			
Crude protein	17.5	U.S.A. August	
Ash	23.1		
Carbohydrate fractions %DM			
Cellulose	8.5		
Reducing sugars	42.8		
Uronic anhydride	4.1		
Major elements %DM			
Sulphur, total	0.90		
Gelidium pristoides MEAL			
Proximate constituents %DM			
Crude protein	21.2	S. Africa 22/6	
Ash	19.8		
Carbohydrate fractions %DM			
Cellulose	6.5		
Reducing sugars	38.7		
Uronic anhydride	2.6		
Major elements %DM			
Sulphur, total	0.74		

Table 1.217 Seaweed: Rhodophyceae (continued)

		Notes	Reference
Gigartina crista MEAL			
Proximate constituents %DM			
Crude protein	15.0	U.S.A.	445
Ash	24.5	August	
Carbohydrate fractions %DM			
Cellulose	4.1		
Reducing sugars	37.4		
Uronic anhydride	2.7		
Major elements %DM			
Sulphur, total	6.38		
Gigartina radula MEAL			
Proximate constituents %DM			
Crude protein	16.9	S. Africa	
Ash	28.3	Sampled 6/6	
Carbohydrate fractions %DM			
Cellulose	1.8		
Reducing sugars	35.2		
Uronic anhydride	1.5		
Major elements %DM			
Sulphur, total	6.08		
Gigartina stellata MEAL			
Proximate constituents %DM		Sampled	
Crude protein	22.5	30th Jan.	
Ash	21.2	Scotland	
Carbohydrate fractions %DM			
Cellulose	2.3		
Reducing sugars	38.9		
Uronic anhydride	3.8		
Major elements %DM			
Sulphur, total	4.74		
Vitamins mg/kg DM			
Carotene	4.0	Collected	398
Carotenoids, total	51.0	3rd March	
Gigartina stririata MEAL			
Proximate constituents %DM			
Crude protein	18.1	S. Africa	445
Ash	28.5	Sampled 10/4	
Carbohydrate fractions %DM			
Cellulose	1.1		
Reducing sugars	36.3		
Uronic anhydride	1.3		
Major element %DM			
Sulphur, total	5.78		
Gracilatia conforvoides MEAL			
Proximate constituents %DM			
Crude protein	23.8	S. Africa	
Ash	31.0	10/6	
Carbohydrate fractions %DM			
Cellulose	3.8		
Reducing sugars	32.0		
Uronic anhydride	1.6		

		Notes	Reference
Major elements %DM			
Sulphur, total	0.94		
Gracilaria foliifera MEAL			
Proximate constituents %DM		Sampled	
Crude protein	16.2	Plymouth	
Ash	38.0	3rd Nov.	
Carbohydrate fractions %DM			
Cellulose	4.6		
Reducing sugars	28.1		
Uronic anhydride	2.4		
Major elements %DM			
Sulphur, total	1.5		
Membranoptera spp MEAL			
Proximate constituents %DM			
Crude protein	23.8	No details	
Ash	33.7		
Carbohydrate fractions %DM			
Cellulose	4.4		
Reducing sugars	28.3		
Uronic anhydride	1.1		
Major elements %DM			
Sulphur, total	3.01		
Phycodrys spp MEAL			
Proximate constituents %DM			
Crude protein	22.5		
Ash	41.4		
Carbohydrate fractions %DM			
Cellulose	3.4		
Reducing sugars	23.2		
Uronic anhydride	0.8		
Major elements %DM			
Sulphur, total	3.54		
Polysiphonia fastigiata MEAL			
Proximate constituents %DM			
Crude protein	30.0	Sampled	
Ash	24.9	4th March	
Carbohydrate fractions %DM			
Cellulose	1.2		
Reducing sugars	35.6		
Uronic anhydride	4.2		
Major elements %DM			
Sulphur, total	3.01		
Porphyra umbilicalis MEAL			
Proximate constituents %DM			
Crude protein	27.5	Sampled	
Ash	21.8	24th April	
Carbohydrate fractions %DM			
Cellulose	3.2		
Reducing sugars	43.2		
Uronic anhydride	3.9		

continued

Table 1.217 Seaweed: Rhodophyceae (continued)

		Notes	Reference
Porphyra umbilicalis (cont'd)			
Major elements %DM			
Sulphur, total	1.74		445
Pterocladia pyramidale MEAL			
Proximate constituents %DM			
Crude protein	18.1	Sampled August	
Ash	28.6		
Carbohydrate fractions %DM			
Cellulose	8.2		
Reducing sugars	31.6		
Uronic anhydride	4.6		
Major elements %DM			
Sulphur, total	1.44		
Ptiloto plumosa MEAL			
Proximate constituents %DM			
Crude protein	25.6	No details	
Ash	34.4		
Carbohydrate fractions %DM			
Cellulose	4.7		
Reducing sugars	30.0		
Uronic anhydride	1.2		
Major elements %DM			
Sulphur, total	1.47		
Rhodochorion floridulum MEAL			
Proximate constituents %DM			
Crude protein	16.2		
Ash	53.3		
Carbohydrate fractions %DM			
Cellulose	7.6		
Reducing sugars	16.7		
Uronic anhydride	1.6		
Major elements %DM			
Sulphur, total	0.17		
Rhodoglossum affine MEAL			
Proximate constituents %DM			
Crude protein	13.1	Sampled August California	
Ash	23.5		
Carbohydrate fractions %DM			
Cellulose	3.0		
Reducing sugars	37.1		
Uronic anhydride	3.1		
Major elements %DM			
Sulphur, total	6.11		
Rhodymenia palmata			
Proximate constituents %DM			
Crude protein	18.1		139
FRESH PROTEIN CONCENTRATE			
Amino acids % of protein N			
Alanine	2.6		
Arginine	5.8		
Aspartic acid	3.0		
Cystine	+		
Amino acids % of protein N (cont'd)			
Glutamic acid	2.6		
Glycine	5.8		
Hystidine	1.0		
Leucine + *iso*leucine	4.7		
Lysine	3.9		
Phenylalanine	1.2		
Proline	1.7		
Serine	3.8		
Threonine	2.1		
Tryptophan	+		
Tyrosine	1.3		
Valine + Methionine	4.6		
MEAL			
Proximate constituents %DM			
Crude protein	21.9		445
	23.4		70
Ether extract	0.3		
Nitrogen-free extract	46.8		
Crude fibre	2.1		
Ash	21.2		445
	27.4		70
Carbohydrate fractions %DM			
Cellulose	2.4		445
Anhydropentose	36.2		
Anhydrohexose	3.5		
Uronic anhydride	3.3		
Major elements %DM			
Calcium	0.72		770
Phosphorus	0.56		
Magnesium	0.39		
Potassium	7.91		
Sodium	2.07		
Chlorine	9.70		
Sulphur, total	0.35		
	0.23		445
Trace elements μg/g DM (except I)			
Barium	21		70
Chromium	34		
Cobalt	2.60		
Copper	48		
Iodine %DM	0.03		
Iron	1355		
Lead	28		
Manganese	110		
Molybdenum	0.83		
Nickel	16.4		
Silver	1.0		
Strontium	90		
Tin	<5		
Titanium	100		
Vanadium	29		
Zinc	200		
Vitamins μg/g DM			
Carotene	37		398
	115		
Carotenoids	108		
	341		
Thiamin	1.5		70
Riboflavin	5.3		
Niacin	28.9		

Table 1.217 Seaweed: Rhodophyceae (continued)

		Notes	Reference
Suhria vittata			
MEAL			
Proximate constituents %DM			
Crude protein	23.8	Sampled 10th April	445
Ash	18.9		
Carbohydrate fractions %DM			
Cellulose	7.9		
Reducing sugars	40.9		
Uronic anhydride	2.9		
Major elements %DM			
Sulphur, total	0.97		

Table 1.218 Seaweed: Unnamed varieties

		Notes	Reference
FRESH PROTEIN CONCENTRATE			
Amino acids g/100 g protein N			
Alanine	5.0	Samples collected 19th March Estimated by paper chrom-atography	139
Arginine	11.8		
Aspartic acid	8.7		
Cystine	+		
Glutamin acid	5.3		
Glycine	5.7		
Histidine	0		
Leucine + *Iso*leucine	8.7		
Lysine	4.1		
Phenylalanine	–		
Proline	+		
Serine	3.3		
Threonine	2.7		
Tryptophan	+		
Tyrosine	0.5		
Valine + Methionine	14.0		

Table 1.219 Metabolisable energy. Oat

(1) Proximate analysis %DM
(2) Digested nutrients %DM
(3) Metabolisable energy, kcal/g dry matter (DM), organic matter (OM) and digested organic matter (DOM)

Description of sample		CP	EE	NFE	CF	OM	Digestibility by	Notes	Reference
Grain, whole:									
var Victory	(1)	9.25	5.78	70.38	10.60	96.01	Leghorn cockerels		366
	(2)	7.16	4.83	54.05	0.93	66.97			
	(3)	DM: 2.75 OM: 2.86 DOM: 4.11							
Not named	(3)	DM: 2.5					Poultry	No data for composition. First 5 values quoted from literature; 6th,"compromise value chosen".	339
		2.7							
		2.7							
		2.8							
		3.4							
		2.7							
Varieties grown in Wales		9.8	5.2	68.9	13.0	96.9	Poultry	Means and ranges for: 82 in 1961 39 in 1962 50 in 1963	370
		(7.2-12.6)	(1.9-8.0)	(63.8-74.2)	(9.0-17.9)	(95.9-97.6)			
	(3)	DM: 2.87 (2.27-3.30)							
	(1)	11.0	5.5	68.5	12.1	97.1		Digested nutrients (or percentage digestibility) not stated. M.E. computed with Bolton's formula.	
		(8.9-13.2)	(3.6-7.6)	(63.9-71.1)	(10.0-15.5)	(96.4-97.8)			
	(3)	DM: 2.96 (2.42-3.43)							
	(1)	11.9	4.9	68.1	12.0	96.9			
		(9.5-14.5)	(0.9-6.8)	(64.1-72.2)	(8.0-15.8)	(95.9- 97.8)			
	(3)	DM: 2.91 (2.26-3.25)							
Huskless Oat:	(1)	14.14	5.40	76.50	1.93	97.97	Leghorn cockerels		366
	(2)	11 62	3.37	69.46	–	84.45			
	(3)	DM: 3.37 OM: 3.44 DOM: 3.99							
Ground whole grain:									
Farm ground: Australian	(1)	7.2	5.2	73.3	9.4	95.1	Poultry	Digested nutrients (or percentage digestibility) not stated for those 3 samples	119
	(3)	DM: 2.8							
Farm ground: Scottish	(1)	9.1	2.7	77.7	8.0	97.5			
	(3)	DM: 3.0							
Sussex ground	(1)	8.6	4.3	74.1	10.2	97.2			
	(3)	DM: 3.1							
Farm ground	(1)	12.90	5.38	66.92	11.52	96.72	Pig		579
	(2)	11.03	4.67	49.72	–	65.42			
	(3)	DM: 2.7 OM: 2.8 DOM: 4.2							
Sussex ground	(1)	13.32	6.47	62.97	13.07	95.83			
	(2)	9.99	5.44	51.13	6.16	72.72			
	(3)	DM: 3.0 OM: 3.1 DOM: 4.1							
Coarse ground	(1)	13.1	6.0	66.8	11.5	97.4	Pig		114
	(2)	10.7	4.5	48.8	1.7	65.7			
	(3)	DM: 2.7 OM: 2.8 DOM: 4.1							
Medium ground	(1)	12.9	6.1	67.2	11.1	97.3			
	(2)	10.7	4.7	48.7	1.1	65.2			
	(3)	DM: 2.7 OM: 2.8 DOM: 4.1							
Fine ground	(1)	12.6	6.4	68.1	10.4	97.5			
	(2)	10.1	5.0	50.1	1.4	66.6			
	(3)	DM: 2.8 OM: 2.8 DOM: 4.1							
Crushed oats:									
Bulk sample	(1)	11.18	5.49	66.85	12.74	96.26	Pig		579
	(2)	7.67	3.50	44.05	–	55.22			
	(3)	DM: 2.3 OM: 2.4 DOM: 4.1							
	(2)	8.83	5.00	53 35	3.25	70.43	Ruminant	Digestibility coefficients from trials with ruminants. Same composition as for pig (assumed)	572
	(3)	DM: 2.9 OM: 3.0 DOM: 4.1							
Oat Offals:									
Oat feedmeal	(1)	9.4	5.7	55.6	24.7	95.4	Sheep		137
	(2)	6.7	2.7	33.0	15.3	57.7			
	(3)	DM: 2.2 OM: 2.3 DOM: 3.8							

Table 1.219 Metabolisable energy. Oat (continued)

Description of sample		CP	EE	NFE	CF	OM	Digestibility by	Notes	Reference
Oat Dust:									
white	(1)	15.9	9.2	54.0	16.3	95.4	Sheep	Proximate analysis mean of two	137
	(2)	10.8	7.8	30.2	3.7	52.5			
	(3)	DM: 2.4	OM: 2.5	DOM: 4.5					
brown (scree)	(1)	13.1	7.0	50.6	23.5	94.2			
	(2)	7.7	5.4	14.8	6.9	34.8			
	(3)	DM: 1.5	OM: 1.6	DOM: 4.4					
Oat Husks:									
Coarsely ground	(1)	2.5	0.9	59.3	33.0	95.7	Pig		114
	(2)	2.0	0.4	46.1	3.3	51.8			
	(3)	DM: 1.9	OM: 2.0	DOM: 3.7					
Medium ground	(1)	2.2	1.0	58.8	34.4	96.4			
	(2)	1.8	0.5	45.3	3.0	50.6			
	(3)	DM: 1.9	OM: 2.0	DOM: 3.7					
Fine ground	(1)	2.2	0.9	58.9	34.6	96.6			
	(2)	1.7	0.5	45.4	1.4	49.0			
	(3)	DM: 1.8	OM: 1.9	DOM: 3.7					

Table 1.220 Metabolisable energy. Wheat

(1) Proximate analysis %DM
(2) Digested nutrients %DM
(3) Metabolisable energy, kcal/g dry matter (DM), organic matter (OM) and digested organic matter (DOM)

Description of sample		CP	EE	NFE	CF	Ash	Digestibility by	Notes	Reference
Whole grain:									
Unnamed	(1)	10.19	1.95	83.84	2.14	1.88	Sheep		538
	(2)	5.56	0.73	70.16	2.14	–			
	(3)	DM: 2.97	OM: 3.03	DOM: 3.78					
	(1)	10.44	2.00	83.31	2.44	1.81			
	(2)	5.75	0.96	71.55	2.44	–			
	(3)	DM: 3.06	OM: 3.11	DOM: 3.80					
Unnamed	(1)	11.5	2.1	82.2	2.5	1.7	Poultry		82
	(2)	9.3	1.1	76.2	–	–			
	(3)	DM: 3.33	OM: 3.38	DOM: 3.84					
Var. Guardsman	(1)	13.8	1.4	81.2	1.7	1.9	Poultry		119
	(3)	DM: 3.69							
Var. Hydrid 46	(1)	13.2	2.0	80.8	1.8	2.2			
	(3)	DM: 3.71							
Unnamed	(3)	DM: 3.67					Poultry	Compromise value from a review	339
Milling offals									
Middlings, coarse (Straight-run pollards)	(1)	18.9	5.0	64.7	7.1	4.3	Poultry		257
	(2)	15.1	4.0	41.1	0.4	–			
	(3)	DM: 2.54	OM: 2.65	DOM: 4.20					
Middling, fine	(1)	9.7	2.6	83.4	2.2	2.1			
	(2)	6.9	2.4	73.8	0.1	–			
	(3)	DM: 3.25	OM: 3.32	DOM: 3.90					
Middlings, coarse (Fine wheat feed)	(1)	18.8	5.8	62.2	8.6	4.6	Sheep		563
	(2)	14.1	3.5	47.4	2.4	–			
	(3)	DM: 2.75	OM: 2.89	DOM: 4.08					
	(2)	14.7	4.0	47.3	1.9	–	Pig		
	(3)	DM: 2.80	OM: 2.93	DOM: 4.12					
Middlings, fine	(1)	20.2	5.9	65.5	4.7	3.7	Poultry		83
	(2)	17.1	5.1	50.2	0.3	–			
	(3)	DM: 3.06	OM: 3.18	DOM: 4.21					
Fine parings	(1)	17.6	5.0	67.4	6.1	3.9	Poultry		119
	(3)	DM: 2.94							
French pollards	(1)	18.6	4.3	62.9	8.4	5.8			
	(3)	DM: 2.22							
Broad bran	(1)	16.6	4.2	60.6	11.9	6.7	Poultry		257
	(2)	10.1	2.2	23.4	1.1	–			
	(3)	DM: 1.54	OM: 1.65	DOM: 4.18					
Coarse bran	(1)	17.3	4.9	60.0	11.9	5.9	Sheep		563
	(2)	12.7	1.0	43.0	3.2	–			
	(3)	DM: 2.34	OM: 2.48	DOM: 3.90					
	(2)	11.8	1.7	38.3	1.4	–	Pig		
	(3)	DM: 2.13	OM: 2.36	DOM: 4.00					
Coarse bran milled	(1)	17.2	5.1	59.9	11.9	5.9			
	(2)	11.9	2.2	40.2	1.9	–			
	(3)	DM: 2.27	OM: 2.41	DOM: 4.02					
Coarse bran	(1)	17.4	5.8	59.9	11.3	5.7	Poultry		260
	(2)	11.9	4.0	25.3	0.5	–			
	(3)	DM: 1.82	OM: 1.93	DOM: 4.37					
	(2)	14 3	4.6	39.4	2.3	–	Rabbit		
	(3)	DM: 2.55	OM: 2.70	DOM: 4.21					
Coarse bran	(1)	15.6	4.9	60.6	11.8	7.1	Poultry		119
	(3)	DM: 1.47							

Table 1.220 Metabolisable energy. Wheat (continued)

Description of sample		CP	EE	NFE	CF	Ash	Digestibility by	Notes	Reference
Fine bran	(1)	17.5	5.1	61.5	10.8	5.1	Sheep		563
	(2)	12.4	3.0	45.5	2.2	–			
	(3)	DM: 2.55	OM: 2.69	DOM: 4.04					
	(2)	13.6	3.4	45.7	1.8	–	Pig		
	(3)	DM: 2.63	OM: 2.78	DOM: 4.09					
Fine bran	(1)	17.3	4.7	63.6	9.3	5.1	Poultry		260
	(2)	12.2	3.5	36.3	0.6	–			
	(3)	DM: 2.24	OM: 2.36	DOM: 4.26					
Wheat and Barley Milling Offals, mixed									
Fine bran	(1)	16.0	5.0	61.6	11.3	6.1	Poultry		260
	(2)	10.0	3.2	25.3	–	–			
	(3)	DM: 1.65	OM: 1.76	DOM: 4.29					
Coarse Millers'	(1)	14.1	4.4	59.1	15.5	6.9	Sheep		563
Offals	(2)	8.8	0.1	40.0	5.1	–			
(10% barley)	(3)	DM: 2.03	OM: 2.18	DOM: 3.76					
	(2)	7.7	2.1	32.9	1.9	–	Pig		
	(3)	DM: 1.79	OM: 1.92	DOM: 4.01					
Fine Millers'	(1)	16.2	5.0	61.2	11.9	5.7	Sheep		
Offals	(2)	11.3	2.8	44.4	3.9	–			
(10% barley)	(3)	DM: 2.49	OM: 2.64	DOM: 4.00					
	(2)	10.3	3.1	39.9	1.9	–	Pig		
	(3)	DM: 2.25	OM: 2.38	DOM: 4.08					

Table 1.221 Metabolisable energy. Barley grain and Distillers' residues

(1) Proximate analysis %DM
(2) Digested nutrients %DM
(3) Metabolisable energy, kcal/g dry matter (DM), organic matter (OM) and digested organic matter (DOM)

Description of sample		CP	EE	NFE	CF	Ash	Digestibility by	Notes	Reference
Grain:	(1)	9.79	1.49	81.47	4.59	2.66	Sheep		538
	(2)	6.69	1.49	73.31	4.59	–			
	(3)	DM: 3.27	OM: 3.36	DOM: 3.80					
	(1)	11.31	1.90	79.51	4.18	3.10			
	(2)	8.05	1.79	71.30	4.18	–			
	(3)	DM: 3.27	OM: 3.37	DOM: 3.83					
	(1)	12.88	2.47	75.64	6.26	2.75	Non-laying hens		82
	(2)	10.34	1.89	62.86	0.59	–			
	(3)	DM: 2.97	OM: 3.05	DOM: 3.92					
	(1)	11.0	1.4	78.7	5.7	3.2	Poultry		119
	(3)	DM: 3.19							
	(1)	11.4	1.1	74.3	5.6	7.6			
	(3)	DM: 2.98							
Unnamed	(3)	DM: 3.11					Poultry	Compromise value from review	339
Unnamed	(3)	DM: 3.32 (3.10–3.56) (1961)					Poultry	ME computed with Bolton's formula; Analyst, 1960, *85*, 189	370
	(3)	DM: 3.47 (3.35–3.59) (1962)							
	(3)	DM: 3.33 (3.09–3.47) (1963)							
Dreg meal	(1)	49.7	14.7	28.7	5.2	98.3	Sheep	Dried extract drained off to leave "distillers' grains" or "draff"	528
	(2)	32.25	14.31	20.00	1.21	–			
	(3)	DM: 3.42	OM: 3.48	DOM: 5.04					
Dried solubles	(3)	DM: 1.81					Poultry	Quoted from Potter and Matterson, 1960	339
	(3)	DM: 3.23						Quoted from Hill, 1961	
	(3)	DM: 2.78						Compromise value chosen	

Table 1.222 Metabolisable energy. Maize and Sorghum

(1) Proximate analysis %DM
(2) Digested nutrients %DM
(3) Metabolisable energy, kcal/g dry matter (DM), organic matter (OM) and digested organic matter (DOM)

Description of sample		CP	EE	NFE	CF	Ash	Digestibility by	Notes	Reference
Maize: Whole grain	(1)	11.61	4.21	81.22	1.63	1.33	Poultry		82
	(2)	10.0	3.7	77.5	0.28	–			
	(3)	DM: 3.59	OM: 3.64	DOM: 3.92					
	(3)	DM: 3.90					Poultry	Compromise value from a survey	339
Maize Meal: yellow	(1)	10.3	3.7	82.4	2.0	1.6	Poultry		119
var. unnamed	(3)	DM: 3.89							
Meal: degermed, cooked	(1)	10.5	1.7	85.4	1.4	1.0	Pig		555
	(2)	9.9	1.4	84.7	1.4	–			
	(3)	DM: 3.73	OM: 3.74	DOM: 3.84					
Maize gluten feed	(1)	25.6	1.1	62.0	10.2	1.1	Poultry		119
	(3)	DM: 2.13							
	(1)	27.9	2.1	56.3	7.8	5.9	Poultry		83
	(2)	24.9	1.2	35.1	0.6	–			
	(3)	DM: 2.54	OM: 2.70	DOM: 4.1					
Maize gluten meal	(3)	DM: 3.56					Poultry	Compromise value from a survey	339
Maize germ meal	(1)	21.0	12.6	47.2	16.5	2.7	Poultry		83
	(2)	16.8	10.7	15.7	4.7	–			
	(3)	DM: 2.36	OM: 2.43	DOM: 4.9					
Maize oil	(3)	DM: 7.8					Poultry		339
Sorghum	(1)	10.30	3.25	82.39	1.30	2.76	Poultry	Dura: white, ground Palestine	85
	(2)	8.27	2.47	73.15	0.11	–			
	(3)	DM: 3.29	OM: 3.38	DOM: 3.91					
	(3)	DM: 3.40					Poultry	Calc. for 90% DM	119
Milo Grain	(3)	DM: 3.37					Poultry		339
	(3)	DM: 3.23							
	(3)	DM: 3.40							
	(3)	DM: 3.33							
	(3)	DM: 3.30							

Table 1.223 Metabolisable energy. Miscellaneous concentrates and residues

(1) Proximate analysis %DM
(2) Digested nutrients %DM
(3) Metabolisable energy, kcal/g dry matter (DM), organic matter (OM) and digested organic matter (DOM)

Description of sample		CP	EE	NFE	CF	Ash	Digestibility by	Notes	Reference
Yeast: dried	(1)	50.67	0.52	37.11	0.52	11.18	Sheep		569
	(2)	43.47	0.21	30.50	–	–			
	(3)	DM: 3.01 OM: 3.39 DOM: 4.06							
Yeast, brewers: dried, unextracted	(1)	44.28	1.06	43.79	0.22	10.65	Pig	*In vivo*	574
	(2)	39.59	–	40.81	0.05	–		Digestibility of CP	
	(3)	DM: 3.21 OM: 3.60 DOM: 3.99						*in vitro* 44.03%	
Yeast: dried	(3)	DM: 2.05					Poultry	Quoted from Potter and Matterson, 1960	339
	(3)	DM: 2.33						Quoted from Hill, 1961	
	(3)	DM: 2.13						Quoted from Fraps and Carlyle, 1942	
	(3)	DM: 2.33						Compromise value chosen	
Penicillium chrysogenum dried felt	(1)	44.43	4.16	29.28	10.28	11.85	Pig		569
	(2)	29.37	1.10	19.21	8.36	–			
	(3)	DM: 2.30 OM: 2.61 DOM: 3.97							
	(1)	44.53	4.68	29.83	9.86	11.10	Sheep		
	(2)	32.91	2.28	24.37	8.62	–			
	(3)	DM: 2.74 OM: 3.09 DOM: 4.03							
Palm kernel extracted meal,	(1)	20.21	2.26	53.75	19.45	4.33	Pig		565
	(2)	12.13	0.58	41.28	11.83	–			
	(3)	DM: 2.47 OM: 2.58 DOM: 3.75							
Sunflower seed var. Hungarian striped	(1)	19.30	28.32	20.65	27.75	3.98	Poultry	Factors for oilseeds Grown in East Anglia	259
	(2)	16.29	26.71	3.51	1.08	–			
	(3)	DM: 3.38 OM: 3.51 DOM: 7.09							
Tapioca flour	(1)	2.06	0 54	92.12	2.83	2.45	Pig		582
	(2)	1.41	0.23	91.18	2.14	–			
	(3)	DM: 3.51 OM: 3.60 DOM: 3.70							
Sago pith meal	(1)	1.94	0.44	88.71	4.97	3.94	Pig		582
	(2)	–	0.20	76.02	–	–			
	(3)	DM: 2.83 OM: 2.95 DOM: 3.71							
Horse Chestnut meal alcohol extracted	(1)	7.35	7.38	74.60	7.99	2.68	Sheep		567
	(2)	–	4.46	53.64	–	–			
	(3)	DM: 2.35 OM: 2.42 DOM: 4.05							
	(2)	1.46	3.14	54.61	1.48	–	Pig		
	(3)	DM: 2.39 OM: 2.46 DOM: 3.94							
water extracted	(1)	7.73	7.76	73.62	8.48	2.41	Sheep		
	(2)	–	5.05	51.39	–	–			
	(3)	DM: 2.32 OM: 2.99 DOM: 4.11							
	(2)	1.76	4.01	53.67	1.31	–	Pig		
	(3)	DM: 2.44 OM: 2.50 DOM: 4.01							
alcohol extracted pressed residue after removal of glucose	(1)	11.17	6.89	52.90	22.88	6.16	Sheep		
	(2)	–	6.26	33.17	–	–			
	(3)	DM: 1.75 OM: 1.86 DOM: 4.43							
	(2)	–	4.95	29.57	–	–	Pig		
	(3)	DM: 1.50 OM: 1.60 DOM: 4.36							
Flax chaff	(1)	10.45	6.27	37.40	37.82	8.06	Sheep		136
	(2)	6.27	4.60	11.80	17.21	–			
	(3)	DM: 1.56 OM: 1.70 DOM: 3.92							
	(1)	11.38	7.54	34.42	38.04	8.62			
	(2)	7.40	5.33	9.02	15.00	–			
	(3)	DM: 1.50 OM: 1.64 DOM: 4.09							

Table 1.224 Metabolisable energy. Milk products

(1) Proximate analysis %DM
(2) Digested nutrients %DM
(3) Metabolisable energy, kcal/g dry matter (DM), organic matter (OM) and digested organic matter (DOM)

Description of sample		CP	EE	NFE	Ash	Digestibility by	Notes	Reference
Whole milk: spray dried	(1)	27.0	29.75	37.0	6.22	Calves	Mean of 2 trials	417
	(2)	24.4	28.6	36.6	–			
	(3)	DM: 5.11	OM: 5.45	DOM: 5.70				
	(1)	27.6	27.3	39.1	6.02	Calves	Mean of 3 trials	418
	(2)	25.83	27.00	38.67	–			
	(3)	DM: 5.10	OM: 5.43	DOM: 5.58				
	(1)	26.8	27.7	39.5	6.00	Calves	Mean of 3 trials	419
	(2)	24.68	26.92	38.24	–			
	(3)	DM: 5.03	OM: 5.35	DOM: 5.60				
Separated milk: spray dried	(1)	38.6	1.46	51.7	8.24	Calves	Mean of 3 trials	418
	(2)	36.28	1.17	50.61	–			
	(3)	DM: 3.61	OM: 3.94	DOM: 4.10				
	(1)	38.4	1.30	52.0	8.34	Calves	Mean of 2 trials	419
	(2)	35.40	0.88	50.49	–			
	(3)	DM: 3.54	OM: 3.86	DOM: 4.08				
Whey: dried	(1)	13.4	0.5	77.2	8.9	–	Of NFE, 61.0 was sugar	120
	(3)	1.89 or 1.90				Poultry	Basis not stated: probably the conventional 90% DM, giving 2.10 or 2.11 kcal/g DM	339
Separated milk: spray dried with lactose added	(1)	26.35	0.97	66.65	6.10	Calves	4 trials with 2 calves	417
	(2)	22.92	0.65	63.85	–			
	(3)	DM: 3.45	OM: 3.68	DOM: 3.95				
Separated milk: spray dried, with hydrogenated palm oil	(1)	25.65	25.85	42.7	5.8	Calves	2 trials [calculation with Axelsson's coefficient for lactose, 3.56, instead of 3.7 for NFE, gives ME values: 4.41, 4.68 and 5.18]	417
	(2)	21.88	22.39	40.95	–			
	(3)	DM: 4.47	OM: 4.75	DOM: 5.25				
Separated milk: spray dried, with unhydrogenated palm oil or palm kernel oil	(1)	27.0	30.1	36.9	6.0	Calves	Palm oil: 2 trials	418
	(2)	25.38	27.42	36.79	–			
	(3)	DM: 4.92	OM: 5.23	DOM: 5.49				
	(1)	27.4	29.5	37.1	6.0	Calves	Palm oil or palm kernel oil: 2 trials	
	(2)	25.10	27.67	36.95	–			
	(3)	DM: 4.93	OM: 5.25	DOM: 5.50				
Separated milk: spray dried, with palm kernel oil	(1)	26.7	27.7	39.7	5.9	Calves	Hydrogenated oil: 2 trials	419
	(2)	22.56	24.29	38.04	–			
	(3)	DM: 4.56	OM: 4.85	DOM: 5.37				
	(1)	25.9	28.6	39.6	5.9	Calves	Unhydrogenated oil: 3 trials	
	(2)	22.20	24.94	38.55	–			
	(3)	DM: 4.62	OM: 4.91	DOM: 5.43				

Table 1.225 Metabolisable energy. Meat and Fish

(1) Proximate analysis %DM
(2) Digested nutrients %DM
(3) Metabolisable energy, kcal/g dry matter (DM), organic matter (OM) and digested organic matter (DOM)

Description of sample		CP*	EE	NFE	Ash	Digestibility by	Notes	Reference
Meat meal:	(1)	71.69	3.17	4.28	20.86	Pig	Fat expelled and extracted	557
	(2)	63.02	2.61	4.28	–			
	(3)	DM: 3.24	OM: 4.09	DOM: 4.63				
	(1)	71.60	11.08	2.44	14.88	Pig	Fat expelled.	
	(2)	67.23	9.86	2.44	–		Pressure expulsion only	
	(3)	DM: 4.03	OM: 4.74	DOM: 5.07				
	(1)	66.38	18.82	2.84	11.96	Pig	Fat drained off	
	(2)	60.34	17.95	2.84	–		*Original protein values	
	(3)	DM: 4.49	OM: 5.10	DOM: 5.53			converted to crude protein	
Whale meat meal:	(1)	92.93	3.66	1.22	2.18	Pig	Cooked, oil expelled, dried	574
	(2)	81.42	3.55	1.22	–			
	(3)	DM: 4.04	OM: 4.13	DOM: 4.69				
Beef tallow:	(3)	DM: 7.02				Pig	Quoted from Potter and Matterson (1960)	339
	(3)	DM: 7.15					Quoted from Hill *et al* (1956)	
Fish meal: White fish	(1)	66.45	2.72	1.52	29.31	Pig	Commercial sample	574
	(2)	63.39	1.09	1.43	–			
	(3)	DM: 3.01	OM: 4.25	DOM: 4.56				
	(3)	DM: 2.7					Quoted from Potter and Matterson (1960)	339
	(3)	DM: 2.0					Quoted from Hill *et al* (1956)	
	(3)	DM: 2.0					Quoted from Fraps and Carlyle (1942)	
	(3)	DM: 2.7					Compromise value chosen	

Table 1.226 Metabolisable energy. Grass and Pasture mixtures

(1) Proximate analysis %DM
(2) Digested nutrients %DM
(3) Metabolisable energy, kcal/g dry matter (DM), organic matter (OM) and digested organic matter (DOM)

Description of sample Date cut		CP	EE	NFE	CF	Ash	Digestibility by	Notes	Reference
Italian Ryegrass									
7 May	(1)	23.7	3.7	46.9	16.1	9.6	Sheep	Cut for silage, young, leafy	351
	(2)	19.4	2.3	40.5	14.1	–			
	(3)	DM: 2.92	OM: 3.23	DOM: 3.83					
Perennial Ryegrass, Timothy, White Clover									
	(1)	9.29	1.92	52.36	29.93	6.50	Sheep	Cut for silage, Stored frozen for the digestibility trials	103
	(2)	5.52	1.42	38.28	21.88	–			
	(3)	DM: 2.40	OM: 2.57	DOM: 3.58					
	(1)	9.91	1.97	53.42	27.72	6.98			
	(2)	6.15	1.46	37.45	21.40	–			
	(3)	DM: 2.38	OM: 2.56	DOM: 3.59					
	(1)	8.65	1.92	54.64	28.26	6.53			
	(2)	5.14	1.42	39.94	20.66	–			
	(3)	DM: 2.41	OM: 2.58	DOM: 3.59					
	(1)	9.50	1.67	53.33	28.92	6.58			
	(2)	5.90	1.24	37.38	22.33	–			
	(3)	DM: 2.38	OM: 2.55	DOM: 3.56					
	(1)	14.64	2.87	45.85	29.64	7.00	Sheep	Cut for silage. Stored frozen for the digestibility trials	104
	(2)	9.75	1.09	33.42	23.92	–			
	(3)	DM: 2.43	OM: 2.62	DOM: 3.57					
	(1)	13.31	2.84	48.12	29.09	6.64			
	(2)	8.70	0.96	35.99	22.22	–			
	(3)	DM: 2.42	OM: 2.60	DOM: 3.57					
	(1)	13.13	2.82	47.67	29.92	6.46			
	(2)	8.74	1.07	34.75	24.15	–			
	(3)	DM: 2.45	OM: 2.61	DOM: 3.56					
	(1)	13.08	2.84	47.39	29.64	7.05			
	(2)	8.55	0.96	35.45	22.64	–			
	(3)	DM: 2.41	OM: 2.59	DOM: 3.57					
Cocksfoot									
19 May	(1)	20.0	3.7	41.3	25.4	9.6	Sheep	Cut for silage	351
	(2)	15.1	1.7	31.2	20.3	–			
	(3)	DM: 2.53	OM: 2.80	DOM: 3.70					
Pasture grass (*Agrostis*, Ryegrass, Cocksfoot, Fescues) herbs, weeds, and clover (red and white)									
20-29 April	(1)	22.66	5.67	50.40	13.20	8.07	Sheep	Grazed by sheep in winter. Mown at monthly intervals. Analyses are means for weekly samples. 1929	585
	(2)	17.98	3.53	44.16	10.25	–			
	(3)	DM: 2.98	OM: 3.24	DOM: 3.92					
14-27 May	(1)	18.88	7.69	46.26	16.85	10.32			
	(2)	14.10	5.47	38.77	12.79	–			
	(3)	DM: 2.84	OM: 3.16	DOM: 3.99					
8-18 June	(1)	18.23	6.79	47.19	18.96	8.83			
	(2)	13.45	3.94	38.27	14.17	–			
	(3)	DM: 2.71	OM: 2.98	DOM: 3.88					
13-26 July	(1)	19.11	5.92	47.49	18.75	8.73			
	(2)	14.13	2.10	35.89	13.04	–			
	(3)	DM: 2.48	OM: 2.71	DOM: 3.80					
11-24 August	(1)	19.04	6:78	47.36	18.51	8.31			
	(2)	14.02	3.07	35.99	13.08	–			
	(3)	DM: 2.56	OM: 2.79	DOM: 3.86					

continued

Table 1.226 Metabolisable energy. Grass and Pasture mixtures (continued)

Description of sample Date cut		CP	EE	NFE	CF	Ash	Digestibility by	Notes	Reference
Pasture grass (*Agrostis* Ryegrass, Cocksfoot, Fescues), herbs, weeds; very little clover									585
15-28 April	(1)	23.52	3.43	45.90	17.72	9.43	Sheep	Same pasture as above, similarly managed. Given farmyard manure, chalk, superphosphate and sulphate of potash in December, and, after the monthly cut, a dressing of sulphate of ammonia.	
	(2)	19.31	1.45	39.96	14.84	–			
	(3)	DM: 2.85	OM: 3.15	DOM: 3.78					
10-23 May	(1)	17.20	4.18	46.13	23.35	9.14			
	(2)	13.08	2.34	38.13	19.82	–			
	(3)	DM: 2.73	OM: 3.00	DOM: 3.72					
7-20 June	(1)	17.24	4.10	45.39	24.29	8.98			
	(2)	12.99	2.00	35.68	19.96	–			
	(3)	DM: 2.61	OM: 2.87	DOM: 3.70					
5-18 July	(1)	17.37	3.91	48.92	20.95	8.85		1930	
	(2)	13.18	1.30	39.00	16.48	–			
	(3)	DM: 2.59	OM: 2.84	DOM: 3.70					
9-22 August	(1)	18.28	4.34	47.00	21.66	8.72			
	(2)	14.20	1.93	39.25	17.96	–			
	(3)	DM: 2.73	OM: 3.00	DOM: 3.73					
6-19 September	(1)	19.47	4.80	46.31	20.26	9.16			
	(2)	15.48	1.77	38.22	16.46	–			
	(3)	DM: 2.70	OM: 2.97	DOM: 3.75					
Pasture, permanent: poor quality									
4 June	(1)	10.62	2.11	47.55	30.43	9.29	Sheep	10–15 cm. height Cut for drying	533
	(2)	6.70	1.18	33.10	21.39	–			
	(3)	DM: 2.22	OM: 2.45	DOM: 3.57					
Pasture, permanent: intensively treated and grazed									
3-12 October	(1)	17.58	2.09	44.86	21.95	13.52	Sheep	Short; little growth Grazed and some cut for drying	
	(2)	13.65	0.81	34.77	17.65	–			
	(3)	DM: 2.45	OM: 2.83	DOM: 3.66					
Permanent pasture: Perennial Ryegrass 45%, *Agrostis* 25%, Rough-stalked meadow grass 13.5%, other grasses 16%, clover 1%, trefoil ½%, herbs, traces.								8-13 cm. height. Basic slag in December; lime in January; sulphate of ammonia in February and some plots further dressings in May or June. Analyses for series of samples but digestibility only for 2nd week in June: applied to 14 May-21 June. A: extra nitrogen B: no extra nitrogen	540
14 May-21 June	(1) A	17.27	4.47	46.01	22.26	9.82	Sheep		
	(2)	12.44	2.64	36.51	18.05	–			
	(3)	DM: 2.62	OM: 2.90	DOM: 3.76					
	(1) B	17.23	4.47	46.53	21.89	9.87			
	(2)	12.30	2.76	35.84	17.07	–			
	(3)	DM: 2.57	OM: 2.85	DOM: 3.77					
Pasture grass with clover								5 weeks' growth at each cut. Farmyard manure in November, and sulphate of ammonia in February, March, April and May.	583
13-23 April	(1)	24.66	5.12	42.68	18.56	8.98	Sheep		
	(2)	18.82	2.93	34.79	15.91	–			
	(3)	DM: 2.79	OM: 3.06	DOM: 3.85					
15-28 May	(1)	15.45	3.63	49.16	23.19	8.57			
	(2)	10.29	1.59	39.03	18.53	–			
	(3)	DM: 2.55	OM: 2.79	DOM: 3.67					
12-21 June	(1)	16.25	4.00	47.19	23.73	8.83			
	(2)	11.10	1.69	35.79	18.93	–			
	(3)	DM: 2.48	OM: 2.72	DOM: 3.68					
10-19 July	(1)	16.72	3.96	49.71	21.75	7.86			
	(2)	12.17	1.50	39.94	17.09	–			
	(3)	DM: 2.61	OM: 2.84	DOM: 3.70					
7-16 August	(1)	18.96	4.12	44.71	23.30	8.91			
	(2)	14.10	1.63	34.25	9.44	–			
	(3)	DM: 2.27	OM: 2.50	DOM: 3.83					
11-20 September	(1)	19.81	3.30	49.19	19.62	8.08			
	(2)	15.04	0.94	40.55	15.96	–			
	(3)	DM: 2.68	OM: 2.92	DOM: 3.70					

Table 1.226 Metabolisable energy. Grass and Pasture mixtures (continued)

Description of sample Date cut		CP	EE	NFE	CF	Ash	Digestibility by	Notes	Reference
Pasture grass: some Ryegrass and Cocksfoot 29 April	(1)	26.04	3.82	44.73	16.73	8.68	Pig	Short, leafy	584
	(2)	17.35	–	31.83	9.52	–			
	(3)	DM: 2.20	OM: 2.41	DOM: 3.75					
29 April	(1)	16.76	3.22	52.62	19.42	7.98		Long, leafy: 4 months' growth Last cut in December	
	(2)	9.22	–	36.24	10.75	–			
	(3)	DM: 2.05	OM: 2.23	DOM: 3.65					
Pasture grass: Fescues, Perennial Ryegrass, Cocksfoot, Yorkshire Fog, Creeping Bent, Rough-stalked meadow grass; very little clover 1-21 March, 1932	(1)	21.41	4.35	52.45	13.76	8.03	Sheep	5 months' growth Sulphate of ammonia after previous cut	586
	(2)	15.33	2.37	42.43	10.47	–			
	(3)	DM: 2.72	OM: 2.96	DOM: 3.85					
3-16 December, 1932	(1)	15.62	2.77	51.75	21.84	8.02			
	(2)	9.76	0.78	39.54	13.04	–			
	(3)	DM: 2.32	OM: 2.52	DOM: 3.68					
2-20 January, 1933	(1)	20.15	3.75	46.37	21.12	8.61			
	(2)	13.12	0.86	35.15	12.95	–			
	(3)	DM: 2.27	OM: 2.49	DOM: 3.70					
1-18 February, 1933	(1)	22.84	3.91	44.47	19.82	8.96			
	(2)	16.35	1.86	33.53	14.07	–			
	(3)	DM: 2.46	OM: 2.70	DOM: 3.79					
Grass, unnamed	(1)	8.34	2.71	53.10	28.87	6.98	Sheep	Cut for conservation	538
	(2)	4.53	1.63	36.15	18.61	–			
	(3)	DM: 2.20	OM: 2.36	DOM: 3.61					
	(1)	8.57	1.89	48.00	33.46	8.08			
	(2)	5.03	1.02	32.94	22.25	–			
	(3)	DM: 2.16	OM: 2.35	DOM: 3.53					
	(1)	8.99	2.04	55.57	26.50	6.90			
	(2)	5.70	1.08	42.67	16.39	–			
	(3)	DM: 2.38	OM: 2.56	DOM: 3.62					
	(1)	10.48	2.14	51.20	24.70	11.48			
	(2)	7.22	1.23	41.93	19.15	–			
	(3)	DM: 2.51	OM: 2.84	DOM: 3.61					
	(1)	10.62	2.11	47.55	30.43	9.29			
	(2)	6.70	1.18	33.10	21.39	–			
	(3)	DM: 2.23	OM: 2.45	DOM: 3.57					
	(1)	10.80	2.53	53.35	25.41	7.91			
	(2)	7.16	0.89	42.11	19.09	–			
	(3)	DM: 2.49	OM: 2.70	DOM: 3.59					
	(1)	13.50	2.66	49.25	26.41	8.18			
	(2)	9.53	1.61	36.50	17.27	–			
	(3)	DM: 2.39	OM: 2.60	DOM: 3.68					
	(1)	12.35	2.36	51.59	24.22	9.48	Sheep	Aftermath	
	(2)	8.76	1.21	40.33	18.48	–			
	(3)	DM: 2.50	OM: 2.76	DOM: 3.63					
	(1)	13.38	3.47	47.86	25.22	10.07			
	(2)	10.84	2.38	39.15	20.80	–			
	(3)	DM: 2.70	OM: 3.01	DOM: 3.69					
	(1)	12.63	2.43	42.10	33.05	9.79	Sheep	"Grazing conditions": advanced stage of maturity	
	(2)	8.84	1.40	31.72	27.69	–			
	(3)	DM: 2.47	OM: 2.73	DOM: 3.54					
	(1)	13.40	2.20	44.15	27.47	12.78			
	(2)	8.69	0.96	30.75	19.76	–			
	(3)	DM: 2.17	OM: 2.49	DOM: 3.59					

continued

Table 1.226 Metabolisable energy. Grass and Pasture mixtures (continued)

Description of sample Date cut		CP	EE	NFE	CF	Ash	Digestibility by	Notes	Reference
Grass, unnamed	(1)	14.74	4.57	47.64	23.69	9.36			
	(2)	10.52	2.82	36.69	18.49	–			538
	(3)	DM: 2.57	OM: 2.83	DOM: 3.75					
	(1)	14.87	4.67	48.25	22.98	9.23			
	(2)	10.71	2.76	38.29	18.63	–			
	(3)	DM: 2.63	OM: 2.90	DOM: 3.74					
	(1)	15.72	3.44	46.06	25.64	9.14	Sheep	Cut for conservation	
	(2)	11.29	1.94	35.66	19.78	–			
	(3)	DM: 2.53	OM: 2.78	DOM: 3.68					
	(1)	16.56	3.29	46.30	22.79	11.06			
	(2)	12.55	1.81	35.43	17.83	–			
	(3)	DM: 2.51	OM: 2.82	DOM: 3.71					
	(1)	16.90	3.28	44.77	25.28	9.77			
	(2)	12.51	1.71	35.03	20.93	–			
	(3)	DM: 2.57	OM: 2.85	DOM: 3.67					
	(1)	17.58	2.09	44.86	21.95	13.52			
	(2)	13.65	0.81	34.77	17.65	–			
	(3)	DM: 2.45	OM: 2.83	DOM: 3.66					
	(1)	17.94	3.08	46.50	22.40	10.08			
	(2)	13.85	1.41	36.05	17.56	–			
	(3)	DM: 2.55	OM: 2.83	DOM: 3.70					
	(1)	15.43	2.65	45.01	26.68	10.23	Sheep	"Grazing conditions":	
	(2)	10.91	1.57	33.82	21.07	–		advanced stage of maturity	
	(3)	DM: 2.45	OM: 2.73	DOM: 3.64					
	(1)	15.78	2.72	45.98	23.32	12.20			
	(2)	11.59	1.19	35.29	18.12	–			
	(3)	DM: 2.42	OM: 2.76	DOM: 3.66					
	(1)	16.01	2.38	45.88	22.18	13.55			
	(2)	12.01	1.02	38.35	19.06	–			
	(3)	DM: 2.57	OM: 2.97	DOM: 3.65					
	(1)	16.08	2.70	43.17	23.08	14.97			
	(2)	11.93	1.09	33.52	18.98	–			
	(3)	DM: 2.39	OM: 2.81	DOM: 3.65					
	(1)	17.99	2.74	42.96	24.88	11.43			
	(2)	13.84	1.69	33.91	21.16	–			
	(3)	DM: 2.59	OM: 2.93	DOM: 3.68					
	(1)	18.39	3.01	42.70	25.21	10.69			
	(2)	14.44	1.65	34.32	18.85	–			
	(3)	DM: 2.57	OM: 2.87	DOM: 3.70					
	(1)	22.46	2.71	40.17	20.98	13.68			
	(2)	18.00	1.35	33.25	18.15	–			
	(3)	DM: 2.64	OM: 3.05	DOM: 3.73					
	(1)	20.58	2.08	45.53	17.70	14.11	Sheep	Cut for conservation	
	(2)	16.58	1.15	37.34	14.12	–			
	(3)	DM: 2.59	OM: 3.02	DOM: 3.75					
	(1)	20.81	3.72	43.77	21.00	10.70			
	(2)	16.85	2.06	35.53	17.63	–			
	(3)	DM: 2.71	OM: 3.04	DOM: 3.76					
	(1)	25.56	4.15	37.67	20.38	12.24			
	(2)	21.06	2.01	29.01	16.91	–			
	(3)	DM: 2.63	OM: 2.99	DOM: 3.81					
Grass, unnamed	(1)	11.77	1.89	54.12	24.63	7.59	Sheep	Grown for hay	
	(2)	8.11	1.09	44.32	19.09	–		Early flowering	537
	(3)	DM: 2.63	OM: 2.84	DOM: 3.62				1932	
	(1)	13.89	2.63	48.38	26.33	8.77			
	(2)	10.28	1.37	37.85	21.80	–		1933	
	(3)	DM: 2.58	OM: 2.83	DOM: 3.64					
	(1)	13.15	2.61	51.08	24.76	8.40			
	(2)	8.96	1.79	38.72	18.57	–		1934	
	(3)	DM: 2.50	OM: 2.72	DOM: 3.68					
	(1)	8.19	2.32	58.33	24.07	7.09			
	(2)	4.64	1.06	46.84	18.29	–		1935	
	(3)	DM: 2.55	OM: 2.74	DOM: 3.60					

Table 1.226 Metabolisable energy. Grass and Pasture mixtures (continued)

Description of sample Date cut		CP	EE	NFE	CF	Ash	Digestibility by	Notes	Reference
Grass, unnamed								Grown for hay	537
	(1)	8.99	2.04	55.57	26.50	6.90	Sheep	Late flowering and mature	
	(2)	5.70	1.07	42.68	16.38	–		1932	
	(3)	DM: 2.38	OM: 2.56	DOM: 3.62					
	(1)	7.55	2.38	52.74	30.57	6.76			
	(2)	4.10	1.43	35.90	19.71	–		1933	
	(3)	DM: 2.19	OM: 2.35	DOM: 3.58					
	(1)	9.65	2.80	53.44	26.80	7.31			
	(2)	6.29	1.60	39.87	18.65	–		1934	
	(3)	DM: 2.41	OM: 2.60	DOM: 3.63					
	(1)	6.42	2.02	54.03	29.42	8.11			
	(2)	2.34	0.91	39.09	17.45	–		1935	
	(3)	DM: 2.09	OM: 2.27	DOM: 3.54					
Pasture grass									
23 May	(1)	13.57	3.69	50.09	25.41	7.24	Calves, 10-12	Stored frozen: Less than	406
	(2)	8.11	1.68	41.52	20.25	–	weeks old	6% clover	
	(3)	DM: 2.60	OM: 2.81	DOM: 3.64					10
Grass, unnamed									
Mid October	(1)	16.02	3.39	42.48	28.63	9.48	Sheep	Cut for silage	160
	(2)	11.58	1.62	30.33	20.27	–		Aftermath	
	(3)	DM: 2.33	OM: 2.58	DOM: 3.66					
	(1)	15.43	3.88	43.86	27.18	9.65			
	(2)	11.16	1.85	31.32	19.24	–			
	(3)	DM: 2.34	OM: 2.59	DOM: 3.68					
	(1)	15.34	3.24	44.16	27.67	9.59			
	(2)	11.09	1.55	31.53	19.59	–			
	(3)	DM: 2.33	OM: 2.58	DOM: 3.66					
Grass, unnamed	(1)	12.47	2.48	44.59	28.78	11.68	Sheep	Aftermath	537
	(2)	6.5	1.3	33.0	22.1	–		1931	
	(3)	DM: 2.24	OM: 2.54	DOM: 3.56					
	(1)	12.35	2.36	51.59	24.22	9.48		1932	
	(2)	8.77	1.21	40.34	18.48				
	(3)	DM: 2.50	OM: 2.76	DOM: 3.63					
	(1)	14.71	3.11	49.23	22.26	10.69			
	(2)	11.91	2.13	40.27	18.36	–		1933	
	(3)	DM: 2.70	OM: 3.02	DOM: 3.72					
Grass, unnamed and									
clover (Wild White)	(1)	15.22	2.70	49.52	23.24	9.32	Sheep	Cut for silage	536
	(2)	11.00	1.46	39.17	18.29	–		Some young, some mature	
	(3)	DM: 2.57	OM: 2.83	DOM: 3.67				Means for 7 samples	
	(1)	13.15	2.83	53.25	22.37	8.40		Means for 7 samples	
	(2)	9.01	1.41	43.24	18.05	–			
	(3)	DM: 2.62	OM: 2.86	DOM: 3.65					
	(1)	15.12	2.99	50.55	22.43	8.91		Means for 9 samples	
	(2)	10.81	1.60	40.34	17.81	–			
	(3)	DM: 2.60	OM: 2.85	DOM: 3.68					
	(1)	10.70	2.61	54.88	24.00	7.81		Means for 4 samples	
	(2)	7.09	0.92	43.30	18.02	–			
	(3)	DM: 2.50	OM: 2.71	DOM: 3.61					
	(1)	15.62	2.65	50.06	22.66	9.01		Means for 6 samples	
	(2)	11.59	1.48	38.90	17.40	–			
	(3)	DM: 2.56	OM: 2.81	DOM: 3.69					
Grass, high proportion								Cut for silage (q.v.)	
of Ryegrass	(1)	13.25	2.16	48.86	29.26	6.47	Sheep	Advanced stage of maturity	105
	(2)	8.04	0.96	35.77	20.22	–			
	(3)	DM: 2.33	OM: 2.49	DOM: 3.59					
	(1)	12.96	2.10	51.55	27.29	6.10			
	(2)	7.87	0.93	37.73	18.86	–			
	(3)	DM: 2.35	OM: 2.51	DOM: 3.60					

Table 1.227 Metabolisable energy, 12 cuts of 4 grasses, primary growth; artificially dried

Reference 514 and 7

Description of sample / Date cut	Stage of growth	Ether solubles	Hexose	Sucrose	Fructosan	Organic acids	Protein	NPN	Cellulose	Hemi-celluloses	Pectin	Lignin	"Not isolated"
		7.8	3.38	3.56	3.76	2.4	4.3	4.3	3.76	3.7	2.4	2.9	3.7
		Metabolisable energy per g digested. Axelsson's coefficients except for Organic acids, for which the general value for available organic acids, from the Report of the FAO Committee on Calorie Conversion Factors and Food Composition Tables (Washington, 1947) was used. To give the correct amount of digested organic matter it was necessary to include "not isolated". Since that fraction was described as "present in the water extract", it was given the energy value of nitrogen-free extract.											
		(1) Composition, %DM		(2) Digested, %DM		(3) ME: kcal or kJ per g dry matter, organic matter and digested organic matter							
Ryegrass S23													
11 May (1)	Young, leafy	9.1	4.6	3.2	6.0	4.2	14.6	3.9	21.3	15.8	2.4	2.7	4.1
(2)		5.91	4.6	3.2	6.0	3.95	11.97	2.93	19.60	14.69	1.73	0.62	4.1
(3)		kcal	DM 3.18		OM: 3.46		DOM 4.01						
		kJ	13.31		14.48		16.78						
25 May (1)	Late leafy	7.6	6.3	2.7	2.8	4.9	11.6	3.6	22.1	18.9	2.1	3.6	5.3
(2)		4.86	6.3	2.7	2.8	4.26	9.05	2.81	19.67	15.88	1.28	0.83	5.3
(3)		kcal	DM 2.98		OM 3.26		DOM 3.94						
		kJ	12.47		13.64		16.48						
1 June (1)	Head emergence	6.5	6.7	2.8	1.8	4.6	10.6	3.2	23.9	19.4	2.2	4.3	6.2
(2)		3.9	6.7	2.8	1.8	4.00	7.53	2.53	20.79	15.33	1.03	0.77	6.2
(3)		kcal	DM 2.85		OM 3.09		DOM 3.89						
		kJ	11.92		12.93		16.28						
29 June (1)	Seed setting	4.7	4.0	2.8	3.8	2.9	6.4	3.2	26.7	25.7	2.2	7.3	4.6
(2)		2.02	4.0	2.8	3.8	1.86	3.20	2.43	19.49	14.39	0.77	–	4.6
(3)		kcal	DM 2.28		OM 2.41		DOM 3.83						
		kJ	9.54		10.08		16.02						
Ryegrass S24													
29 April (1)	Late leafy	8.5	5.1	6.8	6.5	3.8	14.7	3.0	18.5	15.1	2.2	3.1	5.3
(2)		5.61	5.1	6.8	6.5	3.57	12.35	1.98	17.39	13.14	1.56	1.30	5.3
(3)		kcal	DM 3.21		OM 3.47		DOM 3.98						
		kJ	13.43		14.52		16.65						
10 May (1)	20% heads emerged	7.4	5.3	3.8	4.4	4.1	12.1	3.4	20.9	18.7	2.3	4.1	6.5
(2)		3.85	5.3	3.8	4.4	3.65	9.08	1.94	17.97	14.77	1.36	0.78	6.5
(3)		kcal	DM 2.86		OM 3.07		DOM 3.90						
		kJ	11.97		12.84		16.32						
Cocksfoot S37													
27 April (1)	Late leafy	8.6	3.6	4.8	3.6	3.8	18.8	3.8	18.0	14.7	2.4	4.0	6.6
(2)		5.68	3.6	4.8	3.6	3.08	15.79	2.39	16.02	12.50	1.30	1.48	6.6
(3)		kcal	DM 3.11		OM 3.37		DOM 4.05						
		kJ	13.01		14.10		16.95						
16 May (1)	40% heads emerged	6.6	3.2	1.4	0.9	3.8	13.8	3.7	23.8	21.1	2.0	5.6	6.9
(2)		3.83	3.2	1.4	0.9	3.42	10.49	2.11	20.23	16.67	0.84	1.85	6.9
(3)		kcal	DM 2.82		OM 3.04		DOM 3.93						
		kJ	11.80		12.72		16.44						
25 May (1)	90% heads emerged	5.2	3.3	1.9	0.7	3.5	11.1	2.9	27.2	23.1	2.2	6.5	5.8
(2)		2.86	3.3	1.9	0.7	3.05	8.44	1.36	21.49	16.40	0.88	1.23	5.8
(3)		kcal	DM 2.61		OM 2.79		DOM 3.87						
		kJ	10.92		11.67		16.19						
Timothy S48													
1 June (1)	Late leafy	5.8	2.2	2.2	2.7	3.4	9.3	2.0	26.2	24.4	1.4	6.3	7.3
(2)		3.48	2.2	2.2	2.7	2.86	6.70	0.94	21.48	19.28	0.41	1.64	7.3
(3)		kcal	DM 2.77		OM 2.97		DOM 3.89						
		kJ	11.59		12.43		16.28						
15 June (1)	60% heads emerged	4.0	2.4	2.3	0.9	3.5	7.0	1.9	28.4	27.2	1.5	7.3	8.0
(2)		1.80	2.4	2.3	0.9	2.84	4.69	0.76	21.30	19.58	0.41	1.31	8.0
(3)		kcal	DM 2.51		OM 2.66		DOM 3.78						
		kJ	10.50		11.13		15.82						
29 June (1)	All heads emerged	3.0	2.5	1.6	1.2	2.8	5.0	2.0	28.9	29.9	1.6	9.4	6.9
(2)		0.93	2.5	1.6	1.2	2.07	2.70	1.08	18.79	18.84	0.40	1.60	6.9
(3)		kcal	DM 2.19		OM 2.31		DOM 3.73						
		kJ	9.16		9.67		15.61						

Table 1.228 Metabolisable energy. Herbage artificially dried

(1) Proximate analysis %DM
(2) Digested nutrients %DM
(3) Metabolisable energy, kcal/g dry matter (DM), organic matter (OM) and digested organic matter (DOM)

Description of sample Date cut		CP	EE	NFE	CF	Ash	Digestibility by	Notes	Reference
Ryegrass	(1)	6.96	1.98	48.16	35.36	7.54	Sheep	Hay stage	447
	(2)	2.75	0.96	29.75	23.27	–			
	(3)	DM: 1.97	OM: 2.13	DOM: 3.47					
	(1)	5.51	1.76	50.72	34.45	7.56	Sheep	Unthreshed	
	(2)	1.62	0.98	28.03	18.44	–		Mature, saved for seed	
	(3)	DM: 1.72	OM: 1.86	DOM: 3.50					
	(1)	4.97	1.84	49.40	36.27	7.52	Sheep	Threshed	
	(2)	1.15	0.75	25.76	20.16	–		Mature, straw	
	(3)	DM: 1.65	OM: 1.78	DOM: 3.44					
Ryegrass and clover 26 September	(1)	16.36	2.82	43 03	26.00	11.79	Sheep	Second crop	534
	(2)	11.70	0.95	31.73	20.29	–		23–30cm long	
	(3)	DM: 2.34	OM: 2.65	DOM: 3.62					
Perennial Ryegrass, Timothy, White clover 11-13 May	N_0 (1)	9.8	3.2	52.0	26.6	8.4	Sheep	First and second year leys, mixed	372
	(3)	DM: 2.50	DOM: 3.57					Leafy	
22 May	N_0 (1)	8.6	1.8	55.2	25.7	8.7		N_0: no extra nitrogen	
	(3)	DM: 2.29	DOM: 3.56					N_1: 96 'units' N/ha on 10 April	
	N_1 (1)	11.3	1.9	51.4	26.4	9.0		N_2: 190 'units' N/ha on 10 April	
	(3)	DM: 2.41	DOM: 3.77					Metabolisable energy computed by Armstrong's formula from digestible energy:	
	N_2 (1)	12.9	2.6	46.3	28.7	9.5		ME = Y.DE	
	(3)	DM: 2.47	DOM: 3.68					where Y = 79 + 2.1L where L = level of feeding	
Grass, unnamed August-September	(1)	21.70	5.41	43.19	18.38	11.32	Sheep	N dressing	529
	(3)	DM: 2.61	OM: 2.95	DOM: 3.87				Short, leafy	
Late May	(1)	14.55	2.35	49.42	25.39	8.29	Sheep	Permanent grass	530
	(3)	DM: 2.44	OM: 2.66	DOM: 3.63				Early hay: some grasses in head	
Mid June	(1)	10.86	2.40	49.27	28.20	9.27	Sheep	Permanent grass: poor	533
	(3)	DM: 2.43	OM: 2.47	DOM: 3.58				10–15cm height	
6 October	(1)	17.83	2.79	45.92	21.92	11.54	Sheep	Permanent pasture: intensively fertilised,	533
	(3)	DM: 2.48	OM: 2.80	DOM: 3.70				grazed.	
13 October	(1)	16.24	2.20	46.84	21.96	12.76		Short grass	
	(3)	DM: 2.25	OM: 2.58	DOM: 3.68					
No information	(1)	10.86	2.40	49.27	28.20	9.27	Sheep		538
	(2)	6.56	1.33	34.76	20.40	–			
	(3)	DM: 2.26	OM: 2.49	DOM: 3.59					
	(1)	11.33	2.36	48.62	24.39	13.30			
	(2)	6.36	1.29	39.98	19.42	–			
	(3)	DM: 2.42	OM: 2.79	DOM: 3.60					
	(1)	12.47	2.48	44.59	28.78	11.68			
	(2)	6.46	1.28	32.97	22.13	–			
	(3)	DM: 2.24	OM: 2.54	DOM: 3.56					
	(1)	14.55	2.35	49.43	25.38	8.29			
	(2)	9.76	1.20	36.72	19.69	–			
	(3)	DM: 2.44	OM: 2.66	DOM: 3.63					
	(1)	14.82	3.87	46.30	25.72	9.29			
	(2)	10.79	2.62	37.73	21.46	–			
	(3)	DM: 2.69	OM: 2.96	DOM: 3.70					
	(1)	16.24	2.20	46.82	21.96	12.78			
	(2)	9.57	1.23	35.49	14.95	–			
	(3)	DM: 2.25	OM: 2.58	DOM: 3.60					

continued

Table 1.228 Metabolisable energy. Herbage artificially dried (continued)

Description of sample Date cut		CP	EE	NFE	CF	Ash	Digestibility by	Notes	Reference
	(1)	16.25	1.82	45.17	26.37	10.39			
	(2)	10.55	0.66	32.49	21.78	–			538
	(3)	DM: 2.34	OM: 2.61	DOM: 3.57					
	(1)	16.36	2.82	43.03	26.00	11.79			
	(2)	11.70	0.95	31.73	20.29	–			
	(3)	DM: 2.34	OM: 2.65	DOM: 3.62					
	(1)	16.71	3.55	45.59	24.13	10.02			
	(2)	10.58	1.85	32.81	16.43	–			
	(3)	DM: 2.29	OM: 2.55	DOM: 3.71					
	(1)	16.93	3.36	45.39	20.43	13.89			
	(2)	11.64	2.29	37.62	16.31	–			
	(3)	DM: 2.57	OM: 2.98	DOM: 3.74					
	(1)	17.85	3.66	45.41	20.26	12.82			
	(2)	13.48	2.66	36.28	16.18	–			
	(3)	DM: 2.60	OM: 2.98	DOM: 3.79					
No information	(1)	13.8	4.1	52.6	22.1	7.4	Sheep	Coarsely chopped:	77
	(2)	9.22	2.71	45.13	18.32	–		(2) Fed at low level,	
	(3)	DM: 2.81	OM: 3.03	DOM: 3.73				600g per 24 hr.	
	(2a)	8.42	2.35	44.18	16.91	–		(2a) Fed at high level,	
	(3a)	DM: 2.67	OM: 2.88	DOM: 3.72				1500g per 24 hr.	
	(1)	14.2	4.7	51.6	21.3	8.2		Medium ground and	
	(2)	8.79	2.58	42.21	15.91	–		cubed:	
	(3)	DM: 2.60	OM: 2.83	DOM: 3.74				Low level	
	(2a)	8.34	3.34	39.58	12.38	–		High level	
	(3a)	DM: 2.44	OM: 2.66	DOM: 3.84					
	(1)	13.8	4.4	53.0	21.1	7.7		Finely ground and	
	(2)	8.64	2.98	43.46	14.28	–		cubed:	
	(3)	DM: 2.63	OM: 2.85	DOM: 3.79				Low level	
	(2a)	7.56	2.70	39.86	10.53	–		High level	
	(3a)	DM: 2.32	OM: 2.51	DOM: 3.82					
Grass, unnamed and clover									
August	(1)	20.93	5.11	43.36	21.33	9.27	Sheep	5 weeks growth	526
	(2)	16.04	2.64	35.27	15.89	–			
	(3)	DM: 2.66	OM: 2.93	DOM: 3.81					
	(2a)	13.04	1.66	24.69	5.63	–	Rabbit		
	(3a)	DM: 1.77	OM: 1.95	DOM: 3.92					
	(1)	20.76	5.15	43.16	21.56	9.37	Rabbit		
	(2)	12.68	1.16	23.91	5.56	–			
	(3)	DM: 1.68	OM: 1.86	DOM: 3.88					
Clover meal	(1)	9.49	1.62	47.83	33.04	8.02	Sheep		538
	(2)	5.32	0.83	33.33	22.34	–			
	(3)	DM: 2.17	OM: 2.36	DOM: 3.52					
	(1)	13.83	1.42	50.20	25.74	8.81			
	(2)	7.39	0.52	34.47	8.84	–			
	(3)	DM: 1.89	OM: 2.07	DOM: 3.69					
Lucerne meal									
	(1)	13.97	1.98	50.08	24.51	9.46	Sheep		
	(2)	8.74	0.87	36.19	11.30	–			
	(3)	DM: 2.11	OM: 2.33	DOM: 3.70					

Table 1.229 Metabolisable energy. Hay

(1) Proximate analysis %DM
(2) Digested nutrients %DM
(3) Metabolisable energy, kcal/g dry matter (DM), organic matter (OM) and digested organic matter (DOM)

Description of sample Date cut		CP	EE	NFE	CF	Ash	Digestibility by	Notes	Reference
Ryegrass									
	(1)	6.42	2.40	51.15	33.68	6.35	Sheep	Unthreshed	280
	(2)	3.02	1.50	29.68	19.84	–		Mature: grown for seed	
	(3)	DM: 1.92	OM: 2.05	DOM: 3.55					
	A (1)	4.66	2.04	47.53	39.74	6.03	Sheep	Straw after threshing	
	B (1)	4.69	1.78	49.69	37.42	6.42			
	A (2)	0.94	1.15	25.00	24.77	–			
	B (2)	0.87	0.96	26.68	20.76	–			
	A (3)	DM: 1.77	OM: 1.89	DOM: 3.42					
	B (3)	DM: 1.70	OM: 1.82	DOM: 3.45					
Ryegrass S23 22 June	(1)	8.0	1.7	49.2	34.3	6.8	Sheep		382
	(2)	3.57	0.49	29.32	21.54	–			
	(3)	DM: 1.90	OM: 2.04	DOM: 3.46					
Ryegrass 3 June	(1)	7.47	3.05	54.97	27.53	6.98	Sheep	Tripod cured	106
	(2)	5.08	2.23	35.54	20.19	–			
	(3)	DM: 2.29	OM: 2.46	DOM: 3.64					
Ryegrass and other grasses	(1)	7.07	2.50	49.12	31.46	9.85	Sheep	Permanent meadow	448
	(2)	3.44	1.65	31.24	18.31	–			
	(3)	DM: 1.96	OM: 2.18	DOM: 3.59					
	(1)	7.16	3.08	50.13	31.33	8.30			
	(2)	3.17	1.76	30.68	16.79	–			
	(3)	DM: 1.90	OM: 2.07	DOM: 3.62					
Ryegrass and Cocksfoot	(1)	11.57	1.79	47.28	32.17	7.19	Sheep	3rd year ley, little clover	226
	(2)	6.74	0.74	29.6	20.5	–		Sulphate of ammonia, 250 kg/ha 14 days before cutting	
	(3)	DM: 2.05	OM: 2.21	DOM: 3.54					
	(1)	9.67	1.57	49.12	32.48	7.16	Sheep	As above: no fertiliser	
	(2)	4.72	0.65	31.3	21.2	–			
	(3)	DM: 2.02	OM: 2.18	DOM: 3.50					
Ryegrass, perennial and Broad Red Clover									
Early June	A (1)	12.17	1.02	42.74	35.68	8.39	Sheep	Flowering	537
	A (2)	6.6	0.5	25.5	22.2	–			
	A (3)	DM: 1.91	OM: 2.09	DOM: 3.49				A: "Normal" hay	
	B (1)	13.03	1.26	40.85	36.25	8.61	Sheep	B: Piked hay	
	B (2)	7.7	0.5	25.2	24.7	–			
	B (3)	DM: 2.02	OM: 2.21	DOM: 3.47					
	C (1)	11.09	1.12	41.76	37.91	8.12	Sheep	C: "Normal" hay, from the stack, later	
	C (2)	5.1	0.5	24.3	24.1	–			
	C (3)	DM: 1.86	OM: 2.02	DOM: 3.44					
	D (1)	10.82	1.24	40.87	38.97	8.10	Sheep	D: Piked hay, from the stack, later	
	D (2)	5.0	0.6	24.3	26.9	–			
	D (3)	DM: 1.93	OM: 2.10	DOM: 3.41					
Ryegrass and Clover	A (1)	14.3	1.5	49.7	26.0	8.5	Sheep	A: 1st year ley	230
	A (2)	9.2	0.8	36.6	14.3	–		High clover: mean of 5 samples	
	A (3)	DM: 2.23	OM: 2.43	DOM: 3.66				First cut	
	B (1)	10.7	1.4	50.9	29.6	7.4	Sheep	B: 1st year ley	
	B (2)	5.9	0.7	35.8	17.7	–		Medium clover: mean of 11 samples	
	B (3)	DM: 2.16	OM: 2.32	DOM: 3.57					
	C (1)	8.7	1.4	53.0	30.1	6.8	Sheep	C: 1st year ley	
	C (2)	4.6	0.7	37.7	18.5	–		Low clover: mean of 4 samples	
	C (3)	DM: 2.18	OM: 2.34	DOM: 3.55					
	D (1)	14.9	1.5	47.6	27.4	8.6	Sheep	D: 1st year ley	
	D (2)	9.2	0.8	32.0	13.6	–		High clover: 1 sample	
	D (3)	DM: 2.04	OM: 2.23	DOM: 3.66				Second cut	

continued

Table 1.229 Metabolisable energy. Hay (continued)

Description of sample Date cut		CP	EE	NFE	CF	Ash	Digestibility by	Notes	Reference
Ryegrass and Clover									
19 May	A (1)	12.31	2.51	48.88	27.35	8.95	Sheep	A: Tripod cured	106
	A (2)	7.11	1.30	36.35	18.26	–			
	A (3)	DM: 2.28	OM: 2.51	DOM: 3.62					
	B (1)	11.15	2.76	51.01	26.56	8.52	Sheep	B: Ground cured	
	B (2)	6.26	1.62	38.38	19.02	–			
	B (3)	DM: 2.37	OM: 2.59	DOM: 3.63					
Italian Ryegrass and Red Clover									
Mid May	A (1)	17.93	4.11	41.05	27.41	9.50	Sheep	A: Tripod cured	106
	A (2)	13.42	2.66	30.74	20.82	–			
	A (3)	DM: 2.53	OM: 2.79	DOM: 3.73					
	B (1)	14.36	3.67	42.99	31.33	7.65	Sheep	B: Tripod cured	
	B (2)	9.74	2.17	29.13	24.02	–			
	B (3)	DM: 2.36	OM: 2.56	DOM: 3.63					
	C (1)	13.03	3.23	47.00	29.24	7.49	Sheep	C: Tripod cured	
	C (2)	8.17	2.12	33.28	20.73	–			
	C (3)	DM: 2.35	OM: 2.54	DOM: 3.65					
	D (1)	16.68	3.92	43.05	26.88	9.47	Sheep	D: Ground cured	
	D (2)	11.72	2.41	32.55	20.22	–			
	D (3)	DM: 2.48	OM: 2.74	DOM: 3.71					
	E (1)	14.34	2.47	38.99	36.60	7.60	Sheep	E: Ground cured	
	E (2)	8.98	1.21	22.40	27.93	–			
	E (3)	DM: 2.12	OM: 2.29	DOM: 3.50					
	F (1)	11.68	2.51	49.68	29.05	7.08	Sheep	F: Ground cured	
	F (2)	6.77	1.31	32.00	19.35	–			
	F (3)	DM: 2.14	OM: 2.30	DOM: 3.60					
								Data from 3 farms. All had Nitrochalk 500 kg/ha 14 days before cutting	
Ryegrass, Timothy and Clover									
	(1)	10.40	2.21	51.45	31.12	4.82	Sheep	Flowering	516
	(2)	4.67	1.20	34.77	18.75	–			
	(3)	DM: 2.12	OM: 2.23	DOM: 3.58					
	(1)	9.32	2.09	53.37	29.16	6.06			
	(2)	4.66	1.34	36.47	17.90	–			
	(3)	DM: 2.17	OM: 2.31	DOM: 3.60					
	(1)	8.21	5.15	48.10	32.12	6.42			
	(2)	3.20	4.14	28.49	18.46	–			
	(3)	DM: 2.05	OM: 2.20	DOM: 3.78					
	(1)	10.59	2.41	45.74	33.99	7.27			
	(2)	5.26	1.36	26.91	22.31	–			
	(3)	DM: 1.97	OM: 2.13	DOM: 3.54					
Ryegrass, Timothy, Clover and Oat straw									
	(1)	9.81	2.57	49.51	32.53	5.58	Sheep	Preflowering	
	(2)	4.82	1.72	29.71	18.65	–			
	(3)	DM: 1.98	OM: 2.10	DOM: 3.61					
	(1)	14.11	2.02	45.00	32.73	6.14		Flowering	
	(2)	9.23	0.86	25.81	18.97	–			
	(3)	DM: 1.97	OM: 2.10	DOM: 3.59					
Ryegrass, Meadow grass and Clover									
	(1)	10.61	2.47	53.93	26.59	6.40	Sheep	Flowering	
	(2)	6.16	1.64	39.63	17.71	–			
	(3)	DM: 2.37	OM: 2.53	DOM: 3.64					

Table 1.229 Metabolisable energy. Hay (continued)

Description of sample Date cut		CP	EE	NFE	CF	Ash	Digestibility by	Notes	Reference
Ryegrass and Clover	(1)	12.06	2.20	46.33	32.59	6.82	Sheep	Flowering	516
	(2)	6.56	1.36	26.05	18.24	–			
	(3)	DM: 1.88	OM: 2.02	DOM: 3.60					
	(1)	14.47	4.29	41.44	30.13	9.67			
	(2)	8.44	3.35	26.50	19.10	–			
	(3)	DM: 2.16	OM: 2.39	DOM: 3.76					
	(1)	9.33	4.94	46.96	32.60	6.17			
	(2)	4.30	3.53	24.20	18.42	–			
	(3)	DM: 1.89	OM: 2.01	DOM: 3.75					
	(1)	11.86	3.82	57.90	21.28	5.14	Sheep	With a little red clover Preflowering	
	(2)	7.40	2.79	49.86	17.53	–			
	(3)	DM: 2.89	OM: 3.05	DOM: 3.72					
	(1)	12.74	3.78	55.05	20.92	7.51			
	(2)	7.80	2.56	46.77	17.04	–			
	(3)	DM: 2.76	OM: 2.98	DOM: 3.72					
Ryegrass, Brome grass and Clover	(1)	11.81	2.23	45.79	33.45	6.72	Sheep	Flowering	
	(2)	6.39	1.34	27.08	19.47	–			
	(3)	DM: 1.95	OM: 2.09	DOM: 3.58					
Cocksfoot and Timothy: 5% other grasses: 5% Broad Red Clover 26 June	(1)	10.1	1.7	41.4	38.6	8.2	Cows	First cut for hay	40
	(2)	5.32	0.69	23.64	25.78	–			
	(3)	DM: 1.90	OM: 2.07	DOM: 3.44					
Cocksfoot, Timothy, Ryegrass and Clover	(1)	14.69	2.34	42.74	32.40	7.83	Sheep	Flowering	516
	(2)	9.09	1.33	24.56	19.25	–			
	(3)	DM: 1:96	OM: 2.13	DOM: 3.62					
	(1)	9.91	4.31	44.29	35.83	5.66			
	(2)	4.61	3.38	23.41	22.18	–			
	(3)	DM: 1.97	OM: 2.09	DOM: 3.68					
Cocksfoot, Timothy and Clover	(1)	11.13	2.16	50.44	31.19	5.08	Sheep	Flowering	
	(2)	6.68	1.31	34.75	18.47	–			
	(3)	DM: 2.21	OM: 2.33	DOM: 3.61					
Cocksfoot, Clover and Herbs	(1)	11.48	2.40	52.50	28.11	5.51	Sheep	Flowering	
	(2)	6.54	1.19	37.24	17.21	–			
	(3)	DM: 2.25	OM: 2.38	DOM: 3.62					
Cocksfoot and Lucerne	(1)	15.6	2.0	40.0	33.3	9.1	Sheep	Lucerne in mid-flower	230
	(2)	11.0	0.7	26.1	18.4	–			
	(3)	DM: 1.98	OM: 2.18	DOM: 3.52					
Meadow Fescue and Timothy 14-16 June	(1)	8.6	1.3	45.3	37.4	7.4	Steers	No fertiliser treatment (Silage values for same cuts in Silage Table 1.231)	382
	(2)	4.60	0.68	29.58	28.50	–			
	(3)	DM: 2.17	OM: 2.35	DOM: 3.43					
25 May	(1)	11.4	2.2	47.0	30.2	9.4	Sheep		
	(2)	8.39	1.06	36.61	24.49	–			
	(3)	DM: 2.51	OM: 2.77	DOM: 3.56					

continued

Table 1.229 Metabolisable energy. Hay (continued)

Description of sample Date cut		CF	EE	NFE	CF	Ash	Digestibility by	Notes	Reference
Meadow Fescue and Lucerne: equal parts									
23-24 June 1954	(1)	9.6	1.1	42.5	39.8	7.0	Steers	Swath cured	378
	(2)	3.9	0.1	23.1	26.1	–			
	(3)	DM: 1.79	OM: 1.92	DOM: 3.36					
	(1)	10.3	1.4	44.4	36.8	7.1	Steers	Tripod cured	
	(2)	4.4	0.2	26.4	27.1	–			
	(3)	DM: 1.97	OM: 2.12	DOM: 3.39					
21 June 1955	(1)	9.5	1.3	47.1	35.0	7.1	Steers	Swath cured	
	(2)	5.1	0.3	28.2	25.0	–			
	(3)	DM: 2.01	OM: 2.16	DOM: 3.43					
	(1)	10.0	1.5	44.2	36.7	7.6	Steers	Tripod cured	
	(2)	5.6	0.3	27.5	26.0	–			
	(3)	DM: 2.04	OM: 2.20	DOM: 3.43					
Timothy and Clover	(1)	10.72	2.70	43.13	37.60	5.85	Sheep	Preflowering	516
	(2)	4.58	1.7	25.85	19.09	–			
	(3)	DM: 1.84	OM: 1.95	DOM: 3.59					
	(1)	9.21	1.34	49.25	35.26	4.94			
	(2)	3.29	0.36	27.97	22.18	–			
	(3)	DM: 1.85	OM: 1.94	DOM: 3.43					
	(1)	12.23	2.02	47.37	32.48	5.90		Flowering	
	(2)	6.78	0.87	28.12	19.06	–			
	(3)	DM: 1.95	OM: 2.07	DOM: 3.56					
	(1)	10.37	2.61	49.27	31.49	6.26			
	(2)	5.40	1.64	31.26	18.64	–			
	(3)	DM: 2.06	OM: 2.19	DOM: 3.61					
	(1)	9.29	1.93	50.83	31.17	6.78			
	(2)	4.53	1.00	31.12	19.16	–			
	(3)	DM: 1.98	OM: 2.12	DOM: 3.55					
Timothy, Meadow Foxtail and Clover	(1)	8.87	3.09	52.11	31.30	4.63	Sheep	Flowering	
	(2)	3.78	2.17	32.04	19.16	–			
	(3)	DM: 2.07	OM: 2.17	DOM: 3.63					
Timothy, Crested Dogstail and Clover	(1)	11.98	2.83	45.52	33.10	6.57	Sheep	Flowering	
	(2)	6.04	1.94	26.70	20.91	–			
	(3)	DM: 2.01	OM: 2.15	DOM: 3.61					
Grass, unnamed	(1)	10.40	1.54	44.31	34.71	9.04	Sheep	Classified as of good quality	538
	(2)	5.08	0.49	28.59	27.33	–			
	(3)	DM: 2.11	OM: 2.32	DOM: 3.43					
	(1)	12.20	1.86	54.26	23.95	7.73			
	(2)	8.11	0.76	43.36	18.27	–			
	(3)	DM: 2.54	OM: 2.76	DOM: 3.61					
	(1)	12.94	1.97	50.02	26.85	8.22			
	(2)	6.28	0.87	35.22	20.91	–			
	(3)	DM: 2.25	OM: 2.45	DOM: 3.55					
	(1)	13.77	2.41	49.07	26.84	7.91			
	(2)	8.64	1.18	34.60	20.41	–			
	(3)	DM: 2.34	OM: 2.54	DOM: 3.60					
	(1)	14.52	2.41	46.07	28.98	8.02			
	(2)	9.29	1.23	32.05	23.98	–			
	(3)	DM: 2.38	OM: 2.58	DOM: 3.57					

Table 1.229 Metabolisable energy. Hay (continued)

Description of sample Date cut		CP	EE	NFE	CF	Ash	Digestibility by	Notes	Reference
Grass, unnamed	(1)	6.91	1.24	55.02	29.89	6.94	Sheep	From a survey to study quality	449
	(2)	2.55	0.40	33.79	17.95	–			
	(3)	DM: 1.91	OM: 2.05	DOM: 3.50					
	(1)	7.36	1.24	53.77	30.82	6.81			
	(2)	2.66	0.42	35.56	20.25	–			
	(3)	DM: 2.05	OM: 2.20	DOM: 3.48					
	(1)	7.74	1.05	51.12	32.38	7.71			
	(2)	2.95	0.38	30.80	20.74	–			
	(3)	DM: 1.90	OM: 2.06	DOM: 3.46					
Grass, unnamed	(1)	8.11	1.35	47.31	35.21	8.02	Sheep		461
	(2)	3.62	0.29	25.90	21.03	–			
	(3)	DM: 1.75	OM: 1.90	DOM: 3.44					
	(1)	7.92	1.29	48.76	33.68	8.35			
	(2)	3.01	0.31	27.59	19.42	–			
	(3)	DM: 1.74	OM: 1.90	DOM: 3.45					
Grass, unnamed 30 June	A (1)	6.97	1.72	53.33	31.97	6.01	Sheep	Plots laid out and manured since 1897. Some ryegrass, cocksfoot and fescues; little clover.	480
	A (2)	3.26	0.99	32.74	20.39	–			
	A (3)	DM: 2.02	OM: 2.15	DOM: 3.52					
	B (1)	8.14	1.63	57.90	26.64	5.69		A: Farmyard manure	
	B (2)	4.34	0.88	37.64	16.19	–		B: None	
	B (3)	DM: 2.12	OM: 2.25	DOM: 3.59					
	C (1)	8.01	1.77	55.59	28.35	6.28			
	C (2)	4.04	0.99	35.57	18.53	–		C: Basic slag	
	C (3)	DM: 2.10	OM: 2.25	DOM: 3.56					
	D (1)	7.65	1.70	55.69	29.76	5.20		D: Sulphate of ammonia, basic slag and muriate of potash	
	D (2)	3.97	0.96	34.47	18.70	–			
	D (3)	DM: 2.06	OM: 2.18	DOM: 3.55					
Permanent Pasture, mainly meadow grasses, a little Wild White clover	(1)	12.41	1.32	42.09	36.05	8.13	Sheep	Nitrochalk, 375 kg/ha 1 week before cutting	226
	(2)	6.67	0.31	22.9	25.8	–			
	(3)	DM: 1.91	OM: 2.08	DOM: 3.42					
	(1)	10.36	1.32	44.03	36.30	7.99		No extra fertiliser	
	(2)	4.70	0.31	24.0	25.4	–			
	(3)	DM: 1.85	OM: 2.01	DOM: 3.40					
Meadow Hay:	Aa (1)	12.20	1.86	54.26	23.95	7.73	Sheep	Early flowering Aa: sampled when made	537
	(2)	8.11	0.76	43.35	18.27	–			
	(3)	DM: 2.54	OM: 2.75	DOM: 3.61					
	Ab (1)	12.94	1.97	50.02	26.85	8.22		Ab: sampled from stack, later	
	(2)	6.27	0.87	35.21	20.91	–			
	(3)	DM: 2.25	OM: 2.45	DOM: 3.55					
	Ba (1)	13.74	2.02	49.15	27.41	7.68		Ba: sampled when made	
	(2)	8.62	0.99	34.65	20.83	–			
	(3)	DM: 2.52	OM: 2.73	DOM: 3.52					
	Bb (1)	14.19	2.30	47.33	28.19	7.99		Bb: sampled from stack, later	
	(2)	9.08	1.17	32.94	23.33	–			
	(3)	DM: 2.35	OM: 2.56	DOM: 3.55					
	Ca (1)	13.09	2.10	51.36	25.12	8.33		Ca: sampled when made	
	(2)	8.88	0.99	37.85	18.69	–			
	(3)	DM: 2.40	OM: 2.62	DOM: 3.62					
	Cb (1)	14.12	2.71	47.50	26.36	9.31		Cb: sampled from stack, later	
	(2)	9.08	1.53	36.53	19.61	–			
	(3)	DM: 2.44	OM: 2.69	DOM: 3.64					
	Da (1)	9.66	1.87	50.40	30.23	7.84		Da: sampled when made	
	(2)	4.42	0.78	33.21	23.10	–			
	(3)	DM: 2.15	OM: 2.33	DOM: 3.49					
	Db (1)	9.18	1.86	49.68	31.75	7.53		Db: sampled from stack, later	
	(2)	3.79	0.83	32.59	21.88	–			
	(3)	DM: 2.07	OM: 2.24	DOM: 3.50					

continued

Table 1.229 Metabolisable energy. Hay (continued)

Description of sample Date cut		CF	EE	NFE	CF	Ash	Digestibility by	Notes	Reference
Meadow	Aa (1)	6.61	1.13	55.42	30.58	6.26	Sheep	Late flowering, mature	537
Hay (cont'd)	(2)	2.37	0.17	36.74	19.57	–		Aa: sampled when made	
	(3)	DM: 2.22	OM: 2.37	DOM: 3.53					
	Ab (1)	8.31	1.39	52.80	30.30	7.20		Ab: sampled from	
	(2)	4.18	0.15	34.95	19.94	–		stack, later	
	(3)	DM: 2.06	OM: 2.22	DOM: 3.48					
	Ba (1)	7.59	2.08	52.94	30.29	7.10		Ba: sampled when made	
	(2)	3.55	1.08	35.13	19.40	–			
	(3)	DM: 2.27	OM: 2.45	DOM: 3.60					
	Bb (1)	7.88	2.15	53.33	29.36	7.28		Bb: sampled from	
	(2)	3.36	0.98	33.54	18.40	–		stack, later	
	(3)	DM: 2.00	OM: 2.15	DOM: 3.55					
	Ca (1)	9.84	2.01	53.50	27.41	7.24		Ca: sampled when made	
	(2)	6.34	0.92	39.70	18.26	–			
	(3)	DM: 2.49	OM: 2.69	DOM: 3.63					
	Cb (1)	10.06	2.60	52.98	26.48	7.88		Cb: sampled from stack,	
	(2)	5.71	1.30	37.40	18.96	–		later	
	(3)	DM: 2.28	OM: 2.48	DOM: 3.60					
	Da (1)	6.30	1.63	53.25	31.21	7.61		Da: sampled when made	
	(2)	1.77	0.66	35.73	21.38	–			
	(3)	DM: 2.26	OM: 2.45	DOM: 3.53					
	Db (1)	5.92	1.59	52.83	32.38	7.28		Db: sampled from	
	(2)	2.20	0.90	36.03	21.89	–		stack, later	
	(3)	DM: 2.13	OM: 2.30	DOM: 3.50					
Seeds grass (mainly Italian Ryegrass with some Red clover)	(1)	5.98	1.38	52.56	34.12	5.96	Sheep	Means for "traditional" hays for the years 1948-51.	363
	(2)	2.26	0.60	32.05	18.81	–		Ranges also given but not	
	(3)	DM: 1.87	OM: 1.99	DOM: 3.49				number of samples	
	(1)	7.28	1.32	51.61	33.81	5.98	Sheep	Means for hays treated	
	(2)	3.49	0.52	31.28	18.97	–		with N fertilisers, nitrate	
	(3)	DM: 1.90	OM: 2.02	DOM: 3.50				of soda or Nitrochalk: no details given	
Seeds grass	(1)	6.94	1.43	52.30	33.08	6.25	Sheep	1948 1st year : Means for	364
	(2)	3.17	0.71	33.36	18.72	–		hays from 5 areas	
	(3)	DM: 1.97	OM: 2.10	DOM: 3.52					
	(1)	6.99	1.17	49.98	34.63	7.23			
	(2)	3.25	0.48	30.44	20.67	–			
	(3)	DM: 1.90	OM: 2.05	DOM: 3.47					
	(1)	7.36	1.25	47.08	37.18	7.13			
	(2)	3.09	0.47	25.57	21.38	–			
	(3)	DM: 1.74	OM: 1.87	DOM: 3.44					
	(1)	6.07	1.52	52.21	34.11	6.09			
	(2)	2.11	0.70	29.86	17.67	–			
	(3)	DM: 1.76	OM: 1.88	DOM: 3.50					
	(1)	6.58	1.49	53.96	33.60	4.37			
	(2)	2.28	0.67	32.92	18.88	–			
	(3)	DM: 1.92	OM: 2.00	DOM: 3.50					
	(1)	4.59	0.96	58.30	31.01	5.14		1949: Means for hays	
	(2)	1.21	0.31	36.51	15.70	–		from 1 area	
	(3)	DM: 1.88	OM: 1.98	DOM: 3.50					

Table 1.229 Metabolisable energy. Hay (continued)

Description of sample Date cut		CP	EE	NFE	CF	Ash	Digestibility by	Notes	Reference
Seeds grass (cont'd)	(1)	4.84	1.55	54.14	33.83	5.64	Sheep	1950: Means for hays from 2 areas	364
	(2)	1.11	0.64	33.51	17.66	–			
	(3)	DM: 1.85	OM: 1.96	DOM: 3.50					
	(1)	5.13	1.36	53.40	33.42	6.69			
	(2)	1.33	0.55	32.20	16.81	–			
	(3)	DM: 1.78	OM: 1.91	DOM: 3.50					
	(1)	6.07	1.60	51.95	34.94	5.44		1951: Means for hays from 2 areas	
	(2)	2.08	0.74	34.03	19.69	–			
	(3)	DM: 1.98	OM: 2.09	DOM: 3.50					
	(1)	6.70	1.47	50.81	35.37	5.65			
	(2)	2.99	0.75	32.11	20.90	–			
	(3)	DM: 1.98	OM: 2.10	DOM: 3.49					
	(1)	7.6	1.4	51.2	33.6	6.2		1948-51: Means for all hays treated with sodium nitrate or Nitrochalk in amounts varying from 100 to 437 kg/ha	
	(2)	3.72	0.54	31.08	18.98	–			
	(3)	DM: 1.90	OM: 2.03	DOM: 3.50					
Seeds grass and Clover (Ryegrass, Cocksfoot, Timothy, Crested Dogstail and White Clover): 2nd year	(1)	10.80	2.00	47.55	32.30	7.35	Sheep		
	(2)	6.40	0.70	29.44	19.05	–			
	(3)	DM: 1.97	OM: 2.13	DOM: 3.55					

Table 1.230 Metabolisable energy of 47 hays: Reference 5

(1) Proximate analysis %DM
(2) Digested nutrients %DM
(3) Metabolisable energy, kcal and kJ/g dry matter (DM), organic matter (OM) and digested organic matter (DOM)

Sample No.		CP	EE	NFE	CF	Ash
764/65	(1)	8.7	1.0	43.1	37.7	9.5
	(2)	2.6	0.1	16.5	19.9	
	(3)	DM: 1.31	OM: 1.44	DOM: 3.34		
		5.48	6.02	13.97		
765	(1)	8.2	1.3	44.0	38.1	8.4
	(2)	2.6	0.3	22.1	23.8	
	(3)	DM: 1.64	OM: 1.79	DOM: 3.37		
		6.86	7.49	14.10		
798	(1)	10.9	1.6	41.4	37.7	8.4
	(2)	5.6	0.7	23.5	25.9	
	(3)	DM: 1.92	OM: 2.09	DOM: 3.44		
		8 03	8.74	14.39		
606/65A	(1)	8.8	2.2	47.3	34.2	7.5
	(2)	3.9	1.0	24.7	17.4	
	(3)	DM: 1.66	OM: 1.80	DOM: 3.54		
		6.95	7.53	14.81		
591	(1)	6.6	1.6	50.1	34.7	7.0
	(2)	2.8	0.5	30.8	20.7	
	(3)	DM: 1.90	OM: 2.04	DOM: 3.47		
		7.95	8.54	14.52		
592	(1)	11.9	1.8	49.3	28.5	8.5
	(2)	7.6	0.5	37.3	21.3	
	(3)	DM: 2.36	OM: 2.58	DOM: 3.54		
		9.87	10.79	14.81		
593	(1)	8.0	1.8	47.7	34.9	7.6
	(2)	3.6	0.9	30.1	22.5	
	(3)	DM: 1.99	OM: 2.15	DOM: 3.49		
		8.33	9.00	14.60		
603	(1)	12.4	1.8	47.0	31.2	7.6
	(2)	7.1	0.9	30.2	19.3	
	(3)	DM: 2.05	OM: 2.22	DOM: 3.57		
		8.58	9.29	14.94		
604	(1)	9.9	1.6	50.8	29.8	7.9
	(2)	5.6	0.7	35.1	19.1	
	(3)	DM: 2.15	OM: 2.33	DOM: 3.55		
		9.00	9.75	14.85		
605	(1)	11.3	1.8	49.4	29.5	8.0
	(2)	7.7	1.0	34.5	20.0	
	(3)	DM: 2.27	OM: 2.46	DOM: 3.58		
		9.50	10.29	14.98		
631	(1)	7.6	1.6	50.2	32.8	7.8
	(2)	2.8	0.7	32.7	18.7	
	(3)	DM: 1.93	OM: 2.09	DOM: 3.51		
		8.08	8.74	14.69		

Sample No.		CP	EE	NFE	CF	Ash
680	(1)	14.4	2.2	45.3	29.4	8.7
	(2)	10.3	1.0	33.4	23.0	
	(3)	DM: 2.42	OM: 2.65	DOM: 3.58		
		10.13	11.09	14.98		
700	(1)	14.5	2.1	43.6	30.7	9.1
	(2)	9.3	0.5	29.1	22.5	
	(3)	DM: 2.17	OM: 2.39	DOM: 3 53		
		9.08	10.00	14.77		
702	(1)	13.0	1.6	44.0	31.4	10.0
	(2)	8.0	1.0	27.8	24.0	
	(3)	DM: 2.15	OM: 2.39	DOM: 3.53		
		9.00	10.00	14.77		
715	(1)	16.7	2.4	44.9	27.4	8.6
	(2)	12.4	1.3	34.8	23 2	
	(3)	DM: 2.60	OM: 2.84	DOM: 3.62		
		10.88	11.88	15.15		
719	(1)	9.2	1.6	44.8	36.5	7.9
	(2)	4.3	0.5	24.9	21.9	
	(3)	DM: 1.78	OM: 1.93	DOM: 3.45		
		7.45	8.08	14.43		
723	(1)	7.2	1.3	49.8	35.9	5.8
	(2)	3.2	0.1	32.2	21.7	
	(3)	DM: 1.97	OM: 2.09	DOM: 3.44		
		8.24	8.74	14.39		
727	(1)	9.2	1.5	46.8	34 3	8.2
	(2)	4.6	0.5	30.6	22.7	
	(3)	DM: 2.03	OM: 2.21	DOM: 3.47		
		8.49	9.25	14.52		
731	(1)	9.2	1.0	44.7	37.8	7.3
	(2)	4.2	0.0	24.7	25.1	
	(3)	DM: 1.82	OM: 2.00	DOM: 3.37		
		7.61	8.37	14.10		
735	(1)	7.1	1.2	50.6	33.4	7.7
	(2)	3.0	0.4	32.7	18.2	
	(3)	DM: 1.90	OM: 2.06	DOM: 3.50		
		7.95	8.62	14.64		
739	(1)	6.3	1.2	56.8	28.2	7.5
	(2)	2.6	0.6	38.9	15.3	
	(3)	DM: 2.04	OM: 2.21	DOM: 3.56		
		8.54	9.25	14.90		
743	(1)	10.3	1.4	46.4	33.4	8.5
	(2)	5.8	0.6	29.1	20.3	
	(3)	DM: 1.96	OM: 2.14	DOM: 3.52		
		8.20	8.95	14.73		

Sample No.		CP	EE	NFE	CF	Ash
6000	(1)	9.0	1.7	47.4	34.4	7.5
	(2)	4.7	0.6	30.4	25.6	
	(3)	DM: 2.12	OM: 2.29	DOM: 3.45		
		8.87	9.58	14.43		
6001	(1)	8.3	1.3	48.0	34.4	8.0
	(2)	3.7	0.1	31 0	24.1	
	(3)	DM: 2.01	OM: 2.19	DOM: 3.42		
		8.41	9.16	14.31		
6053	(1)	8.7	1.0	48.8	33.2	8.3
	(2)	4.1	0.1	32.0	20.4	
	(3)	DM: 1.96	OM: 2.14	DOM: 3.46		
		8.20	8.95	14.48		
6056	(1)	7.9	1.4	51.5	31.9	7.3
	(2)	4.3	0.1	35.7	21.6	
	(3)	DM: 2.14	OM: 2.31	DOM: 3.47		
		8.95	9.67	14.52		
6057	(1)	7.6	1.4	51.1	31.0	8.9
	(2)	3.8	0.0	35.8	21.5	
	(3)	DM: 2.11	OM: 2.32	DOM: 3.47		
		8.83	9.71	14.52		
6058	(1)	10.9	2.0	46.5	32.8	7.8
	(2)	6.4	0.8	30.8	23.2	
	(3)	DM: 2.15	OM: 2.33	DOM: 3.51		
		9.00	9.75	14.69		
545/66	(1)	9.0	0.8	47.1	34.5	8.6
	(2)	4.5	0.0	29.8	22.1	
	(3)	DM: 1.94	OM: 2.12	DOM: 3.43		
		8.12	8.87	14.35		
549	(1)	8.1	1.2	53.5	29.4	7.8
	(2)	4.0	0.0	35.7	17.0	
	(3)	DM: 1.99	OM: 2.15	DOM: 3.50		
		8.33	9.00	14.64		
550	(1)	8.3	1.2	51.9	31.7	6.9
	(2)	4.2	0.0	34.8	20 3	
	(3)	DM: 2.06	OM: 2.21	DOM: 3.47		
		8.62	9.25	14.52		
557	(1)	11.3	1.3	43.2	34.0	10.2
	(2)	5.4	0.1	21.3	16.9	
	(3)	DM: 1.52	OM: 1.69	DOM: 3.47		
		6.36	7.07	14.52		
592	(1)	7.0	1.0	50 3	33.6	8 1
	(2)	2.7	0.0	28.9	18.1	
	(3)	DM: 1.71	OM: 1.86	DOM: 3.44		
		7.15	7.78	14.39		

For description of hays by Code Number see Table 1.132

Table 1.230 Metabolisable energy of 47 hays: Reference 5 (continued)

Sample No.		CP	EE	NFE	CF	Ash
635	(1)	8.2	2.0	47.3	35.2	7.3
	(2)	3.0	0.9	30.2	22.7	
	(3)	DM: 1.97	OM: 2.13	DOM: 3.48		
		8.24	8.91	14.56		
654	(1)	8.5	1.2	47.8	34.1	8.4
	(2)	4.1	0.7	28.3	19.7	
	(3)	DM: 1.85	OM: 2.02	DOM: 3.50		
		7.74	8.45	14.64		
658	(1)	11.5	1.2	42.5	36.2	8.6
	(2)	5.0	0.2	24.5	24.6	
	(3)	DM: 1.85	OM: 2.02	DOM: 3.41		
		7.74	8.45	14.27		
659	(1)	9.1	1.2	42.9	35.1	11.7
	(2)	3.9	0.2	21.8	19.9	
	(3)	DM: 1.57	OM: 1.77	DOM: 3.42		
		6.57	7.41	14.31		
676	(1)	8.9	1.4	45.4	36.6	7.7
	(2)	5.3	0.5	28.9	24.1	
	(3)	DM: 2.04	OM: 2.20	DOM: 3.46		
		8.54	9.20	14.48		

Sample No.		CP	EE	NFE	CF	Ash
747	(1)	11.2	1.6	46.0	35.5	5.7
	(2)	6.6	0.6	25.5	18.6	
	(3)	DM: 1.81	OM: 1.92	DOM: 3.54		
		7.57	8.03	14.81		
797	(1)	11.5	1.9	43.3	34.9	8.4
	(2)	7.1	0.9	27.1	23.2	
	(3)	DM: 2.05	OM: 2.24	DOM: 3.52		
		8.58	9.37	14.73		
802	(1)	8.8	1.2	48.2	33.8	8.0
	(2)	3.7	0.3	24.4	19.0	
	(3)	DM: 1.64	OM: 1.78	DOM: 3.45		
		6.86	7.45	14.43		
803	(1)	9.7	1.8	51.1	28.8	8.6
	(2)	5.3	0.8	32.3	16.8	
	(3)	DM: 1.97	OM: 2.16	DOM: 3.57		
		8.24	9.04	14.94		
815	(1)	10.6	1.0	54.0	27.5	6.9
	(2)	6.1	0.1	39.0	18.7	
	(3)	DM: 2.26	OM: 2.42	DOM: 3.53		
		9.46	10.13	14.79		

Sample No.		CP	EE	NFE	CF	Ash
502	(1)	8.2	1.2	49 3	34.0	7.3
	(2)	4.6	0.3	33.4	21.9	
	(3)	DM: 2.09	OM: 2.26	DOM: 3.48		
		8.74	9.46	14.56		
630	(1)	7.7	1.7	47.3	35.6	7 7
	(2)	3.5	0.2	27 3	20.9	
	(3)	DM: 1.78	OM: 1.93	DOM: 3 43		
		7.45	8.08	14.35		
632	(1)	9.8	2.0	44.4	37.0	6 8
	(2)	4.9	0.9	24.9	21.5	
	(3)	DM: 1.83	OM: 1.96	DOM: 3.50		
		7.66	8.20	14.64		
633	(1)	10.1	1.7	40.5	41.2	6.5
	(2)	5.5	0.6	20.8	23.0	
	(3)	DM: 1.72	OM: 1.84	DOM: 3.45		
		7.20	7.70	14.43		

For description of hays by Code Number see Table 1.132

Table 1.231 Metabolisable energy. Silage

(1) Proximate analysis %DM
(2) Digested nutrients %DM
(3) Metabolisable energy, Kcal/g dry matter (DM), organic matter (OM) and digested organic matter (DOM)

Description of sample Date cut		CP	EE	NFE	CF	Ash	Digestibility by	Notes	Reference
Ryegrass, S23 17-18 May	(1)	16.7	4.9	35.2	31.9	11.3	Sheep	Early	382
	(2)	12.19	3.53	23.23	26.16	–			
	(3)	DM: 2.42	OM: 2.73	DOM: 3.71					
22 June	(1)	9.3	2.3	41.5	31.9	15.0		Late	
	(2)	5.12	1.20	27.72	25.14	–			
	(3)	DM: 2.07	OM: 2.43	DOM: 3.50					
Ryegrass, unnamed	(1)	14.93	4.70	39.75	31.93	8.69	Sheep	Pit Silage	517
	(2)	8.27	3.68	27.04	26.17	–			
	(3)	DM: 2.40	OM: 2.63	DOM: 3.69					
	(1)	17.98	9.92	29.95	32.72	9.43		Pit Silage	
	(2)	12.17	8.40	20.72	29.26	–			
	(3)	DM: 2.79	OM: 3.08	DOM: 3.96					
	(1)	14.26	9.40	33.27	35.73	7.34		Pit Silage	
	(2)	8.93	8.01	22.85	31.27	–			
	(3)	DM: 2.76	OM: 2.98	DOM: 3.89					
	(1)	15.35	7.26	36.07	32.26	9.06		Barn silage with molasses	
	(2)	9.77	5.93	24.78	27.01	–			
	(3)	DM: 2.58	OM: 2.84	DOM: 3.83					
Italian ryegrass, unnamed 7 May	(1)	24.9	5.3	40.7	17.6	11.5	Sheep	Young, leafy Grass only	351
	(2)	20.7	4.2	32.6	16.0	–			
	(3)	DM: 2.89	OM: 3.26	DOM: 3.93					
	(1)	25.5	5.4	37.5	19.7	11.9		718 kg grass + 192 kg water	
	(2)	21.3	4.4	31.0	18.0	–			
	(3)	DM: 2.93	OM: 3.33	DOM: 3.92					
	(1)	25.1	5.3	39.5	17.9	12.2		909 kg grass + 1 kg soil	
	(2)	20.8	4.2	32.6	16.1	–			
	(3)	DM: 2.90	OM: 3.30	DOM: 3.93					
	(1)	24.7	5.2	38.9	19.4	11.8		Grass + water + soil	
	(2)	21.1	4.3	34.4	16.1	–			
	(3)	DM: 2.97	OM: 3.36	DOM: 3.91					

Table 1.231 Metabolisable energy. Silage (continued)

Description of sample Date cut		CP	EE	NFE	CF	Ash	Digestibility by	Notes	Reference
Ryegrass and Clover									
19 June	(1)	14.83	6.75	34.64	34.74	9.04	Sheep	5 Silages made in small towers, the 4th and 5th with molasses	517
	(2)	10.11	5.66	19.93	26.64	–			
	(3)	DM: 2.39	OM: 2.62	DOM: 3.83					
22 August	(1)	16.82	6.87	34.91	32.08	9.32			
	(2)	11.28	5.41	17.63	20.99	–			
	(3)	DM: 2.17	OM: 2.39	DOM: 3.92					
27 July	(1)	11.12	2.21	45.00	33.37	8.30			
	(2)	6.04	0.74	27.62	25.01	–			
	(3)	DM: 2.06	OM: 2.25	DOM: 3.48					
28 May	(1)	13.87	9.31	37.74	30.25	8.83			
	(2)	8.69	8.01	24.27	23.93	–			
	(3)	DM: 2.59	OM: 2.84	DOM: 3.99					
1 June	(1)	18.60	10.52	33.13	29.48	8.27			
	(2)	13.79	9.32	20.07	23.93	–			
	(3)	DM: 2.76	OM: 3.00	DOM: 4.11					
	(1)	15.01	7.38	30.53	38.23	8.85		3 Barn silages, the 3rd with molasses	
	(2)	10.24	6.59	15.88	31.24	–			
	(3)	DM: 2.45	OM: 2.69	DOM: 3.83					
	(1)	16.26	13.10	28.08	32.37	10.19			
	(2)	9.61	11.46	13.04	23.47	–			
	(3)	DM: 2.47	OM: 2.75	DOM: 4.29					
	(1)	14.82	7.55	40.86	27.63	9.14			
	(2)	9.23	6.51	27.95	21.70	–			
	(3)	DM: 2.57	OM: 2.83	DOM: 3.93					
	(1)	16.77	8.68	35.16	30.91	8.48		3 Pit silages, the 2nd and 3rd with molasses	
	(2)	11.07	7.27	24.49	27.60	–			
	(3)	DM: 2.75	OM: 3.00	DOM: 3.90					
	(1)	12.18	6.09	40.50	30.12	11.11			
	(2)	6.17	4.67	26.63	23.26	–			
	(3)	DM: 2.29	OM: 2.58	DOM: 3.77					
	(1)	14.33	8.04	36.54	31.64	9.45			
	(2)	9.06	6.87	26.38	27.05	–			
	(3)	DM: 2.69	OM: 2.97	DOM: 3.87					
Ryegrass, Cocksfoot, Timothy and Clover									
Between 3 July and 30 September (4 cuts)	(1)	20.72	5.24	39.99	24.85	9.20	Sheep	Long, leafy Molasses 9kg/1000kg of fresh grass NPK fertiliser between cuts	188
	(2)	15.29	3.69	29.29	18.11	–			
	(3)	DM: 2.55	OM: 2.81	DOM: 3.85					
Between 11 July and 11 September (4 cuts)	(1)	19.70	5.30	40.26	25.94	8.80	Sheep	Preflowering Molasses 9kg/1000kg of fresh grass NPK fertiliser between cuts	
	(2)	13.41	3.59	31.24	19.24	–			
	(3)	DM: 2.57	OM: 2.82	DOM: 3.81					
Between 25 July and 26 September (3 cuts)	(1)	17.44	4.50	41.42	27.98	8.66	Sheep	Early flowering Molasses 9kg/1000kg of fresh grass NPK fertiliser between cuts	
	(2)	10.34	2.66	29.88	18.62	–			
	(3)	DM: 2.30	OM: 2.52	DOM: 3.74					
Ryegrass, Timothy, Cocksfoot and Clover									
16 September	(1)	15.10	4.56	40.41	30.07	9.86	Sheep	3 small tower silages, the 2nd and 3rd with molasses	517
	(2)	8.68	1.89	23.81	23.14	–			
	(3)	DM: 2.07	OM: 2.30	DOM: 3.60					
8 June	(1)	21.13	9.94	29.24	31.57	8.12			
	(2)	13.80	8.15	12.73	23.88	–			
	(3)	DM: 2.39	OM: 2.60	DOM: 4.09					
10 June	(1)	18.88	8.67	32.05	30.31	10.09			
	(2)	13.53	7.24	18.47	21.83	–			
	(3)	DM: 2.46	OM: 2.74	DOM: 4.03					

continued

Table 1.231 Metabolisable energy. Silage (continued)

Description of sample Date cut		CP	EE	NFE	CF	Ash	Digestibility by	Notes	Reference
Perennial Ryegrass, Timothy, White Clover	(1)	11.55	3.11	44.65	32.66	8.03	Sheep	Warm fermentation	103
	(2)	6.21	2.03	32.36	23.37	–		Grass, long	
	(3)	DM: 2.30	OM: 2.50	DOM: 3.60					
	(1)	11.76	2.76	43.05	34.27	8.16		Grass, lacerated	
	(2)	7.10	1.99	30.33	24.84	–			
	(3)	DM: 2.30	OM: 2.51	DOM: 3.58					
	(1)	11.25	3.02	43.57	34.09	8.07		Grass, long	
	(2)	5.99	1.98	31.50	24.56	–			
	(3)	DM: 2.29	OM: 2.49	DOM: 3.58					
	(1)	11.56	3.19	43.64	33.16	8.45		Grass, lacerated	
	(2)	6.94	2.31	30.32	24.41	–			
	(3)	DM: 2.31	OM: 2.52	DOM: 3.61					
Perennial Ryegrass, Timothy, White Clover	(1)	16.22	3.81	39.60	31.88	8.49	Sheep	Cold fermentation	104
	(2)	11.58	2.44	30.71	23.75	–		Wilted grass	
	(3)	DM: 2.51	OM: 2.75	DOM: 3.67					
	(1)	14.78	4.27	40.51	32.83	7.61		Unwilted grass	
	(2)	10.67	2.81	31.78	25.19	–			
	(3)	DM: 2.58	OM: 2.80	DOM: 3.67					
	(1)	14.28	3.75	41.02	32.95	8.00		Wilted grass	
	(2)	9.96	2.44	31.63	24.83	–			
	(3)	DM: 2.51	OM: 2.73	DOM: 3.64					
	(1)	15.10	4.12	39.00	33.95	7.83		Unwilted grass	
	(2)	11.10	2.73	30.20	26.30	–			
	(3)	DM: 2.57	OM: 2.79	DOM: 3.65					
Cocksfoot 19 May	(1)	23.2	4.7	29.7	31.6	10.8	Sheep	Grass only	351
	(2)	18.6	2.6	19.0	26.2	–			
	(3)	DM: 2.46	OM: 2.76	DOM: 3.71					
	(1)	20.8	5.6	31.5	32.3	9.8		750 kg grass and 300 kg water	
	(2)	16.9	3.3	21.4	26.6	–			
	(3)	DM: 2.55	OM: 2.83	DOM: 3.74					
	(1)	22.3	5.7	31.5	29.4	11.1		750 kg grass and 22.5 kg molasses in solution	
	(2)	18.5	3.5	23.3	24.7	–			
	(3)	DM: 2.64	OM: 2.97	DOM: 3.78					
	(1)	20.3	5.6	34.5	29.9	9.7		Grass, water and molasses solution as above	
	(2)	15.8	3.3	22.2	25.1	–			
	(3)	DM: 2.49	OM: 2.75	DOM: 3.75					
Cocksfoot and Ryegrass	(1)	20.86	11.58	26.67	29.33	11.56	Sheep	Pit silage	517
	(2)	14.20	9.60	13.15	24.46	–			
	(3)	DM: 2.56	OM: 2.89	DOM: 4.16					
	(1)	19.71	8.73	31.27	30.52	9.77			
	(2)	11.44	7.08	19.87	24.69	–			
	(3)	DM: 2.50	OM: 2.77	DOM: 3.96					
Cocksfoot and Timothy	(1)	9.42	7.11	42.20	34.06	7.21	Sheep	Pit silage	517
	(2)	4.38	5.68	24.18	23.85	–		Grass flowering	
	(3)	DM: 2.22	OM: 2.39	DOM: 3.82					
Cocksfoot and Clover	(1)	11.40	11.21	32.47	36.93	7.99	Sheep	Pit silage	517
	(2)	6.76	9.21	16.77	29.77	–			
	(3)	DM: 2.49	OM: 2.71	DOM: 3.99					
Cocksfoot and Yorkshire Fog (*Holcus lanatus*) with some stemmy docks	(1)	16.47	9.01	26.58	38.75	9.19	Sheep	Pit silage	517
	(2)	11.43	7.44	11.80	29.88	–			
	(3)	DM: 2.37	OM: 2.62	DOM: 3.92					

Table 1.231 Metabolisable energy. Silage (continued)

Description of sample Date cut		CP	EE	NFE	CF	Ash	Digestibility by	Notes	Reference
Cocksfoot, Ryegrass and Clover									
23 June	(1)	12.98	7.63	39.93	30.86	8.60	Sheep	Small tower silage	517
	(2)	8.27	6.47	27.54	22.78	–			
	(3)	DM: 2.54	OM: 2.78	DOM: 3.90					
Meadow Fescue and Timothy									
14-16 June	(1)	10.2	2.5	40.3	38.7	8.3	Steers		382
	(2)	6.05	1.52	29.06	30.38	–			
	(3)	DM: 2.33	OM: 2.55	DOM: 3.48					
25 May	(1)	12.4	3.1	40.4	32.6	11.5	Sheep		
	(2)	9.13	1.82	28.98	26.70	–			
	(3)	DM: 2.38	OM: 2.69	DOM: 3.57					
Timothy and Meadow Fescue	(1)	19.1	4.1	23.2	41.3	12.3	Cows	Unchopped	39
	(2)	12.15	2.07	9.60	33.21	–			
	(3)	DM: 2.00	OM: 2.28	DOM: 3.51					
	(1)	16.6	3.4	32.0	37.1	10.9		Chopped	
	(2)	11.30	1.92	19.39	30.31	–			
	(3)	DM: 2.23	OM: 2.51	DOM: 3.55					
	(1)	17.9	3.5	27.3	40.4	10.9		Lacerated	
	(2)	11.37	1.82	13.79	32.56	–			
	(3)	DM: 2.08	OM: 2.34	DOM: 3.50					
	(1)	17.8	5.1	23.1	43.1	10.9	Steers (Yearling shorthorns)	Unchopped	
	(2)	11.68	3.19	9.22	33.49	–			
	(3)	DM: 2.06	OM: 2.31	DOM: 3.58					
	(1)	16.2	3.3	32.6	37.3	10.6		Chopped	
	(2)	10.55	1.60	19.56	29.36	–			
	(3)	DM: 2.15	OM: 2.41	DOM: 3.53					
	(1)	18.0	3.6	26.9	40.2	11.3		Lacerated	
	(2)	12.17	2.04	13.48	32.08	–			
	(3)	DM: 2.11	OM: 2.38	DOM: 3.53					
Timothy and Lucerne								First Cut	
6-9 June	(1)	17.0	3.2	38.9	31.7	9.2	Cows (dry)	73 kg 2N glycollic acid per 1000 kg green crop	23
	(2)	11.63	1.86	27.54	23.81	–			
	(3)	DM: 2.35	OM: 2.59	DOM: 3.63					
	(1)	17.0	3.6	36.9	32.6	9.9	Cows (dry)	27 kg molasses per 1000 kg green crop	
	(2)	11.73	2.27	25.68	24.48	–			
	(3)	DM: 2.34	OM: 2.60	DOM: 3.65					
	(1)	16.6	2.4	40.8	30.6	9.6	Cows (dry)	No additive Grass wilted	
	(2)	9.63	1.31	29.25	23.29	–			
	(3)	DM: 2.27	OM: 2.51	DOM: 3.58					
Timothy and Lucerne with Barley meal								First cut	
6-9 June	(1)	16.9	2.8	41.7	29.5	9.1	Cows (dry)	50 kg barley meal per 1000 kg chopped herbage	
	(2)	11.49	1.69	30.02	21.71	–			
	(3)	DM: 2.37	OM: 2.60	DOM: 3.65					
Pasture grass, permanent, little clover									
27 May	(1)	15.09	4.17	43.10	27.54	10.10	Sheep	Early hay stage	530
	(2)	10.25	3.20	30.97	21.00	–			
	(3)	DM: 2.45	OM: 2.72	DOM: 3.74				Tower Silo	
28 May	(1)	14.47	4.49	42.33	29.01	9.70			
	(2)	9.76	3.53	30.30	22.71	–			
	(3)	DM: 2.48	OM: 2.74	DOM: 3.73					
Aftermath grass								8-9 weeks' growth	
21 August	(1)	16.82	4.46	39.59	27.76	11.37	Sheep	Old grassland. 531 Hay crop had basic slag and potash.	531
	(2)	5.96	2.86	24.70	22.22	–			
	(3)	DM: 2.04	OM: 2.30	DOM: 3.66					
	(1)	16.29	4.36	43.58	27.36	8.41		Aftermath, nitrogen dressing.	
	(2)	6.55	3.01	27.02	21.26	–		Stack silo	
	(3)	DM: 2.13	OM: 2.33	DOM: 3.69					

continued

Table 1.231 Metabolisable energy. Silage (continued)

Description of sample Date cut		CP	EE	NFE	CF	Ash	Digestibility by	Notes	Reference
Grass, unnamed 23 June	(1)	10.44	2.10	45.02	34.78	7.66	Sheep	Flowering	532
	(2)	4.16	1.34	26.63	23.71	–		Silage used on 8 July	
	(3)	DM: 1.95	OM: 2.12	DOM: 3.50				Pit silo	
Grass, unnamed	(1)	12.55	3.04	47.29	26.87	10.25	Sheep	Tower Silo	538
	(2)	8.46	2.06	37.78	22.22	–			
	(3)	DM: 2.57	OM: 2.86	DOM: 3.64					
	(1)	13.09	4.81	40.43	29.94	11.73			
	(2)	7.75	2.69	27.04	24.35	–			
	(3)	DM: 2.25	OM: 2.55	DOM: 3.64					
	(1)	13.63	3.62	45.37	28.82	8.56			
	(2)	9.51	2.51	34.44	22.95	–			
	(3)	DM: 2.54	OM: 2.78	DOM: 3.67					
	(1)	14.29	4.69	37.87	25.49	17.66			
	(2)	10.30	3.30	29.71	22.95	–			
	(3)	DM: 2.46	OM: 2.99	DOM: 3.72					
	(1)	14.78	4.33	42.71	28.28	9.90			
	(2)	10.01	3.36	30.64	21.85	–			
	(3)	DM: 2.46	OM: 2.73	DOM: 3.74					
	(1)	17.37	3.68	39.79	27.88	11.28			
	(2)	11.01	2.52	29.64	24.37	–			
	(3)	DM: 2.47	OM: 2.79	DOM: 3.66					
	(1)	17.62	3.88	42.02	25.01	11.47		With AIV solution	
	(2)	12.73	2.46	27.89	19.07	–			
	(3)	DM: 2.32	OM: 2.63	DOM: 3.74					
	(1)	18.00	4.91	41.96	23.90	11.23		With AIV solution	
	(2)	13.45	3.15	29.98	17.98	–			
	(3)	DM: 2.45	OM: 2.76	DOM: 3.80					
	(1)	19.94	4.60	37.25	25.14	13.07			
	(2)	14.96	3.13	25.60	19.96	–			
	(3)	DM: 2.41	OM: 2.78	DOM: 3.79					
	(1)	10.44	2.10	45.02	34.78	7.66		Pit silo	
	(2)	4.16	1.30	26.63	23.72	–			
	(3)	DM: 1.95	OM: 2.12	DOM: 3.50					
	(1)	12.06	3.41	49.22	27.10	8.21		With molasses	
	(2)	8.33	2.22	39.40	21.34	–			
	(3)	DM: 2.60	OM: 2.84	DOM: 3.66					
	(1)	12.19	3.59	48.43	26.76	9.03		With molasses and a culture	
	(2)	8.70	2.30	39.13	21.56	–			
	(3)	DM: 2.63	OM: 2.89	DOM: 3.66					
	(1)	12.41	4.12	46.93	27.67	8.87		With whey and a culture	
	(2)	8.58	2.94	37.76	22.11	–			
	(3)	DM: 2.66	OM: 2.92	DOM: 3.68					
	(1)	12.87	3.84	46.88	27.59	8.82		With a solution of dried whey	
	(2)	8.96	2.58	37.09	22.89	–			
	(3)	DM: 2.62	OM: 2.88	DOM: 3.67					
	(1)	13.19	3.75	43.18	30.30	9.58		With acids from a PCl_5 solution	
	(2)	9.35	2.22	32.36	24.08	–			
	(3)	DM: 2.47	OM: 2.73	DOM: 3.63					
	(1)	13.34	4.03	46.29	26.92	9.42		With a solution of dried whey and a culture	
	(2)	9.71	2.95	37.32	22.42	–			
	(3)	DM: 2.68	OM: 2.96	DOM: 3.70					
	(1)	13.41	4.06	47.09	26.21	9.23		With whey and a culture	
	(2)	9.29	2.74	37.19	21.38	–			
	(3)	DM: 2.61	OM: 2.87	DOM: 3.70					
	(1)	13.50	3.64	45.21	28.70	8.95		With whey and a culture	
	(2)	7.32	2.38	31.01	20.65	–			
	(3)	DM: 2.25	OM: 2.47	DOM: 3.66					
	(1)	14.62	3.81	41.58	29.88	10.11		With AIV solution	
	(2)	10.64	2.07	28.98	22.35	–			
	(3)	DM: 2.34	OM: 2.60	DOM: 3.65					

Table 1.231 Metabolisable energy. Silage (continued)

Description of sample Date cut		CP	EE	NFE	CF	Ash	Digestibility by	Notes	Reference
Grass, unnamed (cont'd)	(1)	17.24	4.86	36.82	28.68	12.40		Cold fermentation	538
	(2)	12.28	2.95	27.46	25.56	–			
	(3)	DM: 2.52	OM: 2.87	DOM: 3.69					
	(1)	17.33	4.95	34.99	30.71	12.02		Warm fermentation	
	(2)	11.92	3.09	24.44	26.65	–			
	(3)	DM: 2.43	OM: 2.76	DOM: 3.68					
	(1)	17.39	4.76	39.67	25.80	12.38		With Defu solution and mono-ammonium phosphate	
	(2)	12.30	3.06	29.82	21.79	–			
	(3)	DM: 2.50	OM: 2.85	DOM: 3.73					
	(1)	18.03	4.60	39.65	26.97	10.75		With AIV solution (pH 4)	
	(2)	12.92	2.97	28.56	21.98	–			
	(3)	DM: 2.48	OM: 2.78	DOM: 3.74					
	(1)	18.08	5.00	38.29	26.66	11.97		With Defu solution ($HCl + H_3PO_4$)	
	(2)	13.11	3.35	27.76	22.19	–			
	(3)	DM: 2.50	OM: 2.83	DOM: 3.76					
	(1)	19.06	4.00	39.14	25.98	11.82		With AIV solution	
	(2)	14.37	2.36	30.04	22.42	–			
	(3)	DM: 2.56	OM: 2.91	DOM: 3.71					
	(1)	21.38	5.19	38.68	24.30	10.45		With acids from $POCl_3$ solution	
	(2)	16.36	2.87	27.34	18.90	–			
	(3)	DM: 2.49	OM: 2.78	DOM: 3.80					
	(1)	22.13	5.08	38.73	21.60	12.46		With H_2SO_4	
	(2)	17.71	3.36	27.43	16.63	–			
	(3)	DM: 2.52	OM: 2.88	DOM: 3.87					
	(1)	9.44	2.54	46.50	29.50	12.02		Stack silage With added salt	
	(2)	3.53	1.36	31.52	20.66	–			
	(3)	DM: 2.02	OM: 2.30	DOM: 3.55					
	(1)	9.74	2.78	45.13	33.46	8.89		With added salt	
	(2)	3.19	1.56	28.74	24.76	–			
	(3)	DM: 2.04	OM: 2.24	DOM: 3.50					
	(1)	10.80	3.02	43.71	30.09	12.38		With added salt	
	(2)	5.01	1.66	27.40	22.29	–			
	(3)	DM: 2.01	OM: 2.29	DOM: 3.56					
	(1)	11.81	3.57	38.20	32.15	14.27		With added salt	
	(2)	5.57	1.94	22.29	24.85	–			
	(3)	DM: 1.94	OM: 2.26	DOM: 3.54					
	(1)	12.21	2.08	41.93	34.19	9.59		With added salt	
	(2)	3.03	1.60	23.93	24.98	–			
	(3)	DM: 1.87	OM: 2.06	DOM: 3.48					
	(1)	13.06	3.67	46.88	27.90	8.49		With added salt	
	(2)	7.90	2.52	35.24	21.26	–			
	(3)	DM: 2.46	OM: 2.68	DOM: 3.67					
	(1)	14.10	2.32	41.36	32.53	9.69		With added salt	
	(2)	5.51	1.43	23.55	24.40	–			
	(3)	DM: 1.93	OM: 2.13	DOM: 3.51					
Clover, unnamed	(1)	15.33	1.51	39.40	31.69	12.07	Sheep	Stack silage	
	(2)	5.41	0.73	19.08	16.71	–			
	(3)	DM: 1.48	OM: 1.68	DOM: 3.53					
	(1)	16.29	4.36	43.58	27.36	8.41			
	(2)	6.50	3.00	27.00	21.20	–			
	(3)	DM: 2.13	OM: 2.33	DOM: 3.69					
	(1)	16.82	4.46	39.59	27.76	11.37			
	(2)	5.90	2.90	24.70	22.20	–			
	(3)	DM: 2.04	OM: 2.30	DOM: 3.66					
	(1)	18.02	1.66	40.24	30.76	9.32		With AIV solution	
	(2)	8.49	0.68	23.60	16.04	–			
	(3)	DM: 1.76	OM: 1.94	DOM: 3.60					
Foggage (not described)									
	(1)	18.92	2.86	32.47	31.34	14.41		Stack silage	
	(2)	7.88	1.63	15.97	26.28	–			
	(3)	DM: 1.82	OM: 2.13	DOM: 3.51					

continued

Table 1.231 Metabolisable energy. Silage (continued)

Description of sample Date cut		CP	EE	NFE	CF	Ash	Digestibility by	Notes	Reference
Grass with some white clover	(1)	17.63	7.69	37.18	25.56	11.94	Sheep	No additive	536
	(2)	12.34	5.13	26.73	21.01	–			
	(3)	DM: 2.53	OM: 2.87	DOM: 3.88					
	(1)	14.05	6.46	44.06	25.37	10.06		With molasses, 1-2 kg per 100 kg fresh grass	
	(2)	9.69	4.55	35.29	21.31	–			
	(3)	DM: 2.70	OM: 3.00	DOM: 3.80					
	(1)	16.96	6.06	41.24	25.02	10.72		With AIV solution	
	(2)	12.43	3.90	31.63	20.57	–			
	(3)	DM: 2.61	OM: 2.92	DOM: 3.80					
	(1)	13.27	5.58	47.05	24.69	9.41		With whey to give 1 kg lactose per 100 kg fresh grass	
	(2)	9.32	3.89	37.55	20.39	–			
	(3)	DM: 2.68	OM: 2.96	DOM: 3.77					
	(1)	17.93	6.80	40.26	24.78	10.23			
	(2)	13.21	4.27	29.27	19.97	–			
	(3)	DM: 2.56	OM: 2.86	DOM: 3.84					
Grass, unnamed Mid October	(1)	15.85	9.55	38.62	27.43	8.55	Sheep	Aftermath No additive	160
	(2)	9.86	6.27	22.71	19.78	–			
	(3)	DM: 2.33	OM: 2.54	DOM: 3.97					
	(1)	16.66	6.84	43.24	25.42	7.84		With AIV solution 68 l per 100kg	
	(2)	12.00	3.93	31.09	18.79	–			
	(3)	DM: 2.52	OM: 2.73	DOM: 3.83					
	(1)	16.02	8.28	40.05	25.97	9.68			
	(2)	9.95	5.14	26.43	19.40	–			
	(3)	DM: 2.37	OM: 2.62	DOM: 3.89					
Grass, unnamed June	(1)	13.50	6.27	46.06	24.68	9.49	Sheep	With dried whey to give 1 kg lactose per 100 kg green fodder and lactobacilli	6
	(2)	9.34	4.23	36.39	20.14	–			
	(3)	DM: 2.66	OM: 2.94	DOM: 3.80					
	(1)	13.96	5.76	46.30	24.28	9.70		With a solution of dried whey and lactobacilli	
	(2)	10.16	4.22	37.32	20.23	–			
	(3)	DM: 2.73	OM: 3.03	DOM: 3.80					
	(1)	12.57	5.06	49.05	24.31	9.01		Fresh whey, 680 l per 1000 kg grass and lactobacilli	
	(2)	8.69	3.61	39.49	20.08	–			
	(3)	DM: 2.70	OM: 2.97	DOM: 3.76					
	(1)	13.04	5.23	46.80	25.48	9.45		With dried whey as above	
	(2)	9.08	3.52	37.02	21.15	–			
	(3)	DM: 2.65	OM: 2.92	DOM: 3.74					
	(1)	12.56	4.92	49.14	24.35	9.03			
	(2)	8.97	3.16	39.71	19.63	–			
	(3)	DM: 2.67	OM: 2.94	DOM: 3.74					
Grass, unnamed	(1)	14.65	7.70	39.89	29.38	8.38	Sheep	Lined trench silos No additive	102
	(2)	8.25	6.37	25.69	22.45	–			
	(3)	DM: 2.45	OM: 2.68	DOM: 3.91					
	(1)	14.46	7.11	39.10	31.95	7.38		9 l molasses and 9 l water per 1000 kg grass	
	(2)	8.78	5.99	27.02	24.03	–			
	(3)	DM: 2.54	OM: 2.74	DOM: 3.86					
Grass, unnamed and clover Late May, early June	(1)	14.99	3.71	41.35	33.13	6.82	Lambs	"Low dry matter" silage with molasses. Nitrogen applied in May	327
	(2)	8.58	2.43	25.30	24.15	–			
	(3)	DM: 2.20	OM: 2.36	DOM: 3.63					
September	(1)	15.80	3.64	45.34	24.78	10.44		"High dry matter" silage with molasses. Nitrogen applied again in July	
	(2)	8.51	2.13	30.60	17.67	–			
	(3)	DM: 2.18	OM: 2.43	DOM: 3.69					

Table 1.231 Metabolisable energy. Silage (continued)

Description of sample Date cut		CP	EE	NFE	CF	Ash	Digestibility by	Notes	Reference
Grass, unnamed								At ear emergence	105
	(1)	14.47	2.52	44.71	31.08	7.22	Sheep	Trench silos, lined with 300 gauge polythene film and sealed	
	(2)	9.03	1.38	31.27	22.53	–			
	(3)	DM: 2.31	OM: 2.49	DOM: 3.59					
	(1)	14.07	2.72	45.73	30.45	7.03		Herbage wilted to 50%DM in 24-36h before ensiling	
	(2)	8.80	1.71	32.22	22.72	–			
	(3)	DM: 2.36	OM: 2.54	DOM: 3.61					
Grass, unnamed	(1)	11.52	2.97	42.41	34.51	8.59	Sheep	Farm silage, from top of silo	348
	(2)	5.56	2.07	27.62	26.38	–			
	(3)	DM: 2.19	OM: 2.40	DOM: 3.55					
	(1)	18.00	3.98	32.30	35.69	10.03		Farm silage, bottom of silo	
	(2)	12.80	2.63	16.60	25.10	–			
	(3)	DM: 2.10	OM: 2.33	DOM: 3.67					
	(1)	23.21	6.65	33.11	28.40	8.63		"Good" silage	
	(2)	18.82	4.62	26.02	24.96	–			
	(3)	DM: 2.86	OM: 3.13	DOM: 3.84					
	(1)	23.06	5.55	31.90	28.95	10.54		"Bad" silage	
	(2)	18.61	3.65	24.15	25.42	–			
	(3)	DM: 2.72	OM: 3.04	DOM: 3.78					
	(1)	17.96	2.60	43.78	21.33	14.33		"Good" silage	
	(2)	13.34	1.09	34.76	17.96	–			
	(3)	DM: 2.47	OM: 2.88	DOM: 3.67					
	(1)	17.76	2.30	43.11	22.80	14.03		"Bad" silage	
	(2)	13.60	1.10	34.91	19.61	–			
	(3)	DM: 2.53	OM: 2.94	DOM: 3.66					
	(1)	15.57	3.67	44.16	27.37	9.23		Wilted grass	
	(2)	11.88	2.61	34.14	21.76	–			
	(3)	DM: 2.61	OM: 2.87	DOM: 3.71					
	(1)	14.07	2.54	45.98	28.31	9.10		Fresh grass	
	(2)	10.69	1.57	36.00	22.68	–			
	(3)	DM: 2.57	OM: 2.83	DOM: 3.63					
	(1)	12.15	3.23	43.57	29.31	11.74		Wilted grass	
	(2)	8.16	2.53	34.64	25.50	–			
	(3)	DM: 2.57	OM: 2.91	DOM: 3.63					
	(1)	11.99	3.44	42.73	29.93	11.91		Fresh grass	
	(2)	7.03	2.56	32.22	24.72	–			
	(3)	DM: 2.41	OM: 2.74	DOM: 3.62					
	(1)	13.63	3.15	44.21	30.45	8.56		No additive	
	(2)	9.35	1.91	33.96	24.69	–			
	(3)	DM: 2.52	OM: 2.76	DOM: 3.61					
	(1)	12.88	3.30	44.76	30.35	8.71		With molasses	
	(2)	8.82	1.88	34.74	24.31	–			
	(3)	DM: 2.52	OM: 2.76	DOM: 3.61					
	(1)	15.81	4.50	42.38	26.26	11.05		No additive	
	(2)	11.84	2.70	31.19	20.69	–			
	(3)	DM: 2.47	OM: 2.78	DOM: 3.72					
	(1)	15.97	4.68	40.92	27.47	10.96		With molasses	
	(2)	11.88	3.00	29.87	22.22	–			
	(3)	DM: 2.49	OM: 2.80	DOM: 3.72					
	(1)	14.89	5.17	41.26	28.73	9.95		No additive	
	(2)	10.72	3.77	31.98	22.98	–			
	(3)	DM: 2.61	OM: 2.89	DOM: 3.75					
	(1)	14.74	4.38	41.57	29.20	10.11		With a culture	
	(2)	10.64	2.98	36.30	24.35	–			
	(3)	DM: 2.64	OM: 2.94	DOM: 3.69					
	(1)	14.62	4.87	41.24	29.56	9.71		Lacerated grass	
	(2)	10.18	3.43	30.48	22.97	–			
	(3)	DM: 2.50	OM: 2.77	DOM: 3.73					

continued

Table 1.231 Metabolisable energy. Silage (continued)

Description of sample Date cut		CP	EE	NFE	CF	Ash	Digestibility by	Notes	Reference
Grass, unnamed	(1)	15.09	4.45	39.63	29.91	10.92	Sheep	Lacerated and with a culture	348
(cont'd)	(2)	10.61	2.98	29.13	23.54	–			
	(3)	DM: 2.45	OM: 2.75	DOM: 3.70					
	(1)	11.00	4.36	40.81	34.19	9.64		Chopped grass with molasses	
	(2)	7.02	3.19	28.49	26.77	–			
	(3)	DM: 2.38	OM: 2.63	DOM: 3.64					
	(1)	17.99	5.22	35.56	30.10	11.13		With a culture	
	(2)	13.67	3.97	23.47	24.68	–			
	(3)	DM: 2.45	OM: 2.76	DOM: 3.75					
	(1)	16.68	5.45	37.14	31.26	9.47		With a culture, uncovered	
	(2)	11.11	3.64	23.77	25.29	–			
	(3)	DM: 2.37	OM: 2.62	DOM: 3.72					
	(1)	9.88	2.38	45.55	35.08	7.11		Farm silage, no additive	
	(2)	5.8	1.6	29.1	24.6	–			
	(3)	DM: 2.17	OM: 2.33	DOM: 3.54					

Table 1.232 Metabolisable energy. Lucerne

(1) Proximate composition %DM
(2) Digested nutrients %DM
(3) Metabolisable energy, kcal/g dry matter (DM), organic matter (OM) and digested organic matter (DOM)

Description of sample Date cut		CP	EE	NFE	CF	Ash	Digestibility by	Notes	Reference
Var. Provence									
Mid June	(1)	20.4	2.6	42.3	23.9	10.8	Sheep	4-year old crop. Sown April, 1928.	580
	(2)	15.46	0.66	31.45	11.07	–			
	(3)	DM: 2.20	OM: 2.47	DOM: 3.75					
Early July	(1)	17.4	2.1	39.3	29.7	11.5			
	(2)	12.96	0.18	27.59	12.67	–			
	(3)	DM: 1.96	OM: 2.21	DOM: 3.67					
Early August	(1)	19.3	2.3	39.0	28.5	10.9			
	(2)	14.54	0.46	29.54	10.69	–			
	(3)	DM: 2.06	OM: 2.32	DOM: 3.74					
Late April	(1)	25.3	2.4	38.1	22.1	12.1		1-year old crop. Sown 1.6.1932	581
	(2)	21.25	0.91	30.40	14.09	–			
	(3)	DM: 2.52	OM: 2.86	DOM: 3.78					
May/June	(1)	20.7	2.2	40.7	28.0	8.4		2-year old crop. Sown June, 1931	
	(2)	16.51	0.45	30.77	13.90	–			
	(3)	DM: 2.29	OM: 2.50	DOM: 3.71					
Mid June	(1)	17.1	1.6	41.3	29.9	10.1			
	(2)	12.99	0.22	27.59	13.35	–			
	(3)	DM: 1.98	OM: 2.21	DOM: 3.66					
Early July	(1)	22.4	2.6	39.3	26.1	9.6			
	(2)	17.94	0.51	28.39	12.42	–			
	(3)	DM: 2.22	OM: 2.46	DOM: 3.75					
July/August	(1)	19.3	2.5	39.9	28.8	9.5			
	(2)	15.32	0.48	29.41	13.06	–			
	(3)	DM: 2.16	OM: 2.39	DOM: 3.71					
Mid August	(1)	21.4	2.3	40.3	25.4	10.6			
	(2)	16.67	0.12	29.72	11.94	–			
	(3)	DM: 2.17	OM: 2.43	DOM: 3.72					

Table 1.233 Metabolisable energy. Cereal fodders

(1) Proximate analysis %DM
(2) Digested nutrients %DM
(3) Metabolisable energy, kcal/g dry matter (DM), organic matter (OM) and digested organic matter (DOM)

Description of sample		CP	EE	NFE	CF	OM	Digestibility by	Notes	Reference
Oat: whole plant as fodder									
Var. Ayr Line	(1)	6.9	1.5	48.0	38.2	94.6	Sheep	Cut when grain milky, straw green (1953)	518
Potato: Hay	(2)	2.8	0.6	23.2	22.5	49.1			
	(3)	DM: 1.68	OM: 1.78	DOM: 3.42					
	(1)	7.4	2.7	49.5	35.5	95.1		Cut at late cheesy stage of grain (1953)	
	(2)	4.1	2.0	25.2	20.1	51.4			
	(3)	DM: 1.84	OM: 1.94	DOM: 3.59					
	(1)	5.0	2.4	53.0	34.2	94.6		Cut at flowering stage (1954)	
	(2)	1.5	1.7	26.0	18.0	47.2			
	(3)	DM: 1.68	OM: 1.78	DOM: 3.56					
	(1)	5.1	2.2	49.2	38.1	94.6		Cut when grain milky, straw green (1954)	
	(2)	1.7	1.5	22.4	21.4	47.0			
	(3)	DM: 1.63	OM: 1.72	DOM: 3.47					
	(1)	6.2	1.8	47.2	39.7	94.9		Cut at late cheesy stage of grain (1954)	
	(2)	2.9	1.0	20.3	23.3	47.5			
	(3)	DM: 1.63	OM: 1.72	DOM: 3.43					
Oat straw	(1)	3.1	1.8	46.5	40.9	92.3	Cow		116
	(2)	0.15	0.95	17.50	19.95	38.55			
	(3)	DM: 1.31	OM: 1.42	DOM: 3.39					
Straw pulp	(1)	2.3	–	24.5	69.1	95.9		Mean value for feeding trials used with digestibilities below	33
	(2)	–	–	15.3	51.6	–	Sheep	Digestibility coefficients for sheep and bullock applied to mean values above	224
	(3)	DM: 2.06	OM: 2.15	DOM: 3.08					
	(2)	–	–	17.6	50.7	–	Bullock		
	(3)	DM: 2.12	OM: 2.21	DOM: 3.11					
	(1)	0.36	0.49	16.22	79.82	96.89	Sheep		570
	(2)	–	0.34	6.26	77.32	–			
	(3)	DM: 2.36	OM: 2.43	DOM: 2.98					
	(2)	–	–	10.92	67.93	–	Pig		
	(3)	DM: 2.37	OM: 2.45	DOM: 3.01					

Table 1.234 Metabolisable energy. Kale

(1) Proximate analysis %DM
(2) Digested nutrients %DM
(3) Metabolisable energy, kcal/g dry matter (DM), organic matter (OM) and digested organic matter (DOM)

Description of sample Date cut		CP	EE	NFE	CF	Ash	Digestibility by	Notes	Reference
Marrowstem kale	(1)	9.86	2.20	61.41	14.91	11.62	Sheep		538
	(2)	6.83	1.57	54.78	12.30	–			
	(3)	DM: 2.80	OM: 3.17	DOM: 3.71					
	(1)	11.44	2.03	57.81	16.69	12.03			
	(2)	7.79	1.34	50.55	12.44	–			
	(3)	DM: 2.67	OM: 3.04	DOM: 3.70					
	(1)	11.34	1.95	58.30	16.50	11.91			
	(2)	7.70	1.39	50.63	11.87	–			
	(3)	DM: 2.66	OM: 3.02	DOM: 3.71					
	(1)	12.26	1.98	57.14	16.65	11.97			
	(2)	8.43	1.30	50.41	12.57	–			
	(3)	DM: 2.69	OM: 3.06	DOM: 3.70					
	(1)	22.06	3.45	47.16	11.50	15.83			
	(2)	19.51	2.38	41.00	9.07	–			
	(3)	DM: 2.80	OM: 3.33	DOM: 3.90					
Marrowstem kale 22 September-1 October	(1)	16.19	3.98	46.35	18.32	15.16	Sheep	Unthinned	577
	(2)	13.21	2.38	40.79	12.20	–			
	(3)	DM: 2.62	OM: 3.08	DOM: 3.82					
9-18 January	(1)	14.64	2.81	52.72	18.34	11.49		Unthinned	
	(2)	11.17	1.82	46.55	11.35	–			
	(3)	DM: 2.67	OM: 3.02	DOM: 3.77					
22-31 October	(1)	13.20	2.49	52.24	18.55	13.52		Singled	
	(2)	10.31	1.28	46.75	11.50	–			
	(3)	DM: 2.68	OM: 3.09	DOM: 3.83					
29 January-9 February	(1)	16.01	2.32	51.15	17.88	12.64		Singled	
	(2)	12.62	1.11	45.01	9.69	–			
	(3)	DM: 2.58	OM: 2.95	DOM: 3.76					
Thousand-headed kale 4-26 November	(1)	11.29	2.22	58.07	18.48	9.94	Sheep		
	(2)	8.39	0.79	52.67	10.70	–			
	(3)	DM: 2.68	OM: 2.98	DOM: 3.70					
29 January-9 February	(1)	15.81	2.56	49.07	21.29	11.27			
	(2)	12.55	1.28	42.64	12.62	–			
	(3)	DM: 2.58	OM: 2.91	DOM: 3.74					
Marrowstem kale	(1)	12.28	3.14	45.17	23.57	15.84	Sheep	Silage with AIV solution	538
	(2)	9.54	2.33	38.15	17.76	–			
	(3)	DM: 2.52	OM: 2.99	DOM: 3.72					

Table 1.235 Metabolisable energy. Roots

(1) Proximate analysis %DM
(2) Digested nutrients %DM
(3) Metabolisable energy, kcal/g dry matter (DM), organic matter (OM) and digested organic matter (DOM)

Description of sample Date cut		CP	EE	NFE	CF	Ash	Digestibility by	Notes	Reference
Swedes, var. Victory	(1)	8.5	0.6	73.5	11.2	6.2	Sheep	1960 Mean fresh weight large roots, 1.36 kg	336
	(2)	5.48	0.27	70.12	9.01	–			
	(3)	DM: 3.12	OM: 3.44	DOM: 3.67					
	(1)	7.1	0.6	74.2	12.7	5.4		1960 Mean fresh weight small roots, 0.45 kg	
	(2)	2.95	0.23	68.92	7.82	–			
	(3)	DM: 2.92	OM: 3.45	DOM: 3.66					
	(1)	12.00	0.72	69.77	10.65	6.86		1961 Large	
	(2)	9.13	0.35	67.25	8.52	–			
	(3)	DM: 3.16	OM: 3.39	DOM: 3.70					
	(1)	12.25	0.63	69.10	11.35	6.67		1961 Small	
	(2)	9.09	0.31	65.73	9.03	–			
	(3)	DM: 3.11	OM: 3.33	DOM: 3.69					
Mangolds	(1)	7.61	0.33	70.46	7.41	14.19	Sheep		538
	(2)	4.74	0.33	65.14	7.41	–			
	(3)	DM: 2.85	OM: 3.33	DOM: 3.68					
	(1)	8.16	0.31	68.20	6.82	16.51			
	(2)	4.10	0.31	61.16	6.82	–			
	(3)	DM: 2.66	OM: 3.19	DOM: 3.68					
Potatoes: Cooked	(1)	10.20	0.42	81.79	2.49	5.10	Sheep		538
	(2)	6.92	0.29	77.08	–	–			
	(3)	DM: 3.17	OM: 3.34	DOM: 3.76					
Potatoes: Silage	(1)	11.01	0.38	72.07	3.17	13.37		Steamed and ensiled	
	(2)	7.62	0.34	68.20	2.69	–			
	(3)	DM: 2.96	OM: 3.41	DOM: 3.75					
	(1)	6.00	0.24	78.68	2.63	12.45		Ensiled with AIV solution	
	(2)	1.27	0.08	65.15	0.58	–			
	(3)	DM: 2.49	OM: 2.84	DOM: 3.71					
	(1)	11.47	0.17	81.91	2.11	4.34	Poultry	The fresh and boiled potatoes were from a bulk sample of the variety King Edward; the processed potatoes were from commercial samples. ME was estimated by bomb calorimetry of feed and excreta with a diet of fine bran and potato. Values given here were computed from estimates of ME in kcal/g fresh weight and coefficients of digestibility.	261
Raw	(3)	DM: 0.70	OM: 0.73	DOM: 3.35					
Cooked	(1)	11.32	0.33	82.38	1.81	4.16			
	(3)	DM: 3.25	OM: 3.39	DOM: 4.08					
Dried flakes	(1)	13.31	0.24	78.30	2.48	5.67			
	(3)	DM: 3.58	OM: 3.80	DOM: 4.17					
Dried shreds low temperature	(1)	11.50	0.31	80.44	2.69	5.06			
	(3)	DM: 2.05	OM: 2.16	DOM: 3.91					
Dried shreds high temperature	(1)	11.71	0.17	80.67	2.56	4.89			
	(3)	DM: 3.27	OM: 3.44	DOM: 4.06					
Dried slices	(1)	10.95	0.17	82.32	2.14	4.42			
	(3)	DM: 3.47	OM: 3.63	DOM: 4.08					

Table 1.236 Metabolisable energy. Miscellaneous fodders

(1) Proximate analysis %DM
(2) Digested nutrients %DM
(3) Metabolisable energy, kcal/g dry matter (DM), organic matter (OM) and digested organic matter (DOM)

Description of sample Date cut		CP	EE	NFE	CF	Ash	Digestibility by	Notes	Reference
Sugar beet leaves and tops	(1)	12.44	1.45	54.24	9.52	22.35	Sheep		538
	(2)	9.80	0.78	39.77	7.76	–			
	(3)	DM: 2.18	OM: 2.81	DOM: 3.75					
Lentil Shells	(1)	12.56	0.81	53.99	29.08	3.56			
	(2)	1.49	0.77	37.83	19.58	–			
	(3)	DM: 2.09	OM: 2.17	DOM: 3.51					
Pea Shells	(1)	6.03	0.38	36.18	54.18	3.23			
	(2)	1.09	0.38	29.51	54.09	–			
	(3)	DM: 2.74	OM: 2.83	DOM: 3.22					
	(1)	4.92	0.38	37.77	53.59	3.34			
	(2)	0.59	0.27	31.18	49.50	–			
	(3)	DM: 2.64	OM: 2.73	DOM: 3.23					
Bean Shells	(1)	4.01	0.24	42.15	48.66	4.94			
	(2)	–	0.24	24.22	43.41	–			
	(3)	DM: 2.17	OM: 2.29	DOM: 3.20					
	(1)	4.39	0.67	44.95	46.56	3.43			
	(2)	–	0.67	32.31	37.40	–			
	(3)	DM: 2.33	OM: 2.42	DOM: 3.31					
Sugar beet tops	(1)	11.53	1.71	38.85	11.44	36.47		Silage with AIV solution	
	(2)	7.19	0.60	29.21	9.74	–			
	(3)	DM: 1.72	OM: 2.71	DOM: 3.68					
Oats, peas, beans and tares	(1)	14.38	2.98	33.61	34.99	14.04		Silage with molasses	
	(2)	7.97	2.22	18.98	21.73	–			
	(3)	DM: 1.85	OM: 2.15	DOM: 3.63					
	(1)	12.94	2.59	47.68	27.50	9.29			
	(2)	7.57	1.73	34.86	16.98	–			
	(3)	DM: 2.24	OM: 2.47	DOM: 3.67					

Table 1.237 Metabolisable energy. Uncultivated plants

(1) Proximate analysis %DM
(2) Digested nutrients %DM
(3) Metabolisable energy, kcal/g dry matter (DM), organic matter (OM) and digested organic matter (DOM)

Description of sample Date cut		CP	EE	NFE	CF	Ash	Digestibility by	Notes	Reference
Molinia caerulea (Flying bent)									
3-12 July	(1)	17.85	2.20	47.13	30.56	2.26	Sheep	Herbage wilted	483
	(2)	13.19	–	27.91	20.60	–			
	(3)	DM: 2.20	OM: 2.25	DOM: 3.56					
20-29 July	(1)	15.69	2.22	48.36	30.56	3.17			
	(2)	10.00	–	28.34	20.36	–			
	(3)	DM: 2.07	OM: 2.14	DOM: 3.52					
9-18 August	(1)	14.72	1.77	49.05	31.26	3.20			
	(2)	10.11	–	26.32	19.75	–			
	(3)	DM: 1.98	OM: 2.05	DOM: 3.53					
1-10 September	(1)	12.92	1.92	51.69	30.08	3.39			
	(2)	8.08	–	27.77	17.02	–			
	(3)	DM: 1.87	OM: 1.93	DOM: 3.53					
15-24 September	(1)	11.35	1.85	51.91	31.56	3.33			
	(2)	4.30	–	27.99	17.82	–			
	(3)	DM: 1.74	OM: 1.80	DOM: 3.47					
Nardus stricta (White bent)									
1 July	(1)	9.32	2.27	49.58	31.58	7.25	Sheep	Fresh	479
	(2)	5.38	1.31	28.43	22.11	–			
	(3)	DM: 2.03	OM: 2.19	DOM: 3.54					
13 October	(1)	6.23	1.87	49.24	35.42	7.24			
	(2)	2.18	0.98	22.55	21.65	–			
	(3)	DM: 1.63	OM: 1.76	DOM: 3.45					
Cynosurus cristatus (Crested Dogstail)	(1)	5.25	2.32	49.18	37.50	5.75	Sheep	Late flowering, mature. Grown for seed: unthreshed	447
	(2)	2.37	1.33	26.41	23.49	–			
	(3)	DM: 1.85	OM: 1.96	DOM: 3.47					
	(1)	4.64	2.10	46.15	41.20	5.91	Sheep	Threshed	
	(2)	1.51	1.23	23.27	25.49	–			
	(3)	DM: 1.75	OM: 1.86	DOM: 3.42					
Grasses and Rushes mainly Bents *(Agrostis)*, Sweet Vernal *(Anthoxanthum)* and Yorkshire Fog *(Holcus)*	(1)	7.67	1.97	58.54	25.37	6.45	Sheep	Poorly draining upland	448
	(2)	3.58	1.06	29.24	15.04	–			
	(3)	DM: 1.75	OM: 1.88	DOM: 3.59					
	(1)	9.46	2.49	52.09	30.32	5.64			
	(2)	4.94	1.25	28.14	17.87	–			
	(3)	DM: 1.87	OM: 1.98	DOM: 3.58					
Grasses and Rushes, Bents *(Agrostis)*, Dogstail *(Cynosurus)*, Sweet Vernal *(Anthoxanthum)*, Yorkshire Fog *(Holcus)* 55% and Rushes *(Juncus)* 29%; other species 16%	(1)	8.68	3.01	49.51	32.34	6.46	Sheep	Upland meadow	
	(2)	2.97	1.50	18.52	14.65	–			
	(3)	DM: 1.35	OM: 1.45	DOM: 3.60					
Bents *(Agrostis)* 52% and Ryegrass (35%) with other species (13%)	(1)	11.17	2.23	45.49	32.55	8.56	Sheep	Low lying land, liable to flood	
	(2)	6.31	1.03	24.34	17.87	–			
	(3)	DM: 1.77	OM: 1.94	DOM: 3.57					

Table 1.237 Metabolisable energy. Uncultivated plants (continued)

Description of sample Date cut		CP	EE	NFE	CF	Ash	Digestibility by	Notes	Reference
Bents *(Agrostis)* 48%, Yorkshire Fog *(Holcus)* 15% other grasses (14%), other species (14%) and Rushes (9%)	(1)	9.50	2.86	52.85	26.87	7.92	Sheep	Low lying meadow	448
	(2)	3.66	1.00	23.2	13.97	–			
	(3)	DM: 1.50	OM: 1.63	DOM: 3.58					
Bents *(Agrostis)*, Yorkshire Fog *(Holcus)* and Rushes *(Juncus)*									
Agrostis 14%, *Holcus* 21%, *Juncus* 48% and other species 17%	(1)	9.22	2.15	49.17	31.18	8.28	Sheep		
	(2)	3.74	1.04	22.08	14.41	–			
	(3)	DM: 1.48	OM: 1.61	DOM: 3.59					
Agrostis 14%, *Holcus* 48%, *Juncus* 18% and other species 20%	(1)	10.97	2.54	45.65	32.76	8.08			
	(2)	5.61	1.38	19.68	18.48	–			
	(3)	DM: 1.61	OM: 1.75	DOM: 3.57					
Tall Fescue *(Festuca elatior)* 38%, Yorkshire Fog *(Holcus mollis)* 32%, other species 30%	(1)	8.62	3.46	46.93	30.61	10.38	Sheep	Permanent meadow	
	(2)	3.80	2.09	23.18	16.53	–			
	(3)	DM: 1.66	OM: 1.87	DOM: 3.65					
Pteris aquilina (Bracken)	(1)	14.98	2.22	44.58	31.42	6.80	Sheep	Silage 1938	229
	(2)	0.70	1.70	19.44	8.29	–			
	(3)	DM: 1.12	OM: 1.20	DOM: 3.72					
	(1)	12.78	1.86	49.46	28.81	7.09	Sheep	Silage with molasses, 2% of bracken by weight	
	(2)	0.50	1.24	24.78	9.16	–			
	(3)	DM: 1.30	OM: 1.40	DOM: 3.64					
	(1)	13.22	1.85	48.81	29.21	6.91		Silage 1939	
	(2)	2.30	1.39	26.75	10.90	–			
	(3)	DM: 1.51	OM: 1.63	DOM: 3.66					
July	(1)	14.70	2.97	53.18	22.87	6.28	Sheep	Fresh	365
	(2)	5.56	1.78	34.57	10.45	–			
	(3)	DM: 1.96	OM: 2.09	DOM: 3.74					
August	(1)	10.52	3.04	48.58	32.06	5.80			
	(2)	2.30	2.31	26.27	15.90	–			
	(3)	DM: 1.71	OM: 1.82	DOM: 3.66					
July	(1)	14.47	2.94	51.02	25.34	6.23	Bullocks	Fresh	
	(2)	6.21	1.47	34.90	12.14	–			
	(3)	DM: 2.03	OM: 2.16	DOM: 3.70					
August	(1)	10.24	2.62	50.23	31.26	5.65			
	(2)	1.87	1.89	31.04	14.47	–			
	(3)	DM: 1.80	OM: 1.90	DOM: 3.65					
Late October	(1)	9.06	1.47	47.60	36.67	5.20	Bullock	Hay	
	(2)	1.50	0.19	19.14	4.36	–			
	(3)	DM: 0.91	OM: 0.96	DOM: 3.63					
Mid July	(1)	16.4	1.4	47.2	29.8	5.2	Sheep	Untreated	549
	(2)	7.9	0.7	22.8	13.9	–			
	(3)	DM: 1.64	OM: 1.73	DOM: 3.62					
	(1)	15.4	0.9	47.8	30.4	5.5		Steamed for 1 hour	
	(2)	3.9	0.2	21.0	12.8	–			
	(3)	DM: 1.33	OM: 1.41	DOM: 3.51					

continued

Table 1.237 Metabolisable energy. Uncultivated plants (continued)

Description of sample Date cut		CP	EE	NFE	CF	Ash	Digestibility by	Notes	Reference
Calluna vulgaris (Heather)								Young shoots	
7 February	(1)	6.64	3.54	64.65	21.80	3.37	Sheep	4 years since last burning	11
	(2)	–	–	36.30	4.10	–			
	(3)	DM: 1.46	OM: 1.51	DOM: 3.62					
9 May	(1)	7.15	3.45	65.10	20.77	3.53		90% heather and 10% meadow hay together	
	(2)	1.69	–	39.13	2.94	–			
	(3)	DM: 1.61	OM: 1.66	DOM: 3.67					
	(1)	7.15	3.45	65.10	20.77	3.53		70% heather and 30% meadow hay	
	(2)	2.10	–	42.12	3.57	–			
	(3)	DM: 1.75	OM: 1.82	DOM: 3.67					
	(1)	7.15	3.45	65.10	20.77	3.53		40% heather and 60% meadow hay	
	(2)	1.75	–	42.25	3.76	–			
	(3)	DM: 1.75	OM: 1.81	DOM: 3.66					
17 September	(1)	7.05	3.74	65.14	20.30	3.77	Sheep	Fresh tips, shoots, flowers: some dry shoots.	488
	(2)	1.61	0.49	39.19	5.92	–		Air dried and used with equal weight of meadow hay	
	(3)	DM: 1.73	OM: 1.80	DOM: 3.66				4 years old	
1 October	(1)	6.40	3.84	64.60	21.67	3.49		As above. 10 years old	
	(2)	0.83	0.37	34.60	6.90	–			
	(3)	DM: 1.54	OM: 1.60	DOM: 3.62					
	(1)	7.00	3.70	64.14	21.32	3.84	Sheep	Mainly tips and shoots	461
	(2)	0.94	–	41.67	6.13	–		Air dried and used with equal weight of meadow hay	
	(3)	DM: 1.76	OM: 1.83	DOM: 3.61				3 years old	
	(1)	7.09	3.01	64.66	21.72	3.52		As above. 7 years old.	
	(2)	0.79	0.25	41.91	6.12	–			
	(3)	DM: 1.78	OM: 1.85	DOM: 3.63					
	(1)	7.19	3.37	61.54	24.45	3.45		As above. 14 years old	
	(2)	0.31	0.17	35.35	7.34	–			
	(3)	DM: 1.55	OM: 1.60	DOM: 3.58					

Table 1.238 Metabolisable energy. Seaweed

(1) Proximate analysis %DM
(2) Digested nutrients %DM
(3) Metabolisable energy, kcal/g dry matter (DM), organic matter (OM) and digested organic matter (DOM)

Description of sample Date cut		CP	EE	NFE	CF	Ash	Digestibility by	Notes	Reference
Laminaria digitata Meal	(1)	7.82	0.40	68.47	6.93	16.38	Pig		454
	(2)	2.13	0.09	51.28	3.44	–			
	(3)	DM: 2.10	OM: 2.51	DOM: 3.68					
Laminaria digitata Hydrolysed meal	(1)	7.64	0.36	66.53	7.00	18.47	Pig	Boiled with 5% H_2SO_4 for 1h and neutralised	
	(2)	2.02	–	47.54	2.94	–			
	(3)	DM: 1.93	OM: 2.37	DOM: 3.68					
Laminaria digitata Autumn: frond only	(1)	9.90	1.00	42.35	12.75	34.00	Sheep		451
	(2)	–	–	29.94	6.89	–			
	(3)	DM: 1.31	OM: 1.98	DOM: 3.55					
Laminaria stenophylla	(1)	10.59	2.50	57.54	10.25	19.12	Sheep		
	(2)	–	1.48	38.21	0.43	–			
	(3)	DM: 1.54	OM: 1.91	DOM: 3.84					
Fucus vesiculosus Autumn: dried, ground	(1)	8.40	1.60	61.50	10.50	18.0	Sheep		
	(2)	–	1.43	37.95	–	–			
	(3)	DM: 1.52	OM: 1.85	DOM: 3.85					

Table 1.239 Conversion from Calories to Joules

	0.00	0.01	0.02	0.03	0.04	0.05	0.06	0.07	0.08	0.09
2.0	8.37	8.41	8.45	8.49	8.54	8.58	8.62	8.66	8.70	8.74
2.1	8.79	8.83	8.87	8.91	8.95	9.00	9.04	9.08	9.12	9.16
2.2	9.20	9.25	9.29	9.33	9.37	9.41	9.46	9.50	9.54	9.58
2.3	9.62	9.67	9.71	9.75	9.79	9.83	9.87	9.92	9.96	10.00
2.4	10.04	10.08	10.13	10.17	10.21	10.25	10.29	10.33	10.38	10.42
2.5	10.46	10.50	10.54	10.59	10.63	10.67	10.71	10.75	10.79	10.84
2.6	10.88	10.92	10.96	11.00	11.05	11.09	11.13	11.17	11.21	11.25
2.7	11.30	11.34	11.38	11.42	11.46	11.51	11.55	11.59	11.63	11.67
2.8	11.72	11.76	11.80	11.84	11.88	11.92	11.97	12.01	12.05	12.09
2.9	12.13	12.18	12.22	12.26	12.30	12.34	12.38	12.43	12.47	12.51
3.0	12.55	12.59	12.64	12.68	12.72	12.76	12.80	12.84	12.89	12.93
3.1	12.97	13.01	13.05	13.10	13.14	13.18	13.22	13.26	13.31	13.35
3.2	13.39	13.43	13.47	13.51	13.56	13.60	13.64	13.68	13.72	13.77
3.3	13.81	13.85	13.89	13.93	13.97	14.02	14.06	14.10	14.14	14.18
3.4	14.23	14.27	14.31	14.35	14.39	14.43	14.48	14.52	14.56	14.60
3.5	14.64	14.69	14.73	14.77	14.81	14.85	14.90	14.94	14.98	15.02
3.6	15.06	15.10	15.15	15.19	15.23	15.27	15.31	15.36	15.40	15.44
3.7	15.48	15.52	15.56	15.61	15.65	15.69	15.73	15.77	15.82	15.86
3.8	15.90	15.94	15.98	16.02	16.07	16.11	16.15	16.19	16.23	16.28
3.9	16.32	16.36	16.40	16.44	16.48	16.53	16.57	16.61	16.65	16.69
4.0	16.74	16.78	16.82	16.86	16.90	16.95	16.99	17.03	17.07	17.11
4.1	17.15	17.20	17.24	17.28	17.32	17.36	17.41	17.45	17.49	17.53
4.2	17.57	17.61	17.66	17.70	17.74	17.78	17.82	17.87	17.91	17.95
4.3	17.99	18.03	18.07	18.12	18.16	18.20	18.24	18.28	18.33	18.37
4.4	18.41	18.45	18.49	18.54	18.58	18.62	18.66	18.70	18.74	18.79
4.5	18.83	18.87	18.91	18.95	19.00	19.04	19.08	19.12	19.16	19.20
4.6	19.25	19.29	19.33	19.37	19.41	19.46	19.50	19.54	19.58	19.62
4.7	19.66	19.71	19.75	19.79	19.83	19.87	19.92	19.96	20.00	20.04
4.8	20.08	20.13	20.17	20.21	20.25	20.29	20.33	20.38	20.42	20.46
4.9	20.50	20.54	20.59	20.63	20.67	20.71	20.75	20.79	20.84	20.88
5.0	20.92	20.96	21.00	21.05	21.09	21.13	21.17	21.21	21.25	21.30
5.1	21.34	21.38	21.42	21.46	21.51	21.55	21.59	21.63	21.67	21.71
5.2	21.76	21.80	21.84	21.88	21.92	21.97	22.01	22.05	22.09	22.13
5.3	22.18	22.22	22.26	22.30	22.34	22.38	22.43	22.47	22.51	22.55
5.4	22.59	22.64	22.68	22.72	22.76	22.80	22.84	22.89	22.93	22.97
5.5	23.01	23.05	23.10	23.14	23.18	23.22	23.26	23.30	23.35	23.39
5.6	23.43	23.47	23.51	23.56	23.60	23.64	23.68	23.72	23.77	23.81
5.7	23.85	23.89	23.93	23.97	24.02	24.06	24.10	24.14	24.18	24.23
5.8	24.27	24.31	24.35	24.39	24.43	24.48	24.52	24.56	24.60	24.64
5.9	24.69	24.73	24.77	24.81	24.85	24.89	24.94	24.98	25.02	25.06
7.0	29.29									
7.1	29.71									

BIBLIOGRAPHY

1 Agricultural Research Council: Group on comparison of methods of analysis of mineral elements in plants. Report. London, 1963.

2 Agricultural Research Council: The nutrient requirements of farm livestock. No. 2 Ruminants. Technical Reviews and Summaries. London, 1965.

3 Alder, F.E. The influence of 'Nitro-chalk' on established lucerne leys. J. Brit. Grassland Soc., 1954, **9**, 323-328.

4 Alder, F.E. and Chambers, D.T. Studies on calf management. 1. Preliminary studies of post-weaning grazing. J. Brit. Grassland Soc., 1958, **13**, 13-20.

5 Alderman, G., Collins, F.C., Jones, D.J.C., Morgan, D.E. and Ibbotson, C.F. Recent N.A.A.S. investigations into laboratory methods of estimating the feeding value of hays. Unpublished, 1966.

6 Allen, L.A., Watson, S.J. and Ferguson, W.S. The effect of the addition of various materials and bacterial cultures to grass silage at the time of making on the subsequent bacterial and chemical changes. J. Agric. Sci., 1937, **27**, 294-308.

7 Armstrong, D.G. Evaluation of artificially dried grass as a source of energy for sheep. 2. The energy value of cocksfoot, timothy and 2 strains of rye-grass at varying stages of maturity. J. Agric. Sci., 1964, **62**, 399-416.

8 Armstrong, D.G., Blaxter, K.L. and Waite, R. The evaluation of artificially dried grass as a source of energy for sheep. 3. The prediction of nutritive value from chemical and biological measurements. J. Agric. Sci., 1964, **62**, 417-424.

9 Armstrong, D.G., Cook, H. and Thomas, B. The lignin and cellulose contents of certain grassland species at different stages of growth. J. Agric. Sci., 1950, **40**, 93-99.

10 Armstrong, D.G., Preston, T.R. and Armstrong, R.H. Digestibility of a sample of pasture grass by calves. Nature, 1954, **174**, 1182-1183.

11 Armstrong, D.G. and Thomas, B. The nutritive value of *Calluna vulgaris.* II. A preliminary study of digestibility. J. Agric. Sci., 1953, **43**, 223-228.

12 Armstrong, R.H. Amino-acids in the proteins of herbage. J. Sci. Food Agric., 1951, **2**, 166-170.

13 Armstrong, R.H. and Thomas, B. The availability of calcium in three legumes of grassland. J. Agric. Sci., 1952, **42**, 454-460.

14 Armstrong, R.H., Thomas, B. and Armstrong, D.G. The availability of calcium in three grasses. J. Agric. Sci., 1957, **49**, 446-453.

15 Armstrong, R.H., Thomas, B. and Horner, K. The availability of calcium in three herbs of grassland. J. Agric. Sci., 1953, **43**, 337-342.

16 Armstrong, S.F. and Brandreth, B. Trials of Spratt Archer and Plumage Archer 1924 barleys, 1930-32 (under normal and intensive conditions of manuring). J. Nat. Inst. Agric. Bot., 1934, **3**, 347-359.

17 Ashton, W.M. The chemical composition of the straw of strong and weak-strawed varieties of oats. Welsh J. Agric., 1937, **13**, 144-151.

18 Ashton, W.M. The chemical composition of the grain and straw of varieties of oats bred at the Welsh Plant Breeding Station: comparison with some older varieties. Empire J. Exp. Agric., 1938, **6**, 69-78.

19 Ashton, W.M. and Morgan, D.E. The chemical composition and nutritive value of Welsh hay: comparison with other British hays. I. J. Brit. Grassland Soc., 1952, **7**, 161-171.

20 Ashton, W.M. and Morgan, D.E. The chemical composition and nutritive value of Welsh hay: comparison with other British hays. II. J. Brit. Grassland Soc., 1953, **8**, 131-147.

21 Ashton, W.M. and Sinclair, K.B. Blood serum calcium and magnesium levels in bullocks grazing simple leys and old pastures. J. Brit. Grassland Soc., 1960, **15**, 296-299.

22 Axelsson, J. Fodrets innehall av omsättbar energi. (Metabolisable energy of fodder). Kungl. Lantbruksakad. Tidskr., 1941, **80**, 353-364.

23 Bailey, G.L., Balch, C.C. and Murdoch, J.C. The digestibility and feeding value of a lucerne/timothy sward ensiled in four ways. J. Brit. Grassland Soc., 1955, **10**, 27-34.

24 Bailey, G.L. and Broster, W.H. The fragmentation losses of lucerne hay due to handling. J. Brit. Grassland Soc., 1955, **10**, 191-192.

25 Bailey, G.L. and Broster, W.H. Experiments on the nutrition of the dairy heifer. I. Protein requirements of yearling stock. J. Agric. Sci., 1957, **49**, 435-445.

26 Bailey, G.L., Castle, M.E. and Foot, A.S. Fodder-Beet. III. Feeding trials with dairy cattle. Empire J. Exp. Agric., 1953, **21**, 42-48.

27 Bailey, P.H., Hughes, M. and McDonald, A.N.C. Differential loss of dry matter in the laboratory grinding of dried herbage samples. J. Brit. Grassland Soc., 1957, **12**, 157-165.

28 Baker, C.J.L., Heimberg, M., Alderman, G. and Eden, A. Studies of the composition of sainfoin. J. Agric. Sci., 1952, **42**, 382-394.

29 Baker, F. and Harriss, S.T. Microbial digestion in the rumen (and caecum) with special reference to the decomposition of structural cellulose. Nutrition Abstracts and Reviews, 1947, **17**, 3-12.

30 Baker, H.K. Studies on the root development of herbage plants. III. The influence of cutting treatments on the root, stubble and herbage production of a perennial ryegrass sward. J. Brit. Grassland Soc., 1957, **12**, 197-208.

31 Baker, H.K., Chard, J.R.A. and Hughes, G.P. The production and utilization of winter grass at various centres in England and Wales, 1954-60. I. Management for herbage production. J. Brit. Grassland Soc., 1961, **16**, 185-189.

32 Baker, L.C., Lampitt, L.H., Money, R.W. and Parkinson, T.L. The composition and cooking quality of potatoes from fertilizer trials in the East Riding of Yorkshire. J. Sci. Food Agric., 1950, **1**, 109-113.

33 Balch, C.C., Balch, D.A., Bartlett, S., Bartrum, M.P., Johnson, V.W., Rowland, S.J. and Turner, J. Studies on the secretion of milk of low fat content by cows on diets low in hay and high in concentrates. VI. The effect on the physical and biochemical processes of the reticulo-rumen. J. Dairy Res., 1955, **22**, 270-289.

34 Balch, C.C., Balch, D.A., Bartlett, S., Cox, C.P. and Rowland, S.J. Studies on the secretion of milk of low fat content by cows on diets low in hay and high in concentrates. I. The effect of variations in the amount of hay. J. Dairy Res., 1952, **19**, 39-50.

35 Balch, C.C., Balch, D.A., Bartlett, S., Cox, C.P., Rowland, S.J. and Turner, J. Studies of the secretion of milk of low fat content by cows on diets low in hay and high in concentrates. II. The effect of the protein content of the concentrates. J. Dairy Res., 1954, **21**, 165-171.

36 Balch, C.C., Balch, D.A., Bartlett, S., Hosking, Z.D., Johnson, V.W., Rowland, S.J. and Turner, J. Studies of the secretion of milk of low fat content by cows on diets low in hay and high in concentrates. III. The effect of variations in the amount and physical state of the hay and a comparison of the Shorthorn and Friesian breeds. J. Dairy Res., 1954, **21**, 172-177.

37 Balch, C.C., Balch, D.A., Bartlett, S., Johnson, V.W., Rowland, S.J. and Turner, J. Studies of the secretion of milk of low fat content by cows on diets low in hay and high in concentrates. IV. The effect of variations in the intake of digestible nutrients. J. Dairy Res., 1954, **21**, 305-317.

38 Balch, C.C., Balch, D.A., Johnson, V.W. and Turner, J. Factors affecting the utilization of food by dairy cows. 7. The effect of limited water intake on the digestibility and rate of passage of hay. Brit. J. Nutrition, 1953, **7**, 212-224.

39 Balch, C.C., Murdoch, J.C. and Turner, J. The effect of chopping and lacerating before ensiling on the digestibility of silage by cows and steers. J. Brit. Grassland Soc., 1955, **10**, 326-329.

40 Balch, C.C., Taylor, J. and Thomson, I. The short-term effects of the level of concentrates given to Ayrshire cows and of adding 0.75 lb sodium acetate to the daily diet. J. Dairy Res., 1961, **28**, 5-13.

41 Barber, R.S., Braude, R. and Mitchell, K.G. A comparison of dried skim-milk and white fish-meal as protein supplements for fattening pigs. J. Dairy Res., 1958, **25**, 119-124.

42 Barker, M.G., Hanley, F. and Ridgman, W.J. Studies on lucerne and lucerne-grass leys. I. Summer and autumn management of a lucerne-grass mixture grown on heavy land. J. Agric. Sci., 1955, **46**, 362-376.

43 Barnett, A.J.G. Studies on the digestibility of the cellulose fraction of grassland products. I. The relation between the digestibility of silage cellulose as determined *in vitro* and silage crude fibre digestibility determined by feeding trial. J. Agric. Sci., 1957, **49**, 467-474.

44 Barnett, A.J.G. and Baxter, J. Laboratory studies on the initiation of the silage fermentation by means of whey. J. Brit. Grassland Soc., 1955, **10**, 45-57.

45 Barnett, A.J.G., Dodsworth, T.L. and Miller, T.B. Studies on the composition and nutritive value of three silages made from the same clover-grass crop. I. The crop. J. Brit. Grassland Soc., 1951, **6**, 99-105.

46 Barnett, A.J.G. and Duncan, R.E.B. Volatile fatty acids in laboratory and field silage. J. Sci. Food Agric., 1954, **5**, 120-126.

47 Barnett, A.J.G., McLaren, S.M. and Moore, W. Studies on the ensilage of kale. Empire J. Exp. Agric., 1951, **19**, 139-144.

48 Barnett, A.J.G. and Miller, T.B. Studies on the soluble nitrogen content of dried samples of grass silage. J. Agric. Sci., 1950, **40**, 50-54.

49 Barnett, A.J.G. and Miller, T.B. The laboratory determination of the protein digestibility of silage. J. Agric. Sci., 1951, **41**, 317-321.

50 Barnett, A.J.G., Miller, T.B. and Dodsworth, T.L. Studies on the composition and nutritive value of three silages made from the same clover-grass crop. II. The silage. J. Brit. Grassland Soc., 1952, **7**, 15-20.

51 Barnett, A.J.G. and Reid, R.L. Studies on the production of volatile fatty acids from grass by rumen liquor in an artificial rumen. 1. The volatile acid production from fresh grass. J. Agric. Sci., 1957, **48**, 315-321.

52 Barnett, A.J.G. and Reid, R.L. Studies on the production of volatile fatty acids from grass by rumen liquor in an artificial rumen. II. The volatile fatty acid production from dried grass. J. Agric. Sci., 1957, **49**, 171-179.

53 Bartlett, S. and Broster, W.H. Feeding trials with ammoniated molasses in the diet of young dairy cattle. J. Agric. Sci., 1958, **50**, 60-63.

54 Bartlett, S. and Cotton, A.G. Urea as a protein substitute in the diet of young cattle. J. Dairy Res., 1938, **9**, 263-272.

55 Bartlett, S., Cotton, A.G., Henry, K.M. and Kon, S.K. The influence of various fodder supplements on the production and nutritive value of winter milk. J. Dairy Res., 1938, **9**, 273-309.

56 Bartlett, S., Foot, A.S., Huthnance, S.L. and Mackintosh, J. The protein requirements for milk production. J. Dairy Res., 1940, **11**, 121-135.

57 Barton-Wright, E.C. and Moran, T. The microbiological assay of amino acids. II. The distribution of amino acids in the wheat grain. Analyst, 1946, **71**, 278-282.

58 Bender, A.E. and Doell, B.H. Absence of toxic effects by food yeast on the rat. Brit. J. Nutrition, 1960, **14**, 305-313.

59 Black, W.A.P. The seasonal variation in chemical constitution of some of the sub-littoral seaweeds common to Scotland. III. *Laminaria saccharina* and *Saccorhiza bulbosa*. J. Soc. Chem. Indust., 1948, **67**, 172-176T.

60 Black, W.A.P. The seasonal variation in chemical composition of some of the littoral seaweeds common to Scotland. I. *Ascophyllum nodosum*. J. Soc. Chem. Indust., 1948, **67**, 355-357T.

61 Black, W.A.P. Seasonal variation in chemical composition of some of the littoral seaweeds common to Scotland. Part II. *Fucus serratus, Fucus vesiculosus, Fucus spiralis* and *Pelvetia canaliculata*. J. Soc. Chem. Indust., 1949, **68**, 183-189T.

62 Black, W.A.P. Seaweed as a stockfood. Agriculture, London, 1953-54, **60**, 126-130.

63 Black, W.A.P. The preservation of seaweed by ensiling and bactericides. J. Sci. Food Agric., 1955, **6**, 14-23.

64 Black, W.A.P. Seaweed in animal foodstuffs. 1. Availability and composition. Agriculture, London, 1955-56, **62**, 12-15.

65 Black, W.A.P. and Cornhill, W.J. A method for the estimation of fucosterol in seaweeds. J. Sci. Food Agric., 1951, **2**, 387-390.

66 Black, W.A.P., Dewar, E.T. and Woodward, F.N. Manufacture of algal chemicals. IV. Laboratory-scale isolation of fucoidin from brown marine algae. J. Sci. Food Agric., 1952, **3**, 122-129.

67 Black, W.A.P., Dewar, E.T. and Woodward, F.N. Manufacture of algal chemicals. VII. The isolation of algal carbohydrates by means of charcoal columns. J. Sci. Food Agric., 1955, **6**, 754-763.

68 Black, W.A.P. and Mitchell, R.L. Trace elements in the common brown algae and in sea water. J. Mar. Biol. Assoc., 1951-1952, **30**, 575-584.

69 Black, W.A.P., Richardson, W.D. and Walker, F.T. Chemical and growth gradients of *Laminaria cloustoni* EDM. (*L. hyperborea* FOSL). Econ. Proc. Royal Dublin Soc., 1951-1959, **4**, 137-149.

70 Black, W.A.P. and Woodward, F.N. The value of seaweeds in animal feeding-stuffs as a source of minerals, trace elements, and vitamins. Empire J. Exp. Agric., 1957, **25**, 51-59.

71 Blackman, G.E. The control of annual weeds in cereal crops by dilute sulphuric acid. Emp. J. Exp. Agric., 1934, **2**, 213-227.

72 Blackett, G. and Ravell, J. Fodder beet or swedes . . . for milk? Scot. Agric., 1955-56, **35**, 219-221.

73 Blackett, G.A. The effect of rate and time of application of nitrogen on the yield of winter wheat. Empire J. Exp. Agric., 1957, **25**, 19-23.

74 Bland, B.F. Maize as a silage crop in Eastern England. Agriculture, London, 1953-54, **60**, 311-314.

75 Bland, B.F. and Dent, J.W. Animal preference in relation to the chemical composition and digestibility of varieties of cocksfoot. J. Brit. Grassland Soc., 1964, **19**, 306-315.

76 Blaxter, K.L. Utilization of the metabolisable energy of grass. J. Brit. Grassland Soc., 1964, **19**, 90-99.

77 Blaxter, K.L. and Graham, N. McC. The effect of the grinding and cubing process on the utilization of the energy of dried grass. J. Agric. Sci., 1956, **47**, 207-217.

78 Blaxter, K.L., Graham, N.M. and Wainman, F.W. Some observations on the digestibility of food by sheep, and on related problems. Brit. J. Nutrition, 1956, **10**, 69-91.

79 Blaxter, K.L., Kon, S.K. and Thompson, S.Y. The effect of feeding shark-liver oil to cows on the yield and composition and on the vitamin A and carotene content of the milk. J. Dairy Res., 1946, **14**, 225-230.

80 Blaxter, K.L. and Price, H.A. Experiments on the use of home-grown foods for milk production. V. The protein requirements of growing dairy heifers. J. Agric. Sci., 1946, **36**, 301-309.

81 Boge, G. Amino-acid composition of herring (*Clupea harengus*) and herring meal. Destruction of amino-acids during processing. J. Sci. Food Agric., 1960, **11**, 362-365.

82 Bolton, W. The digestibility of the carbohydrate complex of barley, wheat and maize by adult fowls. J. Agric. Sci., 1955, **46**, 119-122.

83 Bolton, W. The digestibility by adult fowls of wheat fine middlings, maize germ meal, maize gluten feed, soya-bean meal and groundnut meal. J. Sci. Food Agric., 1957, **8**, 132-136.

84 Bolton, W. The determination of digestible carbohydrates in poultry foods. Analyst, 1960, **85**, 189-192.

85 Bondi, A. and Meyer, H. Digestibility trials with poultry. The digestibility of dura, carobs and hide-fleshings. J. Agric. Sci., 1944, **34**, 118-122.

86 Booth, V.H. The α-tocopherol content of forage crops. J. Sci. Food Agric., 1964, **15**, 342-344.

87 Booth, V.H. and Hobson-Frohock, A.H. The α-tocopherol content of leaves as affected by growth rate. J. Sci. Food Agric., 1961, **12**, 251-256.

88 Boruff, C.S., Luthy, P.W. and Van Lanen, J.M. Evaluation of Scotch whisky distiller's feeds in poultry rations. J. Sci. Food Agric., 1964, **15**, 364-369.

89 Boyd, D.A., Church, B.M. and Hills, M.G. Fertilizer use on grassland in England and Wales. J. Brit. Grassland Soc., 1963, **18**, 18-28.

90 Braude, R. Some observations on the need for copper in the diet of fattening pigs. J. Agric. Sci., 1945, **35**, 163-167.

91 Braude, R., Clarke, P.M., Mitchell, K.G Cray, A.S., Franke, A. and Sedgwick, P.H. A comparison of dried skim-milk and white fish meal as protein supplements for fattening pigs. II. On all-meal feeding under commercial conditions. J. Dairy Res., 1958, **25**, 181-190.

92 Braude, R. and Foot, A.S. War-time rations for pigs. Report of experiments with mangolds and biscuit waste, fodder yeast, urea and dried skim milk. J. Agric. Sci., 1942, **32**, 70-84.

93 Braude, R. and Foot, A.S. Pig-feeding experiment using dried *Clostridium* residue. J. Agric. Sci., 1942, **32**, 324-329.

94 Braude, R. and Foot, A.S. Cacao by-products in pig feeding: cocoa-cake meal and de-theobrominized cocoa-cake meal as feeding-stuffs for pigs. Empire J. Exp. Agric., 1942, **10**, 182-188.

95 Braude, R., Henry, K.M. and Kon, S.K. Further studies of the rachitogenic effect of dried yeast in pig diets. Brit. J. Nutrition, 1948-49, **2**, 66-75.

96 Braude, R., Mitchell, K.G., Cray, A.S., Franke, A. and Sedgwick, P.H. A comparison of dried skim-milk and white fish meal as protein supplements. III. For fattening pigs fed whey without restriction under commercial conditions. J. Dairy Res., 1958, **25**, 383-391.

97 Braude, R., Mitchell, K.G. and Robinson, K.L. The value of Australian sorghum for fattening pigs. J. Agric. Sci., 1950, **40**, 84-92.

98 Broster, W.H., Balch, C.C., Bartlett, S. and Campling, R.C. The value of ammoniated sugar beet pulp for dairy cows. J. Agric. Sci., 1960, **55**, 197-202.

99 Broster, W.H., Ridler, B. and Foot, A.S. Levels of feeding of concentrates for dairy heifers before and after calving. J. Dairy Res., 1958, **25**, 373-382.

100 Brown, F. The tocopherol content of farm feeding-stuffs. J. Sci. Food Agric., 1953, **4**, 161-165.

101 Brown, W.O. Dried-grass meal as a riboflavin supplement to animal feeding stuffs with special reference to poultry. J. Sci. Food Agric., 1950, **1**, 219-222.

102 Brown, W.O. and Heaney, I.H. The conservation of grassland herbage in lined trench silos. A comparison of nutrient losses and of feeding value in molassed and ordinary silages. J. Brit. Grassland Soc., 1951, **6**, 91-98.

103 Brown, W.O. and Kerr, J.A.M. Losses in the conservation of grassland herbage in lined trench silos. I. A comparison of long and lacerated silages made by the warm fermentation process. J. Agric. Sci., 1965, **64**, 135-141.

104 Brown, W.O. and Kerr, J.A.M. Losses in the conservation of grassland herbage in lined trench silos. II. Comparison of lacerated silages of low and high dry-matter content made by the cold fermentation process. J. Agric. Sci., 1965, **64**, 143-149.

105 Brown, W.O. and Kerr, J.A.M. Losses in the conservation of heavily-wilted herbage sealed in polythene film in lined trench silos. J. Brit. Grassland Soc., 1965, **20**, 227-232.

106 Brown, W.O. and Robinson, K.L. The composition and digestibility of Northern Irish hays. III. Factors influencing the production of high-quality hays. J. Agric. Sci., 1952, **42**, 362-368.

107 Brown, W.O. and Smyth, V. Losses in the conservation of grassland herbage as molassed and metabisulphite silage in lined trench silos. J. Agric. Sci., 1958, **50**, 307-311.

108 Bunting, E.S. and Willey, L.A. Problems involved in the cultivation of maize for fodder and ensilage. 1. The choice of variety. J. Agric. Sci., 1959, **52**, 95-105.

109 Bunting, E.S. and Willey, L.A. The cultivation of maize for fodder and ensilage. II. The effect of changes in plant density. J. Agric. Sci., 1959, **52**, 313-319.

110 Bunyan, J. and Price, S.A. Studies on protein concentrates for animal feeding. J. Sci. Food Agric., 1960, **11**, 25-37.

111 Burt, A.W.A. The effect of variations in nutrient intake upon the yield and composition of milk. II. Factors affecting rate of eating roughage and responses to an increase in the amount of concentrates fed. J. Dairy Res., 1957, **24**, 296-315.

112 Burt, A.W.A., Bartlett, S. and Rowland, S.J. The use of seaweed meals in concentrate mixtures for dairy cows. J. Dairy Res., 1954, **21**, 299-304.

113 Butler, E.J. and others. The mineral element content of spring pasture in relation to the occurrence of grass tetany and hypomagnesaemia in dairy cows. J. Agric. Sci., 1963, **60**, 329-340.

114 Calder, A.F., Davidson, J., Duckworth, J., Hepburn, W.R., Lucas, I.A.M., Sokarovski, J. and Walker, D.M. Utilisation by pigs of diets containing oats and oat husks ground to different degrees of fineness. J. Sci. Food Agric., 1959, **10**, 682-691.

115 Campling, R.C., Freer, M. and Balch, C.C. Factors affecting the voluntary intake of food by cows. 2. The relationship between the voluntary intake of roughages, the amount of digesta in the reticulo-rumen and the rate of disappearance of digesta from the alimentary tract. Brit. J. Nutrition, 1961, **15**, 531-540.

116 Campling, R.C., Freer, M. and Balch, C.C. Factors affecting the voluntary intake of food by cows. 3. The effect of urea on the voluntary intake of oat straw. Brit. J. Nutrition, 1962, **16**, 115-124.

117 Campling, R.C., Freer, M. and Balch, C.C. Factors affecting the voluntary intake of food by cows. 6. A preliminary experiment with ground, pelleted hay. Brit. J. Nutrition, 1963, **17**, 263-272.

118 Carpenter, K.J. A note on the chemical composition of fish meals used in a hatchability experiment. J. Agric. Sci., 1958, **50**, 113.

119 Carpenter, K.J. and Clegg, K.M. The metabolisable energy of poultry feeding stuffs in relation to their chemical composition. J. Sci. Food Agric., 1956, **7**, 45-51.

120 Carpenter, K.J. and Clegg, K.M. The relative value of three cereals as protein sources for growing chicken. Brit. J. Nutrition. 1957, **11**, 358-364.

121 Carpenter, K.J. and Duckworth, J. The nutritive value of herring 'alkali-reduction' meal for chicks. J. Agric. Sci., 1950. **40**, 44-49.

122 Carpenter, K.J., Ellinger, G.M. and Shrimpton, D.H. The evaluation of whaling by-products as feeding-stuffs. J. Sci. Food Agric., 1955, **6**, 296-304.

123 Carpenter, K.J., Morgan, C.B., Lea, C.H. and Parr, L.J. Chemical and nutritional changes in stored herring meal. 3. Effect of heating at controlled moisture contents on the binding of amino acids in freeze-dried herring press cake and in related model systems. Brit. J. Nutrition, 1962, **16**, 451-465.

124 Castle, M.E. Grassland production and its measurement using the dairy cow. J. Brit. Grassland Soc., 1953, **8**, 195-211.

125 Castle, M.E. The effect of winter weather on the composition of fodder-beet left in the ground. J. Sci. Food Agric., 1955, **6**, 579-581.

126 Castle, M.E. and Drysdale, A.D. Liquid manure as a grassland fertilizer. 1. The response to liquid manure and to dry fertilizer. J. Agric. Sci., 1962, **58**, 165-171.

127 Castle, M.E., Foot, A.S., Hosking, Z.D. and Rowland, S.J. The yield and composition of kale and cabbage with and without nitrogenous top-dressing. J. Agric. Sci., 1957, **48**, 305-314.

128 Castle, M.E., Foot, A.S. and Rowland, S.J. American hybrid maize for silage in the south of England. I. The yield and composition with special reference to plant population and nitrogenous fertilizer. J. Agric. Sci., 1951, **41**, 282-287.

129 Castle, M.E., Foot, A.S. and Rowland, S.J. American hybrid maize for silage in the south of England. II. Composition, nutritive value and losses in the silo. J. Agric. Sci., 1952, **42**, 175-179.

130 Castle, M.E., Foot, A.S. and Rowland, S.J. Fodder-beet. Comparison of Danish fodder-beet and English mangold in the south of England, 1947-9. Empire J. Exp. Agric., 1952, **20**, 1-9.

131 Castle, M.E., Foot, A.S. and Rowland, S.J. Fodder-beet. II. A further comparison of Danish fodder-beet and English mangold in the south of England, 1950 and 1951. Empire J. Exp. Agric., 1952, **20**, 316-325.

132 Castle, M.E. and Reid, D. A comparison of lucerne varieties in south-west Scotland. J. Brit. Grassland Soc., 1960, **15**, 281-286.

133 Cavell, A.J. A rapid method for the determination of nitrogen, phosphorus and potassium in plant materials. J. Sci. Food Agric., 1954, **5**, 195-200.

134 Chambers, W.E. Nutrient composition of the produce of the Broadbalk continuous wheat experiment. I. Changes over seventy years. J. Agric. Sci., 1953, **43**, 473-478.

135 Charles, A.H. Establishment studies. IV. The effect on spring oats of under-sowing with long-term ley mixtures. J. Agric. Sci., 1962, **58**, 243-249.

136 Common, R.H. The feeding value of flax by-products. Empire J. Exp. Agric., 1943, **11**, 229-233.

137 Common, R.H. The composition and digestibility of Northern Irish ryegrass seed and ryegrass seed cleanings. J. Agric. Sci., 1945, **35**, 56-63.

138 Corbett, J.L., Langlands, J.P. and Reid, G.W. Effect of season of growth and digestibility of herbage on intake of grazing dairy cows. Animal Prod., 1963, **5**, 119-129.

139 Coulson, C.B. Plant proteins. V. Proteins and amino-acids of marine algae. J. Sci. Food Agric., 1955, **6**, 674-682.

140 Cowling, D.W. The effect of nitrogenous fertilizer on an established white clover sward. J. Brit. Grassland Soc., 1961, **16**, 65-68.

141 Cowlishaw, S.J. and Alder, F.E. The grazing preferences of cattle and sheep. J. Agric. Sci., 1960, **54**, 257-265.

142 Cowlishaw, S.J., Eyles, D.E., Raymond, W.F. and Tilley, J.M.A. Nutritive value of leaf protein concentrates. I. Effect of addition of cholesterol and amino-acids. J. Sci. Food Agric., 1956, **7**, 768-780.

143 Cox, C.P., Foot, A.S., Hosking, Z.D., Line, C. and Rowland, S.J. The direct evaluation of pasture in terms of the milk production of individually grazed cows. J. Brit. Grassland Soc., 1956, **11**, 107-118.

144 Cullumbine, H., Basnayake, V., Lemottee, J. and Wickramanayake, T.W. Mineral metabolism on rice diets. Brit. J. Nutrition, 1950, **4**, 101-111.

145 Curran, S. Investigation into aphosphorosis and grassland conditions in the midlands of Ireland. J. Dept. Agric. Dublin, 1949, **46**, 60-85.

146 Daji, J.A. The decomposition of green manures in soil. J. Agric. Sci., 1934, **24**, 15-27.

147 David, W.M. Barley: the maltster's point of view. Agriculture, London, 1964, **71**, 117-119.

148 Davidson, J. Procedures for the extraction, separation and estimation of the major fat-soluble pigments of hay. J. Sci. Food Agric., 1954, **5**, 1-7.

149 Davidson, J. The chromogen method for determining the digestibility of dried grass by sheep. J. Sci. Food Agric., 1954, **5**, 209-212.

150 Davies, G.M. The effects of winter and spring defoliation on the subsequent yield of grain and straw of S.147 oats. J. Agric. Sci., 1956, **47**, 363-366.

151 Davies, R.O., Jones, D.I.H. and Milton, W.E.J. Factors influencing the composition and nutritive value of herbage from fescue and *Molinia* areas. J. Agric. Sci., 1959, **53**, 268-285.

152 Davies, R.O. and Morgan, T.B. The yields and composition of lucerne, grass and clover under different systems of management. I. J. Brit. Grassland Soc., 1953, 8, 149-168.

153 Davies, R.P. The chemical composition of the seed and straw of various grasses. Welsh J. Agric., 1939, **15**, 250-260.

154 Davies, W. and Davis, A.G. Lucerne trials at Colesbourne. J. Brit. Grassland Soc., 1951, **6**, 119-124.

155 Davies, W. and Fagan, T.W. Winter keep on temporary leys. Empire J. Exp. Agric., 1938, **6**, 369-376.

156 Davies, W.E. The relative effect of frequency and time of cutting lucerne. J. Brit. Grassland Soc., 1960, **15**, 262-269.

157 Davies, W.E. and Davies, R.O. The yields and composition of lucerne, grass and clover under different systems of management. II. J. Brit. Grassland Soc., 1956, **11**, 127-138.

158 Davies, W.E. and Tyler, B.F. The yields of lucerne varieties in West Wales. J. Brit. Grassland Soc., 1962, **17**, 218-224.

159 Davies, W.E. and Tyler, B.F. The yield and composition of lucerne, grass and clover under different systems of management. IV. Further studies on the effect of frequency of cutting lucerne and lucerne grown with grass. J. Brit. Grassland Soc., 1962, **17**, 306-314.

160 Davies, W.M., Botham, G.H. and Thompson, W.B. Grass silage. A comparison of the changes involved in the ordinary, molasses and A.I.V. processes. J. Agric. Sci., 1937, **27**, 151-161.

161 Dent, J.W. Seasonal yield and composition of lucerne in relation to time of spring cutting. J. Brit. Grassland Soc., 1955, **10**, 330-340.

162 Dent, J.W. The chemical composition of the straw and grain of some varieties of spring oats in relation to time of harvesting, nitrogen treatment and environment. J. Agric. Sci., 1957, **48**, 336-346.

163 Dent, J.W. The composition and yield of the spring oats Radnorshire Sprig, Maldwyn and Sun II, grown for fodder under upland conditions. J. Nat. Inst. Agric. Bot., 1959, **8**, 614-624.

164 Dent, J.W. Seasonal yield and composition of lucerne in relation to the time of cutting of first and second growth. J. Brit. Grassland Soc., 1959, **14**, 262-271.

165 Dent, J.W. Digestibility tests on varieties of fodder crops in trial. J. Nat. Inst. Agric. Bot., 1963, **9**, 395-401.

166 Dent, J.W. and Aldrich, D.T.A. The inter-relationships between heading date, yield, chemical composition and digestibility in varieties of perennial ryegrass, timothy, cocksfoot, and meadow fescue. J. Nat. Inst. Agric. Bot., 1963, **9**, 261-281.

167 Dent, J.W. and Boyd, A.G. Yield and composition of three varieties of spring oats harvested as green fodder and as grain and straw under upland conditions in Scotland. J. Nat. Inst. Agric. Bot., 1964-66, **10**, 180-187.

168 Dent, J.W. (and C.E. Harris) Applications of the two-stage *in vitro* digestibility method to variety testing. (Appendix, C.E. Harris: Comparison of *in vivo* and *in vitro* measurements of the digestibility of fodder crops). J. Brit. Grassland Soc., 1963, **18**, 181-188; 189.

169 Dent, J.W. and Hawkins, R.P. The effect of fertilizers on the yield of red clover seed. J. Agric. Sci., 1962, **59**, 13-23.

170 Dent, J.W. and Zaleski, A. Leafiness and chemical composition of some lucerne strains. J. Brit. Grassland Soc., 1954, **9**, 131-140.

171 Department of Agriculture, Dublin: Report of the Seed Propagation Division, 1939. J. Dept. Agric. Dublin, 1940, **37**, 115-139.

172 Department of Agriculture, Dublin: Report of the Seed Propagation Division, 1940. J. Dept. Agric. Dublin, 1941, **38**, 90-107.

173 Department of Agriculture, Dublin: Report of the Seed Propagation Division, 1941. J. Dept. Agric. Dublin, 1942, **39**, 111-128.

174 Department of Agriculture, Dublin: Report of the Seed Propagation Division, 1942. J. Dept. Agric. Dublin, 1943, **40**, 45-60.

175 Department of Agriculture, Dublin: Report of the Seed Propagation Division, 1943. J. Dept. Agric. Dublin, 1944, **41**, 106-122.

176 Department of Agriculture, Dublin: Report of the Seed Propagation Division, 1944. J. Dept. Agric. Dublin, 1945, **42**, 94-109.

177 Department of Agriculture, Dublin: Report of the Seed Propagation Division, 1945. J. Dept. Agric. Dublin, 1946, **43**, 74-87.

178 Department of Agriculture, Dublin: Report of the Seed Propagation Division, 1946. J. Dept. Agric. Dublin, 1947, **44**, 91-107.

179 Department of Agriculture, Dublin: Report of the Seed Propagation Division, 1947. J. Dept. Agric. Dublin, 1948, **45**, 88-104.

180 Department of Agriculture, Dublin: Report of the Seed Propagation Division, 1948. J. Dept. Agric. Dublin, 1949, **46**, 149-166.

181 Department of Agriculture, Dublin: The value of different liming materials. J. Dept. Agric. Dublin, 1951-52, **48**, 49-67.

182 Department of Agriculture, Dublin: The comparative value of the drill and broadcast methods of phosphate and ground limestone applications for cereals. J. Dept. Agric. Dublin, 1951-52, **48**, 81-113.

183 Deriaz, R.E. Routine analysis of carbohydrates and lignin in herbage. J. Sci. Food Agric., 1961, **12**, 152-160.

184 Dewar, W.A. and McDonald, P. Determination of dry matter in silage by distillation with toluene. J. Sci. Food Agric., 1961, **12**, 790-795.

185 Dodsworth, T.L. Further studies on the fattening value of grass silage and on the effect of the dry-matter percentage of the diet on dry-matter intake in ruminants. J. Agric. Sci., 1954, **44**, 383-393.

186 Dodsworth, T.L. Studies on the starch values of swedes, fodder beet and grass silage and on the complementary value of grass silage and roots when fed together to ruminants. J. Agric. Sci., 1956, **47**, 456-460.

187 Dodsworth, T.L. and Campbell, W.H.McK. Studies on the productivity of first year ley grass when cut at different stages of growth, ensiled and fed to beef cattle. J. Brit. Grassland Soc., 1952, **7**, 151-159.

188 Dodsworth, T.L. and Campbell, W.H.McK. Report on an experiment to compare the fattening values, for beef cattle, of silages made from grass cut at different stages of growth. J. Agric. Sci., 1952, **42**, 395-402.

189 Dougall, B.M. The composition of *Agrostis setacea*, the bristle-leaved or heath bent grass. J. Sci. Food Agric., 1954, **5**, 132-134.

190 Dougall, B.M. The composition and probable feeding value of *Triodia decumbens*. J. Sci. Food Agric., 1954, **5**, 134-136.

191 Dougall, B.M. Carnation sedge and sheep. Agriculture, London, 1955-56, **62**, 348-349.

192 Eastoe, J. and Long, J.E. The amino-acid composition of processed bones and meat. J. Sci. Food Agric., 1960, **11**, 87-92.

193 Eden, A. Some features of the mineral composition of molassed beet pulp. Empire J. Exp. Agric., 1957, **25**, 161-164.

194 Eden, A., Alderman, G., Baker, C.J.L., Nicholson, H.H. and Firth, D.H. The effect of ground water-level upon productivity and composition of Fenland grass. J. Agric. Sci., 1951, **41**, 191-202.

195 Edinburgh and East of Scotland College of Agriculture, Edinburgh. Notes from Boghall. III. Scot. Agric., 1950-51, **30**, 177-179.

196 Eggitt, P.W.R. and Ward, L.D. The chemical estimation of vitamin E activity in cereal products. II. Millers' offals and compound animal foods. J. Sci. Food Agric., 1955, **6**, 329-337.

197 Elliott, F.J. and Thomas, B. On the yields and composition of meadow hay from certain of the Palace Leas plots at Cockle Park. J. Agric. Sci., 1934, **24**, 379-389.

198 Evans, R.E. Studies of the sulphur of pasture grass. I. The cystine content of pasture grass. J. Agric. Sci., 1931, **21**, 806-821.

199 Evans, R.E. Nutrition of the bacon pig. XV. The relative supplemental value of the proteins in dried brewers' yeast and in white-fish meal. J. Agric. Sci., 1952, **42**, 422-437.

200 Evans, R.E. Nutrition of the bacon pig. XVI. The relative supplemental value of the proteins in extracted soya-bean meal and in white-fish meal. J. Agric. Sci., 1952, **42**, 438-453.

201 Evans, R.E. Nutrition of the bacon pig. XVII. The nutritive value of condensed fish solubles. J. Agric. Sci., 1954, **44**, 100-120.

202 Evans, R.E. Nutrition of the bacon pig. XVIII. The influence of dietary penicillin on the growth rate, efficiency of food conversion and the nitrogen retention of the bacon pig. J. Agric. Sci., 1955, **46**, 329-361.

203 Evans, R.E. and Maguire, M.F. The influence of dietary antibiotics on the activity of the cellulose-splitting bacteria in the intestine of the bacon pig. J. Agric. Sci., 1956, **47**, 344-349.

204 Evans, T.A. Renovation of an aged seed stand of S143 cocksfoot. J. Brit. Grassland Soc., 1950, **5**, 287-290.

205 Evans, T.A. Management and manuring for seed production in cocksfoot and timothy. J. Brit. Grassland Soc., 1953, **8**, 245-259.

206 Eyles, D.E. and Cowlishaw, S.J. Simplified rations for poultry at pasture. Empire J. Exp. Agric., 1959, **27**, 158-169.

207 Fagan, T.W. The influence of management on the chemical composition of pastures in winter. Welsh J. Agric., 1936, **12**, 136-139.

208 Fagan, T.W. and Ashton, W.M. The preservation of green fodder. Welsh J. Agric., 1937, **13**, 129-144.

209 Fagan, T.W. and Ashton, W.M. The effect of partial field-drying and artificial drying on the chemical composition of grass. Welsh J. Agric., 1938, **14**, 160-176.

210 Fagan, T.W. and Davies, R.P. The effect of partial field-drying and artificial drying on the chemical composition of grass. Welsh J. Agric., 1939, **15**, 261-271.

211 Fagan, T.W., Jones, E.T. and Ashton, W.M. The yield and chemical composition of cereals under different systems of management. Welsh J. Agric., 1943, **17**, 103-107.

212 Fagan, T.W., Jones, E., Williams, E.E. and Davies, R.O. The influence of liming on the herbage of acidic soils in Merionethshire. Welsh J. Agric., 1940, **16**, 144-153.

213 Fagan, T.W., Jones, E., Williams, E.E. and Davies, R.O. The influence of liming on the herbage of acidic soils in Merionethshire. Welsh J. Agric., 1945, **18**, 73-74.

214 Fagan, T.W. and Milton, W.E.J. A comparison of the nitrogen and mineral content of the pasture, hay and aftermath of four species of grasses grown in a mixture, pure plots and pure drills. Welsh J. Agric., 1933, **9**, 93-109.

215 Fagan, T.W. and Milton, W.E.J. The effect of manures at different altitudes on the nitrogen and mineral content of grass and clover species. Welsh J. Agric., 1934, **10**, 174-189.

216 Fagan, T.W., Phillips, R. and Davies, R.O. The cultivation and composition of kale. Welsh J. Agric., 1943, **17**, 97-101.

217 Fagan, T.W., Phillips, R. and Davies, R.O. Kales and kale silage. Welsh J. Agric., 1945, **18**, 75-78.

218 Fagan, T.W. and Watkins, H.T. The chemical composition of the miscellaneous herbs of pastures. Welsh J. Agric., 1932, **8**, 144-151.

219 Fagan, T.W. and Watkins, H.T. The effect of manures on the nitrogen and mineral content of the produce of contrasting pasture types. Welsh J. Agric., 1932, **8**, 192-196.

220 Fairbairn, C.B. and Thomas, B. The potential nutritive value of some weeds common to north-eastern England. J. Brit. Grassland Soc., 1959, **14**, 36-46.

221 Featherstone, J., Rickaby, C.D. and Cavell, A.J. Variations in the progressive yield and chemical composition of grassland herbage during a period of three months' growth. J. Brit. Grassland Soc., 1951, **6**, 161-166.

222 Ferguson, W.S. Investigations into the intensive system of grassland management by the Agricultural Research Staff of Imperial Chemical Industries, Limited. III. The seasonal variation in the mineral content of pasture with particular reference to drought. J. Agric. Sci., 1931, **21**, 233-240.

223 Ferguson, W.S. Investigations into the intensive system of grassland management by the Agricultural Research Staff of Imperial Chemical Industries, Limited. X. A further study of the mineral content of intensively treated pasture. J. Agric. Sci., 1932, **22**, 251-256.

224 Ferguson, W.S. The digestibility of straw pulp. J. Agric. Sci., 1943, **33**, 174-177.

225 Ferguson, W.S. Marrow-stem kale silage. J. Agric. Sci., 1947, **37**, 77-80.

226 Ferguson, W.S. The effect of late nitrogenous top-dressing on the digestibility of hay. J. Agric. Sci., 1948, **38**, 33-36.

227 Ferguson, W.S. and Armitage, E.R. The chemical composition of bracken *(Pteridium aquilinum)*. J. Agric. Sci., 1944, **34**, 165-171.

228 Ferguson, W.S., Lewis, A.H. and Watson, S.J. The teart pastures of Somerset. I. The cause and cure of teartness. J. Agric. Sci., 1943, **33**, 44-51.

229 Ferguson, W.S. and Neave, O. Bracken ensilage. J. Agric. Sci., 1944, **34**, 172-175.

230 Ferguson, W.S. and Watson, S.J. The composition and nutritive value of seeds hays. J. Agric. Sci., 1944, **34**, 88-92.

231 Field, A.C. Studies on magnesium in ruminant nutrition. 4. Balance trials on sheep given different amounts of grass nuts. Brit. J. Nutrition, 1962, **16**, 99-107.

232 Field, A.C., McCallum, J.W. and Butler, E.J. Studies on magnesium in ruminant nutrition. Balance experiments on sheep with herbage from fields associated with lactation tetany and from control pastures. Brit. J. Nutrition, 1958, **12**, 433-446.

233 Fisher, E.A. Report on the quality – for bread-making purposes – of wheats harvested in 1935 at the headquarters and the sub-stations of the National Institute of Agricultural Botany. J. Nat. Inst. Agric. Bot., 1937, **4**, 147-157.

234 Fisher, E.A. Report on the quality – for bread-making purposes – of wheats harvested in 1937 at the headquarters and the sub-stations of the National Institute of Agricultural Botany. J. Nat. Inst. Agric. Bot., 1938, **4**, 266-274.

235 Fisher, E.A. Report on the quality – for bread-making purposes – of wheats harvested in 1938 at the headquarters and the sub-stations of the National Institute of Agricultural Botany. J. Nat. Inst. Agric. Bot., 1939, **4**, 372-388.

236 Fleming, G.A. Distribution of major and trace elements in some common pasture species. J. Sci. Food Agric., 1963, **14**, 203-208.

237 Food and Agriculture Organisation of the United Nations: Report of Committee on Calorie conversion factors and food composition tables. Washington, D.C., 1947, p. 20.

238 Foot, A.S., Murdoch, J.C. and Rowland, S.J. Marrowstem kale silage. Empire J. Exp. Agric., 1955, **23**, 109-112.

239 Gallagher, E.J. Investigations into the yield and quality of spring wheats grown under Irish conditions – second year's trials. J. Dept. Agric. Fish., Dublin, 1966, **62**, 25-78.

240 Gardner, A.L. A comparison of broadcast and wide-row spaced grasses when managed for foggage production. J. Brit. Grassland Soc., 1958, **13**, 177-186.

241 Gardner, A.L. and Hunt, I.V. Winter utilization of cocksfoot. J. Brit. Grassland Soc., 1955, **10**, 306-316.

242 Gardner, H.W. Nitrogen top dressings of winter wheat. Agriculture, London, 1953-54, **60**, 233-238.

243 Gardner, H.W. Nitrogen top dressings of spring oats. Agriculture, London, 1953-54, **60**, 328-334.

244 Gardner, H.W. and Hutchinson, J.B. Variety and manurial trials with winter oats in Hertfordshire. Agriculture, London, 1951-52, **58**, 208-216.

245 Gardner, R.G. and Mitchell, T.J. Through-circulation drying of seaweed. II. *Laminaria cloustoni* frond. J. Sci. Food Agric., 1953, **4**, 237-245.

246 Garton, G.A. Digestion and absorption of lipids in the ruminant. Proc. Nutr. Soc., 1969, **28**, 131.

247 Grace, N.D. and Richards, E.L. The nutritive value of meat by-products. J. Sci. Food Agric., 1964, **15**, 711-716.

248 Green, J. The distribution of tocopherols during the life-cycle of some plants. J. Sci. Food Agric., 1958, **9**, 801-812.

249 Green, J.O. and Evans, T.A. Manuring and grazing for seed production in S37 cocksfoot and S215 meadow fescue. J. Brit. Grassland Soc., 1956, **11**, 165-173.

250 Greenhalgh, J.F.D., Reid, G.W. and McDonald, I. The indirect estimation of the digestibility of pasture herbage. IV. Regressions of digestibility on faecal nitrogen concentration: effects of different fractions of the herbage upon within- and between-period regressions. J. Agric. Sci., 1966, **66**, 277-283.

251 Greer, E.N., Pringle, W.J.S. and Kent, N.L. The composition of British-grown winter wheat. II. Iron and manganese contents. J. Sci. Food Agric., 1952, **3**, 16-18.

252 Greer, E.N., Ridyard, H.N. and Kent, N.L. The composition of British-grown winter wheat. 1. Vitamin-B_1 content. J. Sci. Food Agric., 1952, **3**, 12-16.

253 ap Griffith, G. and Jones, D.I.H. Note on the pepsin-normal acid digestion for the determination of fibrous residue in herbage. J. Sci. Food Agric., 1965, **16**, 689-690.

254 ap Griffith, G., Jones, D.I.H. and Walters, R.J.K. Specific and varietal differences in sodium and potassium in grasses. J. Sci. Food Agric., 1965, **16**, 94-98.

255 Griffith, M. and Phillips, Ll. Experiments on the wintering of mountain ewes. Welsh J. Agric., 1940, **16**, 206-226.

256 Guttridge, D.G.A. and Lewis, D. Chick bio-assay of methionine and cystine. II. Assay of soyabean meals, ground-nut meals, meat meals, methionine isomers and methionine analogues. Brit. Poultry Sci., 1964, **5**, 193.

257 Halnan, E.T. Digestibility trials with poultry. VII. The digestibility of wheat offals, with a note of the apparent discrepancy between the digestibility coefficients and nutritive values of these products. J. Agric. Sci., 1937, **27**, 126-136.

258 Halnan, E.T. Observations on the value of whale-meat meal as a constituent of chick diets, with a note on the influence of added protein on the efficiency of utilization of the gross energy of a ration. J. Agric. Sci., 1942, **32**, 179-193.

259 Halnan, E.T. Digestibility trials with poultry. IX. The digestibility and metabolizable energy of sunflower seeds. J. Agric. Sci., 1943, **33**, 113-115.

260 Halnan, E.T. Digestibility trials with poultry. X. The effect of war-time changes in milling practice on the composition and nutritive value of fine and coarse wheat bran. J. Agric. Sci., 1944, **34**, 133-138.

261 Halnan, E.T. Digestibility trials with poultry. XI. The digestibility and metabolizable energy of raw and cooked potatoes, potato flakes, dried potato slices and dried potato shreds. J. Agric. Sci., 1944, **34**, 139-154.

262 Hand, C.J.E. and Tyler, C. The effect of feeding different seaweed meals on the mineral and nitrogen metabolism of laying hens. J. Sci. Food Agric., 1955, **6**, 743-754.

263 Hanley, F., Jarvis, R.H. and Whitear, J.D. The effects of time of ploughing and time of drilling on the development and yield of the winter wheat crop following a clover ley. J. Agric. Sci., 1961, **56**, 119-125.

264 Hanley, F., Ridgman, W.J. and Barker, M.G. The effect of leys on soil fertility. I. The effect of management of clover aftermath on the yield of the succeeding wheat crop. J. Agric. Sci., 1957, **49**, 251-257.

265 Harkess, R.D. Studies in herbage digestibility. J. Brit. Grassland Soc., 1963, **18**, 62-68.

266 Harris, C.E. The digestibility of fodder maize and maize silage. Exp. Agric., 1965, **1**, 121-123.

267 Harris, C.E. and Raymond, W.F. The effect of ensiling on crop digestibility. J. Brit. Grassland Soc., 1963, **18**, 204-212.

268 Hart, H.V. (unpublished) Courtesy Res. Assoc. Brit. Flour-Millers (1962).

269 Heathcote, J.G. The protein quality of oats. Brit. J. Nutrition, 1950, **4**, 145-154.

270 Heddle, R.G. Varieties of cocksfoot 1957-1960. Varieties of Italian ryegrass 1961-1962. Edinburgh and East of Scotland College of Agriculture. Technical Bulletin No. 25, 1963.

271 Hemingway, R.G. Effects of salt and other fertilizers on yield and mineral composition of forage crops. I. Turnips. J. Sci. Food. Agric., 1960, **11**, 349-355.

272 Hemingway, R.G. Effects of salt and other fertilisers on yield and mineral composition of forage crops. II. Kale. J. Sci. Food Agric., 1960, **11**, 355-362.

273 Hemingway, R.G. Effects of salt and other fertilisers on yield and mineral composition of forage crops. III. Herbage. J. Sci. Food Agric., 1961(a), **12**, 398-406.

274 Hemingway, R.G. Magnesium, potassium, sodium and calcium contents of herbage as influenced by fertilizer treatments over a three-year period. J. Brit. Grassland Soc., 1961(b), **16**, 106-116.

275 Hemingway, R.G. Copper, molybdenum, manganese and iron contents of herbage as influenced by fertilizer treatments over a three-year period. J. Brit. Grassland Soc., 1962, **17**, 182-187.

276 Henderson, J.L. and Davies, R.O. The yield and composition of mixed cereal-legume crops at different stages of growth. 1. Empire J. Exp. Agric., 1955, **23**, 131-144.

277 Henry, K.M., Kon, S.K., Thompson, S.Y., McCallum, J.W. and Stewart, J. The vitamin D activity of pastures and hays. Brit. J. Nutrition, 1958, **12**, 462-469.

278 Hill-Cottingham, D.G. and Wagner, S.A. An improved method for the determination of nitrogen in plant material using a stable Nessler reagent. J. Sci. Food Agric., 1962, **13**, 669-672.

279 Hirst, E.L., Mackenzie, D.J. and Wylam, C.B. Analytical studies on the carbohydrates of grasses and clovers. IX. Changes in carbohydrate composition during the growth of lucerne. J. Sci. Food Agric., 1959, **10**, 19-26.

280 Holman, W.I.M. and Godden, W. The aneurin (vitamin B_1) content of oats. I. The influence of variety and locality. II. Possible losses in milling. J. Agric. Sci., 1947, **37**, 51-57.

281 Holmes, J.C., Gill, W.D. and Rodger, J.A.B. Rates of nitrogen for winter wheat in South-East Scotland. Scot. Agric., 1959-60, **39**, 137-140.

282 Holmes, J.C., Gill, W.D. and Rodger, J.A.B. The effects of rates and time of application of nitrogenous fertilizer on barley in South-East Scotland. J. Agric. Sci., 1960, **54**, 291-299.

283 Holmes, W. Prickly comfrey. Agriculture, London, 1945-46, **52**, 515-516.

284 Holmes, W. The intensive production of herbage for crop-drying. I. A study of the productivity of two annual crops and two leys and of their responses in yield and chemical composition to applications of nitrogenous manure. J. Agric. Sci., 1948, **38**, 425-436.

285 Holmes, W. The intensive production of herbage for crop-drying. II. A study of the effect of massive dressings of nitrogenous fertilizer and of the time of their application on the yield, chemical and botanical composition of two grass leys. J. Agric. Sci., 1949, **39**, 128-141.

286 Holmes, W. The intensive production of herbage for crop-drying. IV. The effect of massive applications of nitrogen with and without phosphate and potash on the yield of grassland herbage. J. Agric. Sci., 1951, **41**, 70-79.

287 Holmes, W., Arnold, G.W. and Provan, A.L. Bulk feeds for milk production. 1. The influence of level of concentrate feeding in addition to silage and hay on milk yield and milk composition. J. Dairy Res., 1960, **27**, 191-204.

288 Holmes, W. and MacLusky, D.S. The intensive production of herbage for crop-drying. VI. A study of the effect of intensive nitrogen fertilizer treatment on species and strains of grass, grown alone and with white clover. J. Agric. Sci., 1955, **46**, 267-286.

289 Holmes, W., Waite, R., Fergusson, D.L. and MacLusky, D.S. Studies in grazing management. IV. A comparison of close-folding and rotational grazing of dairy cows on intensively fertilized pastures. J. Agric. Sci., 1952, **42**, 304-313.

290 Hood, A.E.M. Nitrogenous manuring of grass-clover leys for conservation, with particular reference to 'late' top-dressing. J. Agric. Sci., 1957, **48**, 145-148.

291 Hood, A.E.M. Soilage for beef cattle. J. Brit. Grassland Soc., 1962, **17**, 264-267.

292 Hughes, B.P. The amino-acid composition of potato protein and of cooked potato. Brit. J. Nutrition, 1958, **12**, 188-195.

293 Hughes, G.P. The utilization of foggage. Agriculture, London, 1948-49, **55**, 98-101.

294 Hughes, G.P. The outwintering of beef stores on grassland. Agriculture, London, 1949-50, **56**, 200-203.

295 Hughes, G.P. The seasonal output of pastures sown with ultra-simple seeds mixture. J. Agric. Sci., 1951, **41**, 203-213.

296 Hughes, G.P. The production and utilization of winter grass. J. Agric. Sci., 1954-55, **45**, 179-201.

297 Hughes, G.P. and Davis, A.G. The development of swards sown with simple mixtures at different rates of seeding under varying systems of management and manuring. J. Brit. Grassland Soc., 1951, **6**, 167-177.

298 Hughes, G.P. and Evans, T.A. Response of an established sheep ley to dressings of sulphate of ammonia. 1. Yields and botanical composition of herbage cut for silage and for hay. 2. Effects on chemical composition of the herbage. Empire J. Exp. Agric., 1951, **19**, 55-64; 65-72.

299 Hughes, G.P. and Eyles, D.E. Extracted herbage leaf protein for poultry feeding. I. Introduction and feeding trial with laying hens. II. The use of leaf protein in chick rations. J. Agric. Sci., 1953, **43**, 136-143; 144-151.

300 Hunt, I.V. Productivity, persistence and response to nitrogen of Italian ryegrass varieties. J. Brit. Grassland Soc., 1962, **17**, 125-129.

301 Hunt, I.V. and Alexander, R.H. The crude-protein content of Italian ryegrass. J. Brit. Grassland Soc., 1961, **16**, 226-229.

302 Hunt, I.V., Alexander, R.H. and Rutherford, A.A. The effect of various manuring practices on the magnesium status of spring herbage. J. Brit. Grassland Soc., 1964, **19**, 224-230.

303 Hunter, H. (with Hartley, H.O.) Relation of ear survival to the nitrogen content of certain varieties of barley (with a statistical study by H.O. Hartley, 496-502). J. Agric. Sci., 1938, **28**, 472-502.

304 Hunter, J.G. The composition of bracken: some major- and trace-element constituents. J. Sci. Food Agric., 1953, **4**, 10-20.

305 Hutchinson, J.B. and Martin, H.F. The chemical composition of oats. I. The oil and free fatty acid content of oats and groats. J. Agric. Sci., 1954-55, **45**, 411-418.

306 Hutchinson, J.B. and Martin, H.F. The chemical composition of oats. II. The nitrogen content of oats and groats. J. Agric. Sci., 1954-55, **45**, 419-427.

307 Ibbotson, C.F. The effect of additions of magnesian limestone and magnesium oxide on silage quality. Empire J. Exp. Agric., 1963, **31**, 259-266.

308 Jarvis, R.H. Studies on lucerne and lucerne-grass leys. 5. Plant population studies with lucerne. J. Agric. Sci., 1962, **59**, 281-286.

309 Jarvis, R.H., Hanley, F. and Ridgman, W.J. The effect of leys on soil fertility. II. The effect of under-sowing with grasses and legumes on the yield of a barley nurse crop. J. Agric. Sci., 1958, **51**, 229-233.

310 Jobson, H.T. and Thomas, B. The composition of gorse *(Ulex europaeus).* J. Sci. Food Agric., 1964, **15**, 652-656.

311 Johnston, M.J. and Waite, R. Studies in the lignification of grasses. 1. Perennial ryegrass (S24) and cocksfoot (S37). J. Agric. Sci., 1965, **64**, 211-219.

312 Jones, C.R. The essentials of the flour-milling process. Proc. Nutr. Soc., 1958, **17**, 7.

313 Jones, D.I.H. Note on the pre-treatment of herbage samples for determination of soluble carbohydrate constituents. J. Sci. Food Agric., 1962, **13**, 83-86.

314 Jones, D.I.H., ap Griffith, G. and Walters, R.J.K. The effect of nitrogen fertilisers on the water-soluble carbohydrate content of grasses. J. Agric. Sci., 1965, **64**, 323-328.

315 Jones, D.J.C. Studies of the chemical composition of kales and rapes. I. The kales. J. Agric. Sci., 1959, **52**, 230-237.

316 Jones, D.J.C. Studies of the chemical composition of kales and rapes. II. The rapes. J. Agric. Sci., 1959, **52**, 238-243.

317 Jones, D.J.C. Studies of the chemical composition of kales and rapes. III. The minor elements. J. Agric. Sci., 1959, **53**, 151-155.

318 Jones, D.J.C. The effect of sulphate of ammonia applications on the sulphur content of various grass and clover mixtures. J. Agric. Sci., 1960, **54**, 188-194.

319 Jones, D.J.C. Studies of the effects of manuring and season on the yield and chemical composition of the cattle cabbage. J. Agric. Sci., 1960, **55**, 165-173.

320 Jones, D.J.C. The effects of singling and advancing season on the composition of thousand-headed kale. J. Agric. Sci., 1962, **58**, 265-275.

321 Jones, D.J.C. The effects of advancing season on the chemical composition of marrow stem kale. J. Agric. Sci., 1965, **65**, 121-128.

322 Jones, E. Studies on the magnesium content of mixed herbage and some individual grass and clover species. J. Brit. Grassland Soc., 1963, **18**, 131-138.

323 Kent, N.L. The quality of oats in North-East Scotland trials, 1952-1955. Scot. Agric., 1957-58, **37**, 98-102.

324 Kent, N.L. The technology of cereals. Pergamon Press, Oxford, 1966.

325 Kent, N.L., Gill, W.D. and Holmes, J.C. The yield and quality of oats in South-East Scotland trials, 1951-56. Scot. Agric., 1958-59, **38**, 126-132.

326 Kent, N.L. and Waterson, H.A. Spring oat trials in the West of Scotland, 1954-59. J. Nat. Inst. Agric. Bot., 1961, **9**, 33-44.

327 Kerr, J.A.M., Brown, W.O. and Morrison, J. The nutritive value of grass silage self-fed to fattening cattle. Animal Prod., 1961, **3**, 321-325.

328 Kimber, D.S. Trials of rye varieties for forage, 1957-63. J. Nat. Inst. Agric. Bot., 1964-66, **10**, 195-203.

329 Knowles, F. and Watkin, J.E. The assimilation and translocation of plant nutrients in wheat during growth. J. Agric. Sci., 1931, **21**, 612-637.

330 Kodicek, E., Braude, R., Kon, S.K. and Mitchell, K.G. The availability to pigs of nicotinic acid in *tortilla* baked from maize treated with lime-water. Brit. J. Nutrition, 1959, **13**, 363-384.

331 Kodicek, E. and Wilson, P.W. The availability of bound nicotinic acid to the rat. 1. The effect of lime-water treatment of maize and subsequent baking into *tortilla.* Brit. J. Nutrition, 1959, **13**, 418-430.

332 Kon, S.K. and Thompson, S.Y. Factors affecting the stability and estimation of carotene in artificially dried grass and hays. J. Agric. Sci., 1940, **30**, 622-638.

333 Laidlaw, R.A. and Reid, S.G. Analytical studies on the carbohydrates of grasses and clovers. I. Development of methods for the estimation of the free sugar and fructosan contents. J. Sci. Food Agric., 1952, **3**, 19-25.

334 Lambert, H.G. Communal grass drying in England and Wales. Results for 1948. Agriculture, London, 1949-50, **56**, 103-106.

335 Lancaster, H.M. Malting quality of spring barleys, 1933-1936. J. Nat. Inst. Agric. Bot., 1938, **4**, 287-292.

336 Lang, R.W. and Holmes, J.C. The effect of plant population and distribution on the yield and quality of swedes. J. Agric. Sci., 1965, **65**, 91-99.

337 Large, R.V., Alder, F.E. and Spedding, C.R.W. Winter feeding of the in-lamb ewe. J. Agric. Sci., 1959, **53**, 102-117.

338 Leat, W.M.F. Studies on pigs reared on diets low in tocopherol and essential fatty acids. Brit. J. Nutrition, 1961, **15**, 259-270.

339 Lewis, D. and Morgan, J.T. Fats and amino acids in broiler rations. 2. Metabolisable energy determination and energy-protein balance. Brit. Poultry Sci., 1963, **4**, 3-11.

340 Line, C., Head, M.J., Rook, J.A.F., Foot, A.S. and Rowland, S.J. Investigations into the use of supplements for the control of hypomagnesaemia in dairy cows during the spring grazing period. J. Agric. Sci., 1958, **51**, 353-360.

341 Linehan, P.A. and Lowe, J. The output of pasture and its measurement. J. Brit. Grassland Soc., 1946, **1**, 7-35.

342 Linehan, P.A., Lowe, J. and Stewart, R.H. The output of pasture and its measurement. II. J. Brit. Grassland Soc., 1947, **2**, 145-168.

343 Linehan, P.A., Lowe, J. and Stewart, R.H. The output of pasture and its measurement. III. J. Brit. Grassland Soc., 1952, **7**, 73-98.

344 Ling, A.W. and Smith, E.L. An investigation into the composition of hay. Journal of the Bath and West and Southern Counties Society. 6th Series. 1939, **14**, 29-60.

345 Mackenzie, D.J. and Wylam, C.B. Analytical studies on the carbohydrates of grasses and clovers. VIII. Changes in carbohydrate composition during the growth of perennial ryegrass. J. Sci. Food Agric., 1957, **8**, 38-45.

346 Macpherson, H.T. Histamine, tryptamine and tyramine in grass silage. J. Sci. Food Agric., 1962, **13**, 29-32.

347 McCance, R.A. and Glaser, E.M. The energy value of oatmeal and the digestibility and absorption of its proteins, fats and calcium. Brit. J. Nutrition, 1948-49, **2**, 221-228.

348 McDonald, P. Studies on the nutritive value of fresh and conserved grass with special reference to silage. Thesis. Univ. Edinburgh, 1956.

349 McDonald, P. and Dewar, W.A. Determination of dry matter and volatiles in silage. J. Sci. Food Agric., 1960, **11**, 566-570.

350 McDonald, P. and Henderson, A.R. Buffering capacity of herbage samples as a factor in ensilage. J. Sci. Food Agric., 1962, **13**, 395-400.

351 McDonald, P., Stirling, A.C., Henderson, A.R. and Whittenbury, R. Fermentation studies on wet herbage. J. Sci. Food Agric., 1962, **13**, 581-590.

352 McNaught, M.L., Smith, J.A.B. and Black, W.A.P. The utilization of carbohydrates of seaweed by rumen microflora *in vitro*. J. Sci. Food Agric., 1954, **5**, 350-352.

353 Mann, H.H. Field studies in green manuring. I. Empire J. Exp. Agric., 1958, **26**, 274-282.

354 Milbourn, G.M., Innes, P. and Holmes, W. The response to fertilizer treatments of winter wheat grown after leys. J. Brit. Grassland Soc., 1963, **18**, 310-317.

355 Miles, D.G., ap Griffith, G. and Walters, R.J.K. The effect of "winter burn" on the chemical composition and *in vitro* dry-matter digestibility of eight grasses. J. Brit. Grassland Soc., 1964, **19**, 75-76.

356 Minson, D.J., Harris, C.E., Raymond, W.F. and Milford, R. The digestibility and voluntary intake of S22 and HI ryegrass, S170 tall fescue, S48 timothy, S215 meadow fescue and Germinal cocksfoot. J. Brit. Grassland Soc., 1964, **19**, 298-305.

357 Minson, D.J., Raymond, W.F. and Harris, C.E. Studies in the digestibility of herbage. VIII. The digestibility of S37 cocksfoot, S23 ryegrass and S24 ryegrass. J. Brit. Grassland Soc., 1960, **15**, 174-180.

358 Mitchell, R.L. Trace elements in some constituent species of moorland grazing. J. Brit. Grassland Soc., 1954, **9**, 301-311.

359 Mitchell, R.L., Reith, J.W.S. and Johnston, I.M. Trace-element uptake in relation to soil content. J. Sci. Food Agric., 1957, **8**, s51-s59.

360 Moon, F.E. The influence of manurial treatment on the carotene content of poor pasture grass, and on the relationship of this constituent to ash and organic fractions. J. Agric. Sci., 1939, **29**, 524-543.

361 Moon, F.E. The composition of grass at various stages of maturity, and the changes occurring during haymaking, with particular reference to carotene-content. Empire J. Exp. Agric., 1939, **7**, 225-234.

362 Moon, F.E. The carotene-contents of some grass and clover species, with a note on pasture weeds. Empire J. Exp. Agric., 1939, **7**, 235-243.

363 Moon, F.E. Improving the nutritive value of Lothians' seeds-hay. Scot. Agric., 1953-54, **33**, 102-105.

364 Moon, F.E. The composition and nutritive value of hay grown in the east of Scotland and the influence of late applications of nitrogenous fertilizers. J. Agric. Sci., 1954, **44**, 140-151.

365 Moon, F.E. and Pal, A.K. The composition and nutritive value of bracken. J. Agric. Sci., 1949, **39**, 296-301.

366 Moon, F.E. and Thomas, B. The digestion of huskless oats by poultry. J. Agric. Sci., 1937, **27**, 458-464.

367 Moran, T. Report on the quality – for bread-making purposes – of wheat harvested in 1939, at the headquarters and sub-stations of the National Institute of Agricultural Botany. J. Nat. Inst. Agric. Bot., 1944, **5**, 9-27.

368 Moran, T. Nutrients in wheat endosperm. Nature, 1945, **155**, 205-206.

369 Moran, T. and Pace, J. A note on the amino acid composition of the protein in heather shoots. J. Agric. Sci., 1962, **59**, 93-94.

370 Morgan, D.E. Variations in the composition of oats and barley grown in Wales. (Proximate constituents, available carbohydrate and "1,000 grain weights"). J. Sci. Food Agric., 1967, **18**, 21-24.

371 Morgan, D.E. Note on variation in the mineral composition of oat and barley grain grown in Wales. J. Sci. Food Agric., 1968, **19**, 393-395.

372 Morgan, D.E. and Jones, M.G.S. The energy value of barn-dried hay. J. Brit. Grassland Soc., 1965, **20**, 174-176.

373 Mortimer, R.G. Feeding maize silage. Agriculture, London, 1960-1961, **67**, 352-356.

374 Murdoch, J.C. The effect of pre-wilting herbage on the composition of silage and its intake by cows. J. Brit. Grassland Soc., 1960, **15**, 70-73.

375 Murdoch, J.C., Balch, D.A., Foot, A.S. and Rowland, S.J. The ensiling of lucerne with addition of formic and glycollic acids, molasses and barley meal, and with wilting. J. Brit. Grassland Soc., 1955, **10**, 139-150.

376 Murdoch, J.C., Balch, D.A., Holdsworth, M.C. and Wood, M. The effect of chopping, lacerating and wilting of herbage on the chemical composition of silage. J. Brit. Grassland Soc., 1955, **10**, 181-188.

377 Murdoch, J.C. and Bare, D.I. The effect of mechanical treatment on the rate of drying and loss of nutrients in hay. J. Brit. Grassland Soc., 1960, **15**, 94-99.

378 Murdoch, J.C., Foot, A.S., Head, M.J., Holdsworth, M.C., Hosking, Z.D. and Line, C. Changes in chemical composition and the loss of nutrients in tripoded and swath-cured hay. J. Brit. Grassland Soc., 1959, **14**, 247-252.

379 Murdoch, J.C. and Holdsworth, M.C. The use of sodium metabisulphite in silage making. J. Brit. Grassland Soc., 1958, **13**, 55-60.

380 Murdoch, J.C. and Holdsworth, M.C. The effect of temperature in the mass on the chemical composition of silage. J. Brit. Grassland Soc., 1960, **15**, 240-245.

381 Murdoch, J.C., Holdsworth, M.C. and Wood, M. The chemical composition and loss of nutrients in silage made with the

addition of sodium metabisulphite and halogenated acetate of glycol. J. Brit. Grassland Soc., 1956, **11**, 16-22.

382 Murdoch, J.C. and Rook, J.A.F. A comparison of hay and silage for milk production. J. Dairy Res., 1963, **30**, 391-397.

383 Nash, M.J. Partial wilting of grass crops for silage. 2. Experimental silages. J. Brit. Grassland Soc., 1959, **14**, 107-116.

384 Nash, M.J. Partial wilting of grass crops for silage. 3. Farm silages. J. Brit. Grassland Soc., 1959, **14**, 177-182.

385 Nash, M.J., McDonald, P., Greenhalgh, J.F.D. and Cunningham, J.M.M. Potato silage. Scot. Agric., 1963-64, **43**, 136-140.

386 Neenan, M. and McCarrick, R.P. Some results from a survey of the types and quantities of silage made in the years 1955-1958. J. Dept. Agric. Dublin, 1960, **57**, 55-67.

387 Nehring, K. European Association for Animal Production, 4th Symposium on Energy Metabolism. Jablonna, near Warsaw. September, 1967.

388 Nicholson, H.H., Firth, D.H., Eden, A., Alderman, G., Baker, C.J.L. and Heimberg, M. The effect of ground water-level upon productivity and composition of Fenland grass (II). J. Agric. Sci., 1953, **43**, 265-274.

389 Nicholson, I.A. The effect of stage of maturity on the yield and chemical composition of oats for haymaking. J. Agric. Sci., 1957, **49**, 129-139.

390 Norman, M.J.T. Intervals of superphosphate application to downland permanent pasture. J. Agric. Sci., 1956, **47**, 157-171.

391 North of Scotland College of Agriculture, Aberdeen Records. Date uncertain.

392 Nowakowski, T.Z. Effects of nitrogen fertilizers on total nitrogen, soluble nitrogen and soluble carbohydrate contents of grass. J. Agric. Sci., 1962, **59**, 387-392.

393 Oldershaw, A.W. Lupins as a light-land crop. Agriculture, London, 1941-42, **48**, 164-168.

394 Oldershaw, A.W. Sweet lupins. Agriculture, London, 1944-45, **51**, 128-134.

395 O'Leary, P. and Brady, J.J. Cereal variety trials at the Department's farms, 1962. J. Dept. Agric. Dublin, 1965, **61**, 112-159.

396 O'Leary, P. and Brady, J.J. Cereal variety trials at the Department's farms, 1963. J. Dept. Agric. Fish. Dublin, 1966, **62**, 198-243.

397 O'Sullivan, T.J.R. Report of the Cereal Station, Ballinacurra, 1961; 1962. J. Dept. Agric. Dublin, 1964, **60**, 68-85; 1965, **61**, 165-208.

398 Owen, E.C. The carotene, carotenoid and chlorophyll contents of some Scottish seaweeds. J. Sci. Food Agric., 1954, **5**, 449-453.

399 Paterson, W.G.R. Some sheep feeding experiments. Trans. Highl. Agric. Soc. Scot., 1934, **46**, 146-169.

400 Paterson, W.G.R. Some further stock-feeding trials. Trans. Highl. Agric. Soc. Scot., 1938, **50**, 38-67.

401 Patterson, J.B.E. A comparison of the feeding-values of grass ensiled by the A.I.V.—process and a ration containing mangolds and hay. Empire J. Exp. Agric., 1935, **3**, 144-152.

402 Pawson, H.C. and Innes, P. Wintering store cattle. Cockle Park trial, 1953-54. Agriculture, London, 1954-55, **61**, 599-604.

403 Pawson, H.C. and Innes, P. Wintering store cattle. Cockle Park trial, 1954-55. Agriculture, London, 1955-56, **62**, 534-539.

404 Pawson, H.C. and Thomas, B. A note on the Melancholy Thistle. J. Minist. Agric., 1937-38, **44**, 424-426.

405 Phillipson, A.T. and Reid, R.S. Thiamine in the contents of the alimentary tract of sheep. Brit. J. Nutrition, 1957, **11**, 27-41.

406 Preston, T.R., Archibald, J.D.H. and Tinkler, W. The digestibility of grass by young calves. J. Agric. Sci., 1957, **48**, 259-265

407 Preston, T.R., Whitelaw, F.G., MacLeod, N.A. and Philip, E.B. The nutrition of the early-weaned calf. VIII. The effect on nitrogen retention of diets containing different levels of fish meal. Animal Prod., 1965, **7**, 53-58.

408 Pritchard, H. and Cawthorne, M. The composition of whale-meat meals of various grades. J. Sci. Food Agric., 1955, **6**, 148-153.

409 Pritchard, H. and Smith, P.A. The composition of feeding meat and bone meal. J. Sci. Food Agric., 1957, **8**, 668-672.

410 Pritchard, H. and Smith, P.A. Composition and amino-acid content of high-grade whale-meat meal. J. Sci. Food Agric., 1960, **11**, 249-252.

411 Pritchard, H. and Wraige, D.R. The value of whale-liver meal and whale-meat meal as sources of the vitamin-B complex in the ration of farm animals. J. Sci. Food Agric., 1952, **3**, 74-77.

412 Pritchard, H. and Wraige, D.R. The vitamin-B group in white fish meal. J. Sci. Food Agric., 1953, **4**, 172-176.

413 Procter, J. and Hood, A.E.M. The close-folding of dairy cows. II. J. Brit. Grassland Soc., 1953, **8**, 239-244.

414 Procter, J., Hood, A.E.M., Ferguson, W.S. and Lewis, A.H. The close-folding of dairy cows. J. Brit. Grassland Soc., 1950, **5**, 243-250.

415 Rahman, H., McDonald, P. and Simpson, D. Effects of nitrogen and potassium fertilisers on the mineral status of perennial ryegrass *(Lolium perenne)*. I. Mineral content. J. Sci. Food Agric., 1960, **11**, 422-428.

416 Ratcliff, E.C. Six ways of haymaking. Agriculture, London, 1947-48, **54**, 63-67.

417 Raven, A.M. and Robinson, K.L. Studies of the nutrition of the young calf. A comparison of starch, lactose, and hydrogenated palm oil, with butterfat, in milk diets. Brit. J. Nutrition, 1958, **12**, 469-482.

418 Raven, A.M. and Robinson, K.L. Studies of the nutrition of the young calf. 2. The nutritive value of unhydrogenated palm oil, unhydrogenated palm-kernel oil and butterfat, as additions to a milk diet. Brit. J. Nutrition, 1959, **13**, 178-190.

419 Raven, A.M. and Robinson, K.L. Studies of the nutrition of the young calf. 3. A comparison of unhydrogenated palm-kernel oil, hydrogenated palm-kernel oil, and butterfat, as constituents of a milk diet. Brit. J. Nutrition, 1960, **14**, 135-146.

420 Raven, A.M. and Thompson, A. The availability of iron in certain grass, clover and herb species. II. Alsike, broad red clover, Kent wild white clover, trefoil and lucerne. J. Agric. Sci., 1959, **53**, 224-229.

421 Raven, A.M. and Thompson, A. The availability of iron in certain grass, clover and herb species. III. Burnet, chicory and narrow-leaved plantain. J. Agric. Sci., 1961, **56**, 229-234.

422 Raymond, W.F. The nutritive value of herbage in The Measurement of Grassland Productivity, p. 156-164. Ed. J.D. Ivins. Butterworth, London, 1959.

423 Raymond, W.F. Biochemistry and animal nutrition. Grassland Research Institute, Ann. Rept. No. 16, 1962-63, p. 61.

424 Raymond, W.F. The digestibility of fodder and herbage crops. J. Nat. Inst. Agric. Bot. 1963, **9**, 392-394.

425 Raymond, W.F. The efficient use of grass. J. Brit. Grassland Soc., 1964, **19**, 81-89.

426 Raymond, W.F. and Harris, C.E. The value of the fibrous residue from leaf protein extraction as a feeding-stuff for ruminants. J. Brit. Grassland Soc., 1957, **12**, 166-170.

427 Raymond, W.F., Harris, C.E. and Kemp, C.D. Studies in the digestibility of herbage. V. The variation, with age, of the ability of sheep to digest herbage, with observations on the effect of season on digestive ability. J. Brit. Grassland Soc., 1954, **9**, 209-220.

428 Raymond, W.F., Harris, C.E. and Kemp, C.D. Studies in the digestibility of herbage. VI. The effect of level of herbage intake on the digestibility of herbage by sheep. J. Brit. Grassland Soc., 1955, **10**, 19-26.

429 Raymond, W.F., Kemp, C.D., Kemp, A.W. and Harris, C.E. Studies in the digestibility of herbage. IV. The use of faecal collection and chemical analysis in pasture studies. (b) Faecal index methods. Appendix: Statistical treatment (Kemp, C.D. and Kemp, A.W.). J. Brit. Grassland Soc., 1954, **9**, 69-78; 79-82.

430 Raymond, W.F., Minson, D.J. and Harris, C.E. Studies in the digestibility of herbage. VII. Further evidence on the effect of level of intake on the digestive efficiency of sheep. J. Brit. Grassland Soc., 1959, **14**, 75-77.

431 Rees, W.J. and Warwick, J. Clover establishment on power-station waste ash. Empire J. Exp. Agric., 1957, **25**, 256-262.

432 Reid, D. Studies on the cutting management of grass-clover swards. III. The effects of prolonged close and lax cutting on herbage yields and quality. J. Agric. Sci., 1962, **59**, 359-368.

433 Reith, J.W.S. The magnesium contents of soils and crops. J. Sci. Food Agric., 1963, **14**, 417-426.

434 Reith, J.W.S. and Williams, E.G. The effectiveness of various liming materials. Empire J. Exp. Agric., 1949, **17**, 265-276.

435 Research Association of British Flour-Millers. Progress Report No. 7, 1953.

436 Research Association of British Flour-Millers. Progress Report No. 9, 1954.

437 Research Association of British Flour-Millers. Progress Report No. 12, 1956.

438 Ridgman, W.J., Hanley, F. and Barker, M.G. Studies on lucerne and lucerne-grass leys. II. The nitrogenous manuring of a lucerne-cocksfoot ley. J. Agric. Sci., 1955, **46**, 441-448.

439 Ridgman, W.J., Hanley, F. and Barker, M.G. Studies on lucerne and lucerne-grass leys. III. The effect of variety of lucerne and strain of grass. J. Agric. Sci., 1956, **47**, 50-58.

440 Robb, W. Oat-breeding in Scotland. Empire J. Exp. Agric., 1934, **2**, 251-257.

441 Roberts, H.W., Kerr, D.H. and Seaton, D. The machair grasslands of the Hebrides. J. Brit. Grassland Soc., 1959, **14**, 223-228.

442 Rook, J.A.F. and Balch, C.C. Magnesium metabolism in the dairy cow. II. Metabolism during the spring grazing season. J. Agric. Sci., 1958, **51**, 199-207.

443 Rook, J.A.F. and Wood, M. Mineral composition of herbage in relation to the development of hypomagnesaemia in grazing cattle. J. Sci. Food Agric., 1960, **11**, 137-143.

444 Rose, T.H. and Dermott, W. Micro-nutrient deficiencies in arable crops in Romney Marsh. I. Emp. J. Exp. Agric., 1960, **28**, 281-293.

445 Ross, A.G. Some typical analyses of red seaweeds. J. Sci. Food Agric., 1953, **4**, 333-335.

446 Russell, E.J. and Watson, D.J. The Rothamsted Field Experiments on barley 1852-1937. 3. The composition and quality of the barley grain. Empire J. Exp. Agric., 1939, **7**, 193-220.

447 Rutledge, W.A. and Common, R.H. The composition and digestibility of Northern Irish hays. I. Unthreshed and threshed ryegrass and crested dogstail hays as saved for seed. J. Agric. Sci., 1947, **37**, 60-63.

448 Rutledge, W.A. and Common, R.H. The composition and digestibility of Northern Irish hays. II. Meadow hays. J. Agric. Sci., 1948, **38**, 28-32.

449 Ryan, J.J. The composition and digestibility of Irish hay. J. Dept. Agric. Dublin, 1947, **44**, 56-61.

450 Scott, R.O. Application of direct photometry to agricultural analysis. J. Sci. Food Agric., 1960, **11**, 584-592.

451 Senior, B.J., Collins, P. and Kelly, M. The feeding value of seaweeds. Econ. Proc. Royal Dublin Soc., 1936-50, **3**, 273-291.

452 Shacklady, C.A. and Newbound, P. Preliminary investigations on ley cropping for large-scale green-crop drying. J. Brit. Grassland Soc., 1951, **6**, 19-28.

453 Shearer, G.D. and McDougall, E.I. Some observations on swayback disease of lambs. J. Agric. Sci., 1944, **34**, 207-212.

454 Sheehy, E.J., Brophy, J., Dillon, T. and O'Muineachain, P. Seaweed *(Laminaria)* as stock food. Econ. Proc. Royal Dublin Soc., 1936-50, **3**, 150-161.

455 Sheehy, E.J., O'Donovan, J., Day, W.R. and Curran, S. Aphosphorosis in cattle in County Offaly. J. Dept. Agric. Dublin, 1948, **45**, 5-28.

456 Shepperson, G. and Grundey, J.K. Recent developments in quick haymaking techniques. J. Brit. Grassland Soc., 1962, **17**, 141-149.

457 Smith, A.M. and Comrie, A. The composition of different kinds of silage. J. Agric. Sci., 1938, **28**, 203-211.

458 Smith, A.M. and Comrie, A. The composition of Lothians' seeds-hay. Scot. Agric., 1948-49, **28**, 67-70.

459 Smith, A.M. and Robb, W. The carotene and protein in oats and barley at different stages of growth. J. Agric. Sci., 1943, **33**, 119-121.

460 Smith, A.M. and Tung Wang. The carotene content of certain species of grassland herbage. J. Agric. Sci., 1941, **31**, 370-378.

461 Smith, A.N. and Thomas, B. The nutritive value of *Calluna vulgaris.* IV. Digestibility at three, seven and fourteen years after burning. J. Agric. Sci., 1956, **47**, 468-475.

462 Smith, E.L. Apple pomace silage. Agriculture, London, 1950-51, **57**, 328-332.

463 Stewart, A.B. and Holmes, W. Manuring of grassland. I. Some effects of heavy dressings of nitrogen on the mineral composition of grassland herbage. J. Sci. Food Agric., 1953, **4**, 401-408.

464 Tayler, J.C. and Rudman, J.E. The production of fattening cattle and extension of autumn grazing following three rates of application of nitrogenous fertilizer to a ryegrass/white clover sward. J. Agric. Sci., 1960, **55**, 75-90.

465 Terry, R.A. and Tilley, J.M.A. The digestibility of the leaves and stems of perennial ryegrass, cocksfoot, timothy, tall fescue, lucerne and sainfoin, as measured by an *in vitro* procedure. J. Brit. Grassland Soc., 1964, **19**, 363-372.

466 Thomas, B. On the composition of common heather *(Calluna vulgaris).* J. Agric. Sci., 1932, **24**, 151-155.

467 Thomas, B. The composition of draw-moss. J. Minist. Agric., 1935-36, **42,** 458-461.

468 Thomas, B. The stool bent or heath rush. J. Minist. Agric., 1936-37, **43,** 262-265.

469 Thomas, B. The composition and feeding value of heather at different periods of the year. J. Minist. Agric., 1936-37, **43,** 1050-1055.

470 Thomas, B. and Armstrong, D.G. The nutritive value of common heather *(Calluna vulgaris).* 1. The preparation of samples of *Calluna vulgaris* for analytical purposes and for digestibility studies. J. Agric. Sci., 1952, **42,** 461-464.

471 Thomas, B. and Boyns, B.M. The composition of grass laid up for winter keep. Empire J. Exp. Agric., 1936, **4,** 368-378.

472 Thomas, B. and Dougall, H.W. The blaeberry. J. Minist. Agric., 1938-39, **45,** 546-552.

473 Thomas, B. and Dougall, H.W. The flying bent. Agriculture, London, 1939-40, **46,** 277-281.

474 Thomas, B. and Dougall, H.W. Yield of edible material from common heather. Scot. Agric., 1947-48, **27,** 35-38.

475 Thomas, B. and Dougall, W. The white bent. Scot. Agric., 1947-48, **27,** 172-175.

476 Thomas, B. and Elliott, F.J. On the yields and composition of pasture grass from the tree field plots at Cockle Park. J. Agric. Sci., 1932, **22,** 736-754.

477 Thomas, B. and Elliott, F.J. A comparison of the yields and composition of pasture grass under intensive and non-intensive manurial treatment. J. Soc. Chem. Indust., 1933, **52,** 182T-185T.

478 Thomas, B., Escritt, J.R. and Trinder, N. The minor elements of common heather *(Calluna vulgaris).* Empire J. Exp. Agric., 1945, **13,** 93-99.

479 Thomas, B. and Fairbairn, C.B. A note on the digestibility of *Nardus stricta.* J. Agric. Sci., 1957, **48,** 413-414.

480 Thomas, B., Holmes, W.B. and Clapperton, J.L. A study of meadow hays from the Cockle Park plots. I. Proximate constituents and digestibility. Empire J. Exp. Agric., 1955, **23,** 25-33.

481 Thomas, B., Holmes, W.B. and Clapperton, J.L. A study of meadow hays from the Cockle Park plots. II. Ash constituents. Empire J. Exp. Agric., 1955, **23,** 101-108.

482 Thomas, B. and Hope, R. The effect of intensive treatment on the yield and quality of grass from an exposed boulder clay pasture. J. Soc. Chem. Indust., 1936, **55,** 146T-151T.

483 Thomas, B. and Ibbotson, C.F. A note on the digestibility of *Molinia coerulea.* J. Agric. Sci., 1947, **37,** 58-59.

484 Thomas, B. and Moon, F.E. A preliminary study of the effects of manurial treatment, and of age, on the carotene-content of grass. Empire J. Exp. Agric., 1938, **6,** 235-245.

485 Thomas, B. and Rogerson, A. A study of some herb-containing swards. Scot. Agric., 1950-51, **30,** 156-160.

486 Thomas, B., Rogerson, A. and Armstrong, R.H. The influence of mineral-rich herbs on the yield and nutritive value of swards. 1. Dry matter production and composition under pasture conditions. J. Brit. Grassland Soc., 1956, **11,** 10-15.

487 Thomas, B., Rogerson, A. and Armstrong, R.H. The influence of mineral-rich herbs on the yield and nutritive value of swards. 2. Yield, composition, digestibility and feeding value under meadow conditions. J. Brit. Grassland Soc., 1956, **11,** 82-85.

488 Thomas, B. and Smith, A.N. The nutritive value of *Calluna vulgaris.* III. Digestibility at four and ten years after burning. Empire J. Exp. Agric., 1948, **16,** 221-230.

489 Thomas, B. and Thompson, A. The ash-content of some grasses and herbs on the Palace Leas hay plots at Cockle Park. Empire J. Exp. Agric., 1948, **16,** 221-230.

490 Thomas, B., Thompson, A., Oyenuga, V.A. and Armstrong, R.H. The ash constituents of some herbage plants at different stages of maturity. Empire J. Exp. Agric., 1952, **20,** 10-22.

491 Thomas, B. and Thompson, F.C. The effect of manurial treatment on the changes in composition which occur during haymaking. J. Soc. Chem. Indust., 1938, **57,** 209T-211T.

492 Thomas, B. and Trinder, N. The ash components of some moorland plants. Empire J. Exp. Agric., 1947, **15,** 237-248.

493 Thomas, B. and Walker, H.F. The B-vitamins in grass. Empire J. Exp. Agric., 1949, **17,** 170-178.

494 Thompson, A. and Raven, A.M. The availability of iron in certain grass, clover and herb species. I. Perennial ryegrass, cocksfoot and timothy. J. Agric. Sci., 1959, **52,** 177-186.

495 Thornburn, C.C. and Willcox, J.S. The caeca of the domestic fowl and digestion of the crude fibre complex. I. Digestibility trials with normal and caecectomised birds. Brit. Poultry Sci., 1965, **6,** 23-31.

496 Thorne, G.N. and Watson, D.J. The effect on yield and leaf area of wheat of applying nitrogen as a top-dressing in April or in sprays at ear emergence. J. Agric. Sci., 1955, **46,** 449-456.

497 Tilley, J.M.A. and Terry, R.A. A two-stage technique for the *in vitro* digestion of forage crops. J. Brit. Grassland Soc., 1963, **18,** 104-111.

498 Tilley, J.M.A., Terry, R.A., Deriaz, R.E. and Outen, G.E. Studies of herbage digestibility using the *in vitro* method. Grassland Research Institute, Ann. Rept. No. 16, 1962-63, p. 64.

499 Tinsley, J. and Nowakowski, T.Z. The composition and manurial value of poultry excreta, straw-droppings composts and deep litter. II. Experimental studies on composts. J. Sci. Food Agric., 1959, **10,** 150-167.

500 Tocher, J.F. Analyses for members during 1929. Trans. Highl. Agric. Soc. Scot., 1930, **42,** 211-216.

501 Tocher, J.F. Analyses for members during 1930. Trans. Highl. Agric. Soc. Scot., 1931, **43,** 197-205.

502 Tocher, J.F. Analyses for members during 1931. Trans. Highl. Agric. Soc. Scot., 1932, **44,** 195-200.

503 Tocher, J.F. Analyses for members during 1932. Trans. Highl. Agric. Soc. Scot., 1933, **45,** 181-187.

504 Tocher, J.F. Analyses for members during 1933. Trans. Highl. Agric. Soc. Scot., 1934, **46,** 285-291.

505 Tocher, J.F. Analyses for members during 1934. Trans. Highl. Agric. Soc. Scot., 1935, **47,** 201-210.

506 Tocher, J.F. Analyses for members during 1935. Trans. Highl. Agric. Soc. Scot., 1936, **48,** 219-226.

507 Tocher, J.F. Analyses for members during 1939. Trans. Highl. Agric. Soc. Scot., 1940, **52,** 154-161.

508 Tocher, J.F. Analyses for members during 1944. Trans. Highl. Agric. Soc. Scot., 1945, **57,** 96-100.

509 Waite, R. and Boyd, J. The water-soluble carbohydrates of grasses. II. Grasses cut at grazing height several times during the growing season. J. Sci. Food Agric., 1953, **4,** 257-261.

510 Waite, R., Fensom, A. and Lovett, S. The amino-acid composition of extracted grass-protein. I. The basic acids. J. Sci. Food Agric., 1953, **4**, 28-33.

511 Waite, R. and Gorrod, A.R.N. The structural carbohydrates of grasses. J. Sci. Food Agric., 1959, **10**, 308-317.

512 Waite, R. and Gorrod, A.R.N. The comprehensive analysis of grasses. J. Sci. Food Agric., 1959, **10**, 317-326.

513 Waite, R., Holmes, W., Campbell, J.I. and Fergusson, D.L. Studies in grazing management. II. The amount and chemical composition of herbage eaten by dairy cattle under close-folding and rotational methods of grazing. J. Agric. Sci., 1950, **40**, 392-402.

514 Waite, R., Johnston, M.J. and Armstrong, D.G. The evaluation of artificially dried grass as a source of energy for sheep. 1. The effect of stage of maturity on the apparent digestibility of ryegrass, cocksfoot and timothy. J. Agric. Sci., 1964, **62**, 391-398.

515 Waite, R. and Sastry, K.N.S. The composition of timothy *(Phleum pratense)* and some other grasses during seasonal growth. Empire J. Exp. Agric., 1949, **17**, 179-187.

516 Walker, D.M. and Hepburn, W.R. The nutritive value of roughages for sheep. I. The relationship between the gross digestible energy and the chemical composition of hays. J. Agric. Sci., 1954-55, **45**, 298-310.

517 Walker, D.M. and Hepburn, W.R. The nutritive value of roughages for sheep. II. The relationship between the gross digestible energy and the chemical composition of silages. J. Agric. Sci., 1956, **47**, 172-186.

518 Walker, H.F. The digestibility of oat hay. J. Agric. Sci., 1959, **53**, 289-295.

519 Walker, T.W., Edwards, G.H.A., Cavell, A.J. and Rose, T.H. The use of fertilizers on herbage cut for conservation. I. Effects on the yield of dry matter, crude protein and botanical composition of herbage cut for silage. J. Brit. Grassland Soc., 1952(a), **7**, 107-130.

520 Walker, T.W., Edwards, G.H.A., Cavell, A.J. and Rose, T.H. The use of fertilizers on herbage cut for conservation. II. Effects on the mineral composition of herbage cut for silage, and correlation of responses to phosphate and potash with soil and crop analyses. J. Brit. Grassland Soc., 1952(b), **7**, 135-150.

521 Walker, T.W., Edwards, G.H.A., Cavell, A.J. and Rose, T.H. The use of fertilizers on herbage cut for conservation. III. Effect of fertilizers and time of application on the yield, chemical and botanical composition of herbage cut for drying. J. Brit. Grassland Soc., 1953, **8**, 45-70.

522 Walsh, T. and O'Donohoe, T.F. Magnesium deficiency in some crop plants in relation to the level of potassium nutrition. J. Agric. Sci., 1945, **35**, 254-263.

523 Walters, W.G.D. Note on the mineral content of some typical North Wales pastures. Welsh Agric. J., 1933, **9**, 109-115.

524 Waterson, H.A. West of Scotland trials of barley varieties, 1957-1960. Scot. Agric., 1961-62, **41**, 122-124.

525 Waterson, H.A. and Kent, N.L. Yield and quality of spring oats in West of Scotland trials. J. Nat. Inst. Agric. Bot., 1956, **7**, 536-550.

526 Watson, C.J. and Godden, W. The comparative digestibility of artificially-dried pasture herbage by sheep and rabbits. Empire J. Exp. Agric., 1935, **3**, 346-350.

527 Watson, D.J. Field experiments on the effect of applying a nitrogenous fertilizer to wheat at different stages of growth. J. Agric. Sci., 1939, **29**, 379-398.

528 Watson, S.J. The digestibility and feeding value of dreg meal. J. Agric. Sci., 1931, **21**, 410-413.

529 Watson, S.J. Investigations into the intensive system of grassland management by the Agricultural Research Staff of Imperial Chemical Industries, Limited. IV. The digestibility and feeding value of artificially dried grass. J. Agric. Sci., 1931, **21**, 414-424.

530 Watson, S.J. Investigations into the intensive system of grassland management by the Agricultural Research Staff of Imperial Chemical Industries, Limited. V. The digestibility and feeding value of grass silage made in a tower, and the digestibility and comparative yield of artificially dried grass obtained from the same source. J. Agric. Sci., 1931, **21**, 425-441.

531 Watson, S.J. Investigations into the intensive system of grassland management by the Agricultural Research Staff of Imperial Chemical Industries, Limited. VI. The digestibility and feeding value of grass silage made in a stack. J. Agric. Sci., 1931, **21**, 452-457.

532 Watson, S.J. Investigations into the intensive system of grassland management by the Agricultural Research Staff of Imperial Chemical Industries, Limited. VII. The digestibility and feeding value of grass silage made in a pit. J. Agric. Sci., 1931, **21**, 469-475.

533 Watson, S.J. and Ferguson, W.S. Investigations into the intensive system of grassland management by the Agricultural Research Staff of Imperial Chemical Industries, Limited. VIII. The comparative digestibility and feeding value of fresh and artificially dried grass. J. Agric. Sci., 1932, **22**, 235-246.

534 Watson, S.J. and Ferguson, W.S. Investigations into the intensive system of grassland management by the Agricultural Research Staff of Imperial Chemical Industries, Limited. IX. The digestibility of artificially dried hay. J. Agric. Sci., 1932, **22**, 247-250.

535 Watson, S.J. and Ferguson, W.S. The chemical composition of grass silage. J. Agric. Sci., 1937, **27**, 1-42.

536 Watson, S.J. and Ferguson, W.S. The losses of dry matter and digestible nutrients in low-temperature silage, with and without added molasses or mineral acids. J. Agric. Sci., 1937, **27**, 67-107.

537 Watson, S.J., Ferguson, W.S. and Horton, E.A. The time of cutting hay, and the losses entailed during haymaking. J. Agric. Sci., 1937, **27**, 224-258.

538 Watson, S.J. and Horton, E.A. Composition, digestibility and nutritive value of samples of grassland products. J. Agric. Sci., 1936, **26**, 142-154.

539 Watson, S.J. and Horton, E.A. Technique of digestibility trials with sheep and its application to rabbits. Empire J. Exp. Agric., 1936, **4**, 25-35.

540 Watson, S.J., Procter, J. and Ferguson, W.S. Investigations into the intensive system of grassland management by the Agricultural Research Staff of Imperial Chemical Industries, Limited. XI. The effect of nitrogen on the yield, composition and digestibility of grassland herbage. J. Agric. Sci., 1932, **22**, 257-290.

541 Whitear, J.D., Hanley, F. and Ridgman, W.J. Studies on lucerne and lucerne-grass leys. VI. Further studies on the effect of systems of grazing management on the persistence of a lucerne-cocksfoot ley. J. Agric. Sci., 1962, **59**, 415-528.

542 Whitehead, D.C. Data on the mineral composition of grassland herbage from the Grassland Research Institute, Hurley, and the Welsh Plant Breeding Station, Aberystwyth. Grassland Research Institute, Tech. Rept. No. 4, Hurley, 1966.

543 Widdowson, F.V. The effects of nitrogen upon three stiff-strawed winter wheat varieties. J. Agric. Sci., 1959, **53**, 17-24.

544 Widdowson, F.V. and Cooke, G.W. Nitrogen fertilizers for spring barley and wheat. J. Agric. Sci., 1958, **50**, 312-321.

545 Widdowson, F.V., Penny, A. and Williams, R.J.B. Applying nitrogen fertilizers for spring barley. J. Agric. Sci., 1961, **56**, 39-45.

546 Widdowson, F.V., Penny, A. and Williams, R.J.B. Experiments with nitrogen and potash on barley. J. Agric. Sci., 1961, **57**, 29-33.

547 Widdowson, F.V., Penny, A. and Williams, R.J.B. Autumn nitrogen for winter wheat. J. Agric. Sci., 1961, **57**, 329-334.

548 Willey, L.A. and Knight, R.L. Russian comfrey. J. Nat. Inst. Agric. Bot., 1962, **9**, 139-144.

549 Williams, D.R. and Evans, R.A. Bracken *(Pteridium aquilinum)*. The effect of steaming on the nutritive value of bracken hay. Brit. J. Nutrition, 1959, **13**, 129-136.

550 Williams, R.D. and Evans, T.W. Chemical composition of various white clovers and of Italian ryegrass. Welsh J. Agric., 1932, **8**, 151-162.

551 Williams, R.J.B., Stojkovska, A., Cooke, G.W. and Widdowson, F.V. Effects of fertilisers and farmyard manure on the copper, manganese, molybdenum and zinc removed by arable crops at Rothamsted. J. Sci. Food Agric., 1960, **11**, 570-575.

552 Wilman, D. The effect of nitrogenous fertilizer on the rate of growth of Italian ryegrass. J. Brit. Grassland Soc., 1965, **20**, 248-254.

553 Wilson, R.F. and Tilley, J.M.A. Amino-acid composition of lucerne and of lucerne and grass protein preparations. J. Sci. Food Agric., 1965, **16**, 173-178.

554 Woodman, H.E. and Eden, A. Nutritive value of lucerne. III. The composition, digestibility and nutritive value of lucerne hay, lucerne meal (English and American) and lucerne leaf meal (American). J. Agric. Sci., 1935, **25**, 50-70.

555 Woodman, H.E. and Evans, R.E. The value of degermed maize meal (cooked) in the nutrition of swine. J. Agric. Sci., 1932, **22**, 670-675.

556 Woodman, H.E. and Evans, R.E. Nutritive value of lucerne. IV. The leaf-stem ratio. J. Agric. Sci., 1935, **25**, 578-597.

557 Woodman, H.E. and Evans, R.E. The composition and digestibility, when fed to pigs, of three grades of meat meal of widely differing fat content. J. Agric. Sci., 1937, **27**, 465-473.

558 Woodman, H.E. and Evans, R.E. The nutrition of the bacon pig. VI. The minimum level of protein intake consistent with quick growth and satisfactory carcass quality (Part III). J. Agric. Sci., 1941, **31**, 232-245.

559 Woodman, H.E. and Evans, R.E. The nutrition of the bacon pig. VII. The chemical composition, digestibility and nutritive value of different types of swill. J. Agric. Sci., 1942, **32**, 85-107.

560 Woodman, H.E. and Evans, R.E. Further investigations of the feeding value of artificially dried potatoes: the composition and nutritive value of potato cossettes, potato meal, potato flakes, potato slices and potato dust. J. Agric. Sci., 1943, **33**, 1-14.

561 Woodman, H.E. and Evans, R.E. The nutrition of the bacon pig. VIII. The value of lawn-grass cuttings in the feeding of bacon pigs. J. Agric. Sci., 1943, **33**, 101-112.

562 Woodman, H.E. and Evans, R.E. The nutrition of the bacon pig. IX. The Lehmann method of pig feeding, with particular reference to the balance of the basal meal and the use of cooked potatoes and molassed beet pulp as the supplemental foods. J. Agric. Sci., 1943, **33**, 155-168.

563 Woodman, H.E. and Evans, R.E. The influence of war-time milling control on the composition, digestibility and nutritive value of the wheaten offals. J. Agric. Sci., 1944, **34**, 35-47.

564 Woodman, H.E. and Evans, R.E. The chemical composition and nutritive value of the pea-canning by-products (green pea pods, pea-pod meal, pea-pod silage and molassed silage from pea haulms with pods). J. Agric. Sci., 1944, **34**, 155-164.

565 Woodman, H.E. and Evans, R.E. The nutrition of the bacon pig. X. The value of extracted palm-kernel meal in the feeding of the bacon pig. J. Agric. Sci., 1945, **35**, 44-55.

566 Woodman, H.E. and Evans, R.E. Nutrition of the bacon pig. XI. The minimum level of protein intake consistent with quick growth and satisfactory carcass quality. (Part IV). J. Agric. Sci., 1945, **35**, 133-149.

567 Woodman, H.E. and Evans, R.E. The horse-chestnut as a source of food for livestock. I. The composition, digestibility and nutritive value of alcohol-extracted horse-chestnut meal, water-extracted horse-chestnut meal and horse-chestnut residue. J. Agric. Sci., 1946, **36**, 29-41.

568 Woodman, H.E. and Evans, R.E. The horse-chestnut as a source of food for livestock. II. The value of alcohol-extracted horse-chestnut meal, water-extracted horse-chestnut meal and horse-chestnut residue as a partial substitute for cereals in the rations of bacon pigs. J. Agric. Sci., 1946, **36**, 42-55.

569 Woodman, H.E. and Evans, R.E. The nutritive value for pigs and ruminants of dried penicillin felt. J. Agric. Sci., 1947, **37**, 81-93.

570 Woodman, H.E. and Evans, R.E. The nutritive value of fodder cellulose from wheat straw. I. Its digestibility and feeding value when fed to ruminants and pigs. J. Agric. Sci., 1947, **37**, 202-210.

571 Woodman, H.E. and Evans, R.E. The nutritive value of fodder cellulose from wheat straw. II. The utilization of cellulose by growing and fattening pigs. J. Agric. Sci., 1947, **37**, 211-223.

572 Woodman, H.E. and Evans, R.E. The chemical composition and nutritive value of ryegrass-seed meal, clover-seed meal, lucerne-seed meal and sainfoin-seed meal. J. Agric. Sci., 1947, **37**, 311-315.

573 Woodman, H.E. and Evans, R.E. The nutrition of the bacon pig. XII. The value of dried grass meal for growing and fattening pigs. J. Agric. Sci., 1948, **38**, 51-63.

574 Woodman, H.E. and Evans, R.E. The composition and nutritive value, when fed to pigs, of whale-meat meal, white-fish meal, feeding-meat meal, extracted decorticated groundnut meal, bean meal and dried yeast. J. Agric. Sci., 1948, **38**, 200-206.

575 Woodman, H.E. and Evans, R.E. Nutrition of the bacon pig. XIII. The minimum level of protein intake consistent with the maximum rate of growth (Part V). J. Agric. Sci., 1948, **38**, 354-365.

576 Woodman, H.E. and Evans, R.E. Nutrition of the bacon pig. XIV. The determination of the relative supplemental values of vegetable protein (extracted, decorticated ground-nut meal) and animal protein (white-fish meal). J. Agric. Sci., 1951, **41**, 102-140.

577 Woodman, H.E., Evans, R.E. and Eden, A. The composition and nutritive value of marrow stem kale and thousand head kale. J. Agric. Sci., 1936, **26**, 212-238.

578 Woodman, H.E., Evans, R.E. and Eden, A. Sheep nutrition. II. Determination of the amounts of grass consumed by sheep on pasturage of varying quality. J. Agric. Sci., 1937, **27**, 212-223.

579 Woodman, H.E., Evans, R.E. and Kitchin, A.W.M. The value of oats in the nutrition of swine. J. Agric. Sci., 1932, **22**, 657-669.

580 Woodman, H.E., Evans, R.E. and Norman, D.B. Nutritive value of lucerne. I. Preliminary studies of yield, composition and nutritive value (season 1932). J. Agric. Sci., 1933, **23**, 419-458.

581 Woodman, H.E., Evans, R.E. and Norman, D.B. Nutritive value of lucerne. II. Investigations into the influence of systematic cutting at three different stages of growth on the yield, composition and nutritive value of lucerne. J. Agric. Sci., 1934, **24**, 283-311.

582 Woodman, H.E., Kitchin, A.W.M. and Evans, R.E. The value of tapioca flour and sago pith meal in the nutrition of swine J. Agric. Sci., 1931, **21**, 526-546.

583 Woodman, H.E. and Norman, D.B. Nutritive value of pasture. IX. The influence of the intensity of grazing on the yield, composition and nutritive value of pasture herbage (Part IV). J. Agric. Sci., 1932, **22**, 852-873.

584 Woodman, H.E. and Norman, D.B. Nutritive value of pasture. X. The utilisation of young grass by swine. J. Agric. Sci., 1934, **24**, 93-103.

585 Woodman, H.E., Norman, D.B. and French, M.H. Nutritive value of pasture. VII. The influence of the intensity of grazing on the yield, composition and nutritive value of pasture herbage (Part III). J. Agric. Sci., 1931, **21**, 267-323.

586 Woodman, H.E. and Oosthuizen, P.M. Nutritive value of pasture. XI. The composition and nutritive value of winter pasturage. J. Agric. Sci., 1934, **24**, 574-597.

587 Woodman, H.E. and Underwood, E.J. Nutritive value of pasture. VIII. The influence of intensive fertilising and composition of good permanent pasture (seasons 1 and 2). J. Agric. Sci., 1932, **22**, 26-71.

588 Woodward, N. Seaweeds as a source of chemicals and stock feed. J. Sci. Food Agric., 1951, **2**, 477-487.

589 Wylam, C.B. Analytical studies on the carbohydrates of grasses and clovers. III. Carbohydrate breakdown during wilting and ensilage. J. Sci. Food Agric., 1953, **4**, 527-531.

590 Zaleski, A. and Dent, J.W. Leafiness, chemical composition and yield of some lucerne varieties. J. Brit. Grassland Soc., 1960, **15**, 21-27.

591 Borthwick, Thos. (Glasgow) Ltd.

592 Scottish Malt Distillers, Borough Briggs, Elgin. 1967.

593 Silcock, R. and Sons Ltd., Liverpool. 1966. Note on definition of Distillery By-Products.

594 Ministry of Agriculture, Fisheries and Food; National Agricultural Advisory Service: "Closed" Conference of Advisory Nutrition Chemists. NC/C/233. September, 1963.

595 United Nations. Dept. of Economic and Social Affairs. Strategy statement on action to avert the protein crisis in the developing countries.
Report of the Panel of Experts on The Protein Problem confronting Developing Countries. United Nations Headquarters, 3-7 May 1971.
United Nations. New York, 1971.

United Nations Publications
Sales No.: E.71.II.A.17

E/5018/Rev.1
St/ECA/144

Index to Tables, Part 1

2S

PART TWO

Field data

Note: Throughout this part, the conventions have been adopted of including the SD when 20 or more samples are present and of replacing the maximum and minimum by the upper and lower decile points if 60 or more samples are represented.

2.1 *AVENA:* OAT, GRAIN

	Mean	*Min.*	*Max.*	*SD*	*No.*
Dry matter %	84.3	73.2	91.5	4.35	753
Proximate constituents % of DM					
Crude protein	11.1	9.1	13.3	1.45	905
Ether extract	5.39	3.96	6.67	1.07	278
	12.5	9.1	16.6	2.44	476
Major elements % of DM					
Calcium	0.14	0.06	0.34	0.06	88
Phosphorus	0.36	0.24	0.45	0.05	88
Magnesium	0.11	0.10	0.12		10
Potassium	0.50	0.42	0.56	0.14	493
Sulphur	0.11	0.01	0.20		2
Trace elements μg per g DM					
Barium	3.52	0.10	18.0	2.68	166
Chromium	0.16	0.07	0.34	0.07	162
Cobalt	0.13	0.01	0.25	0.10	176
Copper	2.90	0.73	9.22	1.14	441
Iron	65	16	439	39.1	192
Lead	1.28	0.55	14.0	2.06	162
Manganese	65	7	317	51.5	411
Molybdenum	0.3	0.03	1.64	0.24	169
Nickel	1.74	0.42	15.0	3.29	167
	0.06	0.03	0.09	0.04	162
	3.55	0.90	29.0	2.73	166
	0.77	0.29	7.60	0.98	162
	2.59	0.64	29.0	3.08	163
	0.08	0.02	0.40	0.07	161
	41	5	214	32.9	163

2.2 OAT, GRAIN, DEHUSKED

	Mean	*Min.*	*Max.*	*SD*	*No.*
Dry matter %	86.4	85.6	88.0		3
Proximate constituents % of DM					
Crude protein	11.2	11.0	11.3		3
Ether extract	5.3	5.2	5.5		3
Crude fibre	3.4	2.6	3.8		3

2.3 OAT, GROATS *

	Mean	*Min.*	*Max.*	*SD*	*No.*
Dry matter %	92.0	91.2	92.8		5
Proximate constituents % of DM					
Crude protein	13.4	13.0	14.1		5
Ether extract	7.3	6.8	8.0		5
Nitrogen-free extract	75.6	74.4	76.3		5
Crude fibre	2.0	0.9	2.4		5
Total ash	1.7	0.9	2.3		5

* This is the kernel of the oat grain

2.4 OAT, PINHEAD MEAL *

	Mean	*Min.*	*Max.*	*SD*	*No.*
Dry matter %	92.1	91.0	93.2		5
Proximate constituents % of DM					
Crude protein	12.7	12.5	13.1		5
Ether extract	7.5	7.0	7.8		5
Nitrogen-free extract	76.2	75.8	77.0		5
Crude fibre	1.8	1.3	2.0		5
Total ash	1.8	1.7	1.9		5

OAT, MEDIUM MEAL *

	Mean	*Min.*	*Max.*	*SD*	*No.*
Dry matter %	92.6	92.1	93.4		5
Proximate constituents % of DM					
Crude protein	13.1	12.8	13.5		5
Ether extract	7.7	7.0	8.4		5
Nitrogen-free extract	75.4	74.7	75.9		5
Crude fibre	1.8	1.2	2.2		5
Total ash	2.0	1.7	2.2		5

OAT, FINE MEAL *

	Mean	*Min.*	*Max.*	*SD*	*No.*
Dry matter %	91.9	91.3	92.5		5
Proximate constituents % of DM					
Crude protein	12.8	12.2	13.7		5
Ether extract	8.2	7.9	8.5		5
Nitrogen-free extract	75.5	74.7	76.1		5
Crude fibre	1.6	1.5	1.7		5
Total ash	1.9	1.6	2.2		5

* Pinhead, Medium and Fine Meal are reasonably well defined grades of oatmeal classification according to their granular size. Most mills produce three grades of meal but particle size and range of particle size vary from mill to mill.

2.5 OAT FLAKE *

	Mean	*Min.*	*Max.*	*SD*	*No.*
Dry matter %	89.7	88.6	90.7		5
Proximate constituents % of DM					
Crude protein	12.6	12.3	13.1		5
Ether extract	8.0	6.5	8.6		5
Nitrogen-free extract	75.8	74.7	77.1		5
Crude fibre	1.8	1.2	2.0		5
Total ash	1.8	1.5	2.0		5

* Product formed from cooked pinhead meal by passing the meal whilst still very hot between two rollers exerting a heavy pressure on the particle and usually referred to as heavy duty roll.

2.6 OAT, CRUSHED

	Mean	Min.	Max.	SD	No.
Dry matter %	81.9	77.2	86.1		9
Proximate constituents % of DM					
Crude protein	14.6	10.0	19.6		9
Ether extract	2.8	1.1	4.6		2
Crude fibre	9.0	4.0	15.2		9
Major elements % of DM					
Calcium	0.39	0.11	1.23		4
Phosphorus	0.57	0.37	1.09		4
Magnesium	0.36				1

2.7 OAT, FLOW MEAL *

	Mean	Min.	Max.	SD	No.
Dry matter %	91.5	87.4	93.0		11
Proximate constituents % of DM					
Crude protein	12.8	11.6	14.0		11
Ether extract	8.64	7.2	10.2		11
Nitrogen-free extract	72.2	71.2	73.1		3
Crude fibre	2.07	0.9	3.8		11
Total ash	2.53	2.4	2.8		3

* The chippings from the groat when it is passed through the cutting process in making Oatmeal. The chippings are screened off.

2.8 OAT DUST *

	Mean	Min.	Max.	SD	No.
Dry matter %	92.3	89.2	94.8		8
Proximate constituents % of DM					
Crude protein	10.6	4.9	15.5		8
Ether extract	6.31	2.9	12.1		8
Nitrogen-free extract	59.6	53.1	67.6		3
Crude fibre	15.9	11.3	29.1		8
Total ash	5.30	4.3	6.0		3

* The material lying between the groat and the husk, which, in the viable grain appears as oat hairs. (The husk and the dust or hairs come off together and are then separated by screening). Usual proportions husk 20% of the grain and the dust 5%.

2.9 OAT BY-PRODUCT *

	Mean	Min.	Max.	SD	No.
Dry matter %	93.7	92.2	94.9	0.93	51
Proximate constituents % of DM					
Crude protein	6.2	4.8	7.4	1.22	51
Ether extract	3.3	2.7	4.2	0.62	51
Nitrogen-free extract	60.8	57.8	63.0	2.51	51
Crude fibre	25.4	22.2	29.2	2.89	51
Total ash	4.2	3.6	4.7	0.73	51

* By-product of oatmeal milling consisting of floury materials, milling materials, scree dust, all finely ground and containing not more than about 27% fibre.

2.10 OAT FEED

	Mean	Min.	Max.	SD	No.
Dry matter %	89.4	88.0	91.2	1.23	60
Proximate constituents % of DM					
Crude protein	3.4	2.5	4.3	0.70	60
Ether extract	1.3	0.7	1.9	0.43	60
Nitrogen-free extract	58.4	56.6	60.2	1.42	60
Crude fibre	32.2	29.9	33.3	1.61	60
Total ash	4.6	4	5.2	0.42	60

2.11 OAT, RAW COCKLE *

	Mean	Min.	Max.	SD	No.
Dry matter %	89.8	86.2	90.9		8
Proximate constituents % of DM					
Crude protein	12.0	11.2	12.5		8
Ether extract	8.4	5.6	12.0		8
Nitrogen-free extract	68.1	67.2	68.7		3
Crude fibre	4.1	1.5	7.6		8
Total ash	4.7	2.6	6.1		3

* The material which separates out on cleaning the oats for removal of other cereals, wheat, barley, etc. and groats themselves. The amount of this material would be about 2½% of the original grain and of this 2½% some 30% would be groats.

2.12 *TRITICUM*, WHEAT, GRAIN, ATLE

	Mean	Min.	Max.	SD	No.
Dry matter %	82.0	81.2	82.8		7
Proximate constituents % of DM					
Crude protein	14.2	13.0	15.6		7
Major elements % of DM					
Calcium	0.06				1
Phosphorus	0.45				1

2.13 WHEAT, GRAIN, ATSON

	Mean	Min.	Max.	SD	No.
Dry matter %	81.2	78.5	83.8		6
Proximate constituents % of DM					
Crude protein	14.6	14.0	15.5		6

2.14 WHEAT, GRAIN, CAPPELLE DESPREZ

	Mean	Min.	Max.	SD	No.
Dry matter %	77.2	70.6	80.5		6
Proximate constituents % of DM					
Crude protein	14.5	12.9	15.5		6

2.15 WHEAT, GRAIN, DOMINATOR

	Mean	Min.	Max.	SD	No.
Dry matter %	83.5	76.9	88.4		10
Proximate constituents % of DM					
Crude protein	12.5	9.7	14.5		10

2.16 WHEAT, GRAIN, ELITE LEPEUPLE

	Mean	Min.	Max.	SD	No.
Dry matter %	85.4	84.0	86.2		7
Proximate constituents % of DM					
Crude protein	11.7	9.8	14.7		7

2.17 WHEAT, GRAIN, HEINES VII

	Mean	Min.	Max.	SD	No.
Dry matter %	82.2	80.5	85.2		7
Proximate constituents % of DM					
Crude protein	13.6	12.3	15.0		7

2.18 WHEAT, GRAIN HYBRID 46

	Mean	Min.	Max.	SD	No.
Dry matter %	80.6	79.3	82.0		10
Proximate constituents % of DM					
Crude protein	13.5	12.2	14.9		10
Major elements % of DM					
Calcium	0.04				1
Phosphorus	0.46				1

2.19 WHEAT, GRAIN, MILFAST

	Mean	Min.	Max.	SD	No.
Dry matter %	83.9	78.3	86.9		9
Proximate constituents % of DM					
Crude protein	12.9	10.4	15.1		9

2.20 WHEAT, GRAIN, MINISTER

	Mean	Min.	Max.	SD	No.
Dry matter %	81.0	79.1	82.4		9
Proximate constituents % of DM					
Crude protein	13.7	12.6	15.2		9

2.21 WHEAT, GRAIN, REDMAN

	Mean	Min.	Max.	SD	No.
Dry matter %	82.2	81.1	83.5		9
Proximate constituents % of DM					
Crude protein	14.9	13.5	16.7		9

2.22 WHEAT, GRAIN, SVENNO

	Mean	Min.	Max.	SD	No.
Dry matter %	81.8	79.3	84.6		6
Proximate constituents % of DM					
Crude protein	15.5	14.2	16.9		6

2.23 WHEAT, GRAIN, TADEPI

	Mean	Min.	Max.	SD	No.
Dry matter %	78.3	75.2	80.8		6
Proximate constituents % of DM					
Crude protein	14.3	12.5	15.1		6

2.24 WHEAT, GRAIN, UNNAMED

	Mean	Min.	Max.	SD	No.
Dry matter %	83.8	73.4	96.4	4.83	525
Proximate constituents % of DM					
Crude protein	13.6	11.6	15.8	1.66	2094
Ether extract	1.98	1.5	2.7		10
Crude fibre	2.8	1.8	5.3	0.50	52
Major elements % of DM					
Calcium	0.06	0.04	0.10		5
Phosphorus	0.42	0.37	0.47	0.04	461
Magnesium	0.15	0.12	0.17		2
Potassium	0.50	0.44	0.56	0.07	495
Trace elements μg per g DM					
Barium	5.73	2.57	8.60		5
Chromium	0.19	0.13	0.25		5
Cobalt	0.04	0.03	0.06		5
Copper	5.26	2.40	9.74		11
Iron	37	24	50		5
Lead	1.28	0.75	3.20		5
Manganese	34	16	65		9
Molybdenum	0.21	0.13	0.40		5
Nickel	0.23	0.16	0.28		5
Silver	0.06	0.05	0.07		5
Strontium	2.68	1.17	5.10		5
Tin	0.60	0.49	0.67		5
Titanium	2.87	0.78	8.60		5
Vanadium	0.22	0.08	0.38		5
Zinc	42	31	52		5

2.25 WHEAT MEAL

	Mean	Min.	Max.	SD	No.
Dry matter %	84.5	83.4	85.5	0.92	47
Proximate constituents % of DM					
Crude protein	14.1	12.5	15.8	1.42	47
Ether extract	1.8	1.4	2.2	0.46	47
Nitrogen-free extract	80.4	77.4	82.6	2.12	47
Crude fibre	2.0	1.3	2.9	0.59	47
Total ash	1.6	1.2	2.5	0.66	47

2.26 WHEAT FEED *

	Mean	Min.	Max.	SD	No.
Dry matter %	85.9	85.2	86.6	0.78	62
Proximate constituents % of DM					
Crude protein	18.3	16.9	19.4	0.94	62
Ether extract	4.0	3.2	4.7	0.80	62
Nitrogen-free extract	63.5	61.0	66.6	2.26	62
Crude fibre	9.1	7.4	10.5	1.20	62
Total ash	5.0	4.4	5.4	0.44	62

* This is the total offal. In addition to the part of the grain which is made into flour, a fairly large proportion of germ is also omitted, or extracted.

2.27 WHEAT, POLLARDS * U.K.

	Mean	Min.	Max.	SD	No.
Dry matter %	88.0	86.1	89.7	1.32	60
Proximate constituents % of DM					
Crude protein	17.4	16.2	18.4	1.03	60
Ether extract	3.73	3.0	4.2	0.44	60
Nitrogen-free extract	63.2	60.8	65.6	2.12	60
Crude fibre	9.9	7.6	12.7	1.78	60
Total ash	5.8	4.9	6.6	0.67	60
Insol. ash	0.37	0.11	0.68	0.20	60
Major elements % of DM					
Calcium	0.26	0.17	0.32		4
Phosphorus	0.90	0.67	1.09		4
Magnesium	0.41	0.29	0.52		8
Chlorine	0.18	0.11	0.23	0.082	60
Trace elements μg per g DM					
Copper	9.3	7.0	14.0		3
Manganese	115	112	119		2
Zinc	98	54	130		9

* Consists of medium type millers offals.

2.28 WHEAT, POLLARDS PELLETS U.K.

	Mean	Min.	Max.	SD	No.
Dry matter %	88.4	87.7	88.9		12
Proximate constituents % of DM					
Crude protein	19.0	18.3	19.6		12
Ether extract	4.0	3.8	4.5		12
Nitrogen-free extract	61.5	59.6	62.8		11
Crude fibre	9.3	8.0	10.3		12
Total ash	6.2	5.6	6.9		11
Carbohydrate fractions % of DM					
Starch	25.5				1

2.29 WHEAT, POLLARDS, GERMANY

	Mean	Min.	Max.	SD	No.
Dry matter %	88.0	87.2	89.0	0.69	51
Proximate constituents % of DM					
Crude protein	16.0	14.2	17.5	1.92	51
Ether extract	3.8	3.1	5.0	0.70	51
Nitrogen-free extract	69.8	66.0	71.2	2.21	51
Crude fibre	6.1	4.2	7.8	1.56	51
Total ash	4.3	3.7	5.0	0.61	51
Insol. ash	0.2	0.1	0.3	0.13	51
Major elements % of DM					
Calcium	0.18	0.09	0.34		8
Phosphorus	0.80	0.57	1.12		8
Magnesium	0.32	0.22	0.45		8
Chlorine	0.06	0.03	0.12	0.06	51

2.30 WHEAT, POLLARDS, NIGERIA

	Mean	Min.	Max.	SD	No.
Dry matter %	89.0	87.5	90.2	0.93	48
Proximate constituents % of DM					
Crude protein	19.0	18.6	19.9	0.66	48
Ether extract	4.0	3.5	4.5	0.37	48
Nitrogen-free extract	61.2	59.8	62.6	1.05	48
Crude fibre	9.6	8.5	10.6	0.84	48
Total ash	6.2	5.8	6.6	0.46	48
Insol. ash	0.4	0.2	0.6	0.13	48
Major elements % of DM					
Calcium	0.29	0.28	0.30		3
Phosphorus	1.06	1.01	1.12		3
Magnesium	0.37				1
Chlorine	0.14	0.11	0.23	0.09	48

2.31 WHEAT, POLLARDS, PLATE

	Mean	Min.	Max.	SD	No.
Dry matter %	89.1	87.9	90.1	0.90	75
Proximate constituents % of DM					
Crude protein	19.4	17.7	20.8	1.16	75
Ether extract	3.6	3.2	3.9	0.31	75
Nitrogen-free extract	60.9	59.3	62.3	1.36	75
Crude fibre	9.7	8.9	10.8	1.01	75
Total ash	6.4	5.5	7.5	0.76	75
Insol. ash	0.43	0.11	0.77	0.23	75
Major elements % of DM					
Chlorine	0.13	0.11	0.22	0.061	63

2.32 WHEAT, COMMON THIRDS *

	Mean	Min.	Max.	SD	No.
Dry matter %	86.1	85.3	87.1	0.69	60
Proximate constituents % of DM					
Crude protein	19.0	17.2	20.6	1.30	60
Ether extract	4.5	3.7	5.3	0.58	60
Nitrogen-free extract	62.2	59.6	65.8	2.31	60
Crude fibre	9.3	7.4	10.8	1.28	60
Total ash	5.0	4.3	5.6	0.56	60

WHEAT, FINE THIRDS *

Dry matter %	86.3	85.4	87.5	0.72	60
Proximate constituents % of DM					
Crude protein	19.1	17.0	20.8	1.41	60
Ether extract	4.3	3.5	5.1	0.61	60
Nitrogen-free extract	69.1	65.5	73.2	2.79	60
Crude fibre	4.2	2.8	5.6	1.04	60
Total ash	3.2	2.4	3.9	0.56	60

* These two grades are reasonably regular throughout the trade as far as names are concerned, but the condition of the two items can vary considerably both from the point of view of protein and fibre. Both products are almost of flour particle size. The Common Thirds being slightly more granular than the Fine Thirds.

2.33 WHEAT, SCREENINGS *

	Mean	Min.	Max.	SD	No.
Dry matter %	87.0	85.9	88.5		18
Proximate constituents % of DM					
Crude protein	12.5	8.4	15.2		18
Ether extract	3.4	2.1	5.6		18
Nitrogen-free extract	67.0	58.7	73.7		18
Crude fibre	13.3	5.8	25.1		18
Total ash	3.9	2.7	6.7		18

* Screenings can vary considerably from mill to mill depending on the types of machines used to clean and extract material other than wheat from the wheat being prepared for milling. Efficiency of different machines can vary and the intensification of cleaning can vary from mill to mill. The screenings consist of a miscellaneous collection of weed seeds, wild oats, normal oats, barley, maize, rye, broken wheats, very small straws and chaff, all of which is ground through a Christie and Norris type grinder. The screens used vary considerably and the product is normally blended in either Common or Fine Thirds so that the foreign matter complies with the Ministry Regulations of under 4% material other than wheat.

2.34 WHEAT, HUSK, CUBED

	Mean	Min.	Max.	SD	No.
Dry matter %	85.1				1
Proximate constituents % of DM					
Crude protein	4.9				1
Ether extract	0.94				1
Crude fibre	22.8				1
Major elements % of DM					
Calcium	1.46				1
Phosphorus	0.35				1

2.35 WHEAT, BRAN, STRAIGHT RUN

	Mean	Min.	Max.	SD	No.
Dry matter %	84.8	83.3	86.6	1.10	30
Proximate constituents % of DM					
Crude protein	17.2	16.2	18.3	0.67	30
Ether extract	3.7	3.1	4.3	0.56	30
Nitrogen-free extract	60.4	58.4	62.3	1.49	30
Crude fibre	12.3	11.3	13.1	0.65	30
Total ash	6.3	5.3	6.8	0.85	30

2.36 WHEAT, BRAN, FLAKE

	Mean	Min.	Max.	SD	No.
Dry matter %	84.6	83.9	85.2		4
Proximate constituents % of DM					
Crude protein	17.2	16.1	18.8		4
Ether extract	2.9	2.4	3.4		4
Nitrogen-free extract	62.7	61.0	64.1		4
Crude fibre	11.8	10.8	12.5		4
Total ash	5.4	5.2	5.6		4

2.36 WHEAT, BRAN, FINE * (continued)

	Mean	Min.	Max.	SD	No.
Dry matter %	84.8	83.6	86.1		7
Proximate constituents % of DM					
Crude protein	17.9	16.1	19.7		7
Ether extract	3.4	2.7	4.0		7
Nitrogen-free extract	61.7	57.8	67.6		7
Crude fibre	11.5	8.7	13.2		7
Total ash	5.5	3.8	6.9		7

* Flake Bran is formed by damping normal straight run bran from the mill and passing this through smooth rolls and then over a sieve. The material or flake larger than the mesh of the sieve is bagged off as flake bran and the fine material passing from the sieve is bagged off and labelled fine bran.

2.37 *HORDEUM:* BARLEY, GRAIN, CARLSBERG

	Mean	Min.	Max.	SD	No.
Dry matter %	82.7	74.0	89.0	3.2	21
Proximate constituents % of DM					
Crude protein	12.2	9.5	18.1	1.8	21

2.38 BARLEY, GRAIN, DOMEN

	Mean	Min.	Max.	SD	No.
Dry matter %	79.3	74.3	83.6	2.72	42
Proximate constituents % of DM					
Crude protein	13.3	11.0	16.4	2.00	42

2.39 BARLEY, GRAIN, EARL

	Mean	Min.	Max.	SD	No.
Dry matter %	81.5	74.3	90.0	3.30	26
Proximate constituents % of DM					
Crude protein	12.6	9.3	16.1	2.12	26

2.40 BARLEY, GRAIN, HERTA

	Mean	Min.	Max.	SD	No.
Dry matter %	81.1	77.1	88.9	2.25	43
Proximate constituents % of DM					
Crude protein	12.5	10.4	15.3	1.40	43
Major elements % of DM					
Calcium	0.07				1
Phosphorus	0.32				1
Magnesium	0.08				1
Trace elements µg per g DM					
Cobalt	0.02				1
Copper	6.5				1

2.41 BARLEY, GRAIN, INGRID

	Mean	Min.	Max.	SD	No.
Dry matter %	80.6	75.4	84.5	3.27	30
Proximate constituents % of DM					
Crude protein	13.1	10.0	15.2	1.85	30

2.42 BARLEY, GRAIN, KENIA

	Mean	Min.	Max.	SD	No.
Dry matter %	85.9	82.3	88.7		9
Proximate constituents % of DM					
Crude protein	9.9	8.7	11.6		9

2.43 BARLEY, GRAIN, MAYTHORPE

	Mean	Min.	Max.	SD	No.
Dry matter %	81.1	76.7	85.9	4.04	80
Proximate constituents % of DM					
Crude protein	12.2	9.8	14.6	1.77	80

2.44 BARLEY, GRAIN, PIONEER

	Mean	Min.	Max.	SD	No.
Dry matter %	82.7	79.5	85.4		11
Proximate constituents % of DM					
Crude protein	10.4	8.7	12.2		11
Ether extract	1.89				1
Crude fibre	5.1				1
Major elements % of DM					
Calcium	0.08				1
Phosphorus	0.34				1
Magnesium	0.09				1
Trace elements µg per g DM					
Copper	3.0				1
Manganese	55				1
Zinc	0.58				1

2.45 BARLEY, GRAIN, PLUMAGE ARCHER

	Mean	Min.	Max.	SD	No.
Dry matter %	81.0	73.4	89.0	4.29	21
Proximate constituents % of DM					
Crude protein	12.9	9.5	16.9	2.37	21

2.46 BARLEY, GRAIN, PROCTOR

	Mean	Min.	Max.	SD	No.
Dry matter %	83.9	78.7	88.8	4.0	254
Proximate constituents % of DM					
Crude protein	11.8	9.6	14.4	1.8	254
Crude fibre	5.3	5.0	5.6		4

2.47 BARLEY, GRAIN, PROVOST

	Mean	Min.	Max.	SD	No.
Dry matter %	80.9	75.1	84.2	3.03	57
Proximate constituents % of DM					
Crude protein	12.6	7.8	15.6	1.67	57

2.48 BARLEY, GRAIN, RIKA

	Mean	Min.	Max.	SD	No.
Dry matter %	81.4	77.1	83.8	3.7	63
Proximate constituents % of DM					
Crude protein	11.9	10.2	14.1	1.4	63
Crude fibre	5.3	4.7	5.9		2
Major elements % of DM					
Calcium	0.07				1
Phosphorus	0.38				1
Magnesium	0.11				1
Trace elements µg per g DM					
Cobalt	0.02				1
Copper	7.0				1

2.49 BARLEY, GRAIN, SPRATT ARCHER

	Mean	Min.	Max.	SD	No.
Dry matter %	83.5	76.7	88.7		14
Proximate constituents % of DM					
Crude protein	10.5	7.9	15.9		14

2.50 BARLEY, GRAIN, UNION

	Mean	Min.	Max.	SD	No.
Dry matter %	83.5	81.7	85.1		6
Proximate constituents % of DM					
Crude protein	11.8	11.2	12.5		6

2.51 BARLEY GRAIN, 1001

	Mean	Min.	Max.	SD	No.
Dry matter %	84.3	79.9	88.7		6
Proximate constituents % of DM					
Crude protein	12.9	11.2	14.8		6

2.52 BARLEY GRAIN, 1015

	Mean	Min.	Max.	SD	No.
Dry matter %	85.6	82.1	89.0		6
Proximate constituents % of DM					
Crude protein	11.5	9.4	13.5		6

2.53 BARLEY, GRAIN, UNNAMED

	Mean	Min.	Max.	SD	No.
Dry matter %	81.6	66.1	93.3	6.05	1079
Proximate constituents % of DM					
Crude protein	11.5	9.5	13.9	1.8	1274
Ether extract	2.03	1.57	2.62	0.23	30
Crude fibre	5.4	4.6	6.7	0.93	142
Major elements % of DM					
Calcium	0.08	0.04	0.12	0.02	35
Phosphorus	0.38	0.30	0.44	0.04	598
Magnesium	0.14	0.13	0.14		2
Potassium	0.35	0.28	0.41		2
Sodium	0.11				1
Trace elements µg per g DM					
Barium	4.8	2.2	9.8		5
Chromium	0.11	0.06	0.25		15
Cobalt	0.03	0.01	0.05	0.008	20
Copper	7.6	0.9	83.0	12.9	36
Iodine	0.11	0.09	0.12		2
Iron	43	33	69		17
Lead	1.04	0.63	1.69		15
Manganese	28	9.9	90	21.0	40
Molybdenum	0.45	0.22	1.36		15
Nickel	0.24	0.08	0.69		15
Silver	0.05	0.05	0.06		15
Strontium	2.1	0.6	3.8		7
Tin	0.65	0.50	0.94		15
Titanium	1.20	0.62	3.20		15
Vanadium	0.05	0.03	0.09		15
Zinc	28	20	39		15

2.54 BARLEY, GROUND

	Mean	Min.	Max.	SD	No.
Dry matter %	82.5	82.0	83.0		2
Proximate constituents % of DM					
Crude protein	9.9	8.8	11.0		2
Ether extract	1.64				1
Nitrogen-free extract	82.1				1
Crude fibre	4.3	3.9	4.7		2
Total ash	2.10				1
Major elements % of DM					
Calcium	0.08				1
Phosphorus	0.36				1

2.55 BARLEY, CRUSHED

	Mean	Min.	Max.	SD	No.
Dry matter %	83.5	81.8	85.1		2
Proximate constituents % of DM					
Crude protein	8.8	8.7	8.9		2
Ether extract	2.2				1
Crude fibre	4.8	4.3	5.3		2

2.56 BARLEY DUST *

	Mean	Min.	Max.	SD	No.
Dry matter %	87.9	87.0	89.0	0.74	40
Proximate constituents % of DM					
Crude protein	12.9	11.0	14.4	1.34	40
Ether extract	4.8	4.0	5.6	0.63	40
Nitrogen-free extract	67.0	63.1	71.8	3.55	40
Crude fibre	11.0	8.3	13.0	2.60	40
Total ash	4.3	3.7	5.2	0.77	40

* In the milling of barley, to produce both Pot and Pearl Barley by passing it over emery stones which rub on the grain to remove the outer skins, an offal is produced which consists of floury matter with a variable proportion of barn fragments through it. This latter product is referred to as barley dust.

2.57 BARLEY RESIDUES, DISTILLERS' WET GRAINS

	Mean	Min.	Max.	SD	No.
Dry matter %	27.3	20.0	31.9	2.93	26
Proximate constituents % of DM and pH					
Crude protein	20.1	17.3	23.2	1.45	25
Ether extract	7.03	5.01	9.19	0.77	25
Crude fibre	18.2	15.4	20.9	1.36	30
pH	4.6				1
Major elements % of DM					
Calcium	0.13	0.05	0.26		6
Phosphorus	0.33	0.18	0.41		6
Magnesium	0.10	0.03	0.18		4
Trace elements μg per g DM					
Copper	27				1
Zinc	15				1

2.58 BARLEY RESIDUES, DISTILLERS' DRY GRAINS

	Mean	Min.	Max.	SD	No.
Dry matter %	90.9	89.5	92.5		7
Proximate constituents % of DM					
Crude protein	19.8	15.6	25.6		7
Ether extract	9.3	7.3	12.2		7
Nitrogen-free extract	49.5	43.5	54.1		7
Crude fibre	18.1	14.5	22.7		7
Total ash	3.34	2.38	4.62		7
Major elements % of DM					
Phosphorus	0.15	0.08	0.21		2
Potassium	0.005	0.004	0.006		2

2.59 DREG *

	Mean	Min.	Max.	SD	No.
Dry matter %	94.1	91.3	96.0	1.52	63
Proximate constituents % of DM					
Crude protein	38.9	31.4	46.5	5.51	63
Ether extract	21.6	14.9	26.8	4.76	62
Crude fibre	12.6				1
Total ash	3.4				1
Carbohydrate fractions % of DM					
Starch	5.3				1
Major elements % of DM					
Calcium	0.28				1
Phosphorus	0.14				1
Chlorine	0.07				1
Trace elements μg per g DM					
Copper	209				1
Iron	433				1

* The product obtained by drying the suspended solids recovered from whisky distillery spent wash after fermentation and distillation and to which no other matter has been added.

2.60 SCOTAFERM *

	Mean	Min.	Max.	SD	No.
Dry matter %	96.0	91.2	97.5	1.63	124
Proximate constituents % of DM					
Crude protein	28.5	26.8	31.3	1.78	124
Ether extract	10.6	9.3	11.7	1.01	124
Nitrogen-free extract	42.5	40.5	49.8	3.54	31
Crude fibre	3.26	1.14	4.8	1.12	92
Insol. ash	0.47	0.21	0.94	0.28	60
Sol. ash	14.3	11.7	17.0	2.37	60
Carbohydrate fractions % of DM					
Sol. sugar	15.9				1
Starch	3.61	3.53	3.69		2
Major elements % of DM					
Calcium	4.85	3.94	5.53	0.64	35
Phosphorus	1.07	0.95	1.28	0.12	35
Magnesium	0.43	0.19	0.48		14
Chlorine	0.36	0.21	0.47	0.11	62
Trace elements µg per g DM					
Copper	60	36	125	34.4	27
Iron	332	238	486		5
Manganese	44	21	71		7
Zinc	92	78	110	10.8	26

* The product obtained by drying by evaporation the spent wash from whisky production after fermentation and distillation and to which no matter other than lime has been added.

2.61 SCOTASOL *

	Mean	Min.	Max.	SD	No.
Dry matter %	94.9	92.3	96.5	1.72	60
Proximate constituents % of DM					
Crude protein	26.5	24.4	28.1	1.50	60
Ether extract	0.31	0.11	0.53	0.23	60
Nitrogen-free extract	46.7	42.1	49.1	2.86	29
Crude fibre	0.26	0.10	0.73	0.29	29
Total ash	26.6	22.6	31.0	3.32	60
Insol. ash	0.72	0.21	1.51	0.44	60
Sol. ash	25.9	21.7	30.7	3.30	60
Major elements % of DM					
Calcium	8.2	3.4	11.1		12
Phosphorus	1.73	1.55	1.82		12
Magnesium	0.55	0.51	0.58		7
Chlorine	0.80	0.42	1.05	0.97	60
Trace elements µg per g DM					
Copper	88	26	131		12
Iron	410	290	580		4
Manganese	55	21	96		11
Zinc	29	20	48		10

* Trade name for dried distillers' solubles "D.D.S." made from malt distilleries.

2.62 BARLEY, DARK GRAINS *

	Mean	Min.	Max.	SD	No.
Dry matter %	94.8	84.5	96.7	2.34	68
Proximate constituents % of DM					
Crude protein	27.7	24.7	29.8	1.89	68
Ether extract	9.4	6.1	11.7	1.77	68
Nitrogen-free extract	54.6	51.9	57.4		8
Crude fibre	11.3	8.6	13.7		8
Total ash	4.7	3.1	5.4		8

* The product obtained by drying a mixture of the spent wash in syrup form from whisky manufacture, after fermentation and distillation, with distillers' grains to which no matter other than lime has been added. The spent wash in syrup form contains 25% dry matter.

2.63 BARLEY, SUPER GRAINS *

	Mean	Min.	Max.	SD	No.
Dry matter %	94.6	92.6	96.3	1.6	68
Proximate constituents % of DM					
Crude protein	25.5	22.4	29.8	3.0	68
Ether extract	5.3	4.0	7.4	1.2	68

* The product obtained by drying the solid residues recovered by screening after fermentation and distillation to which no other matter has been added.

2.64 STIMUFLAV *

	Mean	Min.	Max.	SD	No.
Dry matter %	90.5	88.8	91.4		5
Proximate constituents % of DM					
Crude protein	26.3	24.8	27.4		5
Ether extract	10.5	10.0	11.5		5
Nitrogen-free extract	51.5				1
Crude fibre	8.0	7.4	8.9		3
Total ash	3.6	3.6	3.6		2
Carbohydrate fractions % of DM					
Sol. sugar	17.7				1
Starch	6.6	3.5	9.7		2
Major elements % of DM					
Calcium	0.04	0.03	0.06		2
Phosphorus	0.62	0.61	0.64		2
Chlorine	0.24				1

* A by-product of whisky manufacture

2.65 DRAFT *

	Mean	Min.	Max.	SD	No.
Dry matter %	23.8	21.9	25.3	1.74	62
Proximate constituents % of DM					
Crude protein	19.8	18.0	21.5	1.40	62
Ether extract	8.1	7.4	9.0	0.68	62
Nitrogen-free extract	51.5	49.5	53.3	1.47	62
Crude fibre	17.3	15.7	18.9	1.46	62
Total ash	3.34	2.95	3.69	0.28	62
Major elements % of DM					
Calcium	0.12	0.10	0.16	0.02	62
Phosphorus	0.34	0.27	0.40	0.05	62
Magnesium	0.11	0.08	0.14	0.025	62
Sodium	0.008	0.005	0.011	0.0023	62

* Draff is distillers' wet grains

2.66 DRAFF, SILAGE

	Mean	Min.	Max.	SD	No.
Dry matter %	29.4	28.1	41.1		14
Proximate constituents % of DM and pH					
Crude protein	18.8	15.6	27.3		14
pH	4.2	3.8	7.2		14

2.67 BARLEY, BREWERS' GRAINS, FRESH

	Mean	Min.	Max.	SD	No.
Dry matter %	26.4	19.7	42.2	6.1	176
Proximate constituents % of DM and pH					
Crude protein	20.3	13.5	29.7	3.9	179
Ether extract	7.0	3.3	11.2	2.09	63
Crude fibre	18.9	14.2	22.6	2.5	169
pH	4.3	3.5	5.7	0.68	41
Major elements % of DM					
Calcium	0.48	0.20	1.00	0.21	97
Phosphorus	0.56	0.32	0.76	0.10	96
Magnesium	0.16	0.07	0.47	0.10	43
Potassium	0.03	0.00	0.06	0.019	41
Sodium	0.06	0.01	0.52	0.11	36
Chlorine	0.19	0.00	0.89	0.17	45
Sulphur	0.19				1
Trace elements µg per g DM					
Cobalt	0.37	0.06	0.68		2
Copper	18	12	46	9.75	59
Manganese	42	23	58	8.25	56
Molybdenum	0.85	0.59	1.50		4
Zinc	84	13	121	23.0	43

2.68 BARLEY, BREWERS' GRAINS, DRIED

	Mean	Min.	Max.	SD	No.
Dry matter %	85.6	80.2	90.7		11
Proximate constituents % of DM					
Crude protein	19.2	18.0	19.7		12
Ether extract	6.9	6.2	7.6		9
Nitrogen-free extract	50.4				1
Crude fibre	16.3	13.8	17.3		11
Total ash	3.3				1
Carbohydrate fractions % of DM					
Sol. sugar	12.0				1
Starch	2.25				1
Cellulose	8.0				1
Major elements % of DM					
Calcium	0.47	0.40	0.55		7
Phosphorus	0.57	0.52	0.64		7
Magnesium	0.21	0.09	0.31		4
Potassium	0.04	0.04	0.04		2
Sodium	0.04	0.02	0.05		4
Chlorine	0.05	0.04	0.07		4
Trace elements µg per g DM					
Copper	20	18	23		4
Manganese	50	40	57		4
Zinc	117	100	141		4

2.69 BARLEY, BREWERS' GRAINS, SILAGE

	Mean	Min.	Max.	SD	No.
Dry matter %	30.1	21.9	47.6	3.92	38
Proximate constituents % of DM and pH					
Crude protein	20.5	17.4	25.9	2.51	38
Ether extract	7.2	2.0	9.6		16
Crude fibre	20.4	16.9	25.7	2.7	23
pH	3.9	3.8	4.3		12
Major elements % of DM					
Calcium	0.58	0.39	0.79		9
Phosphorus	0.58	0.32	0.72		9
Trace elements µg per g DM					
Manganese	50				1

2.70 VINEGAR GRAINS

	Mean	Min.	Max.	SD	No.
Dry matter %	25.6	23.8	27.3		2
Proximate constituents % of DM					
Crude protein	30.7	28.4	32.9		2
Ether extract	7.4	7.0	7.8		2
Crude fibre	15.8	15.3	16.2		2
Major elements % of DM					
Calcium	0.16				1
Phosphorus	0.40				1
Magnesium	0.14				1
Potassium	0.04				1
Chlorine	0.02				1
Trace elements µg per g DM					
Copper	12				1
Manganese	30				1
Zinc	110				1

2.71 VINEGAR GRAINS, SILAGE

	Mean	Min.	Max.	SD	No.
Dry matter %	27.6	27.3	27.9		2
Proximate constituents % of DM					
Crude protein	30.5	28.0	33.0		2
Ether extract	7.9	7.0	8.7		2
Crude fibre	15.3	14.4	16.2		2

2.72 MALT CULMS

	Mean	Min.	Max.	SD	No.
Dry matter %	93.1	90.8	96.0	2.15	60
Proximate constituents % of DM					
Crude protein	30.7	25.7	35.5	4.33	60
Ether extract	0.66	0.31	1.05	0.27	60
Nitrogen-free extract	48.8	43.3	52.4	3.99	60
Crude fibre	12.9	10.6	16.6	2.61	60
Total ash	6.88	5.8	8.5	0.95	60
Insol. ash	1.10	0.54	2.41	0.78	60
Sol. ash	5.78	4.7	6.6	0.79	60
Carbohydrate fractions % of DM					
Sol. sugar	16.5				1
Starch	0.59				1
Cellulose	13.4				1
Major elements % of DM					
Calcium	0.43				1
Phosphorus	0.78				1
Chlorine	1.50				1

2.73 *ZEA:* MAIZE, GRAIN

	Mean	Min.	Max.	SD	No.
Dry matter %	87.4	86.7	88.5	0.76	60
Proximate constituents % of DM					
Crude protein	10.5	9.9	11.1	0.48	60
Ether extract	4.35	3.80	5.00	0.47	60
Nitrogen-free extract	80.9	80.1	81.9	0.88	60
Crude fibre	2.48	1.95	2.89	0.42	60
Total ash	1.75	1.38	2.37	0.41	60
Insol. ash	0.21	0.11	0.34	0.22	60
Major elements % of DM					
Calcium	0.064	0.03	0.10	0.028	22
Phosphorus	0.24	0.19	0.28	0.047	22
Magnesium	0.14	0.10	0.34		12
Potassium	0.37	0.23	0.57		15
Sodium	0.04	0.02	0.06		5
Chlorine	0.06	0.03	0.12		10
Trace elements µg per g DM					
Copper	4.7	2.5	6.6		16
Manganese	12.5	4.6	32.0		16
Zinc	32.0	21.6	54.5		16
Vitamins µg per g DM					
β-carotene	4.04	1.8	5.8		7

2.74 PRAIRIE MEAL, MAIZE GLUTEN MEAL *

	Mean	Min.	Max.	SD	No.
Dry matter %	89.1	88.3	90.0	0.65	46
Proximate constituents % of DM					
Crude protein	71.0	67.6	74.1	2.55	46
Ether extract	2.47	1.55	3.99	0.76	46
Nitrogen-free extract	22.5	19.0	27.1	3.21	46
Crude fibre	0.36	0.11	0.67	0.18	46
Total ash	3.64	1.79	6.76	1.89	46
Insol. ash	0.56	0.22	1.14	0.34	46
Major elements % of DM					
Calcium	0.05	0.02	0.10		12
Phosphorus	0.40	0.20	0.48		16
Magnesium	0.04	0.01	0.14		5
Potassium	0.09	0.06	0.12		4
Sodium	0.05	0.02	0.06		4
Chlorine	0.15	0.11	0.33	0.073	41
Trace elements µg per g DM					
Copper	7.3	5.6	9.1		2
Zinc	28	25	32		4

* A by-product, predominantly protein, resulting from the removal of starch and germ from American Yellow Maize, to which no other matter has been added. XANTHOPHYLL CONTENT 150-300 µg/g.

2.75 MAIZE GLUTEN FEED *

	Mean	Min.	Max.	SD	No.
Dry matter %	90.1	88.3	92.1		16
Proximate constituents % of DM					
Crude protein	23.5	20.2	25.7		16
Ether extract	2.6	1.8	3.2		11
Nitrogen-free extract	54.7	53.6	57.4		4
Crude fibre	9.3	8.2	10.4		12
Total ash	9.2	8.1	10.2		4
Major elements % of DM					
Calcium	0.83				1
Phosphorus	1.04				1

* Maize Gluten Feed is equivalent to maize gluten meal, to which bran has been added.

2.76 MAIZE GERM EXPELLER MEAL, OR CAKE *

	Mean	Min.	Max.	SD	No.
Dry matter %	92.1	90.3	95.2		19
Proximate constituents % of DM					
Crude protein	23.2	20.1	26.3		19
Ether extract	10.6	8.6	13.7		19
Nitrogen-free extract	53.9	51.8	55.1		4
Crude fibre	10.5	7.7	13.8		11
Total ash	1.8	1.4	2.1		4

* Ground maize germ from which the oil has been removed in whole or in part.

2.77 MAIZE GERM, EXTRACTED *

	Mean	Min.	Max.	SD	No.
Dry matter %	88.1	86.9	90.4		18
Proximate constituents % of DM					
Crude protein	25.6	24.1	27.2		18
Ether extract	1.9	1.0	3.3		18
Crude fibre	10.3	8.5	11.7		18

* Maize germ meal, solvent extracted.

2.78 MILLET, GRAIN

	Mean	Min.	Max.	SD	No.
Dry matter %	88.0				1
Proximate constituents % of DM					
Crude protein	8.7				1
Crude fibre	11.8				1

2.79 SORGHUM GRAIN

	Mean	Min.	Max.	SD	No.
Dry matter %	87.4	86.2	88.4		3
Proximate constituents % of DM					
Crude protein	12.0	11.1	12.9		3
Ether extract	2.8				1
Nitrogen-free extract	80.1				1
Crude fibre	2.48	2.00	3.05		3
Total ash	1.51				1
Major elements % of DM					
Phosphorus	0.27				1
Potassium	0.44				1
Chlorine	0.04				1

2.80 SORGHUM, POWDERED

	Mean	Min.	Max.	SD	No.
Dry matter %	89.3				1
Proximate constituents % of DM					
Crude protein	9.2				1
Ether extract	3.4				1
Crude fibre	3.2				1

2.81 MILLET, GROUND

	Mean	Min.	Max.	SD	No.
Dry matter %	85.8	84.7	86.8		2
Proximate constituents % of DM					
Crude protein	11.6	9.3	13.9		2
Ether extract	2.4	1.9	2.9		2
Crude fibre	8.6	8.5	8.6		2
Major elements % of DM					
Calcium	0.41	0.41	0.41		2
Phosphorus	0.29	0.29	0.29		2

2.82 MILLET, MEAL

	Mean	Min.	Max.	SD	No.
Dry matter %	87.7	87.5	88.1		3
Proximate constituents % of DM					
Crude protein	14.9	14.1	15.7		3
Ether extract	3.9	2.4	5.5		2
Crude fibre	11.7	11.6	11.8		3
Major elements % of DM					
Calcium	0.84	0.31	1.37		2
Phosphorus	0.43	0.38	0.47		2

2.83 MILO

	Mean	*Min.*	*Max.*	*SD*	*No.*
Dry matter %	88.3	86.9	89.5	0.94	60
Proximate constituents % of DM					
Crude protein	11.9	10.8	13.1	0.87	60
Ether extract	3.17	2.68	3.57	0.36	60
Nitrogen-free extract	80.3	78.7	81.6	1.14	60
Crude fibre	2.59	2.03	3.07	0.44	60
Total ash	2.08	1.59	3.16	0.75	60
Insol. ash	0.25	0.11	0.45	0.14	60
Major elements % of DM					
Calcium	0.13	0.05	0.19	0.091	22
Phosphorus	0.30	0.21	0.40	0.080	21
Chlorine	0.15	0.11	0.34	0.087	60
Trace elements μg per g DM					
Cobalt	7.3	2.25	12.3		2

2.84 MILO GLUTEN FEED PELLETS

	Mean	*Min.*	*Max.*	*SD*	*No.*
Dry matter %	90.4	89.4	92.0		5
Proximate constituents % of DM					
Crude protein	24.3	23.5	25.1		5
Ether extract	3.20	1.52	3.75		5
Nitrogen-free extract	55.4	54.3	56.3		4
Crude fibre	8.8	8.1	9.7		5
Total ash	8.6	7.4	9.1		4

2.85 *ORYZA:* RICE, POLISHED MEAL

	Mean	*Min.*	*Max.*	*SD*	*No.*
Dry matter %	87.1				1
Proximate constituents % of DM					
Crude protein	9.1				1
Crude fibre	1.1				1

2.86 RICE BRAN

	Mean	*Min.*	*Max.*	*SD*	*No.*
Dry matter %	90.9	90.3	91.4	0.46	60
Proximate constituents % of DM					
Crude protein	14.8	13.2	16.4	1.22	60
Ether extract	21.2	17.3	24.6	2.92	60
Nitrogen-free extract	41.7	36.3	48.0	4.46	60
Crude fibre	9.0	6.5	12.4	2.48	60
Total ash	13.3	10.3	16.4	2.11	60
Insol. ash	2.2	0.8	4.7	1.42	60
Major elements % of DM					
Calcium	1.69	0.67	2.63	0.82	55
Phosphorus	1.72	1.18	2.04	0.38	55
Chlorine	0.09	0.03	0.11	0.047	60

2.87 RICE BRAN, EXTRACTED, INDIA

	Mean	*Min.*	*Max.*	*SD*	*No.*
Dry matter %	90.5	89.7	91.5	0.76	60
Proximate constituents % of DM					
Crude protein	13.7	11.2	15.7	1.78	60
Ether extract	0.87	0.33	1.78	0.51	60
Nitrogen-free extract	47.9	42.9	52.3	4.01	60
Crude fibre	18.0	13.8	22.2	3.95	60
Total ash	19.5	17.3	21.7	1.82	60
Insol. ash	11.8	9.4	13.8	1.89	60
Major elements % of DM					
Calcium	0.55	0.28	0.80	0.33	45
Phosphorus	1.41	1.17	1.61	0.25	44
Magnesium	0.84	0.55	1.78		15
Chlorine	0.18	0.11	0.33	0.091	60
Trace elements μg per g DM					
Copper	14.6	5.4	34.6		9
Iron	1992	1544	2379		3
Manganese	266	191	528		10
Zinc	95	59	174		13

2.88 RICE, GRAIN SCREENINGS

	Mean	*Min.*	*Max.*	*SD*	*No.*
Dry matter %	86.7				1
Proximate constituents % of DM					
Crude protein	6.9				1
Ether extract	0.2				1
Major elements % of DM					
Calcium	0.03				1
Phosphorus	0.10				1

2.89 RICE HUSKS

	Mean	*Min.*	*Max.*	*SD*	*No.*
Dry matter %	94.1				1
Proximate constituents % of DM					
Crude protein	4.4				1
Ether extract	6.1				1
Nitrogen-free extract	30.1				1
Crude fibre	37.0				1
Total ash	22.4				1
Insol. ash	19.1				1
Sol. ash	3.3				1
Major elements % of DM					
Calcium	0.16				1
Phosphorus	0.60				1
Chlorine	0.02				1

2.90 RYE GRAIN

	Mean	Min.	Max.	SD	No.
Dry matter %	86.2	85.9	86.8		3
Proximate constituents % of DM					
Crude protein	9.4	8.9	9.8		3
Ether extract	2.04	1.89	2.14		3
Nitrogen-free extract	84.3	83.7	85.3		3
Crude fibre	2.26	1.91	2.64		3
Total ash	1.98	1.82	2.10		3

2.91 SPRING BEAN

	Mean	Min.	Max.	SD	No.
Dry matter %	83.1	81.4	84.4		8
Proximate constituents % of DM					
Crude protein	34.4	33.1	35.6		8
Ether extract	1.6	1.2	1.9		8
Crude fibre	9.1	8.7	9.6		8
Major elements % of DM					
Calcium	0.012				1
Phosphorus	0.48				1
Potassium	1.32				1
Sodium	0.024				1
Trace elements μg per g DM					
Cobalt	0.60				1
Copper	31				1
Iron	120				1
Manganese	23				1
Zinc	31				1
Amino acids % of DM					
Arginine	3.50				1
Cystine	0.35				1
Glycine	1.40				1
Histidine	0.85				1
*Iso*leucine	1.37				1
Leucine	2.40				1
Lysine	2.07				1
Methionine	0.26				1
Phenylalanine	1.43				1
Threonine	1.23				1
Tyrosine	1.20				1
Valine	1.54				1

2.92 WINTER BEAN

	Mean	Min.	Max.	SD	No.
Dry matter %	83.8	81.8	87.6		9
Proximate constituents % of DM					
Crude protein	28.2	26.7	30.1		9
Ether extract	1.8	1.4	2.3		9
Crude fibre	9.3	8.7	10.7		9
Major elements % of DM					
Calcium	0.012	0.012	0.012		2
Phosphorus	0.43	0.37	0.49		2
Potassium	1.35	1.2	1.5		2
Sodium	0.018	0.012	0.024		2
Trace elements μg per g DM					
Cobalt	0.55	0.49	0.61		2
Copper	16	15	18		2
Iron	347	294	400		2
Manganese	16	16	16		2
Zinc	46	38	54		2
Amino acids % of DM					
Arginine	2.25	2.04	2.39		3
Cystine	0.25	0.24	0.26		3
Glycine	1.13	1.12	1.13		3
Histidine	0.69	0.65	0.72		3
*Iso*leucine	1.09	1.04	1.13		3
Leucine	1.99	1.86	2.12		3
Lysine	1.74	1.65	1.80		3
Methionine	0.19	0.14	0.22		3
Phenylalanine	1.14	1.06	1.20		3
Threonine	0.98	0.95	1.00		3
Tyrosine	0.93	0.90	0.96		3
Valine	1.21	1.16	1.23		3

2.93 TIC BEAN

	Mean	Min.	Max.	SD	No.
Dry matter %	86.3	84.1	89.1		7
Proximate constituents % of DM					
Crude protein	30.9	29.4	31.9		7
Ether extract	1.99	1.5	2.2		7
Nitrogen-free extract	54.2	52.9	55.2		3
Crude fibre	8.5	7.4	9.3		4
Total ash	4.4	3.6	5.9		3
Carbohydrate fractions % of DM					
Sol. sugar	6.35				1
Starch	31.9	27.1	36.7		2
Major elements % of DM					
Calcium	0.34	0.12	0.58		5
Phosphorus	0.66	0.57	0.86		5
Chlorine	0.08				1

2.94 *GLYCINE HISPIDA:* SOYA BEAN MEAL, EXTRACTED

	Mean	*Min.*	*Max.*	*SD*	*No.*
Dry matter %	89.2	87.5	90.4	1.20	120
Proximate constituents % of DM					
Crude protein	51.5	49.2	54.2	1.75	120
Ether extract	1.58	0.93	2.32	0.51	120
Nitrogen-free extract	33.9	32.3	35.1	1.50	120
Crude fibre	6.42	4.5	7.6	1.17	120
Total ash	6.57	5.96	7.22	0.49	120
Insol. ash	0.36	0.22	0.57	0.16	60
Major elements % of DM					
Calcium	0.42	0.31	0.50	0.101	25
Phosphorus	0.62	0.48	0.70	0.17	25
Magnesium	0.30	0.27	0.36		8
Chlorine	0.10	0.00	0.22	0.071	60
Trace elements µg per g DM					
Copper	46.4	12.0	171.0		5
Manganese	40.0	37.0	43.0		2
Zinc	76.4	65.0	94.0		12

2.95 *ARACHIS:* GROUNDNUT

	Mean	*Min.*	*Max.*	*SD*	*No.*
Dry matter %	95.3	91.3	96.4	1.02	26
Proximate constituents % of DM					
Crude protein	28.5	27.1	31.2	1.04	26
Ether extract	50.7	44.3	54.4	1.85	26
Nitrogen-free extract	11.2				1
Crude fibre	3.8				1
Total ash	3.44	2.33	5.55		3
Free fatty acid	0.86	0.21	2.31	0.63	26

2.96 GROUNDNUT, EXTRACTED, DECORTICATED, INDIA

	Mean	*Min*	*Max.*	*SD*	*No.*
Dry matter %	91.9	90.8	92.9	0.88	64
Proximate constituents % of DM					
Crude protein	51.4	48.6	56.0	2.28	64
Ether extract	0.72	0.32	1.20	0.33	64
Nitrogen-free extract	29.5	25.9	33.0	2.65	64
Crude fibre	12.0	6.5	16.8	3.90	64
Total ash	6.4	5.7	7.5	0.67	64
Insol. ash	1.4	0.1	2.2	0.52	64
Major elements % of DM					
Calcium	0.33	0.22	0.41		13
Phosphorus	0.54	0.40	0.66		13
Magnesium	0.32	0.27	0.38		17
Potassium	1.42				1
Sodium	0.03				1
Chlorine	0.12	0.11	0.22	0.04	64
Trace elements µg per g DM					
Copper	37	14	82		17
Iron	859				1
Manganese	50	31	67		5
Zinc	65	50	78		13

2.97 GROUNDNUT CAKE, EXTRACTED, DECORTICATED

	Mean	*Min.*	*Max.*	*SD*	*No.*
Dry matter %	91.7	90.4	92.6	1.16	60
Proximate constituents % of DM					
Crude protein	52.8	49.9	56.6	2.14	60
Ether extract	0.83	0.43	1.33	0.39	60
Nitrogeon-free extract	29.3	25.5	33.7	2.78	60
Crude fibre	10.8	5.8	15.6	3.55	60
Total ash	6.27	5.3	7.8	0.94	60
Insol. ash	1.37	0.65	2.60	0.60	60
Sol. ash	4.93	4.3	5.3	0.42	60

2.98 GROUNDNUT CAKE, EXPELLER, DECORTICATED

	Mean	*Min.*	*Max.*	*SD*	*No.*
Dry matter %	93.7	91.1	95.4	1.61	60
Proximate constituents % of DM					
Crude protein	52.0	49.3	53.7	1.97	60
Ether extract	6.8	5.4	9.1	1.40	60
Nitrogen-free extract	28.7	25.0	31.4	2.45	60
Crude fibre	6.4	4.7	8.2	1.63	60
Total ash	6.1	5.2	7.2	0.73	60
Insol. ash	1.6	0.7	2.6	0.64	60
Sol. ash	4.5	4.1	5.0	0.55	60

2.99 GROUNDNUT BRAN

	Mean	Min.	Max.	SD	No.
Dry matter %	86.2				1
Proximate constituents % of DM					
Crude protein	12.2				1
Crude fibre	31.2				1
Major elements % of DM					
Calcium	0.22				1
Phosphorus	0.18				1

2.100 *PISUM:* PEA, SEED

	Mean	Min.	Max.	SD	No.
Trace elements μg per g DM					
Barium	8	2	31		19
Chromium	0.31	0.10	1.03	0.25	20
Cobalt	0.19	0.03	0.69	0.22	20
Copper	9.9	4.1	16.0	2.9	20
Iron	129	40	358	88.0	20
Lead	18.5	0.5	163	38.2	20
Manganese	29	4	96	29.0	23
Molybdenum	1.08	0.22	2.74	0.66	20
Nickel	2.56	0.48	6.70	1.7	20
Silver	0.18	0.06	1.02		17
Strontium	32	1	99		19
Tin	1.2	0.5	2.9	0.57	20
Titanium	7.1	0.6	31.0	7.4	20
Vanadium	0.21	0.03	1.00	0.06	20
Zinc	65	28	129	25.0	20

2.101 *CERATIONIA:* LOCUST BEAN KERNEL

	Mean	Min.	Max.	SD	No.
Dry matter %	85.6				1
Proximate constituents % of DM					
Crude protein	8.2				1

2.102 LOCUST BEAN, KIBBLED, POD + BEAN

	Mean	Min.	Max.	SD	No.
Dry matter %	85.1	80.5	90.4		5
Proximate constituents % of DM					
Crude protein	4.6	4.0	5.0		5
Ether extract	0.6	0.4	0.7		5
Crude fibre	7.7	6.2	8.9		5
Insol. ash	0.6				1
Carbohydrate fractions % of DM					
Sucrose	33.9	18.3	44.4		3
Invert sugar	22.2				1

2.103 *CYAMOPSIS:* GUAR GERM MEAL

	Mean	Min.	Max.	SD	No.
Dry matter %	94.1	92.3	95.4	1.19	64
Proximate constituents % of DM					
Crude protein	43.2	41.2	45.2	1.90	64
Ether extract	5.2	4.7	5.6	0.52	64
Nitrogen-free extract	32.0	30.2	33.8	1.63	60
Crude fibre	13.9	12.3	15.5	1.59	60
Total ash	5.6	5.1	5.9	0.10	64
Insol. ash	0.45	0.22	0.74	0.18	64
Sol. ash	5.13	4.7	5.6	0.31	64
Major elements % of DM					
Calcium	0.44	0.17	0.54	0.12	24
Phosphorus	0.57	0.49	0.62	0.062	25
Magnesium	0.44	0.31	0.57		3
Chlorine	0.11	0.03	0.21	0.049	60
Trace elements μg per g DM					
Copper	14	13	16		4
Manganese	30				1
Zinc	49				1

2.104 *ELAEIS:* PALM, KERNEL

	Mean	Min.	Max.	SD	No.
Dry matter %	92.7	91.8	93.5	0.73	31
Proximate constituents % of DM					
Ether extract	53.2	51.4	54.3	1.20	31

2.105 PALM KERNEL MEAL, EXTRACTED

	Mean	Min.	Max.	SD	No.
Dry matter %	87.0	85.6	88.5	1.06	52
Proximate constituents % of DM					
Crude protein	20.2	19.4	21.0	0.79	52
Ether extract	1.32	0.69	2.16	0.71	52
Nitrogen-free extract	60.7	58.8	61.6		5
Crude fibre	13.7	11.4	15.3	1.34	52
Total ash	4.24	4.07	4.73		5
Insol. ash	0.56	0.45	0.65		5
Sol. ash	3.67	3.50	4.08		5

2.106 PALM KERNEL CAKE, EXPELLER

	Mean	Min.	Max.	SD	No.
Dry matter %	89.7	88.8	91.1	0.93	60
Proximate constituents % of DM					
Crude protein	20.2	19.4	21.4	0.64	60
Ether extract	8.3	7.8	9.0	0.64	60
Nitrogen-free extract	55.6	53.5	57.5	1.35	20
Crude fibre	11.4	9.6	12.8	1.25	60
Total ash	4.15	3.95	4.35	0.13	20
Insol. ash	0.58	0.43	0.95	0.17	20
Sol. ash	3.58	3.25	3.84	0.20	20
Major elements % of DM					
Calcium	0.33	0.26	0.38		6
Phosphorus	0.65	0.57	0.71		5
Magnesium	0.33	0.26	0.38		3
Potassium	0.56				1
Chlorine	0.23	0.00	0.56		11
Trace elements µg per g DM					
Copper	25	25	25		2
Manganese	185				1
Zinc	53	48	55		3

2.107 PALM KERNEL, MOLASSED

	Mean	Min.	Max.	SD	No.
Dry matter %	79.0				1
Proximate constituents % of DM					
Crude protein	12.7				1
Ether extract	0.34				1
Crude fibre	8.0				1
Major elements % of DM					
Calcium	2.52				1
Phosphorus	0.56				1
Sodium	0.79				1
Chlorine	1.21				1

2.108 *GOSSYPIUM:* COTTON CAKE

	Mean	Min.	Max.	SD	No.
Dry matter %	89.3	89.0	89.5		2
Proximate constituents % of DM					
Crude protein	36.1	29.2	43.1		2
Ether extract	5.48	5.17	5.78		2
Crude fibre	20.4	17.3	23.6		2
Major elements % of DM					
Calcium	0.25	0.23	0.26		2
Phosphorus	1.01	0.85	1.17		2
Potassium	1.86	1.82	1.90		2
Chlorine	0.08	0.04	0.12		2

2.109 COTTON CAKE, DECORTICATED, EXTRACTED INDIA

	Mean	Min.	Max.	SD	No.
Dry matter %	92.1	91.3	93.1	0.81	60
Proximate constituents % of DM					
Crude protein	40.6	36.6	44.6	3.13	60
Ether extract	0.64	0.22	1.20	0.38	60
Nitrogen-free extract	35.3	31.6	38.6	2.70	60
Crude fibre	15.5	12.7	18.9	2.56	60
Total ash	7.9	6.7	8.8	0.86	60
Insol. ash	1.6	0.7	2.4	0.76	60
Major elements % of DM					
Calcium	0.48	0.41	0.56	0.07	28
Phosphorus	0.84	0.59	1.08	0.21	28
Magnesium	0.50	0.44	0.56	0.048	29
Sodium	0.039	0.033	0.044		2
Chlorine	0.17	0.11	0.22	0.078	60
Trace elements µg per g DM					
Copper	24	15	30	4.96	54
Manganese	44	27	59	13.4	25
Zinc	87	74	103	11.1	33

2.110 COTTONSEED CAKE, EXPELLER, DECORTICATED

	Mean	Min.	Max.	SD	No.
Dry matter %	91.5	89.9	93.1	1.2	60
Proximate constituents % of DM					
Crude protein	40.5	36.8	43.9	2.90	60
Ether extract	7.8	5.7	9.7	1.56	60
Nitrogen-free extract	31.1	27.4	34.8	3.21	60
Crude fibre	13.8	9.9	17.3	2.86	60
Total ash	6.78	5.97	8.25	0.86	60
Insol. ash	0.44	0.11	0.87	0.35	60
Sol. ash	6.34	5.91	7.10	0.80	60

2.111 COTTONSEED CAKE, EXPELLER, UNDECORTICATED

	Mean	Min.	Max.	SD	No.
Dry matter %	89.7	87.3	92.0	1.68	60
Proximate constituents % of DM					
Crude protein	27.7	25.8	30.1	1.44	60
Ether extract	5.5	4.7	6.4	0.55	60
Crude fibre	23.6	20.3	26.4	2.43	45

2.112 *HELIANTHUS:* SUNFLOWER MEAL, EXTRACTED

	Mean	*Min.*	*Max.*	*SD*	*No.*
Dry matter %	92.1	91.1	93.2	0.89	61
Proximate constituents % of DM					
Crude protein	39.8	36.5	42.6	2.71	61
Ether extract	2.2	1.3	3.1	0.64	61
Nitrogen-free extract	29.4	26.6	31.3	2.22	61
Crude fibre	22.0	18.3	25.9	2.78	61
Total ash	6.6	5.6	7.4	0.71	61
Insol. ash	0.42	0.22	0.76	0.23	61
Major elements % of DM					
Calcium	0.57	0.44	0.76	0.12	21
Phosphorus	0.97	0.86	1.35	0.053	21
Magnesium	0.60	0.51	0.75		8
Chlorine	0.19	0.11	0.22	0.073	59
Trace elements µg per g DM					
Copper	35	27	42		5
Zinc	117	86	169		10

2.113 SUNFLOWER SEED, EXPELLER

	Mean	*Min.*	*Max.*	*SD*	*No.*
Dry matter %	93.5	92.9	94.0		8
Proximate constituents % of DM					
Crude protein	37.6	34.3	41.4		8
Ether extract	10.2	7.5	11.9		8
Nitrogen-free extract	26.8	22.3	31.7		7
Crude fibre	18.0	12.8	27.0		7
Total ash	7.9	5.8	15.3		8

2.114 *LINUM:* LINSEED, GROUND

	Mean	*Min*	*Max.*	*SD*	*No.*
Dry matter %	91.4	91.0	91.8		2
Proximate consistuents % of DM					
Crude protein	24.1	22.4	25.8		2
Ether extract	34.1	32.7	35.5		2
Crude fibre	13.6	7.2	19.9		2
Major elements % of DM					
Calcium	0.28				1
Phosphorus	0.59				1
Sodium	0.03				1
Chlorine	0.04				1

2.115 LINSEED, FLAKES

	Mean	*Min.*	*Max.*	*SD*	*No.*
Dry matter %	93.4				1
Proximate consistuents % of DM					
Crude protein	33.6				1
Ether extract	5.1				1
Crude fibre	9.4				1
Major elements % of DM					
Calcium	0.41				1
Phosphorus	0.96				1
Potassium	1.39				1
Sodium	0.03				1
Chlorine	0.04				1

2.116 LINSEED, CAKE EXPELLER

	Mean	*Min.*	*Max.*	*SD*	*No.*
Dry matter %	89.7	87.5	91.7	1.63	60
Proximate consistuents % of DM					
Crude protein	36.9	35.3	38.9	1.38	60
Ether extract	8.0	6.8	9.7	1.13	60
Nitrogen-free extract	40.3	38.1	42.3	1.81	60
Crude fibre	8.9	8.1	10.0	0.75	60
Total ash	5.73	5.60	6.15	0.47	60
Insol. ash	0.20	0.11	0.34	0.16	60
Sol. ash	5.54	5.03	5.91	0.47	60
Major elements % of DM					
Calcium	0.41	0.34	0.49	0.04	24
Phosphorus	0.86	0.61	0.96	0.06	24
Magnesium	0.58				1
Potassium	1.19	0.83	1.39		5
Chlorine	0.05	0.02	0.06		17
Trace elements µg per g DM					
Barium	6.1				1
Chromium	0.57				1
Cobalt	0.55				1
Copper	25				1
Iron	321				1
Lead	1.3				1
Manganese	41.5	29.5	62.1		7
Molybdenum	0.30				1
Nickel	2.5				1
Silver	0.10				1
Strontium	7.6				1
Tin	1.0				1
Titanium	30				1
Vanadium	0.66				1
Zinc	169				1

2.117 LINSEED PELLETS, EXTRACTED

	Mean	Min.	Max.	SD	No.
Dry matter %	90.1	88.2	91.5	1.67	54
Proximate constituents % of DM					
Crude protein	39.6	37.6	40.7	1.19	54
Ether extract	2.71	1.53	3.69	0.78	54
Nitrogen-free extract	41.3	39.8	43.7	1.68	54
Crude fibre	10.1	9.1	11.6	1.02	54
Total ash	6.32	5.39	7.07	0.58	54

2.118 *BUTYROSPERMUM:* SHEA NUT CAKE EXPELLER

	Mean	Min.	Max.	SD	No.
Dry matter %	93.0	92.4	93.8		14
Proximate constituents % of DM					
Crude protein	13.1	12.3	14.7		14
Ether extract	17.5	15.0	20.7		14
Nitrogen-free extract	54.6	50.5	59.6		14
Crude fibre	9.3	6.8	11.0		14
Total ash	5.5	5.1	5.8		14
Insol. ash	0.76	0.53	0.97		14
Major elements % of DM					
Calcium	0.25	0.14	0.35		7
Phosphorus	0.22	0.19	0.27		8
Magnesium	0.18				1
Potassium	1.81	1.35	2.27		2
Sodium	0.08	0.04	0.11		2
Chlorine	0.18	0.11	0.43		14
Trace elements µg per g DM					
Copper	5.4				1
Zinc	29				1

2.119 SHEA NUT MEAL, EXPELLER, SWEDEN

	Mean	Min.	Max.	SD	No.
Dry matter %	91.7				1
Proximate constituents % of DM					
Crude protein	12.2				1
Ether extract	18.0				1
Crude fibre	12.0				1
Insol. ash	1.9				1
Carbohydrate fractions % of DM					
Sol. sugar	11.6				1
Starch	2.08				1

2.120 *BUTYROSPERMUM:* ILLIPE, EXTRACTED

	Mean	Min.	Max.	SD	No.
Dry matter %	89.7	88.8	90.5		7
Proximate constituents % of DM					
Crude protein	16.9	15.2	18.1		7
Ether extract	0.70	0.44	1.01		7
Nitrogen-free extract	68.9	63.0	74.2		7
Crude fibre	8.21	4.6	10.8		7
Total ash	5.3	4.7	5.9		7
Insol. ash	1.26	0.00	2.00		7
Sol. ash	4.23	3.20	5.70		7
Major elements % of DM					
Calcium	0.39	0.33	0.44		2
Phosphorus	0.29	0.27	0.31		2
Chlorine	0.13	0.11	0.23		7
Trace elements µg per g DM					
Copper	38				1

2.121 *CARTHAMUS:* SAFFLOWER SEED, EXTRACTED, UNDECORTICATED

	Mean	Min.	Max.	SD	No.
Dry matter %	93.5	91.5	95.1		14
Proximate constituents % of DM					
Crude protein	19.6	15.4	24.5		14
Ether extract	0.51	0.00	1.61		14
Nitrogen-free extract	30.2	25.1	35.5		8
Crude fibre	45.2	40.2	48.7		9
Total ash	5.1	3.9	6.0		14
Insol. ash	1.6	0.9	2.8		14
Sol. ash	3.3	2.3	3.9		14
Major elements % of DM					
Calcium	0.41	0.30	0.53		6
Phosphorus	0.50	0.35	0.88		6
Magnesium	0.24	0.22	0.24		4
Chlorine	0.16	0.11	0.22		13
Trace elements µg per g DM					
Copper	26	21	28		5
Iron	480	114	845		2
Manganese	64	51	72		6
Zinc	62	33	77		6

2.122 *SESAMUM:* SESAME CAKE, EXPELLER

	Mean	Min.	Max.	SD	No.
Dry matter %	92.1	91.1	93.0	1.00	60
Proximate constituents % of DM					
Crude protein	45.1	41.0	49.8	3.20	60
Ether extract	8.7	6.4	11.5	1.89	60
Nitrogen-free extract	27.5	24.7	30.0	2.00	60
Crude fibre	5.2	3.8	6.8	1.42	60
Total ash	13.5	12.1	15.5	1.45	60
Insol. ash	2.0	0.9	4.2	1.17	60
Major elements % of DM					
Calcium	2.61	2.24	2.94	0.28	28
Phosphorus	1.16	0.97	1.30	0.12	30
Magnesium	0.62	0.59	0.65		5
Potassium	1.13	1.03	1.23		4
Sodium	0.04	0.02	0.07		4
Chlorine	0.11	0.10	0.22	0.058	60
Trace elements μg per g DM					
Copper	28	15	45		9
Manganese	66	54	96		4
Zinc	108	90	136		8

2.123 *THEOBROMA:* COCOA BEAN MEAL, EXTRACTED

	Mean	Min.	Max.	SD	No.
Dry matter %	94.1				1
Proximate constituents % of DM					
Crude protein	28.4				1
Ether extract	2.2				1
Nitrogen-free extract	52.3				1
Crude fibre	9.7				1
Total ash	7.4				1
Theobromine	2.8				1

2.124 *COCOS:* COPRA

	Mean	Min.	Max.	SD	No.
Dry matter %	96.5	95.9	97.0	0.56	53
Proximate constituents % of DM					
Ether extract	67.4	65.5	69.2	1.61	43
Free fatty acid	3.2	1.9	5.4	1.79	43

2.125 COPRA, EXPELLER

	Mean	Min.	Max.	SD	No.
Dry matter %	90.2	88.0	92.5	1.74	60
Proximate constituents % of DM					
Crude protein	23.6	22.5	24.8	1.08	60
Ether extract	6.72	6.1	7.6	0.65	60
Nitrogen-free extract	51.3	48.9	53.8	2.19	60
Crude fibre	12.6	10.0	14.9	1.93	60
Total ash	5.87	5.3	6.6	0.51	60
Insol. ash	0.37	0.21	0.66	0.22	60
Sol. ash	5.5	4.6	6.1	0.45	60
Major elements % of DM					
Calcium	0.19	0.11	0.37		16
Phosphorus	0.54	0.44	0.64		16
Magnesium	0.37	0.27	0.60		4
Chlorine	1.06	0.9	1.2	0.13	60
Trace elements μg per g DM					
Copper	29	14	39		6
Iron	249				1
Manganese	63	59	68		2
Zinc	65	45	88		6

2.126 *SINAPIS:* MUSTARD SEED MEAL

	Mean	Min.	Max.	SD	No.
Proximate constituents % of DM					
Crude protein	24.7				1
Ether extract	10.0				1
Crude fibre	19.0				1
Major elements % of DM					
Calcium	1.1				1

2.127 *BRASSICA NAPUS:* RAPESEED MEAL, EXTRACTED

	Mean	Min.	Max.	SD	No.
Dry matter %	90.9	89.5	92.3	1.39	60
Proximate constituents % of DM					
Crude protein	41.5	39.9	43.4	1.52	60
Ether extract	1.18	0.77	1.41	0.47	60
Nitrogen-free extract	36.6	32.1	39.4	2.18	37
Crude fibre	12.7	11.6	13.9	0.93	60
Total ash	7.8	7.2	8.2	0.63	37
Insol. ash	0.30	0.11	0.54	0.16	37
Sol. ash	7.5	6.9	7.8	0.57	37

2.128 *BETA VULGARIS:* FODDER BEET, SEED

	Mean	Min.	Max.	SD	No.
Dry matter %	83.3				1
Proximate constituents % of DM					
Crude protein	13.6				1
Ether extract	5.1				1
Crude fibre	38.6				1
Major elements % of DM					
Calcium	1.09				1
Phosphorus	0.32				1
Sodium	0.17				1
Chlorine	0.27				1

2.129 *BRASSICA OLERACEA:* CABBAGE, SEED

	Mean	Min.	Max.	SD	No.
Dry matter %	90.2				1
Proximate constituents % of DM					
Crude protein	22.3				1
Ether extract	21.8				1
Crude fibre	13.9				1
Major elements % of DM					
Calcium	0.60				1
Phosphorus	0.83				1
Sodium	0.04				1
Chlorine	0.06				1

2.130 *SPINACIA OLERACEA:* SPINACH, SEED

	Mean	Min.	Max.	SD	No.
Dry matter %	86.6				1
Proximate constituents % of DM					
Crude protein	13.2				1
Crude fibre	26.0				1
Major elements % of DM					
Sodium	0.07				1
Chlorine	0.11				1

2.131 *BRASSICA NAPOBRASSICA:* SWEDE, UNSPECIFIED SEED

	Mean	Min.	Max.	SD	No.
Dry matter %	92.0				1
Proximate constituents % of DM					
Crude protein	20.8				1
Ether extract	33.6				1
Crude fibre	10.6				1
Major elements % of DM					
Calcium	0.56				1
Phosphorus	0.83				1
Sodium	0.012				1
Chlorine	0.018				1

2.132 *BRASSICA RAPA:* TURNIP SEED

	Mean	Min.	Max.	SD	No.
Dry matter %	90.2				1
Proximate constituents % of DM					
Crude protein	21.0				1
Ether extract	24.7				1
Crude fibre	13.5				1
Major elements % of DM					
Calcium	0.59				1
Phosphorus	0.77				1
Sodium	0.04				1
Chlorine	0.06				1

2.133 *MANIHOT UTILISSIMA:* CASSAVA, ROOT

	Mean	Min.	Max.	SD	No.
Dry matter %	87.6	86.6	88.4		5
Carbohydrate fractions % of DM					
Starch	81.8	78.4	86.5		5

2.134 MANIOC ROOT

	Mean	Min.	Max.	SD	No.
Dry matter %	87.4				1
Carbohydrate fractions % of DM					
Starch	81.7				1

2.135 MANIOC CHIPS

	Mean	Min.	Max.	SD	No.
Dry matter %	88.4	87.5	89.4		8
Proximate constituents % of DM					
Crude fibre	3.62	2.65	4.44		7
Total ash	3.13	2.17	4.10		7
Carbohydrate fractions % of DM					
Starch	76.1	72.6	78.7		7

2.136 MANIOC MEAL

	Mean	Min.	Max.	SD	No.
Dry matter %	88.5	87.8	89.5		12
Proximate constituents % of DM					
Crude fibre	3.2	2.6	4.8		7
Total ash	1.5	0.7	3.4		7
Carbohydrate fractions % of DM					
Starch	80.4	75.3	82.6		12

2.137 TAPIOCA ROOT

	Mean	Min.	Max.	SD	No.
Dry matter %	88.4	87.6	89.4	0.73	60
Proximate constituents % of DM					
Crude protein	2.2				1
Ether extract	0.6				1
Nitrogen-free extract	92.2				1
Crude fibre	3.4	2.7	3.9	0.60	60
Total ash	4.6	3.0	7.6	1.71	60
Carbohydrate fractions % of DM					
Starch	75.4	73.2	77.4	1.88	60

2.138 TAPIOCA MEAL

	Mean	Min.	Max.	SD	No.
Dry matter %	88.0	87.3	89.2		8
Proximate constituents % of DM					
Crude fibre	4.90	2.30	8.92		8
Total ash	5.86	2.68	10.07		8
Carbohydrate fractions % of DM					
Starch	74.1	66.6	81.1		8

2.139 *SAGUS, METROXYLON:* SAGO

	Mean	Min.	Max.	SD	No.
Dry matter %	89.1				1
Proximate constituents % of DM					
Crude protein	0.2				1
Ether extract	0.2				1
Total ash	0.03				1

2.140 APPLE POMACE, DRIED

	Mean	Min.	Max.	SD	No.
Dry matter %	94.8	91.5	96.7		16
Proximate constituents % of DM					
Crude protein	9.4	8.7	10.1		16
Ether extract	3.5	3.0	4.0		16
Nitrogen-free extract	43.5	36.5	51.2		4
Crude fibre	42.6	36.4	47.9		4
Total ash	1.27	0.93	1.59		16
Insol. ash	0.57	0.31	0.73		16
Sol. ash	0.70	0.31	1.06		16
Major elements % of DM					
Phosphorus	0.07	0.04	0.13		5
Magnesium	0.03	0.01	0.10		5
Chlorine	0.11	0.10	0.21		16
Trace elements µg per g DM					
Copper	33	11	43		6
Iron	261				1
Manganese	10	4.1	15.7		5
Zinc	26	15	67		6

2.141 CITRUS PULP, ARTIFICIALLY DRIED

	Mean	Min.	Max.	SD	No.
Dry matter %	89.2				1
Proximate constituents % of DM					
Crude protein	6.6				1
Ether extract	5.3				1
Crude fibre	13.1				1
Major elements % of DM					
Calcium	2.21				1
Phosphorus	0.11				1
Potassium	1.00				1
Sodium	0.02				1
Chlorine	0.04				1

2.142 GRAPESEED MEAL, EXTRACTED

	Mean	Min.	Max.	SD	No.
Dry matter %	88.4	85.8	90.4	1.39	24
Proximate constituents % of DM					
Crude protein	11.4	7.9	14.6	1.47	24
Ether extract	0.74	0.33	1.99	0.39	24
Nitrogen-free extract	30.3	18.4	43.2	5.43	20
Crude fibre	54.0	45.9	69.4	6.05	20
Total ash	3.5	2.2	6.5	0.98	23
Insol. ash	0.43	0.11	1.34	0.29	23
Sol. ash	3.0	2.1	5.3	0.87	23
Major elements % of DM					
Calcium	0.85	0.68	1.10		4
Phosphorus	0.18	0.10	0.25		3
Magnesium	0.10	0.06	0.16		4
Chlorine	0.15	0.11	0.45		12
Trace elements µg per g DM					
Copper	17.0	11.0	26.0		4
Manganese	25.0	17.0	37.0		4
Zinc	35.0	17.0	93.0		7

2.143 GRAPE RESIDUE, DRIED *

	Mean	*Min.*	*Max.*	*SD*	*No.*
Dry matter %	92.6	92.3	92.8		2
Proximate constituents % of DM					
Crude protein	11.0	10.6	11.4		2
Ether extract	6.0	4.8	7.2		2
Nitrogen-free extract	50.3				1
Crude fibre	28.2				1
Total ash	5.4	5.3	5.4		2
Insol. ash	0.27	0.16	0.38		2
Sol. ash	5.08	4.98	5.17		2
Carbohydrate fractions % of DM					
Total sugars	5.4				1
Major elements % of DM					
Calcium	0.82	0.76	0.88		2
Phosphorus	0.21				1
Chlorine	0.03				1

*Residue from wine making.

2.144 BREWERS' YEAST, DRIED

	Mean	*Min.*	*Max.*	*SD*	*No.*
Dry matter %	91.8				1
Proximate constituents % of DM					
Crude protein	42.9				1
Ether extract	1.9				1
Nitrogen-free extract	38.8				1
Crude fibre	2.9				1
Total ash	13.5				1
Major elements % of DM					
Calcium	0.13				1
Phosphorus	1.42				1
Chlorine	0.01				1

2.145 MOLASSES

	Mean	*Min.*	*Max.*	*SD*	*No.*
Dry matter %	73.8	71.9	75.1	1.57	61
Proximate constituents % of DM					
Crude protein	3.77	2.6	4.7	0.87	61
Nitrogen-free extract	86.3	82.3	90.1	2.58	61
Total ash	9.91	7.5	13.4	2.22	61
Carbohydrate fractions % of DM					
Total sugars	70.6	66.8	75.1	3.33	61
Major elements % of DM					
Calcium	1.37	1.36	1.38		2
Phosphorus	0.06	0.04	0.07		2
Chlorine	2.65	1.7	4.2	0.85	61

2.146 MOLASSES, BLACK STRAP

	Mean	*Min.*	*Max.*	*SD*	*No.*
Dry matter %	75.8	73.0	77.7		19
Carbohydrate fractions % of DM					
Sucrose	43.7	29.3	47.9		19
Invert sugars	25.5	17.5	40.4		19
Total sugars	69.1	62.8	76.2		19

2.147 BISCUIT MEAL

	Mean	*Min.*	*Max.*	*SD*	*No.*
Dry matter %	93.0	90.7	95.1	2.16	31
Proximate constituents % of DM					
Crude protein	7.7	5.3	10.7	2.52	31
Ether extract	14.9	11.3	18.9	2.70	31
Nitrogen-free extract	64.4	55.8	75.7		14
Crude fibre	0.75	0.32	3.39		14
Total ash	14.5	2.1	23.4	8.62	31
Insol. ash	0.78	0.11	1.80	0.78	27
Sol. ash	14.2	1.9	21.9	8.08	27
Carbohydrate fractions % of DM					
Sol. sugar	18.8				1
Starch	48.7				1
Major elements % of DM					
Calcium	5.08	0.01	7.68		12
Phosphorus	0.17	0.07	0.54		12
Magnesium	0.15	0.05	0.41		6
Chlorine	0.74	0.22	1.22	0.40	27
Trace elements μg per g DM					
Copper	18	5	80		9
Iron	297	156	438		2
Manganese	77	12	150		8
Zinc	38	15	163		9

2.148 WHEY, DRIED

	Mean	*Min.*	*Max.*	*SD*	*No.*
Dry matter %	94.3	91.9	97.1		7
Proximate constituents % of DM					
Crude protein	13.0	11.5	14.5		7
Ether extract	1.06	0.13	2.07		7
Nitrogen-free extract	76.3	74.2	77.8		5
Total ash	9.2	8.3	10.2		5
Major elements % of DM					
Calcium	0.85	0.67	1.03		2
Phosphorus	0.70	0.69	0.70		2
Chlorine	3.2	2.6	3.7		2

2.149 MEAT MEAL

	Mean	Min.	Max.	SD	No.
Dry matter %	91.3	89.4	93.2		19
Proximate constituents % of DM					
Crude protein	63.0	54.4	69.4		19
Ether extract	5.4	3.7	7.7		19
Nitrogen-free extract	6.4	5.0	9.7		13
Total ash	24.9	17.8	30.7		13
Insol. ash	1.1	0.6	1.7		13
Sol. ash	23.8	17.3	30.1		13
Major elements % of DM					
Calcium	8.0	6.4	11.2		12
Phosphorus	4.1	2.3	5.3		12
Chlorine	1.3	0.4	1.8		19
Trace elements µg per g DM					
Barium	11.0	6.6	13.2		4
Chromium	0.88	0.50	1.33		4
Cobalt	0.14	0.10	0.18		4
Copper	5.1	2.5	8.8		4
Iron	744	368	1180		4
Lead	4.0	2.6	6.6		4
Manganese	21	17	31		4
Molybdenum	0.33	0.10	0.80		4
Nickel	1.74	1.19	2.80		4
Silver	0.22	0.22	0.22		4
Strontium	881	350	1362		4
Tin	2.5	2.2	3.4		4
Titanium	20	14	35		4
Vanadium	1.30	0.64	1.71		4
Zinc	119	106	137		4

2.150 WHALE MEAT MEAL

	Mean	Min.	Max.	SD	No.
Dry matter %	94.0	91.7	95.3	1.12	22
Proximate constituents % of DM					
Crude protein	77.4	73.5	81.7	3.23	22
Ether extract	6.8	4.8	8.9	1.52	22
Total ash	16.5	14.7	18.2	1.33	20
Insol. ash	0.92	0.34	1.97	0.59	20
Sol. ash	15.6	14.0	17.2	1.10	20
Major elements % of DM					
Calcium	4.31	3.40	4.87	0.71	20
Phosphorus	2.17	1.80	2.55	0.24	20
Chlorine	1.87	1.18	2.42	0.12	22

2.151 FISHMEAL, WHITE FISH

	Mean	Min.	Max.	SD	No.
Dry matter %	90.7	89.2	92.5	1.44	60
Proximate constituents % of DM					
Crude protein	70.6	68.8	72.9	1.79	60
Ether extract	5.8	4.6	7.0	0.88	60
Nitrogen-free extract	1.4	0.2	3.5	1.51	60
Total ash	22.2	21.2	24.0	1.15	60
Insol. ash	0.53	0.11	1.44	0.47	60
Sol. ash	21.7	20.2	23.4	1.20	60
Major elements % of DM					
Calcium	6.8	6.2	7.8	0.60	31
Phosphorus	3.8	3.0	4.5	0.46	31
Chlorine	1.08	0.43	1.44	0.39	60

2.152 FISHMEAL, SOUTH AFRICAN PILCHARDS

	Mean	Min.	Max.	SD	No.
Dry matter %	90.4	88.6	91.6	0.97	48
Proximate constituents % of DM					
Crude protein	70.0	66.4	73.8	2.76	48
Ether extract	4.5	2.9	6.7	1.52	48
Nitrogen-free extract	5.38	1.8	9.1	3.01	48
Total ash	20.1	17.9	22.5	1.96	48
Insol. ash	1.02	0.47	1.69	0.49	48
Sol. ash	19.1	17.5	21.5	1.86	48
Major elements % of DM					
Calcium	5.34	4.49	6.55	0.82	48
Phosphorus	3.16	2.80	3.63	0.98	48
Chlorine	2.14	1.22	2.67	0.75	48

2.153 HERRING MEAL

	Mean	Min.	Max.	SD	No.
Dry matter %	91.8	88.7	94.8	1.82	83
Proximate constituents % of DM					
Crude protein	78.2	74.4	81.0	2.53	83
Ether extract	9.6	7.5	11.8	1.80	83
Nitrogen-free extract	2.5	1.0	5.3	2.02	60
Total ash	10.1	8.3	11.3	1.23	60
Insol. ash	0.15	0.10	0.30	0.12	60
Sol. ash	9.9	8.1	11.2	1.22	60
Major elements % of DM					
Calcium	2.08	1.55	2.57	0.40	83
Phosphorus	2.49	1.63	3.10	0.67	83
Chlorine	1.02	0.34	1.64	0.57	83

2.153 HERRING MEAL (continued)

	Mean	Min.	Max.	SD	No.
	Amino acids % of DM				
Alanine	4.85	2.26	5.97		19
Arginine	4.44	3.14	6.15		19
Aspartic acid	7.42	5.90	8.69		19
Cystine	0.80	0.63	1.08		19
Glutamic acid	10.88	8.96	12.91		19
Glycine	4.64	2.95	5.85		19
Histidine	2.05	0.65	4.17		18
*Iso*leucine	3.40	2.95	4.04		19
Leucine	5.77	4.91	7.20		19
Lysine	5.72	4.69	6.58		19
Methionine	2.07	1.75	2.43		18
Phenylalanine	3.00	2.18	4.04		19
Proline	3.24	2.21	4.19		15
Serine	3.25	2.81	4.67		19
Threonine	3.45	2.97	4.24		19
Tryptophan	0.61	0.31	0.74		19
Tyrosine	2.58	2.03	4.16		19
Valine	4.29	3.46	5.37		19
Available lysine	4.84	3.96	5.95		19

2.154 BLOOD MEAL

	Mean	Min.	Max.	SD	No.
Dry matter %	90.4	90.3	90.4		4
	Proximate constituents % of DM				
Crude protein	92.2	91.5	93.2		4
Ether extract	1.30	0.18	2.43		2
Nitrogen-free extract	3.9	3.8	4.0		2
Total ash	2.42	2.06	2.79		2
	Major elements % of DM				
Calcium	0.08	0.01	0.16		2
Phosphorus	0.30	0.08	0.53		2
Chlorine	0.40	0.39	0.40		3

2.155 POULTRY OFFAL MEAL

	Mean	Min.	Max.	SD	No.
Dry matter %	92.6	89.4	95.6	2.01	49
	Proximate constituents % of DM				
Crude protein	68.0	60.6	74.7	6.22	49
Ether extract	24.6	15.2	29.0	5.12	49
Total ash	6.24	4.7	8.0	1.50	44
	Major elements % of DM				
Calcium	1.43	0.41	5.05		6
Phosphorus	0.64	0.29	1.33		6
Chlorine	0.76	0.68	0.83		4

2.156 BONE FLOUR, STEAMED

	Mean	Min.	Max.	SD	No.
Dry matter %	86.5				1
	Proximate constituents % of DM				
Total ash	74.2				1
	Major elements % of DM				
Calcium	26.7				1
Phosphorus	10.0				1

2.157 GREAVES *

	Mean	Min.	Max.	SD	No.
Dry matter %	92.0	90.6	92.8		3
	Proximate constituents % of DM				
Crude protein	47.6	46.8	48.3		3
Ether extract	32.9	32.6	33.2		3
Crude fibre	1.5	1.1	1.7		3
Total ash	17.1	16.3	17.8		3
	Major elements % of DM				
Calcium	3.41	3.29	3.50		3
Phosphorus	1.53	1.49	1.62		3
Chlorine	1.09	1.05	1.12		3

* The sediment from melted tallow "cracklings".

2.158 *DACTYLIS GLOMERATA:* COCKSFOOT, ARTIFICIALLY DRIED

	Mean	*Min.*	*Max.*	*SD*	*No.*
Dry matter %	89.2	85.4	92.4		9
	Proximate constituents % of DM				
Crude protein	15.2	13.2	18.3		10
Crude fibre	24.0	22.7	26.5		4

2.159 COCKSFOOT AND CLOVER, FRESH HERBAGE

	Mean	*Min.*	*Max.*	*SD*	*No.*
Dry matter %	19.6	12.5	39.0	5.0	288
	Proximate constituents % of DM				
Crude protein	18.7	14.7	22.9	3.02	288
Ether extract	3.26	2.88	3.63		2
Crude fibre	19.9	15.8	23.6	3.15	288

2.160 COCKSFOOT AND CLOVER, ARTIFICIALLY DRIED

	Mean	*Min.*	*Max.*	*SD*	*No.*
Dry matter %	85.3				1
	Proximate constituents % of DM				
Crude protein	19.5				1

2.161 COCKSFOOT AND CLOVER, HAY

	Mean	*Min.*	*Max.*	*SD*	*No.*
Dry matter %	82.0	78.9	84.4		3
	Proximate constituents % of DM				
Crude protein	12.3	9.7	14.4		4
Crude fibre	28.8	27.1	30.2		4
	Major elements % of DM				
Calcium	0.91	0.60	1.16		4
Phosphorus	0.31	0.23	0.39		4
	Trace elements µg per g DM				
Copper	6.4				1

2.162 COCKSFOOT AND LUCERNE, HAY

	Mean	*Min.*	*Max.*	*SD*	*No.*
Dry matter %	79.3	70.1	87.2	5.01	189
	Proximate constituents % of DM				
Crude protein	15.2	10.2	19.4	3.23	189
Crude fibre	33.1	28.0	38.4	4.19	181
	Major elements % of DM				
Calcium	1.37	0.76	1.95	0.47	181
Phosphorus	0.30	0.22	0.37	0.08	181

2.163 *LOLIUM MULTIFLORUM:* ITALIAN RYEGRASS, HAY

	Mean	*Min.*	*Max.*	*SD*	*No.*
Dry matter %	80.5	66.5	86.1		6
	Proximate constituents % of DM				
Crude protein	9.7	6.5	13.8		7
Crude fibre	31.5	23.4	39.2		7
	Major elements % of DM				
Calcium	0.51	0.18	0.77		6
Phosphorus	0.24	0.18	0.34		6
	Trace elements µg per g DM				
Cobalt	0.05				1
Copper	4.8				1
Titanium	9.8				1

2.164 ITALIAN RYEGRASS, ARTIFICIALLY DRIED/ BARN DRIED HAY

	Mean	*Min.*	*Max.*	*SD*	*No.*
Dry matter %	87.6	86.3	88.8		3
	Proximate constituents % of DM				
Crude protein	8.5	7.4	12.1		3
Crude fibre	27.1	24.5	32.3		3
	Vitamins µg per g DM				
β-carotene	88				1

2.165 ITALIAN RYEGRASS AND PERENNIAL RYE-GRASS, HAY

	Mean	*Min.*	*Max.*	*SD*	*No.*
	Proximate constituents % of DM				
Crude protein	6.8				1
Crude fibre	31.9				1

2.166 RYEGRASS, PARTLY CURED HAY

	Mean	*Min.*	*Max.*	*SD*	*No.*
Dry matter %	54.7				1
	Proximate constituents % of DM				
Crude protein	12.9				1
Crude fibre	25.0				1
	Major elements % of DM				
Calcium	0.57				1
Phosphorus	0.30				1

2.167 RYEGRASS AND CLOVER, FRESH HERBAGE

	Mean	*Min.*	*Max.*	*SD*	*No.*
Dry matter %	21.0	13.4	32.7	4.78	109
Proximate constituents % of DM					
Crude protein	16.4	10.4	23.1	4.85	433
Crude fibre	19.8	16.3	23.9	3.11	432

2.168 *FESTUCA* SP: FESCUE, HAY

	Mean	*Min.*	*Max.*	*SD*	*No.*
Dry matter %	84.1	82.3	84.8		9
Proximate constituents % of DM					
Crude protein	9.4	8.0	13.0		11
Ether extract	2.8				1
Crude fibre	35.9	32.2	39.9		11
Major elements % of DM					
Calcium	0.54	0.45	0.69		8
Phosphorus	0.25	0.20	0.32		8
Trace elements μg per g DM					
Manganese	41				1
Vitamins μg per g DM					
β-carotene	8	4	12		2

2.169 *PHLEUM PRATENSE:* TIMOTHY, HAY

	Mean	*Min.*	*Max.*	*SD*	*No.*
Dry matter %	86.2	83.1	90.0	1.16	39
Proximate constituents % of DM					
Crude protein	7.4	4.7	9.8	1.07	49
Crude fibre	36.0	32.5	43.8	3.30	49
Major elements % of DM					
Calcium	0.48	0.21	0.75	0.13	24
Phosphorus	0.21	0.15	0.26	0.04	24
Magnesium	0.09				1
Potassium	1.71	1.35	1.95		4
Sodium	0.27	0.26	0.28		2
Sulphur	0.12	0.10	0.13		2
Trace elements μg per g DM					
Cobalt	0.15	0.10	0.18		3
Copper	7.3	5.9	8.7		4
Manganese	43	41	44		4
Molybdenum	1.1	0.7	1.4		2

2.170 TIMOTHY, MEADOWGRASS AND CLOVER

	Mean	*Min.*	*Max.*	*SD*	*No.*
Dry matter %	21.9	19.6	24.0		3
Proximate constituents % of DM					
Crude protein	12.0	10.8	12.8		3
Ether extract	3.0	2.0	3.6		3
Crude fibre	24.0	20.7	26.2		3

2.171 SEEDS GRASS, ARTIFICIALLY DRIED

	Mean	*Min.*	*Max.*	*SD*	*No.*
Dry matter %	89.1	86.8	91.7		7
Proximate constituents % of DM					
Crude protein	12.7	8.3	18.0		7
Crude fibre	17.8	13.2	22.4		2
Major elements % of DM					
Calcium	0.40	0.35	0.44		2
Phosphorus	0.34	0.33	0.35		2

2.172 LEY GRASS, ARTIFICIALLY DRIED

	Mean	*Min.*	*Max.*	*SD*	*No.*
Dry matter %	87.6	82.9	90.5		11
Proximate constituents % of DM					
Crude protein	13.7	9.7	17.8		11
Crude fibre	27.9	25.5	29.8		6
Major elements % of DM					
Calcium	0.51	0.49	0.52		2
Phosphorus	0.31	0.30	0.31		2

2.173 LEY GRASS, HAY

	Mean	*Min.*	*Max.*	*SD*	*No.*
Dry matter %	86.1	81.0	90.7		19
Proximate constituents % of DM					
Crude protein	8.5	5.4	16.9		18
Crude fibre	32.0	27.0	39.0		18
Major elements % of DM					
Calcium	0.58	0.33	1.21		17
Phosphorus	0.24	0.17	0.35		17

2.174 MEADOW GRASS, ARTIFICIALLY DRIED

	Mean	Min.	Max.	SD	No.
Dry matter %	89.5	88.8	90.1		4
Proximate constituents % of DM					
Crude protein	17.5	15.1	19.3		4
Crude fibre	21.4				1
Major elements % of DM					
Calcium	1.14				1
Phosphorus	0.38				1

2.175 MEADOW GRASS AND LEY GRASS, HAY

	Mean	Min.	Max.	SD	No.
Dry matter %	84.2	80.2	86.5		4
Proximate constituents % of DM					
Crude protein	8.7	6.6	9.9		4
Crude fibre	32.0	29.2	35.2		4
Major elements % of DM					
Calcium	0.68	0.53	0.85		4
Phosphorus	0.22	0.20	0.23		4
Trace elements µg per g DM					
Cobalt	0.07	0.07	0.07		2
Copper	4.9	3.8	6.0		2
Titanium	9.8				1

2.176 MEADOW GRASS AND SEEDS GRASS, HAY

	Mean	Min.	Max.	SD	No.
Dry matter %	77.3				1
Proximate constituents % of DM					
Crude protein	10.1				1
Crude fibre	29.8				1
Major elements % of DM					
Calcium	0.86				1
Phosphorus	0.29				1

2.177 PASTURE GRASS, ARTIFICIALLY DRIED

	Mean	Min.	Max.	SD	No.
Dry matter %	90.1	88.2	91.3		6
Proximate constituents % of DM					
Crude protein	12.7	9.4	15.4		7
Crude fibre	24.5	19.6	27.3		4
Major elements % of DM					
Calcium	0.59				1
Phosphorus	0.30				1

2.178 PASTURE GRASS, HAY

	Mean	Min.	Max.	SD	No.
Trace elements µg per g DM					
Barium	20.0	3.1	88.0	14.05	78
Chromium	0.67	0.08	7.6	1.16	87
Cobalt	0.24	0.02	6.30	0.61	235
Copper	6.9	0.1	21.0	3.18	208
Iron	139	35	941	95.04	238
Lead	4.31	0.57	30.0	7.76	87
Manganese	158	16	693	119.6	123
Molybdenum	2.20	0.06	35.0	4.36	199
Nickel	1.82	0.27	7.20	1.32	96
Silver	0.09	0.05	0.50	0.06	87
Strontium	23	3.0	53	9.87	78
Tin	0.85	0.46	2.99	0.49	87
Titanium	13	0.5	69	12.53	91
Vanadium	0.66	0.02	4.70	0.83	87
Zinc	50	15	513	18.73	86

2.179 OAT, WHOLE PLANT AS FODDER

	Mean	Min.	Max.	SD	No.
Dry matter %	25.2	15.6	43.0	4.5	32
Proximate constituents % of DM					
Crude protein	10.4	4.7	27.6	3.9	38
Ether extract	2.58	1.50	7.37	0.84	28
Crude fibre	30.5	11.8	41.5	4.1	32
Major elements % of DM					
Calcium	0.35	0.13	0.80		8
Phosphorus	0.32	0.23	0.38		8
Magnesium	0.14	0.11	0.16		3
Potassium	1.27	0.62	1.97		8
Sulphur	0.15				1
Trace elements µg per g DM					
Barium	20	4.4	84		17
Chromium	0.27	0.08	1.30		19
Cobalt	0.19	0.03	1.30		19
Copper	4.4	0.8	14.0	3.49	128
Iron	113	34	634	121	24
Lead	11	0.6	147		19
Manganese	68	5.0	242	49.9	132
Molybdenum	0.60	0.14	1.78	0.46	20
Nickel	1.56	0.30	7.90	2.2	21
Silver	0.14	0.05	0.70		19
Strontium	15	7	31		17
Tin	1.43	0.50	9.20		19
Titanium	7.6	0.9	52		19
Vanadium	0.22	0.03	1.50		19
Zinc	55	9.3	444	98.6	20

2.180 OAT, LEAF

	Mean	Min.	Max.	SD	No.
	Trace elements µg per g DM				
Barium	11				1
Chromium	0.72				1
Cobalt	0.09				1
Copper	12				1
Iron	65				1
Lead	<0.66				1
Manganese	86				1
Molybdenum	26				1
Nickel	0.26				1
Silver	<0.05				1
Strontium	18				1
Tin	0.61				1
Titanium	1.20				1
Vanadium	0.34				1
Zinc	53				1

2.181 OAT, WHOLE PLANT, ARTIFICIALLY DRIED

	Mean	Min.	Max.	SD	No.
Dry matter %	89.7				1
	Proximate constituents % of DM				
Crude protein	10.5	8.1	12.9		2
Crude fibre	28.7				1

2.182 OAT, HAY

	Mean	Min.	Max.	SD	No.
Dry matter %	83.6	69.1	89.2	6.09	39
	Proximate constituents % of DM				
Crude protein	8.2	6.0	9.7	1.25	49
Ether extract	3.02	2.34	4.25		5
Crude fibre	32.3	26.4	39.1	3.21	46
	Major elements % of DM				
Calcium	0.41	0.23	0.73	0.12	32
Phosphorus	0.29	0.13	0.53	0.09	33
Potassium	0.90	0.80	1.00		3
Sodium	0.43	0.24	0.60		3
	Trace elements µg per g DM				
Cobalt	0.18				1
Copper	4.4	4.3	4.5		2
Manganese	13				1
Molydbenum	0.36				1
	Vitamins µg per g DM				
β-carotene	7.6				1

2.183 WHEAT, WHOLE PLANT AS FODDER

	Mean	Min.	Max.	SD	No.
Dry matter %	28.4	21.7	34.8		9
	Proximate constituents % of DM				
Crude protein	20.9	13.0	32.5		15
Ether extract	2.67	2.56	2.77		5
Crude fibre	11.9	7.3	18.3		9
	Major elements % of DM				
Calcium	0.26	0.15	0.40		8
Phosphorus	0.43	0.37	0.52		11
Magnesium	0.13	0.11	0.14		3
Potassium	2.15	1.24	3.48		10
	Trace elements µg per g DM				
Barium	12	1	28		15
Chromium	1.33	0.12	12.0		15
Cobalt	0.06	0.03	0.21		15
Copper	6.3	3.0	15.0	2.6	33
Iron	228	48	970		16
Lead	1.53	0.71	3.90		15
Manganese	54	4	309	65	29
Molybdenum	0.50	0.08	1.23		16
Nickel	0.40	0.15	0.95		15
Silver	0.10	0.06	0.29		13
Strontium	10.9	3.6	19.0		14
Tin	1.03	0.57	2.86		15
Vanadium	2.82	0.03	29.0		15
Zinc	47	22	193		16

2.184 BARLEY, WHOLE PLANT AS FODDER

	Mean	Min.	Max.	SD	No.
Dry matter %	16.7	7.03	31.5	5.27	101
	Proximate constituents % of DM				
Crude protein	16.9	10.0	26.8	4.3	29
Ether extract	2.89	1.6	3.95		4
Crude fibre	10.4	5.0	16.1		5
	Major elements % of DM				
Calcium	0.61	0.06	0.89		16
Phosphorus	0.36	0.16	0.55	0.14	110
Magnesium	0.10	0.08	0.16		13
Potassium	4.04	2.27	5.67	1.08	150
Sodium	0.18	0.09	0.28	0.09	42
Chlorine	0.08	0.04	0.11		2
	Trace elements µg per g DM				
Barium	6.1	4.5	8.2		9
Chromium	0.27	0.07	1.61		18
Cobalt	0.05	0.02	0.26	0.06	21
Copper	7.4	1.3	53.0	10.5	34
Lead	1.05	0.27	2.78	0.64	20
Manganese	50.0	12.0	158.0	31:0	34

2.184 BARLEY, WHOLE PLANT AS FODDER (continued)

	Mean	Min.	Max.	SD	No.
	Trace elements μg per g DM				
Molybdenum	0.83	0.14	6.20	1.4	20
Nickel	0.31	0.09	1.23	0.29	20
Silver	0.07	0.05	0.24	0.05	20
Strontium	16.9	2.8	36.0		9
Tin	0.56	0.47	0.88	0.12	20
Titanium	3.72	0.91	17.00	4.30	20
Vanadium	0.19	0.05	0.53	0.13	20
Zinc	28.0	17.0	118.0	22.0	20

2.185 MAIZE, WHOLE PLANT AS FODDER

	Mean	Min.	Max.	SD	No.
Dry matter %	19.3	13.0	28.5	3.89	100
	Proximate constituents % of DM				
Crude protein	10.1	7.4	15.5	2.73	115
Ether extract	1.36	0.80	1.84	0.29	64
Crude fibre	25.0	15.9	35.0	5.11	88
	Major elements % of DM				
Calcium	0.56	0.08	0.74		15
Phosphorus	0.28	0.15	0.59	0.08	32
Magnesium	0.46	0.14	0.90		19
Potassium	1.84	0.50	4.90	1.22	32
Sodium	0.04	0.02	0.07		14
Chlorine	0.26	0.02	1.50		11
Sluphur	0.07	0.03	0.13		14
	Trace elements μg per g DM				
Copper	7.6	5.4	9.2		14
Iron	221	110	580		14
Molybdenum	0.3	0.1	0.6		14

2.186 MAIZE COB AS FODDER

	Mean	Min.	Max.	SD	No.
Dry matter %	19.5	10.2	33.9	6.87	129
	Proximate constituents % of DM				
Crude protein	10.1	6.3	12.7	1.67	131
Ether extract	1.63	0.70	4.20	0.57	56
Crude fibre	16.5	10.9	23.2	3.28	131
	Major elements % of DM				
Calcium	0.10	0.04	0.40	0.04	26
Phosphorus	0.33	0.29	0.37	0.02	26
Potassium	1.15	0.79	1.59		12

2.187 MAIZE LEAF AS FODDER

	Mean	Min.	Max.	SD	No.
Dry matter %	22.4	15.8	36.8	5.6	50
	Proximate constituents % of DM				
Crude protein	14.1	6.7	19.1	2.7	50
Ether extract	1.6	0.75	2.20		4
Crude fibre	26.8	23.0	31.9	1.9	50

2.188 MAIZE STEM AS FODDER

	Mean	Min.	Max.	SD	No.
Dry matter %	17.4	12.7	23.0	2.44	127
	Proximate constituents % of DM				
Crude protein	7.5	4.5	11.1	1.77	127
Ether extract	1.07	0.30	2.03	0.36	51
Crude fibre	27.5	23.5	32.4	2.44	126
	Major elements % of DM				
Calcium	0.51	0.28	0.69	0.08	25
Phosphorus	0.24	0.17	0.31	0.03	25
Potassium	1.58	1.33	1.94		10

2.189 RYE, WHOLE PLANT AS FODDER

	Mean	Min.	Max.	SD	No.
Dry matter %	18.7	12.9	30.1	4.73	102
	Proximate constituents % of DM				
Crude protein	21.8	15.3	28.9	5.41	103
Ether extract	2.98	2.61	3.39		8
Crude fibre	20.6	14.4	30.5	7.15	99
	Major elements % of DM				
Calcium	0.51	0.33	1.27		12
Phosphorus	0.43	0.36	0.67		12
Magnesium	0.17				1
Potassium	2.52	1.97	2.78		6
Sodium	0.02	0.02	0.03		6
	Trace elements μg per g DM				
Cobalt	0.08				1
Copper	9.0	8.5	9.5		2
Manganese	14				1

2.190 RYE, HAY

	Mean	*Min.*	*Max.*	*SD*	*No.*
Dry matter %	86.3	84.5	87.8		3
Proximate constituents % of DM					
Crude protein	7.8	6.5	8.7		3
Crude fibre	42.3	38.5	47.5		3
Major elements % of DM					
Calcium	0.36	0.34	0.38		2
Phosphorus	0.28	0.26	0.30		2

2.191 CEREALS, MIXED, WHOLE PLANT AS FODDER

	Mean	*Min.*	*Max.*	*SD*	*No.*
Dry matter %	28.5				1
Proximate constituents % of DM					
Crude protein	9.5				1
Crude fibre	25.2				1

2.192 CEREALS, MIXED, WHOLE PLANT, ARTIFICIALLY DRIED

	Mean	*Min.*	*Max.*	*SD*	*No.*
Dry matter %	86.4				1
Proximate constituents % of DM					
Crude protein	11.1				1
Crude fibre	37.6				1
Vitamins μg per g DM					
β-carotene	6.7				1

2.193 CEREALS, MIXED, HAY

	Mean	*Min.*	*Max.*	*SD*	*No.*
Dry matter %	84.7	83.4	86.0		2
Proximate constituents % of DM					
Crude protein	12.2	10.4	15.2		3
Crude fibre	33.3	25.5	37.6		3
Major elements % of DM					
Calcium	0.72	0.41	1.02		2
Phosphorus	0.34	0.18	0.50		2
Vitamins μg per g DM					
β-carotene	6.7				1

2.194 CEREAL AND LEGUME, WHOLE PLANT AS FODDER

	Mean	*Min.*	*Max.*	*SD*	*No.*
Dry matter %	15.0	12.2	16.9		5
Proximate constituents % of DM					
Crude protein	18.2	14.4	23.3		5
Ether extract	2.42	2.29	2.55		3
Crude fibre	28.7	20.6	34.8		5
Vitamins μg per g DM					
β-carotene	71				1

2.195 CEREAL AND LEGUME, HAY

	Mean	*Min.*	*Max.*	*SD*	*No.*
Dry matter %	82.8	70.0	88.3	4.14	26
Proximate constituents % of DM					
Crude protein	11.8	6.9	18.8	3.2	27
Crude fibre	32.3	25.4	40.3	4.2	27
Major elements % of DM					
Calcium	1.11	0.47	2.36	0.49	27
Phosphorus	0.24	0.13	0.36	0.07	27
Vitamins μg per g DM					
β-carotene	142				1

2.196 *TRIFOLIUM:* CLOVER, ARTIFICIALLY DRIED

	Mean	*Min.*	*Max.*	*SD*	*No.*
Dry matter %	89.3				1
Proximate constituents % of DM					
Crude protein	15.4				1
Crude fibre	30.1				1

2.197 CLOVER MEAL, ARTIFICIALLY DRIED

	Mean	*Min.*	*Max.*	*SD*	*No.*
Dry matter %	94.9	93.7	95.3		6
Major elements % of DM					
Calcium	1.80	1.44	2.28		6
Phosphorus	0.20	0.17	0.22		6
Potassium	1.60	1.02	1.96		6
Trace elements μg per g DM					
Boron	17.2	12.1	19.5		6

2.198 *CERATONIA:* LOCUST BEAN, WHOLE POD

	Mean	Min.	Max.	SD	No.
Dry matter %	86.7	77.4	89.1	1.31	32
Proximate constituents % of DM					
Crude protein	4.69	3.70	5.20	0.57	32
Ether extract	0.36	0.17	0.52	0.15	32
Nitrogen-free extract	84.1	79.8	85.8		16
Crude fibre	7.73	6.68	9.51		16
Total ash	3.12	2.59	4.76		16
Insol. ash	0.26	0.02	0.58		16
Sol. ash	2.85	2.42	4.52		16

2.199 RUNNER BEAN, WHOLE PLANT AS FODDER

	Mean	Min.	Max.	SD	No.
Dry matter %	7.6				1
Proximate constituents % of DM					
Crude protein	24.0				1
Crude fibre	16.6				1
Major elements % of DM					
Calcium	0.29				1
Phosphorus	0.61				1

2.200 PEA, WHOLE PLANT AS FODDER

	Mean	Min.	Max.	SD	No.
Dry matter %	19.9				1
Proximate constituents % of DM					
Crude protein	20.6				1

2.201 PEA HAULM, AS FODDER

	Mean	Min.	Max.	SD	No.
Dry matter %	14.2	14.1	14.5		3
Proximate constituents % of DM					
Crude protein	17.1	14.2	18.9		3
Ether extract	2.15	2.00	2.30		2
Crude fibre	24.7	22.0	28.0		3

2.202 PEA HAULM, HAY

	Mean	Min.	Max.	SD	No.
Dry matter %	76.0				1
Proximate constituents % of DM					
Crude protein	17.6				1
Crude fibre	30.7				1
Major elements % of DM					
Calcium	1.78				1
Phosphorus	0.38				1
Potassium	3.50				1
Sodium	0.22				1

2.203 LUPIN, WHOLE PLANT AS FODDER

	Mean	Min.	Max.	SD	No.
Dry matter %	12.8	6.0	22.4	3.81	53
Proximate constituents % of DM					
Crude protein	20.0	11.2	28.9	3.60	53
Ether extract	2.18	1.70	2.70		3
Crude fibre	28.4	25.0	31.5		4
Major elements % of DM					
Calcium	1.98	1.39	2.62		5
Phosphorus	0.42	0.25	0.53		5
Magnesium	0.20				1
Potassium	2.79	1.03	4.03		5
Sodium	0.31				1
Chlorine	0.33				1

2.204 LUPIN, SWEET, WHOLE PLANT AS FODDER

	Mean	Min.	Max.	SD	No.
Dry matter %	14.0	13.9	14.1		3
Proximate constituents % of DM					
Crude protein	8.9	8.3	9.5		2

2.205 LUPIN, BITTER, WHOLE PLANT AS FODDER

	Mean	Min.	Max.	SD	No.
Dry matter %	11.0				1
Proximate constituents % of DM					
Crude protein	20.0				1

2.206 LEGUMES, MIXED, FRESH HERBAGE

	Mean	Min.	Max.	SD	No.
Dry matter %	27.5	16.2	47.7	5.33	29
Proximate constituents % of DM					
Crude protein	13.3	7.5	22.7	4.6	32
Ether extract	2.46	1.91	2.92		13
Crude fibre	29.9	16.6	40.0	5.18	31
Major elements % of DM					
Calcium	1.23	1.00	1.46		4
Phosphorus	0.32	0.20	0.61		4
Magnesium	0.59				1

2.207 LEGUMES, MIXED, ARTIFICIALLY DRIED

	Mean	*Min.*	*Max.*	*SD*	*No.*
	Proximate constituents % of DM				
Crude protein	14.0	11.6	16.9		3
Crude fibre	27.9	22.7	33.9		3

2.208 LEGUMES, MIXED, HAY

	Mean	*Min.*	*Max.*	*SD*	*No.*
Dry matter %	84.7	76.6	90.5	3.31	46
	Proximate constituents % of DM				
Crude protein	14.5	7.2	19.7	2.4	46
Ether extract	1.7	1.1	3.6	0.6	35
Crude fibre	31.7	22.6	47.3	5.1	46
	Major elements % of DM				
Calcium	1.44	0.52	2.25	0.34	46
Phosphorus	0.27	0.11	0.39	0.06	46
Sodium	0.21	0.09	0.60	0.10	35
Chlorine	0.32	0.13	0.92	0.16	35

2.209 *BRASSICA OLERACEA ACEPHALA:* KALE, MARROWSTEM, WHOLE PLANT

	Mean	*Min.*	*Max.*	*SD*	*No.*
Dry matter %	12.9	9.3	16.3	1.83	98
	Proximate constituents % of DM				
Crude protein	16.1	11.0	25.3	4.62	31
Digestibility of CP	86.4	80.0	92.5		5
Ether extract	2.33				1
Nitrogen-free extract	50.1				1
Crude fibre	18.6	13.0	26.3	4.75	26
Total ash	11.2	7.0	15.5		8
	Major elements % of DM				
Calcium	1.81	1.12	2.91		14
Phosphorus	0.33	0.23	4.31		14
Magnesium	0.16				1
Potassium	3.24				1
	Trace elements μg per g DM				
Cobalt	0.04	0.03	0.05		2
Copper	5.7	2.2	9.0		7
Iodine	0.15	0.08	0.21		2
Manganese	38.0	10.0	126.0		7

2.210 KALE, MARROWSTEM, LEAF

	Mean	*Min.*	*Max.*	*SD*	*No.*
Dry matter %	13.6	8.9	20.3	3.09	143
	Proximate constituents % of DM				
Crude protein	19.5	13.7	25.7	4.64	143
Digestibility of CP	84.7	82.5	88.0	1.50	20
Ether extract	3.7	3.3	4.1		6
Crude fibre	13.1	11.0	15.4	1.82	101
	Major elements % of DM				
Calcium	3.42	2.60	4.05		12
Phosphorus	0.37	0.31	0.43		12
Magnesium	0.22	0.15	0.30		6
Potassium	2.88	2.20	3.90		6
Sodium	0.24	0.10	0.65		6
Chlorine	1.46	1.15	1.80		5
Sulphur	0.50	0.50	0.50		2
	Trace elements μg per g DM				
Copper	4.6	3.5	5.4		5
Manganese	31	16	52		5
Zinc	38	37	40		3

2.211 KALE, MARROWSTEM, STEM

	Mean	*Min.*	*Max.*	*SD*	*No.*
Dry matter %	13.4	7.6	21.4	3.39	150
	Proximate constituents % of DM				
Crude protein	12.5	7.2	19.1	4.11	151
Digestibility of CP	85.7	80.0	95.0	3.2	20
Ether extract	1.37	0.30	2.20		16
Crude fibre	20.6	13.8	29.2	5.81	109
	Major elements % of DM				
Calcium	1.02	0.42	3.09	0.61	22
Phosphorus	0.34	0.21	0.48	0.08	22
Magnesium	0.20	0.12	0.26		16
Potassium	3.88	1.85	5.90		16
Sodium	0.34	0.11	1.11		16
Chlorine	1.29	0.36	1.72		15
Sulphur	0.29	0.08	0.48		12
	Trace elements μg per g DM				
Copper	3.9	2.4	5.6		5
Manganese	19	2.3	79		5
Zinc	26	23	31		3

2.212 KALE, MARROWSTEM, ARTIFICIALLY DRIED

	Mean	Min.	Max.	SD	No.
Dry matter %	88.2				1
Proximate constituents % of DM					
Crude protein	15.1				1
Ether extract	2.4				1
Crude fibre	17.2				1
Major elements % of DM					
Calcium	2.0				1
Phosphorus	0.3				1
Potassium	1.96				1
Sodium	0.73				1
Chlorine	1.13				1

2.213 *BRASSICA OLERACEA FRUTICOSA:* KALE, THOUSAND-HEADED, WHOLE PLANT

	Mean	Min.	Max.	SD	No.
Dry matter %	14.7	11.1	20.7	2.46	44
Proximate constituents % of DM					
Crude protein	20.8	11.4	30.6	4.6	47
Crude fibre	16.9	12.3	22.7	3.2	37
Major elements % of DM					
Calcium	1.81	0.31	2.65		9
Phosphorus	0.45	0.15	1.38		10
Trace elements µg per g DM					
Cobalt	0.08				1
Copper	4.5	3.7	5.5		3
Iodine	0.07				1
Manganese	30	18	43		3

2.214 KALE, THOUSAND-HEADED, LEAF

	Mean	Min.	Max.	SD	No.
Dry matter %	15.8	11.4	29.0	4.14	31
Proximate constituents % of DM					
Crude protein	22.1	16.4	32.4	4.01	32
Ether extract	3.9	3.6	4.3		5
Crude fibre	13.4	10.5	27.5	3.26	31
Major elements % of DM					
Calcium	2.12	1.06	3.58		7
Phosphorus	0.31	0.28	0.34		7
Magnesium	0.13	0.08	0.27		5
Potassium	2.32	2.03	2.75		5
Sodium	0.15	0.14	0.16		5
Chlorine	0.98	0.88	1.20		4
Sulphur	0.62	0.52	0.72		4
Trace elements µg per g DM					
Copper	3.5				1
Manganese	38.0				1

2.215 KALE, THOUSAND-HEADED, STEM

	Mean	Min.	Max.	SD	No.
Dry matter %	19.1	12.6	27.6	4.63	38
Proximate constituents % of DM					
Crude protein	13.8	7.3	18.9	4.1	38
Ether extract	1.12	0.42	2.54		13
Crude fibre	23.9	12.5	42.7	9.67	37
Major elements % of DM					
Calcium	0.86	0.38	1.90		15
Phosphorus	0.21	0.11	0.32		15
Magnesium	0.11	0.07	0.22		13
Potassium	1.20	1.08	3.30		13
Sodium	0.23	0.09	1.06		13
Chlorine	0.43	0.12	0.96		12
Sulphur	0.26	0.09	0.38		12
Trace elements µg per g DM					
Copper	2.6				1
Manganese	11.0				1

2.216 KALE, ASPARAGUS, WHOLE PLANT

	Mean	Min.	Max.	SD	No.
Dry matter %	16.6	15.4	17.8		2
Proximate constituents % of DM					
Crude protein	24.5	21.5	27.5		2
Crude fibre	13.8	12.0	15.6		2

2.217 KALE, ASPARAGUS, LEAF

	Mean	Min.	Max.	SD	No.
Dry matter %	15.1	14.2	16.0		2
Proximate constituents % of DM					
Crude protein	27.3	21.6	32.9		2
Crude fibre	11.2	11.0	11.3		2

2.218 KALE, ASPARAGUS, STEM

	Mean	Min.	Max.	SD	No.
Dry matter %	19.2	18.9	19.5		2
Proximate constituents % of DM					
Crude protein	22.9	21.3	24.6		2
Crude fibre	16.0	13.8	18.1		2

2.219 KALE, COTTAGERS, WHOLE PLANT

	Mean	Min.	Max.	SD	No.
Dry matter %	14.8	13.1	16.5		2
Proximate constituents % of DM					
Crude protein	18.4	17.7	19.0		2
Crude fibre	15.1	13.2	17.0		2

2.220 KALE, COTTAGERS, LEAF

	Mean	Min.	Max.	SD	No.
Dry matter %	13.3	10.9	15.6		2
Proximate constituents % of DM					
Crude protein	21.1	20.3	21.9		2
Crude fibre	10.3	9.9	10.7		2

2.221 KALE, COTTAGERS, STEM

	Mean	Min.	Max.	SD	No.
Dry matter %	17.7	15.8	19.6		2
Proximate constituents % of DM					
Crude protein	14.8	14.1	15.5		2
Crude fibre	21.4	19.8	22.9		2

2.222 KALE, CURLY, WHOLE PLANT

	Mean	Min.	Max.	SD	No.
Dry matter %	15.3	14.1	17.0		4
Proximate constituents % of DM					
Crude protein	20.2	12.4	27.5		4
Crude fibre	16.8	15.4	17.8		4

2.223 KALE, CURLY, LEAF

	Mean	Min.	Max.	SD	No.
Dry matter %	14.7	14.3	15.1		3
Proximate constituents % of DM					
Crude protein	28.8	23.5	37.0		3
Crude fibre	11.7	10.7	12.2		3

2.224 KALE, CYRLY, STEM

	Mean	Min.	Max.	SD	No.
Dry matter %	17.6	14.9	20.7		3
Proximate constituents % of DM					
Crude protein	17.1	12.8	19.8		3
Crude fibre	22.6	20.8	24.3		3

2.225 KALE, HUNGRY GAP, WHOLE PLANT

	Mean	Min.	Max.	SD	No.
Dry matter %	13.8	10.2	17.8		5
Proximate constituents % of DM					
Crude protein	24.8	20.9	29.8		5
Ether extract	4.4				1
Crude fibre	14.9	14.2	15.5		2

2.226 KALE, RAPE, WHOLE PLANT

	Mean	Min.	Max.	SD	No.
Dry matter %	13.2	9.9	17.5		5
Proximate constituents % of DM					
Crude protein	24.8	22.6	28.0		5
Ether extract	4.0				1
Crude fibre	15.0	14.1	15.8		2

2.227 KALE, RAPE, LEAF

	Mean	Min.	Max.	SD	No.
Dry matter %	13.3				1
Proximate constituents % of DM					
Crude protein	32.2				1
Crude fibre	12.8				1

2.228 KALE, RAPE, STEM

	Mean	Min.	Max.	SD	No.
Dry matter %	17.3				1
Proximate constituents % of DM					
Crude protein	19.3				1
Crude fibre	19.0				1

2.229 KALE, UNSPECIFIED, WHOLE PLANT

	Mean	Min.	Max.	SD	No.
Dry matter %	14.7	8.3	30.1	4.78	644
Proximate constituents % of DM					
Crude protein	17.4	10.2	22.7	4.99	775
Ether extract	2.09	1.03	3.32	0.76	29
Crude fibre	19.8	13.9	27.8	5.38	175
Major elements % of DM					
Calcium	1.88	0.99	3.20	0.68	182
Phosphorus	0.33	0.23	0.44	0.08	696
Magnesium	0.16	0.10	0.30	0.07	67
Potassium	2.69	1.97	3.40	0.65	467
Sodium	0.56	0.20	1.25		14
Chlorine	0.58	0.13	1.34		6
Sulphur	0.48	0.29	0.75		13

2.229 KALE, UNSPECIFIED, WHOLE PLANT (continued)

	Mean	Min.	Max.	SD	No.
Trace elements µg per g DM					
Arsenic	0.07				1
Barium	45				1
Chromium	0.2				1
Cobalt	0.09	0.04	0.19		11
Copper	5.1	1.2	10.9	3.8	57
Iodine	1.2	1.1	1.3		2
Iron	68				1
Lead	1.5				1
Manganese	72	2.9	272	92.1	67
Molybdenum	1.7	0.7	3.4		12
Nickel	0.54				1
Silver	0.1				1
Strontium	43				1
Tin	3.0				1
Titanium	4.1				1
Vanadium	0.16				1
Zinc	35	34	36		2

2.230 KALE, UNSPECIFIED, LEAF

	Mean	Min.	Max.	SD	No.
Dry matter %	13.8	8.7	21.7	3.12	673
Proximate constituents % of DM					
Crude protein	21.8	16.4	26.8	4.08	1027
Ether extract	3.66	3.03	4.47		12
Crude fibre	12.7	11.0	14.6	1.45	707
Major elements % of DM					
Calcium	2.66	1.97	3.45	0.60	134
Phosphorus	0.38	0.27	0.49	0.09	305
Magnesium	0.17	0.10	0.31	0.07	109
Potassium	2.42	1.80	3.12	0.52	280
Sodium	0.41	0.06	1.22		7
Chlorine	1.95	1.86	2.04		2
Sulphur	0.71	0.64	0.80		3
Trace elements µg per g DM					
Arsenic	0.1				1
Barium	30				1
Chromium	0.33				1
Cobalt	0.18	0.07	0.28		2
Copper	5.5	3.9	6.4		12
Iron	70				1
Lead	1.17				1
Manganese	47	15	144	23	25
Molybdenum	1.16	0.70	1.97		4
Nickel	0.48				1
Silver	0.06				1
Strontium	62				1
Tin	0.62				1
Titanium	2.96				1
Vanadium	0.20				1
Zinc	41	30	52		2

2.231 KALE, UNSPECIFIED, STEM

	Mean	Min.	Max.	SD	No.
Dry matter %	15.4	8.1	25.0	4.25	658
Proximate constituents % of DM					
Crude protein	13.5	8.1	19.4	4.89	995
Ether extract	1.14	0.88	1.48	0.11	22
Crude fibre	22.6	16.3	29.7	4.66	704
Major elements % of DM					
Calcium	0.87	0.56	1.31	0.26	133
Phosphorus	0.34	0.25	0.43	0.07	297
Magnesium	0.13	0.10	0.16	0.022	106
Potassium	3.00	2.15	3.98	0.65	271
Sodium	0.26	0.03	0.58		8
Chlorine	1.86				1
Sulphur	0.34	0.21	0.43		3
Trace elements µg per g DM					
Barium	47				1
Chromium	0.20				1
Cobalt	0.05	0.03	0.07		2
Copper	4.0	2.4	5.4		13
Iron	28				1
Lead	0.73				1
Manganese	10.1	1.6	15.6	3.3	22
Molybdenum	0.45	0.40	0.50		4
Nickel	0.25				1
Silver	0.06				1
Strontium	62				1
Tin	0.59				1
Titanium	1.57				1
Vanadium	0.07				1
Zinc	22	16	29		2
Vitamins µg per g DM					
β-carotene	6.0				1

2.232 *BRASSICA OLERACEA:* BROCCOLI, LEAF

	Mean	*Min.*	*Max.*	*SD*	*No.*
	Proximate constituents % of DM				
Crude protein	18.1	14.2	21.9		2
	Major elements % of DM				
Calcium	1.77	0.53	3.65		4
Phosphorus	0.35	0.28	0.41		2
Magnesium	0.12	0.08	0.16		4
Chlorine	2.82	2.49	3.28		4
	Trace elements μg per g DM				
Boron	43	26	70		4
Manganese	33	22	46		4

2.233 *BRASSICA OLERACEA:* BRUSSELS SPROUT, WHOLE

	Mean	*Min.*	*Max.*	*SD*	*No.*
Dry matter %	17.3	15.8	18.8		2
	Proximate constituents % of DM				
Crude protein	19.3	15.9	21.4		3
Crude fibre	17.8	16.0	19.0		3
	Major elements % of DM				
Calcium	0.86				1
Phosphorus	0.41				1

2.234 BRUSSELS SPROUT, STALK

	Mean	*Min.*	*Max.*	*SD*	*No.*
Dry matter %	16.6	11.6	21.6		8
	Proximate constituents % of DM				
Crude protein	21.6	15.9	25.2		9
Ether extract	2.9	2.8	3.0		2
Nitrogen-free extract	43.2	38.2	48.2		2
Crude fibre	18.3	13.7	22.1		9
Total ash	12.8	10.2	15.3		2
	Major elements % of DM				
Calcium	0.9	0.68	1.26		4
Phosphorus	0.41	0.24	0.53		4
Potassium	3.24				1
Sodium	0.38	0.31	0.46		2
Chlorine	0.60	0.48	0.71		2

2.235 BRUSSELS SPROUT, DRIED

	Mean	*Min.*	*Max.*	*SD*	*No.*
Dry matter %	90.3				1
	Proximate constituents % of DM				
Crude protein	22.9				1
Ether extract	3.0				1
Crude fibre	12.2				1
	Major elements % of DM				
Calcium	3.97				1
Phosphorus	0.36				1
Potassium	1.84				1
Sodium	0.59				1
Chlorine	0.91				1

2.236 *BRASSICA OLERACEA CAPITATA:* CABBAGE, WHOLE PLANT

	Mean	*Min.*	*Max.*	*SD*	*No.*
Dry matter %	9.9	7.7	13.5	1.85	37
	Proximate constituents % of DM				
Crude protein	19.6	13.1	25.8	3.9	30
Ether extract	2.30	1.86	2.73		2
Crude fibre	11.8	9.6	13.8	1.2	30
	Major elements % of DM				
Calcium	2.22	0.51	4.99		15
Phosphorus	0.39	0.32	0.45		6
Magnesium	0.22	0.10	0.35		10
Potassium	4.95	3.57	5.73		3
	Trace elements μg per g DM				
Barium	18				1
Boron	28	24	31		2
Chromium	0.2				1
Cobalt	0.05				1
Copper	9.6	2.5	21.0		6
Iron	62				1
Lead	0.80				1
Manganese	32	5	65		8
Molybdenum	1.17				1
Nickel	17	8	26		2
Silver	0.1				1
Strontium	26				1
Tin	1				1
Titanium	2.9				1
Vanadium	0.07				1
Zinc	860	440	1280		2

2.237 CABBAGE, ARTIFICIALLY DRIED

	Mean	Min.	Max.	SD	No.
Dry matter %	93.4	85.5	97.7		14
Proximate constituents % of DM					
Crude protein	15.9	10.2	23.4		14

2.238 CABBAGE, LEAF

	Mean	Min.	Max.	SD	No.
Trace elements µg per g DM					
Barium	5.0				1
Chromium	2.16	0.51	5.30		17
Cobalt	0.35	0.08	0.91		17
Copper	5.1				1
Iron	274	54	578		17
Lead	4.2	0.6	12.0		17
Manganese	34				1
Molybdenum	0.44	0.25	0.74		17
Nickel	2.22	1.00	6.00		17
Silver	0.09	0.04	0.16		17
Strontium	10.0				1
Tin	0.92	0.36	1.56		17
Titanium	50	3	128		17
Vanadium	1.04	0.10	2.38		17
Zinc	25	8	58		16

2.239 *BRASSICA OLERACEA:* CAULIFLOWER, WHOLE

	Mean	Min.	Max.	SD	No.
Dry matter %	7.8				1
Proximate constituents % of DM					
Crude protein	22.5	12.8	34.7		5
Crude fibre	15.3				1
Major elements % of DM					
Calcium	0.82	0.16	1.60		5
Phosphorus	0.53	0.32	0.68		5
Magnesium	0.16	0.10	0.20		4
Potassium	3.25	1.95	4.37		4
Trace elements µg per g DM					
Barium	42	33	51		2
Boron	25	22	30		4
Chromium	1.15	0.79	1.51		2
Cobalt	0.27	0.15	0.38		2
Copper	7.0	3.9	10.0		2
Iron	311	197	424		2
Lead	3.6	2.4	4.7		2
Manganese	47	28	66		2
Molybdenum	0.57	0.44	0.69		2
Nickel	11	6.3	16		2
Silver	0.10	0.07	0.12		2
Strontium	55	50	59		2
Tin	1.24	0.75	1.72		2
Titanium	27	8.5	46		2
Vanadium	0.74	0.32	1.16		2
Zinc	827	431	1222		2

2.240 CAULIFLOWER, LEAF

	Mean	Min.	Max.	SD	No.
Dry matter %	10.4	10.1	10.9		5
Major elements % of DM					
Calcium	2.57	1.81	3.48		5
Magnesium	0.24	0.21	0.30		5
Trace elements µg per g DM					
Manganese	32	30	40		5

2.241 *BRASSICA NAPUS:* RAPE, WHOLE PLANT

	Mean	Min.	Max.	SD	No.
Dry matter %	13.4	9.1	24.2	2.6	424
Proximate constituents % of DM					
Crude protein	19.3	12.3	26.0	5.0	319
Crude fibre	9.0	7.4	10.8	0.9	35
Major elements % of DM					
Calcium	1.69	1.00	2.26	0.24	41
Phosphorus	0.27	0.15	0.39	0.09	423
Magnesium	0.25	0.10	0.58		6
Potassium	3.00	1.90	4.19	0.91	216
Sulphur	0.33				1
Trace elements µg per g DM					
Cobalt	0.13	0.02	0.26		5
Copper	4.2	2.0	8.0		13
Iron	91	83	99		2
Manganese	51	18	100		7
Molybdenum	0.51	0.30	0.78		5
Titanium	3.6	3.3	3.8		2

2.242 RAPE AND KALE

	Mean	Min.	Max.	SD	No.
Dry matter %	13.5	11.6	15.3		2
Proximate constituents % of DM					
Crude protein	22.8	19.6	26.0		2
Crude fibre	12.9	12.7	13.1		2
Major elements % of DM					
Calcium	1.84				1
Phosphorus	0.46				1

2.243 *BRASSICA NAPOBRASSICA:* SWEDE, UNSPECIFIED TOPS

	Mean	*Min.*	*Max.*	*SD*	*No.*
Dry matter %	12.8	11.0	15.7	1.25	30
Proximate constituents % of DM					
Crude protein	22.7	17.8	30.2	3.06	30
Crude fibre	10.2	10.2	10.2		2
Major elements % of DM					
Calcium	2.38				1
Phosphorus	0.34	0.20	0.58	0.14	29
Potassium	3.54	2.93	4.08		16
Vitamins μg per g DM					
β-carotene	71				1

2.244 *BRASSICA RAPA:* TURNIP, LEAF

	Mean	*Min.*	*Max.*	*SD*	*No.*
Dry matter %	20.9	14.4	27.7		3
Proximate constituents % of DM					
Crude protein	18.8	11.9	23.1		5
Ether extract	3.2				1
Crude fibre	13.3				1
Major elements % of DM					
Phosphorus	0.35	0.27	0.43		2
Magnesium	0.08	0.08	0.08		2
Potassium	3.09	2.10	4.07		2

2.245 *BETA VULGARIS:* FODDER BEET, WHOLE *PLANT*

	Mean	*Min.*	*Max.*	*SD*	*No.*
Dry matter %	19.8				1
Proximate constituents % of DM					
Crude protein	6.8				1
Ether extract	1.2				1
Crude fibre	6.8				1
Major elements % of DM					
Calcium	0.58				1
Phosphorus	0.13				1
Sodium	0.36				1
Chlorine	0.55				1

2.246 FODDER BEET, TOPS

	Mean	*Min.*	*Max.*	*SD*	*No.*
Dry matter %	11.3	8.1	15.7	2.5	91
Proximate constituents % of DM					
Crude protein	20.5	15.3	26.4	2.9	95
Ether extract	1.8	1.2	2.7		13
Crude fibre	12.3	8.7	16.6	2.4	38
Major elements % of DM					
Calcium	1.51	0.99	2.25		15
Phosphorus	0.32	0.20	0.54	0.11	59
Potassium	3.82				1
Sodium	0.94	0.87	1.01		2
Chlorine	1.45	1.34	1.55		2
Trace elements μg per g DM					
Manganese	145	101	188		4

2.247 FODDER BEET TOPS, ARTIFICIALLY DRIED

	Mean	*Min.*	*Max.*	*SD*	*No.*
Dry matter %	88.3				1
Proximate constituents % of DM					
Crude protein	25.2	15.2	29.9		13
Crude fibre	10.8				1

2.248 *BETA VULGARIS CICLA:* SUGAR BEET, LEAF, ARTIFICIALLY DRIED

	Mean	*Min.*	*Max.*	*SD*	*No.*
Dry matter %	83.1	74.0	87.6		9
Proximate constituents % of DM					
Crude protein	20.8	19.3	22.2		9
Trace elements μg per g DM					
Barium	18	11	34		3
Chromium	1.3	0.9	1.7		3
Cobalt	0.36	0.29	0.42		3
Copper	10	8	11		3
Iron	513	390	610		3
Lead	3.0	2.5	3.3		3
Manganese	74	59	198		3
Molybdenum	0.66	0.54	0.75		3
Nickel	1.61	1.12	2.00		3
Silver	0.24	0.16	0.34		3
Strontium	29	10	40		3
Tin	1.8	1.4	2.4		3
Titanium	55	40	71		3
Vanadium	1.5	1.3	2.0		3
Zinc	56	55	57		3

2.249 SUGAR BEET, PETIOLE, ARTIFICIALLY DRIED

	Mean	Min.	Max.	SD	No.
Trace elements µg per g DM					
Barium	34	17	50		2
Chromium	1.8	1.2	2.3		2
Cobalt	0.39	0.29	0.48		2
Copper	7.1	5.9	8.3		2
Iron	488	440	535		2
Lead	3.1	2.7	3.5		2
Manganese	86	67	104		2
Molybdenum	0.46	0.23	0.68		2
Nickel	1.57	0.96	2.18		2
Silver	0.23	0.16	0.30		2
Strontium	41	28	54		2
Tin	26	2.3	49		2
Titanium	54	35	73		2
Vanadium	1.3	0.9	1.7		2
Zinc	35	30	40		2

2.250 SUGAR BEET, TOPS

	Mean	Min.	Max.	SD	No.
Dry matter %	12.5	9.0	16.3	1.04	34
Proximate constituents % of DM					
Crude protein	19.5	12.2	27.4	3.9	65
Crude fibre	13.8	8.2	26.3		4
Major elements % of DM					
Calcium	1.33	0.85	2.82	0.53	36
Phosphorus	0.37	0.19	0.67	0.13	53
Magnesium	0.58	0.11	1.25	0.27	49
Potassium	4.27	1.45	7.47	1.42	51
Trace elements µg per g DM					
Barium	31	20	42		4
Copper	13	6	25		9
Manganese	141	25	538		15
Nickel	2.0	0.5	4.3		5
Zinc	146	27	260		7

2.251 SUGAR BEET TOPS, ARTIFICIALLY DRIED

	Mean	Min.	Max.	SD	No.
Dry matter %	90.6	86.3	94.0		4
Proximate constituents % of DM					
Crude protein	12.9	10.2	14.1		4
Ether extract	1.6	1.2	1.8		3
Crude fibre	12.8	10.8	16.7		3
Major elements % of DM					
Calcium	2.41	1.36	3.07		3
Phosphorus	0.22	0.20	0.26		3
Potassium	1.05	0.82	1.28		2
Sodium	1.41	0.61	2.20		3
Chlorine	2.18	0.94	3.40		3

2.252 *BETA VULGARIS RAPA:* BEETROOT, LEAF

	Mean	Min.	Max.	SD	No.
Trace elements µg per g DM					
Barium	112	74	150		4
Chromium	0.7	0.4	1.0		4
Cobalt	0.61	0.28	0.98		4
Copper	12	10	14		4
Iron	195	158	254		4
Lead	5.0	1.9	6.7		4
Manganese	859	207	1620		4
Molybdenum	0.83	0.36	1.30		4
Nickel	2.44	1.16	3.80		4
Silver	0.23	0.11	0.34		4
Strontium	33	28	39		4
Tin	1.0	1.0	1.0		4
Titanium	22	12	44		4
Vanadium	0.66	0.51	0.91		4
Zinc	583	410	730		4

2.253 BEETROOT, TOPS

	Mean	Min.	Max.	SD	No.
Dry matter %	17.8				1
Proximate constituents % of DM					
Crude protein	19.1	11.9	24.4		6
Crude fibre	6.2				1
Major elements % of DM					
Phosphorus	0.41	0.10	0.83		5
Magnesium	0.37	0.19	0.87		5
Potassium	4.02	1.52	6.02		5
Trace elements µg per g DM					
Manganese	618	294	952		5

2.254 *BETA VULGARIS:* MANGOLD, TOPS

	Mean	Min.	Max.	SD	No.
Dry matter %	9.9	7.5	15.5	1.92	36
Proximate constituents % of DM					
Crude protein	21.9	15.3	32.5	3.53	67
Ether extract	0.16				1
Crude fibre	10.8	8.8	13.2	1.0	35
Major elements % of DM					
Calcium	2.20	1.19	3.46	1.02	31
Phosphorus	0.30	0.09	0.57	0.08	35
Magnesium	1.43	0.04	2.34	0.58	30
Potassium	2.11	0.40	5.14	1.36	35
Trace elements μg per g DM					
Copper	10				1
Manganese	276	10	800		8

2.255 *DAUCUS SATIVUS:* CARROT, TOPS

	Mean	Min.	Max.	SD	No.
Dry matter %	16.2	14.5	17.0		3
Proximate constituents % of DM					
Crude protein	14.3	11.6	16.3		4
Major elements % of DM					
Calcium	2.45	2.36	2.59		3
Phosphorus	0.15	0.13	0.20		3
Magnesium	0.14	0.07	0.20		4
Trace elements μg per g DM					
Barium	24	13	37		4
Chromium	0.7	0.5	0.9		4
Cobalt	0.11	0.08	0.15		4
Copper	7.0	5.0	9.1		4
Iron	131	98	158		4
Lead	2.7	1.4	3.9		4
Manganese	113	29	245		5
Molybdenum	0.51	0.21	0.87		4
Nickel	2.08	0.69	3.50		4
Silver	0.07	0.07	0.07		4
Strontium	16	11	18		4
Tin	0.7	0.7	0.7		4
Titanium	9	5	14		4
Vanadium	0.32	0.22	0.45		4
Zinc	52	35	70		4

2.256 *APIUM GRAVEOLENS:* CELERY

	Mean	Min.	Max.	SD	No.
Dry matter %	8.8				1
Proximate constituents % of DM					
Crude protein	18.0	15.8	20.3		4
Crude fibre	12.1	11.6	12.5		2
Major elements % of DM					
Calcium	2.65	1.26	3.71		3
Phosphorus	0.29	0.26	0.31		3
Magnesium	0.25	0.23	0.26		2
Potassium	3.02	1.97	4.07		2
Sulphur	0.21				1
Trace elements μg per g DM					
Cobalt	0.18				1
Copper	4.3				1
Manganese	22	19	24		2
Molybdenum	0.4				1

2.257 CELERY, ARTIFICIALLY DRIED

	Mean	Min.	Max.	SD	No.
Dry matter %	84.4	81.7	87.1		2
Proximate constituents % of DM					
Crude protein	16.6	12.5	20.6		2
Ether extract	2.3	1.3	3.2		2
Crude fibre	6.2	5.1	7.2		2
Major elements % of DM					
Calcium	1.52	0.99	2.05		2

2.258 *LINUM USITATISSIMUM:* FLAX

	Mean	Min.	Max.	SD	No.
Dry matter %	18.5	18.0	19.3		3
Proximate constituents % of DM					
Crude protein	17.1	14.3	20.3		3
Ether extract	3.1	2.6	3.5		3
Crude fibre	36.4	32.4	42.2		3
Major elements % of DM					
Calcium	1.07				1
Phosphorus	0.34				1
Sodium	0.27				1
Chlorine	0.41				1
Potassium	1.23				1
Trace elements μg per g DM					
Copper	20	20	20		2
Zinc	48	40	56		2

2.259 FLAX MEAL

	Mean	Min.	Max.	SD	No.
Dry matter %	88.6	87.2	89.5		4
Proximate constituents % of DM					
Crude protein	11.6	9.6	13.6		4
Ether extract	6.7	5.5	7.5		4
Crude fibre	29.3	26.0	33.4		4
Major elements % of DM					
Calcium	2.02				1
Phosphorus	0.26				1
Sodium	0.26				1
Chlorine	0.39				1

2.260 FLAX CHAFF

	Mean	Min.	Max.	SD	No.
Dry matter %	85.9	81.6	97.5		10
Proximate constituents % of DM					
Crude protein	10.0	6.4	13.0		10
Ether extract	7.9	2.4	11.4		10
Crude fibre	38.0	35.0	42.3		10
Major elements % of DM					
Calcium	0.89	0.66	1.20		6
Phosphorus	0.20	0.13	0.33		6

2.261 *LACTUCA SATIVA:* LETTUCE, LEAF

	Mean	Min.	Max.	SD	No.
Trace elements μg per g DM					
Chromium	5.09	0.55	11.0		8
Cobalt	0.74	0.09	1.45		8
Iron	1060	140	2170		8
Lead	10.5	1.4	24.0		8
Molybdenum	0.67	0.15	1.51		8
Nickel	2.87	0.86	5.40		8
Silver	0.29	0.05	0.54		8
Tin	2.9	0.51	5.35		8
Titanium	208	18	444		8
Vanadium	6.21	0.41	13.0		8
Zinc	36	25	49		8

2.262 *SINAPIS* SP: MUSTARD

	Mean	Min.	Max.	SD	No.
Dry matter %	11.2				1
Proximate constituents % of DM					
Crude protein	19.7	15.5	22.5	2.5	21
Ether extract	2.3	1.8	2.4		3
Crude fibre	17.3	8.1	28.3		3
Major elements % of DM					
Calcium	3.14				1
Phosphorus	0.36	0.31	0.38		19
Potassium	4.67	3.76	5.64		19
Sodium	0.56				1
Chlorine	0.87				1
Trace elements μg per g DM					
Copper	24	9	39		5
Zinc	264	152	297		5

2.263 MUSTARD, BRAN

	Mean	Min.	Max.	SD	No.
Dry matter %	91.7				1
Proximate constituents % of DM					
Crude protein	18.1				1
Ether extract	10.1				1
Crude fibre	26.2				1
Major elements % of DM					
Calcium	1.09				1
Phosphorus	0.30				1
Sodium	0.02				1
Chlorine	0.04				1

2.264 *SOLANUM TUBEROSUM:* POTATO, HAULM

	Mean	Min.	Max.	SD	No.
Dry matter %	7.0				1
Proximate constituents % of DM					
Crude protein	28.6				1
Ether extract	2.9				1
Nitrogen-free extract	40.8				1
Crude fibre	17.7				1
Total ash	10.0				1
Major elements % of DM					
Calcium	2.04				1
Magnesium	0.27				1
Potassium	6.90				1
Trace elements μg per g DM					
Barium	13				1
Chromium	0.65				1
Cobalt	0.20				1
Copper	16				1
Iron	185				1
Lead	2.7				1
Manganese	196				1
Molybdenum	0.47				1
Nickel	1.08				1
Strontium	88				1
Tin	1.0				1
Titanium	21				1
Vanadium	0.77				1
Zinc	26				1

2.265 POTATO LEAF

	Mean	Min.	Max.	SD	No.
	Proximate constituents % of DM				
Crude protein	29.9	25.0	33.7		4
Total ash	15.1	12.8	16.5		6
	Major elements % of DM				
Calcium	0.83	0.58	1.13		3
Phosphorus	0.34	0.13	0.87		5
Magnesium	0.20	0.06	0.70		8
Potassium	4.17	2.07	6.06		9
	Trace elements μg per g DM				
Manganese	352	190	570		5

2.266 *ALLIUM ASCALONICUM:* SHALLOT, LEAF

	Mean	Min.	Max.	SD	No.
	Trace elements μg per g DM				
Barium	21	19	24		4
Chromium	0.6	0.4	0.7		4
Cobalt	0.09	0.08	0.10		4
Copper	5.9	4.8	7.0		4
Iron	124	103	159		4
Lead	3.6	2.1	4.7		4
Manganese	30	26	31		4
Molybdenum	1.12	0.41	1.93		4
Nickel	0.69	0.41	0.91		4
Silver	0.36	0.20	0.60		4
Strontium	23	21	25		4
Tin	0.07	0.07	0.07		4
Titanium	11	8	14		4
Vanadium	0.32	0.25	0.41		4
Zinc	17	8	29		4

2.267 *SPINACIA* SP: SPINACH, LEAF

	Mean	Min.	Max.	SD	No.
	Major elements % of DM				
Calcium	1.55	1.41	1.68		2
Magnesium	0.41	0.38	0.44		2
Potassium	8.6	8.3	9.0		2
	Trace elements μg per g DM				
Boron	52	50	54		2
Iron	1398	689	2106		2
Manganese	63	44	100		4

2.268 *LYCOPERSICUM:* TOMATO, WHOLE PLANT

	Mean	Min.	Max.	SD	No.
	Proximate constituents % of DM				
Crude protein	15.2	12.2	17.4		4
	Major elements % of DM				
Calcium	0.21	0.16	0.27		4
Phosphorus	0.40	0.24	0.63		10
Magnesium	0.22	0.14	0.42		10
Potassium	4.35	4.07	4.90		4
	Trace elements μg per g DM				
Aluminium	150	96	238		6
Barium	9.1	5.3	15.0		6
Boron	19	13	27		6
Chromium	1.56	1.28	2.01		6
Cobalt	2.05	0.52	6.60		6
Copper	9	7	13		6
Iron	195	60	327		10
Lead	6.6	5.2	9.7		6
Manganese	775	303	1568		6
Molybdenum	2.2	0.5	4.3		6
Nickel	2.05	1.27	3.20		6
Silver	0.12	0.10	0.18		6
Strontium	45	22	65		6
Tin	1.19	1.00	1.75		6
Titanium	17	14	20		6
Vanadium	1.26	1.05	1.54		6
Zinc	144	98	211		6

2.269 TOMATO, LEAF

	Mean	Min.	Max.	SD	No.
	Proximate constituents % of DM				
Crude protein	28.5	20.5	34.3		8
	Major elements % of DM				
Calcium	3.0	2.77	3.62		6
Phosphorus	0.33	0.19	0.50		8
Magnesium	0.61	0.09	1.28		6
Potassium	5.18	3.90	6.22		6
	Trace elements μg per g DM				
Manganese	780	780	780		2

2.270 TOMATO, AERIAL PARTS, ARTIFICIALLY DRIED

	Mean	*Min.*	*Max.*	*SD*	*No.*
	Major elements % of DM				
Phosphorus	0.42	0.24	0.76		4
Magnesium	0.21	0.13	0.28		4
	Trace elements μg per g DM				
Aluminium	275	205	370		4
Barium	17	13	30		4
Boron	19.8	14	29		4
Chromium	1.62	1.45	1.90		3
Cobalt	3.81	0.72	5.40		3
Copper	13	9	20		4
Iron	267	231	300		4
Lead	4.01	3.14	5.10		3
Manganese	991	391	1540		4
Molybdenum	1.76	0.55	4.10		3
Nickel	1.76	1.09	2.21		3
Silver	0.15	0.11	0.19		3
Strontium	71	50	117		4
Tin	1.49	1.14	2.13		3
Titanium	16	14	17		3
Vanadium	1.22	1.07	1.42		3
Zinc	111	97	145		4

2.271 *HELIANTHUS TUBEROSUS:* TOPINAMBUR*, LEAF

	Mean	*Min.*	*Max.*	*SD*	*No.*
Dry matter %	20.8	13.5	27.2		4
	Proximate constituents % of DM				
Crude protein	20.3	11.9	27.4		4
Ether extract	1.32	1.00	1.63		2
Nitrogen-free extract	39.3				1
Crude fibre	13.2	12.7	13.6		3
Total ash	25.9				1
Sol. ash	15.3	14.0	16.6		2
Insol. ash	3.30	2.90	3.71		2
	Major elements % of DM				
Calcium	2.85	2.00	4.86		4
Phosphorus	0.39	0.29	0.55		4
Potassium	5.85	5.39	6.31		2
Magnesium	0.35	0.30	0.39		3
	Trace elements μg per g DM				
Copper	14.2	11.2	16.0		3
Manganese	460				1

* Topinambur is synonymous with artichoke.

2.272 TOPINAMBUR, STEM

	Mean	*Min.*	*Max.*	*SD*	*No.*
Dry matter %	15.8	9.7	26.4		7
	Proximate constituents % of DM				
Crude protein	9.7	3.6	15.6		7
Ether extract	1.10	0.70	1.42		5
Nitrogen-free extract	42.7	39.1	46.4		4
Crude fibre	35.8	24.3	49.5		7
Total ash	11.4	7.2	13.9		4
Sol. ash	12.1	11.4	12.8		2
Insol. ash	0.63	0.56	0.70		2
	Major elements % of DM				
Calcium	0.87	0.69	1.01		4
Phosphorus	0.23	0.12	0.36		4
Potassium	5.48	5.15	5.81		2
Magnesium	0.13	0.10	0.15		3
	Trace elements μg per g DM				
Copper	14.2	11.2	16.0		3
Manganese	150				1

2.273 *RAPHANUS SATIVUS:* RADISH, TOPS

	Mean	*Min.*	*Max.*	*SD*	*No.*
	Major elements % of DM				
Phosphorus	0.46				1
Magnesium	0.12				1
	Trace elements μg per g DM				
Barium	12	10	17		3
Chromium	2.47	2.24	2.70		2
Cobalt	0.20	0.17	0.22		2
Copper	9.1	8.6	9.5		3
Iron	287	281	295		3
Lead	5.0	4.6	5.4		2
Manganese	34	25	43		3
Molybdenum	1.09	0.91	1.27		2
Nickel	1.88	1.83	1.93		2
Silver	0.15	0.14	0.15		2
Strontium	57	50	63		3
Tin	1.32	1.24	1.39		2
Titanium	30	30	31		2
Vanadium	0.94	0.91	0.96		2
Zinc	125	67	187		3

2.274 BEETROOT, ROOT

	Mean	*Min.*	*Max.*	*SD*	*No.*
Dry matter %	15.0				1
	Proximate constituents % of DM				
Crude protein	16.3				1
Crude fibre	11.0				1

2.275 FODDER BEET, ROOT

	Mean	Min.	Max.	SD	No.
Dry matter %	18.4	11.4	26.5	3.49	194
Proximate constituents % of DM					
Crude protein	7.6	5.3	10.8	1.5	171
Ether extract	1.6	0.2	7.7	2.4	45
Crude fibre	5.8	4.7	7.4	1.1	117
Major elements % of DM					
Calcium	0.26	0.11	0.88	0.30	50
Phosphorus	0.18	0.09	0.37	0.11	65
Magnesium	0.28				1
Potassium	1.17				1
Sodium	0.20	0.09	0.27		8
Chlorine	0.27	0.14	0.43		8
Trace elements μg per g DM					
Manganese	81	68	94		5

2.276 FODDER BEET, ROOT, ARTIFICIALLY DRIED

	Mean	Min.	Max.	SD	No.
Proximate constituents % of DM					
Crude protein	8.3	7.8	9.2		12
Ether extract	0.12	0.06	0.19		12
Crude fibre	5.9	4.9	8.0		12

2.277 *BETA* SP: BEET, UNSPECIFIED, ROOT

	Mean	Min.	Max.	SD	No.
Dry matter %	11.9	8.8	16.8	2.3	23
Proximate constituents % of DM					
Crude protein	10.6	8.0	13.3	1.4	24
Major elements % of DM					
Phosphorus	0.27	0.18	0.33	0.05	24
Potassium	3.95	2.36	5.32	0.78	24

2.278 SUGAR BEET, ROOT

	Mean	Min.	Max.	SD	No.
Dry matter %	20.2	17.7	23.8	1.38	28
Proximate constituents % of DM					
Crude protein	6.6	5.0	10.1	1.1	34
Ether extract	0.68	0.10	1.29		7
Crude fibre	4.5	3.7	5.3		9
Major elements % of DM					
Calcium	0.28	0.13	0.98	0.32	25
Phosphorus	0.17	0.11	0.31	0.05	29
Magnesium	0.15	0.07	0.45	0.07	28
Potassium	2.43	1.41	6.98	1.48	28
Trace elements μg per g DM					
Barium	18	14	23		4
Chromium	0.62	0.28	0.92		4
Cobalt	0.17	0.06	0.31		4
Copper	8.5	5.0	11.4		4
Iron	273	1.07	488		4
Lead	2.4	1.2	4.1		4
Manganese	39	27	51		6
Molybdenum	0.41	0.25	0.73		4
Nickel	0.56	0.25	0.92		4
Silver	0.01	0.1	0.1		4
Strontium	6	5	8		4
Tin	1.0	0.7	1.4		4
Titanium	23	9	48		4
Vanadium	1.0	0.3	1.5		4
Zinc	41	28	56		4

2.279 SUGAR BEET, PULP

	Mean	Min.	Max.	SD	No.
Dry matter %	17.9	10.7	25.9		12
Proximate constituents % of DM					
Crude protein	11.7	9.8	15.9		11
Ether extract	0.96	0.74	1.18		2
Nitrogen-free extract	57.9	49.4	66.4		2
Crude fibre	21.2	17.0	31.4		8
Total ash	7.53	3.73	15.71		8
Major elements % of DM					
Calcium	0.81	0.47	1.03		6
Phosphorus	0.22	0.09	0.57		4
Magnesium	0.16	0.10	0.23		2

2.280 SUGAR BEET, PULP, ARTIFICIALLY DRIED

	Mean	*Min.*	*Max.*	*SD*	*No.*
Dry matter %	87.4	81.2	91.0	3.16	59
Proximate constituents % of DM					
Crude protein	10.6	8.9	13.6	1.42	63
Ether extract	0.5	0.4	0.8	0.12	27
Crude fibre	14.3	10.8	19.5	2.1	53
Major elements % of DM					
Calcium	0.60	0.35	0.93	0.19	52
Phosphorus	0.09	0.06	0.30	0.09	51
Magnesium	0.13	0.10	0.18	0.02	22
Potassium	2.21	1.92	2.49	0.23	21
Sodium	0.30	0.17	0.47		17
Chlorine	0.52	0.38	0.85	0.13	21
Sulphur	0.11	0.07	0.17		4
Trace elements µg per g DM					
Cobalt	1.3				1
Copper	25	11	38	9.3	32
Manganese	51	29	87	12.3	29
Molybdenum	0.8	0.5	1.7		7
Zinc	32	23	40	4.1	20

2.281 SUGAR BEET, PULP, CUBES

	Mean	*Min.*	*Max.*	*SD*	*No.*
Dry matter %	90.0				1
Proximate constituents % of DM					
Crude protein	10.4				1
Ether extract	1.7				1
Crude fibre	15.0				1
Major elements % of DM					
Calcium	0.74				1
Phosphorus	0.14				1
Potassium	1.75				1
Sodium	0.24				1
Chlorine	0.14				1
Trace elements µg per g DM					
Cobalt	1.3				1
Copper	21				1

2.282 SUGAR BEET, SWEEPINGS, ARTIFICIALLY DRIED

	Mean	*Min.*	*Max.*	*SD*	*No.*
Dry matter %	87.2				1
Proximate constituents % of DM					
Crude protein	10.2				1
Ether extract	0.5				1
Crude fibre	18.4				1
Major elements % of DM					
Calcium	1.00				1
Phosphorus	0.11				1
Sodium	0.38				1
Chlorine	0.58				1

2.283 SUGAR BEET, TAILINGS

	Mean	*Min.*	*Max.*	*SD*	*No.*
Dry matter %	13.2	10.0	16.3		2
Proximate constituents % of DM					
Crude protein	11.3	7.2	15.3		2
Ether extract	0.75	0.50	1.00		2
Crude fibre	14.2	12.0	16.3		2
Major elements % of DM					
Calcium	2.77				1
Phosphorus	0.66				1
Sodium	0.16				1
Chlorine	0.24				1

2.284 MANGOLD, ROOT

	Mean	*Min.*	*Max.*	*SD*	*No.*
Dry matter %	10.9	8.4	13.8	1.42	186
Proximate constituents % of DM					
Crude protein	10.0	7.3	13.5	2.13	182
Digestibility of CP	80.7	71.0	86.0		4
Ether extract	0.25				1
Crude fibre	7.3	5.6	11.7	1.47	58
Major elements % of DM					
Calcium	0.29	0.12	2.79	0.18	50
Phosphorus	0.21	0.07	0.37	0.11	131
Magnesium	0.53	0.13	1.02		8
Potassium	3.86	1.93	6.01	0.75	91
Sodium	0.99	0.29	2.08		9
Chlorine	1.51	0.45	3.10		9
Sulphur	0.03				1
Trace elements µg per g DM					
Cobalt	0.09				1
Copper	9.4	9.4	9.5		3
Manganese	39	25	55		5

2.285 CARROT, ARTIFICIALLY DRIED

	Mean	Min.	Max.	SD	No.
Dry matter %	87.8	86.5	90.0		3
Proximate constituents % of DM					
Crude protein	8.2	6.5	9.8		3
Ether extract	1.15	1.0	1.3		2
Crude fibre	7.4	6.2	8.7		2
Carbohydrate fractions % of DM					
Starch	11.1				1
Major elements % of DM					
Calcium	0.43				1
Phosphorus	0.35				1
Vitamins µg per g DM					
β-carotene	<11.6				1
Trace elements µg per g DM					
Barium	38	27	48		2
Chromium	3.79	1.27	6.30		2
Cobalt	4.33	2.46	6.20		2
Copper	43	19	66		8
Iron	2172	533	3810		2
Lead	32	18	45		2
Manganese	244	55	433		2
Molybdenum	1.42	1.14	1.70		2
Nickel	54	42	66		2
Silver	0.80	0.12	1.47		2
Strontium	38	32	44		2
Tin	8.62	2.24	15.0		2
Titanium	207	31	382		2
Vanadium	5.64	1.28	10.0		2
Zinc	396	153	643		8

2.286 CARROT MEAL, EXTRACTED

	Mean	Min.	Max.	SD	No.
Dry matter %	90.0	74.1	98.5		1
Proximate constituents % of DM					
Crude protein	8.3				1
Ether extract	1.0				1
Crude fibre	6.2				1

2.287 POTATO TUBER

	Mean	Min.	Max.	SD	No.
Dry matter %	19.8	10.4	27.2	3.54	165
Proximate constituents % of DM					
Crude protein	10.3	8.4	11.6	1.36	549
Ether extract	0.41	0.37	0.44		2
Crude fibre	2.5	1.9	4.3	0.5	43
Major elements % of DM					
Calcium	0.11	0.09	0.12		6
Phosphorus	0.23	0.17	0.28	0.06	513
Magnesium	0.16	0.06	0.20		4
Potassium	1.96	1.55	2.38	0.36	509

2.288 POTATO, ARTIFICIALLY DRIED

	Mean	Min.	Max.	SD	No.
Dry matter %	91.0				1
Proximate constituents % of DM					
Crude protein	10.7	8.4	13.8	1.4	33
Ether extract	0.4				1
Crude fibre	2.6				1
Major elements % of DM					
Calcium	0.11	0.07	0.14	0.02	33
Phosphorus	0.24	0.18	0.29	0.03	33
Potassium	2.25	1.96	2.69	0.17	32

2.289 TOPINAMBUR (ARTICHOKE), TUBER

	Mean	Min.	Max.	SD	No.
Dry matter %	18.3	16.3	24.2		4
Proximate constituents % of DM					
Crude protein	13.9	10.1	16.1		4
Ether extract	2.4	0.5	5.5		3
Crude fibre	4.4	4.1	4.8		4

2.290 TURNIP, ROOT

	Mean	Min.	Max.	SD	No.
Dry matter %	10.4	6.7	15.1	3.38	466
Proximate constituents % of DM					
Crude protein	11.4	8.1	15.6	3.19	410
Ether extract	1.18	0.58	2.57		11
Nitrogen-free extract	74.6	65.4	81.5		8
Crude fibre	9.2	2.1	13.0		10
Ash	6.36	3.89	9.20		8
Major elements % of DM					
Calcium	0.61	0.41	0.82		4
Phosphorus	0.30	0.15	0.42	0.22	455
Magnesium	0.12				1
Potassium	1.94	1.41	2.45	0.37	291
Sulphur	0.18				1
Trace elements μg per g DM					
Barium	24	3.4	150	31.9	23
Chromium	0.27	0.08	0.98	0.25	23
Cobalt	0.20	0.02	0.52	0.16	29
Copper	5.9	1.1	54.0	9.8	29
Iron	93	21	640	135	28
Lead	0.99	0.55	2.41	0.53	23
Manganese	35	6	196	49	25
Molybdenum	0.20	0.09	0.67	0.14	27
Nickel	2.25	0.33	16	3.9	24
Silver	0.07	0.04	0.17	0.04	23
Strontium	43	11	224	62	23
Tin	0.74	0.44	1.60	0.34	23
Titanium	12	0.5	112	25.4	24
Vanadium	0.28	0.02	2.07	0.49	23
Zinc	36	10	280	58	20

2.291 RADISH, ROOT

	Mean	Min.	Max.	SD	No.
Trace elements μg per g DM					
Barium	10.2	9.7	11.0		3
Chromium	1.44	1.14	1.64		3
Cobalt	0.12	0.10	0.15		3
Copper	3.7	3.0	5.0		3
Iron	149	136	168		3
Lead	2.14	1.79	2.56		3
Manganese	11.6	9.7	13.0		3
Molybdenum	0.42	0.34	0.55		3
Nickel	1.51	1.24	1.97		3
Silver	0.10	0.10	0.10		3
Strontium	24	20	30		3
Tin	1.19	0.99	1.56		3
Titanium	21	19	22		3
Vanadium	0.42	0.36	0.49		3
Zinc	71	28	104		3

2.292 COCKSFOOT, SILAGE

	Mean	Min.	Max.	SD	No.
Dry matter %	23.4	16.2	31.1	5.74	182
Proximate constituents % of DM and pH					
Crude protein	12.7	8.4	17.8	2.82	180
Crude fibre	32.9	25.7	40.6	3.96	58
pH	4.5	3.7	5.6	0.5	169
Major elements % of DM					
Phosphorus	0.22				1

2.293 COCKSFOOT AND LUCERNE, SILAGE

	Mean	Min.	Max.	SD	No.
Dry matter %	22.8	11.0	48.3	5.87	870
Proximate constituents % of DM and pH					
Crude protein	16.0	12.6	19.4	2.79	870
pH	4.96	4.20	5.50	0.51	870

2.294 RYEGRASS AND CLOVER, SILAGE

	Mean	Min.	Max.	SD	No.
Dry matter %	23.1	12.3	48.8	5.52	541
Proximate constituents % of DM and pH					
Crude protein	13.9	9.9	18.5	3.18	544
Crude fibre	24.0				1
pH	4.3	3.4	5.0	0.48	544

2.295 FESCUE, SILAGE

	Mean	Min.	Max.	SD	No.
Dry matter %	17.7	16.3	20.1		3
Proximate constituents % of DM and pH					
Crude protein	10.9	8.7	14.8		3
Crude fibre	33.2	31.7	34.6		2
pH	5.0	4.5	5.7		3

2.296 TIMOTHY, SILAGE

	Mean	Min.	Max.	SD	No.
Dry matter %	26.9	18.0	35.3		4
Proximate constituents % of DM and pH					
Crude protein	13.2	10.7	15.7		4
Crude fibre	30.3	29.3	31.2		2
pH	4.4	4.1	4.8		3

2.297 MEADOW GRASS, SILAGE

	Mean	Min.	Max.	SD	No.
Dry matter %	23.9	14.3	44.4	5.06	516
Proximate constituents % of DM and pH					
Crude protein	11.9	9.2	14.7	2.18	508
Digestibility of CP	51.7	45.7	58.8	11.44	434
Crude fibre	33.4	28.7	38.6	4.01	434
pH	4.4	3.8	5.1	0.47	417
Major elements % of DM					
Phosphorus	0.25	0.24	0.25		2
Sulphur	0.08				1
Trace elements μg per g DM					
Cobalt	0.11	0.10	0.12		2
Copper	10.2	7.4	13.4		4
Molybdenum	1.0				1

2.298 MEADOW GRASS, PASTURE GRASS AND SEEDS GRASS, SILAGE

	Mean	Min.	Max.	SD	No.
Proximate constituents % of DM and pH					
Crude protein	13.2	8.3	20.7	2.9	30
Crude fibre	31.3				1
pH	4.5	3.6	5.7	0.57	30

2.299 GRASS AND CLOVER, SILAGE

	Mean	Min.	Max.	SD	No.
Proximate constituents % of DM and pH					
Crude protein	14.4	10.1	18.8	3.31	821
Crude fibre	29.6				1
pH	4.4	3.8	5.3	0.56	821

2.300 GRASS AND LUCERNE, SILAGE

	Mean	Min.	Max.	SD	No.
Dry matter %	22.9	11.3	61.0	9.9	478
Proximate constituents % of DM and pH					
Crude protein	16.0	12.0	19.7	2.96	541
Crude fibre	33.8				1
pH	4.82	4.00	5.50	0.55	541

2.301 OAT, SILAGE

	Mean	Min.	Max.	SD	No.
Dry matter %	21.7	11.3	33.7	5.06	126
Proximate constituents % of DM and pH					
Crude protein	11.0	7.3	15.2	2.89	131
Crude fibre	34.8	26.0	40.6	3.6	20
pH	4.5	3.8	5.2	0.5	131
Major elements % of DM					
Calcium	0.58				1
Phosphorus	0.35				1
Sulphur	0.12				1
Trace elements μg per g DM					
Copper	7.6				1
Molybdenum	0.9				1

2.302 WHEAT, SILAGE

	Mean	Min.	Max.	SD	No.
Dry matter %	22.5	17.5	27.3		6
Proximate constituents % of DM and pH					
Crude protein	16.9	11.3	21.5		6
pH	4.5	3.7	5.3		6

2.303 BARLEY, SILAGE

	Mean	Min.	Max.	SD	No.
Dry matter %	19.3	10.8	25.9		7
Proximate constituents % of DM and pH					
Crude protein	13.9	12.4	15.6		9
Ether extract	2.71	1.01	4.40		2
Crude fibre	32.7	31.7	33.8		4
pH	4.6	4.5	4.7		7

2.304 MAIZE, SILAGE

	Mean	Min.	Max.	SD	No.
Dry matter %	17.8	13.2	24.1	2.53	131
Proximate constituents % of DM and pH					
Crude protein	11.3	8.4	14.8	2.06	248
Ether extract	2.75	1.40	5.42	0.71	39
Crude fibre	29.0	21.1	35.3	4.27	156
pH	4.10	3.63	4.84	0.41	235

2.305 RYE, SILAGE

	Mean	Min.	Max.	SD	No.
Dry matter %	21.2	14.3	35.2	6.82	38
Proximate constituents % of DM and pH					
Crude protein	12.9	7.3	20.1	4.5	38
Ether extract	4.8	4.7	4.9		2
Crude fibre	34.8	24.2	46.0		10
pH	4.8	3.9	5.8	0.53	36

2.306 SORGHUM, SILAGE

	Mean	Min.	Max.	SD	No.
Dry matter %	20.6				1
Proximate constituents % of DM and pH					
Crude protein	11.0				1
pH	4.7				1

2.307 CEREALS, MIXED, SILAGE

	Mean	Min.	Max.	SD	No.
Dry matter %	22.7	17.9	27.3		14
Proximate constituents % of DM and pH					
Crude protein	10.7	7.7	16.5		14
Crude fibre	26.8				1
pH	4.3	3.7	4.8		14

2.308 CEREAL AND LEGUME, SILAGE

	Mean	Min.	Max.	SD	No.
Dry matter %	21.8	13.5	30.2	5.25	459
Proximate constituents % of DM and pH					
Crude protein	13.1	8.9	16.9	3.15	462
pH	4.54	3.80	5.30	0.57	462

2.309 MASHLAM *

	Mean	Min.	Max.	SD	No.
Dry matter %	22.9	10.6	30.0	4.62	696
Proximate constituents % of DM and pH					
Crude protein	10.3	5.4	20.3	3.9	696
pH	4.5	4.0	5.4	0.40	696

* Mashlam is oat, bean and pea silage.

2.310 OAT AND TARES, SILAGE

	Mean	Min.	Max.	SD	No.
Dry matter %	21.0	14.0	28.6	5.49	595
Proximate constituents % of DM and pH					
Crude protein	12.9	9.1	16.6	3.12	595
pH	4.57	3.80	5.30	0.55	595

2.311 BEAN, SILAGE

	Mean	Min.	Max.	SD	No.
Dry matter %	22.1	15.5	25.8		9
Proximate constituents % of DM and pH					
Crude protein	15.5	10.5	18.3		9
pH	4.4	3.7	5.1		9

2.312 PEA, SILAGE

	Mean	Min.	Max.	SD	No.
Dry matter %	21.7	14.5	31.0	3.75	691
Proximate constituents % of DM and pH					
Crude protein	15.2	11.6	19.3	3.04	695
Ether extract	6.25				1
Crude fibre	16.7				1
pH	4.1	3.6	4.7	0.44	695

2.313 LEGUME SILAGE

	Mean	Min.	Max.	SD	No.
Proximate constituents % of DM and pH					
Crude protein	17.9	7.2	24.6	3.1	57
pH	4.7	3.6	5.7	0.7	57
Major elements % of DM					
Calcium	1.7				1
Phosphorus	0.17				1
Trace elements µg per g DM					
Manganese	280				1

2.314 KALE, MARROWSTEM, SILAGE

	Mean	Min.	Max.	SD	No.
Dry matter %	14.2	10.9	17.7		3
Proximate constituents % of DM and pH					
Crude protein	11.0	10.1	11.9		3
Crude fibre	34.3	30.0	40.0		3
pH	4.2	3.7	4.6		3

2.315 KALE, UNSPECIFIED, SILAGE

	Mean	*Min.*	*Max.*	*SD*	*No.*
Dry matter %	15.5	11.8	20.7	2.04	87
Proximate constituents % of DM and pH					
Crude protein	14.0	9.5	19.6	2.76	87
Ether extract	4.48	2.42	9.23		19
Crude fibre	30.0	24.8	34.8	2.5	30
pH	4.1	3.6	5.2	0.38	87
Major elements % of DM					
Calcium	2.7	2.66	2.74		2
Phosphorus	0.32	0.28	0.35		2
Magnesium	0.13	0.13	0.13		2
Potassium	2.32	2.19	2.44		2
Sodium	0.38	0.38	0.38		2
Trace elements μg per g DM					
Copper	5.0	4.7	5.2		2
Manganese	80				1

2.316 CABBAGE, SILAGE

	Mean	*Min.*	*Max.*	*SD*	*No.*
Dry matter %	13.8	11.0	16.5		2
Proximate constituents % of DM and pH					
Crude protein	17.5	14.6	20.3		2
pH	4.3	4.2	4.4		2

2.317 RAPE, SILAGE

	Mean	*Min.*	*Max.*	*SD*	*No.*
Dry matter %	14.5	11.5	16.8		5
Proximate constituents % of DM and pH					
Crude protein	15.2	12.7	19.1		5
pH	4.5	3.8	5.2		5

2.318 RAPE AND KALE, SILAGE

	Mean	*Min.*	*Max.*	*SD*	*No.*
Dry matter %	18.1	15.5	21.8		3
Proximate constituents % of DM and pH					
Crude protein	14.5	13.3	16.6		3
pH	4.5	3.9	5.2		3
Major elements % of DM					
Calcium	1.56				1
Phosphorus	0.17				1

2.319 FODDER BEET, TOPS, SILAGE

	Mean	*Min.*	*Max.*	*SD*	*No.*
Dry matter %	16.1	15.0	17.8		5
Proximate constituents % of DM and pH					
Crude protein	15.9	12.9	18.7		5
pH	4.9	3.8	5.4		5

2.320 SUGAR BEET, TOPS, SILAGE

	Mean	*Min.*	*Max.*	*SD*	*No.*
Dry matter %	21.4	13.3	31.4	5.8	90
Proximate constituents % of DM and pH					
Crude protein	12.8	7.8	18.1	3.9	90
Crude fibre	17.0	13.2	21.4		5
pH	4.3	3.8	5.0	0.5	90

2.321 SUGAR BEET, PULP, SILAGE

	Mean	*Min.*	*Max.*	*SD*	*No.*
Dry matter %	15.5	10.8	21.0		8
Proximate constituents % of DM and pH					
Crude protein	13.2	9.9	21.0		8
Crude fibre	27.3	25.0	29.6		2
pH	3.8	3.4	4.4		8

2.322 SUGAR BEET, TOPS AND PULP, SILAGE

	Mean	*Min.*	*Max.*	*SD*	*No.*
Dry matter %	17.2	14.8	21.5		3
Proximate constituents % of DM and pH					
Crude protein	10.1	9.4	11.2		3
pH	3.7	3.5	4.0		3

2.323 POTATO, TUBER, SILAGE

	Mean	*Min.*	*Max.*	*SD*	*No.*
Dry matter %	24.8	16.8	35.0	6.6	31
Proximate constituents % of DM and pH					
Crude protein	10.9	5.6	16.2	2.5	30
Crude fibre	4.1	2.9	6.0		10
pH	4.0	3.5	4.8	0.5	24
Major elements % of DM					
Calcium	0.06				1
Phosphorus	0.24				1

2.324 COCKSFOOT, STRAW

	Mean	Min.	Max.	SD	No.
Dry matter %	86.8	84.1	88.5		6
Proximate constituents % of DM					
Crude protein	5.8	5.0	7.2		7
Crude fibre	44.2	42.2	48.7		7
Major elements % of DM					
Calcium	0.44	0.38	0.54		6
Phosphorus	0.12	0.09	0.18		6

2.325 ITALIAN RYEGRASS, STRAW

	Mean	Min.	Max.	SD	No.
Dry matter %	82.2				1
Proximate constituents % of DM					
Crude protein	5.1				1
Crude fibre	37.3				1
Major elements % of DM					
Calcium	0.49				1
Phosphorus	0.19				1

2.326 RYEGRASS, STRAW

	Mean	Min.	Max.	SD	No.
Dry matter %	84.9	83.2	86.7		7
Proximate constituents % of DM					
Crude protein	5.9	5.0	7.5		10
Ether extract	1.27				1
Crude fibre	35.8	33.1	37.6		10
Major elements % of DM					
Calcium	0.51	0.44	0.57		3
Phosphorus	0.16	0.15	0.18		3

2.327 FESCUE, STRAW

	Mean	Min.	Max.	SD	No.
Proximate constituents % of DM					
Crude protein	5.3	5.2	5.4		3
Crude fibre	39.1	38.7	39.3		3
Major elements % of DM					
Calcium	0.50	0.45	0.55		2
Phosphorus	0.13	0.10	0.15		2

2.328 TIMOTHY, STRAW

	Mean	Min.	Max.	SD	No.
Dry matter %	86.0	80.4	89.5		3
Proximate constituents % of DM					
Crude protein	4.6	3.0	6.6		4
Ether extract	1.55				1
Crude fibre	38.6	37.7	39.4		4
Major elements % of DM					
Calcium	0.41	0.39	0.43		2
Phosphorus	0.14	0.12	0.16		2

2.329 SEEDS GRASS, STRAW

	Mean	Min.	Max.	SD	No.
Dry matter %	83.5				1
Proximate constituents % of DM					
Crude protein	5.1	3.6	6.6		2
Crude fibre	45.5	38.1	52.8		2
Major elements % of DM					
Calcium	0.45				1
Phosphorus	0.15				1

2.330 OAT, STRAW

	Mean	Min.	Max.	SD	No.
Dry matter %	79.3	59.2	90.5	12.4	109
Proximate constituents % of DM					
Crude protein	3.35	1.94	6.12	1.24	132
Ether extract	1.40	0.87	2.03		13
Crude fibre	41.3	24.8	49.1	6.3	123
Major elements % of DM					
Calcium	0.41	0.25	0.59	0.08	39
Phosphorus	0.14	0.07	0.26	0.05	37
Magnesium	0.07	0.06	0.08		4
Potassium	0.97	0.83	1.10		2
Sodium	0.15	0.09	0.21		2

2.330 OAT, STRAW (continued)

	Mean	Min.	Max.	SD	No.
Trace elements µg per g DM					
Barium	29	6.7	54	17.9	142
Chromium	0.21	0.05	0.40	0.25	147
Cobalt	0.08	0.02	0.16	0.06	160
Copper	2.2	1.03	3.41	1.00	392
Iron	49	12	700	64.0	175
Lead	1.72	0.39	3.30	1.32	147
Manganese	69	9	157	67.3	361
Molybdenum	0.30	0.06	0.60	0.21	159
Nickel	0.29	0.16	0.45	0.15	148
Silver	0.04	0.02	0.05	0.015	147
Strontium	17	9.5	27	6.8	142
Tin	0.35	0.26	0.52	0.14	147
Titanium	3.41	2.24	4.30	1.81	148
Vanadium	0.21	0.06	0.40	0.15	147
Zinc	29	3.4	425	72.1	148

2.331 OAT, STRAW AND GRAIN

	Mean	Min.	Max.	SD	No.
Dry matter %	89.0				1
Proximate constituents % of DM					
Crude protein	4.6				1
Crude fibre	34.4				1
Major elements % of DM					
Calcium	0.23				1
Phosphorus	0.37				1

2.332 WHEAT, STRAW

	Mean	Min.	Max.	SD	No.
Dry matter %	68.2	43.3	89.9	14.0	114
Proximate constituents % of DM					
Crude protein	2.66	1.90	3.59	0.74	202
Crude fibre	43.8	36.5	47.8	3.9	20
Major elements % of DM					
Calcium	0.26	0.25	0.26		2
Phosphorus	0.10	0.03	0.15	0.04	184
Trace elements µg per g DM					
Barium	36	9.3	63		2
Chromium	0.17	0.14	0.19		2
Cobalt	0.03	0.02	0.03		2
Copper	2.7	2.0	3.6		7
Iron	31	23	38		2
Lead	1.34	0.17	1.51		2
Manganese	15				1
Molybdenum	0.37	0.28	0.45		2
Nickel	0.15	0.10	0.20		2
Silver	0.06	0.05	0.06		2
Strontium	18	6.3	30		2
Tin	0.51	0.47	0.55		2
Titanium	1.52	1.24	1.79		2
Vanadium	0.25	0.19	0.31		2
Zinc	13	11	15		2

2.333 BARLEY, STRAW

	Mean	Min.	Max.	SD	No.
Dry matter %	63.2	39.9	89.6	12.8	235
Proximate constituents % of DM					
Crude protein	3.9	2.0	5.8	1.3	387
Ether extract	1.45	1.21	1.64		5
Crude fibre	43.1	35.6	48.7	2.6	31
Major elements % of DM					
Calcium	0.44	0.30	0.67		16
Phosphorus	0.10	0.06	0.15	0.03	368
Potassium	1.29	0.74	1.85	0.41	346
Sodium	0.24				1
Trace elements µg per g DM					
Barium	28	14	41		3
Chromium	0.19	0.08	0.23		11
Cobalt	0.03	0.02	0.09		14
Copper	2.8	1.2	8.4	2.0	25
Iron	55	26	173		15
Lead	4.18	0.32	7.20		12
Manganese	84	15	528	129	22
Molybdenum	0.47	0.13	1.12		12
Nickel	0.22	0.11	0.39		12
Silver	0.04	0.03	0.05		12
Strontium	11	6.5	16		3
Tin	0.42	0.26	0.49		12
Titanium	2.93	2.37	3.50		11
Vanadium	0.20	0.10	0.27		12
Zinc	16	4.9	33		12

2.334 CEREALS, MIXED, STRAW

	Mean	Min.	Max.	SD	No.
Dry matter %	89.6	89.2	90.0		2
Proximate constituents % of DM					
Crude protein	7.0	2.9	11.1		2
Crude fibre	45.3	39.9	50.6		2
Major elements % of DM					
Calcium	0.58				1
Phosphorus	0.29				1

2.335 CLOVER, STRAW

	Mean	Min.	Max.	SD	No.
Dry matter %	84.5	81.6	86.6		6
Proximate constituents % of DM					
Crude protein	13.0	10.3	15.7		6
Crude fibre	33.0	31.4	36.4		5
Major elements % of DM					
Calcium	1.37	1.21	1.52		4
Phosphorus	0.29	0.27	0.31		4

2.336 PEA, STRAW

	Mean	Min.	Max.	SD	No.
Dry matter %	79.5	77.2	81.8		2
Proximate constituents % of DM					
Crude protein	10.35	10.2	10.5		2
Crude fibre	33.65	29.6	37.7		2
Major elements % of DM					
Calcium	2.22	2.15	2.29		2
Phosphorus	0.13	0.13	0.13		2

2.337 PEA AND BEAN, STRAW

	Mean	Min.	Max.	SD	No.
Proximate constituents % of DM					
Crude protein	12.6	7.3	18.4		7
Crude fibre	34.7	25.4	51.9		7
Major elements % of DM					
Calcium	1.73	1.27	2.74		7
Phosphorus	0.20	0.09	0.27		7

2.338 TREFOIL, STRAW

	Mean	Min.	Max.	SD	No.
Dry matter %	85.3	82.7	87.2		3
Proximate constituents % of DM					
Crude protein	8.0	6.8	9.2		3
Crude fibre	40.4	37.7	43.2		3
Major elements % of DM					
Calcium	0.95	0.94	0.95		2
Phosphorus	0.13	0.12	0.13		2

2.339 LEGUMES, MIXED, STRAW

	Mean	Min.	Max.	SD	No.
Dry matter %	86.2	85.3	86.7		3
Proximate constituents % of DM					
Crude protein	9.3	8.2	10.4		4
Crude fibre	38.5	38.3	39.7		4
Major elements % of DM					
Calcium	2.66	2.15	3.17		2
Phosphorus	0.14	0.13	0.15		2

2.340 *CYNOSURUS CRISTATUS*: CRESTED DOGSTAIL

	Mean	Min.	Max.	SD	No.
Trace elements μg per g DM					
Barium	5.7	5.2	6.5		4
Chromium	0.12	0.08	0.14		4
Cobalt	0.03	0.03	0.03		4
Copper	3.2	1.69	6.50	0.97	24
Iron	29	25	35		4
Lead	0.70	0.65	0.72		4
Manganese	102	65	150	22	24
Molybdenum	0.65	0.28	1.53		4
Nickel	0.49	0.45	0.56		4
Silver	0.06	0.05	0.06		4
Strontium	9.5	8.9	10.0		4
Tin	0.54	0.50	0.57		4
Titanium	1.61	1.49	1.73		4
Vanadium	0.03	0.03	0.04		4
Zinc	27	19	34		4

2.341 *MOLINIA CAERULEA*: FLYING BENT, HAY

	Mean	Min.	Max.	SD	No.
Dry matter %	89.0	88.8	89.2		2
Proximate constituents % of DM					
Crude protein	12.8	10.8	14.8		2
Ether extract	2.6				1
Crude fibre	30.3				1

2.342 *CALLUNA VULGARIS:* HEATHER

	Mean	*Min.*	*Max.*	*SD*	*No.*
Dry matter %	40.9	30.5	53.9	6.47	21
Proximate constituents % of DM					
Crude protein	8.1	5.8	9.1	1.35	21
Ether extract	3.4	3.3	3.5		2
Nitrogen-free extract	60.6	60.5	60.8		2
Crude fibre	25.0	22.0	30.0	3.07	21
Total ash	3.2	3.2	3.2		2
Trace elements µg per g DM					
Barium	15	11	18		11
Chromium	1.1	0.9	1.6		11
Cobalt	0.22	0.02	0.43	0.12	29
Copper	8.1	5.2	14.0	2.3	21
Iron	121	47	258	54	28
Lead	34	15	64		11
Manganese	400	267	660		13
Molybdenum	0.17	0.05	0.32	0.09	21
Nickel	2.4	1.3	4.4		11
Silver	0.09	0.06	0.16		11
Strontium	4	3	6		11
Tin	1.0	0.7	1.5		11
Titanium	12	5	22		11
Vanadium	1.3	1.0	2.1		11
Zinc	56	37	75		11

2.343 HEATHER, SHOOTS

	Mean	*Min.*	*Max.*	*SD*	*No.*
Dry matter %	57.8				1
Proximate constituents % of DM					
Crude protein	8.5				1
Digestible CP	1.6				1
Ether extract	2.3				1
Crude fibre	24.8				1

2.344 GRASS AND HEATHER, HILL PASTURE

	Mean	*Min.*	*Max.*	*SD*	*No.*
Dry matter %	33.4	30.9	35.8		2
Proximate constituents % of DM					
Crude protein	8.5	8.0	8.9		2
Ether extract	2.9	2.6	3.1		2
Nitrogen-free extract	57.8	57.5	58.2		2
Crude fibre	27.2	25.5	28.8		2
Total ash	3.62	2.94	4.30		2
Major elements % of DM					
Calcium	0.16				1
Phosphorus	0.11				1
Magnesium	0.21				1
Potassium	0.44				1
Sodium	0.06				1

2.345 SEAWEED MEAL

	Mean	*Min.*	*Max.*	*SD*	*No.*
Dry matter %	83.2	77.6	88.9		2
Proximate constituents % of DM					
Crude protein	5.93	5.75	6.10		2
Major elements % of DM					
Calcium	1.07	0.98	1.15		2
Phosphorus	0.11	0.10	0.11		2
Magnesium	0.92				1
Potassium	2.78				1

Index to Tables, Part 2

Typesetting by Martin Dawson, Aberdeen & Glasgow

Printed in Great Britain by
Robert MacLehose & Co. Ltd, Printers to the University of Glasgow